Study Guide

to accompany

JACALYN NEWMAN
University of Pittsburgh

ED DZIALOWSKI
University of North Texas

LINDSAY GOODLOE
Cornell University

BETTY MCGUIRE
Cornell University

NANCY GUILD
University of Colorado at Boulder

Sinauer Associates, Inc.

W. H. Freeman and Company

Cover photograph © Dr. Merlin D. Tuttle/Photo Researchers, Inc.

Study Guide to accompany *Life: The Science of Biology,* **Ninth Edition**

Address editorial correspondence to:
Sinauer Associates, Inc.
23 Plumtree Road
Sunderland, MA 01375 U.S.A.
Fax: 413-549-1118
Internet: www.sinauer.com; publish@sinauer.com

Address orders to:
MPS/W.H. Freeman & Co. Order Department
16365 James Madison Highway, U.S. Route 15
Gordonsville, VA 22942 U.S.A.
Examination copy information: 1-800-446-8923
Orders: 1-888-330-8477

ISBN 978-1-4292-3569-3
Printed in U.S.A.

4 3 2 1

To the Student

Biology is an incredibly exciting field of study, but in order to appreciate new discoveries and discussions, it is necessary to have a firm grasp of the underlying concepts and ideas. Your textbook is designed to give you a comprehensive overview of important biological phenomena. It will also serve as a resource to you in future studies. Together with your instructor, your textbook will provide you with invaluable information for beginning your study of biology.

This Study Guide is designed to supplement, not replace, your textbook and your instructor. It was written for you, the student, in language that you can understand, but it does emphasize proper usage of biological terminology. Important concepts and ideas have been synthesized into short, easy-to-read summaries that provide an overview of the biological concepts discussed in your textbook. Each Study Guide chapter includes four review elements: The Big Picture, Common Problem Areas, Study Strategies, and Important Concepts—these can help you preview a chapter before reading it, check your understanding, and review the chapter later.

Each Study Guide chapter also includes a series of questions, which have been grouped into three categories. New for the Ninth Edition Study Guide, Diagram Exercises present questions in a format that allows you to review your factual and conceptual knowledge of each chapter visually. In some chapters, diagrams are provided and you are asked to label them or answer questions based on them, while in other chapters, you are asked to create your own diagrams based on a question or series of questions. Knowledge and Synthesis Questions are designed to determine if you have retained information from a chapter, and if you can put together various concepts in order to answer questions. Application Questions ask you to apply the knowledge you have gleaned from a chapter to answer questions that are more open-ended. The latter type of questions require you to have assimilated several concepts and to think beyond what you have just read. Your instructor may ask questions similar to these on exams, or may use an entirely different approach to assess your knowledge, but these questions will be a good check of how well you understand the material.

Brief answers to the questions are provided at the end of each chapter. These answers are not exhaustive, but are instead designed to point to the correct concepts in the textbook. Because of the nature of many of the Application Questions, your answers should be more expansive than the short explanations given in this Study Guide.

Strategies for Studying Biology

Each individual has his or her own unique study pattern. However, there are some successful study strategies that are universal. We recommend that you first preview a chapter in your textbook. In this initial preview, it is important to note the organization of the chapter and the main points, and to go over the chapter summary at the end. By referring to the Study Guide at this point, you will further understand the organization of the material. If you follow these steps, you might have some specific questions in your mind that you will expect your reading to answer.

Reading a textbook is an active process. We recommend that you always have a pencil and paper available. Jotting notes in margins also serves as an excellent mental trigger when it comes time to review. As you read each section of your textbook, see if you can summarize it in your own words. Compare your summaries to those provided at the end of each section and at the end of the chapter. You want to assure yourself that the main points you are noting match those that the author has selected. Refer back to those questions that came up as you previewed the chapter. You should be able to answer your own questions by the time you have finished your reading.

As you are reading, take time to review all the figures and tables. Your textbook makes use of figures to illustrate points, describe pathways, and give visual representation to complex topics. Often these figures are more helpful than the paragraphs of written explanation. In places where

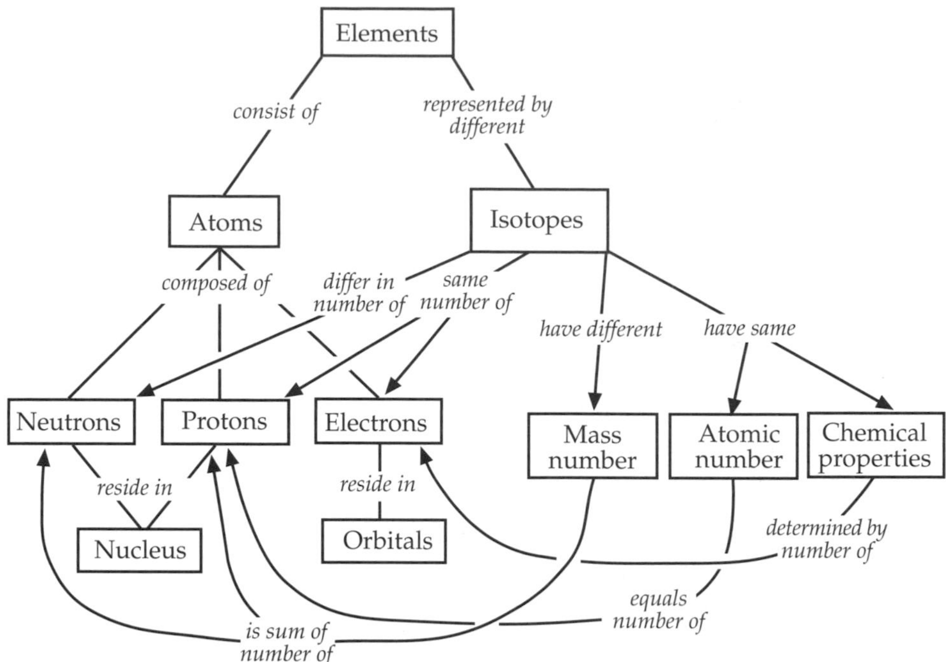

figures might be beneficial to you but are not provided in the textbook, try to draw them yourself. This is especially important when attempting to understand structures and pathways. You will find that the Study Guide also refers you to specific tutorials and activities in the Life, Ninth Edition BioPortal, which is available online at yourBioPortal.com. Each of these visual tools will aid in your understanding of the material. In addition, BioPortal includes many other study and review resources, some of which your instructor may assign. (Many of these resources are also available on the textbook's free Companion Website, at www.thelifewire.com.)

Once you have read the material, summarized it in your own words, and reviewed all of the figures and tables, it is time to do another quick review. Scan the summary in your textbook again and read over the important concepts in this Study Guide. We recommend further that you construct a "concept map" of the main points of a chapter. This will assure you that you see how concepts are interrelated and that you understand the organization of the material. An example of a concept map is shown below. Main concepts are in boxes, and arrows are used to connect concepts and show relationships.

Finally, when you are comfortable with chapter content, move to the questions section of this Study Guide. We recommend working through all of the questions before checking your answers. Any questions that you do not understand or that you answer incorrectly should be flagged as material you need to review. Remember that your goal is to understand the material completely, not just to answer these particular questions.

Experiments have suggested that the average attention span of a reader is approximately 12 minutes. You cannot expect to sit down and master an entire chapter in one sitting. Break up your study into short segments of approximately 30 minutes each. This will give you time to get down to business, maximize your attention, and learn without becoming frustrated or drained. At the end of this time, move to some other activity or area of study. When you return to biology, you will find your mind is ready to absorb more.

This entire process should take place well in advance of your exams. You cannot learn even one chapter adequately the day before an exam. Mastering biology requires daily study and review. If you keep up with learning the concepts as your course proceeds, you will find that reviewing the textbook summaries and the Study Guide, along with the online review tools, is adequate preparation. The key to learning biology well is slow and diligent daily work.

Doing Well on Biology Exams

Your instructors are your best resource for doing well on exams. They are there to assist you in learning the material that is presented in your textbook. They are experts in their fields and understand how the concepts presented are vital to your biology education. Follow their lead in preparing for exams. Exam performance is directly linked to classroom attendance and daily study. Sit front and center in your lecture hall and pay close attention to what your instructor writes and provides in presentations or overhead displays. Your instructor will give you guidance about the most important points to study. You should take careful notes and ask for clarification when necessary.

You should always compare your classroom and lecture notes to your textbook and the Study Guide. Mark in the books those points your instructor emphasizes. Were there specific questions that the instructor brought up in class? Do these correspond to the textbook end-of-chapter questions or to those in the Study Guide? If so, chances are you will see them again on the exam. This practice should be part of your daily study regimen.

Approach the exam itself with confidence. Read the directions carefully and read the questions completely. One of the biggest mistakes students make on exams has to do with not following instructions. We recommend that you read carefully through the entire exam even before you begin to answer the first question. You may find that early questions are answered partially in later ones, and that an overview of the whole exam helps to trigger your memory of the material.

If your exam contains multiple choice questions, treat each of the answer options as if it were a separate true/false question. In other words, mentally fill in the question with each answer and ask yourself if the resulting statement is true or false. Be sure to read each of the options, even if the first one strikes you as the correct one. You may be dealing with a question that has multiple correct answers or asks you to select "all of the above" or "none of the above." If you are unsure about how to answer a question, begin by ruling out answers that you know are incorrect. By eliminating some possibilities, your chance of selecting the right one is greatly improved! If you narrow the choices down to two possible answers and just can't decide between them, go with your gut response. You just may be right. And when you go over your answers, change your response only if you know you answered a question incorrectly. If you aren't sure, it may be best to leave the answer as-is. Your first instincts are often correct.

Essay questions require you not only to think through the concepts, but also to decide how you will present them. Organized, concise essay answers are always preferred to rambling answers with little direction. Take a look at the point value of each question. This is often an indication of how many "points" you need to make in your discussion. A 10-point question rarely can be answered with a single sentence. Look carefully at what your instructor is asking. If you are required to "discuss" a concept or a problem, do not present a list or some scattered phrases. Write in complete sentences and paragraphs unless you are asked specifically to list or itemize the material. Be sure that you address all points in the question. Essay questions frequently have multiple parts to test your understanding of the links between various concepts.

Always be aware of the time you have to complete your exam. Answer the questions you know swiftly, and allow yourself time to go back and really think about those you are struggling with. Be sure to concentrate on the questions with the highest point value. Those are the questions testing the most critical concepts and they are the ones that will have the most influence on your grade.

Once you think you have completed the exam, take a few minutes to go back over it. If there are any questions about which you are still unclear, be sure to ask your instructor for clarification.

Learn from Your Mistakes

After your instructor returns your exam or posts the answers, review the mistakes you have made. Take this opportunity to clarify misunderstandings and re-learn material when necessary. You will find this helpful not only for the final exam, but also as you approach upper-level courses. Material reviewed, corrected, and re-learned is more likely to be retained in the long run.

Get Help before It Is Too Late

Many students come to college without adequate study habits and/or are ill prepared for the rigor of biology at the college level. But this is a problem that can be solved fairly easily. Many colleges and universities have academic centers that assist with locating tutors, teaching study habits, and acquainting you with a variety of other resources. Make good use of these facilities. They are there for you. Also, for many additional study tips and strategies, see the *Survival Skills* document that is available within BioPortal.

Good luck in your study of introductory biology!

Contents in Brief

Contents in Brief (continued)

1 Studying Life

The Big Picture

- Biology is the study of living things. Living things are composed of cells, which contain genetic material, require nutrients for energy, maintain a constant internal environment, and interact with one another. All available evidence points to an origin of life some 4 billion years ago. Natural selection has led to the current host of organisms inhabiting the planet. There are specific characteristics of life, and the evolutionary roots of these characteristics can be traced through time.
- All life on Earth is organized in the tree of life based on molecular evidence. When studying organisms from an evolutionary perspective, scientists are concerned with evolutionary relatedness and common ancestors.
- Science follows a specific hypothesis–prediction method. The necessity of a testable and falsifiable hypothesis sets science apart from other methods of inquiry. Control of variables and repeated testing lend credibility to the method.

Common Problem Areas

- Many students have trouble remaining open-minded enough to begin to understand evolutionary biology. Neither your instructors nor this book are asking you to set aside your beliefs. We are asking you to learn science. We encourage you to examine the evidence supporting evolution without prejudice, as you would any other scientific subject.
- Refrain from learning the hypothesis–prediction method merely as a series of steps. Think about how each step in the process follows from the preceding step. Two characteristics of science set it apart from other modes of inquiry. One is the setting of hypotheses. The other is the continual process by which it is carried out. With each conclusion leading to a new set of hypotheses, science has the capacity to alter our overall base of knowledge.

Study Strategies

- Many students find that understanding the hypothesis–prediction approach makes understanding some of the evidence behind evolution easier. If you find you are having trouble, work on that section first.
- Drawing a timeline or examining the calendar model of Figure 1.8 will help you understand the chronology of the major evolutionary events discussed in your textbook.
- Go to yourBioPortal.com to review the following tutorial and activities:

  Animated Tutorial 1.1: The Scientific Method

  Web Activity 1.1 The Hierarchy of Life

  Web Activity 1.2 The Major Groups of Organisms

Important Concepts

Biology is the study of living things.

- Living things are made up of cells. They can be unicellular (composed of one cell) or multicellular (composed of multiple cells). Cell theory states that cells are the basic structural and physiological units of living things and that they can work alone or together to make up more complex organisms. This theory also states that all cells are made of the same materials, come from other cells, are the sites of the chemical reactions of life, and contain the genetic material that is passed from cell to cell. While viruses lack a cellular structure, they must parasitize cellular organisms to carry out their functions.
- Charles Darwin proposed that all living things are related to one another and that species have evolved through the process of natural selection. A species can be defined as a group of organisms that are similar and can successfully breed and produce offspring. Natural selection works because individuals vary in their traits, and the characteristics of traits possessed by some of the individuals may provide them with a better chance of survival and reproduction. Natural selection can produce adaptations, which are structural, physiological, or behavioral traits that enhance the chance of survival of an organism in a given environment.
- The genome of a cell contains all the genetic information in the form of DNA. Segments of DNA make up genes, which are used to produce the proteins neces-

sary for life. All the cells in an organism contain the same genome, but they express different genes.

- All cells require nutrients for survival. These nutrients are used to carry out cellular work, such as the work of building the cell or mechanical work, such as the moving of molecules from one location to another. Metabolism or metabolic rate is a measure of all the chemical transformations and work done in a cell or organism.
- Cells continuously carry out multiple chemical reactions and thus must carefully regulate their internal environment. Multicellular organisms must also regulate the extracellular fluid that bathes each cell. This requires the specialization of the cells making up multicellular organisms. A group of similar cells that work together are known as a tissue. A number of different tissues can be organized into an organ, and multiple organs work together to form an organ system.
- Ecology is the study of how different species interact with one another. Organisms from one species living together make up a population. The many populations of different species in an area make up a community. Communities and their abiotic environment make up ecosystems.
- Biologists use model species to study biological processes.

Certain evolutionary results had to occur before others.

- All species on Earth share a common ancestor. The study of the evolution of life on Earth involves both the fossil record and modern molecular methods for comparing genomes.
- Life arose approximately 4 billion years ago. The random aggregation of complex chemicals led to the existence of the first biological molecules. These initial simple molecules led to molecules that could reproduce themselves and act as templates for larger, more complex molecules.
- The next step in the origin of life probably involved the enclosure of these molecules in a membrane enclosure, allowing all the necessary complex biological molecules to interact. Early cells were simple, ocean-living prokaryotes with no internal membrane-enclosed compartments.
- Metabolism, which is the sum of all chemical reactions that occur within a cell, requires a source of energy. The ability to carry out photosynthesis arose approximately 2.7 billion years ago. Photosynthesis involves converting light energy from the sun into usable chemical energy, with oxygen as a by-product. Large numbers of photosynthetic organisms increased atmospheric oxygen levels, allowing for the evolution of aerobic metabolism. Photosynthetic organisms also contributed to the insulating ozone layer that shields Earth from radiation and modifies temperature. This shield eventually made life on land possible.
- Eukaryotic cells have intracellular membrane-enclosed compartments known as organelles that carry out specific cellular functions. These compartments are thought to have arisen when larger prokaryotic cells engulfed and assimilated smaller cells.
- Multicellularity, which evolved about 1 billion years ago, made it possible for cells to specialize and for organism size to increase.

All organisms are descendants of a single unicellular organism.

- Different species come about when two groups of organisms in one species are isolated from each other so that they can no longer mate and produce viable offspring.
- The tens of millions of different species on Earth today resulted from the splitting of populations called speciation events. By means of phylogenetic trees constructed according to similarities in genome sequences, systematists study the evolutionary relationship among all organisms.
- Organisms are referred to by their genus and species names (e.g., humans are *Homo sapiens*).
- Scientists are determining the branching pattern of the tree of life based on the molecular evidence found in organisms. Knowledge about a species' placement on the tree of life reveals a great deal about its biology. Information about a new species is provided by comparisons with closely related species.
- All life is placed into one of three major domains: Archaea, Bacteria, and Eukarya.
- Plantae, Fungi, and Animalia are three major groups of the Eukarya. These three groups evolved from different protists.

Biologists use observation and experimentation to investigate life.

- Biologists have always studied life by means of observation. As we move forward, many new tools are being developed that help us observe life. Many of these new tools allow scientists to quantify their observations.
- The scientific method, or hypothesis–prediction method, is the basis of most scientific investigations. This method involves making observations and asking questions. These questions lead to the formation of hypotheses, which are provisional answers to the questions. Predictions are made in regard to the hypotheses, and then they are tested.
- A hypothesis can be tested by either comparative or controlled experiments. A controlled experiment involves isolating variables of interest while keeping other variables that may influence the outcome as steady as possible. Observations are then made to test the hypothesis. A comparative experiment involves gathering data from multiple groups or conditions to examine the patterns found in nature.

- Statistical methods are used to determine if the results of an experiment are significant. Typically, these statistical tests start with a null hypothesis stating that there are no differences.
- Science depends on a hypothesis that is testable and that can be rejected by direct observation and experiments. Observations must also be reproducible and quantifiable for a conclusion to be considered scientific.

Biological science has an impact on every human being and is the basis for many public policy decisions.

- The environmental stresses of population growth, the emergence of agricultural and medical technologies, and other modern challenges have increased the need for everyone to understand biological information.

Test Yourself

Diagram Exercise

The study of biology is advanced by using the scientific method. Create a flow chart showing the steps involved in the scientific method.

Textbook Reference: *1.3 How Do Biologists Investigate Life? p. 14*

Knowledge and Synthesis Questions

1. Life arose on Earth approximately _______ years ago.
   a. 4 billion
   b. 4 million
   c. 4,000
   d. 1.5 billion
   e. 400,000
   Textbook Reference: *1.2 How Is All Life on Earth Related? p. 10*
2. Which of the following is the feature or component of organisms that allows for life in such a wide variety of environments on Earth?
   a. Prokaryotic cells
   b. Eukaryotic cells
   c. Homeostasis
   d. Adaptation
   e. Model systems
   Textbook Reference: *1.1 What Is Biology? p. 6*
3. Which of the following is *not* a characteristic of most living organisms?
   a. Regulation of internal environment
   b. One or more cells
   c. Ability to produce biological molecules
   d. Ability to extract energy from the environment
   e. None of the above
   Textbook Reference: *1.1 What Is Biology? p. 2*
4. Photosynthesis was a major evolutionary milestone because
   a. photosynthetic organisms contributed oxygen to the environment, which led to the evolution of aerobic organisms.
   b. photosynthesis led to conditions that allowed life to arise on land.
   c. photosynthesis is the only metabolic process that can convert light energy to chemical energy.
   d. photosynthesis provides food for other organisms.
   e. All of the above
   Textbook Reference: *1.2 How Is All Life on Earth Related? pp. 10–11*
5. The fact that all cells come from preexisting cells is a component of _______ theory.
   a. animal
   b. genetic
   c. cell
   d. plant
   e. adaptation
   Textbook Reference: *1.1 What Is Biology? p. 4*
6. A group of cells that work together to carry out a similar function is known as a(n)
   a. tissue.
   b. organ system.
   c. unicellular organism.
   d. protein.
   e. a gene.
   Textbook Reference: *1.1 What Is Biology? p. 8*
7. Which of the following does *not* contribute to adaptation in the wild?
   a. Artificial selection
   b. Genetic drift
   c. Natural selection
   d. Sexual selection
   e. All of the above contribute to adaptation in the wild.
   Textbook Reference: *1.1 What Is Biology? pp. 5–6*
8. Which of the following is *not* a domain on the tree of life?
   a. Archaea
   b. Plantae
   c. Eukarya
   d. Bacteria
   e. All of the above are domains.
   Textbook Reference: *1.2 How Is All Life on Earth Related? p. 12*
9. The information needed to produce proteins is contained in
   a. nutrients.
   b. tissues.
   c. evolution.
   d. organs.
   e. genes.
   Textbook Reference: *1.1 What Is Biology? p. 6*
10. Evolution is
    a. not important to the study of biology.
    b. the change in the genetic makeup of a population through time.
    c. the change in protein expression of a population through time.
    d. not influenced by natural selection.
    e. None of the above

Textbook Reference: *1.1 What Is Biology? p. 5*

11. In a model experiment, researchers subjected frogs to various levels of atrazine while keeping all other variables constant. This is an example of a _______ experiment.
    a. controlled
    b. repeated
    c. laboratory
    d. comparative
    e. None of the above
    Textbook Reference: *1.3 How Do Biologists Investigate Life? p. 15*
12. For a hypothesis to be scientifically valid, it must be _______ and it must be possible to _______ it.
    a. testable; prove
    b. testable; reject
    c. controlled; prove
    d. controlled; reject
    e. testable; control
    Textbook Reference: *1.3 How Do Biologists Investigate Life? p. 14*
13. Eukaryotic cells differ from prokaryotic cells in that they have
    a. genes.
    b. proteins.
    c. organelles.
    d. membranes.
    e. All of the above
    Textbook Reference: *1.2 How Is All Life on Earth Related? p. 11*
14. In the names of organisms, the _______ is placed first and the _______ is placed second.
    a. species; genus
    b. genus; domain
    c. domain; genus
    d. genus; species
    e. domain; species
    Textbook Reference: *1.2 How Is All Life on Earth Related? p. 12*
15. Metabolism refers to
    a. natural selection.
    b. the chemical transformations and work of a cell.
    c. communities.
    d. mutations in DNA.
    e. cellular structure.
    Textbook Reference: *1.1 What Is Biology? p. 7*

Application Questions

1. Discuss how the process of scientific inquiry is different from other forms of inquiry. Include in your discussion a description of the hypothesis–prediction approach.
   Textbook Reference: *1.3 How Do Biologists Investigate Life? p. 16*
2. Look over a recent newspaper to see how many articles are directly related to biology. Select one article and discuss how the researchers followed or did not follow the hypothesis–prediction approach.
   Textbook Reference: *1.3 How Do Biologists Investigate Life? p. 13*
3. What is one example from your own experience of biology's influence on public policy?
   Textbook Reference: *1.4 How Does Biology Influence Public Policy? pp. 17–18*
4. Scientists interested in human biology typically perform experiments with other model systems. Why do scientists use model systems in this way?
   Textbook Reference: *1.1 What Is Biology? p. 9*
5. The study of biology can be organized from the most basic unit, the molecule, up to the biosphere. Describe how each level is connected with the level below it.
   Textbook Reference: *1.1 What Is Biology? p. 8*

Answers

Diagram Exercise Answer

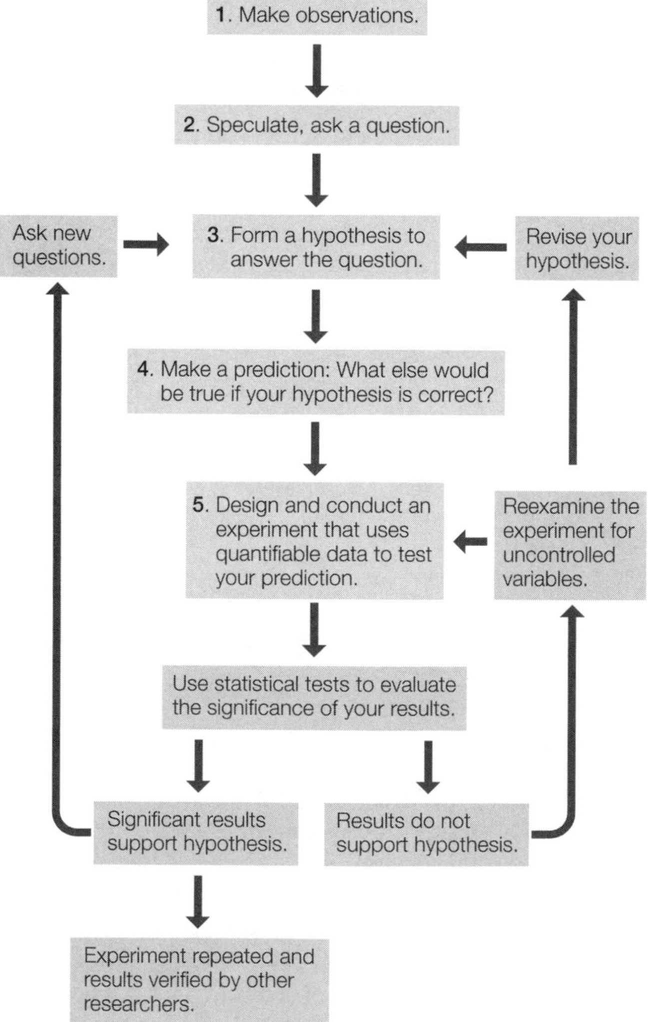

Knowledge and Synthesis Answers

1. **a.** Available evidence places the beginning of life at about 3.8 billion years ago.
2. **d.** Adaptations are the differences found in organisms that allow them to live in an environment.
3. **e.** Most living organisms are composed of cells, have the ability to make biological molecules, can regulate their internal environment, and get energy from the external environment.
4. **d.** Photosynthesis caused the accumulation of oxygen in the atmosphere and contributed to the formation of an ozone layer. The presence of oxygen made the evolution of aerobic organisms possible, and the ozone layer shielded Earth from harmful radiation. These two phenomena contributed to the evolution of terrestrial life.
5. **c.** The cell theory states that all life is based on the cell.
6. **a.** Multicellular organisms have tissues that are formed from many similar cells.
7. **a.** Adaptations arise from genetic drift, natural selection, and sexual selection. Artificial selection can produce adaptations, but they occur through human selection and not in the wild.
8. **b.** Plantae is a kingdom found in the domain Eukarya.
9. **e.** Genes are specific sequences of DNA that contain the information used to make proteins.
10. **b.** Evolution is a change in the frequency of genes within a population over time and can occur through natural selection. This is the major unifying principle of biology.
11. **a.** For experiments to be scientifically valid, they must be controlled.
12. **b.** Scientific hypotheses are set apart from mere conjecture by being testable and falsifiable.
13. **c.** Both eukaryotic and prokaryotic cells contain genes, proteins, and some sort of membrane. Only the eukaryotes have organelles.
14. **d.** An example is *Homo sapiens.*
15. **b.** The metabolism, or metabolic rate, of an organism is the sum of all the chemical transformations and work that it performs.

Application Answers

1. The process of scientific inquiry is different because a hypothesis must be testable, and it must be possible to reject it. The hypothesis–prediction approach begins with observations that lead to questions. From the questions, hypotheses are formed that are probable explanations for the observed phenomena. Predictions are formed from the hypotheses and tested. Conclusions are drawn from the tests. These conclusions may, in turn, lead to additional hypotheses.
2. Often an article reporting on science news does not provide enough information for the reader to know whether a particular study followed the hypothesis–prediction approach. It pays to be a smart reader and consumer in this respect.
3. Many public policies rely on sound biological findings. Examples include medical issues, environmental quality issues, and ecology of local populations of wild animals.
4. Model systems are useful in the study of biology because all organisms have evolved from a common ancestor. Therefore, cellular pathways in a zebrafish are very similar to those found in the human. Model systems are valuable because in many cases they can be manipulated experimentally.
5. The cell is composed of many different types of molecules. Cells with the same function and coordination are grouped together to form tissues. Several tissue types work together to form a functioning organ. A complex multicellular organism is composed of organs and organ systems. Many organisms of the same species living together make up a population. A community encompasses all of the populations within a given area. An ecosystem is composed of many communities in the same geographical area. All of the ecosystems on Earth make up the biosphere.

2 Small Molecules and the Chemistry of Life

The Big Picture

- Understanding the chemical building blocks of all matter is essential to understanding the biology of organisms. Atoms, which contain protons, neutrons, and electrons, form elements or combine to form molecules. Molecules may be held together with covalent or ionic bonds and stabilized in three-dimensional conformations by hydrogen bonds. Chemical reactions change reactants to products without creating or destroying matter.
- Water has unique properties that make it biologically important. An essential property is its ability to act as a solvent. How a substance ionizes in water determines if it is acidic or basic. Acids have a high concentration of H^+, whereas bases have a low concentration of H^+. The pH scale allows us to compare the concentrations of H^+ in a variety of solutions. Weak acids acting as buffers are necessary for an organism to maintain homeostasis.

Common Problem Areas

- pH tends to confuse many people. Remember that it is a concentration of H^+ ions. Those that have a higher concentration are acidic and have a low pH, and those that have a lower concentration are basic with a high pH. Also remember that pH is measured on a logarithmic scale. This means that if you decrease the pH from 8 to 7, you increase the concentration of H^+ ions by a factor of 10.
- Learning quantitative terminology and how solutions are made is often difficult. Completing practice problems is the best way to learn this material.
- People are sometimes overwhelmed by the periodic table. Consult Figure 2.2 in the textbook to make sure you understand the components and how they are arranged. Do not try to memorize the table, but keep it handy as a reference.
- Use Table 2.1 to understand the differences among the various types of bonds and interactions.

Study Strategies

- The best way to learn the quantitative terms and pH material is to work on many practice problems. Also take advantage of any laboratory experience in which you make solutions. Every good biologist makes many solutions over time.
- Use the figures in your book to help you visualize what atoms and molecules look like and how they interact.
- Be careful not to get lost in terminology. Think about what each term means, and focus on understanding the concept rather than memorizing the term.
- Go to yourBioPortal.com to review the following tutorial and activity:

  Animated Tutorial 2.1 Chemical Bond Formation

  Web Activity 2.1 Electron Orbitals

Important Concepts

All matter is composed of the same chemical elements.

- Atoms combine to form all matter. The nucleus of atoms contains a defined number of positively charged protons and neutral neutrons. Negatively charged electrons move in their outer shells. An element is made up of only one type of atom. There are more than 100 elements, and they are grouped according to certain characteristics in the periodic table.
- Six elements compose the majority of every living organism: carbon, hydrogen, oxygen, nitrogen, phosphorus, and sulfur.
- The number of protons in an element determines its type and is known as its atomic number. Neutrons are found in every element except hydrogen. The total number of protons and neutrons equals the mass number of the atom.
- Isotopes of an element have the same number of protons but differ in the number of neutrons present. Radioisotopes are unstable isotopes that give off energy. This decay eventually causes the atom to

change to a new element. Radioisotopes are important in the medical field.

- Protons and neutrons have mass and contribute to atomic weight. Atomic weights on a periodic chart are averages of all isotopes of an element.

Biologists are interested in chemical reactions and the interactions of electrons.

- Interactions between atoms involve the associations of their electrons. Electrons continuously orbit the nucleus of an atom in a defined space.
- The orbital of an atom is the space in which an electron is found at least 90 percent of the time. These patterns of orbitals compose a series of electron shells or energy levels, each with a specific number of electrons. The first shell (*s* orbital) can contain up to two electrons. The second shell can have as many as four orbitals (*s* orbital and three *p* orbitals), each containing two electrons. The number of electrons in the outermost shell determines how the atom interacts with other atoms. Many biologically important atoms, such as carbon and nitrogen, are stable when they have eight electrons in the outermost shell. This phenomenon is referred to as the octet rule.
- Two or more linked atoms compose a molecule resulting in stabilization of the outer electron shell of the atoms.

Chemical bonds hold elements together to make molecules.

- Atoms share pairs of electrons to stabilize their outer shells in covalent bonding. This type of bond is very stable and strong and can be broken only with a great deal of energy. All molecules have a three-dimensional shape, and the interactions between a given pair of atoms always have the same length, angle, and direction. A molecule's shape affects how it behaves.
- Molecules containing two or more elements in a fixed ratio make a compound. The molecular weight of a compound is the sum of the atomic weights of all atoms in the compound.
- Multiple covalent bonds may exist. Although a single covalent bond involves one pair of shared electrons, double bonds involve two pairs of shared electrons. Triple bonds share three pairs, but are rare.
- Covalent bonding between atoms of the same element results in equal sharing of electrons. The attractive force an atom exerts on electrons is known as electronegativity. Two atoms that have the same electronegativity form nonpolar covalent bonds.
- Bonds between different elements generally result in polar covalent bonds with unequal sharing of electrons. Unequal sharing of electrons results in partial (δ) charges in molecules because one nucleus is more electronegative than the other, attracting the electrons more strongly. One atom of a molecule may be partially negative, whereas the other is partially positive. This balance of partial charges results in a polar molecule with a δ^- pole and δ^+ pole.
- Ions are formed when atoms lose or gain electrons, resulting in a net positive or negative charge. Cations are positively charged and have fewer electrons than protons. Anions are negatively charged and have more electrons than protons.
- More than one electron can be gained or lost from an atom or molecule. A group of covalently bonded atoms may also gain or lose electrons to form complex ions.
- Ionic bonds form between ions of opposite charge. Ionic bonds are not as strong as covalent bonds and they also attract partial charges of polar molecules.
- Partial charges allow hydrogen bonding of molecules. This occurs often between molecules of water where the positive hydrogen is attracted to the negative oxygen. This is a weak bond that is easily broken because no electrons are shared, but this type of bond is important in stabilizing the three-dimensional shape of large molecules such as proteins and DNA.
- Polar molecules are hydrophilic, or "water loving," because of the partial charges. Nonpolar molecules are hydrophobic, or "water hating," and they have hydrophobic interactions. Attractions of nonpolar molecules are enhanced by van der Waals forces. Polar molecules tend to aggregate with other polar molecules, and nonpolar molecules aggregate with other nonpolar molecules.

Chemical reactions occur when atoms combine and form new bonding partners.

- Chemical reactions involve reactants that are altered to produce products. Matter is neither created nor destroyed, but changed. Energy is the capacity to do work. Some reactions produce energy and release it as heat, whereas others require input of energy.
- Chemical bonds hold potential energy. When these bonds are broken, release of this energy is important to living systems.

Water is a biologically important molecule with many unique and crucial properties.

- Water is a polar molecule with the ability to form hydrogen bonds. Due to the four pairs of electrons in the outer shell of oxygen, water has a tetrahedral shape.
- Ice floats because of the less-dense pattern of hydrogen bonding that results when water molecules are frozen. Ice insulates ponds and lakes in the winter, allowing life to continue beneath the frozen layer.
- As ice melts, breaking the hydrogen bonds requires the input of lots of heat energy. In contrast, heat energy is lost during the freezing of water.
- Large bodies of water help moderate temperature on land and in the atmosphere. The amount of energy needed to raise the temperature of 1 gram of something by 1°C is known as the specific heat. Liquid water

has a high specific heat, which provides it with a high heat capacity and the ability to moderate temperatures.

- Water has a high heat of vaporization that requires the input of heat energy to break the hydrogen bonds and go from a liquid to a gaseous state. This makes evaporating water an effective coolant.
- The polar nature of water and the formation of hydrogen bonds contribute to the cohesive strength of water and its surface tension, both of which are biologically important.
- Solutions are formed when a substance is dissolved in water or another liquid. An aqueous solution has water as the solvent.

Quantitative terms are important for the understanding of both biology and chemistry.

- Concentration is the amount of a substance in a given amount of solution.
- A mole is the amount of a substance (in grams) that is numerically equal to its molecular weight.
- Avogadro's number relates the number of molecules of any substance to its weight and is 6.02×10^{23} molecules per mole. A 1 molar (1 *M*) solution is one mole of substance dissolved in water to make 1 liter.

Acids and bases are substances that dissolve in water and are biologically important.

- Acids dissolve in water and release H^+ ions. Bases dissolve in water and attach to H^+ ions.
- Acids and bases do not all ionize in the same way. Some ionize easily and completely and are called strong acids or bases. Others ionize only partially or reversibly and are called weak acids or bases.
- When strong acids and bases are ionized, the reaction is irreversible. Ionization of weak acids and bases can be reversible reactions.
- Water is a very weak acid that requires two water molecules for ionization into the base hydroxide ion and the acid hydronium ion. Though ionization of water is uncommon, it is biologically important.
- Solutions are referred to as acidic or basic, whereas compounds and ions are referred to as bases or acids.
- The pH of a solution refers to the concentration of H^+ ions and is measured as the $\log_{10}$ of the H^+ concentration, $pH = -\log_{10}[H^+]$. Lower pH indicates a higher H^+ concentration or a more acidic solution. A higher pH indicates a lower H^+ concentration or a more basic solution.
- Homeostasis is the maintenance of a constant internal environment. A buffer is a mixture of a weak acid and its corresponding base. Buffers prevent fluctuations in the pH of a solution and play an important role in maintaining pH homeostasis.

Test Yourself

Diagram Exercise

Draw the electron shells and place electrons in the appropriate shells based on the atomic numbers of the elements carbon ($_6$C), nitrogen ($_7$N), sodium ($_{11}$Na), chlorine ($_{17}$Cl), argon ($_{18}$Ar), and oxygen ($_8$O).
Textbook Reference: *2.1 How Does Atomic Structure Explain the Properties of Matter? p. 24*

Knowledge and Synthesis Questions

1. The atomic number of an element refers to the number of _______ in an atom.
   a. protons and neutrons
   b. protons
   c. electrons
   d. neutrons
   e. atomictrons
   Textbook Reference: *2.1 How Does Atomic Structure Explain the Properties of Matter? p. 22*
2. Which of the following statements concerning electrons is *false*?
   a. Electrons orbit the nucleus of an atom in defined orbitals.
   b. The outer shell of all atoms must contain eight electrons.
   c. An atom may have more than one valence shell.
   d. Electrons are negatively charged particles.
   e. All of the above are true.
   Textbook Reference: *2.1 How Does Atomic Structure Explain the Properties of Matter? pp. 23–24*
3. The element with which of the following atomic numbers would be most stable?
   a. 1
   b. 3
   c. 12
   d. 15
   e. 18
   Textbook Reference: *2.1 How Does Atomic Structure Explain the Properties of Matter? p. 25*
4. What is the difference between an element and a molecule?
   a. Molecules may be composed of different types of atoms, whereas elements are always composed of only one type of atom.
   b. Molecules are composed of only one type of atom, whereas elements are composed of different types of atoms.
   c. Molecules are elements.
   d. Molecules always have larger atomic weights than elements.
   e. Molecules do not have electrons, whereas elements do.
   Textbook Reference: *2.1 How Does Atomic Structure Explain the Properties of Matter? p. 25*

5. The strongest chemical bonds occur when
   a. two atoms share electrons in a covalent bond.
   b. two atoms share electrons in an ionic bond.
   c. hydrogen bonds are formed.
   d. van der Waals forces are in effect.
   e. there are hydrophobic interactions.

   Textbook Reference: *2.2 How Do Atoms Bond to Form Molecules? p. 26*

6. You have discovered that a molecule is hydrophilic. What else do you know about this molecule?
   a. It cannot form hydrogen bonds.
   b. It is a polar molecule.
   c. It is a nonpolar molecule.
   d. It has a partial positive region and a partial negative region.
   e. Both b and d

   Textbook Reference: *2.2 How Do Atoms Bond to Form Molecules? p. 29*

7. The stability of the three-dimensional shape of many large molecules is dependent on
   a. covalent bonds.
   b. ionic bonds.
   c. hydrogen bonds.
   d. van der Waals attractions.
   e. hydrophobic interactions.

   Textbook Reference: *2.2 How Do Atoms Bond to Form Molecules? p. 27*

8. The molecular weight of glucose is 180. If you added 180 grams of glucose to a 0.5 liter of water, what would be the molarity of the resulting solution? (See Figure 2.2 for the periodic table.)
   a. 18
   b. 1
   c. 9
   d. 2
   e. 0.5

   Textbook Reference: *2.4 What Makes Water So Important for Life? p. 33*

9. Why does ice float in water?
   a. Ice is less dense than water.
   b. There are no hydrogen bonds in ice.
   c. Ice is denser than water.
   d. Water has a higher heat capacity than ice.
   e. Ice has more covalent bonds than water.

   Textbook Reference: *2.4 What Makes Water So Important for Life? pp. 31–32*

10. Cola has a pH of 3; blood plasma has a pH of 7. The hydrogen ion concentration of cola is _______ than the hydrogen ion concentration of blood plasma.
    a. 4 times greater
    b. 4 times smaller
    c. 400 times greater
    d. 10,000 times greater
    e. 30,000 times greater

    Textbook Reference: *2.4 What Makes Water So Important for Life? p. 34*

11. If solution A has a pH of 2 and solution B has a pH of 8, which of the following statements is true?
    a. A is basic and B is acidic.
    b. A is acidic and B is basic.
    c. A is a base and B is an acid.
    d. A has a greater $[OH^-]$ than B.
    e. None of the above

    Textbook Reference: *2.4 What Makes Water So Important for Life? p. 34*

12. One mole of glucose ($C_6H_{12}O_6$) weighs
    a. 180 grams.
    b. 42 atomic mass units.
    c. 96 grams.
    d. 342 grams.
    e. 6.02 grams.

    Textbook Reference: *2.4 What Makes Water So Important for Life? p. 33*

13. The role of a buffer is to
    a. allow the pH of a solution to vary widely.
    b. make a solution basic.
    c. maintain pH homeostasis.
    d. disrupt pH homeostasis.
    e. make a solution more acidic.

    Textbook Reference: *2.4 What Makes Water So Important for Life? pp. 34–35*

14. Which of the following statements about water is true?
    a. Water has a low heat of vaporization.
    b. Water has a high specific heat.
    c. When water freezes, it gains energy from the environment.
    d. All of the above
    e. None of the above

    Textbook Reference: *2.4 What Makes Water So Important for Life? p. 32*

15. Which of the following statements about chemical reactions is true?
    a. The bonding partners of atoms remain constant.
    b. All reactions release energy as they proceed.
    c. The bonding partners of atoms change.
    d. All reactions consume energy as they proceed.
    e. None of the above

    Textbook Reference: *2.3 How Do Atoms Change Partners in Chemical Reactions? p. 30*

Application Questions

1. Calcium has an atomic number of 20. Draw structures for Ca and Ca^{2+}. What is the difference between these structures? Why is the most common ion of lithium Li^+?

   Textbook Reference: *2.2 How Do Atoms Bond to Form Molecules? p. 28*

2. Nitrogen atoms can form triple bonds with each other. What is a triple bond? How many electrons are shared between two N atoms?

   Textbook Reference: *2.2 How Do Atoms Bond to Form Molecules? p. 27*

3. Water is a polar molecule. This property contributes to cohesion and surface tension. Draw six water molecules. In your drawing, indicate how hydrogen bonding between molecules contributes to cohesion and surface tension. (Be sure to include appropriate covalent bonds in each molecule.)
 Textbook Reference: *2.2 How Do Atoms Bond to Form Molecules? p. 29*
4. Rank the following solutions in order from the most acidic to the most basic: lemon juice, pH = 2; Mylanta, pH = 10; Sprite, pH = 3; drain cleaner, pH = 15; seawater, pH = 8. Of the preceding, which has the highest concentration of H^+ ions? Which has the lowest concentration of H^+ ions?
 Textbook Reference: *2.4 What Makes Water So Important for Life? p. 34*
5. If you have 12 moles of a substance, how many molecules do you have of that substance? Suppose the substance has a molecular weight of 342. How many grams of that substance would you have to dissolve in a liter of water to make a 12 *M* solution?
 Textbook Reference: *2.4 What Makes Water So Important for Life? p. 33*

Answers

Diagram Exercise Answer

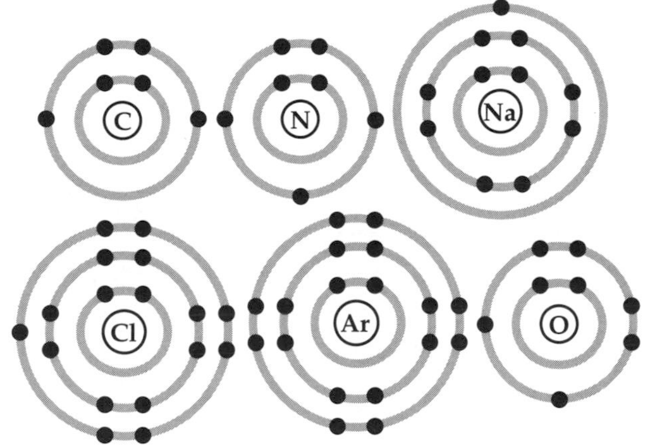

Knowledge and Synthesis Answers

1. **b.** The atomic number refers to the number of protons in an atom. The atomic weight can vary with the isotope and is the total of the number of protons and neutrons in an atom.
2. **b.** The first energy shell requires only two electrons to be full; subsequent shells may hold eight or more electrons.
3. **e.** The element with atomic number 18 (Argon) is the most stable because its outermost electron shell is full.
4. **a.** An element is always composed of only one type of atom. Molecules contain many types of atoms that are bonded together.
5. **a.** Covalent bonds, resulting from the sharing of a pair (or more) of electrons, form the strongest types of bonds. Ionic bonds, in which electrons are donated or received, are the second strongest. Hydrogen bonds are the weakest bonds and result from the attraction of partial charges to one another. Van der Waals forces are not bonds but attractions between nonpolar molecules. Hydrophobic interactions involve nonpolar substances and water.
6. **e.** Hydrophilic molecules are polar and have a partial positive and a partial negative pole. These molecules are said to be hydrophilic because they easily form hydrogen bonds with water.
7. **c.** Hydrogen bonds, though weak individually, are quite effective in large numbers and are responsible for maintaining the structural integrity of many large molecules such as proteins and DNA.
8. **d.** The molarity of the solution is 2. A 1 *M* solution is made by adding 180 grams to 1 liter of water. Adding the same amount to half a liter would result in a solution with twice the molarity.
9. **a.** When water forms ice, it becomes more structured. However, the resulting form is less dense than water in the liquid state, allowing ice to float on water.
10. **d.** pH is measured on a logarithmic scale. Each increase or decrease in pH value is by a factor of 10. Therefore, a difference between pH 3 and pH 7 would be a difference of $10 \times 10 \times 10 \times 10$, or 10,000.
11. **b.** A solution with a pH of 2 is acidic, and one with a pH of 8 is basic. The lower the pH value, the higher the concentration of H^+ ions.
12. **a.** Each carbon atom weighs 12 atomic mass units, each hydrogen atom is 1 atomic mass unit, and each oxygen is 16 atomic mass units. Adding these yields a molecular weight of 180 atomic mass units. Because one mole is equal to the atomic weight in grams, one mole of glucose equals 180 grams.
13. **c.** Buffers are solutions that contain a weak acid (carbonic acid) and the corresponding base (bicarbonate ions). Buffers help maintain the pH of a solution when either an acid or base is added.
14. **b.** Water has both a high heat of vaporization and a high specific heat. In addition, when water freezes it gives off a lot of energy to the environment.
15. **c.** Chemical reactions involve the combining or changing of bonding partners of the reactants to produce the products. During this process matter is neither created or destroyed. These reactions can either release energy or require energy.

Application Answers

1.

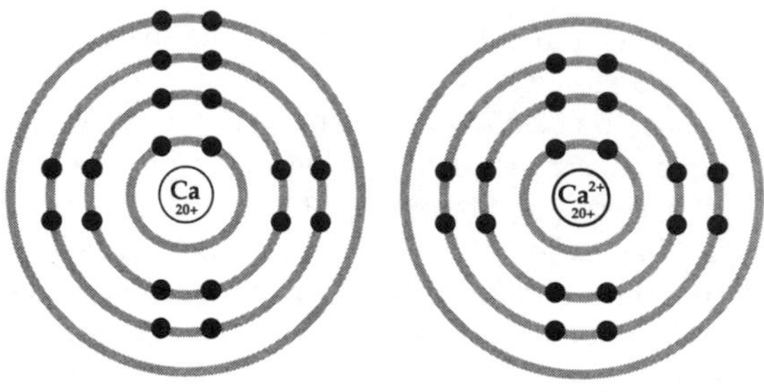

The difference between these structures is that calcium (Ca) has two electrons in its outer shell. Calcium ion (Ca^{2+}) has lost its two outer electrons and therefore has a positive charge of 2 because it has two more protons than electrons. Lithium has only one electron in its outer shell; when that electron is lost, the ion gains a positive charge of only 1.

2. Three pairs of electrons are shared between two N atoms in a triple bond.

3.

Hydrogen bond

The partially positive hydrogens of one water molecule are attracted to the partially negative oxygens of another molecule of water. This attraction tends to cause water molecules to "stick" together, creating surface tension.

4. Lemon juice, pH = 2 (highest concentration of H^+ ions); Sprite, pH = 3; seawater, pH = 8; Mylanta, pH = 10; drain cleaner, pH = 15 (lowest concentration of H^+ ions).

5. You would multiply Avogadro's number (6.02×10^{23}) times 12 to determine the number of molecules. This gives you 7.224×10^{24} molecules of this substance. A 12 *M* solution would be made by multiplying the formula weight of the molecule times 12 and adding that many grams of the substance to one liter of water. The formula weight of this molecule is 342, so you would need 4104 g.

3 Proteins, Carbohydrates, and Lipids

The Big Picture

- Macromolecules provide for specific biological activities and contribute to the cell's energy and information storage, structure, catalytic activity, and other functions. There are four basic classes of biologically important macromolecules: proteins, carbohydrates, nucleic acids, and lipids. All macromolecules are synthesized from smaller molecules through condensation and can be broken down through hydrolysis.
- Macromolecules all have regular primary structures held by covalent bonds, but many, especially proteins, can be folded into complex shapes held by ionic or hydrogen bonds or attractive forces.
- The side chains of amino acids contain functional groups that are important in determining the structure and function of a protein.

Common Problem Areas

- Many students are overwhelmed by the different types of macromolecules of biological systems. Remember that each macromolecule is synthesized from smaller components that give each class of molecule its specific function.

Study Strategies

- Make a table of macromolecule characteristics. This will help you see the patterns of similarity and the differences.
- Avoid the temptation to memorize structures. If you understand condensation and hydrolysis and know the basic components of the macromolecules, memorizing the structures of the large macromolecules is unnecessary.
- Go to yourBioPortal.com to review the following tutorial and activities:

  Animated Tutorial 3.1 Macromolecules

  Web Activity 3.1 Functional Groups

  Web Activity 3.2 Features of Amino Acids

  Web Activity 3.3 Forms of Glucose

Important Concepts

Molecules have specific properties that enhance their biological activity.

- Functional groups are small groups of atoms that confer specific properties to the molecules they are attached to. In chemical structures, the functional group is often shown bonded to an R (for "remainder") to indicate that the functional group can be attached to a wide variety of carbon skeletons.
- Isomers are molecules that have the same number and kinds of atoms but in which the atoms are arranged differently. In structural isomers, the order and binding of the atoms differ. In optical isomers, the carbon has four different groups or atoms attached to it in differing orders.

Macromolecules are very large polymers with a wide range of biological functions.

- All organisms are composed of four major biological macromolecules: proteins, nucleic acids, carbohydrates, and lipids. In all life forms the specific types of macromolecules share similar roles that are dependent on the chemical properties of their component monomers. Macromolecules provide energy storage, structural support, protection, catalysis, transport, defense, regulation, movement, and information storage.
- Polymers are made by the covalent bonding of monomers in a process called condensation (see Figure 3.4A). Polymers may be broken down into their constituent monomers through hydrolysis (see Figure 3.4B). The size and structure of each polymer is related to its biological function.

Proteins, which are polymers of amino acids, function in structural support, protection, transport, catalysis, defense, regulation, and movement.

- Proteins have many functions in an organism, including enzymatic activity, defense, hormonal regulation, storage, structural support, transport, and genetic regulation. Proteins range in size from small proteins with 50 to 60 amino acids to large proteins with more than 4,600 amino acids. Proteins may be made of single chains of amino acids or multiple interacting chains.

- Proteins are composed of 20 amino acids, each of which has a central carbon atom with a hydrogen atom, amino group, and carboxyl group attached to it.
- The shape and structure of proteins are influenced by the properties of the side chains of the constituent amino acids. Five amino acids have charged side chains, which attract water; five have polar, hydrophilic side chains that form hydrogen bonds; seven have hydrophobic, nonpolar side chains; and three have special hydrophobic functions.
  - The side chain of cysteine, –SH, can form disulfide bridges with other cysteines and influences protein chain folding.
  - Glycine is small and fits into folds of proteins.
  - Proline has limited bonding ability and often contributes to looping in proteins.
- Amino acids are chained or polymerized together during condensation. The resulting bonds between the amino and carboxyl groups are called peptide bonds or linkages (see Figure 3.6). Peptide chains always begin with the N terminus and end with the C terminus.
- The C–N peptide linkages are relatively rigid, limiting the extent of protein folding. Hydrogen bonds are favored within the folded proteins, helping maintain the form and function of proteins.

Proteins have four levels of structural organization.

- A protein's primary structure refers to the basic sequence of amino acids. The primary structure is held together by covalent bonds. All higher levels of structure are due to the specific amino acid sequence in the primary structure of the protein.
- Secondary structure refers to regular, repeated spatial patterns of the amino acid chain stabilized by hydrogen bonds. The α helix consists of a right-handed coiling of a single polypeptide chain. The β pleated sheet is formed from two or more hydrogen-bonded chains. A single chain can also fold over and form hydrogen bonds in a similar manner.
- Tertiary structure refers to an amino acid chain that is bent and folded into a more complex pattern. This structure is stabilized by disulfide bridges, hydrophobic interactions, van der Waals forces, ionic interactions, and hydrogen bonds. The sequence of R groups in the primary structure is responsible for this level of folding.
- Quaternary structure refers to the three-dimensional interactions of multiple protein subunits.

Protein shape and surface chemistry influence function.

- The ability of proteins to bond noncovalently to other molecules allows them to carry out their specific function in the cell.
- Binding sites in proteins have characteristic shapes, allowing for selective and specific interactions with other molecules. The particular orientation and position of chemical groups within the binding site and their ability to interact with other molecules are a result of the protein's three-dimensional shape and the properties of its constituent amino acids.
- Because a protein's three-dimensional structure is stabilized by relatively "weak" bonds, environmental conditions such as temperature, pH changes, or altered salt concentrations can change the shape of the protein into an inactive form. This is called denaturation.
- Chaperones are a class of proteins that prevent other proteins from denaturing by inhibiting inappropriate bonding. Heat shock proteins (HSPs) are a group of chaperones that are found in most eukaryotic cells.

Sugar polymers called carbohydrates are a primary energy source and make up the carbon skeleton in biological systems.

- Carbohydrates have the general formula $C_n(H_2O)_n$. They are used to store energy, transport energy, or as carbon skeletons.
- Monosaccharides are simple sugar monomers that are used in the synthesis of complex carbohydrates. More complex arrangements include disaccharides (2 monosaccharides), oligosaccharides (3–20 monosaccharides), and polysaccharides (large, complex carbohydrates).
- Simple sugars (monomers) are biologically important as an energy source (glucose) and as a structural backbone for RNA and DNA (ribose and deoxyribose, two pentoses).
- Simple sugars typically have three to six carbon atoms and may exist as different isomers.
- Dehydration synthesis between monomers results in glycosidic linkages. Oligosaccharides are composed of several monosaccharides that have been bound by glycosidic linkages.
- Polysaccharides are very large chains of monomers that provide energy storage or structural support. The specific structure of polysaccharides contributes to their function. For instance, starch, glycogen, and cellulose are all chains of glucose, but the glycosidic bonds are in different orientations and yield very different chemical properties.
- Added functional groups, such as a carboxyl group, give carbohydrates altered functions. Sugar phosphates and amino sugars, such as chitin, have significant biological importance.

Lipids are chemically diverse but water insoluble, and provide many biological functions.

- Lipids are insoluble in water due to many nonpolar covalent bonds. Lipids aggregate because of their nonpolar nature and are held together by van der Waals forces. They play a wide role in the biology of organisms; they provide energy storage, play structural roles, help plants capture light, play regulatory roles as hormones and vitamins, provide electrical and thermal insulation, and repel water.

- Fats and oils, composed of fatty acids and glycerol, are also known as triglycerides or simple lipids, and function primarily in the storage of energy. At room temperature, fats are solid and oils are liquid.
- Each triglyceride contains a glycerol bound to three fatty acids with carbon atoms that are all single-bonded (saturated fatty acids) or contain double bonds (unsaturated fatty acids). During condensation, three fatty acids are covalently linked to the glycerol by ester linkages. The characteristics of the carbon bonds of fatty acids (i.e., whether they are saturated or unsaturated) influences the shape of the molecule and determines its melting point. The greater the saturation of the fatty acid chains, the higher the melting point.
- Phospholipids form cellular membranes. In phospholipids, the hydrophobic fatty acid of a typical lipid is replaced with a hydrophilic phosphate group. This allows phospholipids to be amphipathic: they have a hydrophilic "water-loving" head and hydrophobic "water-hating" tails. In an aqueous environment, the hydrophobic tails of the phospholipids tend to aggregate, with the phosphate heads facing out. This bilayer effect allows for the establishment of a hydrophobic inside surrounded by an aqueous environment.
- Carotenoids and steroids are lipids that do not have the glycerol–fatty acid structure. Carotenoids trap light energy during photosynthesis. Steroids serve as signaling molecules and are synthesized from cholesterol.
- Some vitamins are lipids and are not synthesized in the human body. Vitamins A, D, E, and K are all lipid-derived.
- The nonpolar nature of lipids makes them water repellent. Wax coatings on hair, feathers, and leaves are all lipid-derived.

Test Yourself

Diagram Exercise

Draw a phospholipid and a bilayer. What characteristics of phospholipids make them perfectly suited for membranes? What do you think might happen if phospholipids did not form a bilayer? How might they arrange themselves in an aqueous environment?
Textbook Reference: *3.4 What Are the Chemical Structures and Functions of Lipids? p. 56*

Knowledge and Synthesis Questions

1. Which of the following statements about polymers is *false?*
   a. Polymers are synthesized from monomers during condensation.
   b. Polymers are synthesized from monomers during dehydration.
   c. Polymers consist of at least two types of monomers.
   d. Both a and c
   e. Both b and c
   Textbook Reference: *3.1 What Kinds of Molecules Characterize Living Things? p. 39*

2. A macromolecule with many hydrogen and peptide bonds is most likely a
   a. carbohydrate.
   b. lipid.
   c. protein.
   d. nucleic acid.
   e. vitamin.
   Textbook Reference: *3.2 What Are the Chemical Structures and Functions of Proteins? p. 44*

3. An α helix is an example of the _______ level of protein structure.
   a. primary
   b. secondary
   c. tertiary
   d. quaternary
   e. hepternary
   Textbook Reference: *3.2 What Are the Chemical Structures and Functions of Proteins? p. 46*

4. Which of the following statement about isomers is true?
   a. They all have different chemical formulas but the same arrangement.
   b. They are found only in proteins.
   c. They can only be structural.
   d. They all have the same chemical formula but different arrangements.
   e. None of the above
   Textbook Reference: *3.1 What Kinds of Molecules Characterize Living Things? p. 40*

5. Cellulose and starch are composed of the same monomers but have structural and functional differences. Which of the following is the characteristic that accounts for those differences?
   a. Different types of glycosidic linkages
   b. Different numbers of glucose monomers
   c. Different types of bonds holding them together
   d. A linear shape in one versus a ring shape in the other
   e. None of the above
   Textbook Reference: *3.3 What are the Chemical Structures and Functions of Carbohydrates? p. 53*

6. Which of the following statements about proteins is *false?*
   a. Enzymes are proteins.
   b. They are part of the phospholipid bilayer.
   c. Some hormones are proteins.
   d. They are structural components of the cell.
   e. All of the above are true of proteins.
   Textbook Reference: *3.2 What Are the Chemical Structures and Functions of Proteins? p. 42*

7. A disulfide bridge is formed by
   a. two cysteine side chains.
   b. two glycerol linkages.
   c. two proline side chains.
   d. condensation.
   e. hydrolysis.
   Textbook Reference: *3.2 What Are the Chemical Structures and Functions of Proteins? p. 44*

8. Triglycerides are synthesized from ______ and _______.
   a. glycerol; amino acids
   b. amino acids; cellulose
   c. steroid precursors; starch
   d. cholesterol; glycerol
   e. fatty acids; glycerol
   Textbook Reference: *3.4 What Are the Chemical Structures and Functions of Lipids? p. 54*
9. Proteins consists of amino acids linked together by
   a. noncovalent bonds.
   b. peptide bonds.
   c. phosphodiester bonds.
   d. van der Waals forces.
   e. Both a and b
   Textbook Reference: *3.2 What Are the Chemical Structures and Functions of Proteins? p. 44*
10. Which of the following characteristics distinguishes carbohydrates from other macromolecule types?
    a. Carbohydrates are constructed of monomers that always have a ring structure.
    b. Carbohydrates never contain nitrogen.
    c. Carbohydrates consist of a carbon bonded to hydrogen and a hydroxyl group.
    d. Carbohydrates contain glycerol.
    e. None of the above
    Textbook Reference: *3.3 What Are the Chemical Structures and Functions of Carbohydrates? pp. 49–50*
11. Which of the following statements about carbohydrates is *false*?
    a. Monomers of carbohydrates have six carbon atoms.
    b. Monomers of carbohydrates are linked together during dehydration.
    c. Carbohydrates are energy-storage molecules.
    d. Carbohydrates can be used as carbon skeletons.
    e. All of the above are true.
    Textbook Reference: *3.3 What Are the Chemical Structures and Functions of Carbohydrates? p. 50*
12. The R groups of amino acids located on the surface of protein molecules in the interior of biological membranes would be
    a. hydrophobic.
    b. hydrophilic.
    c. polar.
    d. able to form disulfide.
    e. electrically charged.
    Textbook Reference: *3.2 What Are the Chemical Structures and Functions of Proteins? p. 43; 3.4 What Are the Chemical Structures and Functions of Lipids? p. 55*
13. The characteristic of phospholipids that allows them to form a bilayer is their
    a. hydrophilic fatty acid tail.
    b. hydrophobic head.
    c. hydrophobic fatty acid tail.
    d. hydrophilic glycogen acid tail.
    e. All of the above
    Textbook Reference: *3.4 What Are the Chemical Structures and Functions of Lipids? p. 55*
14. A five-carbon sugar is known as a
    a. glutamine.
    b. glucose.
    c. hexose.
    d. pentose.
    e. None of the above
    Textbook Reference: *3.3 What Are the Chemical Structures and Functions of Carbohydrates? p. 50*
15. Which of the following statements about lipids is *false*?
    a. Lipids are a major component of the phospholipid bilayer.
    b. Lipids provide waterproofing for the surfaces of organisms.
    c. Steroid hormones are lipids.
    d. A number of vitamins are lipids.
    e. All of the above statements are true.
    Textbook Reference: *3.4 What Are the Chemical Structures and Functions of Lipids? pp. 56–57*

Application Questions

1. Differentiate between primary, secondary, tertiary, and quaternary structures as they relate to proteins. If a protein is immersed in an unfavorable pH solution, which structures are most likely to disassociate first, and why?
   Textbook Reference: *3.2 What Are the Chemical Structures and Functions of Proteins? pp. 45–47*
2. Dietary guidelines encourage people to stay away from saturated fats. What is meant by the term "saturated fat"? Why is this type of fat of more concern than unsaturated fats in the diet? What is the structural difference between saturated and unsaturated fats?
   Textbook Reference: *3.4 What Are the Chemical Structures and Functions of Lipids? pp. 54–55*
3. Complex carbohydrates should be a mainstay of one's diet. What properties of carbohydrates make them excellent food sources?
   Textbook Reference: *3.3 What Are the Chemical Structures and Functions of Carbohydrates? pp. 52–53*
4. Discuss how a protein's three-dimensional structure makes it perfect for acting as a carrier and receptor molecule. In what ways are proteins uniquely suited for this function compared to other macromolecules?
   Textbook Reference: *3.2 What Are the Chemical Structures and Functions of Proteins? p. 48*
5. How do the different chemical properties of amino acid R groups contribute to the final three-dimensional shape of the molecule?
   Textbook Reference: *3.2 What Are the Chemical Structures and Functions of Proteins? pp. 43–44*
6. Suppose that you have isolated a protein with the following amino acid sequence: RSCFLA. Refer to the diagrams in Table 3.1 of the textbook, and draw this protein. In your drawing, label the N terminus and the C terminus and show all peptide linkages. How many water molecules are generated in the synthesis of this protein?
   Textbook Reference: *3.2 What Are the Chemical Structures and Functions of Proteins? p. 43*

7. Consider the following triglycerides (***A*** and ***B***).

$$\begin{array}{ll}
CH_3(CH_2)_{16}-\overset{O}{\overset{\|}{C}}-O-CH_2 & CH_3(CH_2)_3-\overset{O}{\overset{\|}{C}}-O-CH_2 \\
CH_3(CH_2)_{16}-\overset{O}{\overset{\|}{C}}-O-CH_2 & CH_3\text{-}HC{=}CH\text{-}CH_2-\overset{O}{\overset{\|}{C}}-O-CH_2 \\
CH_3(CH_2)_{16}-\overset{O}{\overset{\|}{C}}-O-CH_2 & CH_3\text{-}HC{=}CH\text{-}CH_2-\overset{O}{\overset{\|}{C}}-O-CH_2 \\
\qquad\qquad A & \qquad\qquad B
\end{array}$$

a. In ***B***, circle the remnant of the glycerol portion of the triglyceride.

b. Which triglyceride (***A*** or ***B***) is probably a solid at room temperature? Explain your answer.

c. Which triglyceride (***A*** or ***B***) is probably derived from a plant? Explain your answer.

d. How many water molecules result from the formation of triglyceride **B** from glycerol and three fatty acids?

Textbook Reference: *3.4 What Are the Chemical Structures and Functions of Lipids? p. 55*

Answers

Diagram Exercise Answer

Phosphatidylcholine

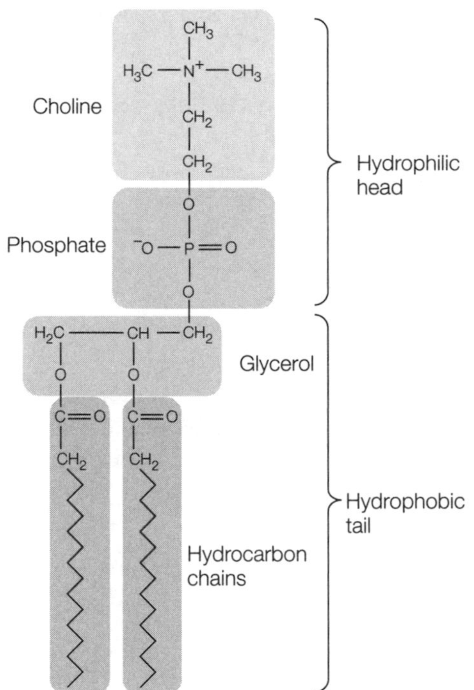

Phospholipid bilayer

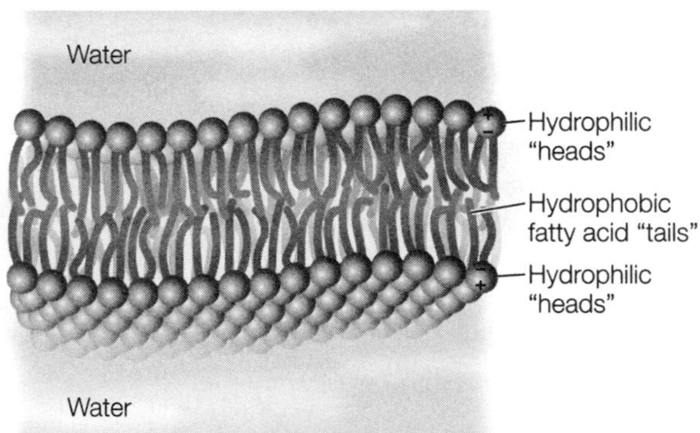

The hydrophilic head and the hydrophobic tail of phospholipids allow them to have an "inside" that resists an aqueous environment and an "outside" that can reside in such an environment. When they exist as a bilayer, the hydrophobic tails aggregate. If they did not exist in two layers, the tails would still try to aggregate.

This would result in a spherical aggregation of phospholipids called a micelle in which the tails are arranged toward the center of the sphere, away from the aqueous environment, and the heads are immersed in the aqueous environment.

Knowledge and Synthesis Answers

1. **e.** Polymers are synthesized via condensation and broken down via hydrolysis. They may be made of a single type of monomer or from multiple types of monomers.
2. **c.** Proteins consist of amino acids joined by peptide bonds. Higher levels of structure are stabilized by hydrogen bonds.
3. **b.** α helices are examples of secondary structure and are maintained by hydrogen bonds between the amino acid residues.

4. **d.** Proteins, carbohydrates, or lipids can have the same chemical formula, but the isomer molecules are arranged differently.
5. **a.** Starch has β-glycosidic linkages, whereas cellulose has α-glycosidic linkages. This difference accounts for structural and functional differences between the two macromolecules.
6. **b.** Proteins do not make up the phospholipid bilayer, which is composed of phospholipids. Proteins do perform all of the other functions.
7. **a.** Disulfide bridges are formed when the –SH groups of two cysteines interact.
8. **e.** Triglycerides are formed from one glycerol and three fatty acid molecules.
9. **b.** The peptide bond that links amino acids to form proteins is a type of covalent bond.
10. **c.** Carbohydrates always have carbon atoms bonded to hydrogen atoms and hydroxyl groups. They may have a variety of other associate molecules in addition to these.
11. **a.** Carbohydrate monomers may have different numbers of carbon atoms. Five-carbon monomers (pentoses) and six-carbon monomers (hexoses) are both common.
12. **a.** The interior of the plasma membrane is hydrophobic; therefore, an embedded protein would have hydrophobic residues.
13. **c.** Phospholipids are composed of a hydrophilic head and a hydrophobic tail. When they are placed in water, the hydrophobic tails come together in the interior of the bilayer, surrounded by the hydro-philic heads facing outward.
14. **d.** Five-carbon sugars are known as pentoses and form the backbones for RNA and DNA.
15. **e.** Lipids play a role in all of the functions listed.

Application Answers

1. See Figure 3.7 in the textbook for diagrams of primary, secondary, tertiary, and quaternary structures. The quaternary structure is the least stable and breaks down first in unfavorable conditions. Protein structures continue to denature by unfolding the tertiary structures, then the secondary structures. Primary structure is maintained by covalent bonds and is the last to break down. Disruptions in tertiary and quaternary structures are often reversible. Disruptions in primary or secondary structures are generally irreversible.
2. Saturated fats contain only single bonds and are "saturated" with hydrogen, which allows fat molecules to pack together densely. This characteristic is easily seen in that most saturated fats are solid at room temperature. Unsaturated fats contain double bonds that affect the shape of the molecule and are not "saturated" with hydrogen. This characteristic keeps them from packing together tightly, and they tend to be liquid at room temperature. Saturated fats are dangerous because of this "packing" ability.
3. Complex carbohydrates are easily broken down into glucose monomers, which provide nearly all cellular energy. By storing glucose monomers in large carbohydrates, the osmotic strain on any given cell is reduced without sacrificing availability of energy.
4. The three-dimensional nature of proteins allows them to form binding sites. These binding sites are uniquely shaped to interact with other molecules.
5. The size of the R group, the charge of the R group, and any special binding properties all contribute to the final orientation of a protein molecule.
6. Five water molecules are generated.

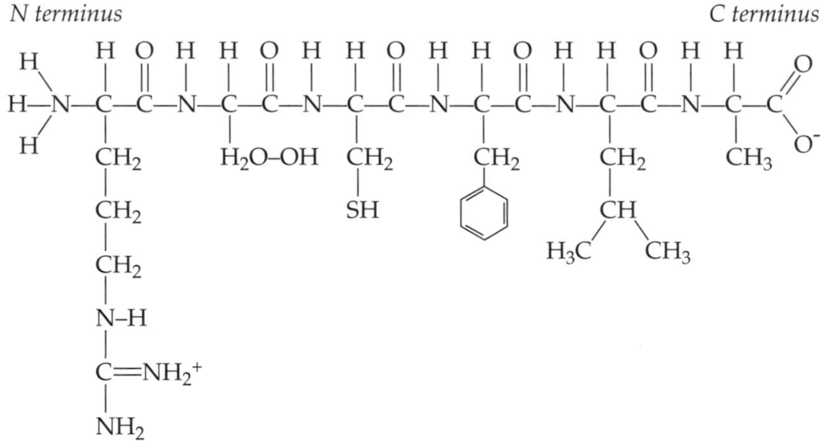

N terminus *C terminus*

```
      H O H  H O H  H O H  H O H  H O H  H    O
H     | || |  | || |  | || |  | || |  | || |   //
 \N—C—C—N—C—C—N—C—C—N—C—C—N—C—C—N—C—C
H/  |        |       |       |       |       |   \O-
    CH2    H2O-OH   CH2     CH2     CH2     CH3
    |                |      (phenyl)  |
    CH2              SH             CH
    |                              /  \
    CH2                          H3C   CH3
    |
    N-H
    |
    C=NH2+
    |
    NH2
```

7. a.

$$CH_3(CH_2)_3\text{—}\overset{O}{\overset{\|}{C}}\text{—O—}CH_2$$
$$CH_3\text{–}HC\text{=}CH\text{–}CH_2\text{—}\overset{O}{\overset{\|}{C}}\text{—O—}CH$$
$$CH_3\text{–}HC\text{=}CH\text{–}CH_2\text{—}\overset{O}{\overset{\|}{C}}\text{—O—}CH_2$$

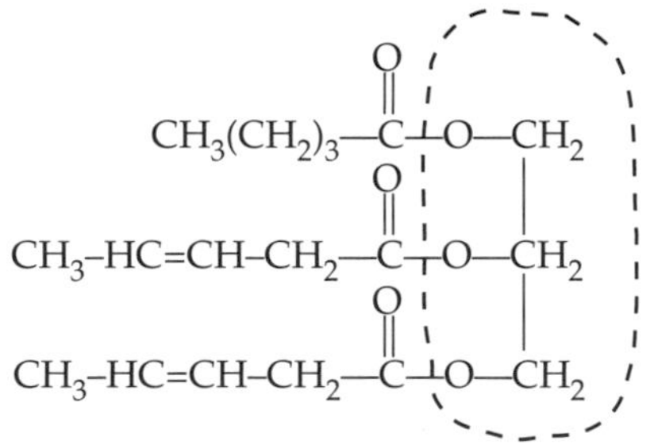

b. Triglyceride ***A*** is probably solid at room temperature. Its fatty acid chains are saturated (no double bonds) and relatively long; both characteristics of solid, animal-derived triglycerides.

c. Triglyceride ***B*** is probably derived from a plant. Its fatty acid chains are unsaturated (double bonds) and relatively short; both characteristics of liquid, plant-derived triglycerides.

d. Three water molecules will result. One water molecule results for each of the three fatty acids added to glycerol by a condensation reaction.

4 Nucleic Acids and the Origin of Life

The Big Picture

- Sequences of nucleic acids are found as either single-stranded RNA or double-stranded DNA and contain the genetic information needed to code for the proteins required for life. The "central dogma" of molecular biology states that DNA can reproduce itself exactly and that the information stored in DNA is transcribed into RNA and then translated into specific proteins.
- The origin of small molecules required for life first occurred in water. The first theory for the origin of life involves meteorites containing the molecules required for life. Alternatively, the atmospheric environment of early Earth may have been conducive to the formation of amino acids, nucleic acid bases, and carbon sugars. Larger polymers, such as proteins and nucleic acids, may have evolved in concentrated droplets with the help of an external energy source for polymerization of reactions or as nucleic polymers with catalytic functions. The cell may have evolved from a simple protocell.

Common Problem Areas

- Students are sometimes confused by the different theories regarding the origin of life. Remember that the monomers required for life had to arise first, and this was followed by the ability to replicate and carry out metabolic processes.

Study Strategies

- Use the figures in your textbook to help you understand the structural relationship between the different nucleotides, DNA, RNA, and proteins.
- Go to yourBioPortal.com to review the following tutorials and activities:

  Animated Tutorial 4.1 Pasteur's Experiment

  Animated Tutorial 4.2 Synthesis of Prebiotic Molecules

  Web Activity 4.1 Nucleic Acid Building Blocks

  Web Activity 4.2 DNA Structure

Important Concepts

The nucleic acids (DNA and RNA) store, transmit, and use genetic information.

- The primary role of nucleic acids is to store and transmit hereditary information.
- Nucleic acids are made up of nucleotide monomers. Each nucleotide monomer consists of a pentose (five-carbon) sugar, a phosphate group, and a nitrogenous base. DNA has one fewer oxygen atoms in the ribose sugar than RNA and is thus called deoxyribonucleic acid. DNA and RNA also differ in one of their nitrogenous bases.
- The backbone of DNA consists of alternating sugars and phosphates with the bases projecting outward. In both RNA and DNA, the nucleotides are joined by phosphodiester linkages between the nucleotide with the phosphate group. Most RNA is a single strand of nucleotides, whereas DNA is a double strand. The two strands in DNA are stabilized by hydrogen bonds and run in opposite directions.
- Nucleotide bases are capable of complementary base pairing. Only four types of nucleotides are found in DNA: two purines (adenine and guanine) and two pyrimidines (thymine and cytosine). In DNA, adenine and thymine pair, and cytosine and guanine pair. In RNA, thymine is replaced by uracil. Complementary base pairing is facilitated by the size and shape of the molecules, the geometry of the sugar–phosphate backbone, and hydrogen bonding. DNA is a double helix with the two complementary strands paired and twisted together.
- The sequences of bases in DNA provide a means of storing information in the cell. The "central dogma" of molecular biology states that DNA can replicate and be transcribed into RNA, which in turn is translated into specific polypeptides. Bond pairs for DNA (A-T and G-C) are replaced with their appropriate RNA counterpart (A-U and G-C) during transcription.
- The genome contains a complete set of an organism's DNA. Genes are the specific portions of the DNA that encode for proteins and are transcribed only when the cell requires the specific proteins.

- Similarities in base sequences of DNA can help determine evolutionary relationships among organisms because it is a conserved and regular molecule. Closely related species tend to have greater similarity in base sequences compared to distantly related species.
- Nucleotides have additional biological functions. ATP and GTP serve as energy sources. cAMP serves as a messenger in many cellular processes.

Two alternative theories exist for the beginning of life on Earth.

- In 1668, the Italian physician Francesco Redi conducted experiments with flies that showed that life does not arise by spontaneous generation. Louis Pasteur conducted similar experiments to show that microorganisms arise from other microorganisms.
- Approximately 4.6 billion years ago, the solar system was formed from an exploding star, producing our sun and other bodies known as planetesimals. The first signs of life on Earth date back 4 billion years. Water high in the atmosphere condensed on the surface of Earth as it cooled, and simple chemical reaction might have led to life. But comets and meteorites hitting Earth would have released large amounts of energy, and likely destroyed any initial molecules necessary for life.
- One theory of the origin of life on Earth is based on the idea that water and some molecules characteristic of life may have come from space. In 1969, a meteorite containing purines, pyrimidines, sugars, and ten amino acids was found in Murchison, Australia. A meteorite from Mars found in Antarctica, ALH 84001, contained water, simple carbon-containing molecules, and crystals of magnetite (iron oxide mineral made by living organisms on Earth).
- Another theory of the origin of life on Earth, known as chemical evolution, states that the conditions on Earth billions of years ago led to the emergence of biological molecules, either physically or chemically.
  - Originally, the atmosphere of Earth had very little oxygen, which began to accumulate only as single-celled organisms began to carry out photosynthesis about 2.8 million years ago.
  - Stanley Miller and Harold Urey produced a primitive atmosphere containing hydrogen gas, ammonia, methane gas, and water, and by means of electrical charges passed through these gasses (the so-called hot experiment) were able to produce the initial building blocks of life. This experiment provided support for the chemical origin of life, since all five nucleotide bases, 17 of 20 amino acids, and 3- to 6-carbon sugars were produced. While life relies on only the L-isomers, the amino acids produced were a mixture of D- and L-isomers, and it is possible that rocks on Earth segregated the two amino acid isomers, resulting in the elimination of D-amino acids. Additional experiments adding carbon dioxide, nitrogen, hydrogen sulfide, and sulfur dioxide have produced additional molecules required for life.
- Stanley Miller was also able to produce amino acids and nucleotide bases from ammonia gas, water vapor, and cyanide in a cold environment. These initial molecules were placed in a sealed tube, frozen for 25 years, and reopened. It is thought that the new molecules form within pockets of liquid water within the ice.

Formation of polymers may have been the result of chemical evolution.

- Two theories regarding the appearance of genetic material and proteins on Earth take opposite sides on the question.
  - According to the metabolism first theory, life began from chemical changes in droplets of water that concentrated their contents. The mineral pyrite could have served as an energy source for polymerization reactions and the eventual formation of nucleic acids and proteins. From this, nucleic acid replication and enzymatic function could have evolved.
- The second theory suggests that nucleic acids were the first to evolve. From these initial random polymers, the ability to replicate and produce proteins arose. Problems with this model include the inability to reproduce nucleic acid polymers in prebiotic simulations and the fact that DNA is not self-catalytic. However, some RNAs, such as ribozymes, have the ability to act as a catalyst. RNA contains genetic code, making it a potential precursor for the evolution of DNA.

Evolution of the first cells involved the formation of a cell membrane.

- The evolution of the cell membrane provided a barrier between the internal and external environments of a cell. It is possible that the amphipathic nature of fatty acids may have formed a lipid bilayer protocell that trapped water and other molecules.
- Large molecules such as DNA and RNA cannot cross this protocell bilayer, but single nucleic acids and sugars can.
- Thus, if a self-replicating RNA strand were to be in the protocell, it could produce new polynucleotide chains using external nucleic acids.

Test Yourself

Diagram Exercise

Two different models have been proposed for the origin of living cells: the chemical (metabolism first) model and the replicator first model. Diagram the two proposed pathways, beginning your diagram with the prebiotic synthesis of the monomers.

Textbook Reference: *4.3 How Did the Large Molecules of Life Originate? p. 70*

Knowledge and Synthesis Questions

1. DNA utilizes the bases guanine, cytosine, thymine, and adenine. In RNA, _______ is replaced by _______.
   a. adenine; arginine
   b. thymine; uracil
   c. cytosine; uracil
   d. cytosine; arginine
   e. cytosine; thymine
   Textbook Reference: *4.1 What Are the Chemical Structures and Functions of Nucleic Acids? p. 61*
2. The pairing of purines with pyrimidines to create a double-stranded DNA molecule is called
   a. complementary base pairing.
   b. phosphodiester linkages.
   c. antiparallel synthesis.
   d. dehydration.
   e. the genome.
   Textbook Reference: *4.1 What Are the Chemical Structures and Functions of Nucleic Acids? p. 63*
3. The end product of transcription is _______ and the end product of translation is _______.
   a. proteins; DNA
   b. proteins; RNA
   c. RNA; DNA
   d. DNA; RNA
   e. RNA; proteins
   Textbook Reference: *4.1 What Are the Chemical Structures and Functions of Nucleic Acids? p. 64*
4. Which of the following is *not* a nucleotide?
   a. Adenosine triphosphate
   b. Guanosine triphosphate
   c. Cyclic adenosine monophosphate
   d. Thymine
   e. All of the above are nucleotides.
   Textbook Reference: *4.1 What Are the Chemical Structures and Functions of Nucleic Acids? p. 64*
5. An individual human's genome
   a. consists only of the genes that code for specific proteins.
   b. is identical for every individual.
   c. is the complete set of the individual's DNA.
   d. is the complete set of the individual's RNA.
   e. None of the above
   Textbook Reference: *4.1 What Are the Chemical Structures and Functions of Nucleic Acids? p. 64*
6. Francesco Redi's experiment with flies disproved the concept of
   a. the genome.
   b. translation.
   c. the "central dogma."
   d. spontaneous generation.
   e. the protocell.
   Textbook Reference: *4.1 What Are the Chemical Structures and Functions of Nucleic Acids? p. 64*
7. DNA replication, transcription, and translation are all part of the concept of
   a. hydrolysis.
   b. the "central dogma."
   c. the genome.
   d. nucleic acids.
   e. the double helix.
   Textbook Reference: *4.1 What Are the Chemical Structures and Functions of Nucleic Acids? p. 64*
8. The double-helix formation of DNA is caused by
   a. ionic bonds.
   b. covalent bonds.
   c. hydrogen bonds.
   d. hydrophobic side chains.
   e None of the above
   Textbook Reference: *4.1 What Are the Chemical Structures and Functions of Nucleic Acids? p. 64*
9. Life is thought to have originated
   a. on land.
   b. in the air.
   c. underground.
   d. in water.
   e. All of the above
   Textbook Reference: *4.2 How and Where Did the Small Molecules of Life Originate? p. 65*
10. Which of the following were *not* synthesized in the experiments conducted by Miller and Urey?
    a. Uracil
    b. Amino acids
    c. 6-carbon sugars
    d. 3-carbon sugars
    e. 5-carbon sugars
    Textbook Reference: *4.2 How and Where Did the Small Molecules of Life Originate? p. 68*
11. The theory that life may have come from outside of Earth assumes that _______ brought life to Earth.
    a. protocells
    b. the moon
    c. volcanic eruptions
    d. meteorites
    e. All of the above
    Textbook Reference: *4.2 How and Where Did the Small Molecules of Life Originate? p. 66*
12. The first biological catalyst may have been similar to
    a. a ribozyme.
    b. DNA.
    c. a protein.
    d. glucose.
    e. a lipid.
    Textbook Reference: *4.3 How Did the Large Molecules of Life Originate? p. 71*
13. Prebiotic, water-filled structures that are defined by a lipid bilayer are known as
    a. lipases.
    b. protocells.
    c. genomes.
    d. neocells.
    e. lipocells.

Textbook Reference: 4.4 *How Did the First Cells Originate? p. 72*

14. In the origin of life, _______ may have evolved first, followed by _______.
    a. proteins; amino acids
    b. DNA; RNA
    c. RNA; DNA
    d. RNA; nucleic acids
    e. All of the above appeared at the same time.

    Textbook Reference: *4.3 How Did the Large Molecules of Life Originate? p. 70*

15. In DNA and RNA, nucleotides are joined by
    a. phosphodiester linkages.
    b. hydrogen bonds.
    c. peptide linkages.
    d. glycosidic linkages.
    e. All of the above

    Textbook Reference: *4.1 What Are the Chemical Structures and Functions of Nucleic Acids? p. 61*

Application Questions

1. What properties of a DNA structure make it well suited for its function as an informational molecule?

   Textbook Reference: *4.1 What Are the Chemical Structures and Functions of Nucleic Acids? pp. 63–64*

2. In the diagram below, use the base-pairing rules for DNA and RNA to label the strand of RNA (right) that is complementary to the single strand of DNA (left), where C = cytosine, G = guanine, A = adenine, T = thymine, and U = uracil. Then circle and label an example of a nucleotide and a nucleoside, showing the double-stranded DNA and RNA hybrid. Based on the orientation of the sugar molecules, label the four ends of the molecule as 3′ or 5′, showing the double-stranded DNA and RNA hybrid.

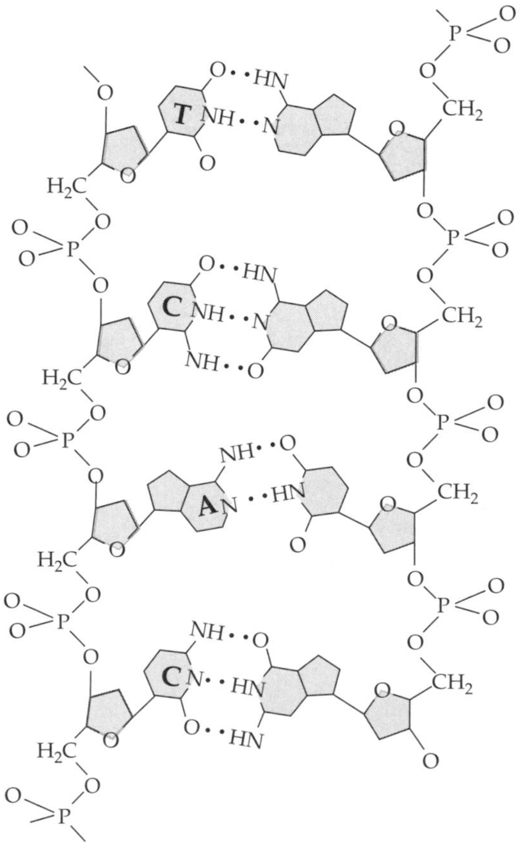

   Textbook Reference: *4.1 What Are the Chemical Structures and Functions of Nucleic Acids? pp. 62–64*

3. If you were to transcribe both strands of the DNA sequence below, what would be the two RNA sequences?

   AAGCGTC
   TTCGCAG

   Textbook Reference: *4.1 What Are the Chemical Structures and Functions of Nucleic Acids? p. 64*

4. What is the "central dogma" of molecular biology?

   Textbook Reference: *4.1 What Are the Chemical Structures and Functions of Nucleic Acids? p. 64*

5. One of the first steps leading to a cell that could reproduce may have been the formation of protocells. What characteristics may have allowed protocells to be the precursors to the living cell?

   Textbook Reference: *4.4 How Did the First Cells Originate? pp. 72–73*

Answers

Diagram Exercise Answer

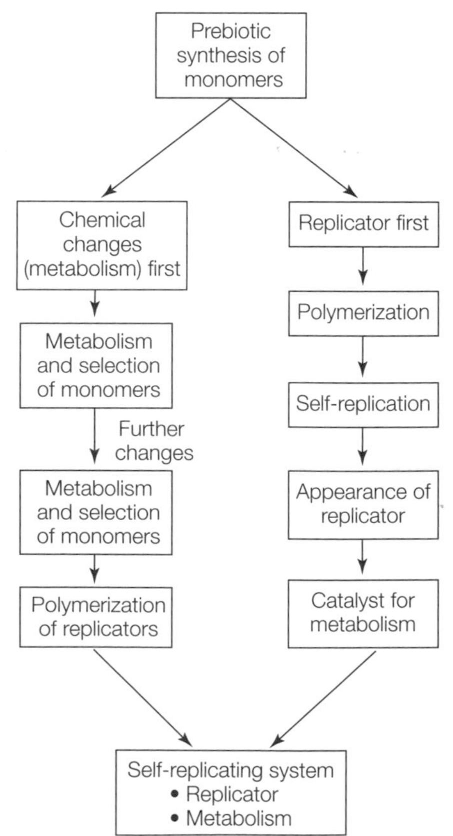

Knowledge and Synthesis Answers

1. **b.** RNA and DNA differ by one oxygen molecule in their ribose sugar and in the substitution of uracil (RNA) for thymine (DNA).
2. **a.** Complementary base pairing results from the attraction of charges and the ability of purines to pair with pyrimidines through hydrogen bond formation.
3. **e.** Transcription involves making RNA from DNA, and translation involves making proteins from RNA.
4. **e.** These are all examples of nucleotides. ATP and GTP both serve as energy sources. cAMP is involved in signal transduction of hormones and the nervous system.
5. **c.** A person's genome is the complete set of that individual's DNA, including the portions that code for genes and those that do not code for genes.
6. **d.** Redi was an Italian scientist who did experiments with flies and meat to show that life comes from other life, rather than arising by spontaneous generation.
7. **b.** The "central dogma" has two major ideas. First, DNA can replicate. Second, DNA is used as a template for transcription to make RNA, which then is used to make proteins.
8. **c.** The double-helix structure of DNA is due to the hydrogen bonding.
9. **d.** Life is thought to have originated in water.
10. **e.** The hot chemistry experiments conducted by Miller and Urey made the five nucleotide bases, 17 of 20 amino acids, and the 3- and 6-carbon sugars.
11. **d.** Molecules required for life may have hitched a ride on meteorites that come from outside Earth.
12. **a.** Ribozymes are catalytic RNA molecules and may have been the first catalytic molecules.
13. **b.** These prebiotic cells are known as protocells, and similar structures may have been involved in trapping water and other molecules necessary for the evolution of life.
14. **c.** The replicator first theory for the origin of life suggests that RNA may have evolved first, followed by DNA. This is plausible because RNA is both catalytic and stores information.
15. **a.** The sugar of one nucleotide and the phosphate of the next are joined by phosphodiester linkages in both DNA and RNA.

Application Answers

1. DNA structure is highly regular and conserved, which makes it usable by all cells, while the different bases allow it to encode specific genetic information.
2. 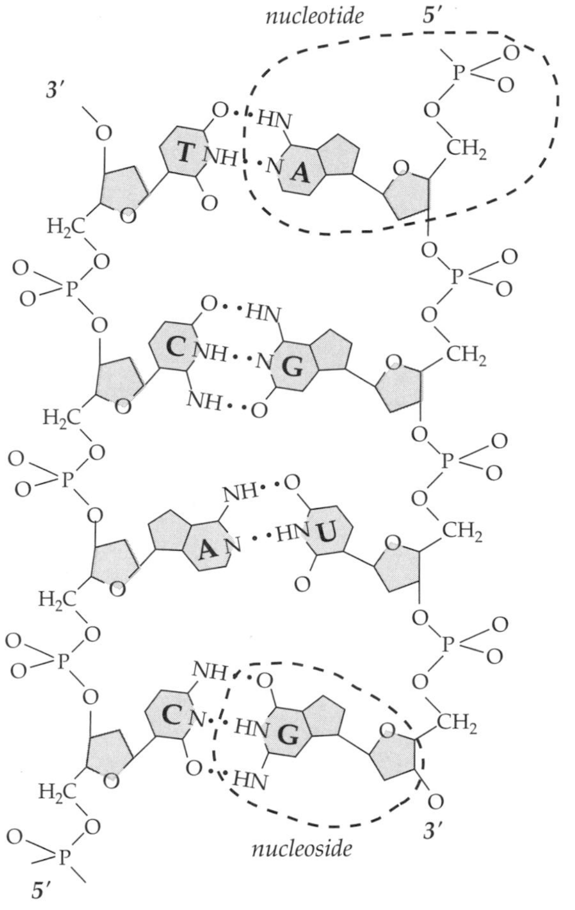
3. RNA sequence for top: UUCGCAG; RNA sequence for bottom: AAGCGUC.
4. The "central dogma" of molecular biology is the concept that DNA contains the genetic material necessary for life that can be replicated exactly. This genetic information is transcribed into RNA, which then produces proteins during translation. These proteins are based upon the specific nucleotide sequences contained in the DNA and subsequent RNA.
5. Protocells are lipid bilayers that allow for organization of molecules in an environment distinct from the external environment. Experiments have shown that sugars and nucleotides can enter these protocells, while larger molecules such as DNA and RNA cannot leave. In the presence of a self-replicating nucleic acid strand, they will produce polynucleotide chains. Thus, structures similar to protocells could have been the precursors to living cells.

5 Cells: The Working Units of Life

The Big Picture

- All living things are composed of cells, and all cells come from preexisting cells. There are two basic cell forms: the small prokaryotic cell with no organelles and the eukaryotic cell with many membrane-enclosed organelles.
- The organelles of eukaryotic cells have a variety of functions, including information storage and transmission, modification and packaging of cellular products, and energy transformation. See Figure 5.7 of the textbook to help you understand the organization, structure, and function of each component.
- The cytoskeleton of the cell provides support, structure, and protection. Components of the cytoskeleton also provide movement of the cell and within the cell.

Common Problem Areas

- Students often are overwhelmed by the number of organelles and their diverse functions in a eukaryotic cell. Remember that each organelle has a distinct location within the cell and that organelle's structure is related to its function.

Study Strategies

- Terminology is the most difficult part of this material. The terminology is easier to learn if you understand the concepts before you memorize the terms.
- Use Figure 5.7 while learning the organelles and their components. Visualizing the jobs each must do and how their functions relate to the structure makes learning easier.
- Go to yourBioPortal.com to review the following tutorials and activities:

  Animated Tutorial 5.1 The Golgi Apparatus

  Animated Tutorial 5.2 Eukaryotic Cell Tour

  Web Activity 5.1 The Scale of Life

  Web Activity 5.2 Know Your Techniques

  Web Activity 5.3 Lysosomal Digestion

Important Concepts

The cell is the building block of life.

- The cell theory states that the cell is the basic unit of life, that all organisms are composed of cells, and that all cells come from preexisting cells.
- Cell size is limited by surface area-to-volume ratio. Very large cells are not feasible because as volume increases, surface area increases at a slower rate. A cell's capacity for chemical activity is related to the cell volume. However, cells with large volumes are unable to take up enough material across their surface to support the metabolic activity of the volume.
- Because cells are small, either light microscopes or electron microscopes must be used to visualize them. Light microscopes, which use lenses and light, can resolve objects as small as 0.2 μm. Electron microscopes, which use electron beams, can resolve objects as small as 0.2 nm. Staining techniques are used to assist in visualizing cellular components.
- All cells are surrounded by a plasma membrane composed of a phospholipid bilayer with many embedded and protruding proteins. The plasma membrane and the associated proteins are selectively permeable, permitting some small molecules to pass through but preventing others from doing so. The membrane allows for maintenance of the internal cellular environment and is responsible for communication and interactions with other cells.

Cells are classified as either prokaryotic or eukaryotic.

- Prokaryotic cells do not have internal membrane compartments and are generally smaller than eukaryotic cells. Prokaryotes are one-celled organisms found only in domains Archaea and Bacteria.
- Prokaryotic cells have a plasma membrane and a nucleoid containing the genetic material that is not membrane-enclosed. The cytoplasm of the cell consists of the liquid cytosol and insoluble particles that act as subcellular machinery (i.e., ribosomes for protein synthesis).

- Some prokaryotic cells have special features:
  - Cell walls: This rigid structure is outside the cell membrane and is composed of peptidoglycans. Cell wall type is frequently used as a characteristic to identify bacteria.
  - Outer membrane: This membrane is rich in polysaccharides and located outside the cell wall.
  - Capsule: This slime layer is outside the wall and outer membrane. Though it provides protection, it is not necessary for the cell's survival.
  - Internal membranes: Photosynthetic prokaryotes have stacks of folded membranes on which photosynthesis is carried out. Other prokaryotes have membrane folds, which function in energy reactions and cell division.
  - Flagella: These tiny protein machines contain the protein flagellin and cause motion of the cell.
  - Pili: These are protein extensions that assist in adherence.
  - Cytoskeleton: Actin-like proteins may help the cell maintain its shape.
- Eukaryotic cells are larger than prokaryotic cells and are distinguished by their membrane-enclosed internal components such as the nucleus, mitochondrion, endoplasmic reticulum, Golgi apparatus, lysosome, vacuole, and chloroplast.
  - Each organelle (membrane-enclosed compartment) has its own internal environment uniquely suited to its function. The membrane assists in regulating that environment. Organelles and their functions can be studied by microscopy or by cell fractionation and organelle isolation.

Some organelles store and process information.

- Ribosomes found in the cytoplasm, on the surface of the endoplasmic reticulum, and in the mitochondria act as information transcription centers and guide the synthesis of proteins from nucleic acid blueprints.
- The nucleus stores DNA and is the site of DNA duplication and regulation of DNA transcription. The nucleolus region of the nucleus is involved in ribosome assembly and RNA synthesis.
- The nucleus is bounded by a double membrane called the nuclear envelope. Small openings in the nuclear envelope called nuclear pores allow passage of RNA and ribosomes to the cell cytoplasm. Proteins can move through the nuclear pores if they contain a short sequence of amino acids known as the nuclear localization signal (NLS). The outer membrane of the nuclear envelope is continuous with the endoplasmic reticulum.
- Inside the nucleus, DNA and proteins combine to make chromatin. At cell division, chromatin condenses to form chromosomes. Chromatin is organized on the nuclear matrix and the nuclear lamina. The protein nuclear lamina helps maintain nuclear shape.

The endomembrane system functions in manufacturing and packaging cellular products.

- The endomembrane system consists of endoplasmic reticulum (ER), Golgi apparatus, lysosomes, and vesicle shuttles.
- The endoplasmic reticulum is a complex of membrane sacs throughout the cell and is continuous with the nuclear membrane. The ER has a large surface area due to the large amount of folding of the membrane.
- The endoplasmic reticulum is classified into two types based on the presence or absence of attached ribosomes.
  - Rough endoplasmic reticulum (RER) is studded with active ribosomes. Because ribosomes actively aid in the synthesis of new proteins, the RER is important for storage, transport, and modification of proteins. Many of the cell's membrane bound-proteins are produced in the RER. Carbohydrate groups are added to proteins to make glycoproteins in the RER.
  - Smooth endoplasmic reticulum (SER) also functions in protein modification and transport, but its more important functions are in the modification of chemicals taken in by the cell, such as drugs and pesticides, and as the site of hydrolysis of glycogen and synthesis of steroids and lipids.
- The Golgi apparatus is an organelle composed of flattened membrane stacks called cisternae and vesicles. It contributes to further modification, packaging, and concentrating of proteins. In plants, it is the site of polysaccharide synthesis.
- The three different regions of the Golgi have different enzymes and functions. The *cis* region is closest to the nucleus or RER. The *trans* region is closest to the cell surface. Proteins are released from the ER in a vesicle and are transported to the *cis* region, where the vesicle membrane fuses with the Golgi membrane and the contents are released into the Golgi. Vesicles containing proteins pinch off of the *trans* Golgi for transport to the plasma membrane or lysosomes.
- Lysosomes are "digestion centers" within a cell that break down proteins, polysaccharides, nucleic acids, and lipids into their monomer components. They contain a host of powerful enzymes in a slightly acidic environment that break down engulfed molecules and cellular waste. In the process of phagocytosis, the cell takes up material in a phagosome, which fuses with a primary lysosome to make a secondary lysosome. Through the process of autophagy, lysosomes digest organelles such as mitochondria, breaking them down to monomers for reuse in new organelles.

Mitochondria and plastids function in energy transformation.

- During cellular respiration, mitochondria convert potential chemical energy stored in glucose into adenosine triphosphate (ATP), an energy form readily usable by the cell.

- Mitochondria have two membranes—an outer smooth membrane and a highly folded internal membrane. The folds are called cristae, and the remaining internal space is called the matrix. The matrix holds ribosomes and DNA. Protein complexes used during cellular respiration are embedded in the cristae.
- Plastids are found in plants and protists, but not in animal cells. Plant cells have several different types of plastids, each with unique functions.
- Chloroplasts are the site of photosynthesis where light energy is converted to chemical energy. Chloroplasts are green because they contain the photosynthetic pigment chlorophyll.
- Chloroplasts have two membranes. The innermost membrane contains circular compartments, or thylakoids, that are folded into stacks called grana and hold chlorophyll and enzymes for photosynthesis. The liquid content of the chloroplast is called stroma and contains ribosomes and DNA.
- Other plastid types include chromoplasts, which contain pigments involved in flower color, and starch-storing leucoplasts.

Other specialized organelles are found in some but not all cells.

- Peroxisomes function to break down harmful peroxide by-products.
- Glyoxysomes are found in young plants, and convert lipids to carbohydrates.
- Vacuoles are found in plants and protists and have various functions, depending on the organism, in storage of waste products, structural support, reproduction, food storage, or water regulation.

The cytoskeleton provides support, shape, and movement for the cell.

- Microfilaments, composed of the protein actin, stabilize cell shape and assist with contraction of the cell. These filaments are involved in muscle cell contraction, cell division, cytoplasmic streaming, and cell shape. In muscle cells the protein myosin interacts with actin to produce muscle contraction.
- Intermediate filaments are found only in multicellular organisms and function in stabilizing structure and resisting tension. These make up the nuclear lamina, form desmosomes, and anchor the nucleus.
- Microtubules are long hollow tubes of the dimer protein tubulin that contribute to the rigidity of the cell and act as a framework of movement for motor proteins. They have a very specific structure that can be quickly added to or reduced. Motor proteins use microtubules as tracks from one area of the cell to another.
- Cilia and flagella of eukaryotes are powered by microtubules. Though cilia and flagella differ in size and function, they have the same basic structure: a "9 + 2" arrangement of microtubules (see Figure 5.20).
- Movement of cilia and flagella occurs when the microtubules slide past one another. The sliding is caused by an ATP-driven shape change in molecules of dynein bound to the microtubules (see Figure 5.21). Motor proteins called kinesin are involved in the transfer of vesicles along microtubules within the cell.
- Centrioles are also made up of microtubules and are involved in the formation of the mitotic spindle during cell division.

Extracellular structures are outside the plasma membrane and provide support, protection, and anchoring for the cell.

- Plants have semirigid cell walls composed of cellulose, which function to support and protect the cell. Plasmodesmata are small holes in the cell walls that allow connections between plant cells.
- The collagen- and glycoprotein-containing extracellular matrix of some animal cells functions to hold cells together, filter materials, orient cell movement, and assist with chemical signaling. Some cells, such as bone cells, secrete an elaborate and rigid matrix.

Eukaryotic cells evolved through membrane folding and the endosymbiotic theory.

- The endomembrane system may have originated from folding of the plasma membrane. The interactions of the ER and nuclear envelope support this theory.
- Mitochondria and chloroplasts have their own DNA and are enclosed in a double membrane. They also reproduce within the cell. In other words, they act very much like cells within cells, but they are still under nuclear control.
- The endosymbiotic theory suggests that a small photosynthetic prokaryote was engulfed and retained by a larger cell, with each getting mutual benefit from the other (endosymbiosis). Mitochondria may have originated as an engulfed respiring prokaryote (see Figure 5.26).

Test Yourself

Diagram Exercise

The role of a certain cell in an organism is to secrete a protein. Create a flowchart in which you trace the production of that protein from the nucleus through all necessary organelles to the point of release from the cell. ***Textbook Reference:*** *5.3 What Features Characterize Eukaryotic Cells? pp. 89–91*

Knowledge and Synthesis Questions

1. A mass of cells is found in the sediment surrounding a thermal vent in the ocean floor. The salinity in the area is quite high. Microscopic examination of the cells

reveals no evidence of membrane-enclosed organelles. This cell would be classified as a
a. eukaryotic cell.
b. prokaryotic cell.
c. member of domain Archaea or Bacteria.
d. Both a and c
e. Both b and c
Textbook Reference: *5.2 What Features Characterize Prokaryotic Cells? p. 82*

2. Centrifugation of a cell results in the rupture of the cell membrane and the compacting of the contents into a pellet in the bottom of the centrifuge tube. Bathing this pellet with a glucose solution yields metabolic activity, including the production of ATP. One of the contents of this pellet is most likely which of the following?
a. Cytosol
b. Mitochondria
c. Lysosomes
d. Golgi bodies
e. Thylakoids
Textbook Reference: *5.3 What Features Characterize Eukaryotic Cells? p. 92*

3. Eukaryotic cells are thought to be derived from prokaryotic cells that underwent phagocytosis without digestion of the phagocytized cell. This mutualistic relationship is explained by the _______ theory.
a. endosymbiotic
b. cell
c. evolutionary
d. parasite
e. prokaryotic
Textbook Reference: *5.5 How Did Eukaryotic Cells Originate? p. 101*

4. Though science fiction has produced stories like "The Blob," we do not see very many large single-celled organisms. Which of the following tends to limit cell size?
a. The difficulty in maintaining a continuous large membrane
b. The difficulty of reproduction in a large cell
c. Surface area-to-volume ratios
d. All of the above
e. None of the above
Textbook Reference: *5.1 What Features Make Cells the Fundamental Units of Life? pp. 78–79*

5. Microscopes are used to resolve images that cannot be seen with the unaided eye. Electron microscopes use _______ to resolve images, whereas light microscopes use _______ to resolve images.
a. light and lenses; diffraction of electron beams
b. diffraction of electron beams; light and lenses
c. lasers; light and lenses
d. light and lenses; lasers
e. None of the above
Textbook Reference: *5.1 What Features Make Cells the Fundamental Units of Life? pp. 80–81*

6. What is the correct cellular function of the RER?
a. DNA synthesis
b. Photosynthesis
c. Cellular respiration
d. Protein synthesis
e. mRNA degradation
Textbook Reference: *5.3 What Features Characterize Eukaryotic Cells? p. 89*

7. Photosynthesis occurs in the
a. chloroplast.
b. mitochondria.
c. Golgi apparatus.
d. nucleus.
e. RER.
Textbook Reference: *5.3 What Features Characterize Eukaryotic Cells? p. 93*

8. Lysosomes are involved in
a. DNA synthesis.
b. the breakdown of phagocytized material.
c. protein folding.
d. pigment production.
e. cell membrane production.
Textbook Reference: *5.3 What Features Characterize Eukaryotic Cells? p. 91*

9. The packaging of proteins to be used outside the cell occurs in the
a. nucleus.
b. SER.
c. Golgi apparatus.
d. chromoplast.
e. nuclear pore.
Textbook Reference: *5.3 What Features Characterize Eukaryotic Cells? p. 90*

10. Which of the following organelles is *not* enclosed in a double membrane?
a. Nucleus
b. Chloroplast
c. Mitochondrion
d. RER
e. All of the above have double membranes.
Textbook Reference: *5.3 What Features Characterize Eukaryotic Cells? pp. 88, 92–93*

11. Movement of cells in both prokaryotes and eukaryotes is accomplished by which of the following structures?
a. Cilia
b. Pili
c. Dynein
d. Cell membranes
e. Flagella
Textbook Reference: *5.3 What Features Characterize Eukaryotic Cells? p. 98*

12. Which of the following statements about mitochondria and chloroplasts is true?
a. Animal cells produce chloroplasts.
b. Both mitochondria and chloroplasts may be found in the same cell.

c. Mitochondria and chloroplasts are not found in the same cell.
d. In certain conditions, chloroplasts can revert to mitochondria.
e. None of the above

Textbook Reference: *5.3 What Features Characterize Eukaryotic Cells? pp. 92–93*

13. Which of the following statements best describes ribosomes?
    a. Ribosomes guide protein synthesis.
    b. Ribosomes are found only in the nucleus or on the RER.
    c. There are no ribosomes in the mitochondria.
    d. Ribosomes are the site of photosynthesis.
    e. All of the above

    Textbook Reference: *5.3 What Features Characterize Eukaryotic Cells? p. 90*
14. Nuclear DNA exists as a complex of proteins called _______ that condenses into _______ during cellular division.
    a. chromosomes; chromatin
    b. chromatids; chromosomes
    c. chromophors; chromatin
    d. chromatin; chromosomes
    e. None of the above

    Textbook Reference: *5.3 What Features Characterize Eukaryotic Cells? p. 89*
15. Rough endoplasmic reticulum and smooth endoplasmic reticulum differ
    a. only by the presence (RER) or absence (SER) of ribosomes.
    b. both in the presence (RER) or absence (SER) of ribosomes and in their function.
    c. only in microscopic appearance.
    d. only in their function.
    e. None of the above

    Textbook Reference: *5.3 What Features Characterize Eukaryotic Cells? p. 89*

Application Questions

1. Explain how microtubules and dynein function to make cilia and flagella move.

   Textbook Reference: *5.3 What Features Characterize Eukaryotic Cells? pp. 97–98*
2. Compare and contrast prokaryotic and eukaryotic cells.

   Textbook Reference: *5.2 What Are the Characteristics of Prokaryotic Cells? 5.3 What Features Characterize Eukaryotic Cells? p. 82; 84*
3. Explain the significance of organelles. What are the costs and benefits of having large compartmentalized cells?

   Textbook Reference: *5.3 What Features Characterize Eukaryotic Cells? p. 84*
4. What is the primary function of a cell membrane? What characteristics of membranes allow them to contribute to metabolic activity?

   Textbook Reference: *5.1 What Features Make Cells the Fundamental Units of Life? p. 79*
5. The organelles that contain their own DNA are all enclosed in double membranes. Relate this observation to the endosymbiotic theory.

   Textbook Reference: *5.5 How Did Eukaryotic Cells Originate? p. 102*
6. There are structural similarities between mitochondria and chloroplasts. If we can assume that form follows function, what would be the explanation for the similarities between these two organelles?

   Textbook Reference: *5.3 What Features Characterize Eukaryotic Cells? pp. 92–93*

Answers

Diagram Exercise Answer

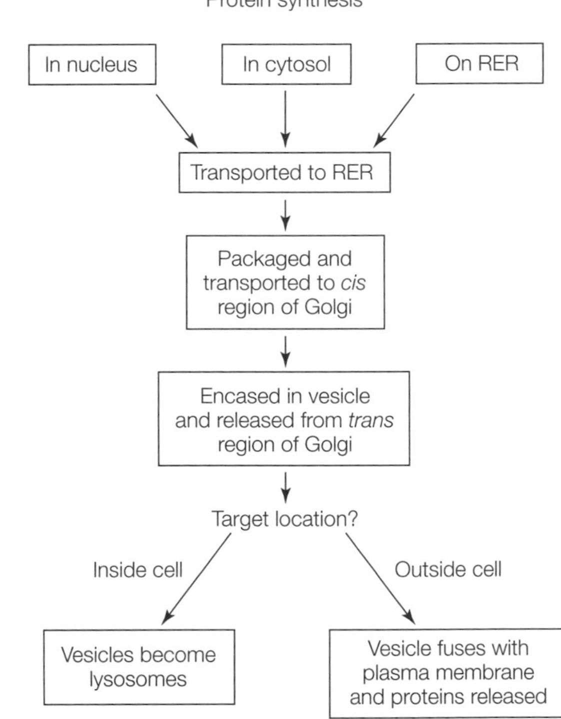

Knowledge and Synthesis Answers

1. **e.** Several characteristics suggest that this is a prokaryote. It survives in high salinity and high heat, although the sure indication is that it contains no membrane-enclosed organelles. Prokaryotes are in the domain Archaea and Bacteria.
2. **b.** The pellet is undergoing cellular respiration, a function that occurs in the mitochondria. You can also assume that if the single membrane of the cell itself is ruptured, other organelles enclosed in single membranes would be ruptured as well.
3. **a.** The endosymbiotic theory explains the presence of DNA in mitochondria and chloroplasts as well as the presence of two membranes around these organelles.

4. **c.** As volume increases, the surface area available for exchange does not increase proportionally. Eventually the surface is not large enough for maintenance of the metabolic activity of the cell.
5. **b.** In electron microscopy, a concentrated beam of electrons is focused on an object, allowing resolution of structures as small as 2 nm. Light microscopy, using lights and lenses, can only resolve objects down to approximately 0.2 μm.
6. **d.** The RER is the site of protein synthesis.
7. **a.** The chloroplasts are the organelles involved in photosynthesis.
8. **b.** Lysosomes are organelles that contain digestive enzymes used to break down macromolecules taken in by phagocytosis.
9. **c.** The Golgi apparatus packages proteins for both internal and external use.
10. **d.** The nucleus, the mitochondria, and the chloroplasts are the only organelles enclosed in double membranes.
11. **e.** Though the flagella are of different structures, they serve the same role in prokaryotes and eukaryotes.
12. **b.** Mitochondria and chloroplasts may be found in the same cell. Almost all eukaryotic cells contain mitochondria.
13. **a.** Ribosomes, found in the nucleus, the cytosol, and in organelles such as the mitochondria, RER, and chloroplasts, complex with RNA to guide protein production.
14. **d.** The complex of proteins and DNA is called chromatin. Chromatin takes the form of chromosomes only during cell reproduction.
15. **b.** Both the structure and the function of RER and SER differ.

Application Answers

1. Dynein molecules bind to pairs of microtubules in the flagella or cilia. With the addition of cellular energy, the dynein molecules undergo a conformational change that results in microtubules sliding past one another, resulting in a whiplike motion of the flagella.
2. Prokaryotic cells are small in size, have no membrane-enclosed organelles, are found only in domains Archaea and Bacteria, and have DNA in a nucleoid; Eukaryotic cells are 10 or more times greater in size, have membrane-enclosed organelles, are found in all domains other than Archaea and Bacteria, and have DNA in a nucleus.
3. Organelles allow different metabolic environments to exist in the same cell. This partitioning of jobs allows for greater specialization but comes at an energy cost. Eukaryotic cells are more energy expensive.
4. A cell membrane exists to form an inside and an outside of a cell. The presence of an inside and an outside allows for the establishment of different environments. In addition, membranes hold integral proteins with a variety of chemical properties and activities. This allows for the enzymatic activity associated with membranes. Stacks of membranes, such as those in mitochondria and chloroplasts, increase the amount of chemical activity in an area.
5. See Figure 5.26 for a description of the origin of double membranes from endosymbiosis.
6. Both mitochondria and chloroplasts are involved in energy-transformation activities that require many enzymes. The stacking or folding of membranes provides enzymatic activity centers for these reactions.

6 Cell Membranes

The Big Picture

- Cellular membranes are a dynamic composition of a phospholipid bilayer, integral and peripheral proteins, and carbohydrates. The nature of the constituent phospholipids allows the formation of a barrier that is semipermeable. Small hydrophilic molecules can traverse the membrane via simple diffusion, water may cross the membrane through aquaporins by osmosis, small charged ions may pass through protein channels, and some other molecules may be shepherded through by carrier proteins. Transport of molecules against their concentration gradient is active and requires energy input, either directly from ATP or as coupled to ATP-driven transport. Larger substances depend on endocytosis and exocytosis for transport into and out of the cell.
- In multicellular organisms, adhesion and cell-to-cell connections are important for structure and communication. Tight junctions allow multiple cells to form a sheet that directs movement of substances in a single direction. Desmosomes enhance structural support of sheets of cells that are prone to abrasion. Gap junctions allow communication by connecting adjacent cells.

Common Problem Areas

- In the study of membranes, the processes of diffusion and osmosis are the most difficult to understand. It is very easy to get the terminology confused, especially the terms " hypertonic" and " hypotonic." A consideration of the Latin roots of the words is helpful. "Hyper-" generally means excess, and "hypo-" generally means "less than." " Tonic" refers to solute concentration. Therefore, "hypertonic" means excess solutes, and "hypotonic" means fewer solutes.
- Secondary active transport also tends to be confusing. Remember that secondary active transport does not use ATP directly, but is tightly coupled to ion transport that does require ATP.

Study Strategies

- When studying the fluid mosaic model, think about the properties of the constituent molecules. This will make understanding the membrane's structure much easier. Draw a cartoon of the plasma membrane and all the potential components.
- For the study of diffusion and osmosis, draw diagrams of the movement of water and solutes. Diagrams will help you visualize what is happening across a membrane.
- Go to yourBioPortal.com to review the following tutorials and activities:

  Animated Tutorial 6.1 Passive Transport

  Animated Tutorial 6.2 Active Transport

  Animated Tutorial 6.3 Endocytosis and Exocytosis

  Interactive Tutorial: Lipid Bilayer: Temperature Effects on Composition

  Web Activity 6.1 The Fluid-Mosaic Model

  Web Activity 6.2 Animal Cell Junctions

Important Concepts

Biological membranes have a specific content and structure.

- Biological membranes are composed of lipids, proteins, and carbohydrates. Lipids provide the structure and barrier functions for the membrane. Proteins are involved in creating channels and transporting materials across the lipid barrier. Carbohydrates found on the outside of the plasma membrane are involved in signaling and adhesion. The fluid mosaic model describes how lipids interact to produce a fluid membrane on which the proteins and carbohydrates float.
- A biological membrane is mostly composed of lipids, with phospholipids as the most abundant component. Phospholipids have both hydrophilic ("water-loving") heads and hydrophobic ("water-hating") tails. They arrange themselves into a bilayer with the hydrophobic tails touching and the heads extending into the aqueous environment inside and outside the cell. This arrangement allows for the fluid movement of the two layers on top of each other and sealing of any disruptions in the membrane.

- Though the basic structure of a bilayer is always the same, the inner and outer halves differ in lipid composition and thus have slightly different properties. Phospholipids differ in their length, degree of unsaturation, and degree of polarity. Cholesterol can make up to 25 percent of the lipid content in the membrane of animals.
- The amount of cholesterol present and the degree of fatty acid saturation (membrane kinks) influences the fluidity of the membrane. Increases in cholesterol and fatty acid saturation make the membrane less fluid. Decreases in temperature result in a less fluid membrane. Organisms have the ability to change the lipid composition of their membranes in response to a change in temperature.
- Proteins may be embedded in or extend across membranes. Regions or domains of a membrane protein with hydrophobic amino acid side chains tend to be found in the hydrophobic environment of the membrane. The regions of a membrane protein containing hydrophilic amino acid side chains extend out from the membrane. The phospholipids and proteins are independent of each other, allowing for movement throughout the membrane. This happens for the most part because the constituents interact only noncovalently.
- Proteins that penetrate the phospholipid bilayer are called integral proteins. Special types of integral proteins, transmembrane proteins, span the entire bilayer. Those not embedded are referred to as peripheral proteins. Proteins are distributed in membranes asymmetrically "as needed," and the numbers of proteins and their placement vary greatly based on cell type. The "inside" and "outside" of a membrane often have different properties due to the different characteristics of the transmembrane proteins on the two sides.
- Some membrane proteins may be anchored to cytoskeletal components or bound to lipid rafts. Such proteins have very specific functions.
- Membranes are in a constant state of flux. Phospholipids are produced on the surface of the smooth endoplasmic reticulum and distributed to the Golgi as vesicles. From the Golgi they move to the plasma membrane.
- Membrane carbohydrates serve as recognition sites for other cells and molecules. Carbohydrates are frequently bound to lipids or proteins, forming glycolipids or glycoproteins, respectively. Because of the nature of carbohydrates, they contribute to cell adhesion.

Cells must be able to recognize and adhere to other cells to form tissues, organs, and organisms.

- Cells have the ability to recognize and adhere to other cells, allowing for the grouping of cells in an organism. Cell adhesions are specific to the organism and to the cell type. A sponge is an animal composed of many cells that exhibit species-specific cell adhesion.
- A cell's ability to recognize and adhere to other cells is due to the presence of recognition proteins on the cell surface. Often the proteins on two cells that adhere to each other are the same; these molecules and the bonds they form are referred to as homotypic. Occasionally the two cells have different proteins, but complementary binding sites. In this case the molecules and the binding are called heterotypic.

Cells must make connections to other cells to communicate and stabilize tissues.

- Cell junctions are sites where cell recognition proteins from two cells form a connection. The three main types of cell junctions are tight junctions, desmosomes, and gap junctions.
  - Tight junctions are found specifically in epithelial cells and function to prevent substances from moving through the spaces between cells. They also restrict migration of membrane proteins and phospholipids that would otherwise be free to float around in the "fluid" plasma membrane. Tight junctions help tissues move materials in specific directions.
  - Desmosomes are structural connections between cells that hold them together. Desmosomes have plaques on the plasma membrane that attach to special cell adhesion molecules. Desmosome plaques are also attached to keratin fibers of the cytoskeleton that cross the cell membrane. The connections thus have integrity and are not easily broken.
  - Gap junctions facilitate communication between cells. They are made of protein connexons that span adjacent plasma membranes and act as channels.

Transport of substances across membranes can be a passive process.

- Plasma membranes are not permeable to all substances, but are selectively permeable. Passive transport is the movement of a substance across a lipid bilayer through channel proteins or by carrier molecules. Passive transport does not require an input of energy; it can occur either by simple diffusion or facilitated diffusion.
- Diffusion is the net movement of a substance from an area of greater concentration to an area of lesser concentration. It is a random process that moves toward equilibrium.
- A membrane is said to be permeable to those substances that can pass through it and impermeable to those that cannot. If a substance can pass through a membrane, it will diffuse until concentrations on either side of the membrane are equal. At equilibrium, the molecules of the substance continue to move across the membrane, but the net movement of molecules in both directions is equal.
- The rate of diffusion depends on the size of the diffusing substance, the temperature of the solution, and the concentration gradient. Diffusion within small areas

such as single cells may occur rapidly, but diffusion occurs more slowly with increasing distance.

- In simple diffusion, small molecules pass through a membrane. The more lipid-soluble the molecules are, the faster they diffuse across the membrane. In contrast, charged and polar molecules do not readily pass through a membrane due to the formation of many hydrogen bonds with water and the hydrophobic nature of the internal layer of the membrane.
- Osmosis is the diffusion of water across a membrane, typically through channels in the membrane. Water will move across a membrane to equalize solute concentrations on either side of the membrane if the solute cannot move. Water moves from areas of low solute concentration to high solute concentration in an attempt to equalize solute concentrations on both sides of the membrane.
  - Solute concentrations separated by a membrane are classified as isotonic, hypertonic, or hypotonic. Isotonic solutions have the same solute concentrations on both sides of a membrane. A hypertonic solution has a higher solute concentration than the solution on the other side of the membrane. A hypotonic solution has a lower solute concentration than the solution with which it is being compared. Environmental and cellular solute concentrations dictate the direction of osmosis in living cells.
- The pressure within cells, called turgor pressure, changes according to the amount of water taken up by osmosis. Turgor pressure can build up only if there is a cell wall to limit cell expansion.

Facilitated diffusion makes use of proteins that span the membrane.

- Facilitated diffusion is the passive movement of a substance across a membrane with the help of membrane-bound proteins that act as channels or carriers.
- A substance may cross the membrane by facilitated diffusion through protein channels running through the plasma membrane. The pores of these proteins have polar amino acids that allow polar molecules and ions to cross the membrane.
- The best studied channels are ion channels. Most ion channels are either ligand-gated or voltage-gated, allowing the passage of ions to be controlled depending on the cellular environment. Movement of ions through a voltage-gated channel depends on both the concentration gradient for the ion and the distribution of electrical charge across the membrane.
- All living cells have an unequal distribution of ions between their inside and outside. This unequal separation of ions creates a charge difference across the membrane, known as a membrane potential. The imbalance is typically a K^+ imbalance, with a larger amount of K^+ inside than outside. The membrane potential can be calculated with the Nernst equation. The membrane potential of a cell is typically around –70 mV.
- Ion channels are very specific for the ions that they allow to pass through their pore. For example, a K^+ channel has a highly polar oxygen at the channel opening. This oxygen pulls K^+ ions away from water, allowing them to move through the channel.
- Specific channels called aquaporins allow water to cross the membrane by osmosis. Water can also diffuse through ion channels.
- Facilitated diffusion may be aided by carrier proteins that bind a substance and transport it across the membrane. Diffusion in this case is not only limited by the concentration gradient, but also by the number of available membrane carrier proteins. When all of the carrier proteins are bound to the substance, they are saturated, limiting the rate of diffusion.

Active transport is directional, works against a concentration gradient, and requires an input of energy.

- Primary active transport requires the direct participation of ATP to move ions against their concentration gradient. Three types of protein transporters carry out active transport. Uniport transporters move a single solute across a membrane in one direction. Symport transporters move two solutes in the same direction across a membrane. Antiport transporters move two solutes in opposite directions.
- Primary active transport directly uses energy from the hydrolysis of ATP to drive the transporter. The sodium–potassium pump is an important antiporter found in animal cells that uses ATP to transport two K^+ ions into the cell and three Na^+ ions out of the cell.
- Secondary active transport uses ATP indirectly by coupling solute transport with the ion concentration gradient established by primary transport. It typically moves amino acids and sugars across the membrane with the help of an ion like Na^+.

Endocytosis and exocytosis move large molecules into and out of cells.

- Endocytosis is the process by which a cell brings large substances into the cell. This is accomplished by the cell membrane, which folds around the substance to form an endocytotic vesicle. See Figure 6.18.
  - Large substances and even entire cells are engulfed in the process of phagocytosis. Once inside the cell, the vesicle fuses with a lysosome for digestion. The cell takes up liquids in small vesicles from the outside in the process of pinocytosis.
  - Animal cells use receptor-mediated endocytosis to capture specific macromolecules such as cholesterol from the environment. Receptors for specific macromolecules cluster together on the cell surface in coated pits containing the protein clathrin. Upon binding of the specific molecule to the receptors, the

coated pit invaginates to form a vesicle. The resulting vesicle becomes clathrin coated until it is well inside the cell, where it loses its coat and fuses with a lysosome. Animal cells take up cholesterol by the process of receptor-mediated endocytosis.

- Exocytosis moves materials in vesicles out of the cell. Binding proteins found on the surface of the cell-produced vesicles bind with receptor proteins of the cytoplasmic side of the cell membrane. The vesicle membrane fuses with the cell membrane, and the contents of the vesicle are released outside the cell membrane.

Membranes have other functions.

- Membranes have additional biological functions. Though the primary function of membranes is to act as barriers and binding sites, they also have roles in information processing, energy transformation, and the organization of chemical reactions. In order to carry out these functions, membranes are necessarily dynamic. They are continually being formed, modified, and degraded.

Test Yourself

Diagram Exercise

Diagram a cell membrane and label the phospholipid bilayer, integral proteins, peripheral membrane proteins, and carbohydrates. Describe the fluid mosaic model with reference to your diagram.
Textbook Reference: *6.1 What Is the Structure of a Biological Membrane? p. 107*

Knowledge and Synthesis Questions

1. Which of the following statements regarding cellular membranes is *false*?
   a. The hydrophobic nature of the phospholipid tails limits the migration of polar molecules across the membrane.
   b. Integral proteins and phospholipids move fluidly throughout the membrane.
   c. Membrane phospholipids flip back and forth from one side of the bilayer to the other.
   d. Glycolipids and glycoproteins serve as recognition sites on the cell membrane.
   e. All of the above are true.
   Textbook Reference: *6.1 What Is the Structure of a Biological Membrane? pp. 106–110*
2. Which of the following contributes to differences in the two sides of the cell membrane?
   a. Differences in peripheral proteins
   b. Different domains expressed on the ends of integral proteins
   c. Differences in phospholipid types
   d. Differences in the carbohydrates attached to membrane proteins
   e. All of the above
   Textbook Reference: *6.1 What Is the Structure of a Biological Membrane? pp. 108–109*
3. Which of the following cell membrane components serve as recognition signals for interactions between cells?
   a. Cholesterol
   b. Glycolipids or glycoproteins
   c. Phospholipids
   d. Carrier proteins
   e. All of the above
   Textbook Reference: *6.1 What Is the Structure of a Biological Membrane? p. 110*
4. Which of the following types of junctions are responsible for communication between cells?
   a. Tight junctions
   b. Desmosomes
   c. Gap junctions
   d. Active transporters
   e. None of the above
   Textbook Reference: *6.2 How Is the Plasma Membrane Involved in Cell Adhesion and Recognition? p. 113*
5. You are monitoring the diffusion of a molecule across a membrane. Which of the following will result in the fastest rate of diffusion?
   a. An internal concentration of 5 percent and an external concentration of 60 percent
   b. An internal concentration of 60 percent and an external concentration of 5 percent
   c. An internal concentration of 35 percent and an external concentration of 40 percent
   d. An internal concentration of 50 percent and an external concentration of 50 percent
   e. Both a and b
   Textbook Reference: *6.3 What Are the Passive Processes of Membrane Transport? p. 114*
6. What will happen if a red blood cell with an internal salt concentration of about 0.85 percent is placed in a saline solution that is 4 percent?
   a. The red blood cell will lose water and shrivel.
   b. The red blood cell will gain water and burst.
   c. The turgor pressure in the cell will increase greatly.
   d. The turgor pressure in the cell will decrease greatly.
   e. The cell will remain unchanged.
   Textbook Reference: *6.3 What Are the Passive Processes of Membrane Transport? p. 115*
7. In which of the following is solution X hypotonic relative to solution Y?
   a. Solution X has a greater solute concentration than solution Y.
   b. Solution X has a lower solute concentration than solution Y.
   c. Solution X and solution Y have the same solute concentration.
   d. All of the above
   e. None of the above
   Textbook Reference: *6.3 What Are the Passive Processes of Membrane Transport? p. 116*

8. Which of the following statements regarding osmosis is *false*?
   a. Osmosis refers to the movement of water along a concentration gradient.
   b. In osmosis, water moves to equalize solute concentrations on either side of the membrane.
   c. The movement of water across a membrane can affect the turgor pressure of some cells.
   d. If osmosis occurs across a membrane, then diffusion is not occurring.
   e. During osmosis water is moving through membrane channels.

   Textbook Reference: *6.3 What Are the Passive Processes of Membrane Transport? pp. 115–116*

9. Channel proteins allow ions that would not normally pass through the cell membrane to pass through via the channel. What property of the channel proteins is responsible for this?
   a. A pore of polar amino acid groups
   b. A pore of hydrophobic amino acid groups
   c. A pore of Ca^{2+}
   d. All of the above
   e. None of the above

   Textbook Reference: *6.3 What Are the Passive Processes of Membrane Transport? p. 117*

10. Which of the following limits the movement of molecules by means of carrier-mediated facilitated diffusion?
    a. The concentration gradient
    b. The availability of carrier molecules
    c. Temperature
    d. Both a and b
    e. All of the above

    Textbook Reference: *6.3 What Are the Passive Processes of Membrane Transport? p. 119*

11. Active transport differs from passive transport in that active transport
    a. requires energy.
    b. never requires direct input of ATP.
    c. moves molecules with a concentration gradient.
    d. Both b and c
    e. Both a and c

    Textbook Reference: *6.4 What Are the Active Processes of Membrane Transport? p. 120*

12. Single-celled animals, such as amoebas, engulf entire cells for food. This manner of "eating" is called
    a. exocytosis.
    b. endocytosis.
    c. facilitative transport.
    d. active transport.
    e. osmosis.

    Textbook Reference: *6.5 How Do Large Molecules Enter and Leave a Cell? p. 122*

13. Many cells have a sodium–potassium pump. In order to function, sodium–potassium pumps require
    a. ATP.
    b. a channel protein.
    c. no concentration gradient.
    d. ADP.
    e. All of the above

    Textbook Reference: *6.4 What Are the Active Processes of Membrane Transport? p. 120*

14. Bacterial cells are often found in very hypotonic environments. Which of the following characteristics keeps them from taking in too much water from their environment?
    a. The presence of a cell wall, which allows for a buildup of turgor pressure, preventing additional water from entering the cell
    b. The presence of a cell wall, which allows for a buildup of tonic pressure, preventing additional water from entering the cell
    c. The capacity of the cell to expel water as quickly as it takes it up
    d. The presence of an active water pump
    e. None of the above

    Textbook Reference: *6.3 What Are the Passive Processes of Membrane Transport? p. 116*

15. The rate of diffusion can be affected by
    a. temperature.
    b. molecule size.
    c. the concentration gradient.
    d. the electrical charge.
    e All of the above

    Textbook Reference: *6.3 What Are the Passive Processes of Membrane Transport? pp. 114, 117*

Application Questions

1. The cells that make up a few animals, such as sponges, can be mechanically separated but then will re-form themselves over time. How are the separated cells able to reorganize themselves?

   Textbook Reference: *6.2 How Is the Plasma Membrane Involved in Cell Adhesion and Recognition? p. 111*

2. Cells have the ability to take in large molecules by endocytosis and secrete them to the environment by exocytosis. Describe each process and explain why both are important for the cell.

   Textbook Reference: *6.5 How Do Large Molecules Enter and Leave a Cell? pp. 123–124*

3. A marathon runner has just arrived in the emergency room with severe dehydration, and the physician must decide which type of solution to pump into his veins: pure water, 0.9 percent saline, or 1.5 percent saline. In order to be certain, blood samples are treated with each solution and observed under a microscope. Describe what is likely to happen to the blood cells when exposed to each solution. (Hint: Blood cells are approximately 0.9 percent saline.) Which solution should the physician choose for rehydrating the runner?

   Textbook Reference: *6.3 What Are the Passive Processes of Membrane Transport? p. 116*

4. Compare and contrast active and passive transport.
Textbook Reference: *6.4 What Are the Active Processes of Membrane Transport? p. 120*
5. Barrier formation is only one function of the cell membrane. Describe some other functions of the membrane and discuss how the membrane is suited for those functions.
Textbook Reference: *6.6 What Are Some Other Functions of Membranes? pp. 124–125*

Answers

Diagram Exercise Answer

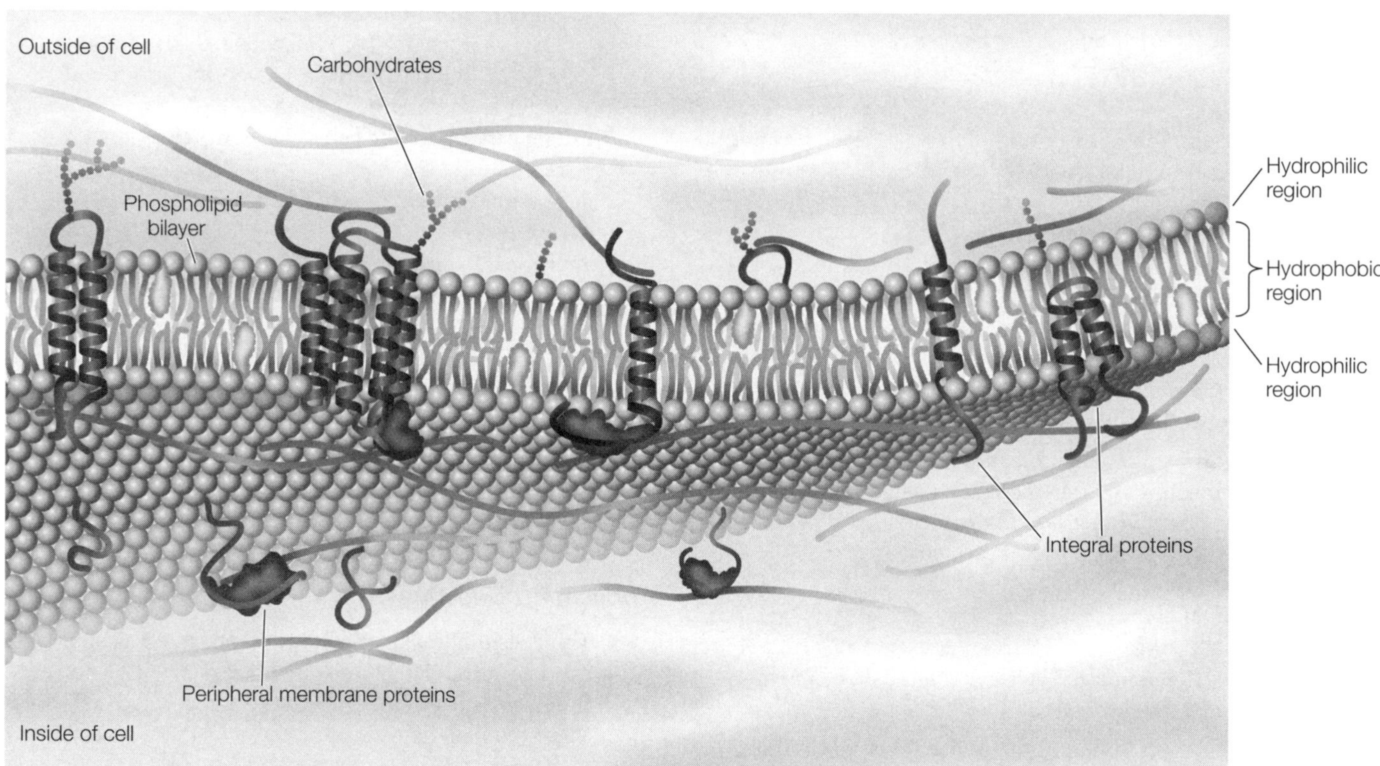

The fluid mosaic model states that the phospholipid bilayer allows the embedded proteins to float freely through the bilayer.

Knowledge and Synthesis Answers

1. **c.** Because of the hydrophobic tails and hydrophilic head of the phospholipids, it is impossible for them to flip back and forth from one side of the membrane bilayer to the other.
2. **e.** The cell membrane is asymmetric and has different properties and functions on the cytoplasmic side versus the extracellular side. These properties arise from differences in the constituents of the membrane.
3. **b.** Both glycolipids and glycoproteins serve as recognition signals.
4. **c.** Gap junctions are involved in chemical and electrical signaling between cells.
5. **e.** Diffusion may take place in either direction across a membrane and always follows a concentration gradient. The larger the gradient, the faster diffusion will occur.
6. **a.** The cell will lose water as solute concentrations on both sides of the membrane equalize.
7. **b.** A solution that has a lower solute concentration than another is hypotonic in comparison to the other one.
8. **d.** Diffusion and osmosis are not mutually exclusive and may take place at the same time.
9. **a.** The charged or polar lining of the channel proteins allows passage of polar and charged molecules.

10 **d.** Anything that affects the rate of diffusion will affect carrier-mediated facilitated diffusion. Carrier-mediated facilitated diffusion also relies on the availability of carrier molecules.

11. **a.** Active transport works against a concentration gradient and requires energy to do so. That energy does not always have to be directly supplied in the form of ATP.

12. **b.** Cells carry out cellular eating by phagocytosis, which is a type of endocytosis.
13. **a.** Sodium–potassium pumps are forms of primary active transport and require energy in the form of ATP.
14. **a.** Turgor pressure limits osmosis, and once a cell is turgid, no more water may be taken on.
15. **d.** Temperature, molecule size, molecule charge, and concentration gradients all affect the rate at which diffusion takes place.

Application Answers

1. The cells in the sponge have the ability to recognize and adhere to other cells from their original body. This cell recognition is tissue-specific (so that the tissues re-form correctly) and species-specific (so that only cells of one species adhere to one another). This ability to adhere and recognize other cells to re-form the sponge is due to specific glycoproteins on the plasma membrane that bind to one another and result in adherence.
2. Cells take up large particles, foreign cells, and food sources by endocytosis, in which the plasma membrane of the cell surrounds the particle to form an endocytotic vesicle. Cells secrete substances such as undigested material, digestive enzymes, neurotransmitters, and material for plant wall construction by exocytosis. During exocytosis, the membrane of a secretory vesicle fuses with the plasma membrane and the contents are released to the outside of the cell.
3. In pure water, the blood cells will take on water through osmosis, swell, and eventually rupture. In 0.9 percent saline, the cells should neither gain nor lose a significant amount of water. In a 1.5 percent saline solution, the cells should lose water and shrivel. In order to rehydrate the runner, a solution isotonic to the patient's blood cells, 0.9 percent, should be infused into his bloodstream. A hypotonic solution would end up rupturing the patient's cells.
4. The main difference between active and passive transport is that active transport goes against a concentration gradient and requires energy, whereas passive transport diffuses passively and does not require energy.
5. Membranes function in processing energy transformation and in the organization of chemical reactions. Integral and peripheral proteins contribute to these functions. The membrane serves as a holding site for the catalytic enzymes associated with these processes.

7 Cell Signaling and Communication

The Big Picture

- Signal transduction is the means by which cells receive information from the environment or other cells and react to those signals. Transduction is a highly regulated series of events that depends on the binding of a signal ligand to a receptor protein. The signal binding must cause a change in the shape of the receptor protein, which causes a responder protein to initiate events in the cell that change its function. The effects of signals are often mediated by secondary messengers.

Common Problem Areas

- It's easy to get overwhelmed by the different examples of signal transduction. Try to focus on particular details of the systems. For instance, distinguish between direct and indirect signal transduction. Compare plasma membrane receptors with cytoplasmic receptors. List the three kinds of secondary messengers for signal transduction. Then, pick one of the signal transduction examples and create a table that includes the signal, the receptor, the transduction (responders and amplification), and the effect in the cell. Expand your table to include other examples of signal transduction.
- G protein action frequently gives students trouble. Remember that the G protein interacts with the receptor, binds GTP, and then interacts with the effector protein.
- Gap junctions and plasmodesmata are physical connections between adjacent cells, and although they are similar in function, they are fundamentally different in structure.

Study Strategies

- Create a flow chart or a diagram of the signal transduction pathways.
- Make a list each of the different secondary signals and give an example of how each signal is activated and what it affects.
- Review the figures of the chapter to clarify the different examples of signal transduction.
- Go to yourBioPortal.com to review the following tutorials and activities:

  Animated Tutorial 7.1 Signal Transduction Pathway

  Animated Tutorial 7.2 Signal Transduction and Cancer

  Web Activity 7.1 Signal Transduction

  Web Activity 7.2 Concept Matching

Important Concepts

Cells respond to specific environmental signals by changing their cellular function. This process of receiving a signal and communicating it to the cell is called signal transduction.

- Cells receive signals from their environment (chemicals, light, temperature, touch, or sound) and from other cells (usually chemicals).
- Multicellular organisms receive signals from their environment, from other cells, or from extracellular fluid.
- Autocrine signals are local signals that affect the cells that make them (see Figure 7.1A).
- Paracrine signals are local signals that affect nearby cells (see Figure 7.1A).
- Hormones are circulatory signals that travel through the circulatory system and affect distant cells (see Figure 7.1B).
- To transmit a signal, a cell must receive it and respond to it, and that response must have some effect on the function of the cell.
- Signal transduction includes the following elements:
  - A receptor binds the signal, altering its conformation.
  - A responder is activated in response to the conformational change of the receptor (e.g., a kinase enzyme adds a phosphate to another protein).
  - The signal is amplified (e.g., a single enzyme will repeatedly catalyze the addition of a phosphate to its target protein as long as the signal is bound to the receptor).
  - The signal transduction pathway is activated further (e.g., transcription factor activation causes gene products to be expressed and to alter the cell's activity).

- The response of *E. coli* to osmotic change provides one example of signal transduction (see Figure 7.3). An environmental signal (a change in solute concentration in the inner membrane space) causes a conformational change in a receptor protein (EnvZ), which is a transmembrane protein in the bacterium's plasma membrane. Signal transduction then proceeds along the following pathway: EnvZ picks up a phosphate from ATP and transfers it to OmpR (a protein in the cytoplasm). The OmpR protein changes shape so that it can bind to the promoter for the gene that encodes the protein OmpC, thereby increasing production of OmpC protein. The OmpC protein blocks pores in the outer membrane, preventing solutes from entering the intermembrane space.

Receptors are specific. Not all cells have the same receptors, so a given signal can have different effects (or no effect) on different cell types.

- Binding to the receptor requires a fit of the ligand to the receptor binding site. Once the ligand binds with the receptor, the receptor must undergo a conformational change to have an effect.
- The ligand is not changed in the binding process, and binding of the signal to the receptor is reversible and follows the law of mass action. Binding sites may be inhibited by competing chemicals.
- There are two classes of receptors: plasma membrane receptors, which bind large and/or polar ligands that cannot cross the plasma membrane, and cytoplasmic receptors, which bind small nonpolar ligands that diffuse across the plasma membrane.
- There are three main types of plasma membrane receptors: ion channel receptors, protein kinase receptors, and G protein-linked receptors.
  - Ion channel receptors: Ligand binding causes a conformational change to open "gates" that allow ions (Na^+, K^+, Ca^+, or Cl^-) to pass through (see Figure 7.6). These receptors respond to sensory stimuli and chemical ligands. An example is the acetylcholine receptor found on muscle cells. When two molecules of acetylcholine bind to the receptor, the channel in the receptor opens and sodium flows through.
  - Protein kinases: Ligand binding stimulates the transfer of a phosphate group from ATP to a target protein (see Figure 7.7).
  - G protein-linked receptors: This term applies to a group of receptors, each of which is composed of a single protein with seven transmembrane-spanning domains. Ligand binding to the transmembrane receptor protein changes its shape so that a G protein on the cytoplasmic side can bind. When the G protein is activated (by binding to the cytoplasmic side of the G-linked receptor), it binds GTP and activates an effector protein, changing cellular function. After binding the effector protein, GTP is hydrolyzed to GDP, causing the G protein to be inactivated until the receptor binds its ligand again (see Figure 7.8). The G protein can either inhibit or stimulate a cellular response.
- Cytoplasmic receptors are located in the cell cytoplasm. Ligand binding causes a conformational change in the receptor that allows passage of the receptor-signal complex into the cell nucleus where it can function as a transcription factor (see Figure 7.9).

Signal transduction may be direct (occurring at ligand binding) or indirect (requiring second messenger molecules).

- Direct signal transduction is a function of the receptor at the plasma membrane (see Figure 7.10A).
  - The phosphorylation of a target protein by a protein kinase is an example of direct signal transduction (see Figure 7.12). In a protein kinase cascade, several different kinases are activated after the initial signal is bound. Protein kinase cascades can result in the transcriptional activation of genes. They are useful signal transducers because at each step the signal is amplified, the signal is communicated to the nucleus, different target proteins can provide variation, and specificity is made possible by the multiple steps in the pathway.
- Indirect signal transduction utilizes a second messenger to mediate the response between receptor binding and cellular response (see Figure 7.10B).
- Secondary messengers are released to the cytoplasm after ligand binding and function to amplify the signal. They act as cofactors or allosteric regulators of target enzymes and allow the cell to have multiple responses to a single signal.
  - cAMP is a second messenger that can affect ion channels in the cell or activate protein kinases. The membrane-bound effector protein adenylyl cyclase is responsible for making cAMP from ATP. Adenylyl cyclase is activated by a G protein.
  - Inositol trisphosphate (IP_3) and diacylglycerol (DAG) are lipid end products of the hydrolysis of PIP2 (phosphatidyl inositol-bisphosphate) by phospholipase C. They can interact with protein kinase C or Ca^{2+} channels in the endoplasmic reticulum.
  - Calcium ions frequently act as secondary messengers to activate protein kinase C, control other channels, and stimulate exocytosis. Calcium levels are controlled by ion channels and membrane pumps.
  - The gas nitric oxide, which is produced by the endothelial cells of blood vessels, stimulates the formation of cGMP. This can initiate a protein kinase cascade that causes the smooth muscles to relax.
- Signal transduction pathways are highly regulated to ensure their proper function. Nitric oxide breaks down rapidly. Ca^{2+} concentration is controlled by membrane pumps and ion channels. cAMP levels are affected by phosphodiesterase, which breaks down cAMP.

Function in a pathway is dependent on the synthesis and breakdown of the enzymes involved and the activation or inhibition of these enzymes.

- These signaling transduction pathways have a number of ways in which they can bring about the response in the cell. Some signal transduction pathways open ion channels (as seen in sensing a particular odor). Others alter enzyme function through either inhibition or activation (as seen with epinephrine stimulation). Signal transduction pathways also influence the expression of genes by regulating transcription.

Signals are often transduced between cells.

- Cell-to-cell signaling in animal cells can be facilitated by gap junctions. Gap junctions are protein-lined (connexon protein) channels that allow the passage of small molecules, including signal molecules and ions between adjacent cells.
- Plasmodesmata are membrane-lined channels connecting adjacent plant cells. They are filled with tubules derived from the endoplasmic reticulum called desmotubules, which allow small metabolites and ions to move between plant cells. The plant and plant viruses can stimulate the synthesis of "movement proteins" to increase pore size and permit proteins, mRNAs, and viruses to move through plasmodesmata.

Test Yourself

Diagram Exercise

Diagram the Ras signaling pathway by means of a flow chart. Be sure to identify the signal, the receptor, and the mechanisms involved.
Textbook Reference: *7.3 How Is a Response to a Signal Transduced through the Cell? pp. 136–137*

Knowledge and Synthesis Questions

1. Which of the following does *not* occur during signal transduction?
   a. Binding of ligand to receptor
   b. Conformational change to the receptor protein
   c. Conformational change of the signal
   d. Alteration of cellular activity
   e. Conformational change in the ligand
   Textbook Reference: *7.1 What Are Signals, and How Do Cells Respond to Them? pp. 130–132*
2. *E. coli* cells respond to osmotic changes by
   a. changing the permeability of the membrane.
   b. changing the DNA that is transcribed.
   c. phosphorylating OmpR.
   d. binding the ligand to EnvZ.
   e. increasing metabolism.
   Textbook Reference: *7.1 What Are Signals, and How Do Cells Respond to Them? p. 131*
3. What do G protein-linked receptors and protein kinases have in common?
   a. Ligand binding
   b. Conformational change once the ligand is bound
   c. Amplification of the signal
   d. All of the above
   e. None of the above
   Textbook Reference: *7.2 How Do Signal Receptors Initiate a Cellular Response? pp. 135–136*
4. Signal binding to a receptor differs from enzyme–substrate binding in that
   a. in signal binding the signal is not altered during the process.
   b. the single-binding process is reversible.
   c. in signal binding, cell receptors can become saturated by signal molecules.
   d. signal binding can be inhibited by an antagonist.
   e. None of the above
   Textbook Reference: *7.2 How Do Signal Receptors Initiate a Cellular Response? p. 132*
5. Which of the following statements about secondary messengers is true?
   a. They amplify the signal.
   b. They bind to the active site of the receptor.
   c. They result in multiple effects from a single signal.
   d. Both a and c
   e. All of the above
   Textbook Reference: *7.3 How Is a Response to a Signal Transduced through the Cell? pp. 137–139*
6. Caffeine is a stimulant that works because it acts as a(n) _______ to the adenosine receptors in a person's brain, and stimulates _______ in that person's heart and liver that increases blood flow and blood glucose.
   a. effector; a pathway
   b. inhibitor; a cascade pathway
   c. inhibitor; a ligand
   d. signal; inhibitors
   e. pathway; a ligand
   Textbook Reference: *7.2 How Do Signal Receptors Initiate a Cellular Response? p. 133*
7. Cytoplasmic receptors bind
   a. small signals that can diffuse through the plasma membrane.
   b. secondary messengers, such as cAMP.
   c. hydrophilic molecules.
   d. ligands.
   e. All of the above
   Textbook Reference: *7.2 How Do Signal Receptors Initiate a Cellular Response? p. 135*
8. Which of the following statements about protein kinase cascades is true?
   a. Amplification can occur at each step in the path.
   b. Information at the plasma membrane is communicated to the nucleus.
   c. The multiple steps allow for specificity of the process.
   d. Different targets can produce variation in the cellular response.
   e. All of the above
   Textbook Reference: *7.3 How Is a Response to a Signal Transduced through the Cell? pp. 136–137*

9. Which of the following is *not* part of the IP_3 / DAG pathway?
   a. The binding of a hormone to a receptor
   b. The production of IP_3 by a G protein
   c. Activation of phospholipase C
   d. The opening of Ca^{2+} channels
   e. All of the above are part of the pathway.

   Textbook Reference: *7.3 How Is a Response to a Signal Transduced through the Cell? p. 139*

10. Which of the following statements about receptors is true?
    a. Receptors are only found on the surface of cells.
    b. Receptors are specific to the signal ligand.
    c. Most receptors can bind with many signal ligands.
    d. All receptors can act as ion channels.
    e. All of the above

    Textbook Reference: *7.2 How Do Signal Receptors Initiate a Cellular Response? pp. 132–134*

11. Gap junctions and plasmodesmata differ in that
    a. gap junctions are connected by protein tubules called connexons, whereas plasmodesmata are connected by extensions of the plant's plasma membrane.
    b. gap junctions allow much larger molecules to pass through them.
    c. gap junctions have no real physical connection but are rather the space between adjacent cell membranes.
    d. one is of animal origin and the other is of plant origin.
    e. gap junctions are connected by desmotubules and plasmodesmata are connected by connexons.

    Textbook Reference: *7.5 How Do Cells Communicate Directly? pp. 144–145*

12. Paracrine signals
    a. act on the cells that made them.
    b. move through the blood and act on cells far from their source.
    c. act on cells that are near to those that secrete them.
    d. do not act through receptors.
    e. require large concentrations of the signaling molecule to function.

    Textbook Reference: *7.1 What Are Signals, and How Do Cells Respond to Them? p. 130*

13. Which of the following signals results in the direct activation of an enzyme?
    a. Acetylcholine binding to its receptor
    b. Insulin binding to its receptor
    c. Cortisol binding to its receptor
    d. Fertilization of an egg by a sperm cell
    e. All of the above

    Textbook Reference: *7.2 How Do Signal Receptors Initiate a Cellular Response? pp. 133–135*

14. Which of the following represents the correct order of signal transduction?
    a. Binding of signal, release of secondary messenger, alteration of receptor conformation, alteration of cellular function
    b. Binding of signal, release of secondary messenger, alteration of receptor conformation, transcription of gene
    c. Binding of signal, activation of target protein by responder, alteration of receptor conformation, release of secondary messenger; transcription of gene
    d. Binding of signal, alteration of receptor conformation, alteration of cellular function, release of second messenger
    e. Binding of signal, alteration of receptor conformation, activation of target protein by responder, alteration of cellular function

    Textbook Reference: *7.1 What Are Signals, and How Do Cells Respond to Them? pp. 130–132*

15. Which of the following is *not* a response to a signal binding its receptor?
    a. A channel opening
    b. A G protein's exchange of GDP for GTP
    c. Increase in intracellular concentration of nitrous oxide
    d. The diffusion of solutes through the porous outer membrane of *E. coli*
    e. Activation of a protein kinase

    Textbook Reference: *7.2 How Do Signal Receptors Initiate a Cellular Response? pp. 133–136; 7.3 How Is a Response to a Signal Transduced through the Cell? p. 140*

Application Questions

1. Based on your knowledge of prokaryotic and eukaryotic cell structure and function, explain how signal transduction might differ in prokaryotes and eukaryotes?

   Textbook Reference: *7.1 What Are Signals, and How Do Cells Respond to Them? pp. 130–131*

2. Describe how protein kinases and G protein-linked receptors may interact in a signal transduction cascade.

   Textbook Reference: *7.3 How Is a Response to a Signal Transduced through the Cell? pp. 136–137*

3. Discuss the role of secondary messengers in a signaling pathway. How are they different from signals and receptors? What roles do secondary messenger and signals have in common?

   Textbook Reference: *7.3 How Is a Response to a Signal Transduced through the Cell? pp. 137–138*

4. Can a receptor act as a signal? Explain your answer.

   Textbook Reference: *7.3 How Is a Response to a Signal Transduced through the Cell? p. 137*

5. Trace how changes in osmotic pressure in the plasma membrane of *E. coli* result in changes in gene transcription. What would be the outcome if OmpR could not bind to the *ompC* promoter?

   Textbook Reference: *7.1 What Are Signals, and How Do Cells Respond to Them? pp. 130–132*

Answers

Diagram Exercise Answer

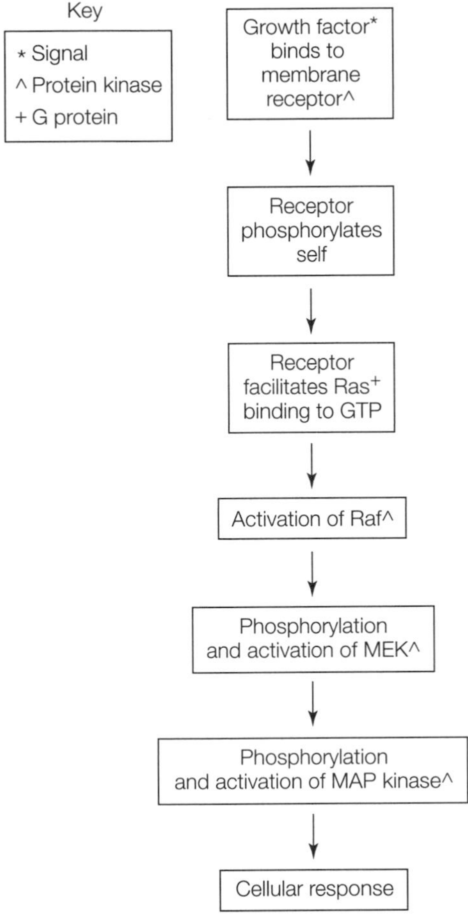

Knowledge and Synthesis Answers

1. **c.** In order for signal transduction to take place, the ligand must bind to the receptor, the receptor must undergo a conformational change, and the activity of the cell must be altered.
2. **a.** The ultimate goal of any signaling pathway is an alteration in the function of the cell in response to the signal.
3. **d.** All types of receptors bind ligands, undergo conformational change, and amplify the signal.
4. **a.** Substrates are altered in enzyme–substrate complexes. Signals are not altered when they bind to receptors.
5. **d.** Secondary messengers both amplify the signal and have multiple effects within a cell.
6. **b.** Caffeine and adenosine bind to the same receptor protein. The binding of caffeine prevents adenosine from binding in the brain. In other organs, a series of cascades begin in response to caffeine binding.
7. **a.** To bind to a cytoplasmic receptor, the ligand must be able to pass through the plasma membrane.
8. **e.** Amplification, regulation, specificity, and variation are all roles of protein kinase cascades.
9. **b.** IP_3 and DAG are made when phospholipase C hydrolyzes PIP2.
10. **b.** Receptors interact only with specific ligands to bring about a response in the cell.
11. **a.** Both plasmodesmata and gap junctions provide physical connections between the cytoplasms of adjacent cells.
12. **c.** Autocrine signals affect the cells that make them and paracrine cells affect nearby cells. A receptor in the plasma membrane binds hydrophilic signals.
13. **b.** Insulin binds and activates a protein kinase receptor. Acetylcholine leads to a channel opening, cortisol binds to a cytoplasmic receptor and activates transcription, and fertilization leads to an increase in intracellular calcium concentrations.
14. **e.** Signals must bind their receptors, the conformation of the receptor is altered, a responder activates a target protein, and cell function changes.
15. **d.** The diffusion of solutes through the porous outer membrane of *E. coli* is a signal; the other answer options are responses to signals.

Application Answers

1. Signal transduction in prokaryotes is limited to only one cell; there is no nuclear membrane to cross, and cascades tend to be simpler, with fewer intermediate steps.
2. G protein-linked receptors frequently expose the protein kinase activities of effector molecules.
3. Secondary messengers function to amplify and spread signals. They do not bind to receptors and do not act like signals in the cascade. Signals result in a change in cell function; secondary messengers are part of the pathway that results in the change in cell function.
4. A receptor can act as a signal for the next step in a given cascade. In a protein kinase cascade, the receptor is a kinase that then sends a signal to the next step of the cascade.
5. Refer to Figure 7.3 in the textbook. If OmpR could not bind to the *ompC* promoter, then the *ompC* gene would not be transcribed and translated, and the *E. coli* cell would be unable to prevent further solute entry into the inner membrane space.

8 Energy, Enzymes, and Metabolism

The Big Picture

- Cells need energy to carry out their functions. Metabolism is a series of energy-transferring reactions that fuel the processes of the cell. All reactions either give off energy (exergonic) or require energy to proceed (endergonic). Energy may be stored in chemical bonds (potential energy) and released when the bonds are broken to do work (kinetic energy). ATP serves as the energy shuttle for many metabolic processes.
- Enzymes aid biological reactions by lowering the activation energy required to start the reaction. Each enzyme has a specific three-dimensional conformation that interacts specifically with the substrate. Interaction between the enzyme and substrate results in an optimal orientation for a reaction to take place. Enzymes are proteins and are affected by temperature, pH, reactants, activators, and inhibitors.
- Metabolism is regulated by enzymes. Most metabolic pathways are under allosteric control. Enzyme complexes allow for interactions and regulation of adjacent active sites. Often the final product of a specific pathway regulates the commitment step of the pathway itself.

Common Problem Areas

- The chemistry and physics involved in this material are often overwhelming, and you may question why you are learning this in biology. Remember that biological processes are based on physical and chemical properties. To understand the function of organisms, it is necessary to understand the basics of chemistry and physics as they pertain to energy transfer.
- Allosteric regulation can be confusing because the terms "allosteric regulation" and "allosteric enzyme" are related but slightly different. Remember that an allosteric enzyme is an enzyme with more than one binding site: an active site (for binding and acting on substrates) and one or more allosteric sites. When allosteric regulators are bound to the allosteric site, they control whether or not the enzyme can bind substrate.

Study Strategies

- With this chapter, the best strategy is to take "small bites." You may find the concepts to be very unfamiliar; therefore, make sure you understand each section before proceeding. This chapter contains a large amount of terminology. Making a terminology/vocabulary list may be helpful.
- When thinking about the equation $\Delta G = \Delta H - T\Delta S$ it may be helpful to convert the concepts into familiar terms. For example, your net pay is equal to your gross pay minus income taxes and other deductions. So the energy in the raw materials (ΔH) is reduced by the entropy "tax" ($T\Delta S$) the universe imposes on all chemical reactions, giving you a final amount of energy that can be used to do work (ΔG).
- When studying the energy equations, be sure that you understand how any change in either free energy, unusable energy, or absolute temperature will affect the total energy of the system. Also, focus on what a larger or smaller ΔG value means in terms of how reactions proceed and whether they require or release energy.
- When thinking of enzyme-mediated reactions, be sure to consider that altering anything on the left side of the equation is going to lead to changes in the amount of product formed. Always remember that enzymes are conserved across the equation.
- Go to yourBioPortal.com to review the following tutorials and activities:

  Animated Tutorial 8.1 Enzyme Catalysis

  Animated Tutorial 8.2 Allosteric Regulation of Enzymes

  Web Activity 8.1 ATP and Coupled Reactions

  Web Activity 8.2 Free Energy Changes

Important Concepts

Energy is the capacity to do work.

- No living cell manufactures energy. The energy needed for life must be obtained from an environmental energy source and transformed to a usable form.

- Energy transformations are linked to chemical transformations in cells.
- Kinetic energy is energy in motion and does work. Potential energy is stored energy. Energy can be stored biologically in chemical bonds of fatty acids and other molecules. Breaking the bonds converts the energy to kinetic energy.
- Metabolism is all of the chemical reactions occurring in a living organism. Anabolic reactions or anabolism link simple molecules to create complex molecules and store energy in the resulting bonds. Catabolic reactions or catabolism break down complex molecules and release stored energy.
- First law of thermodynamics: Energy is neither created nor destroyed. Energy can be converted from one form to another.
- Second law of thermodynamics: Not all energy can be used. When energy is converted from one form to another, some becomes unusable.
  - Total energy (enthalpy, H) = usable energy (free energy, G) + unusable energy (entropy, S) × absolute temperature (T).
- We cannot measure absolute energy; we can only measure the change (Δ) in energy. We determine the change in usable energy as follows: $\Delta G = \Delta H - T\Delta S$.
  - If ΔG is negative, free energy is released, and the reaction is exergonic.
  - If ΔG is positive, free energy is required, and the reaction is endergonic.
- With each conversion, entropy (S) tends to increase; therefore, processes tend toward disorder and randomness.
- All biochemical reactions must either release ($-\Delta G$) or take up ($+\Delta G$) usable energy.
- Reactions that are exergonic in one direction are endergonic in the other. The direction of the reaction depends on concentrations of reactants versus products. The point at which the forward reaction occurs at the same rate as the reverse reaction is termed chemical equilibrium.
- The further toward completion a reaction's equilibrium lies, the more free energy is given off (larger $-\Delta G$). A larger $+\Delta G$ means equilibrium favors the reverse reaction (equilibrium falls toward the reactant). A ΔG near zero indicates that both products and reactants have free energy.

ATP is the primary energy currency of living cells.

- ATP (adenosine triphosphate) is needed to capture, transfer, and store free energy (usable energy) in cells to do work. Additionally, ATP can donate a phosphate group to different molecules by phosphorylation.
- Hydrolysis of ATP to ADP and P_i releases a large amount of free energy ($\Delta G = -7.3$ kcal/mol). The equilibrium falls toward ADP production.
- The formation of ATP is endergonic, requiring as much energy as is released by ATP hydrolysis. This results in an "energy-coupling cycle" of the endergonic and exergonic reactions. ATP is formed when coupled with the exergonic reactions of cellular respiration. ATP is hydrolyzed to fuel endergonic reactions like protein synthesis (see Figure 8.6).
- ***For review, go to Diagram Exercise 1.***

Enzymes catalyze biochemical reactions by lowering activation energy.

- Catalysts speed up reactions without being consumed in the process. Most biological catalysts are protein enzymes. A protein enzyme acts as scaffolding upon which the reaction takes place.
- Energy barriers exist between reactants and products that slow down reactions. Energy must be added to get past this barrier; it may be thought of it as a little "shove" to get the reaction going. This is called activation energy (E_a) (see Figure 8.8).
- For reactions to take place, the reactants must be slightly destabilized and turned into "transition-state species" with higher free energy than that of either the reactants or the products. This is accomplished with the help of a catalyst.
- Enzymes are highly specific biological catalysts that act on specific reactants or substrates. They have specific substrate-binding areas on their surfaces called active sites. The enzyme's specificity comes from the shape of its active site. The names for enzymes typically end in the suffix –ase. Only specific substrates can fit in and bind with an enzyme's active site. Once bound, the site is referred to as an enzyme–substrate complex (ES).
- The enzyme alters the conformation of the substrate, thus lowering the activation energy for the reaction. From this, the product is formed, and the enzyme remains unchanged. Such reactions can be represented as follows:

$$E + S \rightarrow ES \rightarrow E + P$$

- Enzymes do not alter the equilibrium of a reaction. Both the forward and reverse reactions are accelerated.
- ***For review, go to Diagram Exercise 2.***

The structure of an enzyme determines its function.

- Enzymes speed up reactions by orienting substrates for maximum chemical interactions, by inducing strain on the substrate, or by adding charges to substrates.
- Enzymes are very large macromolecules with small active sites for binding small substrates.
- Binding of small molecules to the active sites depends on H-bonds, charge interactions, and hydrophobic interactions.
- Binding of substrate by an enzyme can change the enzyme's shape. The enzyme's shape can be altered by the substrate to produce an "induced fit." The large size of an enzyme helps it position the correct amino acids

at the active site and helps regulate shape and allow for the induced fit.

- Cofactors, coenzymes, and prosthetic groups all contribute to an enzyme's activity. Cofactors are inorganic ions such as zinc, copper, or iron. Coenzymes are organic molecules that bind in the active site and can be thought of as a co-substrate. Coenzymes are chemically changed during the reaction and regenerated in other pathways. An example would be ATP and ADP and energy carrier coenzymes. Prosthetic groups are permanently bound to the enzyme.
- The rate of a reaction increases as the substrate concentration increases, until all enzyme active sites are occupied. At that point, no amount of additional substrate will increase the reaction rate because no more enzyme is available for catalysis. Enzyme efficiency is measured in turnover number, or how fast an enzyme can convert substrate to product and free up its active site.

Homeostasis requires that enzyme activity be regulated.

- All life must maintain stable internal conditions, or homeostasis, and this is accomplished by the regulation of enzymes. The chemical reactions occurring in an organism are organized into specific metabolic pathways that are catalyzed by specific enzymes at each step.
- Inhibitors are substances that bind with enzymes to inhibit their function. Irreversible inhibition occurs when an inhibitor forms a covalent bond to an enzyme and permanently destroys the active site.
- Reversible inhibition occurs when an inhibitor binds to and alters the enzyme active site but the inhibition is reversible, depending on the concentration of the inhibitor and substrate. Some reversible inhibitors are called competitive inhibitors because they compete with the substrate for the active site. Others are called noncompetitive inhibitors because they bind elsewhere but alter the active site so that the substrate cannot bind, or the rate of binding is reduced.

Allosteric enzymes are regulated by other molecules.

- Allostery is the regulation of an enzyme by a molecule that binds to the enzyme at a site other than the active site, resulting in a change in the enzyme shape. Noncompetitive inhibitors are an example of inhibitors that work allosterically.
- Most allosteric enzymes exist naturally in an active form and an inactive form, and they can switch back and forth between the two forms. When an allosteric regulator binds to the enzyme, the enzyme becomes locked in either its active form or inactive form.
- Most allosteric enzymes are proteins with quaternary structures in which each subunit may have a different regulatory job. Catalytic subunits contain the active site for a specific substrate, whereas the regulatory subunit contains the site for the activator or inhibitor.
- Allosteric interactions help control metabolism. The first step in an enzyme-mediated pathway is referred to as the commitment step. Once initiated, the pathway is followed to completion. Frequently, the final product allosterically inhibits the enzyme of the commitment step, preventing overproduction of the final product. This process is called end-product or feedback inhibition.
- *For review, go to Diagram Exercise 3.*

Enzyme function is influenced by the cellular environment.

- Changes in pH influence charges of amino and other groups on the enzyme or substrate. This may change the folding of the enzyme or its ability to interact with the substrate and drastically alter its catalytic ability. Enzymes typically have an ideal pH at which they function best.
- Increases in temperature may aid in reduction of activation energy by adding kinetic energy; however, this may also result in denaturing of enzymes. Each enzyme tends to have an optimal temperature. Organisms may produce different forms of the same enzyme, called isozymes, which have different optimal temperatures.
- *For review, go to Diagram Exercises 4–5.*

Test Yourself

Diagram Exercises

1. What is the result of the hydrolysis of ATP? Draw the products using the reactants structures given in the diagram below as a guide.

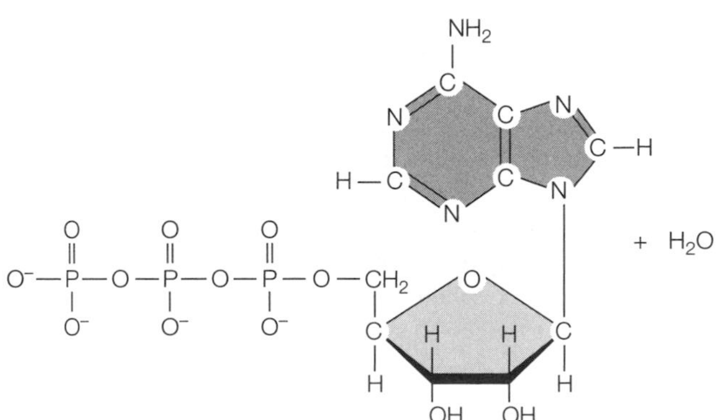

Textbook Reference: *8.2 What Is the Role of ATP in Biochemical Energetics? p. 154, Figure 8.5*

2. Label the graph below with the following: Activation energy for catalyzed reaction, activation energy (E_a) for uncatalyzed reaction, ΔG for catalyzed reaction, ΔG for uncatalyzed reaction, free energy of reactants, free

energy of products, least stable state on graph, most stable state on graph.

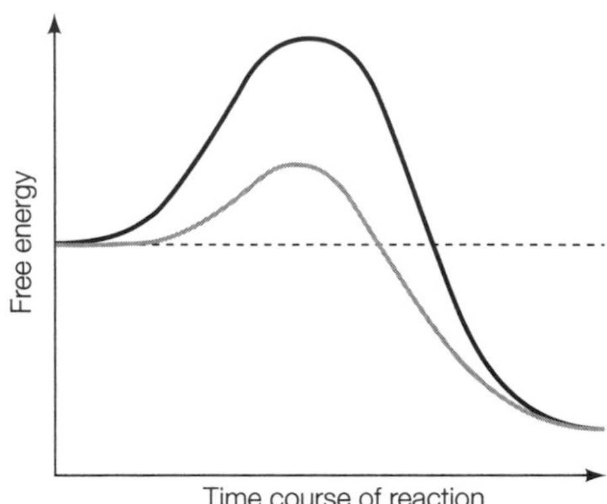

Textbook Reference: *8.3 What Are Enzymes? p. 157, Figure 8.10*

3. Using the following structures as a guide, draw the enzyme–substrate interaction in the presence of a competitive inhibitor, noncompetitive inhibitor, and allosteric activator.

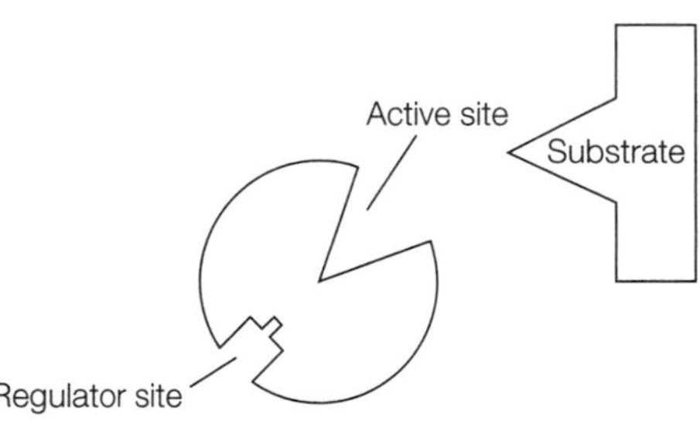

Textbook Reference: *8.5 How Are Enzyme Activities Regulated? p. 163, Figures 8.16, 8.17*

4. *Escherichia coli* is a common bacteria found naturally in the human gut and in waterways contaminated with mammalian fecal matter. Given this information about *E. coli,* draw the predicted enzyme activity profile of an enzyme naturally found in *E. coli.* Some reference temperatures: Water freezes at zero °C, rainbow trout prefer a stream temperature of 16°C, room temperature is about 23°C, normal human body temperature is 37°C, the highest temperature recorded at Death Valley is 56.7°C, and water boils at 100°C.

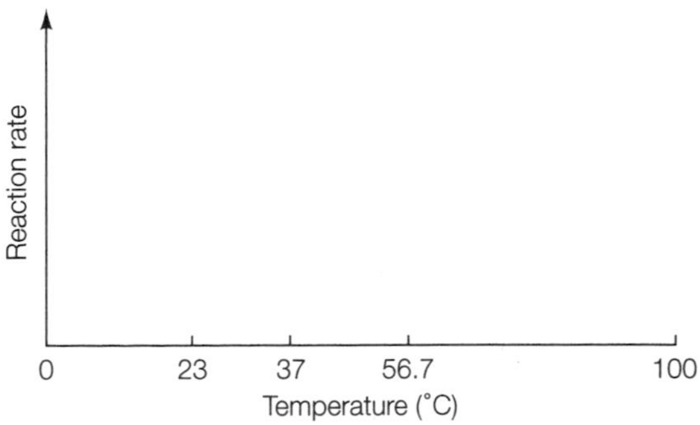

Textbook Reference: *8.5 How Are Enzyme Activities Regulated? p. 165, Figure 8.20*

5. Enzyme X has an optimal pH of 9.0 and is completely inactive at ph 7.0 and ph 11.0. Draw the predicted enzyme activity and determine if this enzyme prefers an acidic, neutral, or basic environment.

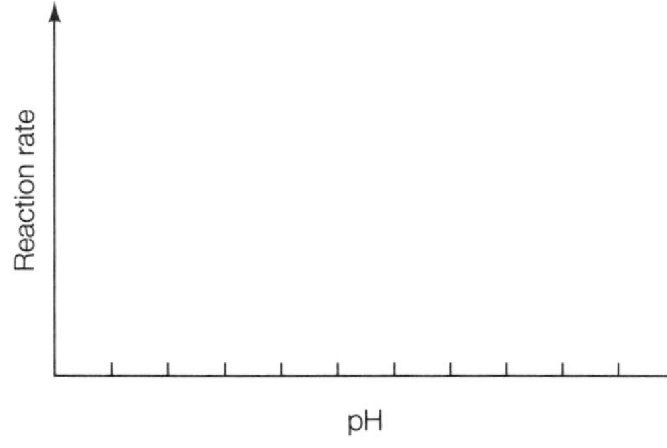

Textbook Reference: *8.5 How Are Enzyme Activities Regulated? p. 165, Figure 8.20*

Knowledge and Synthesis Questions

1. ATP is necessary for the conversion of glucose to glucose 6-phosphate. Splitting ATP into ADP and P_i releases energy into what form used by the cell?
   a. Potential
   b. Kinetic
   c. Entropic
   d. Enthalpic
   e. Heat

   Textbook Reference: *8.2 What Is the Role of ATP in Biochemical Energetics? p. 154*
2. Before ATP is split into ADP and P_i, it holds what type of energy?
   a. Potential
   b. Kinetic
   c. Entropic
   d. Enthalpic
   e. Physical

   Textbook Reference: *8.1 What Physical Principles Underlie Biological Energy Transformations? p. 149*
3. Which of the following statements concerning energy transformations is true?
   a. Increases in entropy reduce usable energy.
   b. Energy may be created during transformation.
   c. Potential energy increases with each transformation.
   d. Increases in temperature decrease total amount of energy available.
   e. Decreases in entropy reduce usable energy.

   Textbook Reference: *8.1 What Physical Principles Underlie Biological Energy Transformations? p. 151*
4. A reaction has a ΔG of –20 kcal/mol. This reaction is
   a. endergonic, and equilibrium is far toward completion.
   b. exergonic, and equilibrium is far toward completion.
   c. endergonic, and the forward reaction occurs at the same rate as the reverse reaction.
   d. exergonic, and the forward reaction occurs at the same rate as the reverse reaction.

e. of an indeterminate nature according to the information supplied.
Textbook Reference: *8.1 What Physical Principles Underlie Biological Energy Transformations? p. 152*

5. ATP hydrolysis is
   a. endergonic.
   b. exergonic.
   c. chemoautotrophic.
   d. anabolic.
   e. None of the above
   Textbook Reference: *8.2 What Is the Role of ATP in Biochemical Energetics? p. 155*
6. Enzymes are biological catalysts and function by
   a. increasing free energy in a system.
   b. lowering activation energy of a reaction.
   c. lowering entropy in a system.
   d. increasing temperature near a reaction.
   e. altering the equilibrium of the reaction.
   Textbook Reference: *8.3 What Are Enzymes? p. 157*
7. Which of the following contributes to the specificity of enzymes?
   a. Each enzyme has a wide range of temperature and pH optima.
   b. Each enzyme has an active site that interacts with many substrate.
   c. Substrates themselves may alter the active site slightly for optimum catalysis.
   d. Enzymes are more active at higher temperatures.
   e. All of the above
   Textbook Reference: *8.4 How Do Enzymes Work? p. 158*
8. Coenzymes and cofactors, as well as prosthetic groups, assist enzyme function by
   a. stabilizing three-dimensional shape.
   b. assisting with the binding of enzyme and substrate.
   c. maintaining active sites in an active configuration.
   d. reversibly binding to the enzyme to regulate the enzyme's activity.
   e. a, b, and c only
   Textbook Reference: *8.4 How Do Enzymes Work? p. 160*
9. Which of the following are characteristics of enzymes?
   a. They are consumed by the enzyme-mediated reaction.
   b. They are not altered by the enzyme-mediated reaction.
   c. They raise activation energy.
   d. They can be composed of RNA or proteins.
   e. They are only rarely regulated.
   Textbook Reference: *8.3 What Are Enzymes? p. 156*
10. Ascorbic acid, found in citrus fruits, acts as an inhibitor to catecholase, the enzyme responsible for the browning reaction in fruits such as apples, peaches, and pears. One explanation for the inhibiting function of ascorbic acid could be its similarity, in terms of size and shape, to catechol, the substrate of the browning reaction. If this explanation is correct, then this inhibition is most likely an example of _______ inhibition.
    a. competitive
    b. indirect
    c. noncompetitive
    d. allosteric
    e. feedback
    Textbook Reference: *8.5 How Are Enzyme Activities Regulated? p. 162*
11. Refer to question 10. Suppose further studies indicate that ascorbic acid is not similar to catechol in size and shape but that the pH of the ascorbic acid solution is altering the protein folding of catecholase. If this is true, then this inhibition is most likely an example of
    a. competitive inhibition.
    b. enzyme denaturation.
    c. noncompetitive inhibition.
    d. allosteric regulation.
    e. feedback inhibition.
    Textbook Reference: *8.5 How Are Enzyme Activities Regulated? pp. 161–162*
12. Metabolism is organized into pathways that are linked in which of the following ways?
    a. All cellular functions feed into a central pathway.
    b. All steps in the pathway are catalyzed by the same enzyme.
    c. The product of one step in the pathway functions as the substrate in the next step.
    d. Products of the pathway accumulate and are secreted from the cell.
    e. Different substrates are acted on by the same enzyme.
    Textbook Reference: *8.5 How Are Enzyme Activities Regulated? p. 161*
13. Which of the following represents an enzyme-catalyzed reaction? (E = enzyme, P = product, S = substrate)
    a. $E + P \rightarrow E + S$
    b. $E + S \rightarrow E + P$
    c. $E + S \rightarrow P$
    d. $E + S \rightarrow E$
    e. $P + S \rightarrow E$
    Textbook Reference: *8.3 What Are Enzymes? p. 157*
14. In the pathway $A + B \rightarrow C + D$, enzyme X facilitates the reaction. If compound D inhibits enzyme X, you would conclude that
    a. enzyme X is an allosteric inhibitor of the above reaction.
    b. compound D is an allosteric stimulator of the above reaction.
    c. compound D is a competitive inhibitor of the above reaction.
    d. enzyme X is subject to feedback stimulation.
    e. compound D is a coenzyme in the above reaction.
    Textbook Reference: *8.5 How Are Enzyme Activities Regulated? pp. 161–163*
15. You are studying a new species never before studied. It lives in acidic pools in volcanic craters where temperatures often reach 100°C and normally stay

above 90°C. You determine that it has a surface enzyme that catalyzes a reaction leading to its protective coating, and you decide to study this enzyme in the laboratory. Under what conditions would you most likely find optimal activity of this enzyme?
a. 0°C
b. 37°C
c. 55°C
d. 95°C
e. 105°C
Textbook Reference: *8.5 How Are Enzyme Activities Regulated? p. 165*

16. Enzymes alter the
a. ΔG value of the reaction.
b. activation energy of the reaction.
c. equilibrium of the reaction.
d. rate of the reaction.
e. Both a and b
Textbook Reference: *8.3 What Are Enzymes p. 157*

17. You fill two containers with identical amounts of reactants A and B and enzymes 1–4. In the reactions shown below, if product D inhibits enzyme 2 and product F is an allosteric stimulator of enzyme 1, what will be the final result if you add extra product D to the second container? (Assume that both containers are given enough time for the reactions to go to completion.)

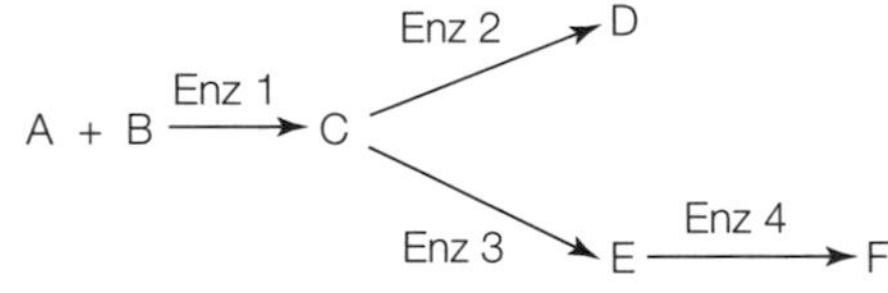

a. The concentration of product C will increase and there will be no change in product F concentration compared to the first container.
b. The concentration of reactants A and B will increase relative to the first container.
c. The concentration of product F will increase in the second container because more of D is converted back to C.
d. The concentration of products E and F will both increase in the second container, since D inhibits enzyme 2.
e. The concentration of product F will increase relative to the first container, since enzyme 2 will have been inhibited from converting as much of C into D.
Textbook Reference: *8.5 How Are Enzyme Activities Regulated? pp. 163–164*

18. You are given an unlabeled enzyme and told to add a compound to the container that will irreversibly bind to the enzyme and increase its function. You ask for information about the enzyme, but your instructor simply hands you a list of possible compounds. Based on what you have learned about the enzyme partners below, which one is the best choice?
a. Coenzyme A
b. Zinc (Zn^{2+})
c. Flavin
d. ATP
e. NAD
Textbook Reference: *8.4 How Do Enzymes Work? p. 160, Table 8.1*

Application Questions

1. It is estimated that approximately 90 percent of energy that passes between levels in a food web is "lost" at each level. Explain the first law of thermodynamics and discuss why this apparent loss of energy does not contradict the law.
Textbook Reference: *8.1 What Physical Principles Underlie Biological Energy Transformations? p. 150*

2. Amylase is a digestive enzyme that breaks down starch and is secreted in the mouth of humans. Amylase functions well in the mouth but ceases to function once it hits the acidic stomach environment. Explain why amylase does not function in the stomach.
Textbook Reference: *8.5 How Are Enzyme Activities Regulated? p. 164*

3. The ultimate goal of metabolism is to drive ATP synthesis. ATP is considered the energy currency of the cell. Discuss how ATP couples endergonic and exergonic reactions and why it is so important in cellular functions.
Textbook Reference: *8.2 What Is the Role of ATP in Biochemical Energetics? p. 154*

4. Figure 8.17 shows the behavior of an allosteric enzyme that has binding sites for a negative regulator. Describe the behavior of an enzyme with binding sites for a positive regulator instead of a negative regulator.
Textbook Reference: *8.5 How Are Enzyme Activities Regulated? p. 163*

5. Explain how substrate concentration affects the rate of an enzyme-mediated reaction.
Textbook Reference: *8.5 How Are Enzyme Activities Regulated? p. 163*

6. Explain how free energy, total energy, temperature, and entropy are related.
Textbook Reference: *8.1 What Physical Principles Underlie Biological Energy Transformations? p. 151*

7. Use a graph to explain how temperature affects enzyme activity.
Textbook Reference: *8.5 How Are Enzyme Activities Regulated? p. 165*

8. You decide to purchase a new water heater and start looking at the energy efficiency ratings. You find one unit that is labeled as 100 percent energy efficient, and the salesperson says that the more efficient the appliance is, the more money you will save. However, you don't trust that the store is providing accurate information, and you do not buy the product. Was this decision correct?
Textbook Reference: *8.1 What Physical Principles Underlie Biological Energy Transformations? p. 151*

Answers

Diagram Exercise Answers

1.

2.

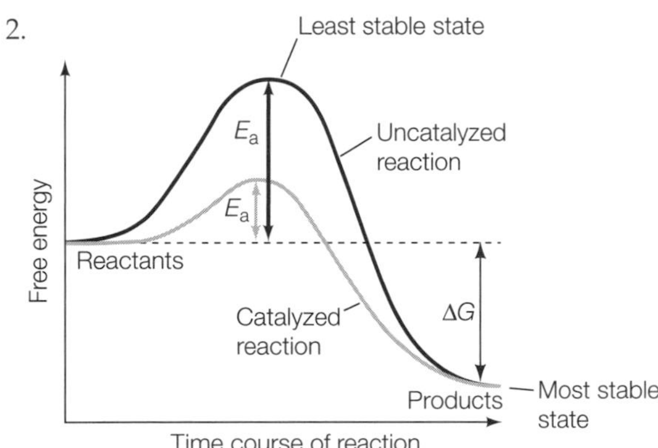

3.

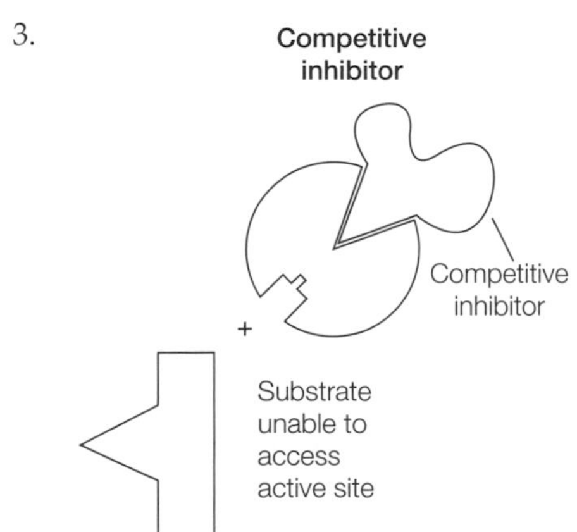

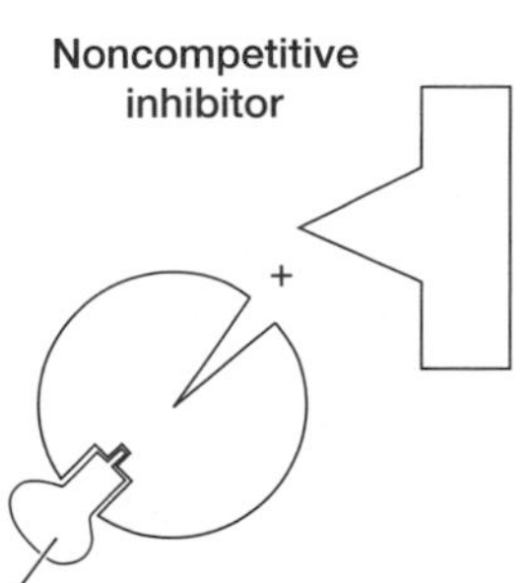

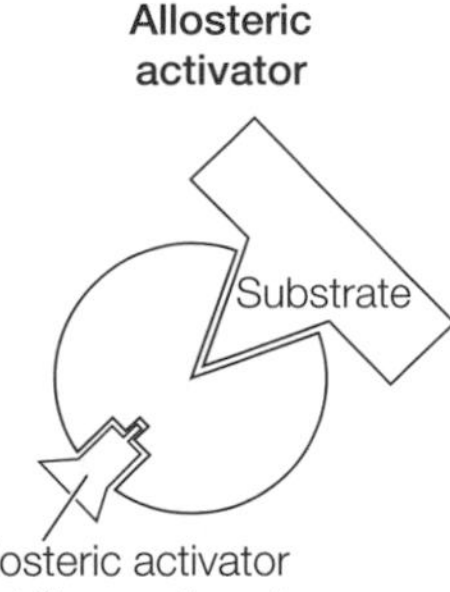

4.

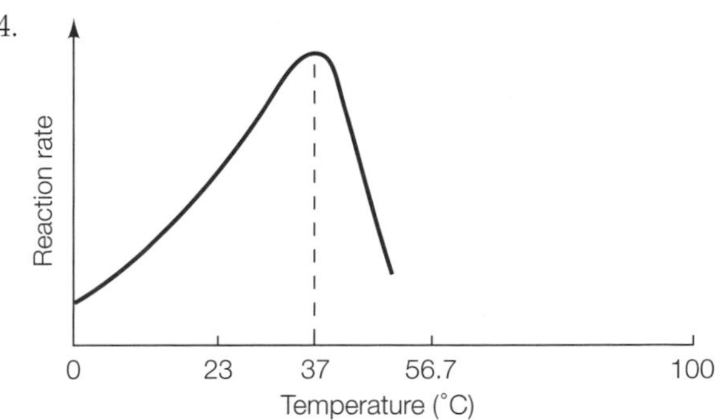

E. coli's optimal temperature is 37°C, matching its natural environment in the human gut. The enzyme's activity will increase slowly up to the optimal temperature, then rapidly drop off as the enzyme denatures at temperatures above the optimal.

5.

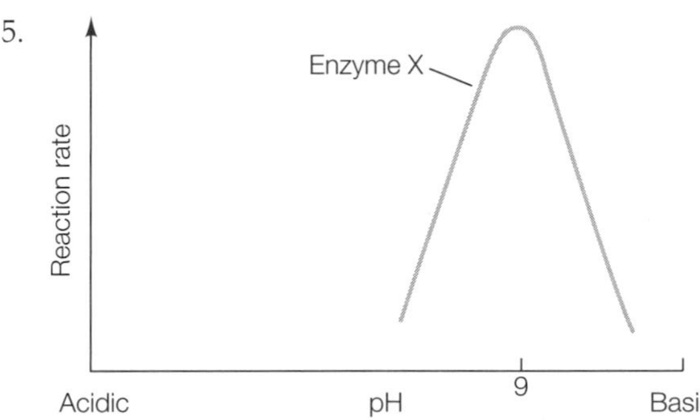

The enzyme prefers a basic environment. Your graph should be symmetrical around a peak at pH 9.0.

Knowledge and Synthesis Answers

1. **b.** The released energy is available to do work; therefore, it is kinetic energy. Some energy is lost in the form of entropy, but it will not be used by the cell to do work.
2. **a.** Potential energy is energy held within chemical bonds that may be converted to working kinetic energy.
3. **a.** Total energy = Free energy + entropy × temperature. Any increase in entropy is necessarily going to reduce free energy.
4. **b.** A negative ΔG indicates an exergonic reaction with energy being liberated. A large ΔG indicates that equilibrium lies toward completion.
5. **b.** ATP hydrolysis is exergonic, resulting in a ΔG of –7.3 kcal/mol.
6. **b.** Enzymes reduce activation energy and speed up reactions.
7. **d.** Enzymes are specific to particular substrates that may actually "adjust" the fit of the active site. They also function in specific, narrow optimum ranges of pH and temperature.
8. **e.** Cofactors, coenzymes, and prosthetic groups assist with the maintenance of an enzyme's three-dimensional shape and the conformation of the active osthetic groups are permanently bound to the enzyme.

9. **b.** Enzymes are not consumed or altered in any way during an enzyme-mediated reaction, and they function to lower the activation energy of a reaction. Ribozymes (composed of RNA) are catalysts, but are not true enzymes.
10. **a.** Competitive inhibitors compete for the active site with the substrate.
11. **b.** Destroying the three-dimensional structure of an enzyme, or denaturing it, is usually irreversible.
12. **c.** Within a given pathway, the products of the preceding step act as substrates for subsequent steps.
13. **b.** Substrate is converted to product and the enzyme is unchanged.
14. **a.** Products of a reaction that inhibit the enzymes via feedback inhibition are allosteric inhibitors.
15. **d.** Enzymes typically work at a maximal rate at a particular temperature or within a range of temperatures. This optimal temperature tends to be correlated with the body temperature of the organism.
16. **b.** The free energy levels of the reactants and products are not changed by enzymes. Only the activation energy is altered. See Figure 8.10.
17. **e.** The addition of more D to the second container will reduce the activity of Enzyme 2. The pathway from A + B going all the way to F will be the predominant reaction that takes place, leading to a greater final concentration of F in the second container. The first container will convert some of C into D, and thus the level of intermediate E and final product F will be lower than in the second container. (Hint: Draw a diagram of the experiment.)
18. **c.** Of the five compounds listed, only one is a prosthetic group, which is irreversibly bound to its target enzyme.

Application Answers

1. The first law of thermodynamics states that energy cannot be created or destroyed, but that it may be converted from one form to another. In the transfer between levels of a food web, approximately 90 percent of the energy is converted to unusable heat energy. There is a net loss of usable energy during each conversion, but the total amount of energy (usable and unusable) remains the same.
2. The pH optimum of amylase is approximately 7. At that pH, the protein has the three-dimensional shape to allow starch to bind to its active site and catalyze its hydrolysis. When it is at the stomach pH (approximately 2), the protein is denatured, and its three-dimensional shape and active site are lost; therefore, it can no longer catalyze the reaction.
3. The conversion of ATP to ADP and P_i releases approximately 7.3 kcal/mol of energy. This energy release fuels (endergonic) reactions in the cell. Equilibrium of the reaction is far to the right and favors the formation of ADP. In the converse, the formation of ATP from ADP and P_i is energy intensive and can be coupled to highly exergonic reactions within the cell. Thus, ATP functions as an energy shuttle between endergonic and exergonic reactions. The small size of the molecule and its ubiquitousness allow it to be available and move freely within the cell.
4. A positive regulator would stabilize the enzyme in its active form. In the absence of the regulator, the enzyme would alternate between its inactive and active forms. When it encountered substrate, it would bind it only if it happened to be in its active form. When bound to the regulator, the enzyme would be fixed in its active form, and it would bind substrate at a greater rate.
5. Increasing substrate concentration will result in an increased rate of reaction until all available active sites are occupied. At that point, no amount of substrate increase will increase the rate of reaction (see Figure 8.13).
6. Total energy = free energy + unusable energy × absolute temperature.
7. See Figure 8.21.
8. The decision was correct. An appliance with 100 percent energy efficiency is not possible. The second law of thermodynamics indicates that every time energy is transformed, some is lost in the form of entropy. An appliance with no energy lost to entropy therefore does not exist.

9 Pathways that Harvest Chemical Energy

The Big Picture

- Metabolic processes occur in pathways and are regulated by enzymes. These pathways may be further controlled by compartmentalization into organelles in eukaryotic cells. Because metabolic processes are energy-transferring reactions, ATP and NAD are necessary for shuttling energy and electrons between steps of the pathways. Redox reactions are the basis of metabolism. As one compound is oxidized, another is reduced.
- Glucose provides cellular energy. It may be metabolized in the absence of oxygen through glycolysis and fermentation. In the presence of oxygen, it is metabolized through glycolysis and cellular respiration in the form of the citric acid cycle, electron transport, and oxidative phosphorylation. Compared to aerobic respiration, anaerobic respiration produces significantly less cellular energy (in the form of ATP) from the same amount of glucose.
- Each step of the catabolic pathway is regulated by the products of subsequent pathways and depends on the presence of oxygen as the final electron acceptor. This level of control is achieved through allosteric regulation.

Common Problem Areas

- The biggest mistake you can make in studying glucose metabolism is to focus on memorizing the pathways without understanding what is happening. When you begin your study, do not focus on the pathways themselves. Focus on understanding the beginning and end products of each pathway and why the pathway does what it does. Once you conceptually understand why the pathway is present, move on to learning the pathway itself.
- Students often fail to see how regulation of the pathways occurs. By understanding how the presence or absence of intermediates affects the entire pathway, you can better understand the pathways themselves.

Study Strategies

- Remember that ATP and NAD are energy currencies. They move energy from pathway to pathway. In order to understand the pathways completely, think about the roles of ATP and NAD in redox reactions.
- Do not memorize pathways. Make sure you *understand* the pathways and how they are connected. Also understand how changes in one pathway can alter another pathway.
- This chapter is very visual in nature. You should spend a significant amount of time with the diagrams.
- Go to yourBioPortal.com to review the following tutorials and activities:

  Animated Tutorial 9.1 Electron Transport and ATP Synthesis

  Animated Tutorial 9.2 Two Experiments Demonstrate the Chemiosmotic Mechanism

  Web Activity 9.1 Energy Pathways in Cells

  Web Activity 9.2 Glycolysis and Fermentation

  Web Activity 9.3 The Citric Acid Cycle

  Web Activity 9.4 Respiratory Chain

  Web Activity 9.5 Energy Levels

  Web Activity 9.6 Regulation of Energy Pathways

Important Concepts

Glucose is the energy source most often used by living organisms.

- All living organisms require a source of energy to survive. Most organisms use glucose ($C_6H_{12}O_6$) as a metabolic fuel source. Cells oxidize glucose for energy: $C_6H_{12}O_6 + 6\ O_2 \rightarrow 6\ CO_2 + 6\ H_2O$ + energy. The energy is typically captured in ATP and is used to carry out cellular work.
- The metabolism of glucose involves up to three metabolic processes. In the first pathway, glycolysis, glucose is converted to pyruvate with a small net energy release. In the second pathway cellular respiration, the pyruvate from glycolysis is converted to CO_2 and energy in the form of ATP. Cellular respiration

occurs only in the presence of O_2 and is thus an aerobic process. In the absence of O_2, pyruvate is converted to lactic acid or ethanol with a small net energy release in the process of fermentation. Both fermentation and glycolysis occur in the absence of O_2, making them anaerobic processes.

- Complete conversion of glucose to CO_2 in the presence of O_2 releases 686 kcal/mol of energy. Incomplete breakdown through fermentation produces substantially less energy.
- ***For review, go to Diagram Exercise 1.***

Redox reactions are the electron transferring mechanism of metabolism.

- Energy is transferred from compound to compound through transfer of electrons in redox reactions.
- The gain of electrons or hydrogen atoms is called reduction, and the loss of electrons or hydrogen atoms is called oxidation. Oxidation and reduction are always coupled. In order for a material to lose an electron or a hydrogen atom, another material must accept it.
- Oxidizing agents accept electrons and become reduced. In metabolism, O_2 is the oxidizing agent.
- Reducing agents donate electrons and become oxidized. In metabolism, glucose is the reducing agent.
- The net ΔG of a redox reaction is negative; therefore, energy is liberated as heat.
- ***For review, go to Diagram Exercise 2.***

Just as ATP is an energy shuttle in metabolism, NAD is an electron shuttle in the redox reactions of metabolism.

- NAD (nicotinamide adenine dinucleotide) may exist as either NAD^+ or NADH + H^+ and thus acts as an electron carrier.
- Two electrons are transferred in the reduction of NAD^+ to NADH + H^+. The oxidation of NADH by O_2 is exergonic and liberates 52.4 kcal/mol. FAD (flavin adenine dinucleotide) is another electron carrier used in glucose metabolism.

Glycolysis is the process of converting glucose to pyruvate and usable energy.

- Glycolysis is a fundamental metabolic pathway occurring in the cytosol of the cell. It occurs during both aerobic and anaerobic cell respiration.
- During glycolysis, glucose is converted to two molecules of pyruvate, with a net production of two ATP and two NADH.
- Though glycolysis results in a net gain of two ATP, the initial steps are endergonic and require energy input.
- Refer to Figure 9.5 in your text to follow each of the ten enzyme-catalyzed reactions involved in the conversion of glucose to two molecules of pyruvate. The first five steps are endergonic and require energy input. Subsequent steps are exergonic.
- The following are key points to remember about glycolysis:
  - Glycolysis begins with a 6-C sugar that is ultimately converted to two 3-C pyruvate molecules.
  - During the initial three energy-intensive steps of glycolysis, two phosphate groups from two ATP molecules are added to the 6-C sugar to produce the 5-C fructose-1,6-biphosphate (FBP). During steps 4 and 5, the carbon ring is opened and split into two 3-C phosphorylated molecules: glyceraldehyde 3-phosphate (G3P).
  - The subsequent energy-producing steps occur when the phosphate groups are transferred to ADP to make ATP, and NAD^+ is reduced.
- The conversion of glyceraldehyde 3-phosphate (G3P) to 1,3-bisphosphoglycerate (BPG) during step 6 is significant because it is an oxidation process, and energy is stored in two NADH + H^+ after this conversion. The cell has only small amounts of NAD, so it must be recycled for glycolysis to continue.
- During steps 7 and 10, the phosphate groups on BPG (step 7) and phosphoenolpyruvate (PEP, step 10) are transferred to ADP to produce ATP in the process of substrate-level phosphorylation. Phosphorylation is the term used to describe the addition of a phosphate group to a molecule.
- During glycolysis, two ATPs are used and four ATPs are produced, resulting in a net gain of two ATPs.
- ***For review, go to Diagram Exercise 3.***

Aerobic conditions allow the oxidation of pyruvate to acetate, which is combined with coenzyme A to form acetyl CoA.

- Pyruvate oxidation to acetyl CoA occurs on the inner mitochondrial membrane with the help of the pyruvate dehydrogenase complex.
- CO_2 is liberated, and NADH + H^+ and acetyl CoA are produced.

The citric acid cycle oxidizes acetate to CO_2 and forms $FADH_2$, NADH + H^+, and ATP.

- The citric acid cycle consists of eight reactions. Figure 9.7 details the reactions of the citric acid cycle. In the initial step, citrate is formed by the combination of acetyl CoA and oxaloacetate. During this step, CoA is liberated and recycled for combination with another pyruvate.
- The concentrations of the intermediate molecules involved in the citric acid cycle are maintained at a constant or steady state.
- In the third and fourth reactions of the cycle, one CO_2 molecule and one NADH + H^+ are formed at each step. Reaction 5 involves the conversion of GDP + P_i into GTP, which is then used to produce ATP. Reaction 6 yields one $FADH_2$. The last reaction in the cycle results in the formation of one NADH + H^+ and oxaloacetate, which reenters the cycle at the beginning. So one turn

of the cycle produces two CO_2 molecules, three NADH + H^+, one $FADH_2$, and one ATP.

- Remember that each glucose molecule yields two acetyl CoA that enter the cycle, so the yield-per-glucose is doubled.
- The majority of the citric acid cycle takes place in the mitochondrial matrix.
- The NADH produced by glycolysis and the citric acid cycle must be reoxidized during fermentation in the absence of oxygen or oxidative phosphorylation in the presence of oxygen.
- ***For review, go to Diagram Exercise 4.***

Energy is liberated from reduced NAD^+ and FAD through electron transport in the respiratory chain.

- The process of oxidative phosphorylation involves the passing of electrons through a series of membrane-associated electron carriers in the mitochondria.
- The respiratory chain is composed of four enzyme complexes (I, II, III, IV) plus cytochrome *c* and ubiquinone (Q). The first two complexes shuttle the electrons of NADH + H^+ and $FADH_2$ to Q. The third complex moves electrons from Q to cytochrome *c*. The final complex passes the electrons on to O_2, the ultimate electron acceptor, resulting in H_2O as a by-product. Figures 9.8 and 9.8 detail the respiratory chain.
- The purpose of the electron shuttling is to pump protons (H^+) across the mitochondrial membrane against a concentration gradient. This movement of electrons by these proton pumps results in the establishment of a proton gradient across the inner mitochondria membrane. This also polarizes the membrane by creating a charge differential across the membrane. These coupled effects are called the proton-motive force.

ATP is produced as protons diffuse across the mitochondrial membrane.

- An ATP synthase in the inner mitochondrial membrane couples the movement of protons with the synthesis of ATP in the chemiosmotic mechanism. ATP synthase is a protein made of two parts. The F_0 unit spans the membrane and acts as the channel for H^+, and the F_1 unit is the site of ATP synthesis. (Chemiosmosis is not limited to the inner mitochondrial membrane and is also found in chloroplasts and some bacterial membranes; this concept will be explained again in later chapters.)
- Because the ATP synthase reaction may go in either direction, ATP is removed from the mitochondrial matrix immediately to favor ATP formation. Also, the proton gradient is maintained to favor ATP synthesis.
- If the proton flow is uncoupled from ATP synthase, the resulting energy is lost as heat. This phenomenon is seen in thermoregulation.
- ***For review, go to Diagram Exercise 5.***

In the absence of O_2, ATP is generated via fermentation.

- Fermentation occurs in the cell cytosol. Fermentation reduces pyruvate (or a metabolite) and returns NAD for use in glycolysis in the absence of O_2. This allows the cell to continue to produce small amounts of ATP by glycolysis.
- There are two types of fermentation based on the final end products produced. In lactic acid fermentation, pyruvate serves as the electron acceptor and lactic acid is produced. In yeast and some plant cells, alcoholic fermentation occurs with ethyl alcohol as the end product.

Fermentation and cellular respiration have different energy yields.

- Fermentation yields a net total of two ATP for every one glucose molecule, regardless of whether it occurs through lactic acid fermentation or alcoholic fermentation.
- Aerobic respiration yields a total of 32 ATP for every one glucose molecule. Two ATP are produced in glycolysis, two from the citric acid cycle, and 28 from the respiratory chain.
- The general formula for aerobic cellular respiration is:

$$C_6H_{12}O_6 + 6\ O_2 \rightarrow 6\ O_2 + 6\ H_2O + 32\ ATP$$

Glycolysis and cellular respiration occur while other metabolic processes are occurring.

- Components of the metabolic pathways are also components of other metabolic processes, including catabolism and anabolism.
- Catabolism is the breaking down of molecules to release energy. Polysaccharides, lipids, and proteins all feed into the metabolic pathways at different points. Polysaccharides are broken down into glucose and enter at glycolysis. Lipids are broken down and can enter at glycolysis or the citric acid cycles as acetyl CoA. Proteins enter glycolysis and the citric acid cycle as amino acids.
- Just as intermediates can be broken down (catabolized) to release energy, they can also be synthesized in anabolism for storage or use by the cell. Gluconeo-genesis is the anabolic process by which glucose is produced from the intermediates of glycolysis.
- Metabolism, both anabolism and catabolism, is regulated by enzymes to maintain stable concentrations of intermediates for metabolic homeostasis.

Metabolic enzymes are allosterically controlled to regulate the production of ATP.

- Metabolism responds to both positive and negative feedback provided by intermediates and end products of the process.
- Each process has a specific control point for regulation. The control point for glycolysis is phosphofructokinase,

which is inhibited by ATP. This allows glycolysis to speed up during fermentation and slow down during cellular respiration.

- The main control point for the citric acid cycle is isocitrate dehydrogenase. NADH + H^+ and ATP inhibit the enzyme to slow down the process, and NAD^+ and ADP act as activators.
- If the citric acid cycle slows due to abundant ATP, glycolysis slows as well.
- If excess acetyl CoA is produced and ATP is abundant, acetyl CoA can be shuttled to fatty acid synthesis for storage.

Test Yourself

Diagram Exercises

1. In the diagram below, name the processes bracketed, enter the products in the blanks indicated, and indicate how many ATP molecules are generated.

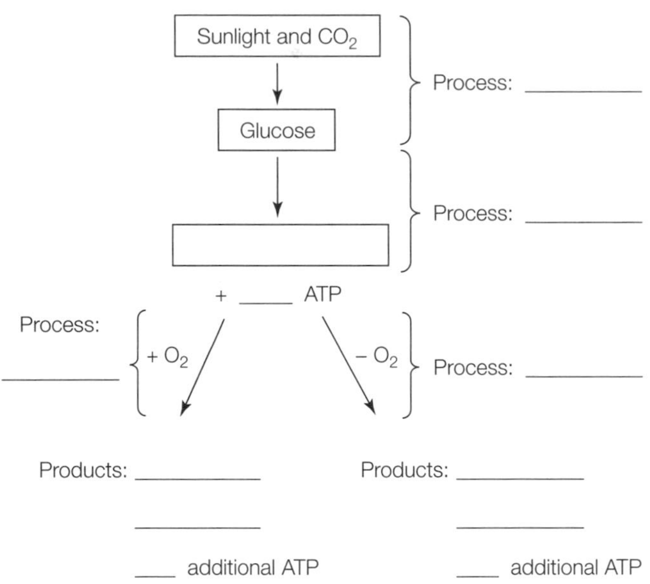

Textbook Reference: *9.1 How Does Glucose Oxidation Release Chemical Energy? p. 170, Figure 9.1 and p. 182, Figures 9.11 and 9.12*

2. In the diagram below, label each compound as being either oxidized or reduced. Circle the two lowest energy compounds. Which is the oxidizing agent and which is the reducing agent?

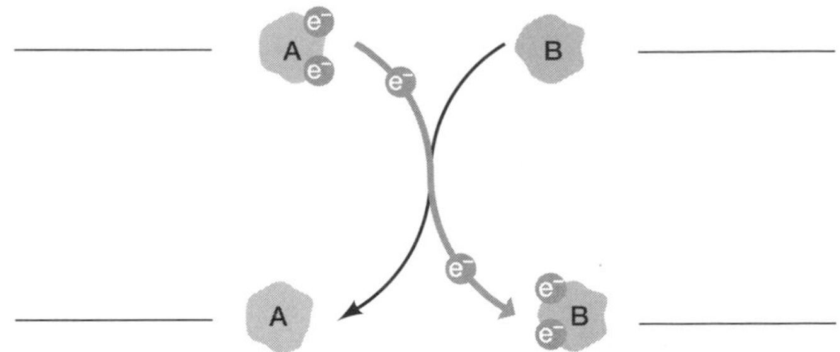

Textbook Reference: *9.1 How Does Glucose Oxidation Release Chemical Energy? p. 170, In-text art*

3. In the diagram below, fill in the numbers of reactants and products on each side of the equation.

1 Glucose { + ____ ADP; + ____ NAD^++ H^+; + ____ P_i; + ____ ATP } → ____ Pyruvate { + ____ ATP; + ____ NADH; + ____ H_2O; + ____ ADP; + ____ P_i }

Cross out anything appearinig on both sides of the equation to get the net reation shown below.

1 Glucose { + ____ ADP; + ____ NAD^++ H^+; + ____ P_i } → ____ Pyruvate { + ____ ATP; + ____ NADH; + ____ H_2O }

Textbook Reference: *9.2 What Are the Aerobic Pathways of Glucose Metabolism? p. 173, Figure 9.5*

4. In the diagram below, label all of the structures and indicate the necessary reactants and products for the different steps of the pathway.

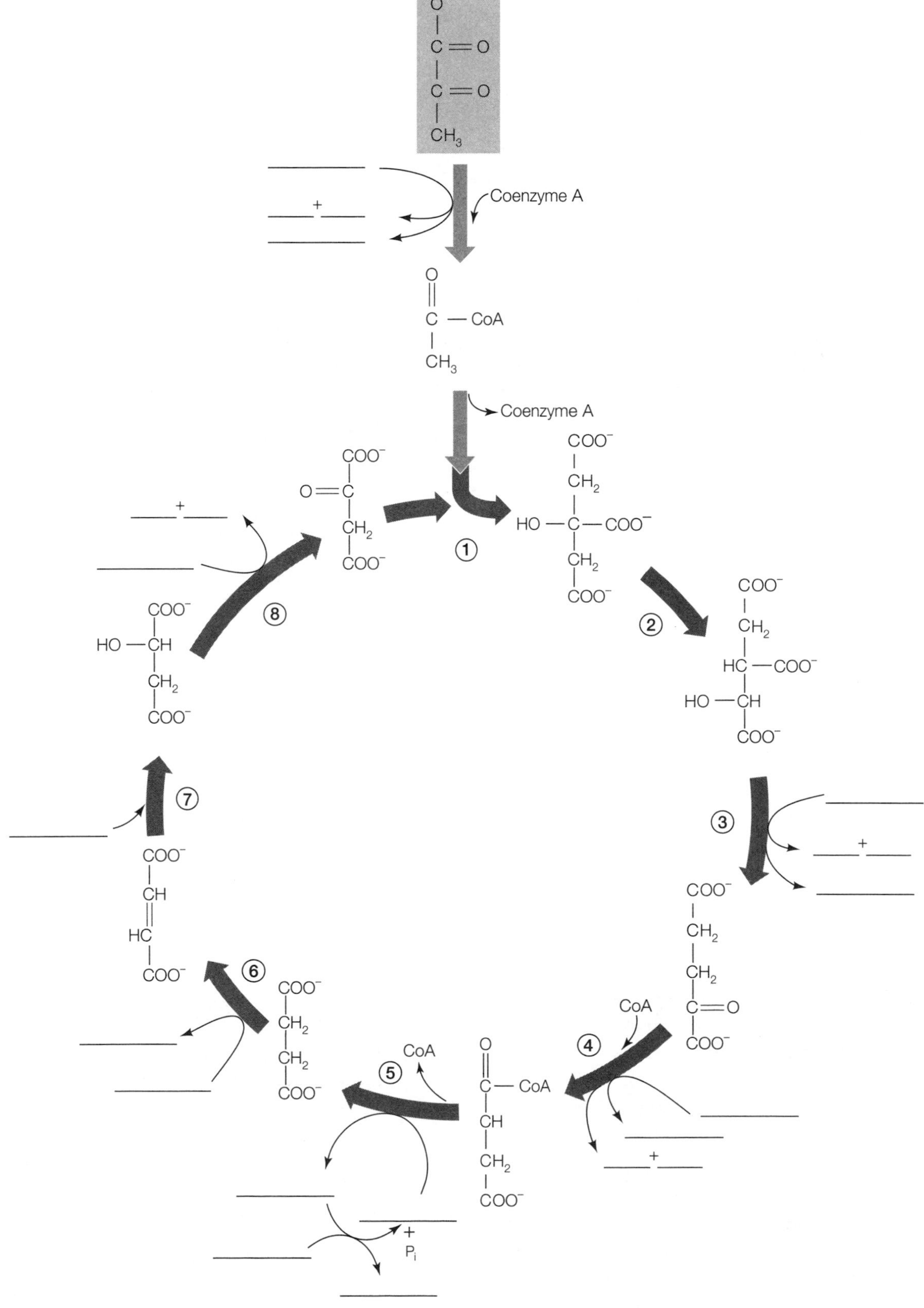

Textbook Reference: *9.2 What Are the Aerobic Pathways of Glucose Metabolism? p. 176, Figure 9.7*

5. Label the diagram below of the respiratory chain and chemiosmosis in the mitochondria. Be sure to label the mitochondrial parts as well.

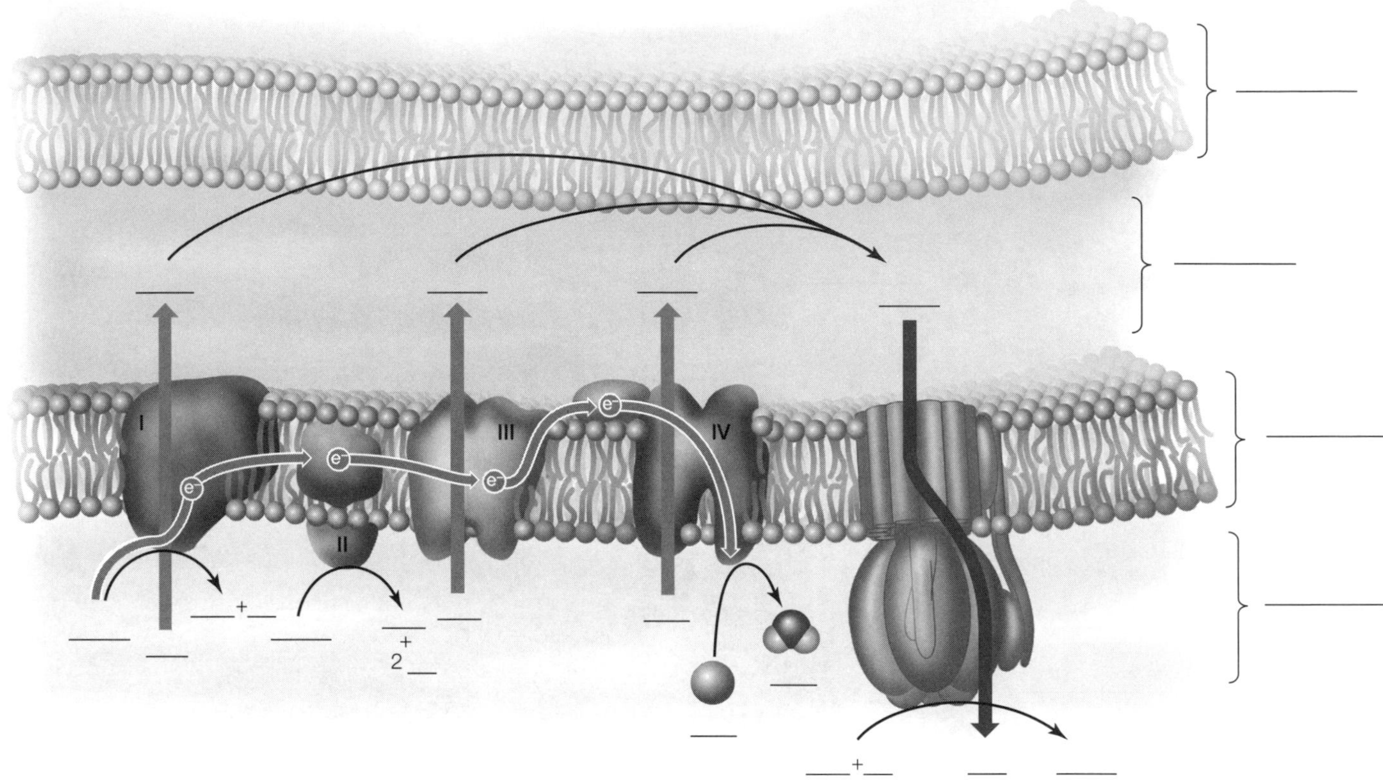

Textbook Reference: 9.3 *How Does Oxidative Phosphorylation Form ATP? p. 179, Figure 9.9*

Knowledge and Synthesis Questions

1. Which of the following cellular metabolic processes is active in *all* cells, regardless of the presence or the absence of oxygen?
   a. The citric acid cycle
   b. Electron transport
   c. Glycolysis
   d. Fermentation
   e. Pyruvate oxidation
   Textbook Reference: *9.1 How Does Glucose Oxidation Release Chemical Energy? p. 170, Figure 9.1*
2. Which of the following statements regarding glycolysis is *false*?
   a. A 6-C sugar is broken down to two 3-C molecules.
   b. Two ATP molecules are consumed.
   c. Glycolysis requires oxygen.
   d. A net sum of two ATP molecules is generated.
   e. Glycolysis occurs in the cytosol.
   Textbook Reference: *9.2 What Are the Aerobic Pathways of Glucose Metabolism? p. 173, Figure 9.5*
3. During which process is most ATP generated in the cell?
   a. Glycolysis
   b. The citric acid cycle
   c. Electron transport coupled with chemiosmosis
   d. Fermentation
   e. Pyruvate oxidation
   Textbook Reference: *9.4 How Is Energy Harvested from Glucose in the Absence of Oxygen? p. 183, Figure 9.13*
4. One purpose of the electron transport chain is to
   a. cycle NADH + H^+ back to NAD^+.
   b. use the intermediates from the citric acid cycle.
   c. break down pyruvate.
   d. increase the number of protons in the mitochondrial matrix.
   e. consume excess ATP.
   Textbook Reference*: 9.3 How Does Oxidative Phosphorylation Form ATP? p. 178*
5. Cellular respiration is allosterically controlled. Which of the following act(s) as inhibitors at the various control points?
   a. ATP
   b. NADH
   c. Both ATP and NADH
   d. ADP
   e. Both ADP and NADH
   Textbook Reference: *9.5 How Are Metabolic Pathways Interrelated and Regulated? p. 186*

6. Which of the following describes the role of the inner mitochondrial membrane?
   a. It acts as an anchor for the membrane-associated enzymes of cellular respiration.
   b. It allows for the establishment of a proton gradient.
   c. It separates the mitochondria's environment from that of the cytosol.
   d. It anchors enzymes and allows for the establishment of the proton gradient, but it is not involved in separating the contents of the mitochondria from the cytosol.
   e. It anchors enzymes, allows for the establishment of the proton gradient, and is involved in separating the contents of the mitochondria from the cytosol.

   Textbook Reference: *9.3 How Does the Oxidative Phosphorylation Form ATP? p. 179, Figure 9.9*

7. In the following redox reaction, _______ is oxidized and _______ is reduced.

   Glyceraldehyde 3-phosphate (G3P) + NAD^+ + H^+ + P_i → 1,3-Bisphosphoglycerate (BPG) + NADH
   a. G3P; NAD^+
   b. BPG; NADH + H^+
   c. G3P; NADH + H^+
   d. NAD^+; NADH + H^+
   e. The equation does not show a redox reaction.

   Textbook Reference: *9.2 What Are the Aerobic Pathways of Glucose Metabolism? p. 173*

8. Which of the following statements about redox reactions is true?
   a. Oxidizing agents accept electrons.
   b. Oxidizing agents donate electrons.
   c. A molecule that accepts electrons is said to be oxidized.
   d. A molecule that donates electrons is said to be reduced.
   e. Oxidizing agents accept electrons and are reduced in the process.

   Textbook Reference: *9.1 How Does Glucose Oxidation Release Chemical Energy? p. 170*

9. Cyanide poisoning inhibits aerobic respiration at cytochrome *c* oxidase. Which of the following is *not* a result of cyanide poisoning at the cellular level?
   a. Reduction of oxygen to water
   b. Cessation of ATP synthesis in the mitochondria because electron transport is never completed
   c. Switching of cells to anaerobic respiration and fermentation if possible
   d. Continuation of glycolysis as long as NAD^+ is available
   e. Less acidic pH of the intermenbrane space

   Textbook Reference: *9.3 How Does Oxidative Phosphorylation Form ATP? p. 179, Figure 9.9*

10. Which of the following is correctly matched with its catabolic product?
    a. polysaccharides → amino acids
    b. lipids → glycerol and fatty acids
    c. proteins → glucose
    d. polysaccharides → glycerol and fatty acids
    e. nucleic acids → monosaccharides

    Textbook Reference: *9.5 How Are Metabolic Pathways Interrelated and Regulated? p. 184*

11. The main function of cellular respiration is the
    a. conversion of energy stored in the chemical bonds of glucose to an energy form that the cell can use.
    b. recovery of NAD^+ from NADPH.
    c. conversion of kinetic to potential energy.
    d. creation of energy in the cell.
    e. elimination of excess glucose from the cell.

    Textbook Reference: *9.1 How Does Glucose Oxidation Release Chemical Energy? p. 170*

12. Which of the following statements concerning the synthesis of ATP in the mitochondria is *false*?
    a. ATP synthesis cannot occur without the presence of ATP synthase.
    b. The proton-motive force is the establishment of a charge and concentration gradient across the mitochondrial membrane.
    c. The proton-motive force drives protons back across the membrane through channels established by the ATP synthase channel protein.
    d. The ATP synthase protein is composed of two units.
    e. The intermembrane space is more basic than the mitochondrial matrix.

    Textbook Reference: *9.3 How Does Oxidative Phosphorylation Form ATP? p. 179, Figure 9.9*

13. Which of the following does *not* occur in the mitochondria of eukaryotic cells?
    a. Fermentation
    b. Oxidative phosphorylation
    c. Citric acid cycle
    d. Electron transport chain
    e. Creation of a proton gradient

    Textbook Reference: *9.1 How Does Glucose Oxidation Release Chemical Energy? p. 172, Table 9.1*

14. The largest change in free energy during glycolysis occurs at which reaction?
    a. Reaction 2: G6P → F6P
    b. Reaction 5: DAP → G3P
    c. Reaction 6: G3P → BPG + NADH
    d. Reaction 7: BPG → 3PG + ATP
    e. Reaction 8: 3PG → 2PG

    Textbook Reference: *9.2 What Are the Aerobic Pathways of Glucose Metabolism? p. 173, Figure 9.5*

15. Which of the following is recycled and reused in cellular metabolism?
    a. ADP
    b. NAD^+
    c. FAD
    d. P_i
    e. All of the above

    Textbook Reference: *9.2 What Are the Aerobic Pathways of Glucose Metabolism? pp. 172–173*

16. For each molecule of glucose, how many ATPs are synthesized in fermentation?
   a. 0
   b. 1
   c. 2
   d. 3
   e. 4
   Textbook Reference: *9.2 What Are the Aerobic Pathways of Glucose Metabolism? pp. 172–173*
17. If additional malate is added to a cell undergoing cellular respiration, there will be
   a. an increase in CO_2 production but no increase in ATP synthesis.
   b. an increase in CO_2 production and a decrease in ATP synthesis.
   c. an increase in both the CO_2 production and ATP synthesis.
   d. a decrease in both the CO_2 production and ATP synthesis.
   e. no change in the rates of CO_2 production or ATP synthesis.
   Textbook Reference: *9.2 What Are the Aerobic Pathways of Glucose Metabolism? pp. 176–177*

Application Questions

1. Why is oxygen necessary for aerobic respiration?
   Textbook Reference: *9.3 How Does Oxidative Phosphorylation Form ATP? p. 177*
2. Glycolysis yields two molecules of pyruvate, two ATP, and two NADH + H^+, regardless of whether oxygen is present or not. What are the fates of these molecules in the absence of oxygen? What would happen if NADH + H^+ was not recycled?
   Textbook Reference: *9.4 How Is Energy Harvested from Glucose in the Absence of Oxygen? pp. 181–182*
3. Explain how the proton-motive force drives chemiosmosis.
   Textbook Reference: *9.3 How Does Oxidative Phosphorylation Form ATP? p. 179*
4. The fate of acetyl CoA differs according to how much ATP is present in the cell. Explain what happens to acetyl CoA when ATP is limited, and compare that to what happens when acetyl CoA is abundant. How do these processes help regulate metabolism?
   Textbook Reference: *9.5 How Are Metabolic Pathways Interrelated and Regulated? p. 184*
5. Cellular respiration occurs simultaneously with many other cellular processes. Describe, in general, how cellular respiration interacts with other cellular metabolic events.
   Textbook Reference: *9.5 How Are Metabolic Pathways Interrelated and Regulated? p. 184*
6. Compare and contrast energy yields from aerobic respiration and fermentation.
   Textbook Reference: *9.4 How Is Energy Harvested from Glucose in the Absence of Oxygen? p. 182*
7. One of the by-products of aerobic cellular respiration is carbon dioxide. Assuming you begin with labeled glucose, trace the fate of that molecule until carbon dioxide is released.
   Textbook Reference: *9.2 What Are the Aerobic Pathways of Glucose Metabolism? p. 173, Figure 9.5, and p. 176, Figure 9.7*
8. Identify the controlling steps of glycolysis, the citric acid cycle, and electron transport. Which regulators affect each of these steps?
   Textbook Reference: *9.5 How Are Metabolic Pathways Interrelated and Regulated? p. 186, Figure 9.16*
9. Cyanide kills by inhibiting cytochrome oxidase in the mitochondria so that oxygen can no longer be utilized and the electron transport chain halts. However, many cells in the human body are capable of lactic acid fermentation. Since cyanide does not inhibit glycolysis and fermentation, what could explain cyanide's lethal affect on humans?
   Textbook Reference: *9.4 How Is Energy Harvested from Glucose in the Absence of Oxygen? p. 181*
10. Where do the bubbles in beer come from?
   Textbook Reference: *9.4 How Is Energy Harvested from Glucose in the Absence of Oxygen? p. 182*
11. When a person consumes a packet of pure sugar and burns it for energy, where does the carbon in the sugar ultimately go? Is the same true of the carbons in fat molecules when a person loses weight?
   Textbook Reference: *9.3 How Does Oxidative Phosphorylation Form ATP? p. 176, Figure 9.7*

Answers

Diagram Exercise Answers

1.

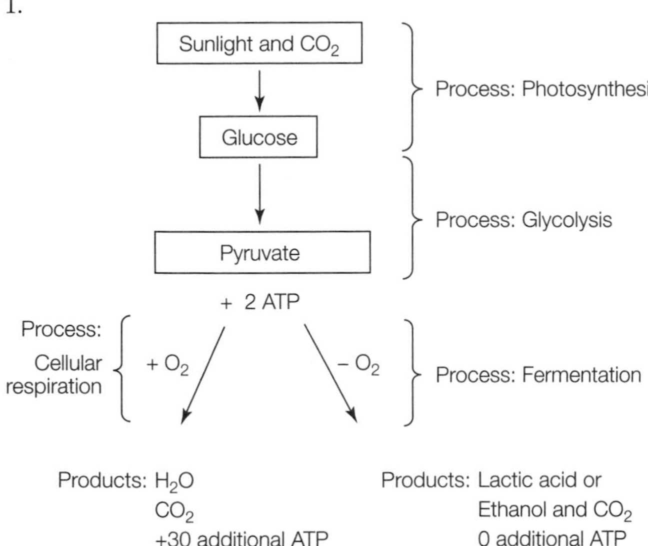

2.

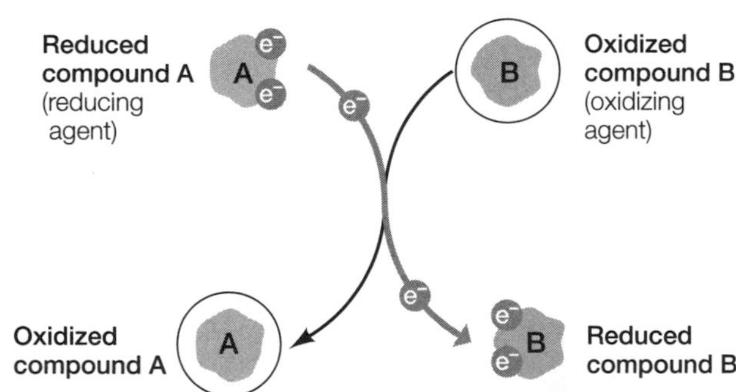

3.

1 Glucose { + ~~4~~ ADP (becomes 2 ADP) + 2 NAD^+ + H^+ + 4 P_i (becomes 2 P_i) + ~~2 ATP~~ } → 2 Pyruvate { + ~~4~~ ATP (becomes 2 ATP) + 2 NADH + 2 H_2O + ~~2 ADP~~ + ~~2 P_i~~ }

Net Reaction:

1 Glucose { + 2 ADP + 2 NAD^+ + H^+ + 2 P_i } → 2 Pyruvate { + 2 ATP + 2 NADH + 2 H_2O }

4.

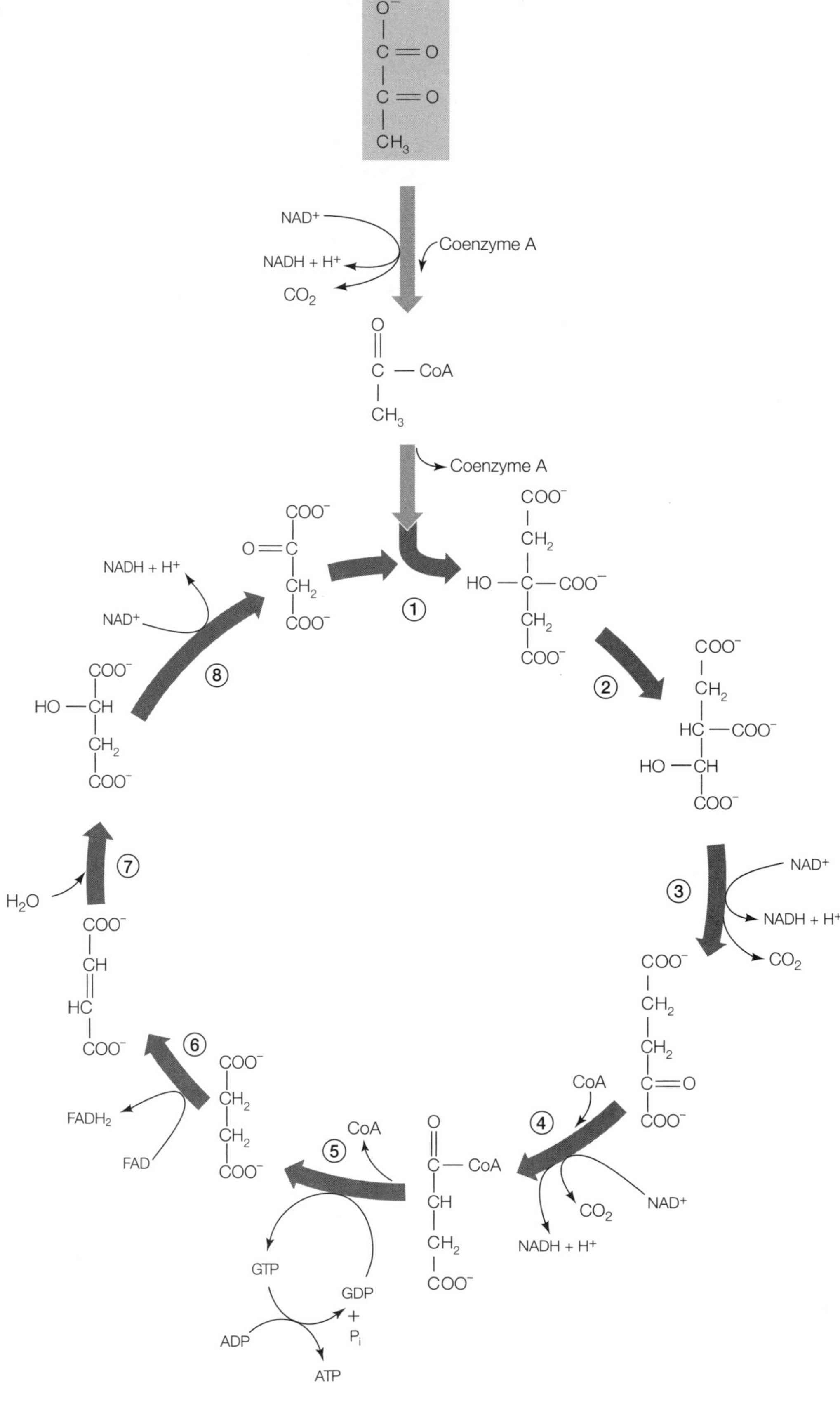

NAD+
NADH + H+
CO2
Coenzyme A
C — CoA
CH3
Coenzyme A
NADH + H+
NAD+
1
2
3
4
5
6
7
8
HO — C — COO−
HC — COO−
HO — CH
NAD+
NADH + H+
CO2
CoA
CO2
NAD+
NADH + H+
C— CoA
CoA
GTP
GDP
+
Pi
ADP
ATP
FADH2
FAD
H2O
HO — CH

5.

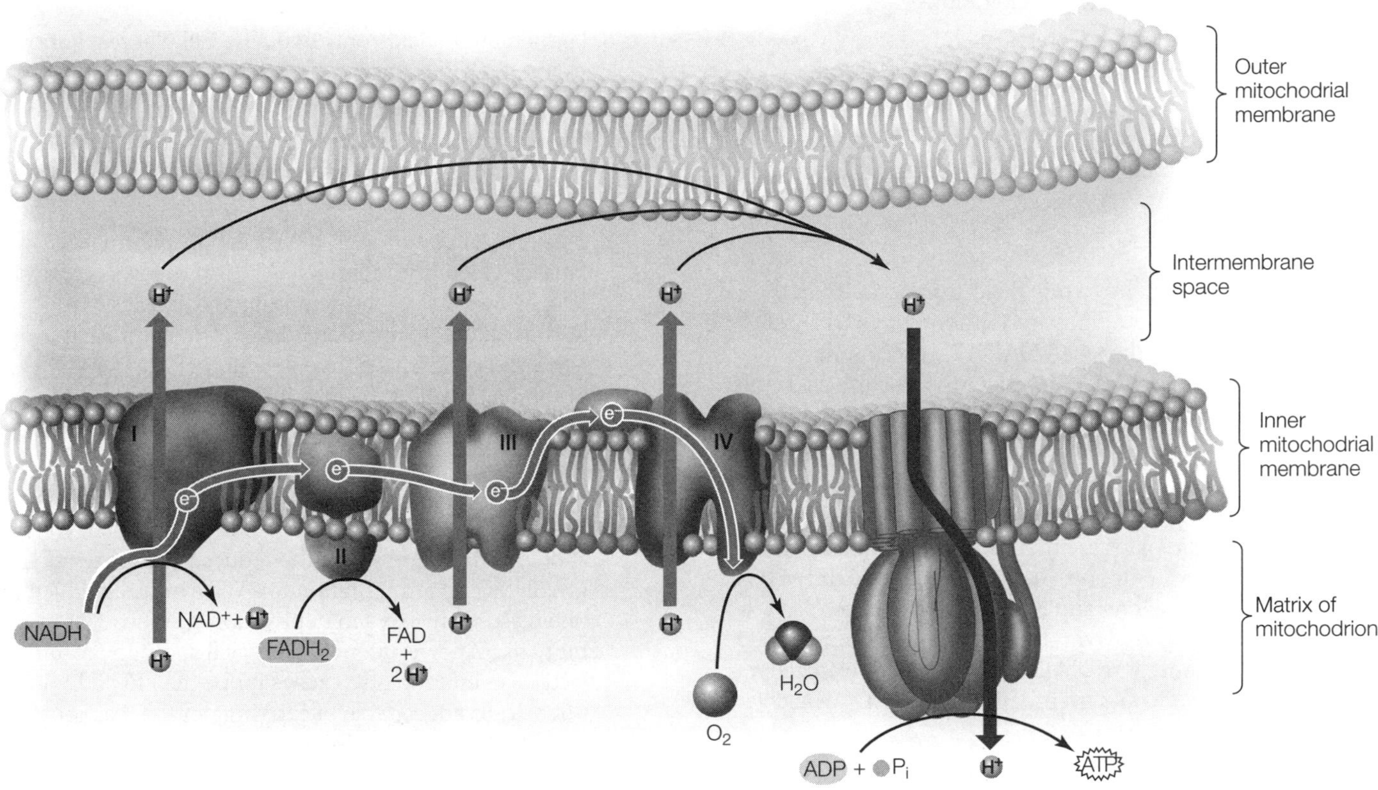

Knowledge and Synthesis Answers

1. **c.** Glycolysis proceeds during both fermentation and cellular respiration. Only in cellular respiration is oxygen needed as the terminal electron acceptor of the pathway.
2. **c.** During glycolysis, 6-C glucose is broken down into two 3-C pyruvate molecules. In the process, four total ATP are produced, but two are consumed, leaving a net production of two ATP molecules. No oxygen is required in glycolysis.
3. **c.** Most of the ATP produced during cellular respiration is produced by electron transport and chemiosmosis coupled in oxidative phosphorylation.
4. **a.** The electron transport chain is responsible for oxidizing NADH + H^+ back to NAD^+.
5. **c.** Both ATP and NADH allosterically control metabolism. ATP controls both phosphofructokinase and isocitrate dehydrogenase, which are commitment steps for glycolysis and the citric acid cycle, respectively. NADH controls isocitrate dehydrogenase.
6. **c.** The mitochondrial membrane is necessary for the anchoring of proteins as well as the establishment of a barrier across which a gradient can be established.
7. **a.** A molecule is oxidized when it loses electrons or protons and is reduced when it gains electrons or protons. In this reaction, G3P donates electrons and therefore is oxidized, while NAD^+ accepts them and thus is reduced.
8. **e.** Oxidizing agents accept electrons and cause oxidation of another molecule. Reducing agents donate electrons and cause the reduction of another molecule.
9. **a.** Cyanide stops aerobic cellular respiration because cytochrome *c* oxidase loses the ability to reduce oxygen to water. Those cells that can switch to anaerobic respiration (fermentation) do so and use their remaining glucose more quickly. Cells that cannot make that switch die.
10. **b.** Lipids are broken down into glycerol and fatty acids; polysaccharides are broken down into glucose; proteins are broken down into amino acids. Nucleic acids are converted into some amino acids and fed into the citric acid cycle.
11. **a.** Cellular respiration is the cell's way of converting potential energy in the chemical bonds of glucose to potential energy that the cell ultimately can use.
12. **a.** Substrate level phosphorylation occurs in step 5 of the citric acid cycle in the mitochondria. The intermembrane space is more acidic than the matrix.
13. **a.** Fermentation occurs in the cytosol, whereas all the other processes occur in the mitochondria of eukaryotic cells.

14. **c.** The largest change in free energy occurs in reaction 6, with more than 100 kcal/mol released.
15. **e.** ADP, NAD^+, FAD, and P_i are all recycled and reused in the process of cellular respiration.
16. **a.** Fermentation regenerates NAD^+ so that glycolysis can continue.
17. **c.** Additional malate will increase the carbon compounds cycling through the citric acid cycle, resulting in an increase in the products of the citric acid cycle, including ATP and CO_2.

Application Answers

1. Oxygen acts as the terminal electron acceptor in the electron transport pathway. Without it, NADH + H^+ cannot be cycled back to NAD^+. The accumulated NADH + H^+ acts as an inhibitor to the citric acid cycle and effectively shuts it down. Therefore, in the absence of oxygen, a cell can only undergo glycolysis.
2. In the absence of oxygen, pyruvate is either reduced to lactate (in lactic acid fermentation) or it is metabolized and its metabolites are reduced to ethyl alcohol (in alcoholic fermentation). In either case, NADH + H^+ is the reducing agent, and it is oxidized back to NAD^+ in the process. The two molecules of ATP would be used as cellular energy. If NADH + H^+ was not oxidized to NAD^+, there would eventually be no NAD^+ available for glycolysis.
3. The proton-motive force results in a concentration and charge gradient across the mitochondrial membrane. For that gradient to equalize, the protons must flow through a channel protein. If this channel protein has an associated ATP synthase, ATP is generated as protons flow through.
4. If ATP is limited, acetyl CoA enters the citric acid cycle, and cellular respiration utilizes it to produce ATP. If ATP is abundant, acetyl CoA is shuttled to fatty acid synthesis, thus storing the energy in chemical bonds.
5. Consult Figure 9.14 to see where different metabolic pathways in the cell interact.
6. Fermentation yields only two ATP. Cellular respiration yields 32 ATP.
7. Follow Figures 9.5 and 9.7. Carbon dioxide is liberated when pyruvate is oxidized to acetate, isocitrate is oxidized to α-ketoglutarate, and α-ketoglutarate is oxidized to succinyl CoA.
8. This control point for glycolysis is phosphofructokinase, which is inhibited by ATP. This allows glycolysis to speed up during fermentation and slow down during cellular respiration. The control point for the citric acid cycle is isocitrate dehydrogenase. NADH + H^+ and ATP inhibit the enzyme, and NAD^+ and ADP are activators. Electron transport is controlled by the amount of NADH + H^+ fed in and by NADH-Q reductase.
9. Based on what you have learned in this chapter, you should be thinking about the reduced efficiency of glycolysis and fermentation for ATP synthesis, which is one reason why oxygen deprivation is so deadly to humans. (And, as later chapters will show, the resulting lactic acid buildup also causes problems.) In addition, not all cells are capable of carrying out the reactions of fermentation. Brain cells, for example, will simply die in the absence of oxygen.
10. The bubbles in beer are bubbles of CO_2 released during the fermentation of pyruvate into ethyl alcohol. See Figure 9.12.
11. The carbon in the sugar is exhaled in the form of CO_2. When someone loses weight, the carbon from the fat molecules is also exhaled in the form of CO_2.

10 Photosynthesis: Energy from Sunlight

The Big Picture

- Photosynthesis allows organisms with the appropriate pigments and metabolic processes to convert light energy from the sun to chemical energy that can be used in the cell or stored. This conversion process forms the basis of all food webs and is vital to life on Earth. Photosynthetic organisms fix CO_2 into sugars. This process requires water and light energy and produces the by-products O_2 and water.
- The process of photosynthesis occurs in two steps. The first step utilizes light energy to produce ATP and NADPH with the help of the electron transport system. The movement of electrons through the electron transport system generates cellular energy in the form of ATP that can be used in the fixing of CO_2. The fixing of CO_2 occurs in the Calvin cycle and is mediated by an enzyme called rubisco, and the reactions are CO_2 concentration dependent.
- Rubisco can also act as an oxygenase and allow a plant to go through a metabolically expensive process called photorespiration. Plants that grow in conditions favoring photorespiration have evolved a wide variety of systems to avoid photorespiration.

Common Problem Areas

- There is the temptation to learn pathways without understanding why they occur. Be sure to focus on why the plant has a particular pathway before attempting to learn the pathway itself.
- It is very common to think that the light reactions happen in the light and the Calvin cycle occurs in the dark. This is not the case! The light reactions must occur simultaneously with the Calvin cycle to supply the needed energy in the form of ATP and NADPH + H^+.
- Remember that plants respire as well as photosynthesize and that photorespiration and respiration are different processes. All plant cells go through cellular respiration, which many students forget as they study Chapters 9 and 10. Cellular respiration takes the energy stored in photosynthesis and makes it available to drive other cellular processes. Photorespiration is energy expensive and is of little benefit to the plant. Photosynthesis occurs in specialized photosynthetic cells at the same time as respiration.
- Be very careful not to confuse the pathways for glycolysis, fermentation, and cellular respiration from Chapter 9 with the pathways of photosynthesis in Chapter 10. There are connections between the two pathways (see Figure 10.20) but they are not interchangeable. Read questions carefully and identify which set of pathways you are being asked about.

Study Strategies

- The process of photosynthesis is easier to understand when you visualize what is happening. Use the figures in the book to understand where energy and carbon are flowing. Remember that the plant is taking light energy and converting it to chemical energy that can be stored. In the process it uses CO_2 and gives off O_2. Pay attention to where energy is flowing and how this process proceeds.
- The properties of chlorophyll and light itself greatly affect how well a plant photosynthesizes. Take some time to understand the properties of light and how pigments interact with light.
- Rubisco is the most abundant protein on the planet. Focus on how the relative concentrations of O_2 and CO_2 dictate how the enzyme functions. Also understand that C_4 and CAM plants have modifications to avoid photorespiration.
- Go to yourBioPortal.com to review the following tutorials and activities:

  Animated Tutorial 10.1 The Source of the Oxygen Produced by Photosynthesis

  Animated Tutorial 10.2 Photophosphorylation

  Animated Tutorial 10.3 Tracing the Pathway of CO_2

  Web Activity 10.1 The Calvin Cycle

  Web Activity 10.2 C_3 and C_4 Leaf Anatomy

Important Concepts

The photosynthetic production of O_2 by green plants is necessary for most organisms to obtain the energy for life.

- Photosynthesis uses CO_2, water, and light to produce carbohydrate and O_2 in the reaction:

$$6\ CO_2 + 12\ H_2O \rightarrow C_6H_{12}O_6 + 6\ O_2 + 6\ H_2O$$

- In plants, water for photosynthesis must be acquired by the roots and transported to the leaves, where photosynthesis takes place. CO_2, O_2, and water vapor are exchanged through openings or stomata in the leaf's surface. Light from the sun is required for photosynthesis to occur. Photosynthesis in eukaryotes occurs in chloroplasts.
- Photosynthesis occurs in a two-pathway process. The light reactions produce ATP from light energy. The light-independent reactions use ATP and NADPH + H^+ produced by the light reactions to trap CO_2 and produce sugars that act as an energy store. The light-independent pathways are the Calvin cycle, C_4 photosynthesis, and crassulacean acid metabolism. *Neither* pathway operates in the absence of light.

The interaction of light and pigment is fundamental to photosynthesis.

- Light is a form of electromagnetic radiation that comes in discrete packets called photons and has wavelike properties. Photons can be scattered, transmitted, or absorbed. The properties of light depend on its wavelength. For light to be biologically available, it must be absorbed by the receptive molecule and contain enough energy to carry out chemical work. When a molecule absorbs a proton, it goes from a grounded state to an excited state.
- Pigments are molecules that absorb wavelengths in the visible spectrum. The pigment chlorophyll appears green because blue and red light are absorbed and green light is reflected.
- Compounds have unique absorption spectra, depending on which wavelengths of light are absorbed (Figure 10.6). An action spectrum can be measured that relates biological activity to particular wavelengths.
- In plants, photosynthesis relies on chlorophyll *a*, chlorophyll *b*, and accessory pigments. Chlorophylls are more abundant, but they can only absorb photons with blue or orange-red wavelengths. Accessory pigments, such as carotenoids and phycobilins, absorb photons that chlorophylls cannot, and transfer energy to the chlorophylls.

In photosynthesis, a pigment molecule in the excited state passes the absorbed energy along.

- Once a pigment absorbs light, two things can happen: the energy can be released as fluorescence, or it can be passed to another pigment.
- Photosynthetic antennae systems function to pass the light energy from one pigment to another until the reaction center is reached, where light energy is converted to chemical energy. The excitation energy is passed from one pigment that absorbs at a shorter wavelength to another pigment that absorbs at a longer wavelength. In plants, the reaction center is chlorophyll *a*. Excited chlorophyll acts as a reducing agent, resulting in electron transport.
- Noncyclic electron flow results in the production of ATP and NADPH + H^+. Water is oxidized to form O_2, H^+, and electrons. Electrons move from water to chlorophyll, through electron carriers, and ultimately to NADP$^+$. Release of free energy drives ATP synthesis through chemiosmosis.
- Noncyclic electron flow requires two photosystems in the thylakoid membrane to continuously absorb light. Photosystem II takes slightly higher energy (680 nm) than photosystem I (700 nm). These two photosystems interact to pass electrons in a model called the Z scheme.
- Figure 10.10 shows the flow of electrons in the Z scheme from the splitting of water to the reduction of NADP$^+$. The electron transport pumps protons actively into the thylakoid compartment. This proton gradient powers the formation of ATP. Thus, noncyclic flow produces equal amounts of ATP and NADPH + H^+.
- The light-independent reactions require more ATP than NADPH + H^+. Cyclic electron flow supplies the additional ATP by producing ATP without producing NADPH + H^+.
- Cyclic electron transport involves only photosystem I and cycles back to the same chlorophyll molecule (see Figure 10.11). The amount of NADPH + H^+ present regulates whether cyclic or noncyclic electron flow occurs. If NADPH + H^+ is abundant, electrons are accepted by plastoquinone, which pumps two protons back across the thylakoid membrane.
- ATP synthesis in either cyclic or noncyclic electron flow is produced via chemiosmosis. Just as in the mitochondria, a proton-motive force coupled to ATP synthase is established in the chloroplast as electrons are passed along the transport chain. This results in the photophosphorylation of ADP to ATP.
- ***For review, go to Diagram Exercise 1.***

The Calvin cycle incorporates CO_2 into sugars.

- In the Calvin cycle, ATP and NADPH from the light reactions are used to convert CO_2 to sugars for storage. Because ATP and NADPH are energy-rich coenzymes that cannot be stockpiled, the Calvin cycle reactions occur only in the light when ATP and NADPH can be made.
- The Calvin cycle occurs in the stroma of the chloroplasts.

- The cycle was revealed using radiolabeled CO_2 and observing where it was incorporated. Following this process, researchers revealed the pathway by which CO_2 is converted to sugar. See Figure 10.13 to understand how this was done.
- CO_2 is combined with ribulose 1,5-bisphosphate (RuBP), which is then split into two molecules of 3-phosphoglycerate (3PG) by an enzyme called rubisco. Rubisco is the world's most abundant protein, and it regulates the Calvin cycle.
- ATP and NADPH + H^+ from the light reactions are then used to convert the fixed CO_2 in 3PG into carbohydrate (glyceraldehyde 3-phosphate, or G3P).
- Finally, ATP is used to regenerate RuBP so that additional CO_2 may be fixed.
- The resulting carbohydrate (G3P) is converted to starch or sucrose, either for storage or for conversion to glucose and fructose for cellular fuel.
- Light stimulates the Calvin cycle by inducing pH changes that result in the activation of Calvin cycle enzymes and by reducing disulfide bonds on four of the Calvin cycle enzymes.
- ***For review, go to Diagram Exercise 2.***

Rubisco also mediates the process of photorespiration.

- Rubisco can act as an oxygenase and fix O_2 at the expense of CO_2 to produce phosphoglycolate. Phosphoglycolate enters membrane-enclosed peroxisomes and is converted to glycine. The glycine is then converted into glycerate and CO_2 in the mitochondrion.
- This process is known as photorespiration. ATP and NADPH from the light reaction are used in photorespiration, but CO_2 is released instead of being used to make carbohydrate.
- Whether CO_2 or O_2 is fixed depends on relative concentrations of CO_2 and O_2. Excess O_2 (abundant, for example, on hot, dry days) forces photorespiration.

Photorespiration is metabolically expensive, so plants have evolved mechanisms to avoid it.

- To avoid photorespiration, the level of CO_2 around rubisco must be high. This is difficult to achieve because O_2 levels in the air are much higher than CO_2 levels. If hot conditions force the closing of stomata, CO_2 levels are quickly depleted.
- Plants such as roses, wheat, and rice that fix CO_2 to 3PG are called C_3 plants and are very sensitive to CO_2/O_2 levels. Photorespiration occurs frequently and limits the range of conditions under which C_3 plants can grow.
- Many tropical plants such as corn, sugarcane, and other tropical grasses are C_4 plants. They fix CO_2 to the acceptor phospho-enolpyruvate (PEP) using the enzyme PEP carboxylase to form oxaloacetate. PEP carb-oxylase has no oxygenase activity and fixes CO_2 at very low levels. The resulting four-carbon compound diffuses to the bundle sheath cells in the interior of the leaf, releases CO_2, and is recycled. The released CO_2 goes through the Calvin cycle including rubisco as normal. The early fixation process is a CO_2-concentrating mechanism around rubisco.
- Cacti, pineapples, and other succulents undergo crassulacean acid metabolism (CAM) that separates CO_2 fixation and the Calvin cycle. Plants that carry out CAM open their stomata only at night and store CO_2 for use during the day when light is present. CO_2 is fixed as oxaloacetate and converted to malic acid. CO_2 is stored in malic acid until it is transferred to the chloroplasts or photosynthetic cells when light is available.

Metabolic pathways in plants involve photosynthesis and respiration.

- Because plants are autotrophic, they can synthesize all necessary molecules to survive from CO_2, H_2O, phosphate, sulfate, and NH_4. All cells, whether photosynthetic or not, go through respiration to produce the ATP necessary for cellular processes, and they do so in the light and the dark.
- The Calvin cycle and the respiratory pathways are closely linked. G3P can be converted to pyruvate and enter respiration, or it can enter the gluconeogenic pathway to form sucrose. In order for plants to grow, the energy stored must exceed the energy used in respiration.
- Figure 10.20 illustrates how plant metabolism is interconnected.
- ***For review, go to Diagram Exercise 3.***

Test Yourself

Diagram Exercises

1. Label the photosystems, pathways, reactants, and products in the diagram below. Include both electron pathways in your answer.

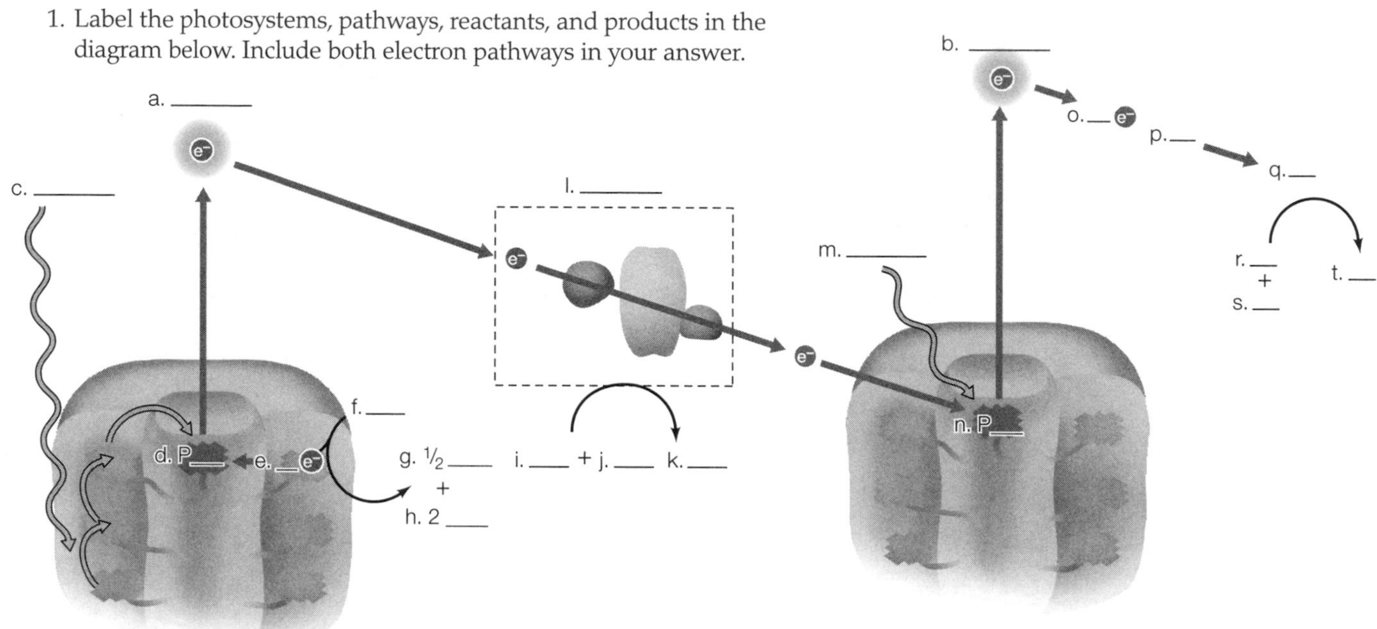

a. ______
b. ______
c. ______
d. P___
e. ___
f. ___
g. 1/2___
+
h. 2 ___
i. ___ + j. ___ k. ___
l. ______
m. ______
n. P___
o. ___
p. ___
q. ___
r. ___
+
s. ___
t. ___

Textbook Reference: *10.2 How Does Photosynthesis Convert Light Energy into Chemical Energy? p. 197*

2. Label the stages, reactants, and products in the diagram below. Which enzyme is most important in this system?

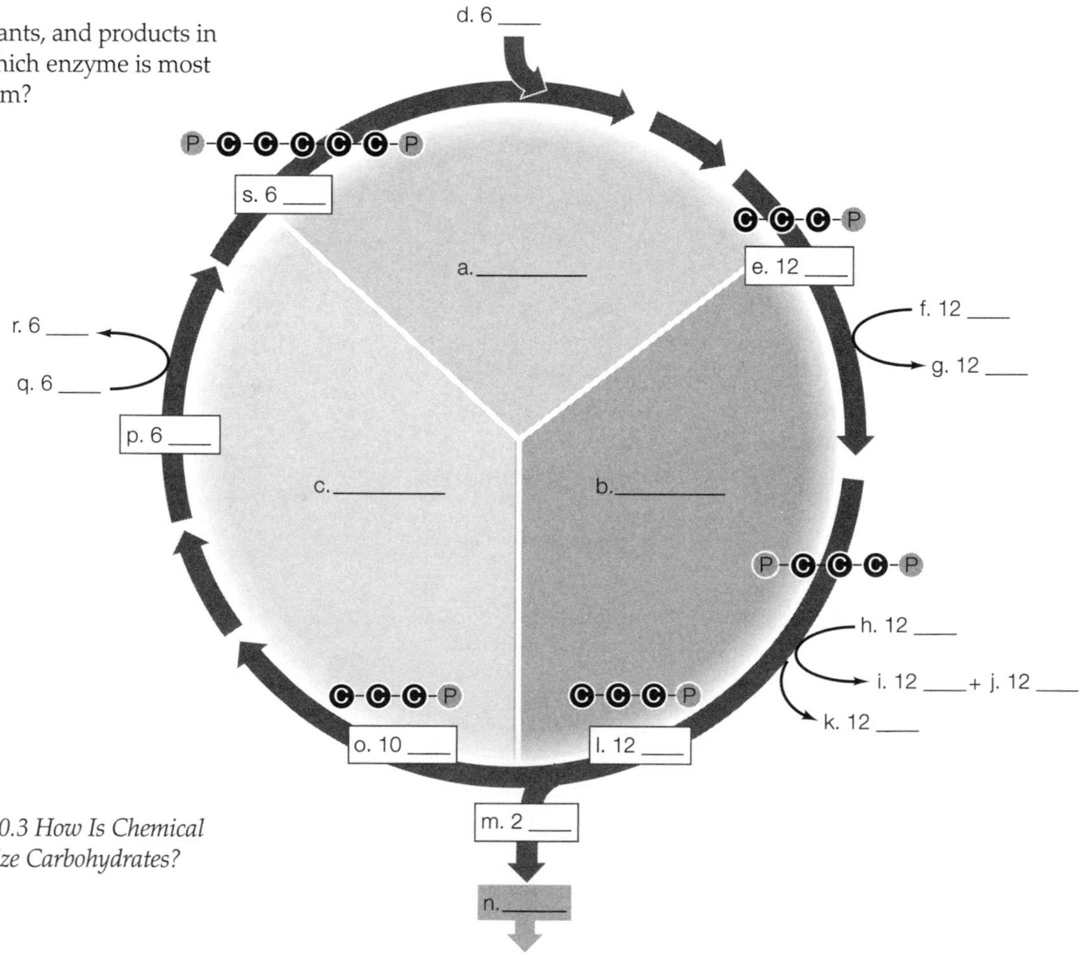

d. 6 ___
s. 6 ___
a. ______
e. 12 ___
f. 12 ___
g. 12 ___
r. 6 ___
q. 6 ___
p. 6 ___
c. ______
b. ______
h. 12 ___
i. 12 ___ + j. 12 ___
k. 12 ___
o. 10 ___
l. 12 ___
m. 2 ___
n. ___

Textbook Reference: *10.3 How Is Chemical Energy Used to Synthesize Carbohydrates? p. 201*

3. Label the cycles, pathways, reactants, and products in the diagram below. As you work, focus on the connections between the two systems. Note how the products of one system provide the raw materials for the other system.

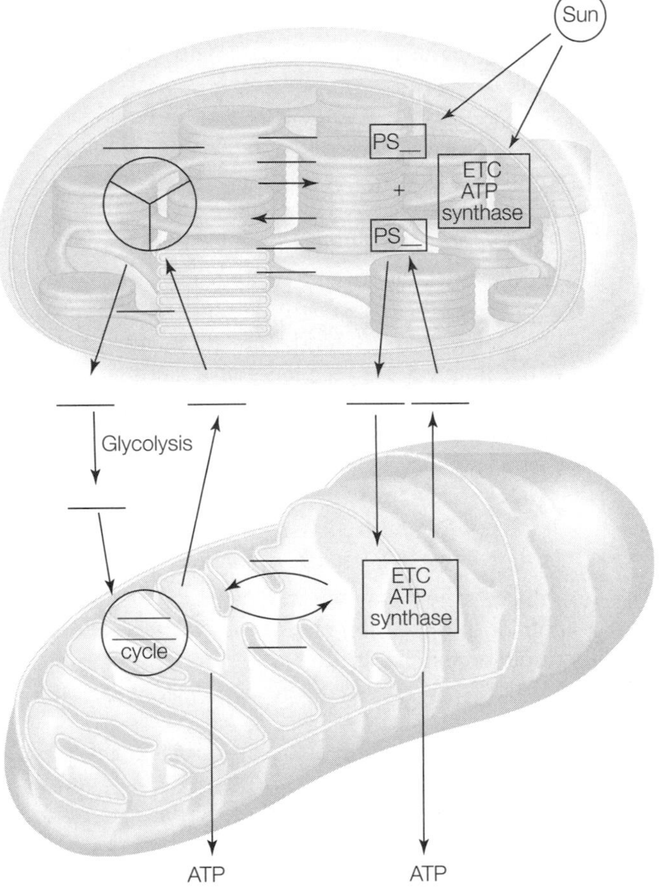

Textbook Reference: *10.5 How Does Photosynthesis Interact with Other Pathways? p. 206*

Knowledge and Synthesis Questions

1. The main function of photosynthesis is the
   a. consumption of CO_2.
   b. production of ATP.
   c. conversion of light energy to chemical energy.
   d. production of starch.
   e. production of O_2.
   Textbook Reference: *10.1 What Is Photosynthesis? p. 190*
2. Which of the following best represent the components that are necessary for photosynthesis to take place?
   a. Mitochondria, accessory pigments, visible light, water, and CO_2
   b. Chloroplasts, accessory pigments, visible light, water, and CO_2
   c. Mitochondria, chlorophyll, visible light, water, and O_2
   d. Chloroplasts, chlorophyll, visible light, water, and CO_2
   e. Chlorophyll, accessory pigments, visible light, water, and O_2
   Textbook Reference: *10.1 What Is Photosynthesis? Figure 10.3, p. 192*
3. Chlorophyll is suited for the capture of light energy because
   a. certain wavelengths of light raise it to an excited state.
   b. in its excited state it gives off electrons.
   c. its structure allows it to attach to thylakoid membranes.
   d. it can transfer absorbed energy to another molecule.
   e. All of the above
   Textbook Reference: *10.2 How Does Photosynthesis Convert Light Energy into Chemical Energy? p. 194*
4. Plants give off O_2 because
   a. O_2 results from the incorporation of CO_2 into sugars.
   b. they do not respire; they photosynthesize.
   c. water is the initial electron donor, leaving O_2 as a photosynthetic by-product.
   d. electrons moving down the electron chain bind to water, releasing O_2.
   e. O_2 is synthesized in the Calvin cycle.
   Textbook Reference: *10.2 How Does Photosynthesis Convert Light Energy into Chemical Energy? pp. 196–197*
5. Cyclic and noncyclic electron flow are used in plants to
   a. meet the ATP demands of the Calvin cycle.
   b. produce excess NADPH + H^+.
   c. synthesize proportional amounts of ATP and NADPH + H^+ in the chloroplast.
   d. consume the products of the Calvin cycle.
   e. produce O_2 for the atmosphere.
   Textbook Reference: *10.3 How Is Chemical Energy Used to Synthesize Carbohydrates? p. 200*
6. Which of the following statements concerning the light reactions of photosynthesis is true?
   a. Photosystem I cannot operate independently of photosystem II.
   b. Photosystems I and II are activated by different wavelengths of light.
   c. Photosystems I and II transfer electrons and create proton equilibrium across the thylakoid membrane.
   d. Photosystem I is more significant than Photosystem II.
   e. Oxygen gas is a product of Photosystem I.
   Textbook Reference: *10.2 How Does Photosynthesis Convert Light Energy into Chemical Energy? pp. 196–197*
7. ATP is produced during the light reactions via
   a. CO_2 fixation.
   b. chemiosmosis.
   c. reduction of water.
   d. glycolysis.
   e. noncyclic electron flow from photosystem I.
   Textbook Reference: *10.2 How Does Photosynthesis Convert Light Energy into Chemical Energy? p. 197*

8. Because of the properties of chlorophyll, plants need adequate _______ light to grow properly.
   a. green
   b. blue and red
   c. infrared
   d. ultraviolet
   e. blue and blue-green
   Textbook Reference: *10.2 How Does Photosynthesis Convert Light Energy into Chemical Energy? p. 194, Figure 10.6*
9. Which of the following statements concerning the Calvin cycle is *false*?
   a. Light energy is not required for the cycle to proceed.
   b. CO_2 is assimilated into sugars.
   c. RuBP is regenerated.
   d. It uses energy stored in ATP and NADPH + H^+.
   e. All of the above are false.
   Textbook Reference: *10.3 How Is Chemical Energy Used to Synthesize Carbohydrates? p. 201*
10. Which of the following statements concerning rubisco is true?
   a. Rubisco is a carboxylase.
   b. Rubisco preferentially binds to O_2 over CO_2.
   c. Rubisco is absent from C_4 and CAM plants.
   d. Rubisco catalyzes the splitting in water to release O_2.
   e. Rubisco is more allosterically regulated by CO_2.
   Textbook Reference: *10.3 How Is Chemical Energy Used to Synthesize Carbohydrates? p 200; 10.4 How Do Plants Adapt to the Inefficiencies of Photosynthesis? p. 202*
11. Which of the following begins the Calvin cycle that results in the entire pathway being carried out under environmental conditions?
   a. 3PG is reduced to G3P using ATP and NADPH + H^+.
   b. RuBP is regenerated.
   c. CO_2 and RuBP join forming 3PG.
   d. G3P is converted into glucose and fructose.
   e. Any of the above; as a cycle, it can start at any point.
   Textbook Reference: *10.3 How Is Chemical Energy Used to Synthesize Carbohydrates? p. 200*
12. The Calvin cycle results in the production of
   a. glucose.
   b. starch.
   c. rubisco.
   d. G3P.
   e. ATP.
   Textbook Reference: *10.3 How Is Chemical Energy Used to Synthesize Carbohydrates? p. 200*
13. Which of the following statements regarding photorespiration is true?
   a. Photorespiration is a metabolically expensive pathway.
   b. Photorespiration is avoided when CO_2 levels are low.
   c. Photorespiration increases the overall CO_2 that is converted to carbohydrates.
   d. Photorespiration increases by 75% the net carbon that is fixed.
   e. Photorespiration is most common in C_4 plants.
   Textbook Reference: *10.4 How Do Plants Adapt to the Inefficiencies of Photosynthesis? pp. 202–203*
14. The fixation of CO_2 by PEP carboxylase functions to
   a. concentrate O_2 for use in photosynthetic cells.
   b. allow plants to close stomata without the occurrence of photorespiration.
   c. allow plants to photosynthesize in the dark.
   d. reduce water loss by the plant.
   e. All of the above
   Textbook Reference: *10.4 How Do Plants Adapt to the Inefficiencies of Photosynthesis? p. 204*
15. CAM plants differ from C_4 plants in that
   a. photosynthesis can occur at night in CAM plants.
   b. CO_2 is stored in CAM plants as malic acid.
   c. the stomata of CAM plants close during periods that favor photorespiration.
   d. CAM plants use PEP carboxylase to fix CO_2.
   e. the Calvin cycle is only found in C_4 and C_3 plants, not in CAM plants.
   Textbook Reference: *10.4 How Do Plants Adapt to the Inefficiencies of Photosynthesis? pp. 203–204*
16. Which of the following statements regarding the relationship between photosynthesis and cellular respiration in plants is true?
   a. Photosynthesis occurs in specialized photosynthetic cells.
   b. Cellular respiration occurs in specialized respiratory cells.
   c. Cellular respiration and photosynthesis can occur in the same cell.
   d. Photosynthesis is limited to specialized plant cells and cellular respiration does not occur in plant cells.
   e. Both a and c
   Textbook Reference: *10.5 How Does Photosynthesis Interact with Other Pathways? pp. 205–206*
17. Photosynthesis occurs
   a. in all plant cells.
   b. only in photosynthetic plant cells.
   c. only in plant cells lacking mitochondria.
   d. only in the stroma.
   e. only in the thylakoid membrane.
   Textbook Reference: *10.1 What Is Photosynthesis? p. 192*
18. Activities such as amino acid synthesis and active transport in plant cells are powered by
   a. the light-dependent and light-independent reactions of photosynthesis.
   b. ATP from the light reactions of photosynthesis.
   c. ATP from fermentation.
   d. ATP from glycolysis and cellular respiration.
   e. All of the above
   Textbook Reference: *10.5 How Does Photosynthesis Interact with Other Pathways? p. 206*

Application Questions

1. Plants consume CO_2 and give off O_2. How is this possible if plants must also undergo cellular respiration?
   Textbook Reference: *10.5 How Does Photosynthesis Interact with Other Pathways? pp. 205–206*
2. Why do plants undergo both the light reactions of photosynthesis and the Calvin cycle? Why don't they simply use the ATP produced in the light reactions of photosynthesis to drive cellular processes?
   Textbook Reference: *10.3 How Is Chemical Energy Used to Synthesize Carbohydrates? p. 200*
3. Why do plants undergo both photosynthesis and cellular respiration, even in the daytime? Why don't they simply use the ATP produced in the light reactions of photosynthesis to drive cellular processes?
   Textbook Reference: *10.5 How Does Photosynthesis Interact with Other Pathways? pp. 205–206*
4. Rubisco has both carboxylase and oxygenase activities. These processes compete with each other. What determines which function the enzyme has? What conditions favor photorespiration? What conditions favor photosynthesis?
   Textbook Reference: *10.4 How Do Plants Adapt to the Inefficiencies of Photosynthesis? pp. 202–203*
5. Compare and contrast C_3, C_4, and CAM plants.
   Textbook Reference: *10.4 How Do Plants Adapt to the Inefficiencies of Photosynthesis? p. 205*
6. The Calvin cycle was once referred to as the "dark" reactions of photosynthesis. Why is this a misnomer?
   Textbook Reference: *10.1 What Is Photosynthesis? p. 193*
7. Explain the differences between cyclic and noncyclic electron flow. Why are both processes necessary?
   Textbook Reference: *10.2 How Does Photosynthesis Convert Light Energy into Chemical Energy? pp. 197–198*
8. How do accessory pigments enhance photosynthetic activity in plants?
   Textbook Reference: *10.2 How Does Photosynthesis Convert Light Energy into Chemical Energy? p. 194*
9. Why are plants green?
   Textbook Reference: *10.2 How Does Photosynthesis Convert Light Energy into Chemical Energy? p. 194*

Answers

Diagram Exercise Answers

1. a. Photosystem II
   b. Photosystem I
   c. Photon
   d. 680
   e. 2
   f. H_2O
   g. O_2
   h. H^+
   i. ADP
   j. P_i
   k. ATP
   l. Electron transport
   m. Photon
   n. 700
   o. 2
   p. Fd
   q. $NADP^+$ reductase
   r. $NADP^+$
   s. H^+
   t. NADPH
2. The most important enzyme in this system is rubisco, which is active in the first stage.
   a. Carbon fixation
   b. Reduction and sugar production
   c. Regeneration of RuBP
   d. CO_2
   e. 3PG
   f. ATP
   g. ADP
   h. NADPH
   i. $NADP^+$
   j. H^+
   k. P_i
   l. G3P
   m. G3P
   n. Sugars
   o. G3P
   p. RuMP
   q. ATP
   r. ADP
   s. RuBP
3.

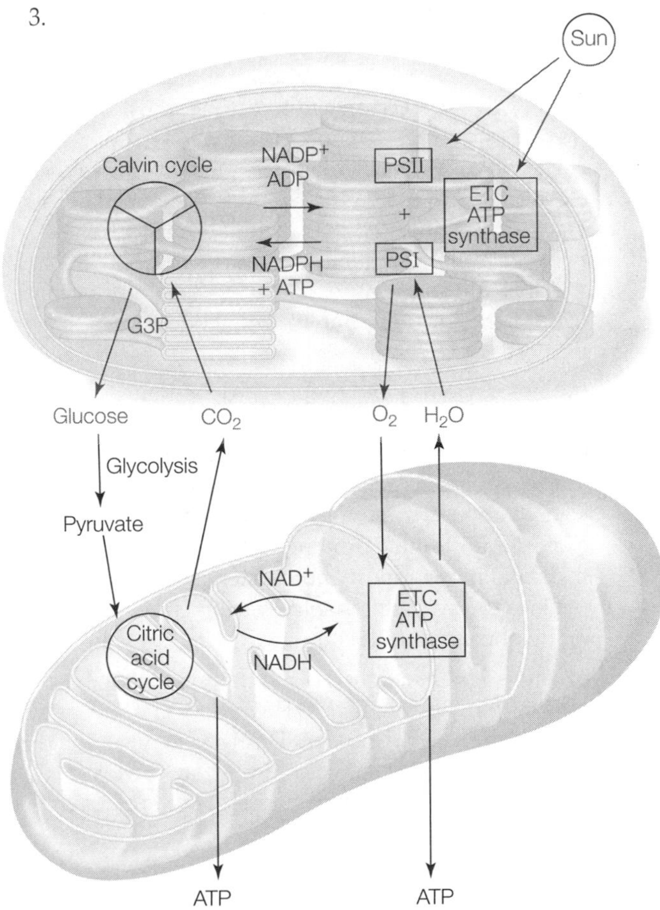

Knowledge and Synthesis Answers

1. **c.** Photosynthetic organisms, including but not limited to plants, are the only life forms capable of trapping light energy and converting it to chemical energy. Because of this they form the basis of many of Earth's food chains.

2. **d.** Chloroplasts are the site of the photosynthetic reactions; chlorophyll is excited by photons of light and serve as reaction centers for the photosystems; visible light is necessary to excite chlorophyll and accessory pigments; water is the initial electron donor for the pathway; and CO_2 is necessary to make precursor molecules for energy storage.
3. **e.** The"tails"of chlorophyll molecules are associated with the thylakoid membranes of the chloroplasts. This close membrane association assists with establishing the proton-motive force that will drive ATP synthesis. When excited by light, the chlorophyll moves into an excited state and passes electrons to acceptor molecules. This begins to set up the proton gradient across the membrane that will drive ATP synthesis.
4. **c.** Water is split at photosystem II to donate electrons to the reaction center. The resulting protons are moved across the membrane to establish the proton-motive force, and O_2 is given off as a by-product.
5. **a.** ATP is required at higher levels in the Calvin cycle than NADPH + H^+ is; therefore, there must be a mechanism for producing additional ATP. Cyclic electron flow provides that mechanism. If noncyclic electron flow were to be sped up to meet ATP needs, an excess of NADPH + H^+ would result. Shifting between cyclic and noncyclic flow balances ATP/NADPH + H^+ ratios. Oxygen gas is a by-product of the light reactions, but its production is not the purpose of the reactions.
6. **b.** Photosystems I and II operate depending on whether electron flow is cyclic or noncyclic. Activity is controlled by the ATP levels in the chloroplast. Photosystem II is activated by light of a higher energy level than photosystem I. Both photosystems transfer electrons and create proton gradients across the thylakoid membranes; photosystem I does this via the cyclic pathway. Water is split by a structure embedded in the photosystem II complex.
7. **b.** In the light reactions, ATP synthesis occurs when protons flow through an ATP synthase channel protein in the thylakoid membrane. This is a chemiosmotically driven process. Photosystem II is always involved, while Photosystem I participates via cyclic electron transport only.
8. **b.** Chlorophyll and accessory pigments absorb light in the blue and red wavelengths of visible light. Green light is reflected; therefore, plants appear green. (Accessory pigments allow energy from additional wavelengths to be absorbed as well.)
9. **a.** Light energy is required for the Calvin cycle to proceed. ATP synthesis is dependent on light energy, and the Calvin cycle is dependent on ATP.
10. **a.** Rubisco, the most abundant enzyme on Earth, has both oxygenase and carboxylase activities. It is present in C_3, C_4, and CAM plants and binds CO_2 with greater affinity than O_2.
11. **c.** The first step of the Calvin cycle is the fixation of CO_2 into 3PG. This is the regulatory step, and it requires ATP and NADPH + H^+. While it is true that the Calvin cycle is a cycle, there is a net consumption of CO_2 for the purpose of building carbohydrates.
12. **d.** The Calvin cycle produces only G3P, which can then be metabolized into storage products like sugars and starch.
13. **a.** Photorespiration uses as much ATP as photosynthesis, but results in no energy gains for the plant and reduces net carbon fixation by 25 percent compared with the Calvin cycle. If CO_2 is abundant, rubisco acts as a carboxylase rather than an oxygenase.
14. **b.** Plants do not photosynthesize in the dark. PEP carboxylase allows the fixation of CO_2 at low concentrations in the leaf so that it can be sent to rubisco for the Calvin cycle.
15. **b.** CAM plants functionally store CO_2 as malic acid.
16. **e.** Photosynthesis occurs only in plant cells that have the necessary structures, but cellular respiration occurs in *every* living plant cell that has mitochondria and O_2.
17. **b.** Photosynthesis is limited to photosynthetic plant cells. There are many plant cells that are not exposed to light or that lack chloroplasts; these cells rely on cellular respiration.
18. **d.** Plant cells have mitochondria (see Figure 5.7) and rely on the processes of glycolysis and cellular respiration to provide ATP for cellular activities. Photosynthesis converts light energy into potential energy stored in chemical form, but that energy must then be made usable by the cells. Plant cells release this stored energy via the catabolic reactions covered in Chapter 9 (see Figure 9.14).

Application Answers

1. Plant cells undergo cellular respiration in all living cells. Therefore, all living cells consume O_2. Photosynthesis occurs in specialized cells that consume both CO_2 and O_2. Because atmospheric O_2 levels are high, excess O_2 is available for the plant to utilize; therefore, O_2 continues to be emitted from the plant.
2. The light reactions of photosynthesis produce ATP. ATP cannot be stored for use later (such as when light is not available); therefore, there has to be a mechanism for that energy to be stored. The Calvin cycle stores the energy in the chemical bonds of G3P, which can be incorporated into carbohydrates for longer-term storage.
3. Though photosynthesis produces all the necessary energy for a plant, a plant cannot be continuously photosynthetically active. Therefore, a plant stores energy in carbohydrates. Cellular respiration is necessary to break down stored carbohydrates.
4. Whether rubisco acts as a carboxylase or an oxygenase depends on the relative ratio of O_2 to CO_2. At higher CO_2 levels, it acts as a carboxylase. At low CO_2 levels,

it acts as an oxygenase. Photorespiration is favored during hot, dry weather, which forces the closing of stomata and leads to increases in O_2 levels within the leaf. Photosynthesis is favored when stomata can remain open and light intensity is optimal.

5. Refer to Table 10.1 in your book.
6. Light is required for both the light reactions of photosynthesis and the Calvin cycle. The Calvin cycle depends on the ATP generated during the light-dependent reactions.
7. Noncyclic electron flow involves both photosystems I and II. It results in equal amounts of ATP and NADPH being synthesized (see Figure 10.10). However more ATP than NADPH is required for the Calvin cycle (see Figure 10.15). To provide the additional ATP, Photosystem I sends electrons to the electron carrier ferredoxin in electron transport chain driving ATP synthesis (see Figure 10.11). This cyclic pathway provides the necessary ATP for the Calvin cycle to regenerate RuBP.
8. Accessory pigments allow utilization of light in many wavelengths of the visible spectrum that could not be used by chlorophyll alone. The energy absorbed is channeled to the reaction centers of the photosystems.
9. The primary pigments in plants are chlorophylls. Chlorophylls absorb blue and orange-red wavelengths of light and reflect green light, thus making plants appear green. See the absorption spectra and action spectra of chlorophyll in Figure 10.6.

11 The Cell Cycle and Cell Division

The Big Picture

- In the presence of an internal or external environmental signal, cells undergo a specific set of steps to replicate their DNA, segregate those newly replicated chromosomes, and divide their cytoplasm into two daughter cells. These steps ensure that the daughter cells receive a complete set of genetic instructions.
- Diploid organisms produce haploid gametes, which fuse during fertilization to form new diploid organisms.
- During the generation of these gametes, crossing over between sister chromatids and random segregation of those chromatids create genetically diverse gametes, leading to increased species diversity and survival.
- Cancer cells lose their ability to respond to cell cycle signals.

Common Problem Areas

- Be sure you can distinguish the differences between mitosis and meiosis. In mitosis, the replicated chromosomes are segregated to the daughter cells, resulting in each daughter cell's having two sets of chromosomes (diploid), like its parent cell. After every DNA replication, there is one cell division. In meiosis, the replicated chromosomes undergo two cell divisions (for every one round of DNA replication), resulting in cells that have only one set of chromosomes (haploid). Furthermore, the homologous chromosomes pair in prophase of meiosis I (which is not the case in mitosis), and crossing over occurs.

Study Strategies

- Make several sequential drawings that show a cell with four chromosomes and what those chromosomes look like after DNA replication and during each phase of mitosis. Repeat these sequential drawings for meiosis I and II, using different colors for each homologous chromosome. In your drawings be sure to include crossing over in prophase I of meiosis.
- Draw a picture of the cell cycle and indicate the points at which the cyclin–Cdk complexes act during that cell cycle. Refer to Figure 11.6 as well as the information in your textbook.
- List the five phases of mitosis and describe what happens in each phase. Repeat this process for meiosis.
- Go to yourBioPortal.com to review the following tutorials and activities:

  Animated Tutorial 11.1 Mitosis

  Animated Tutorial 11.2 Meiosis

  Web Activity 11.1 The Mitotic Spindle

  Web Activity 11.2 Images of Mitosis

  Web Activity 11.3 Sexual Life Cycle

  Web Activity 11.4 Images of Meiosis

Important Concepts

Cell division is a multistep process.

- Cell division requires a reproductive signal either from inside or outside the cell, subsequent DNA replication, segregation of the newly replicated chromosomes to opposite poles of the cell, and the division of the cytoplasm (cytokinesis) to form two daughter cells.

Prokaryotes divide by fission.

- Prokaryotic cells divide by increasing in size, replicating their DNA (often a single circular chromosome), and dividing into two new cells through the process of binary fission.
- Environmental signals and nutrient concentrations influence the decision to divide and the rate of division in prokaryotes.
- DNA replication in prokaryotic cells requires a number of different proteins that form a replication complex through which the DNA is threaded. Special sites on the chromosome facilitate the initiation (the *ori* site) and the termination (the *ter* site) of replication. Segregation of the newly replicated chromosomes occurs with the help of special proteins, ATP hydrolysis, and the prokaryotic cytoskeleton.
- Cytokinesis is caused by the pinching in of the plasma membrane, and division is completed by the formation of the cell wall between the two cells.

Eukaryotic cells divide by mitosis or meiosis.

- Cell division in multicellular eukaryotes occurs in response to the needs of the entire organism.
- Eukaryotic cells have more than one chromosome, and those chromosomes are linear.
- Eukaryotic cells replicate their chromosomes in response to a reproductive signal, segregate those chromosomes (which are closely associated with each other as sister chromatids) during mitosis, and divide their cytoplasm during cytokinesis.
- Meiosis is the process of nuclear division in cells involved in sexual reproduction. In cells that produce gametes (eggs and sperm), nuclear division occurs via meiosis, which generates diversity.

The signaling pathways for eukaryotic cell division are highly controlled.

- The period between cell divisions is called the cell cycle and includes mitosis, cytokinesis, and interphase. Interphase is divided into three subphases: G1 (or Gap 1), S (or DNA synthesis), and G2 (or Gap 2). The M phase includes mitosis and cytokinesis.
  - G1 is the most variable phase of the cell cycle. In this phase, the decision and subsequent preparation for DNA synthesis occurs.
  - The G1-to-S transition (called the restrictive point, R) is the point at which the commitment to DNA replication and subsequent cell division occurs.
  - S is the phase of the cell cycle in which all the chromosomes are replicated; each replicated chromosome consists of two identical sister chromatids.
  - During G2, the cell prepares for mitosis (M). During the M phase, the cell segregates the newly replicated chromosomes to opposite poles of the cell, and nuclear division occurs.

Specific signals trigger events in the cell cycle.

- Cdk, a *c*yclin-*d*ependent *k*inase that phosphorylates other proteins in the cell, is catalytically active when it is bound to another protein, cyclin.
- Different cyclin–Cdk protein complexes act as checkpoints in the cell cycle: during G1 before the cell enters S, during S phase to stimulate DNA replication, and after S to initiate the transition from G2 to M.
- Other proteins, such as retinoblastoma protein (RB), are regulated by cyclin–Cdk complexes.
- Growth factors are external chemical signals that stimulate cell division, often by activating cylcin–Cdk complexes.

Eukaryotic chromosomes are composed of protein and DNA.

- Eukaryotic chromosomes are composed of protein and DNA, collectively called chromatin. Cohesin is a protein that holds the sister chromatids together along their length.
- When eukaryotic chromosomes are replicated, the daughter chromosomes, termed sister chromatids, are still joined at the centromere, and condensin proteins make the chromosome more compact.
- In a eukaryotic nucleus, DNA molecules are wrapped around beadlike particles of protein called nucleosomes. Nucleosomes are composed of eight histone molecules, two from each class. Histones are very basic proteins.
- The nucleosomes are further condensed into a coil that twists into another larger coil. During mitosis, more compaction occurs, and the sister chromatids become visible (see Figures 11.8 and 11.9). Interphase chromosomes are less densely packed.

Mitosis proceeds through a number of phases.

- During mitosis a single nucleus gives rise to two nuclei that are genetically identical to the parent nucleus.
- Centrosomes determine the plane of cell division. Centrosome duplication occurs during S phase. Centrosomes may contain centrioles, which consist of two hollow tubes of microtubules positioned at right angles to each other.
- Duplicated centrosomes migrate to opposite poles of the cell during the G2-to-M transition.
- Microtubules grow from a microtubule-organizing center, the centrosome. The microtubules form a spindle along which the chromosomes will move. Later in the cell cycle, some of these microtubules (the kinetochore microtubules) attach to the chromosomes and help separate them to opposite poles of the cell.
- Mitosis can be divided into five phases: prophase, prometaphase, metaphase, anaphase, and telophase.
- During prophase, the mitotic spindle forms and includes polar and kinetochore microtubules. The newly replicated chromosomes condense so that the sister chromatids are visible. Kinetochores assemble at the centromere of each newly replicated chromosome, and microtubules attach to the kinetochore. Polar microtubules form the framework of the spindle (see Figure 11.10).
- Prometaphase signals the disappearance of the nuclear envelope and the nucleoli and the attachment of kinetochore microtubules to the kinetochores of each newly replicated chromosome.
- Metaphase signals the positioning of all of the centromeres at the equatorial (metaphase) plate.
- At the beginning of anaphase the cohesion proteins are hydrolyzed by the protease separase. Separase is activated by an anaphase-promoting complex (APC) which is a cyclin–Cdk complex.
- During anaphase, the chromosomes move to opposite poles of the spindle, assisted by dynein motors acting at the kinetochores and by the shortening of the microtubules at the poles (see Figure 11.11).

- At telophase, the poles break down, the chromosomes begin to uncoil, and the nuclear envelope forms, resulting in two identical nuclei.

Cytokinesis results in two distinct cells.

- During cytokinesis in animal cells, the cell membrane contracts due to the interaction of actin and myosin microfilaments at the cell membrane, forming a contractile ring (see Figure 11.13A).
- In plant cell cytokinesis, a cell plate forms between the newly segregated chromosomes. This cell plate is derived from Golgi vesicles located at the equatorial region, which fuse to form a plasma membrane. The plasma membrane secretes plant cell-wall materials into the cell plate to complete cell division.
- On completion of cytokinesis, there are two distinct cells, each with a full complement of chromosomes.

Cells reproduce asexually or sexually.

- There are two kinds of reproduction. The first is asexual (or vegetative reproduction) in which the replicated chromosomes are separated by mitosis. After cytokinesis the offspring are genetically identical (clones) to the parent. The second kind of reproduction is sexual, in which gametes from two different parents fuse to form a zygote. This fusion is called fertilization and produces offspring that are different from both parents.
- Somatic cells from diploid organisms contain pairs of homologous chromosomes, one set of chromosomes coming from each parent. Homologous chromosomes are similar in size and appearance and bear corresponding genetic information.
- Gametes each contain a single set of chromosomes (n) resulting from meiosis (and thus are haploid), and the zygote contains two sets of chromosomes ($2n$), one set from each parent, and is diploid.
- Mature organisms can exist primarily in the diploid or haploid state or can alternate between haploid and diploid states during their life cycles (see Figure 11.15).
- Haplontic organisms are haploid as mature organisms. They produce spores that fuse to form a zygote, which is diploid. The zygote then undergoes meiosis to become haploid again. Those cells divide mitotically to become the mature organism, which can be single-celled or multicellular.
- Most plants and some protists have mature forms that alternate between diploid and haploid stages (see Figure 11.15).
- Diplontic organisms are diploid in their mature state; only their gametes are haploid.
- Fusion of gametes to form a zygote increases diversity because the zygote's genetic makeup is derived from two different parents, each contributing one haploid set of randomly selected chromosomes.
- Mitotic chromosomes can be characterized by their shape, number, size, and centromere position. Each organism has a distinctive set of chromosomes called a karyotype.

Meiosis proceeds through a number of phases.

- Meiosis is distinguished from mitosis by *two* nuclear divisions (meiosis I and II), but only *one* round of DNA replication, resulting in a reduction in the number of chromosomes from diploid to haploid (see Figure 11.17).
- The purpose of meiosis is to reduce the chromosome number from diploid to haploid, to ensure haploid products have a complete set of chromosomes, and to promote genetic diversity in gametes and in the species.
- Each haploid cell has one complete set of chromosomes, and the combinations of alleles on the chromosomes of each gamete are different. The advantage of creating gametes with different combinations of alleles is genetic diversity in the future zygote, potentially increasing the resultant organism's ability to survive.
- The phases of meiosis are: prophase, prometaphase, metaphase, anaphase, and telophase.
- In prophase of meiosis I, the newly replicated chromosomes pair with their homologs (synapsis). Because there are four chromatids (two from each homolog), these paired chromosomes are called tetrads or bivalents.
- Crossing over occurs between nonsister chromatids on homologous chromosomes during prophase at special sites called chiasmata (see Figure 11.18). The resulting recombinant chromosomes contribute to an increase in the genetic variation in the products of meiosis.
- During anaphase of meiosis I, each pole receives one member of the homologous pair.
- Unlike in mitosis, no DNA replication occurs before meiosis II begins, and the number of chromosomes on the equatorial plate is half the number in the mitotic nucleus (even though each chromosome still consists of two chromatids). Unlike the chromatids in mitosis, each chromatid is different from its sister chromatid, due to crossing over in meiosis I.
- As in mitosis, segregation of each chromatid occurs in anaphase II, and each of the resulting four nuclei is haploid.
- Crossing over in prophase I and independent assortment in meiosis I (of homologs) and in meiosis II (of chromatids) results in genetic diversity in the daughter cells.

Failures in meiosis can lead to aneuploidy or polyploidy.

- Aneuploidy, an abnormal number of chromosomes, occurs in cells whose chromosomes fail to segregate in meiosis I (both homologs fail to separate and go to the same pole) or meiosis II (both chromatids fail to separate and go to one pole; see Figure 11.21).

- One of the causes of nondisjunction in meiosis I may be the breakdown of cohesins, which orient chromosomes at the equatorial plate during metaphase I. In the absence of cohesins, homologs may line up at random and wind up in the wrong daughter cell.
- Aneuploid gametes can produce trisomies as in Down syndrome, in which the individual has an extra chromosome 21. Monosomic zygotes have one copy of a homologous chromosome.
- Translocations, another chromosome abnormality, occur when part of one chromosome breaks away and becomes attached to another chromosome.
- Polyploids are cells that have increased numbers of complete sets of chromosomes. Polyploids with an even number of sets of chromosomes ($4n$, $6n$, $8n$) can undergo meiosis successfully, whereas odd-number polyploids ($3n$, $5n$, etc.) cannot, because each homolog needs to pair with its partner.

Cell death occurs by necrosis or apoptosis.

- Cell death can occur in cells in two ways: by necrosis (cells are damaged by toxins or starved of oxygen or essential nutrients) or by apoptosis (programmed cell death). Apoptosis is a normal process during development and is used to eliminate unneeded tissue. Apoptosis also occurs in older cells where genetic damage may be more prevalent.
- Signals for cell death include a lack of mitotic signals and recognition of DNA damage, or changes in a receptor protein that activates signal transduction.
- Signal transduction can lead to the activation of caspases, which are proteases that hydrolyze target molecules leading to cell death.

Cancer is caused by unregulated cell division.

- Cancer cells do not respond to cell division controls.
- Cancer cells divide continuously, forming tumors.
- Benign tumors resemble the tissue they came from and can remain localized, but often they need to be removed.
- Malignant tumors do not look like the parent tissues and metastasize, spreading to other areas of the body (see Figure 11.23).
- Oncogenes encode proteins that are positive regulators of cell division (such as HER2). Tumor suppressors (such as RB) prevent the cell cycle from proceeding through cell division. In cancer cells, these regulators do not function normally, leading to uncontrolled cell division.
- Cell death occurs by necrosis or apoptosis in response to environmental signals.
- Cancer cells lose their ability to respond to positive regulators of apoptosis.
- Radiation and cancer drugs target the cell cycle (See Figure 11.25).

Test Yourself

Diagram Exercise

Diagram normal meiosis, meiosis with a nondisjunction in meiosis I, and meiosis with a nondisjunction in meiosis II. Label the gametes that would result from each meiotic event. What kind of zygotes would be produced from these gametes?
Textbook Reference: *11.5 What Happens during Meiosis? pp. 226–229*

Knowledge and Synthesis Questions

1. Which of the following statements about mitosis is true?
   a. Cytokinesis follows mitosis.
   b. DNA replication is completed prior to the beginning of this phase.
   c. The chromosome number of the resulting cells is the same as that of the parent cell.
   d. The daughter cells are genetically identical to the parental cell.
   e. All of the above
   Textbook Reference: *11.3 What Happens during Mitosis? p. 217*
2. Which of the following statements about meiosis is true?
   a. The chromosome number in the resulting cells is halved.
   b. DNA replication occurs before meiosis I and meiosis II.
   c. The homologs do not pair during prophase I.
   d. The daughter cells are genetically identical to the parental cell.
   e. The chromosome number of the resulting cells is the same as that of the parent cell.
   Textbook Reference: *11.5 What Happens during Meiosis? p. 227*
3. Which of the following statements about kinetochores on mitotic chromosomes is true?
   a. They are located at the centromere of each chromosome.
   b. They are the sites where microtubules attach to separate the chromosomes.
   c. They are organized so that there is one per sister chromatid.
   d. Kinetochore microtubules from opposite poles attach to each sister chromatid.
   e. All of the above
   Textbook Reference: *11.3 What Happens during Mitosis? pp. 217–219*
4. Which of the following statements about the mitotic spindle is true?
   a. It is composed of polar and kinetochore microtubules, both of which attach to chromosomes.
   b. It is composed of actin and myosin microfilaments.
   c. It is composed of kinetochores at the metaphase plate.
   d. It is composed of microtubules, which help separate the chromosomes to opposite poles of the cell.

e. It originates only at the centrioles in the centrosomes.
Textbook Reference: *11.3 What Happens during Mitosis? pp. 216–217*

5. Imagine that there is a mutation in the Cdk gene such that its gene product is nonfunctional. What kind of effect would this mutation have on a mature red blood cell?
   a. The cyclin that bound to this Cdk would not be phosphorylated.
   b. There would be no effect, because mature red blood cells do not enter the cell cycle.
   c. The cell would be unable to replicate its DNA.
   d. The cell would not be able to enter G1.
   e. The cell would be unable to reproduce itself.
   Textbook Reference: *11.2 How Is Eukaryotic Cell Division Controlled? pp. 214–215*
6. Imagine that there is a mutation in the Cdk gene such that its gene product is nonfunctional. What kind of effect would this mutation have on a mammalian white blood cell?
   a. The cell would be unable to replicate its DNA.
   b. The cell would be unable to enter mitosis.
   c. The cell would be unable to reproduce itself.
   d. The cell would not be able to phosphorylate its associated cyclin.
   e. All of the above
   Textbook Reference: *11.2 How Is Eukaryotic Cell Division Controlled? pp. 214–215*
7. Which of the following statements about DNA replication and cytokinesis in *Escherichia coli* is true?
   a. DNA replication occurs in the nucleus.
   b. Cytokinesis is facilitated by microfilaments of actin and myosin.
   c. DNA replication occurs during the S phase of the cell cycle.
   d. Cell reproduction is initiated by reproductive signals, which result in DNA replication, DNA segregation, and cytokinesis.
   e. The *E. coli* chromosome is linear.
   Textbook Reference: *11.1 How Do Prokaryotic and Eukaryotic Cells Divide? pp. 210–211*
8. Which of the following statements about chromatids is true?
   a. They are replicated chromosomes still joined together at the centromere.
   b. They are identical in mitotic chromosomes.
   c. They undergo recombination in mitosis.
   d They are identical in meiotic chromosomes.
   e. Both a and b
   Textbook Reference: *11.3 What Happens during Mitosis? pp. 215–216*
9. Histones are positively charged because
   a. the majority of the ions in the nucleus of the cell are negatively charged.
   b. histones interact with acidic residues of proteins found in the nucleus.
   c. the basic side chains of histone proteins interact with the negatively charged DNA.
   d. histones have a majority of acidic residues in their protein sequence.
   e. the pH of the nucleus needs to be increased.
   Textbook Reference: *11.3 What Happens during Mitosis? pp. 216–217*
10. Chromosome movement during anaphase is the result of
    a. the hydrolysis of ATP by dynein.
    b. molecular motors at the kinetochores that move the chromosomes toward the poles.
    c. molecular motors at the centrosome that pull the microtubules toward the poles.
    d. shortening of the microtubules at the centrosome that pull the chromosomes toward the poles.
    e. a, b, and d
    Textbook Reference: *11.3 What Happens during Mitosis? p. 219*
11. Programmed cell death (apoptosis)
    a. occurs in cells that have been deprived of essential nutrients.
    b. occurs only in cells that have damaged DNA.
    c. is a natural process during development.
    d. is signaled by the initiation of mitosis.
    e. is well controlled in cancer cells.
    Textbook Reference: *11.6 In a Living Organism, How Do Cells Die? pp. 229–230*
12. If the *ori* site on the *E. coli* chromosome is deleted,
    a. nothing will happen.
    b. replication will start but not be able to continue.
    c. replication will not start.
    d. replication will initiate at another *ori* site on the chromosome.
    e. the chromosome will be replicated but the cell will not be able to divide.
    Textbook Reference: *11.1 How Do Prokaryotic and Eukaryotic Cells Divide? p. 211*
13. Chiasmata
    a. are sites where nonsister chromatids can exchange genetic material during meiosis.
    b. are sites where sister chromatids can exchange genetic material during meiosis.
    c. increase genetic variation among the products of meiosis.
    d. increase genetic variation among the products of mitosis.
    e. Both a and c
    Textbook Reference: *11.5 What Happens during Meiosis? p. 226*
14. The difference between asexual and sexual reproduction is that
    a. asexual reproduction occurs only in bacteria, whereas sexual reproduction occurs in plants and animals.

b. asexual reproduction results from meiosis, whereas sexual reproduction results from mitosis.
c. asexual reproduction results in an organism that is identical to the parent, whereas sexual reproduction results in an organism that is not identical to either parent.
d. asexual reproduction results from the fusion of two gametes, whereas sexual reproduction produces clones of the parent organism.
e. asexual reproduction occurs only in haplontic organisms, whereas sexual reproduction occurs only in diplontic organisms.

Textbook Reference: *11.4 What Role Does Cell Division Play in a Sexual Life Cycle? pp. 221–222*

15. A chromatid is
   a. a chromosome before it has undergone DNA replication.
   b. one of the pairs of homologous chromosomes.
   c. a homologous chromosome.
   d. a newly replicated bacterial chromosome.
   e. one-half of a newly replicated eukaryotic chromosome.

   Textbook Reference: *11.3 What Happens during Mitosis? pp. 215–216*

Application Questions

1. How is cell division different in prokaryotic cells and eukaryotic cells?
   Textbook Reference: *11.1 How Do Prokaryotic and Eukaryotic Cells Divide? pp. 210–211*
2. By using a chemical that inhibits cytokinesis, you have created peaches that are tetraploid. How many sets of chromosomes do these peaches have? (What is the ploidy of these chromosomes?) Will these peaches produce gametes that are fertile? What if the peaches were triploid?
   Textbook Reference: *11.5 What Happens during Meiosis? p. 229*
3. How does cytokinesis differ in animal and plant cells?
   Textbook Reference: *11.3 What Happens during Mitosis? p. 220*
4. Describe how two meters of DNA in a typical human cell can fit into the nucleus, which is 5 μm in diameter.
   Textbook Reference: *11.3 What Happens during Mitosis? p. 217, Figure 11.9*
5. Describe two ways that the genetic diversity of organisms is increased during meiosis.
   Textbook Reference: *11.5 What Happens during Meiosis? pp. 226–227*

Answers

Diagram Exercise Answer

Normal meiosis:

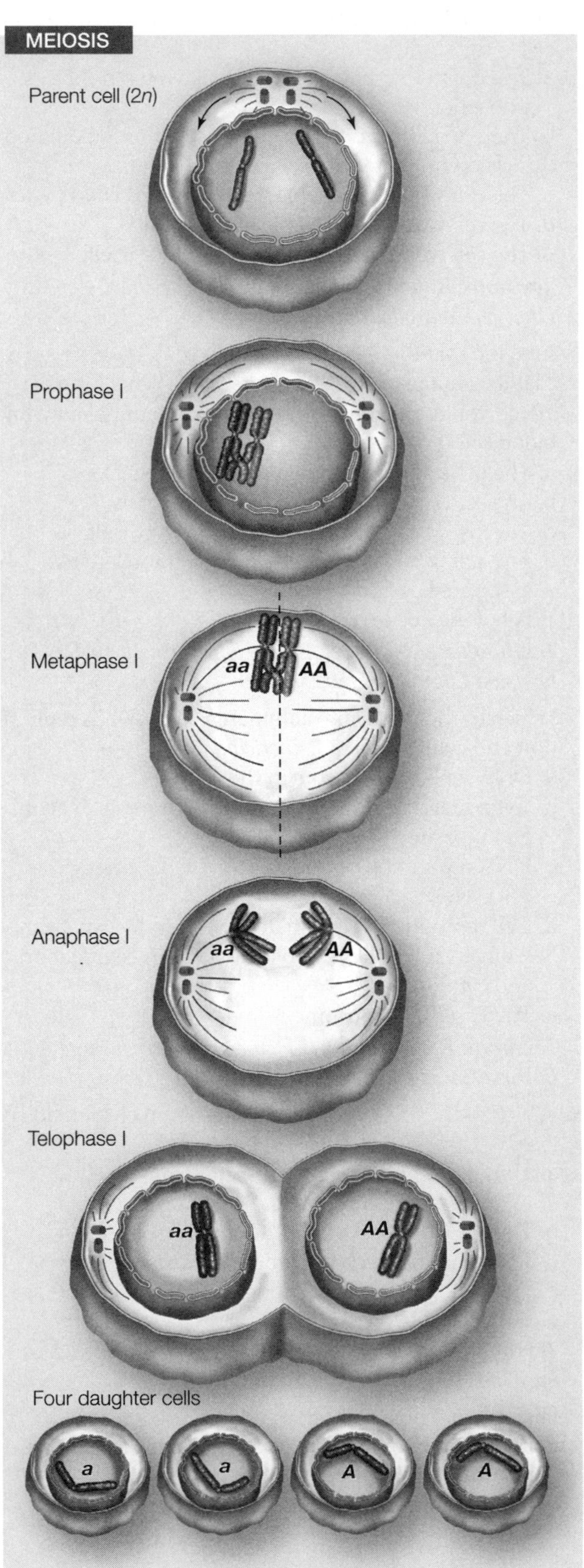

Meiosis with a nondisjunction in meiosis I:

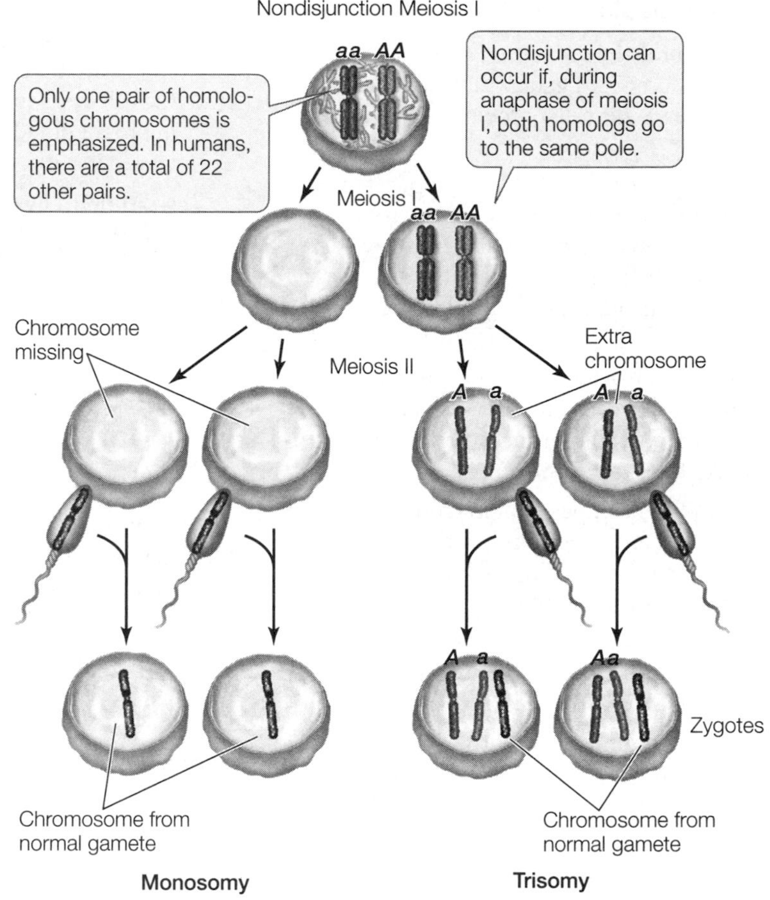

Meiosis with a nondisjunction in meiosis II:

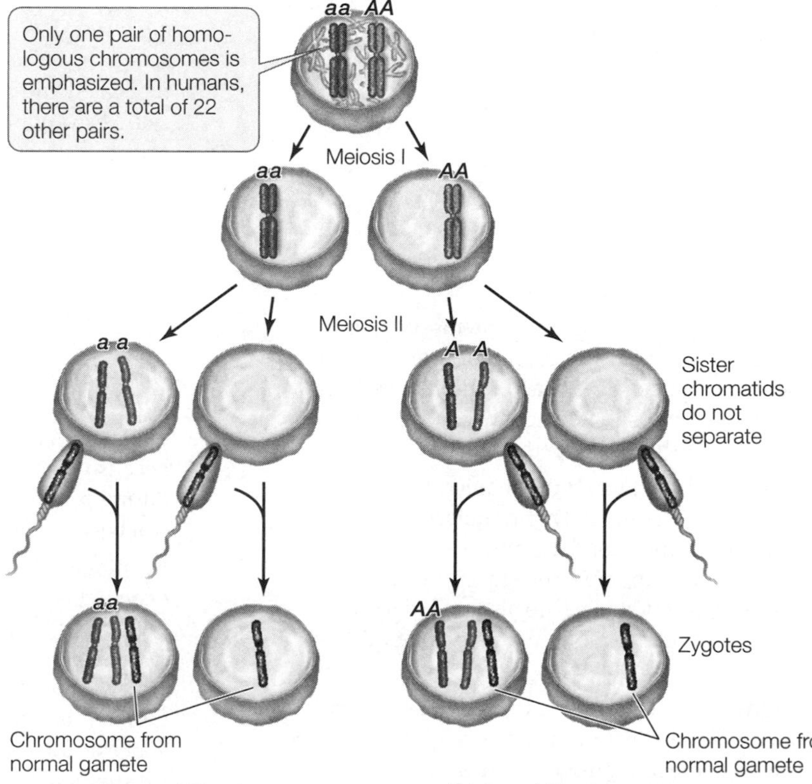

Zygotes would either have one too many chromosomes (trisomy) or would be missing a chromosome (monosomy).

Knowledge and Synthesis Answers

1. **e.** Mitosis occurs after DNA replication and results in cells with the same number of genetically identical chromosomes as the parent cell. Cytokinesis follows mitosis.
2. **a.** Meiosis occurs after one round of DNA replication. Homologous chromosomes pair during prophase I of meiosis, and after meiosis II the resulting cells have half the number of chromosomes as the parent cell. Those chromosomes are not genetically identical to the parental cells.
3. **e.** Kinetochores, one per sister chromatid, are assembled at the centromere of each chromosome and are the sites in which microtubules from opposite poles attach to segregate the chromosomes.
4. **d.** The mitotic spindle is composed of microtubules, not actin and myosin filaments. The spindle originates from the centrosome, which may or may not have centrioles, and only the kinetochore microtubules attach to the chromosomes.
5. **b.** Many cells, such as red blood cells, muscle cells, and nerve cells, lose their ability to divide as they mature.
6. **e.** Cyclin-Cdk's affect the transition into DNA replication into mitosis and are required for cell division. Functional Cdk is require to phosphorylate its associated cyclin.
7. **d.** *Escherichia coli* is a prokaryote. It lacks a nucleus and does not undergo the cell cycle seen in eukaryotes. It has a circular chromosome, and does not synthesize actin or myosin proteins. Cytokinesis in *E. coli* is a result of a reproductive signal that causes the DNA to be replicated and segregated and finally causes the cell to divide.
8. **e.** Chromatids are highly condensed, newly replicated chromosomes, which will be segregated to the daughter cells. After DNA replication, chromatids are still attached to each other at the centromere. Meiotic sister chromatids are different from each other due to recombination (crossing over) in prophase of meiosis I. Mitotic sister chromatids are identical.
9. **c.** The positive charges on histone proteins are due to the large number of basic amino acid residues found in these proteins. These positive charges interact with the negatively charged phosphate sugar backbone of DNA during assembly of the DNA on the nucleosome.
10. **e.** Chromosomes are attached to the microtubules at their kinetochores. There are dynein molecular motors at the kinetochores (but not the centrosome) which hydrolyze ATP and help move the chromosomes to opposite poles. Chromosomes are also pulled toward the poles by the shortening of the kinetochore microtubules.
11. **c.** Programmed cell death occurs during the development of many organisms (for instance, tadpoles lose their tails to become adult frogs). One of the stimuli for programmed cell death is DNA damage, but it is not the only cause of death. Necrosis (cell death that is not programmed) occurs when cells have been deprived of essential nutrients. The initiation of mitosis is part of the cell cycle, in which cells reproduce, and is not a step in programmed cell death. Apoptosis is not well controlled in cancer cells.
12. **c.** Without the origin of replication (there is only one in *E. coli*), there would be no site for the replication proteins to bind to initiate DNA replication, so DNA synthesis would not start.
13. **e.** Chiasmata are sites where nonsister chromatids can exchange genetic material during meiosis, which increases genetic variation in the gametes (the products of meiosis).
14. **c.** Asexual reproduction, which results from mitosis, produces cells that are identical to the parent and can occur in plants. Sexual reproduction can occur in haplontic organisms (such as fungi) and results in an organism that is not genetically identical to either parent.
15. **e.** A chromatid is one-half of a newly replicated eukaryotic chromosome, and is connected to the other (sister) chromatid at the centromere.

Application Answers

1. In most prokaryotic cells there is only one circular chromosome. As the cell enlarges to prepare for division, the newly replicated daughter chromosomes are separated at opposite sides of the cell. During fission, the cell membrane pinches in, and cell wall components are synthesized between the daughter cells. In eukaryotic cells, there are more chromosomes, and they are linear. The cell undergoes a sequential set of steps called the cell cycle, in which the chromosomes are replicated and then separated to opposite poles of the cell. Microtubules are used to segregate the chromosomes equally into the daughter cells, and actin filaments and myosin cause the cell membrane to form a contractile ring and separate to form two daughter cells.
2. Peaches that are tetraploid have four sets of chromosomes. Because there are an even number of chromosomes ($4n$), each replicated homologous chromosome will be able to find a replicated homolog to pair with at meiosis and will produce fertile gametes. These gametes will be diploid. Triploid cells will not be fertile because one of the three homologs will not find its pair during prophase of meiosis I, and the single homologs will be segregated randomly into the daughter cells.
3. In animal cells, cytokinesis results from the interaction of actin filaments and myosin, which causes the cell membrane to pinch in and divide the cytoplasm into two cells. In plant cells, a cell plate forms between the newly segregated chromosomes, and Golgi vesicles fuse at that site to form the new cell membranes. Cell wall components are then secreted between the plasma membranes to complete cytokinesis.

4. See Figures 11.8 and 11.9.
5. Genetic diversity is increased during crossing over of prophase I of meiosis so that each gamete has chromosomes with different combinations of alleles. During meiosis, each homologous chromosome is randomly segregated to one of the two poles, resulting 2^{23} different possible combinations of homologous chromosomes per gamete.

12 Inheritance, Genes, and Chromosomes

The Big Picture

- Genes are units of inheritance that are passed down to the progeny by the fusion of gametes to form a new organism.
- Meiosis generates gametes that contain only one allele for each gene, and those alleles are segregated independently (unless they are linked).
- By studying the expression of particular phenotypes in successive generations, one can begin to understand the alleles that gave rise to those phenotypes (i.e., rare recessive, rare dominant, sex-linked).
- Exceptions to Mendel's rules also help explain the interactions of gene products and the interaction of alleles that lead to the observed phenotype.
- Bacteria can transfer genes to other recipient cells during conjugation.

Common Problem Areas

- It is easy to be overwhelmed by all the exceptions to Mendel's rules. Initially, make sure you understand Mendel's laws of segregation and independent assortment. Examine the exceptions to these rules by thinking about the molecular role of these alleles. For example, what is the explanation for a codominant phenotype at the molecular level, such as the AB blood type in humans? Both gene products are synthesized by the cell and modify the cell surface glycoproteins, producing both A and B antigens. Or what is the molecular explanation for epistasis? The enzyme required to produce an initial substrate necessary for pigmentation is missing, even though the other gene products (further down the biochemical pathway) for producing a particular color are present in the cell.

Study Strategies

- Review the figures of meiosis in Chapter 11 to see how Mendel's laws of independent assortment and segregation apply. Draw the chromosomes with different alleles (*S* on one chromosome, *s* on the other) and follow how those alleles are segregated to the gametes during meiosis.
- Use a Punnett square to predict the outcome of different monohybrid and dihybrid crosses. Use the probability rules to predict the genotypes and phenotypes of these same crosses and compare the two methods.
- After reviewing Mendel's laws of independent assortment and allele segregation, make a list of the exceptions to those laws with specific examples for each exception.
- Review the linkage problem in Figure 12.22 to familiarize yourself with allele loci on a chromosome and how recombination frequencies affect the genotypes of the gametes.
- Go to yourBioPortal.com to review the following tutorials and activities:

  Animated Tutorial 12.1 Independent Assortment of Alleles

  Animated Tutorial 12.2 Alleles That Do Not Sort Independently

  Interactive Tutorial: Pedigree Analysis

  Web Activity 12.1 Homozygous or Heterozygous?

  Web Activity 12.2 Concept Matching I

  Web Activity 12.3 Concept Matching II

Important Concepts

Genes are units of inheritance.

- Mendel studied the inheritance of characters (such as flower color) and traits (a particular feature, such as a red, purple, or white color) as they were passed (heritable traits) from one generation of pea plants to another.
- He used plants that were true-breeding (which meant that the observed trait must be the only form for many generations). To test if plants were true-breeding, Mendel allowed these plants to self-fertilize for many successive generations to determine if the trait was consistently expressed in every generation.
- In genetic crosses, the two individuals in the initial cross are the parents (the P generation) and their offspring are the first filial generation (the F_1 generation). The F_2 (second filial generation) are the offspring from a cross of the F_1 generation.

- Monohybrid crosses are matings between two true-breeding parents, each expressing one different trait (i.e., a cross between one true-breeding parent with a spherical seed and one true-breeding parent with a wrinkled seed). Mendel found that in these crosses, the F_1 generation expressed only one trait (spherical), and in the F_2 generation both the spherical and the wrinkled traits were expressed in ratios of 3:1 (spherical to wrinkled). He termed the F_1 trait (the spherical seed) dominant and the F_2 trait (the wrinkled seed) recessive.
- Mendel's theory stated that discrete particles of inherited traits existed in pea plants in pairs (one from each parent, diploid) and that these particles separated from each other and were assorted independently during gamete formation. Each of the gametes received one of these particles (haploid), whereas the zygote received two particles for a trait, one from each parent. This unit of inheritance is a gene. The totality of all genes in an organism is its genome.
- Alleles are different forms of the same gene. The gene for seed shape in peas has a spherical allele and a wrinkled allele.
- True-breeding parents are homozygous for that trait and have two copies of the same allele, one on each homologous chromosome.
- Heterozygous individuals have two different alleles for a gene, one on each homologous chromosome. For example, the F_1 generation of a monohybrid cross between pea plants with spherical seeds and pea plants with wrinkled seeds is heterozygous for seed shape, having one spherical allele and one wrinkled allele.
- The phenotype of an organism is the physical appearance of that organism (i.e., having seeds that are spherical or wrinkled).

The outcome of genetic crosses can be predicted.

- Mendel's theories have been restated as laws. Mendel's first law, the law of segregation, states that when any individual produces gametes, alleles separate so that each gamete receives only one allele for each gene.
- Punnett squares can be utilized to analyze the potential outcome of a genetic cross. All of the possible gametes from one parent are lined up on one side of the Punnett square, and all of the gametes of the other parent are lined up on the other side. Genotypes of the progeny are predicted by placing both alleles of the gametes in the square (see Figures 12.4, 12.6, and 12.7).
- Genes are located at particular sites on chromosomes called loci (locus, singular).
- Test crosses are used to determine if a given individual is heterozygous or homozygous for a dominant allele. That individual is crossed with an individual homozygous for the recessive trait (the test cross). If the individual from this test cross is homozygous for that trait, then all of the offspring will express the dominant trait. If the test individual is heterozygous, then approximately one-half of the offspring from this cross will express the dominant trait and one-half will express the recessive trait (see Figure 12.6).
- Mendel's second law, the law of independent assortment, states that all of the alleles of different genes assort independently of one another during gamete formation. When the F_1 generation that is heterozygous for two traits—for example, seed shape and seed color—is crossed (a dihybrid cross), all the alleles are expressed in the F_2 generation. In the F_2 progeny from this dihybrid cross, both yellow and green seed color are observed as well as wrinkled and spherical seeds. Review the Punnett square in Figure 12.7.

Probability calculations can be used to predict the outcome of genetic crosses.

- The product rule states that the probability of two independent events happening together (a joint probability) is equal to the product of the probability of each of those individual events (see Figure 12.9).
- The sum rule is used to predict the probability of an event that can occur in two or more different ways. The probability that the event will occur in one of these ways is equal to the sum of the individual probabilities (see Figure 12.9).

Pedigrees can be used to predict how traits are inherited from generation to generation.

- If a trait is due to a rare dominant allele, (1) the affected person has an affected parent; (2) on average, one-half of the offspring of an affected parent will be affected; and (3) the phenotype is seen equally in both sexes (see Figure 12.10A).
- If a trait is due to a rare recessive allele, (1) the affected individual can have unaffected parents; (2) on average, one-fourth of the children from unaffected parents express the trait; and (3) the phenotype occurs equally in both sexes (see Figure 12.10B).

Alleles interact in different ways.

- New alleles arise by mutation.
- "Wild type" refers to traits that occur in most individuals in nature, whereas "mutant" refers to traits that are different from wild type.
- A polymorphic trait is a trait that has many different forms, with each individual phenotype present in less than 99 percent of the population.
- Because of random mutations, multiple alleles for a given gene may exist in a group of individuals (for example, rabbit fur color; see Figure 12.11).
- Incomplete dominance is common in nature. It is a situation in which neither of the two alleles of a gene is dominant and individuals that are heterozygotic for a particular trait appear as intermediate phenotypes of the parents. For example, a cross between a red-flowered plant and a white-flowered plant yields F_1 plants with pink flowers (see Figure 12.12).

- In codominance, both alleles are expressed in the individual (for example, blood types in humans) (see Figure 12.13).
- A pleiotropic allele is a single allele that can have more than one effect in an individual. For example, in the Siamese cat, a single allele can affect both pigmentation and crossed eyes.

Genes interact in different ways.

- In a situation of epistasis, the phenotypic expression of one gene is affected or masked by another gene. For example, an allele for yellow coat color in Labrador retrievers can mask another allele for brown or black color (see Figure 12.14).
- The superiority of individuals that are heterozygous for a particular trait is known as "hybrid vigor," or heterosis. For example, hybrid varieties of corn produce much better yields than their homozygous parents (see Figure 12.15).
- Environmental conditions such as light, temperature, or nutrition can affect the expression of a particular trait. Both penetrance and expressivity are related to environmental effects (see Figure 12.16).
    - Penetrance is the proportion of individuals within a group with a particular genotype that actually show the expected phenotype.
    - Expressivity is the degree a particular genotype is expressed in an individual.
- Variation of phenotypes within a population is called quantitative, or continuous, variation. More than one gene (quantitative trait loci) can affect a particular phenotype, with each allele intensifying or diminishing the phenotype (for example, human height). These traits can be affected by the environment.

Genes on the same chromosome are linked.

- Different alleles on the same chromosome do not assort independently, but rather are linked, as first noted by Morgan in his studies on *Drosophila* (see Figure 12.18).
- Genes can be exchanged between chromatids of a homologous pair during prophase I of meiosis.
    - Linkage groups are the full set of loci on a given chromosome.
    - Recombination frequencies between alleles that are linked (due to crossing over during prophase of meiosis I, see Figure 12.19) can be used to map positions of genes on a chromosome. Recombination frequencies are expressed in map units and are equal to the percent of recombinant progeny divided by the total progeny. A map unit (also referred to as a centimorgan, cM) corresponds to a recombination frequency of 0.01 (see Figures 12.20, 12.21, 12.22, and 12.23).

Sex is determined in different ways in different species.

- Monoecious organisms (earthworms, pea plants, corn, etc.) produce both female and male gametes.
- Dioecious organisms (some plants and most animals) produce only male or only female gametes, resulting in two separate sexes.
- In many animals including humans, sex is determined by one or two sex chromosomes. The other chromosomes are called autosomes.
- In mammals, there are two X chromosomes in females and one X chromosome and one Y chromosome in males. Males produce two kinds of gametes, which carry either an X or a Y chromosome.
- Nondisjunction of the sex chromosomes produces gametes that can result in Turner syndrome (female, XO) or Klinefelter syndrome (male, XXY).
- The *SRY* (*s*ex-determining *r*egion on the *Y* chromosome) determines maleness in males. *SRY* is responsible for the primary sex determination in males and causes the embryo to develop sperm-producing tissue, the testes. Males without the *SRY* gene develop as females. The *DAX1* gene, which encodes the anti-testis factor, is found on the X chromosome and inhibits the development of testis tissue. In XY individuals (males), the *SRY* gene product inhibits the effect of *DAX1*, and testes develop.
- Secondary sex determination, the outward manifestation of male and female characteristics, is determined by autosomal and X-linked genes that control the actions of hormones such as testosterone or estrogen.
- In *Drosophila*, sex is determined by the ratio of X chromosomes to autosomes: XX is female (as is XXY), and XY is male (as is XO).

Genes on sex chromosomes are inherited in special ways.

- In mammals and other organisms, including some insects, females have two copies of each X-linked gene, whereas males only have one copy of the X-linked genes.
- Because males are hemizygous for X-linked traits, all their X-linked traits will be expressed.
- If an allele is located on the X chromosome, reciprocal crosses do not have the same result. For instance, a cross between a homozygous red-eyed female *Drosophila* and a white-eyed male *Drosophila* will generate all red-eyed flies. The reciprocal cross between a white-eyed female (who must be homozygous, because the white-eyed allele is recessive) and a red-eyed male will produce an F_1 generation in which all female flies have red eyes and all male flies have white eyes (see Figure 12.23).
- X-linked recessive traits show the following characteristics:

- The phenotype is much more common in males than in females.
- Males with X-linked traits can pass those traits only to their daughters.
- Females who are carriers of an X-linked trait (heterozygous) pass that trait on average to one-half of their sons and one-half of their daughters. Because sons always receive the X chromosome from their mother, one-half of those sons, on average, will express the X-linked trait. Daughters who receive the X-linked trait from their mothers will be heterozygous carriers (assuming they received the X chromosome from their father that carried the dominant allele).

- X-linked traits can skip a generation; a trait that is expressed in the grandfather will be carried by his daughter and may be expressed in his grandson.
- There are several dozen genes in humans that are Y-linked. These traits are always passed from father to son.

Cytoplasmic inheritance differs from the Mendelian pattern.

- Mitochondria and plastids contain genomes that are passed to the zygote from the gamete that makes the largest cytoplasmic contribution (in humans, the egg).
- These organelles are generally so numerous that genes for cytoplasmic traits are polyploid.
- Genes in organelles mutate faster than those in the nucleus, leading to multiple alleles for many nonnuclear genes.
- In plants, mutations in plastid genes affect proteins that assemble chlorophyll molecules into photosystems, resulting in white rather than green leaves (see Figure 12.25).
- Mutations in mitochondrial genes affect the electron transport chain and the synthesis of ATP. These mutations have a strong effect on tissues that require high ATP concentrations, particularly the nervous and muscular systems and kidneys.

Prokaryotes exchange genes by conjugation.

- *E. coli* has a single circular chromosome which is 1 μM in circumference and carries a few thousand genes.
- Some bacteria mate by transferring part of their chromosome through a conjugation tube to a recipient bacterial cell. Those bacteria must contact the recipient cell through a projection called a sex pilus (see Figure 12.26A).
- Genes that are transferred during the mating process recombine with the recipient cell's chromosome.
- Many bacterial cells harbor plasmids, which are small circular DNA molecules that replicate independently.
- Some plasmids harbor genes with metabolic capabilities, genes for conjugation, or genes for antibiotic resistance.

Test Yourself

Diagram Exercise

1. a. Diagram two separate pairs of chromosomes each bearing a different allele; for instance, *Ss* and *Yy*, as seen in the dihybrid cross with seed shape (*Ss*) and seed color (*Yy*). Show how these alleles assort inde pendently during meiosis to produce haploid gam etes (review Figure 12.8).

   b. Draw a diagram starting with the same parent (*SsYy*) but assume that *S* and *Y* are linked and are 20 map units apart. For this parent, the *S* and *Y* alleles are on one chromosome and the *s* and *y* alleles are on the other chromosome. Draw the gametes you would expect if there is a crossover between *S* and *Y*.

   Textbook Reference: *12.1 What Are the Mendelian Laws of Inheritance? pp. 242–245*

Knowledge and Synthesis Questions

1. Hemophilia is a trait carried by the mother and passed to her sons. The allele for hemophilia, therefore,
   a. is carried on one of the mother's autosomal chromosomes.
   b. is carried on the Y chromosome.
   c. can be carried on the X or Y chromosome.
   d. is on the X chromosome and can only be inherited by the son if the mother is a carrier (heterozygous).
   e. is carried in the mitochondrial genome because sons inherit this allele from their mothers.

   Textbook Reference: *12.1 What Are the Mendelian Laws of Inheritance? pp. 246–247*

2. Originally, genetic inheritance was thought to be a function of the blending of traits from the two parents. Which exception to Mendel's rules is an example of blending?
   a. X linkage
   b. Polygenic inheritance
   c. Incomplete dominance
   d. Codominance
   e. Pleiotropism

   Textbook Reference: *12.2 How Do Alleles Interact? p. 249*

3. True-breeding plants
   a. produce the same offspring when crossed for many generations.
   b have no mutations.
   c. result from a monohybrid cross.
   d. result from a dihybrid cross.
   e. result from crossing over during prophase I of meiosis.

   Textbook Reference: *12.1 What Are the Mendelian Laws of Inheritance? p. 239*

4. What is the probability that a cross between a true-breeding pea plant with spherical seeds and a true-breeding pea plant with wrinkled seeds will produce F_1 progeny with spherical seeds?
   a. ½
   b. ¼
   c. 0

d. $\frac{1}{8}$
e. 1
Textbook Reference: *12.1 What Are the Mendelian Laws of Inheritance? p. 241, Figure 12.3*

5. What is the pattern of inheritance for a rare recessive allele?
a. Every affected person has an affected parent.
b. Unaffected parents can produce children who are affected.
c. Affected parents do not produce affected children.
d. Unaffected mothers have affected sons and daughters who are carriers.
e. None of the above
Textbook Reference: *12.1 What Are the Mendelian Laws of Inheritance? pp. 246–248*

6. What is the pattern of inheritance for a rare dominant allele?
a. Every affected person has an affected parent.
b. Unaffected parents can produce children who are affected.
c. Affected parents do not produce affected children.
d. Unaffected mothers have affected sons and daughters who are carriers.
e. None of the above
Textbook Reference: *12.1 What Are the Mendelian Laws of Inheritance? pp. 246–248*

7. What is the pattern of inheritance for a sex-linked allele?
a. Every affected person has an affected parent.
b. Unaffected parents can produce children who are affected.
c. Affected parents do not produce affected children.
d. Unaffected mothers have affected sons and daughters who are carriers.
e. None of the above
Textbook Reference: *12.4 What Is the Relationship between Genes and Chromosomes? p. 257*

8. Penetrance and expressivity are related to
a. the increased expression of a particular trait when a hybrid species is formed.
b. quantitative traits that diminish or intensify a particular phenotype.
c. the influence of environment on the expression of a particular genotype.
d. the expression of one gene masking the effects of another gene.
e. the expression of a dominant phenotype in a heterozygote.
Textbook Reference: *12.3 How Do Genes Interact? p. 252*

9. Sex determination in humans and *Drosophila* is similar because
a. females are hemizygous.
b. males have one X chromosome and females have two X chromosomes.
c. all males from both species always have one Y chromosome.
d. secondary sex characteristics are determined by genes on the X chromsome.
e. the ratio of X chromosomes to sets of autosomes determines maleness or femaleness.
Textbook Reference: *12.4 What Is the Relationship between Genes and Chromosomes? pp. 257–258*

10. Linked genes are genes that
a. assort independently.
b. segregate equally in the gametes during meiosis.
c. always contribute the same trait to the zygote.
d. are found on the same chromosome.
e. recombine during mitosis.
Textbook Reference: *12.4 What Is the Relationship between Genes and Chromosomes? p. 253*

11. Cytoplasmic inheritance
a. results from polygenic nuclear traits.
b. is determined by nuclear genes.
c. is the result of gametes contributing equal amounts of cytoplasm to the zygote.
d. is determined by genes on DNA molecules in mitochondria and chloroplasts.
e. follows Mendel's law of segregation.
Textbook Reference: *12.5 What Are the Effects of Genes Outside the Nucleus? pp. 259–260*

12. Epistasis is
a. the degree to which a particular genotype is expressed in an individual.
b. the proportion of individuals within a group with a particular genotype that show the expected phenotype.
c. a situation in which a heterozygotic individual expresses an intermediate phenotype of the parents.
d. a situation in which one gene masks the expression of another gene.
e. a situation in which both alleles are expressed equally.
Textbook Reference: *12.3 How Do Genes Interact? p. 250*

13. Quantitative traits are traits
a. that are affected by the environment.
b. that affect the same physical characteristic.
c. in which each allele intensifies or diminishes the phenotype.
d. All of the above
e. None of the above
Textbook Reference: *12.3 How Do Genes Interact? p. 252*

14. A test cross
a. is used to determine if an organism that is displaying a dominant trait is heterozygous or homozygous for that trait.
b. is used to determine if an organism that is displaying a recessive trait is heterozygous or homozygous for that trait.
c. causes the loss of hybrid vigor.
d. results in an F_2 generation with a phenotypic ratio of $\frac{3}{4}$ dominant to $\frac{1}{4}$ recessive.

e. results in the same alleles being transferred from generation to generation.

Textbook Reference: *12.1 What Are the Mendelian Laws of Inheritance? p. 242*

15. An individual has a karyotype that is XX but is phenotypically male. What could explain this result?
    a. The Y chromosome was not visible in the karyotype.
    b. Sex determination is determined by the autosomes and not the X chromosomes.
    c. A translocation has occurred, placing the *SRY* gene on one of the X chromosomes.
    d. The DAX 1 protein is overproduced.
    e. A nondisjunction event has resulted in Kleinfelter syndrome.

    Textbook Reference: *12.4 What Is the Relationship between Genes and Chromosomes? p. 257*

16. A bacterial cell's genotype can be altered by
    a. the introduction of a plasmid that carries some of the bacteria's genes.
    b. mating with a bacterial cell with the same genotype.
    c. homologous recombination with human DNA.
    d. the forming of a conjugation tube with another bacterial cell.
    e. the transferring of genetic material from a different strain of bacteria.

    Textbook Reference: *12.6 How do Prokaryotes Transmit Genes? pp. 260–261*

Application Questions

1. Draw a pedigree for three generations in which the grandfather has red–green color blindness and his daughter is a carrier. This daughter has four sons. Predict how many of the sons will be color-blind.

   Textbook Reference*: 12.4 What Is the Relationship between Genes and Chromosomes? pp. 246–248, p. 259*

2. Draw a sample pedigree with three generations in which the paternal grandfather has a rare dominant autosomal trait. What is the probability that one of his children will have the disease? What is the probability that one of his grandchildren will have the disease?

   Textbook Reference: *12.1 What Are the Mendelian Laws of Inheritance? pp. 246–248*

3. Draw a sample pedigree with three generations in which the maternal grandmother and paternal grandfather are carriers of a rare recessive autosomal trait. What is the probability that one of their children will be carriers of this trait? What is the probability that a grandchild will have the disease?

   Textbook Reference: *12.1 What Are the Mendelian Laws of Inheritance? pp. 246–248*

4. Cytoplasmic traits in certain species of trees are passed from the male plant to all of its progeny. Compare this observation to cytoplasmic inheritance in humans.

   Textbook Reference: *12.5 What Are the Effects of Genes Outside the Nucleus? pp. 259–260*

5. Suppose you are a genetics counselor who is working with a 21-year-old pregnant woman who has just discovered that her father has Huntington's chorea, a rare dominant autosomal trait. This disease usually develops in middle age, so people carrying this trait do not find out they have this genetic disorder until midlife. What are the chances that the child she is carrying will develop the disease? (Assume that her husband's family has no history of the disease.) What is the chance that she has Huntington's chorea?

   Textbook Reference: *12.1 What Are the Mendelian Laws of Inheritance? pp. 246–248*

Answers

Diagram Exercise Answer

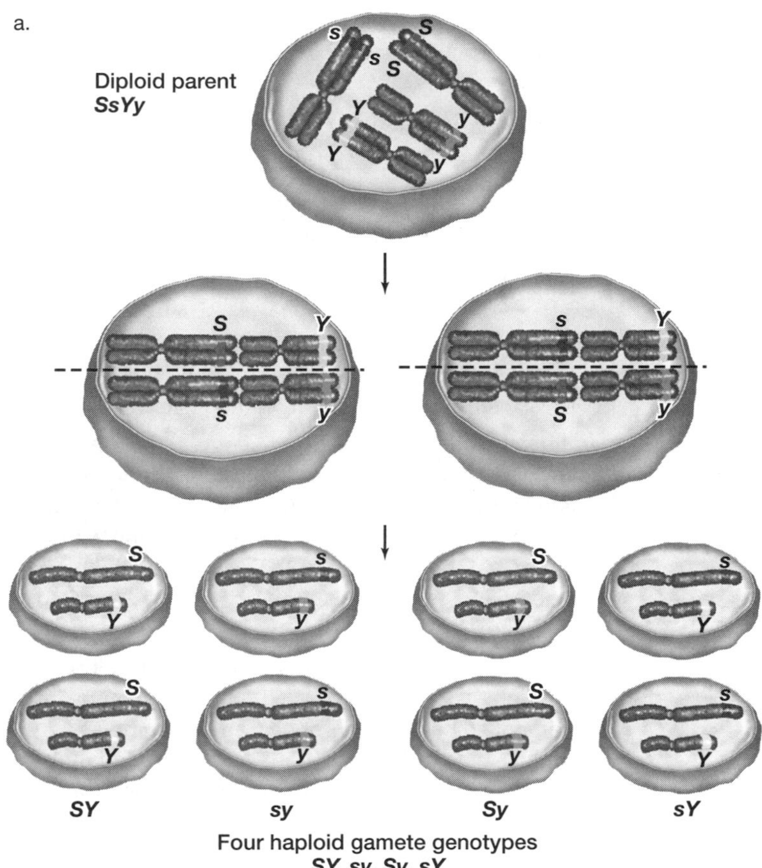

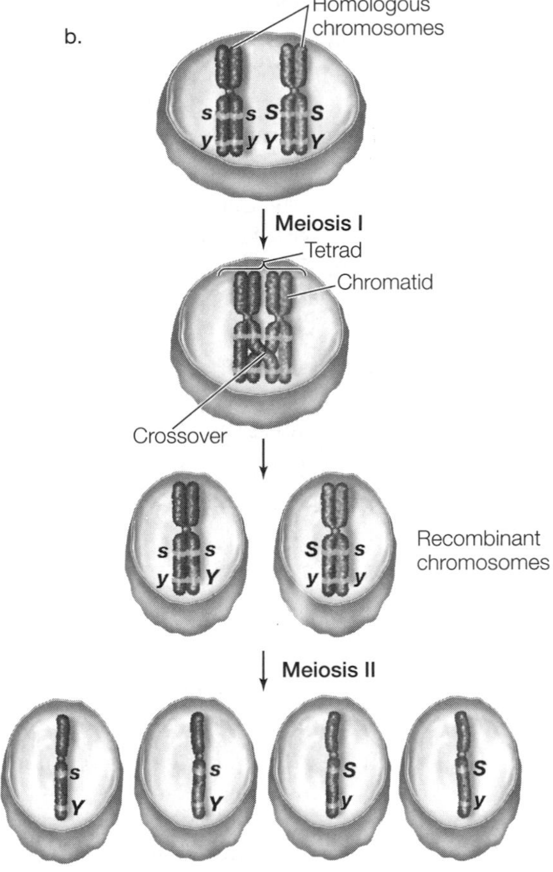

Knowledge and Synthesis Answers

1. **d.** Hemophilia is an X-linked trait and can only be inherited by the son from his mother's X chromosome (and not her mitochondrial chromosome). The father contributes the Y chromosome to his son (not his X chromosome) and thus cannot pass any of his X-linked alleles to his son.
2. **c.** Incomplete dominance results in the progeny's expressing an intermediate form of the two parental alleles. (In a cross between red-flowered plants and white-flowered plants, the expression of pink-flowered plants would be a "blend" of the parental traits.) Co-dominance is not an example of blending because both alleles are fully expressed in the individual.
3. **a.** Monohybrid and dihybrid crosses produce heterozygous individuals; true-breeding individuals are always homozygous.
4. **e.** This is an example of a monohybrid cross. All of the F_1 progeny would have spherical seeds. (The F_1 generation would all have the genotype *Ss,* producing the phenotype of spherical seeds because the spherical allele, *S,* is dominant to the wrinkled allele, *s.*)
5. **b.** Rare recessive alleles can be carried by both parents but not expressed in those parents. If the parents are heterozygous for this allele (*Aa*), their children will have a one-fourth probability of expressing that recessive allele (*aa*). If both parents are affected (*aa*), their children will also be affected (*aa*).
6. **a.** If an allele is dominant, every affected individual has at least one dominant allele. An affected individual must have received that allele from one of his or her parents. Because the allele is dominant, that parent must also be affected.
7. **d.** The most common sex-linked alleles are X-linked and are passed from a mother to her son (because the mother always donates one of her X chromosomes to her son, and the father always donates the Y chromosome to his son). Daughters can also receive the X-linked allele from their mothers, but the father donates the other X chromosome, so daughters can be carriers.
8. **c.** Penetrance and expressivity are related to the effects the environment has on a particular phenotype. Answer **a** refers to hybrid vigor, answer **b** refers to quantitative traits, answer **d** refers to epistasis, and answer **e** refers to expression of a dominant allele.

9. **b.** In these species, females have two X chromosomes and males have one X chromosome. The alleles for secondary sex characteristics are found on the X chromosome and the autosomes.
10. **d.** Linked genes by definition are on the same chromosome and thus do not sort independently, do not contribute the same trait to the zygote, and do not recombine during mitosis or segregate equally to the gametes during meiosis.
11. **d.** The genes on the mitochondria and chloroplast chromosomes which are cytoplasmically inherited (unlike nuclear genes) are passed on to all of the progeny from the gamete that contributes the majority of the cytoplasm.
12. **d.** Answer **a** refers to expressivity, answer b refers to penetrance, answer **c** refers to incomplete dominance, and answer **e** refers to codominance.
13. **d.** Quantitative traits are traits that are affected by the environment and can either diminish or intensify one phenotype.
14. **a.** A test cross is used to determine if an organism that is expressing a dominant trait is homozygous or heterozygous for that trait. A ratio of ¾ dominant to ¼ recessive in the F_2 generation results from a monohybrid cross. True-breeding individuals continue to express the same alleles generation after generation.
15. **c.** The *SRY* (sex-determining region) gene has been moved via a translocation to the X chromosome, and in the presence of the SRY protein, the XX individual has developed sperm-producing testes. SRY also inhibits the expression of the *DAX 1* gene, which encodes a male inhibitor. If *DAX 1* was overproduced during embryonic development, a female would be produced. Individuals with Klinefelter syndrome are XXY.
16. **e.** A bacterial cell's genotype changes when new genetic information is introduced either on a plasmid or by conjugation (which requires a conjugation tube and the transfer of DNA). If conjugation occurs, those genes need to recombine into the recipient cell's chromosome to be maintained. Transferring the same genetic information on a plasmid or mating with a bacterial cell with the same genotype will not change the genotype of the recipient bacteria. Human DNA cannot recombine with a bacterial chromosome unless there are identical (homologous) DNA sequences in both DNA molecules.

Application Answers

1. See Figure 12.24, generations II, III, and IV. One-half of her sons could be color-blind.
2. See Figure 12.10A. One-half of his children could get the disease; one-fourth of his grandchildren could get the disease.
3. See Figure 12.10B, generations III, and IV. One-half of the children of these grandparents could be carriers. One-sixteenth of the children could have the disease.
4. In humans, the gamete with the largest cytoplasmic contribution is the egg, so cytoplasmic inheritance is passed from the female parent to all her children. In certain tree species, the male gamete contributes the majority of the cytoplasm to the zygote, so all the mitochondria and chloroplasts in the zygote are inherited from the male parent.
5. There is a 50 percent chance that she will develop Huntington's chorea. Because the trait is an autosomal dominant allele, one-half of her father's gametes will contain the homologous chromosome carrying that allele and one-half of his gametes will contain the homologous chromosome that carries the wild-type allele. If she received the Huntington's allele, her child has a 50 percent chance of receiving this allele from her. The product rule is used to predict the probability that her child will inherit the Huntington's allele: ½ (the probability that she has the Huntington's allele) × ½ (the probability her child will inherit this allele from her) = ¼ (the probability her child has the allele). Her child has a 25 percent chance of carrying the Huntington's chorea allele and thus of developing the disease.

13 DNA and Its Role in Heredity

The Big Picture

- Knowing that DNA is the genetic material has allowed scientists to understand how hereditary information is passed on at the molecular level and how mutations can alter that hereditary information.
- Advances in our knowledge of DNA replication provide new technologies for understanding genes, their function, their expression, and genetic relatedness in different organisms.

Common Problem Areas

- Students frequently get lost in the history of discovery. Do not focus on *who* and *when,* but instead on *how* the experiments provided evidence to prove that DNA is the genetic material.
- Understanding DNA replication is a highly visual process. Use the figures in your textbook to "see" what is occuring. Make your own diagrams of these processes.
- 5′ and 3′ ends are often confused when parental (template) strands and daughter (newly synthesized) strands are compared. On both strands, the 3′ end corresponds to the site (or former site) of a hydroxyl group (—OH), and the 5′ end corresponds to the site (or former site) of a phosphate tail.
- Take advantage of laboratory activities that simulate or allow you to experience sequencing or PCR. Nearly all molecular research labs utilize one or both of these techniques.

Study Strategies

- Focus on how scientific evidence proved that DNA is the genetic material and how Meselson and Stahl demonstrated that DNA replication is semiconservative.
- Draw a picture of the replication fork. Position the primers on the fork, and then draw in the leading and lagging strands. Label all of the 3′ and 5′ ends. Draw a box to indicate where ligase will seal up the ends.
- Make a list of all the proteins involved in DNA replication. Describe the function of each protein.
- Go to yourBioPortal.com to review the following tutorials and activity:

  Animated Tutorial 13.1 DNA Replication, Part 1: Replication of a Chromosome and DNA Polymerization

  Animated Tutorial 13.2 The Meselson–Stahl Experiment

  Animated Tutorial 13.3 DNA Replication, Part 2: Coordination of Leading and Lagging Strand Synthesis

  Web Activity 13.1 The Replication Complex

Important Concepts

DNA is the genetic material.

- By the 1920s, Feulgen's staining techniques showed that DNA is present in the nucleus and the chromosomes, that different organisms show varying amounts of DNA in their nuclei, and that proportional amounts of DNA are found in somatic and germ cells.
- Griffith's transformation experiments, done around the same time, indicated that some transforming principle (later identified as DNA) could cause a heritable change in living cells (see Figure 13.1).
- In 1944 Avery, MacLeod, and McCarty used purified DNA to demonstrate that the transforming factor containing the genetic information is DNA (see Figure 13.2).
- In 1952, Hershey and Chase, using a virus that infected bacteria, demonstrated definitively that DNA carries the hereditary information (see Figures 13.3 and 13.4).
- Transformation in eukaryotes (termed transfection) using a genetic marker was the final proof that DNA is the genetic material.

A series of experiments elucidated the structure of DNA.

- In the early 1950s, Franklin and Wilkins prepared X-ray crystallographic images, which were critical to understanding the structure of DNA.
- In 1950, Chargaff observed that the relative ratios of pyrimidines and purines, the nitrogenous bases of the nucleotides of DNA, are consistent with A = T and G = C (Chargaff's rules).
- In 1953, Watson and Crick coupled results from previous experimentation and model-building and revealed the three-dimensional, double-helical structure of DNA.

The form of DNA is tied to its function.

- DNA is a right-handed double-stranded helix formed by two antiparallel DNA strands.
- The two strands are aligned in the helix with the 5′ end of one strand pairing with the 3′ end of its complementary strand.
- The sugar–phosphate backbones of the polynucleotide chains are on the outside of the helix, and the nitrogenous bases point toward the center of the helix.
- Base pairs are complementary; they consist of a pyrimidine pairing with a purine via hydrogen bonding (A pairs with T, G pairs with C). Hydrophobic interactions between the base pairs help them stack on top of one another in the center of the molecule, stabilizing it and maintaining a constant diameter.
- The DNA helix has a major and minor groove (see Figure 13.9), and the bases in these grooves provide distinct surfaces for protein recognition.
- The specific base sequences in DNA are used to store genetic information. Altering the base sequence in the DNA by mutation may cause a change in the genetic information. Precise replication of the DNA molecule is accomplished by means of complementary base pairing. The expression of genetic information in the DNA results in specific phenotypes.

DNA must be able replicate itself in order to be heritable.

- All of the following must be present for replication to take place: (1) a template strand; (2) all four deoxyribonucleoside triphosphates (dATP, dCTP, dGTP, and dTTP); (3) a primer base-paired to the template DNA; and (4) DNA polymerase.
- Meselson and Stahl's experiments demonstrated that DNA replication is semiconservative: Each parental strand of the DNA duplex serves as a template for the newly synthesized (daughter) DNA, and each new DNA molecule is composed of one parental (template) and one newly-replicated (daughter) strand (see Figure 13.11).
- During replication, DNA is unwound, and the new strand is synthesized by complementary base pairing of each incoming nucleotide to the parental strand via hydrogen bonds. New nucleotides are always added at the 3′ (—OH) end of the elongating strand via covalent (phosphodiester) bonds (see Figure 13.12).
- Replication complexes assemble at origins (*ori*) on the DNA to begin replication.
- Replication complexes are composed of many proteins: Helicases unwind parental DNA, single-stranded binding proteins prevent parental strands from reassociating, primases generate RNA primers, DNA polymerases elongate daughter strands from the 3′ OH end of those primers, and DNA ligases seal the nicks in the sugar–phosphate backbone.
- All chromosomes have at least one origin of replication where the replication complex binds.
- New daughter strands must begin with a short RNA primer, formed with the parental strand as a template. Once a primer is synthesized, DNA polymerase then adds deoxyribonucleotides to the 3′ (—OH) end of the primer.
- There are several different DNA polymerases in cells; all replicate DNA, some remove primers, and some are involved in DNA repair.
- Replication proceeds in both directions from the replication origin, forming replication forks. Both leading-strand and lagging-strand synthesis occur at replication forks (see Figure 13.18).

Replication proceeds in a specified direction.

- Nucleotides can be added only to the 3′ (—OH) end of a strand. Thus, the parental strand is read in the 3′-to-5′ direction, and the daughter strand elongates in the 5′-to-3′ direction.
- The leading strand elongates continuously in the "right" direction (5′ to 3′) as the replication fork opens up.
- The lagging strands (Okazaki fragments) are synthesized in a direction opposite to opening of the replication fork (5′ to 3′). This process requires many primers, and the resulting Okazaki fragments that are generated are ligated (covalently joined by DNA ligase) to form a continuous strand. RNA primers are removed prior to this ligation by DNA polymerase I (see Figure 13.17).
- The sliding clamp (proliferating cell nuclear antigen PCNA in mammals) increases the efficiency of and rate of DNA replication by keeping DNA polymerase tightly associated with the newly replicated strand (see Figure 13.18).
- In eukaryotes, DNA is threaded through a stationary replication complex that is attached to chromatin.
- Most eukaryotes have many origins of replication for their large linear chromosomes.
- Telomeres are repetitive sequences found at the ends of eukaryotic chromosomes, which, with the protein telomerase, help maintain the integrity of those ends during each replicative cycle. These telomeres are gradually lost during each replicative cycle, and this loss can eventually lead to cell death.
- Telomerase is continually expressed in cancer cells.

Errors can occur in DNA synthesis.

- DNA polymerase proofreads and repairs base mismatches during replication (see Figure 13.22A).
- Mismatch repair systems scan for possible errors in newly synthesized DNA (see Figure 13.22B).
- Excision repair mechanisms replace abnormal bases with functional bases (see Figure 13.22C).

Advanced techniques in molecular biology have allowed us to make multiple copies of DNA from a single strand and to determine the sequence of bases in DNA.

- The polymerase chain reaction (PCR) allows the rapid amplification of short DNA sequences into many identical copies (see Figure 13.23). PCR requires primers, dNTPs, a DNA polymerase that is heat-tolerant, the appropriate buffers and salts, and template DNA. The three steps in PCR amplification are denaturation, annealing, and polymerization.
- DNA sequencing is now a highly automated process in which modified bases are mixed with the normal substrates for DNA replication to generate a mixture of DNA fragments. These fragments are used to determine the base sequence of the DNA.

Test Yourself

Diagram Exercise

Diagram the replication complex and label the following: helicase, single-stranded DNA binding protein, DNA polymerase, primase, the leading and lagging strands, and the leading and lagging template strands. Be sure to label the 5′ and 3′ ends of the parental and newly replicated DNA. What would happen to DNA replication if the genes encoding each of these proteins were mutated such that the proteins were nonfunctional?

Textbook Reference: *13.3 How Is DNA Replicated? pp. 280–283, Figure 13.15*

Knowledge and Synthesis Questions

1. Griffith's experiments showing the transformation of R strain pneumococcus bacteria to S strain pneumococcus bacteria in the presence of heat-killed S strain bacteria provided evidence that
   a. an external factor was affecting the R strain bacteria.
   b. DNA was definitely the transforming factor.
   c. S strain bacteria could be reactivated after heat killing.
   d. S strain bacteria required special genes to be pathogenic.
   e. All of the above

   Textbook Reference: *13.1 What Is the Evidence that the Gene Is DNA? pp. 267–268*

2. Experiments by Avery, MacLeod, and McCarty supported DNA as the genetic material by showing that
   a. both protein and DNA samples provided the transforming factor.
   b. DNA has to be destroyed by DNase in order to transform the R bacteria.
   c. DNA is not complex enough to be the genetic material.
   d. only samples with DNA provided transforming activity.
   e. even though DNA is molecularly simple, it provides adequate variation to act as the genetic material.

   Textbook Reference: *13.1 What Is the Evidence that the Gene Is DNA? p. 269, Figure 13.2*

3. Hershey and Chase used radioactive ^{35}S and ^{32}P in experiments to provide evidence that DNA is the genetic material. These experiments pointed to DNA because
   a. progeny viruses retained ^{32}P but not ^{35}S.
   b. the presence of ^{32}P in progeny viruses indicated that DNA was passed on.
   c. the absence of ^{35}S in progeny viruses indicated that proteins were not passed on.
   d. ^{32}P indicated where the DNA label was localized.
   e. All of the above

   Textbook Reference: *13.1 What Is the Evidence that the Gene Is DNA? p. 270, Figure 13.4*

4. Chargaff observed that the amount of _______ was roughly equal to the amount of _______ in all tested organisms.
   a. purines; pyrimidines
   b. A; T
   c. A + T; G + C
   d. A + G; T + C
   e. a, b, and d

   Textbook Reference: *13.2 What Is the Structure of DNA? p. 273*

5. Watson and Crick's model allowed them to visualize
   a. the molecular bonds of DNA.
   b. the sugar and phosphate component of the DNA molecule's surface.
   c. how the purines and pyrimidines fit together in a double helix.
   d. the antiparallel design of two strands of the DNA double helix.
   e. All of the above

   Textbook Reference: *13.2 What Is the Structure of DNA? pp. 273–275*

6. A fundamental requirement for the functioning of genetic material is that it must be
   a. conserved among all organisms with very little variation.
   b. passed intact from one species to another.
   c. accurately replicated.
   d. found outside the nucleus.
   e. replicated accurately over many millions of years of evolution.

   Textbook Reference: *13.2 What Is the Structure of DNA? p. 273*

7. Evidence of the semiconservative nature of DNA replication came from
   a. DNA staining techniques.
   b. X-ray crystallography.
   c. DNA sequencing.
   d. density gradient studies using "heavy" nucleotides.
   e. None of the above

   Textbook Reference: *13.3 How Is DNA Replicated? pp. 276–278*

8. The primary function of DNA polymerase is to
   a. add nucleotides to the growing daughter strand.
   b. seal nicks along the sugar–phosphate backbone of the daughter strand.
   c. unwind the parent DNA double helix.
   d. generate primers to initiate DNA synthesis.
   e. prevent reassociation of the denatured parental DNA strands.

   Textbook Reference: *13.3 How Is DNA Replicated? pp. 279–280, Figures 13.13 and 13.15*

9. When the lagging daughter strand of DNA is synthesized, unreplicated gaps are formed on the parental DNA. Lagging strand synthesis fills these gaps by
   a. synthesizing short Okazaki fragments in a 5′-to-3′ direction.
   b. synthesizing multiple short RNA primers to initiate DNA replication.
   c. using DNA polymerase I to remove RNA primers from Okazaki fragments.
   d. filling in those gaps with new strands of complementary DNA as the replication fork proceeds.
   e. All of the above

   Textbook Reference: *13.3 How Is DNA Replicated? pp. 281–282*

10. RNA primers are necessary in DNA synthesis because
    a. DNA polymerase is unable to initiate replication without an origin.
    b. the DNA polymerase enzyme can catalyze the addition of deoxyribonucleotides only onto the 3′ (—OH) end of an existing strand.
    c. RNA primase is the first enzyme in the replication complex.
    d. primers mark the sites where helicase has to unwind the DNA.
    e. All of the above

    Textbook Reference: *13.3 How Is DNA Replicated? p. 279, Figure 13.13*

11. Proofreading and repair occur
    a. at any time during or after synthesis of DNA.
    b. only before DNA methylation.
    c. only in the presence of DNA polymerase.
    d. only in the presence of an excision repair mechanism.
    e. only during replication.

    Textbook Reference: *13.4 How Are Errors in DNA Repaired? pp. 285–286*

12. Thirty percent of the bases in a sample of DNA extracted from eukaryotic cells are adenine. What percentage of cytosine is present in this DNA?
    a. 10 percent
    b. 20 percent
    c. 30 percent
    d. 40 percent
    e. 50 percent

    Textbook Reference: *13.2 What Is the Structure of DNA? p. 272*

13. Which of the following represents a bond between a purine and a pyrimidine (in the correct order)?
    a. C–T
    b. G–A
    c. G–C
    d. T–A
    e. A–G

    Textbook Reference: *13.2 What Is the Structure of DNA? p. 273*

14. Which of the following statements about DNA replication is *false*?
    a. Okazaki fragments are synthesized as part of the leading strand.
    b. Replication forks represent areas of active DNA synthesis on the chromosomes.
    c. Error rates for DNA replication are reduced by proofreading of the DNA polymerase.
    d. Ligases and polymerases function in the vicinity of replication forks.
    e. The sliding clamp protein increases the rate of DNA synthesis.

    Textbook Reference: *13.3 How Is DNA Replicated? pp. 279–284, 13.4 How Are Errors in DNA Repaired? p. 285*

15. The PCR technique
    a can amplify only very small samples of DNA.
    b. amplifies several random DNA sequences within a genome.
    c. requires synthetic primers to flank the regions of interest.
    d. is accomplished in three sequential steps: annealing, denaturation, and replication.
    e. generates DNA molecules that all have variable sequences.

    Textbook Reference: *13.5 How Does the Polymerase Chain Reaction Amplify DNA? pp. 286–287*

16. Which of the following would *not* be found in a DNA molecule?
    a. Purines
    b. Ribose sugars
    c. Phosphates
    d. Sulfur
    e. Nitrogenous bases

    Textbook Reference: *13.2 What Is the Structure of DNA? pp. 273–275, Figure 13.9*

17. If a high concentration of a particular nucleotide lacking a hydroxyl group at the 3′ end is added to a PCR reaction,
    a. no additional nucleotides will be added to a growing strand containing that nucleotide.
    b. strand elongation will proceed as normal.
    c. nucleotides will be added only at the 5′ end.
    d. *T. aquaticus* DNA polymerase will become nonfunctional.
    e. the primer will be unable to anneal with the DNA template.

    Textbook Reference: *13.5 How Does the Polymerase Chain Reaction Amplify DNA? pp. 286–287*

18. Meselson and Stahl were trying to determine if DNA replication was semiconservative, conservative, or dispersive by labeling *E. coli* DNA with a regimen of heavy nitrogen (H) for one round of replication and then transferring these cells to light nitrogen (L) for two more rounds of replication. Which of the following statements would *not* be true within the context of this experiment?
   a. If DNA replication were conservative, no DNA molecules of intermediate density (H-L) would have been seen.
   b If DNA replication were dispersive, only DNA molecules that were of intermediate density (H-L) would have been seen.
   c. If DNA replication were semiconservative, the DNA molecules that were made would all continue to be heavy density (H-H).
   d. If DNA replication were semiconservative, the DNA molecules would consist of one parental strand base-paired to one newly replicated strand.
   e. If DNA replication were semiconserative, a higher proportion of DNA molecules from future divisions would have been low density (L-L).

   Textbook Reference: *13.3 How Is DNA Replicated? pp. 276–278*

19. The telomeres at the ends of linear chromosomes allow
   a. the 5′ ends of the chromosomes to undergo recombination.
   b. the gaps left by primer removal of lagging strands to be repaired by telomerase.
   c. DNA repair enzymes to recognize those ends and remove them.
   d. normal cells to divide continuously.
   e. DNA breaks to be examined at cell division checkpoints.

   Textbook Reference: *13.3 How Is DNA Replicated? pp. 283–286*

Application Questions

1. Given the following parent strand sequence, what would the daughter strand sequence look like?

   5′ – G C T A A C T G T G A T C G T A T A A G C T G A – 3′

   Textbook Reference: *13.2 What Is the Structure of DNA? p. 273*

2. Diagram the double helix. Be sure to label those properties that make it most suited as the genetic material.

   Textbook Reference: *13.2 What Is the Structure of DNA? p. 275*

3. Diagram a replication fork as it would be seen in a replicating segment of DNA. In your diagram, label the 5′ and 3′ ends of each parent strand and daughter strand. Indicate which new strand is the leading strand and which is the lagging strand of the daughter DNA.

   Textbook Reference: *13.3 How Is DNA Replicated? pp. 280–281*

4. Based on your diagram in Question 3, construct a flowchart of DNA replication. Divide your chart into three parts: initiation, elongation, and termination of replication. Indicate the roles of helicase, DNA polymerase, single-stranded DNA binding proteins, nucleotides, parental (template) DNA strands, DNA ligase, RNA primase, RNA primers, and the sliding clamp protein.

   Textbook Reference: *13.3 How Is DNA Replicated? pp. 280–283*

5. Explain the difference between conservative and semiconservative models of DNA replication. What results supported the semiconservative model? What would the results have looked like if the conservative model of DNA replication had been accurate? Are there any other potential hypotheses?

   Textbook Reference: *13.3 How Is DNA Replicated? pp. 276–278*

6. Explain the role of Okazaki fragments in the synthesis of the lagging strand.

   Textbook Reference: *13.3 How Is DNA Replicated? p. 281*

7. Differentiate between proofreading, mismatch repair, and excision repair. Which of these repair mechanisms is responsible for repairing a mutation that occurs in an adult cell from overexposure to the sun? Explain your answer.

   Textbook Reference: *13.4 How Are Errors in DNA Repaired? pp. 285–286*

8. Explain how PCR amplifies a particular sequence of DNA.

   Textbook Reference: *13.5 How Does the Polymerase Chain Reaction Amplify DNA? pp. 286–287*

Answers

Diagram Exercise Answer

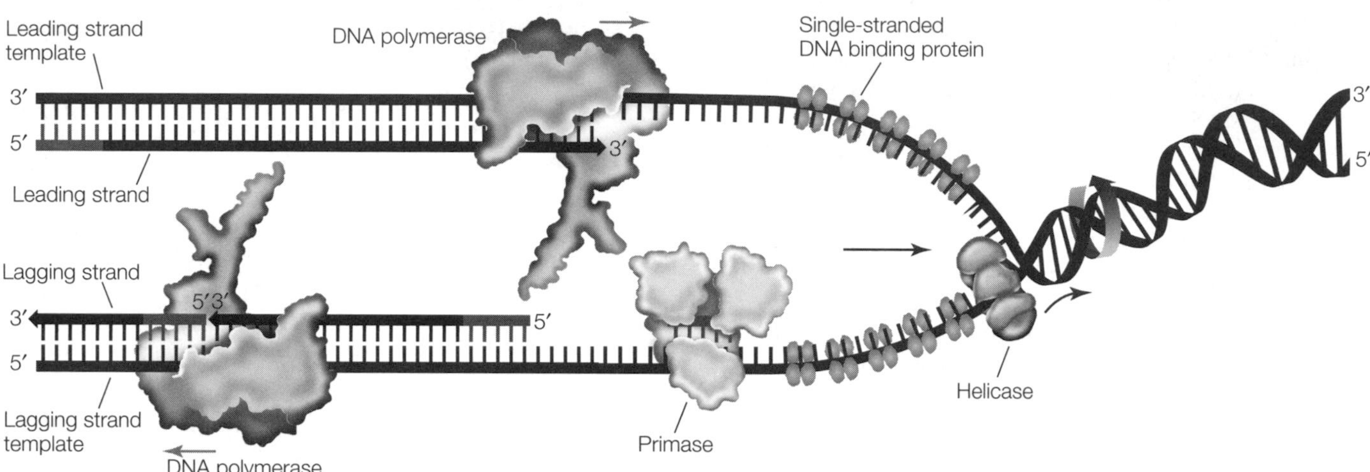

If helicase was nonfunctional, replication would not start because the DNA strands would not be unwound.

If single-stranded DNA binding protein was nonfunctional, the single strands of DNA would not be stabilized, and replication could not start.

If primase was nonfunctional, DNA polymerase could not initiate. DNA polymerase requires a primer with a 3′ OH group base-paired to the template DNA to initiate replication.

If DNA polymerase was nonfunctional, replication could not occur, even though the DNA was unwound and stabilized in an "open" configuration.

Knowledge and Synthesis Answers

1. **a.** The experiments showed only that something was causing the R strain pneumococcus to develop the S strain virulence. Additional experiments were necessary to provide evidence that the transforming factor was indeed DNA.
2. **d.** These researchers were able to isolate nearly pure DNA samples. It was only these samples that provided transformation activity.
3. **e.** Only DNA incorporates radioactive ^{32}P. Radioactive ^{35}S is incorporated into proteins. The fact that progeny viruses retained radioactive ^{32}P and not radioactive ^{35}S indicated that DNA is heritable and protein is not.
4. **e.** Chargaff found that in most DNA sampled, the amount of A equaled the amount of T and the amount of G equaled the amount of C. It follows that the amount of purines (A + G) equals the amount of pyrimidines (T + C). The same does not hold for the amount of A + T versus the amount of G + C.
5. **e.** Model-building by Watson and Crick created a three-dimensional visualization of the size, bond angles, base pairings, and overall structure of the DNA molecule. The information contributing to their model came from a variety of experiments performed by other scientists.
6. **c.** Replication of DNA is a fundamental requirement for its function as the genetic material. DNA must be correctly replicated in each cell of an organism and passed from parent to offspring or from cell to daughter cell. DNA is not passed from one species to another. There is variation in DNA sequences from different organisms. DNA is found outside the nucleus in eukaryotic cells (specifically in the mitochondria, and in the chloroplasts of plants), but this DNA is inherited in a non-Mendelian fashion (see Chapter 12). For evolution to occur, there must be some mistakes in DNA replication over time.
7. **d.** Density gradient labeling allowed Meselson and Stahl to track parental versus daughter strands of DNA. Their research showed that DNA replication is a semiconservative process, with each newly synthesized DNA molecule containing one parental (template) strand base-paired to one daughter (newly synthesized) strand.
8. **a.** DNA polymerase adds nucleotides to an existing nucleotide strand, ligase seals nicks, helicase unwinds the DNA, primase generates primers, and a single-stranded DNA binding protein prevents the reassociation of the parental strands.

9. **e.** Okazaki fragments are small segments of newly synthesized DNA that have been added to short RNA primers. The RNA is removed from these small fragments by DNA polymerase I and replaced with DNA. The remaining DNA fragments are ligated (covalently bound) to form a continuous, newly synthesized DNA strand.
10. **b.** DNA polymerase cannot initiate synthesis of a nucleotide strand; it can only add onto an existing strand of RNA primer or DNA. Primers provide DNA polymerase with the 3′ (—OH) end required for the addition of deoxyribonucleotides. The origin is also required for DNA synthesis, but it is the site where the replication complex is initially assembled. Primase is not the first enzyme in the replication complex.
11. **a.** The mismatch repair mechanism operates before the newly synthesized DNA strand is methylated, but for the integrity of DNA to be maintained, repair mechanisms must be active during synthesis, modification, and utilization of DNA.
12. **b.** If 30 percent of DNA is adenine, then according to Chargaff's rule, 30 percent must be thymine. The remaining 40 percent of the DNA is cytosine and guanine. Because the ratio of cytosine to guanine must be equal, the percentage of cytosine in this DNA must be 20 percent.
13. **c.** Guanine is a purine, and its paired pyrimidine is cytosine.
14. **a.** Okazaki fragments are involved in synthesis of the lagging strand.
15. **c.** PCR amplifies specific DNA sequences from small or large samples of DNA and requires three sequential steps: denaturation, annealing, and replication. The amplified DNA molecules do not have variable DNA sequences.
16. **d.** Sulfur is a constituent of many protein molecules, but it is not found in DNA.
17. **a.** A hydroxyl group at the 3′ position of a nucleotide is necessary for the binding of any additional nucleotides. If this hydroxyl group was absent, no other nucleotides could be added to a growing strand.
18. **c.** DNA replication is semiconservative, and DNA molecules would be of intermediate density (H-L) after one round of replication in heavy nitrogen and one round in light nitrogen.
19. **b.** Telomerase binds the telomeres at the ends of the chromosomes, protects them from being degraded or recombined with other chromosomes, and ensures that these ends will be replicated after primer removal. Telomeres are not associated with cell division checkpoints or continuous cell division.

Application Answers

1. 3′–CGATTGACACTAGCATATTCGACT–5′
2. See Figures 13.8 and 13.9 in the textbook.

3.–4. See Figures 13.15 and 13.16 in the textbook.

5. In the conservative model of DNA replication, the parental DNA remains intact, and a newly synthesized molecule consists of two newly replicated daughter strands. If this were indeed DNA's method of replication, Meselson and Stahl would not have seen the intermediate density band in the first generation of replication. They would have seen a dense band corresponding to the parental DNA (H-H) and a light band corresponding to the daughter DNA (L-L). Because they saw an intermediate band, they knew that one strand was heavy and one was light; therefore, replication was semiconservative, with each new molecule consisting of one parental strand and one daughter strand. A third hypothesis was the dispersive model, according to which each new molecule contains bits and pieces of both old and new strands. This model was rejected because examination of the second round of replication in light nitrogen showed that the result was an H-L band and an L-L band. The L-L band could not have been generated by dispersive replication; all of the bands would again be of intermediate density (H-L).
6. See Figure 13.17 in the textbook.
7. Proofreading occurs during synthesis of the DNA molecule, mismatch repair occurs after synthesis, and excision repair removes any abnormalities in "mature" DNA. DNA damage due to UV exposure is generally repaired via excision repair mechanisms.
8. Refer to Figure 13.22 in the textbook.

14 From DNA to Protein: Gene Expression

The Big Picture

- Every protein and RNA in the cell has a DNA blueprint (the gene sequence) that specifies the amino acid or nucleotide sequence of that gene product. Those sequences determine how that gene product will fold up three-dimensionally and ultimately function in the cell.
- Alterations in those gene products, which are caused by mutations in the genes, help us understand how those proteins and RNAs function in the cell.

Common Problem Areas

- Students often try to understand every detail of gene expression without comprehending the larger picture: the way genetic information is accessed in the DNA and expressed in the cell so that the cell can function. Familiarize yourself with the details of the central dogma, but then take a step back and understand that all of these details describe the steps required to synthesize gene products.
- Students often confuse transcription and translation. Be able to distinguish between the two processes, and for each one carefully review the template, the product, and the sites of initiation and termination.

Study Strategies

- Draw diagrams of the processes of transcription and translation that include initiation, elongation, and termination. Label the components that play a role in each process. Be sure to orient the 5′ and the 3′ ends of the nucleic acid, and the N and the C terminus of the protein.
- Choose a hypothetical gene that encodes a peptide of four amino acids and draw a flowchart showing how that gene in the chromosome is made into protein. Now follow the details of gene expression: the synthesis of the RNA from the DNA, and the synthesis of the protein from the mRNA. Be sure to include the start and stop signals in each of these processes. Then alter one of those nucleotides in the DNA sequence, and transcribe and translate the gene. Is the protein sequence changed? Repeat this process using a different mutation in the DNA.
- Review the experiments that revealed the presence of introns in the primary RNA transcript and describe how those introns are removed during splicing.
- Go to yourBioPortal.com to review the following tutorials and activities:

  Animated Tutorial 14.1 Transcription

  Animated Tutorial 14.2 Deciphering the Genetic Code

  Animated Tutorial 14.3 RNA Splicing

  Animated Tutorial 14.4 Protein Synthesis

  Web Activity 14.1 Eukaryotic Gene Expression

  Web Activity 14.2 The Genetic Code

Important Concepts

Genes code for proteins.

- In the early twentieth century, Archibald Garrod studied a rare disease called alkaptonuria, in which patients accumulate homogentisic acid. He concluded the disease was genetically determined and that these patients lacked an essential enzyme.
- Beadle and Tatum studied special mutant (auxotroph) and wild-type (prototroph) strains of *Neurospora*, a bread mold. Auxotrophs cannot grow on minimal media without the addition of an external mineral, vitamin, or nutrient, because they are missing the enzymes needed to synthesize these nutrients. Prototrophs, also known as "wild-type" strains, can grow on minimal media because in these cells, all the enzymes needed to synthesize these nutrients are functional.
  - For the amino acid arginine, for example, each of these enzymes defines a step in the biochemical pathway of arginine synthesis.
  - Supplying mutant strains with chemical intermediates in the arginine pathway allowed Beadle and Tatum to determine which enzymes are nonfunctional (see Figure 14.1) in the auxotroph. They also analyzed individual cells for enzyme activity.
- Beadle and Tatum concluded from their experiments that one gene encodes one enzyme. Because enzymes

can consist of several different polypeptides, this idea has been modified to the conclusion that one gene encodes one polypeptide.

- Genes also code for some RNAs (such as ribosomal RNA, transfer RNA, and regulatory RNA) that do not become translated into polypeptides.

Information flows from genes to proteins.

- Genes are transcribed into complementary RNA molecules. RNA sequence is translated into the amino acid sequence of a protein.
- The central dogma of molecular biology states that information flows from DNA, which is transcribed into RNA, which is translated into protein.

RNA differs from DNA.

- Unlike DNA, which consists of two polynucleotide strands, RNA is generally a single polynucleotide strand. In RNA, the sugar molecule is ribose rather than deoxyribose, as in DNA.
- The bases A, G, and C in DNA also occur in RNA, but the T in RNA is replaced by U. In RNA, G still base-pairs with C, but A base-pairs with U.
- A strand of RNA can base-pair with a single strand of DNA. A strand of RNA may also fold over to base-pair with itself.
- Three types of RNA are involved in protein synthesis: messenger RNA (mRNA), transfer RNA (tRNA), and ribosomal RNA (rRNA).
  - Messenger RNA is complementary to one DNA strand of the gene, and the information sequence consists of three sequential base units called codons.
  - Transfer RNA (or tRNA) translates the nucleic acid (in the messenger RNA) into proteins, using its anticodon to read the codon in the mRNA.
  - Ribosomal RNA (rRNA) catalyzes peptide bond formation and provides a structural framework for the ribosome.
- Some viruses (including the tobacco mosaic virus, the poliovirus, and the influenza virus) use RNA as their genome and during infection make more RNA (using viral RNA as a template). Other viruses, known as retroviruses (like HIV), make a DNA copy of their RNA using reverse transcriptase and use that DNA to synthesize more RNA. Retroviruses can also insert that DNA copy into the host's chromosome. These viruses are exceptions to the central dogma.

DNA is transcribed into RNA.

- The enzyme RNA polymerase synthesizes RNA using one strand of the DNA as a template. The resulting RNA molecules can be messenger RNAs (which are used to synthesize proteins), or they can become transfer RNA, ribosomal RNA, or regulatory RNA (such as microRNA).
- RNA polymerase initiates transcription at special sequences on the DNA called promoters. When RNA polymerase binds the promoter, it unwinds the DNA. The promoter tells the RNA polymerase where to start transcription, which strand of DNA to transcribe, and in which direction to move on the DNA to begin synthesizing RNA.
- RNA polymerase unwinds about 10 base pairs in the DNA as it reads and synthesizes the complementary RNA strand.
- New RNA is synthesized in a 5′-to-3′ direction, antiparallel to the template DNA strand. New nucleotides are added to the 3′ end of the growing strand. No primer is required for initiation of RNA synthesis.
- RNA polymerase does not proofread, resulting in transcription errors of one for every 10^4 to 10^5 bases.
- RNA polymerase terminates transcription at special sequences in the DNA.

The information for translation is in the genetic code.

- The genetic code provides the key to "reading" mRNA sequences.
  - Genetic information in the messenger RNA is read one codon (three bases) at a time.
  - There are four possible bases that can be used in a codon, so there are 4^3, or 64, different codons.
  - Sixty-one codons specify specific amino acids, so the genetic code is redundant, meaning that there is more than one codon for many of the 20 amino acids.
  - Translation starts with the AUG codon, which also specifies the amino acid methionine.
  - Three codons—the stop codons UAG, UAA, and UGA—are used to terminate translation.
  - The genetic code for nuclear genes is almost identical in all living cells. There are slight differences in the genetic code of one group of protists and in mitochondria and chloroplasts.
  - The genetic code was first determined by experiments in which artificial RNAs were added to reaction mixtures containing all of the necessary components needed for protein translation (see Figure 14.5). The resulting polypeptides were analyzed to determine the protein-synthesizing instructions for these RNA sequences. A simple UUU mRNA caused the tRNA carrying phenylalanine to bind to the ribosome.

Introns play an important role in eukaryotic transcription and RNA processing.

- Each eukaryotic gene has its own promoter.
- Eukaryotic RNA polymerase requires other molecules to help it recognize promoters.
- Terminators are special sequences in eukaryotic DNA that signal the end of transcription.
- Eukaryotic genes contain noncoding regions called introns interspersed with coding regions (exons).

Introns must be removed from the pre-mRNA transcripts by RNA processing to ensure that all the exons are adjacent to each other for translation.

- Introns were discovered by means of nucleic acid hybridization. Mature mRNAS were hybridized to denatured DNA containing the gene for that mRNA (see Figure 14.8 and 14.9). Introns in the DNA were observed to be single-stranded loops that did not hybridize to the mature mRNA. When unprocessed pre-mRNA was used in these experiments, no DNA loop was observed, indicating that intron removal occurred during mRNA maturation.
- Some exons encode functional domains of a protein.

Modifications of pre-mRNA occur in the nucleus of eukaryotic cells.

- A cap of modified GTP (a G cap) is added to the 5′ end of eukaryotic mRNA to facilitate binding of mRNA to the ribosome in the cytoplasm and to protect the mRNA from ribonuclease degradation in the cytoplasm.
- The pre-mRNA is cut by an enzyme that recognizes the sequence A A U A A A at the 3′ end of the mRNA, and a poly A tail (100–300 nucleotides) is added to the 3′ end.
- The poly A tail assists in the transport of the mRNA from the nucleus to the cytoplasm and increases the mRNA stability in the cytoplasm.
- Introns are "spliced," or removed, from pre-mRNAs by small nuclear ribonucleoprotein particles (snRNPs).
- snRNPs bind consensus sequences at the 3′ and 5′ boundaries of the introns by means of complementary base pairing.
- Other proteins bind the snRNPs to form the spliceosome, which uses the energy of ATP hydrolysis to cut the RNA, release the intron, and rejoin the RNA to form the mature mRNA (see Figure 14.11).
- Some forms of β-thalassemia (in which there is an inadequate amount of β-globin, resulting in severe anemia) are the result of a mutation in the consensus sequence, causing inappropriate splicing of β-globin RNA and a nonfunctional gene product.
- Mature RNA is transported from the nucleus to the cytoplasm through the nuclear pores.

mRNA is translated into proteins.

- Transfer RNA (tRNA) is the link between a codon and its amino acid. Each tRNA binds (is "charged") to a particular amino acid; they base-pair with codons in the mRNA and interact with ribosomes.
  - tRNA is a small RNA molecule (75–80 nucleotides) that base-pairs with itself to form a characteristic shape found in all tRNAs (see Figure 14.12).
  - The anticodon of the tRNA (found at about the midpoint in the molecule) base-pairs with the codon of the mRNA in an antiparallel fashion. This base pairing occurs on the ribosome during translation.
  - Base pairing between the anticodon and the codon is always complementary (A pairs with U, C pairs with G) in the first two positions. At the third position of the codon (and the first position of the anticodon), base pairing does not have to be exact. This "wobble" position allows the cell to use one tRNA to base-pair with up to four different codons.
  - An amino acid is covalently attached to the 3′ end of the tRNA; thus the tRNA (with an anticodon of AAA) has arginine covalently attached to its 3′ end. The enzymes that catalyze the covalent attachment of the amino acids to the correct tRNAs are tRNA synthase (see Figure 14.13).
- The ribosome is the site in the cell where protein translation occurs.
  - Ribosomes are composed of a large subunit and a small subunit; both subunits consist of different ribosomal proteins and RNA molecules (ribosomal RNA) (see Figure 14.14).
  - The important sites on a ribosome are the A site (where the anticodon of the charged tRNA binds the codon of the mRNA), the P site (where the tRNA with the growing polypeptide chain is located), and the E site (where the tRNA exits from the ribosome).
- Translation takes place in three steps: initiation, elongation, and termination.
  - Initiation of translation requires (1) the small subunit of the ribosome; (2) the initiator tRNA charged with methionine; (3) the mRNA with the start codon AUG positioned correctly on the ribosome; and (4) the initiation factors (see Figure 14.15).
  - In prokaryotes, the rRNA of the small subunit of the ribosome base-pairs with a complementary sequence (the Shine–Dalgarno sequence) upstream from the start codon (AUG).
  - In eukaryotes, the small subunit of the ribosome loads onto the 5′ cap of the mRNA and moves along the RNA until it finds the first AUG.
- When all these components are in place, the large subunit of the ribosome joins the small subunit of the ribosome, and the incoming charged tRNA is positioned at the A site. The anticodon of this tRNA base-pairs with the codon on the mRNA that is at the A site of the ribosome.
- The large subunit of the ribosome catalyzes the formation of a peptide bond between the amino acid on the tRNA at the P site and the amino acid on the tRNA at the A site. This peptidyl transferase activity can be attributed to the RNA (not to any of the proteins) in the large ribosomal subunit.
- Once the peptide bond is formed, the ribosome shifts down the messenger RNA so that the next codon is placed at the A site. The tRNA that was in the P site is

moved to the E site and will exit the ribosome. The tRNA that was at the A site is shifted to the P site.

- During elongation, the messenger RNA is read 5′ to 3′, with each new codon being moved to the A site during elongation so that the next incoming charged tRNA can base-pair with the codon in the mRNA (see Figure 14.16).
- Termination occurs when a stop codon is moved to the A site of the ribosome. Release factor binds the stop codon, and peptidyl transferase hydrolyzes the bond between the amino acid (which will be the C terminus) that is covalently attached to the tRNA in the P site. The polypeptide is released from the ribosome, and the small and large subunits of the ribosome dissociate (see Figure 14.17).
- In prokaryotes, one messenger RNA is usually bound by many translating ribosomes, forming a polysome (see Figure 14.18).

Changes and modifications occur alter translation.

- Signal sequences within the amino acid sequence of some proteins in eukaryotic cells serve as addresses for the protein's final cellular destination.
  - In the case of nuclear, mitochondrial, plastid, or peroxisomal proteins, these signal sequences can be recognized by docking proteins, which form pores in the membrane of their destination.
  - Alternatively, they can be recognized by signal recognition particles in the cytoplasm that direct the partially translated protein to the endoplasmic reticulum for further addressing and completion of translation (see Figure 14.19). The signal recognition particle binds the docking protein in the endoplasmic reticulum and protein translation continues into the ER through a receptor pore.
- Proteins synthesized in the endoplasmic reticulum may (1) be sent to the Golgi, lysosomes, or the plasma membrane; (2) be secreted; or (3) stay in the endoplasmic reticulum and be modified further. Proteins that lack signal sequences remain in the cytoplasm.
- Protein modification after translation includes proteolysis, glycosylation, and phosphorylation.
  - Proteolysis is the process of cutting proteins into smaller pieces. The removal of the signal sequence from a protein in the ER is an example of proteolysis.
  - Glycosylation is the addition of sugars to proteins. Glycosylation in the Golgi apparatus marks proteins as destined for lysosomes, the plasma membrane, or the vacuole (in plants).
  - Phosphorylation is the addition of phosphate groups to proteins and is catalyzed by protein kinases. Phosphorylation can alter the tertiary structure of a protein.

Test Yourself

Diagram Exercise

Draw a diagram of a eukaryotic cell and show where in the cell the gene is transcribed and translated. In your diagram, indicate how this particular gene product is targeted to a compartment in the cell such as the endoplasmic reticulum. Create a step-wise list of all the important proteins and enzymes required for this process. Imagine that the signal sequence has been deleted from this gene without deleting the start codon. How will this deletion affect gene expression? How will this deletion affect the targeting of the protein to the endoplasmic reticulum?

Textbook Reference: *14.6 What Happens to Polypeptides after Translation? pp. 310–313*

Knowledge and Synthesis Questions

1. Transcription in prokaryotic cells
   a. occurs in the nucleus, whereas translation occurs in the cytoplasm.
   b. is initiated at a start codon with the help of initiation factors and the small subunit of the ribosome.
   c. is initiated at a promoter and uses only one strand of DNA (the template strand) to synthesize a complementary RNA strand.
   d. is terminated at stop codons.
   e. is initiated at an *ori* site on the chromosome.

   Textbook Reference: *14.3 How Is the Information Content in DNA Transcribed to Produce RNA? pp. 296–298*

2. Which of the following statements about RNA polymerase is *false*?
   a. It synthesizes mRNA in a 5′-to-3′ direction, reading the DNA strand 3′ to 5′.
   b. It synthesizes mRNA in a 3′-to-5′ direction, reading the DNA strand 5′ to 3′.
   c. It binds at the promoter and unwinds the DNA.
   d. It does not require a primer to initiate transcription.
   e. It uses only one strand of DNA as a template for synthesizing RNA.

   Textbook Reference: *14.3 How Is the Information Content in DNA Transcribed to Produce RNA? pp. 296–298*

3. Translation of messenger RNA into protein occurs in a _______ direction, and from _______ terminus to _______ terminus.
   a. 3′-to-5′; N; C
   b. 5′-to-3′; N; C
   c. 3′-to-5′; C; N
   d. 5′-to-3′; C; N
   e. 3′-to-5′; C; C

   Textbook Reference: *14.5 How Is RNA Translated into Proteins? p. 307*

4. If codons were read two bases at a time instead of three bases at a time, how many different possible amino acids could be specified?
   a. 16
   b. 64

c. 8
d. 32
e. 128
Textbook Reference: *14.3 How Is the Information Content in DNA Transcribed to Produce RNA? pp. 298–299*

5. Translate the following mRNA:
3′ – G A U G G U U U U A A A G U A – 5′
a. NH_2 met—lys—phe—leu—stop COOH
b. NH_2 met—lys—phe—trp—stop COOH
c. NH_2 asp—gly—phe—lys—val COOH
d. NH_2 met—gly—phe—lys—val COOH
e. NH_2 asp—gly—phe—lys—stop COOH
Textbook Reference: *14.3 How Is the Information Content in DNA Transcribed to Produce RNA? p. 299, Figure 14.6*

6. What would happen if a mutation occurred in DNA such that the second codon of the resulting mRNA was changed from UGG to UAG?
a. Translation would continue and the second amino acid would be the same.
b. Nothing. The ribosome would skip that codon and translation would continue.
c. Translation would continue, but the reading frame of the ribosome would be shifted.
d. Translation would stop at the second codon, and no functional protein would be made.
e. Translation would continue, but the second amino acid in the protein would be different.
Textbook Reference: *14.3 How Is the Information Content in DNA Transcribed to Produce RNA? p. 299, Figure 14.6*

7. If the following synthetic RNA were added to a test tube containing all the components necessary for protein translation to occur, what would the amino acid sequence be?
5′– A U A U A U A U A U A U – 3′
a. Polyphenylalanine
b. Isoleucine–tyrosine–isoleucine–tyrosine
c. Isoleucine–isoleucine–isoleucine–isoleucine
d. Tyrosine–tyrosine–tyrosine–tyrosine
e. Aspargine–aspargine–aspargine–aspargine
Textbook Reference: *14.3 How Is the Information Content in DNA Transcribed to Produce RNA? p. 299, Figure 14.6*

8. Which part of the tRNA base-pairs with the codon in the mRNA?
a. The 3′ end, where the amino acid is covalently attached
b. The 5′ end
c. The anticodon
d. The start codon
e. The promoter
Textbook Reference: *14.5 How Is RNA Translated into Proteins? pp. 304–305*

9. Peptidyl transferase is an
a. enzyme found in the nucleus of the cell that assists in the transfer of mRNA to the cytoplasm.
b. enzyme that adds the amino acid to the 3′ end of the tRNA.
c. enzyme found in the large subunit of the ribosome that catalyzes the formation of the peptide bond in the growing polypeptide.
d. RNA molecule that is catalytic.
e. Both c and d
Textbook Reference: *14.5 How Is RNA Translated into Proteins? p. 307*

10. Termination of translation requires
a. a termination signal, RNA polymerase, and a release factor.
b. a release factor, initiator tRNA, and ribosomes.
c. initiation factors, the small subunit of the ribosome, and mRNA.
d. elongation factors and charged tRNAs.
e. a stop codon positioned at the A site of the ribosome, peptidyl transferase, and a release factor.
Textbook Reference: *14.5 How Is RNA Translated into Proteins? p. 308*

11. If the DNA encoding a nuclear signal sequence were placed in the gene for a cytoplasmic protein, the protein would
a. be modified in the Golgi.
b. be directed to the lysosomes.
c. be directed to the nucleus.
d. be directed to the cytoplasm.
e. stay in the endoplasmic reticulum.
Textbook Reference: *14.6 What Happens to Polypeptides after Translation? p. 311*

12. Auxotrophs are mutant strains that
a. can grow on a minimal medium.
b. require the addition of an essential nutrient to grow on a minimal medium.
c. cannot make any enzymes.
d. behave like wild-type strains.
e. can grow only if arginine is added to the growth medium.
Textbook Reference: *14.1 What Is the Evidence that Genes Code for Proteins? pp. 292–293*

13. The central dogma of molecular biology states that _______ is transcribed into _______, which is (are) translated into _______.
a. a gene; polypeptides; a gene product
b. protein; DNA; RNA
c. DNA; mRNA; tRNA
d. DNA; RNA; protein
e. RNA; DNA; protein
Textbook Reference: *14.2 How Does Information Flow from Genes to Proteins? pp. 294–295*

14. A gene product can be
a. an enzyme.
b. a polypeptide.
c. RNA.
d. microRNA.
e. All of the above
Textbook Reference: *14.1 What Is the Evidence that Genes Code for Proteins? p. 294*

15. The enzyme that catalyzes the synthesis of RNA is
    a. peptidyl transferase.
    b. DNA polymerase.
    c. tRNA synthase.
    d. ribosomal RNA.
    e. RNA polymerase.

    Textbook Reference: *14.5 How Is RNA Translated into Proteins? p. 307*

16. A mutation occurs such that a spliceosome cannot remove one of the introns in a gene. What effect will this have on that gene?
    a. It will have no effect; the gene will be transcribed and translated into protein.
    b. Transcription will terminate early and the protein will not be made.
    c. Transcription will proceed, but translation will stop at the site where the intron remains.
    d. Translation will continue, but a nonfunctional protein will be made.
    e. Translation will continue and will skip the intron sequence.

    Textbook Reference: *14.3 How Is the Information Content in DNA Transcribed to Produce RNA? pp. 296–298*

Application Questions

1. What would be the effect of a deletion of the DNA encoding the targeting sequence for that gene product? (Imagine that this protein was targeted to go to the endoplasmic reticulum and the signal sequence was removed as a result of this deletion.)

   Textbook Reference: *14.6 What Happens to Polypeptides after Translation? p. 311*

2. What would happen if the tRNA synthase for tryptophan added a phenylalanine to the tryptophan tRNAs instead of tryptophan?

   Textbook Reference: *14.5 How Is RNA Translated into Proteins? p. 305*

3. Suppose that two different mutant strains of a bacterium are unable to grow on a minimal medium without the addition of the amino acid lysine. Explain how this phenotype might be caused by different mutations in each strain, perhaps on the same gene and perhaps in two different genes.

   Textbook Reference: *14.1 What Is the Evidence that Genes Code for Proteins? pp. 292–293*

4. Suppose that two auxotrophic mutants that had been isolated are able to grow when fed the same biochemical intermediate. According to the experiments of Beadle and Tatum, the mutations in each of these auxotrophs should be in the same gene, because they were blocked at the same step in a biochemical pathway. Yet, these two auxotrophs had mutations that mapped in different genes. How do you explain this?

   Textbook Reference: *14.1 What Is the Evidence that Genes Code for Proteins? pp. 292–293*

Answers

Diagram Exercise Answer

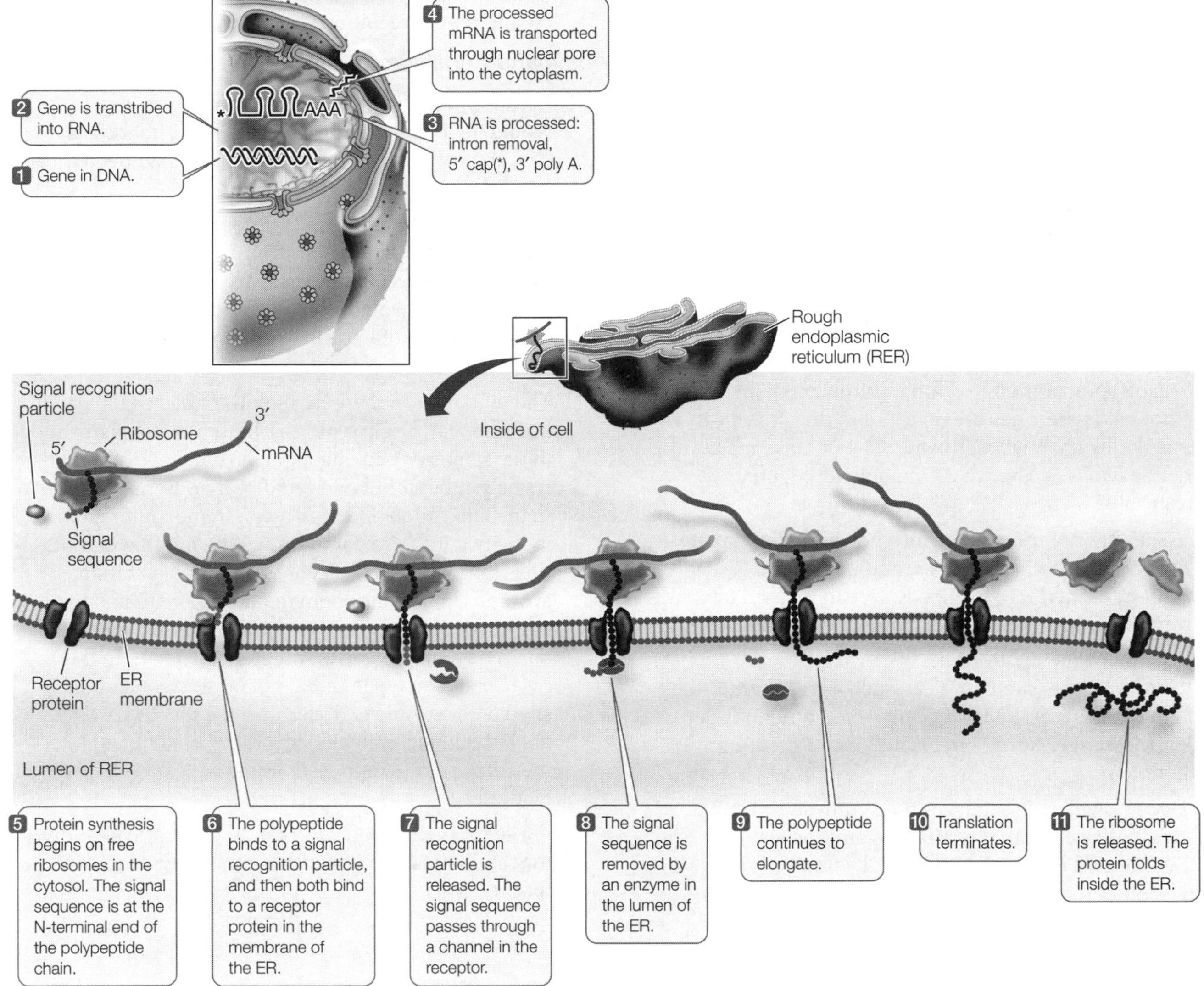

Deleting the signal sequence will not affect the transcription or translation of this gene, but it will impact the targeting of this gene product. During translation of this mRNA, there will be no signal peptide made and the signal recognition particle will be unable to bind. The mRNA will continue to be translated in the cytoplasm and remain there.

Knowledge and Synthesis Answers

1. **c.** Answer **a** describes transcription and translation in eukaryotic cells; answers **b** and **d** describe translation. Answer **e** refers to the *ori* site, where replication of the circular chromosome starts.
2. **b.** RNA polymerase binds at a promoter, unwinds the DNA, synthesizes mRNA in a 5′-to-3′ (not 3′-to-5′) direction, and does not require a primer to synthesize the RNA.
3. **b.** Translation of messenger RNA occurs 5′ to 3′, and the polypeptide is synthesized from the N terminus to the C terminus.
4. **a.** Four possible bases read two at a time would yield 4^2, or 16, different codons.
5. **b.** See the codon table (Figure 14.6). Recall that translation occurs in the 5′-to-3′ direction.
6. **d.** UAG is a stop codon, so translation would terminate at that site.
7. **b.** See the codon table (Figure 14.6).

8. **c.** Neither the 3′ end nor the 5′ end of the tRNA is part of the anticodon. The promoter is a DNA sequence, to which RNA polymerase binds to initiate transcription. The start codon is found in the mRNA.
9. **e.** Peptidyl transferase is the enzyme that catalyzes the formation of the peptide bond, and it is located in the large subunit of the ribosome. Its catalytic activity is due to ribosomal RNA found in the large subunit of the ribosome.
10. **e.** Termination of translation requires a stop codon positioned at the A site of the ribosome, peptidyl transferase, and a release factor. Peptidyl transferase hydrolyzes the last amino acid attached to the tRNA in the P site, creating the C terminus.
11. **c.** The nuclear sequence would direct this protein to the nucleus.
12. **b.** Auxotrophs cannot grow on minimal medium and are not wild type. They are deficient in one enzyme in a particular biochemical pathway, but can make many other enzymes. Answer **e** is true only for arginine auxotrophs.
13. **d.** Genes are not transcribed into polypeptides, protein is not used to synthesize DNA, and messenger RNAs are not translated into tRNAs. RNA can be used to synthesize DNA using reverse transcriptase, but DNA cannot be utilized to make protein.
14. **e.** Gene products can be RNAs (such as rRNA, tRNA, and microRNA) as well as enzymes and other polypeptides. Messenger RNA is translated into a gene product, protein.
15. **e.** DNA polymerase catalyzes the synthesis of DNA, tRNA synthase covalently attaches amino acids to tRNAs, and ribosomal RNA (peptidyl transferase in the large subunit) catalyzes the formation of the peptide bond during translation.
16. **d.** When an intron fails to be removed, that noncoding sequence is retained in the RNA within the coding sequence. When this RNA is translated, the protein will likely be nonfunctional due to the insertion of a noncoding sequence within the coding sequence.

Application Answers

1. Because the protein lacked its targeting sequence, it would no longer be moved to the mitochondria and would remain in the cytoplasm after it had been translated.
2. If the tRNA synthetase for tryptophan added phenylalanine to the tryptophan tRNAs, whenever a tryptophan codon was read by these tryptophan tRNAs, phenylalanine would be added to the polypeptide. This would create proteins that were nonfunctional, and the cell would die.
3. The mutations in these two strains of bacteria apparently interfere with lysine synthesis. Both mutations might be in the same gene coding for an enzyme necessary for lysine synthesis, but one could be a nonsense mutation in the fifth codon and the other a frame-shift mutation in the twenty-third, for example (the number of mutations that can disable a gene is enormous). If lysine synthesis in this bacterium requires more than one enzyme (as is likely), the two mutations could be in different genes coding for different enzymes. In this case, the phenotypes would not be strictly identical; it should be possible to distinguish the two by trying to grow them on minimal media to which different intermediates in the synthesis of lysine have been added (see Figure 12.1).
4. These two genes must encode different polypeptides that are both subunits for the same enzyme in this biochemical pathway.

15 Gene Mutation and Molecular Medicine

The Big Picture

- Some diseases in humans can be directly related to a mutations in the genome, causing the production of an abnormal gene product. Other molecular diseases are affected by the individual's environment. Understanding the molecular alterations in a particular gene product and its effect can lead to therapeutic approaches, new diagnostic tools, and preventative measures in health care.

Common Problem Areas

- RFLP mapping, oligonucleotide hybridization, and allele-specific cleavage analyses are similar techniques used in genetic screening to detect mutant alleles. Individual differences in DNA sequences (including simple base changes and DNA deletions, as well as a difference in the number of repeated sequences) can be utilized to distinguish individuals and to map genes. Review Figures 15.15, 15.20, and 15.21 to distinguish these screening techniques.
- Many different mutations can occur in a gene that will alter the function of the gene product. Some of those mutations may have more deleterious effects than others. Review the different kinds of mutations and how they impact gene function and the health of the individual.

Study Strategies

- Make a list of all the kinds of mutations discussed in this chapter. Describe how those mutations can result in a nonfunctional gene product and give examples of diseases that are caused by those mutations.
- Make a list of the molecular techniques that are used to screen for abnormal genes. Select a few of the human diseases that are discussed in this chapter and describe the techniques that would be useful to screen for those diseases.
- Describe some of the techniques that are used to treat genetic diseases.
- Go to yourBioPortal.com to review the following tutorials and activity:

  Animated Tutorial 15.1 Separating Fragments of DNA by Gel Electrophoresis

  Animated Tutorial 15.2 DNA Testing by Allele-Specific Cleavage

  Interactive Tutorial: Gene Mutations

  Web Activity 15.1 Allele-Specific Cleavage

Important Concepts

Mutations are heritable changes in genetic information.

- Mutations can occur in germline cells and thus be passed on to progeny, or they can occur in somatic cells.
- Silent mutations are mutations that do not affect protein function. They can occur in noncoding DNA or in coding DNA where, because of the redundancy of the codons, the meaning of the codon is not altered by the mutation.
- Loss of function mutations lead to a loss of protein function and are almost always recessive.
- Gain of function mutations lead to altered protein function and usually shows dominant inheritance to the wild-type allele.
- Conditional mutants are a class of mutations that are observable only under restrictive conditions, such as high temperature. The inability of certain mutants to grow at high temperature can be related to the unstable tertiary structure of the protein whose gene sequence has been altered.
- All mutations are alterations in DNA sequence and can be classified into two categories: point mutations and chromosomal mutations.
- Point mutations are changes of one or two base pairs and include silent mutations, missense mutations, nonsense mutations, and frame-shift mutations.
  - Silent mutations are base changes that do not alter the protein function because they do not change the meaning of the codon. For example, the change of an A to a C in the third position of a codon would change a CCA to a CCC, but both codons specify proline (see Figure 15.2).

- Missense mutations are single-base changes that change one codon to another (e.g., GAU, which codes for asparagine, to GUU, which codes for valine) (see Figure 15.2).
- Nonsense mutations are single-base changes that change a codon to a stop codon (e.g., UGG, which codes for tryptophan, to UAG, stop). Nonsense mutations result in a shortened polypeptide (see Figure 15.2).
- Frame-shift mutations are deletions or insertions of one or two bases and alter the reading frame of the genetic message (see Figure 15.2).

- Chromosomal mutations are extensive changes in the chromosome and include larger deletions, duplications, inversions, or translocations (see Figure 15.4).

Mutations can be spontaneous or induced.

- Spontaneous mutations occur without any outside influence and can be caused by tautomerization of the bases, deamination of the bases, errors during DNA replication, and errors during meiosis (leading to unequal crossovers or unequal separation of the chromosomes).
- Induced mutations are caused by chemicals and radiation, which alter the bases, add groups to the bases, or break the backbone of the DNA.
- Hot spots, which are usually sites on the DNA containing methylated cytosines, are more prone to mutation than average. Loss of the amino group on a cytosine converts that base to uracil, which is removed from the DNA by repair enzymes. Loss of an amino group on a methylated cytosine converts that base to thymine, which is retained in the DNA, resulting in a base change at that site. Repair of the GT base pair may result in a correction to the original GC base pair or the introduction of a new AT mutation in the DNA at that site.
- Mutagens can be natural (e.g., aflatoxins in plants) or artificial (e.g., nitrites in preserved foods).
- Mutations for the most part are deleterious, but some mutations may benefit the organism by creating an altered gene product that improves survival.
- Mutations in germ line cells are carried on to the next generation. Somatic mutations can lead to cancer (see Chapter 11).

DNA molecules and mutations can be analyzed.

- Bacteria defend themselves against viral infection by synthesizing enzymes known as restriction endonucleases. Restriction endonucleases cut viral DNA but leave their own DNA unharmed. Bacteria use their own enzymes (methylases) to add a methyl group to the bases of the DNA, making their DNA unable to be cut by their own restriction enzymes.
- Restriction endonucleases cut the backbone of the DNA at specific recognition sites (restriction sites) between the 3′ OH of one nucleotide and the 5′ PO_4 of the next nucleotide. For example, *Eco*RI is a restriction endonuclease that cuts the sequence 5′ – GAATTC – 3′ and its complementary strand, 3′ – CTTAAG – 5′, between the G and the A on each strand. This sequence is a palindrome; it reads the same in both directions.
- *Eco*RI cuts, on average, about 1 in every 4,000 base pairs in prokaryotic genomes, but sizes of the DNA fragments vary, depending on the actual location of these *Eco*RI sites in the chromosome.
- Hundreds of purified restriction enzymes are used in research laboratories to manipulate DNA.

DNA fragments can be separated by means of gel electrophoresis.

- A gel is a porous molecular sieve that allows smaller DNA molecules to move through it faster than larger DNA molecules. When an electric field is applied to the gel, the negatively charged phosphates of the DNA are pulled toward the positively charged electrode (see Figure 15.8).
- Restriction digests of a particular DNA sample will reveal the number of DNA fragments, their sizes, and their abundance.

DNA fingerprinting uses electrophoresis and restriction digestion.

- Genes that are used for DNA fingerprinting are highly polymorphic and include single nucleotide polymorphisms (SNPs) or short tandem repeats (STRs) (see Figure 15.9).
- DNA fingerprinting is used in forensic cases to identify suspects and to determine relatedness between skeletal remains and living relatives (e.g., the identification of the Romonovs; see Figure 15.10).

A DNA barcode project has been proposed to classify all organisms.

- Scientists have proposed a DNA barcode using a short sequence from a highly conserved gene (cytochrome oxidase) to classify all organisms on Earth. This project has the potential to help us understand evolution and species diversity and detect undesirable microbes in food.

Many diseases in humans are caused by a specific genetic defect that leads to defective proteins.

- Proteins function in cells in many ways: as enzymes, receptors, transporters, and structures. Mutations in genes encoding these proteins can cause many diseases.
- Phenylketonuria (PKU) is caused by a mutation in the gene encoding an enzyme (phenylalanine hydroxylase) in the biochemical pathway for the synthesis of tyrosine (see Figure 15.12).
  - When this enzyme is nonfunctional, a buildup of phenylalanine in the blood occurs. As phenylalanine accumulates, it is converted to phenylpyruvic acid, which in high concentrations prevents normal brain

development in infants. Patients with PKU also have light-colored skin and hair because the skin pigment melanin is made from tyrosine (see Figure 15.12).

    - The molecular alterations in many patients with PKU is a change in the 408th amino acid (from arginine to tryptophan) or a change in the 280th amino acid (from glutamate to lysine) of phenylalanine hydroxylase.

- Many proteins are polymorphic. Protein sequencing has revealed a 30 percent variation in amino acid sequences for the same protein in different individuals. Some mutations in these genes have no effect, producing functional gene products.
- In sickle-cell disease, which is homozygous in 1 in 655 African-Americans, abnormal alleles produce an abnormal protein (β-globin).
    - The abnormal hemoglobin protein aggregates in the red blood cells, causing the cells to sickle. Sickle cells block narrow capillaries and damage highly vascularized tissues, resulting in organ damage, especially at low blood oxygen concentrations.
    - The abnormal allele in sickle-cell disease is caused by a mutation that results in a change at amino acid position number 6 (from a glutamic acid to a valine) in the β-globin gene. This changes the structure of hemoglobin, reducing its oxygen-carrying capacity and resulting in anemia.
- Hemoglobin is easy to isolate and to study. Hundreds of β-globins with alterations in their primary amino acid sequence have been isolated. Some alterations have no effect, whereas others are quite severe. Hemoglobin C disease results from a change at position 6 of β-globin, which changes the glutamic acid to a lysine. The result is anemia that is not as severe as sickle-cell disease (see Figure 15.13).

Most common genetic diseases result from altered proteins.

- Familial hypercholesterolemia (FH), cystic fibrosis, Duchenne muscular dystrophy, and hemophilia are all examples of human disease in which abnormal proteins are formed due to specific mutations in the genes encoding those proteins.

Prion diseases are disorders of protein conformation.

- Transmissible spongiform encephalopathy (TSE) is a degenerative brain disease in which the brain develops holes and the brain tissue appears spongelike. This disease has been identified in sheep and goats (scrapie), in cows ("mad cow" disease), and in humans.
    - TSE is transmitted from one animal to another by the eating of infected tissue from the diseased animal.
    - The infectious agent is a nonfunctional conformation of a membrane protein (PrP^{sc}), or prion (proteinaceous infective particle). Altered proteins pile up as fibers in the brain tissue, causing cell death.
    - The prion (PrP^{sc}) seems to induce a misfolding in the naturally occurring protein (PrP^{c}) so that it is converted to the abnormal, misfolded form.
- Prions apparently play a role in blocking the synthesis of β-amyloid. β-amyloid plaques accumulate in patients with Alzheimer's.

Most diseases are multifactorial.

- Although some diseases are the result of a mutant allele that produced a single dysfunctional protein, most diseases are multifactorial, caused by many genes and proteins interacting with the environment. Up to 60 percent of people may be affected by diseases that are genetically influenced.

Identifying the genes that cause disease has been accomplished in a variety of ways.

- Using reverse genetics, DNA variations are used to identify the proteins causing particular diseases.
- Genetic markers and family pedigrees are compared to determine if a particular marker is different in affected versus unaffected individuals. In order for a marker to be useful, it has to have allelic polymorphisms.
- RFLPs (*r*estriction *f*ragment *l*ength *p*olymorphisms) are due to mutations in restriction sites. Restriction digests from different DNA samples are probed with a radioactively labeled single-stranded DNA probe to generate RFLP patterns (see Figure 15.15).
- Single nucleotide polymorphisms (SNPs) are genetic markers that can be used to distinguish individual variations in chromosomes. They can be detected using direct sequencing, PCR, or mass spectrometry.
- If a particular RFLP or SNP is positioned adjacent to a mutated gene, it can be used as a landmark to isolate the gene because the marker and the disease gene are inherited together. Once a linked region is identified, the DNA region can be searched for other STRs, SNPs, or candidate genes. Comparison of these DNA sequences in affected and unaffected individuals can help determine if the candidate gene is mutated in affected individuals only.

Different kinds of mutations can cause human disease.

- Disease-causing mutations may involve a singe base pair, a long stretch of DNA, multiple segments of DNA, or even an entire chromosome.
- Some human diseases have the same point mutation (as seen in sickle-cell anemia), while others can have a wide variety of different mutations. For example, there are 500 different mutations that have been isolated in different patients with PKU.
- Larger mutations include deletions, which can vary in size. Small deletions generally have less severe consequences than large deletions, which can remove more than one gene. Duchenne muscular dystrophy is due to a deletion of the dystrophin gene on the X chromosome.

- Chromosomal abnormalities include the gain or loss of one or more chromosomes, loss of a piece of a chromosome, translocations, and abnormal constrictions of chromosomes.
- Expanding triplet repeats are found in fragile-X syndrome (a repeat of CGG in the *FMR1* gene), in myotonic dystrophy (a repeat of CTG), and in Huntington's disease (a repeat of CAG). Triplet repeats are more extensive in affected individuals (200–2000 in fragile-X) and are thought to expand during DNA replication due to slippage of DNA polymerase in those regions. Triplet repeats can be found within or outside the coding sequence of a gene. Premutated individuals carry a moderate number of repeats (55–200) and show no symptoms (see Figure 15.18). Normal individuals carry a small number of repeats (6–54).

Genetic screening is used to identify people who might be carriers of a particular disease.

- Genetic screening can be done prenatally, on newborns, or on asymptomatic individuals who have close relatives with a genetic disease. DNA testing can be used to look for mutations responsible for human disease. DNA extracted from several cells can be PCR amplified, and the resulting sequences are analyzed.
- In PKU screening, phenylalanine concentrations are assayed in blood samples from newborns (see Figure 15.19). Individuals who test positive for PKU are fed special diets low in phenylalanine to prevent abnormal brain development.
- Preimplantation screening for mutant alleles for *in vitro* fertilized human embryos can be done at the eight-cell stage. One of these embryonic cells is removed, and its DNA is analyzed for the disease gene. Once a normal genotype has been confirmed, the seven-cell embryo can be implanted in the mother and will develop normally. Fetuses can also be screened with chorionic villus sampling or by amniocentesis.
- Once the DNA has been isolated, different kinds of genetic screens can be used. Allele-specific cleavage differences can detect changes in restriction endonuclease sites in the DNA containing the gene of interest (see Figure 15.20). Oligonucleotides that are specific for the mutant allele and the normal allele can hybridized with DNA isolated from individuals who are being tested (see Figure 15.21).

Treating genetic diseases includes modifying the phenotype and gene therapy.

- There are three ways to treat genetic disease: restrict the substrate of the deficient enzyme, inhibit a harmful metabolic reaction, or supply the normal version of the protein.
  - For individuals with PKU, treatment involves restricting the substrate (phenylalanine) for the missing enzyme (phenylalanine hydroxylase), which prevents the buildup of phenylpyruvic acid.
  - For individuals with chronic myelogenous leukemia, drugs have been developed that target the protein responsible for cell proliferation.
  - Patients with hemophilia A are given the missing gene product, which is a clotting factor, in purified form.
- In gene therapy, the new gene is inserted into the affected individual. The gene must be taken up efficiently, precisely inserted into host DNA, and appropriately expressed (see Figure 15.20).
- *Ex vivo* gene therapy is a technique in which the gene of interest is inserted into the patient's cells and then those cells are reintroduced into the patient. This technique has been used to replace the adenosine deaminase gene using viral DNA to infect the deficient cells (see Figure 15.23).
- Another approach to gene therapy involves inserting the gene directly into the cells in which the defect exists.

Test Yourself

Diagram Exercise

Put the following sequence on one of the DNA strands: 5′ – ATGAAATTTTTG – 3′. Consider ATG the start codon for the peptide that is made from this gene.

Then, for each of the following, write the new DNA sequence and translate that sequence into amino acids:

a. For the second codon, alter one of the bases to make it a silent mutation.
b. For the fourth codon, alter one of the bases to make it a nonsense mutation.
c. Pick any codon and change it to a missense codon.
d. Insert an A after the first A in the sequence and translate the resulting peptide.
e. Invert the sequence AATTT and translate the resulting peptide.

Textbook Reference: *15.1 What Are Mutations? pp. 318–320*

Knowledge and Synthesis Questions

1. Fragile-X syndrome is caused by
   a. a single base change in the DNA.
   b. the changing of a valine into a glutamic acid.
   c. triplet expansion.
   d. a chromosomal translocation.
   e. nondisjunction of the X chromosome in meiosis.
   Textbook Reference: *15.4 What DNA Changes Lead to Genetic Diseases? p. 333*
2. A nonfunctional membrane protein is responsible for
   a. hemophilia.
   b. sickle-cell disease.
   c. Duchenne muscular dystrophy.
   d. cystic fibrosis.
   e. PKU.

Textbook Reference: *15.3 How Do Defective Proteins Lead to Diseases? p. 329*

3. Expanding triplet repeats
   a. occur during DNA replication due to slippage of DNA polymerase.
   b. are caused by errors in DNA synthesis with reverse transcriptase.
   c. can be identified using RFLP mapping.
   d. occur in individuals with a normal number of triplet codons.
   e. Both a and c

   Textbook Reference: *15.4 What DNA Changes Lead to Genetic Diseases? pp. 333–334*

4. In sickle-cell disease,
   a. a clotting factor in the blood is nonfunctional.
   b. the sixth amino acid is changed from a valine to a glutamic acid.
   c. the sixth amino acid is changed to a stop codon.
   d. hemoglobin builds up in the red blood cells.
   e. the structure of β-globin is altered, and the hemoglobin protein forms aggregates in the red blood cells.

   Textbook Reference: *15.1 What Are Mutations? p. 318*

5. Metabolic disorders are
   a. caused by mutations in centromeres.
   b. the result of mutations in genes encoding enzymes that are required to synthesize particular compounds in the cell (such as an amino acid).
   c. due to abnormal membrane proteins that transport chloride ions.
   d. caused by prions.
   e. All of the above

   Textbook Reference: *15.6 How Are Genetic Diseases Treated? p. 337*

6. Prions
   a. cause scrapie in sheep.
   b. are caused by abnormally folded proteins that interfere with normal brain-cell function.
   c. cause "mad cow" disease.
   d. are transmitted through the ingestion of infected tissue.
   e. All of the above

   Textbook Reference: *15.3 How Do Defective Proteins Lead to Diseases? pp. 329–330*

7. Human genetic disease can be caused by
   a. an autosomal recessive trait.
   b. translocations.
   c. a deletion in a chromosome.
   d. a mutant allele that is dominant to the wild-type allele.
   e. All of the above

   Textbook Reference: *15.1 What Are Mutations? p. 320*

8. Mutations that result in human disease include
   a. expanding triplet repeats that occur during DNA synthesis.
   b. point mutations that do not change the amino acid sequence of the gene.
   c. prion diseases such as "mad cow" in humans.
   d. mutations in restriction sites in the DNA.
   e. Both a and d

   Textbook Reference: *15.4 What DNA Changes Lead to Genetic Diseases? p. 333*

9. Genetic screening
   a. has been used to treat embryos carrying a mutant allele.
   b. requires that gene sequences are the same in affected and unaffected individuals.
   c. has been used to identify fathers who are carriers for X-linked alleles.
   d. has been used to identify mutant alleles in embryos.
   e. requires dissimilarity between the alleles being tested and those that are linked to a particular disease.

   Textbook Reference: *15.5 How Is Genetic Screening Used to Detect Diseases? pp. 334–337*

10. For gene therapy to be most effective, genes should
    a. be delivered to the germ-line cells using a viral vector.
    b. be inserted into the host DNA chromosome at random sites.
    c. be expressed in the host cells in response to the appropriate environmental signals.
    d. be similar to the defective gene.
    e. have antibiotic resistance markers.

    Textbook Reference: *15.6 How Are Genetic Diseases Treated? p. 338*

11. Allele-specific oligonucleotides
    a. should be tightly linked to the disease allele.
    b. can be used to induce spontaneous mutations.
    c. can detect a change in a restriction endonuclease pattern.
    d. are used to detect mutations in protein.
    e. a, c, and d

    Textbook Reference: *15.5 How Is Genetic Screening Used to Detect Diseases? pp. 335–336*

12. Conditional mutations
    a. are mutations that are expressed in wild-type cells.
    b. are deletion mutations.
    c. are analyzed under permissive and restrictive conditions.
    d. produce gene products that are nonfunctional under all conditions.
    e. can be analyzed only under restrictive conditions.

    Textbook Reference: *15.1 What Are Mutations? p. 318*

13. Gel electrophoresis
    a. causes DNA to be pulled through the gel toward the negative end of the field.
    b. causes larger DNA fragments to move more quickly through the gel than smaller DNA fragments.
    c. is used to identify and isolate DNA fragments.
    d. is required for PCR reactions.
    e. is used in allele-specific oligonucleotide hybridization.

Textbook Reference: *15.2 How Are DNA Molecules and Mutations Analyzed? p. 324*

14. Gain of function mutations
    a. are dominant mutations that are expressed in wild-type cells.
    b. are dominant mutations that are expressed in mutant cells.
    c. are the cause of continuous division in cancer cells.
    d. can be analyzed only under restrictive conditions.
    e. are expressed only with the appropriate environmental signals.

    Textbook Reference: *15.1 What Are Mutations? p. 318*

15. Which of the following mutations would still allow protein X to be functional?
    a. A missense mutation in the third codon of gene *X*
    b. A silent mutation in the third codon of gene *X*
    c. A deletion of the first three codons of gene *X*
    d. A frame-shift mutation in the third codon of gene *X*
    e. A nonsense mutation in the third codon of gene *X*

    Textbook Reference: *15.1 What Are Mutations? pp. 318–320*

16. Which of the following chromosomal mutations would still allow protein X to be functional?
    a. Deletion of the last 100 codons of gene *X*
    b. A duplication of gene *X*
    c. An inversion of the last 100 codons of gene *X*
    d. A translocation of the last 100 codons of gene *X* to another chromosome
    e. None of the above

    Textbook Reference: *15.1 What Are Mutations? pp. 318–320*

17. Which of the following mutations would probably be the most deleterious?
    a. A missense mutation in the second codon
    b. A frame-shift mutation in the second codon
    c. A nonsense mutation in the last codon
    d. A silent mutation in the second codon
    e. A missense mutation in the last codon

    Textbook Reference: *15.1 What Are Mutations? pp. 317–320*

Application Questions

1. Phenylketonuria (PKU) and Duchenne muscular dystrophy (MD) are both molecular diseases that are caused by mutations in particular genes, yet PKU is easier to treat than MD. Why?

   Textbook Reference: *15.6 How Are Genetic Diseases Treated? p. 337*

2. Two individuals have a mutation in gene *X* but at different sites. The mutation affects the first individual adversely, and the second individual experiences no effect. Explain this observation.

   Textbook Reference: *15.1 What Are Mutations? pp. 318–320, Figures 15.1, 15.2*

3. Bacterial restriction endonucleases provide a type or "immunity" to viral infections. Describe how these endonucleases protect the bacterial cell from viral infections.

   Textbook Reference: *15.1 What Are Mutations? pp. 324–325*

4. When a gene that has been mutated is expressed, the resulting protein may misfold and be nonfunctional. How is a misfolded protein (due to a mutation) different from a misfolded prion protein?

   Textbook Reference: *15.3 How Do Defective Proteins Lead to Diseases? pp. 329–330*

5. You have characterized over 100 mutations in a particular gene that are linked to a certain disease. The mutations you have identified are all due to a single base change in one codon of the gene encoding protein X. Which DNA technique would you use to analyze patients who might have this disease?

   Textbook Reference: *15.3 How Do Defective Proteins Lead to Diseases? pp. 327–329*

Answers

Diagram Exercise Answer

5′– ATGAAATTTGG – 3′ met-lys-phe-trp

a. 5′– ATGAA**G**TTTGG – 3′ met-lys-phe-trp
b. 5′– ATGAAATTT**A**G – 3′ met-lys-phe-stop
c. 5′– ATG**C**AATTTGG – 3′ met-gln-phe-trp
d. 5′– **A**ATGAAATTTGG – 3′ asn-glu-ile-leu
e. 5′– ATGA**TTTAA**TGG – 3′ met-ile-stop

Knowledge and Synthesis Answers

1. **c.** Fragile-X syndrome is caused by a triplet expansion on the X chromosome, not a single base change, a translocation, or an alteration of protein sequence. Nondisjunction of the X chromosome would result in XO, also known as Turner's syndrome.

2. **d.** The chloride transporter is a membrane protein that, when nonfunctional, causes cystic fibrosis. Hemophilia is caused by a defect in a clotting factor. Sickle-cell disease is caused by a defect in β-globin. PKU is caused by a defect in an enzyme in a biochemical pathway. Duchenne muscular dystrophy is caused by a defect in the muscle protein dystrophin.

3. **e.** Expanding triplet repeats are thought to be due to slippage of DNA polymerase during DNA replication, and result in an increasing number of triplet repeats in those cells. That increase in the DNA sequence can be distinguished from the DNA sequence of a normal individual by means of RFLP analysis. Reverse transcriptase is an enzyme from HIV that copies RNA into DNA.

4. **e.** In sickle-cell disease, the sixth amino acid is changed from a glutamic acid to a valine (not from a valine to a glutamic acid, and not to a stop codon). The resulting change causes the hemoglobin protein to form aggregates in the red blood cell. The concentration of hemo-

globin in the red blood cell is not altered in sickle-cell disease.

5. **b.** Some metabolic disorders are caused by mutations in genes that encode enzymes in a biochemical pathway (such as the pathway for the synthesis of phenylalanine). Mutations in centromeres and genes for chloride transporters do not result in metabolic disorders. Prions are not the result of a mutation; they are the result of the abnormal folding of a protein.
6. **e.** Prions result from abnormally folded proteins that interfere with normal brain-cell function. "Mad cow" disease and scrapie are caused by prion formation. They are thought to be passed from organism to organism by the eating of infected tissue.
7. **e.** All these mutations and chromosome alterations can cause human disease.
8. **e.** Expanding triplet repeats cause a variety of diseases, including fragile-X syndrome, Huntington's chorea, and myotonic dystrophy. Point mutations that do not alter the amino acid sequence of a protein are silent mutations and do not cause disease. Prion diseases are not the result of a mutation in a gene, but are due to the abnormal folding of a particular protein. Mutations in a gene coding sequence or regulatory sequence could alter a restriction site.
9. **c.** Genetic screening has been used to analyze *in vitro* fertilized embryos (at the eight-cell stage) but cannot be used to treat those embryos. To be useful, the alleles being tested have to be different in affected versus unaffected individuals, and the specific alleles that are linked to a disease trait should be the same. Fathers are not carriers for X-linked diseases; they express all their X-linked alleles.
10. **c.** The most effective conditions for gene therapy are for genes to reside permanently at the appropriate site in the host chromosome in the individual's cells and be expressed under the correct environmental signals. The gene should be identical to the wild-type copy of the gene and does not require antibiotic resistance markers. Viral delivery of the gene to germ-line cells will not affect the diseased somatic cells.
11. **e.** Allele-specific oligonucleotides can detect specific changes in the DNA, including mutations, disease alleles, and changes in restriction sites. To be useful, they should be tightly linked to the disease allele. They cannot induce spontaneous mutations.
12. **c.** Conditional mutations are used to compare gene expression under permissive and restrictive conditions. Conditional mutations are not expressed in mutant cells under restrictive conditions but are expressed under permissive conditions. Under restrictive conditions they produce nonfunctional gene products. They cannot be deletions because the gene product has to be functional at the permissive temperatures.
13. **c.** DNA fragments migrate toward the positive end of the electric field, with the smallest fragments migrating the fastest. PCR reactions do not require gel electrophoresis, although their products are analyzed by gel electrophoresis. Allele-specific oligonucleotide hybridization does not require gel electrophoresis.
14. **b.** Gain of function mutations are dominant and are expressed in mutant cells. They do not respond to the appropriate environmental signals and do not have a restrictive condition under which they are expressed. Some gain of function mutations are seen in cancer cells, but mutations in tumor supressors are also responsible for unregulated cell division in cancer cells.
15. **b.** A silent mutation (no change in the amino acid sequence) would result in no change in protein function. All of the other mutations would alter the amino acid sequence. A missense mutation would change the amino acid sequence at that codon and could result in a nonfunctional protein. A deletion would remove the first three amino acids of the protein. A frame shift would cause all of the codons after that mutation to be out of frame. A nonsense mutation would terminate translation at that codon.
16. **c.** A duplication of gene *X* might allow the protein to be functional. All of the rest of these chromosomal mutations would remove large regions of the coding sequence or would position part of that sequence in the opposite direction (inversion), which would produce a nonfunctional protein X.
17. **b.** The mutation that is potentially the most dangerous is the frame shift in the second codon. All of the other codons would be out of register and no functional protein could be made. A missense mutation in the second codon could be disastrous or minimal, depending on how crucial that second amino acid was to the folding and functioning of the protein. A nonsense or missense mutation at the end of the coding sequence would have less effect on protein function because almost every amino acid in the protein would have been synthesized correctly. Again, this would depend on how important that last codon was to the proper folding and functioning of the protein. A silent mutation would have no effect on protein function because the amino acid inserted at that codon would be the same.

Application Answers

1. Phenylketonuria is a metabolic disease that results from a deficient enzyme, phenylalanine hydroxylase. Patients with PKU can build up harmful levels of phenylalanine in their systems and can be treated by restricting their intake of phenylalanine. Muscular dystrophy is caused by a defect in a structural protein, dystrophin, and results in dysfunctional muscle tissue. This structural dysfunction cannot be reversed with the application of any drug.

2. The mutation in gene *X* in the first individual must have occurred in an essential region of the gene that is required for its function. The mutation in gene *X* in the second individual may be a silent mutation, or it may be in a region that is nonessential for the function of that protein.
3. Restriction endonucleases will cut DNA at specific sites and can degrade the viral DNA once it has infected the bacterial cell. Bacteria modify their own DNA so the restriction endonucleases cannot cut their own DNA.
4. The misfolding of a prion protein is due to the presence of another misfolded prion in its environment and not due to a mutation in the gene sequence that results in a misfolded protein.
5. Since all of the mutations you isolated occur in the same codon, you could use a set of SNPs for that codon. SNPs identify single base changes in the DNA.

16 Regulation of Gene Expression

The Big Picture

- Prokaryotes regulate gene expression transcriptionally at promoters and posttranscriptionally by mRNA degradation, translational regulation, protein hydrolysis, and the inhibition of protein function.
- Prokaryotic cells use operons to coordinately regulate the expression of several genes. In response to environmental signals (the presence of lactose or the absence of tryptophan, for example) the repressor protein releases the operator site, leaving the promoter accessible to RNA polymerase to initiate transcription. In bacteriophage λ, other proteins bind promoters and control whether lysis or lysogeny will occur.
- Eukaryotes regulate gene expression with transcription factors, repressors, and activators, and they coordinate that regulation by placing regulatory sequences in front of different genes. Histone modifications also affect gene expression. Posttranscriptional regulation includes alternate splicing, modification of the 5′ cap, control of translational initiation, and protein degradation.
- Gene expression in eukaryotes can occur at transcriptional and posttranscriptional levels but can be regulated epigenetically. Eukaryotic gene expression requires that the DNA sequences be made accessible through chromatin remodeling before transcription can begin.

Common Problem Areas

- It is easy to confuse inducible operons (the *lac* operon) with repressible operons (the *trp* operon) because they both use repressors to regulate gene expression. Review the environmental conditions that must exist for each repressor to bind its operator site and what causes each repressor to release the operator site. (In the presence of lactose, the *lac* repressor cannot bind the *lac* operator site; in the presence of tryptophan, the *trp* repressor binds the *trp* operator site.) Review gene expression (Chapter 14) to see how, once transcription is initiated, proteins are produced that will act on those environmental signals.
- It may seem puzzling at first that some viruses insert their chromosome into the host chromosome, because the goal of viral infection seems to be rapid multiplication and infection of adjacent cells. However, lysogeny allows the viral genome to persist while environmental resources are abundant. When resources are depleted or cell damage occurs, the virus can enter the lytic cycle to infect other cells.

Study Strategies

- Gene expression often occurs in response to environmental signals. Make a list of the environmental signals that a prokaryote receives and then outline the steps initiated by the cell in response to those signals. Include the following signals: lactose in the cell, glucose and lactose in the cell, high levels of glucose in the cell, high or low levels of tryptophan in the cell, bacteriophage λ infection under poor or rich environmental conditions.
- Outline the different animal virus life cycles, from the start of infection through multiplication of virus to viral release (review Figures 16.5 and 16.6). Compare these life cycles to the lytic and lysogenic life cycles of bacteriophage (see Figure 16.3).
- Gene regulation can occur at many steps in a eukaryotic cell. Starting with chromatin remodeling, list the steps that must occur for a gene to be made into functional protein in the cytoplasm of the cell.
- Go to yourBioPortal.com to review the following tutorials and activities:

  Animated Tutorial 16.1 The *lac* Operon

  Animated Tutorial 16.2 The *trp* Operon

  Animated Tutorial 16.3 Initiation of Transcription

  Web Activity 16.1 Eukaryotic Gene Expression Control Points

  Web Activity 16.2 Concept Matching

Important Concepts

Genes can be negatively or positively regulated.

- Gene expression can be negatively regulated by a repressor that inhibits transcription or positively regulated by an activator protein that stimulates transcription.

Some bacteriophage are lytic, and some alternate between lytic and lysogenic life cycles.

- Outside the host cell, viruses exist as individual particles called virions, which consist of nucleic acid and proteins.
- Viruses are acellular, do not regulate transport in and out of cell membranes, and perform no metabolic function. They develop and reproduce only inside the cells of specific hosts.
- Some bacteriophage (the virulent viruses) are lytic. During infection they take over the host machinery to synthesize new viral particles and lyse the host cell.
- During the early phase of the lytic cycle, the phage injects its nucleic acid into the host cytoplasm, and viral genes are transcribed and translated. These early gene products shut down host transcription, degrade host DNA, and stimulate viral genome replication and viral gene transcription. Late gene products include viral capsid proteins and proteins that lyse the cell at the end of the lytic cycle. The lytic cycle takes about 30 minutes and produces hundreds of bacteriophage per cell (see Figure 16.4).
- During the lytic cycle, some bacteriophage can package host DNA and transfer it to the next bacterial cell. This process is called transduction.
- Other bacteriophage (the temperate viruses) can alternate between lytic and lysogenic life cycles.
- Lysogeny occurs when the viral genome inserts itself into the host genome and the viral genome is replicated as part of the host chromosome. The virus is called a prophage at this stage. When the cell encounters stressful environmental conditions, lysogenic viruses respond with a change in gene expression, resulting in excision from the host chromosome and completion of the lytic cycle.
- Lysogenic phage have a genetic switch that senses conditions in the host. Two proteins (Cro and cI) compete for two promoters in lambda (λ), a lysogenic phage. When the Cro protein binds those promoters, lytic genes are activated. When the cI protein binds those promoters, genes involved in lysogeny are activated (see Figure 16.5).

Eukaryotic viruses are diverse.

- Animal viruses include RNA and DNA viruses and retroviruses.
- DNA viruses can be double-stranded or single-stranded and can be both lytic and lysogenic (e.g., herpes and papillomaviruses).
- RNA viruses are usually single-stranded (like influenza). Their genome is translated by the host machinery, and some of those gene products are used to replicate the RNA genome.
- Retroviruses are viruses whose genome is RNA (e.g., HIV). Once the host is infected, that RNA is used as a template to synthesize double-stranded DNA, which is integrated into the host genome.
  - Human immunodeficiency virus (HIV) is an enveloped virus. It is enclosed in a plasma membrane derived from the host cell.
  - HIV is a retrovirus that copies its RNA genome into DNA using reverse transcriptase. The DNA copy of the viral genome is then integrated into the host cell's chromosome, forming a provirus (see Figure 16.6). The provirus resides permanently in the host chromosome.
  - When activated, the provirus is transcribed as mRNA and translated into protein by the host cell's translation machinery. The viral protein tat (*t*rans*a*ctivation of *t*ranscription) prevents termination, allowing viral gene transcription by the host RNA polymerase.
  - Viruses are assembled at the host cell membrane, and mature virions bud from the cell membrane.
  - Viral proteins activate a switch in response to environmental factors, which can determine if temperate phage (such as lambda, λ) will make a "decision" to lyse or lysogenize.

Prokaryotes conserve energy by making proteins only when they need them.

- Prokaryotes conserve energy by making proteins only when they need them. They regulate gene expression by (1) downregulating transcription of mRNAs; (2) hydrolyzing mRNA before its translation; (3) preventing mRNA from being translated by the ribosome; (4) hydrolyzing the protein after it is made; and (5) inhibiting the function of the protein.
- There are three proteins involved in lactose uptake: β-galactoside permease, β-galactosidase, and β-galactoside transacetylase.
  - β-galactoside permease transports lactose into the cell.
  - β-galactosidase breaks down lactose into glucose and lactose.
  - β-galactoside transacetylase transfers acetyl groups to certain β-galactosides.
- These structural genes are expressed in the presence of the inducer, lactose (see Figure 16.8).
- Genes can be expressed in response to specific environmental conditions (inducible) or constitutively (all the time).
- Enzyme activity can be repressed by end-product feedback (also known as feedback inhibition) or by regulating gene expression (see Figure 16.9).

Regulation of the *lac* operon.

- In prokaryotic cells, one promoter can control more than one gene.
- Transcription in prokaryotes can be regulated by proteins that bind operator sites on the DNA.

- The structural genes for the metabolism of lactose are adjacent to one another in the *lac* operon and are all transcribed from one promoter (see Chapter 14 for a review of transcription). There are three structural genes which code for three enzymes necessary for the cell to use lactose (see Figure 16.11).
- The *lac* operon is an inducible operon.
- Operons are units of transcription in prokaryotes that consist of structural genes and the DNA sequences that control their transcription. The control sequences include the promoter (which is bound by RNA polymerase) and the operator (which is bound by the *lac* repressor).
- When RNA polymerase binds the promoter of the operon, all three structural genes are transcribed.
- The operator is a DNA sequence immediately adjacent to the promoter in the operon. If a repressor molecule is bound to the operator, it prevents RNA polymerase from binding to the promoter.
- Operons can be inducible or repressible and can be regulated by a repressor. Operons can also be regulated by an activator protein.
  - The repressor for the *lac* operon can bind the operator site of the *lac* operon and the inducer, lactose. When the repressor binds the inducer, it cannot bind the operator, and transcription of the *lac* genes will occur (see Figure 16.11). The repressor regulates the expression of the lactose genes by binding the operator site on the DNA.
  - In summary, inducible operons have the following features: the operator and promoter are bound by a regulatory protein (the repressor) to control transcription; in the absence of the inducer, the operon is off; the repressor prevents transcription of the operon; and adding the inducer changes the tertiary structure of the repressor so transcription of the operon occurs.
- The *trp* operon is a repressible operon. This means that its repressor is *not* bound to the operator unless another molecule, the corepressor (tryptophan, in this case), first binds to the repressor. When the corepressor binds to the repressor, the repressor becomes active and binds to the operator, preventing transcription of the structural genes.
- In inducible systems (like that of the *lac* operon), the substrates of these pathways bind the repressor, allowing transcription to occur. The presence of the substrate thus turns *on* production of the machinery necessary to catabolize the substrate.
- In repressible systems (like that of the *trp* operon), the product of the metabolic pathway (the corepressor) interacts with the regulatory protein, enabling it to bind the promoter and block transcription. The presence of the product thus turns *off* the transcriptional machinery necessary to produce more of that product.

Transcription can be positively regulated by increasing promoter efficiency.

- The cAMP receptor protein (CRP), when bound to cAMP, will cause the transcription of certain operons to increase by helping RNA polymerase bind the promoter more efficiently (see Table 16.1 for an outline of positive and negative regulation of transcription).
- Glucose reduces the levels of cellular cAMP, resulting in lower transcription efficiency of the *lac* operon, which is termed catabolite repression. This type of gene regulation occurs when the presence of a preferred energy source (glucose) represses other catabolic pathways.

The expression of eukaryotic genes is precisely regulated.

- Eukaryotic promoters have two important sequences, a recognition sequence and a TATA box.
- Initiation of transcription in eukaryotic cells requires regulatory proteins called transcription factors. These are proteins that bind to the promoter and form a transcription complex to which RNA polymerase II can bind in order to initiate transcription.
- TFIID, a transcription factor, binds the TATA box, changing the shape of the DNA and the transcription factor. Other transcription factors then bind the promoter, and RNA polymerase binds to initiate transcription.
- Other short sequences bind regulatory proteins that affect transcription positively (enhancers, which bind activator proteins) and negatively (silencers, which bind repressors).
- Regulatory sequences can be located close to or far away from the gene they affect.
- The transcription of any eukaryotic gene is determined by the combination of transcription factors, repressors, and activators.

Proteins that bind DNA have characteristic protein motifs.

- Activators, transcription factors, regulatory proteins, and repressors have characteristic protein motifs.
- DNA protein-binding motifs include helix-turn-helix, leucine zipper, zinc finger, and helix-loop-helix (see Figure 16.16).
- These proteins can interact with the DNA by fitting into the major or minor groove, using amino acid side chains that can project into the interior of the helix. Some of these may form hydrogen bonds with the bases.

Gene regulation is a coordinated process.

- Coordinated gene regulation is achieved by the positioning of regulatory sequences that bind the same transcription factors near the promoters of each of those genes. Proteins bind to these regulatory sequences and stimulate RNA synthesis.

- In plants, regulatory elements known as stress response elements (SREs) are found near the promoters of genes that respond to drought (see Figure 16.17).

Histone protein modifications affect transcription.

- The term "epigenetics" refers to changes in the expression of a gene or genes that occur without any change in the gene sequence. This process includes DNA methylation and chromosomal protein alterations.
  - In methylation, cytosine residues are methlyated by DNA methyl transferase, usually at CpG islands. Maintenance methylases catalyze the formation of 5-methylcytosine on the replicated DNA strand; demethylases remove the methyl group from the cytosine. Repressor proteins bind methylated regions on DNA.
  - Demethylation occurs in male and female genomes upon fertilization and in tumor suppressor genes of cancer cells.
- The DNA in eukaryotic chromosomes is associated with nucleosomes and other chromosomal proteins, which can inhibit the initiation and elongation steps of transcription. Chromatin remodeling must occur to make the DNA accessible to transcription complexes.
- Nucleosomes become disassembled and then reassembled with the help of acetylation, deacetylation, methylation, and phosphorylation of histones (see Figure 16.19).
- Epigenetic changes induced by the environment can be inherited.
- DNA methylation patterns differ in male and female gametes, leading to genomic imprinting (see Figure 16.20). Examples of genomic imprinting include Angelman Syndrome (male imprinting for that gene is inherited and the female region for that gene is deleted) and or Prader-Willi (female imprinting for that gene is inherited and the male region for that gene is deleted).
- The genotype for the sex chromosomes in mammals is XY in males and XX in females. However, the expression of X-linked genes (the gene dosage) is the same in both sexes, due to X-inactivation in females.
  - Early in embryonic development, one of the two X chromosomes is randomly inactivated, producing a highly condensed heterochromatic chromosome called a Barr body (see Figure 16.21).
  - The number of Barr bodies is equal to the number of X chromosomes minus one; an XXX female has two Barr bodies and normal females (XX) have one Barr body.
  - Most of the DNA of the inactive X is heavily methylated, with methyl groups added to the 5′ carbon of cytosine. However, transcription of the *XIST* gene occurs on the inactive X chromosome.
  - The *XIST* RNA (also known as an interference RNA) binds the X chromosome from which it was transcribed and inactivates it (see Figure 16.21).

Different RNAs can be made from the same eukaryotic gene after transcription.

- Different RNAs can be made from the same gene by alternate splicing (see Figure 16.22).
- MicroRNAs are small RNAs (about 22 bp) that bind specific mRNAs and block their translation.
- MicroRNAs are transcribed as longer precursors in the cell and are then cleaved by dicer protein to produce small double-stranded microRNAs. Those RNAs are converted to single-stranded RNA by a protein complex and inhibit translation of mRNAs, targeting them for degradation.

Translation of mRNA can be regulated.

- Messenger RNAs in the cytoplasm are not always translated, because other factors regulate the accessibility of these mRNAs to the translational machinery. If the G of the 5′ cap is not modified (as in the egg cells of the tobacco hornworm moth), the mRNA is not translated. After fertilization, this G cap is modified, and the mRNA is translated.
- Ferritin is a storage protein for iron. The mRNA necessary for making ferritin is present in cells at steady levels, regardless of the concentration of iron ions in the cell. When the iron level in the cell is low, a repressor protein binds to ferritin mRNA, preventing its translation into ferritin. As the iron level rises in the cell, some iron ions bind to the repressor protein, altering its shape so that it can no longer bind the ferritin mRNA. This permits translation of ferritin from the ferritin mRNA; the new ferritin is then available for storage of iron ions.

Protein degradation is regulated.

- Proteins that are targeted for degradation are first linked to ubiquitin through a lysine residue. Subsequently, more ubiquitin chains attach, forming a polyubiquitin chain. This polyubiquitin complex binds to a proteasome, which has ATPase activity.
- When a protein–ubiquitin complex enters the proteasome, ubiquitin is removed, the protein is unfolded, and three proteases digest the protein, using ATP hydrolysis (see Figure 16.24).

Test Yourself

Diagram Exercise

Diagram a gene in both a prokaryotic cell and a eukaryotic cell. Include the types of sequences that are important for transcriptional regulation, the location of those sequences, and the proteins that bind those regions.

Textbook Reference: *16.3 How Is Eukaryotic Gene Transcription Regulated? p. 354*

Knowledge and Synthesis Questions

1. Viruses consist of
   a. a protein core and a nucleic acid capsid.
   b. a cell wall surrounding nucleic acid.
   c. RNA and DNA enclosed in a membrane.
   d. a nucleic acid core surrounded by a protein capsid, and in some cases, a membrane.
   e. a nucleic acid core surrounded by a cell membrane.
   Textbook Reference: *16.1 How Do Viruses Regulate Their Gene Expression? p. 343*
2. Lytic bacterial viruses
   a. infect the cell, replicate their genomes, and lyse the cell.
   b. infect the cell, replicate their genomes, transcribe and translate their genes, and lyse the cell.
   c. infect the cell, replicate their genomes, transcribe and translate their genes, package those genomes into viral capsids, and lyse the cell.
   d. infect the cell, transcribe and translate their RNA, replicate their genomes, package those genomes into viral capsids, and lyse the cell.
   e. insert their chromosome into the host chromosome.
   Textbook Reference: *16.1 How Do Viruses Regulate Their Gene Expression? pp. 343–345*
3. Animal viruses that integrate their DNA into the host chromosome
   a. have DNA as their genome.
   b. are prophages.
   c. copy their RNA genome into DNA using reverse transcriptase.
   d. replicate their genome using RNA polymerase.
   e. can undergo only a lytic infection cycle.
   Textbook Reference: *16.1 How Do Viruses Regulate Their Gene Expression? p. 343*
4. An operon
   a. is regulated by a repressor binding at the promoter.
   b. has structural genes that are all transcribed from same promoter.
   c. has several promoters, but all of the structural genes are related biochemically.
   d. is a set of structural genes that are all under the same translational regulation.
   e. is transcribed when RNA polymerase binds the operator.
   Textbook Reference: *16.2 How Is Gene Expression Regulated in Prokaryotes? p. 349*
5. If the gene encoding the *lac* repressor is mutated so that the repressor can no longer bind the operator, will transcription of that operon occur?
   a. Yes, because the repressor transcriptionally activates the *lac* genes.
   b. Yes, but only when lactose is present.
   c. No, because RNA polymerase is needed to transcribe the genes.
   d. Yes, because RNA polymerase will be able to bind the promoter and transcribe the operon.
   e. No, because cAMP levels are low when the repressor is nonfunctional.
   Textbook Reference: *16.2 How Is Gene Expression Regulated in Prokaryotes? pp. 350–351, Figure 16.11*
6. If the gene encoding the *trp* repressor is mutated such that it can no longer bind tryptophan but can still bind the operon, will transcription of the *trp* operon occur?
   a. Yes, because the *trp* repressor can bind the *trp* operon and block transcription only when it is bound to tryptophan.
   b. No, because this mutation does not affect the part of the repressor that can bind the operator.
   c. No, because the *trp* operon is repressed only when tryptophan levels are high.
   d. Yes, because the *trp* operon can allosterically regulate the enzymes needed to synthesize the amino acid tryptophan.
   e. No, because the repressor will be continuously bound to the operator.
   Textbook Reference: *16.2 How Is Gene Expression Regulated in Prokaryotes? p. 351*
7. Transcriptional regulation in prokaryotes can occur by
   a. a repressor binding an operator and preventing transcription.
   b. an activator binding upstream from a promoter and positively affecting transcription.
   c. different promoter sequences binding RNA polymerase more tightly, resulting in more effective transcriptional initiation.
   d. the control of promoter efficiency.
   e. All of the above
   Textbook Reference: *16.2 How Is Gene Expression Regulated in Prokaryotes? pp. 348–350*
8. For the bacteriophage λ, the "decision" to become a prophage is made
   a. if environmental resources (e.g., food) for the host are limited.
   b. when the Cro protein binds the promoter.
   c. when the cI protein binds the promoter.
   d. when bacterial lysis occurs.
   e. at the end of the lytic infection.
   Textbook Reference: *16.1 How Do Viruses Regulate Their Gene Expression? p. 346*
9. Imagine that the TATA box for gene *X* becomes highly methylated. How will this affect the expression of gene *X*?
   a. There will be no effect.
   b. Gene *X* will be transcribed but not translated.
   c. Gene *X* will be transcribed if the transcription factors receive the appropriate environmental signal.
   d. Gene *X* will not be transcribed or translated.
   e. Gene *X* will be transcribed if the histones become acetylated.
   Textbook Reference: *16.4 How Do Epigenetic Changes Regulate Gene Expression? pp. 356–357*

10. Imagine that gene *X* is moved to a part of the chromosome where the histones are highly acetylated. How will this affect its expression?
    a. There will be no effect.
    b. It will be transcribed but not translated.
    c. It will be transcribed if the transcription factors receive the appropriate environmental signal.
    d. It will not be transcribed or translated.
    e. It will be transcribed if the histones become deacetylated.
    Textbook Reference: *16.4 How Do Epigenetic Changes Regulate Gene Expression? pp. 357–358*
11. Which of the following is an example of regulation of eukaryotic transcription?
    a. Iron binding the repressor protein for the ferritin mRNA and increasing ferritin expression
    b. Proteostome breakdown of protein–ubiquitin complexes
    c. MicroRNAs binding their target mRNA and causing its degradation
    d. Alternate splicing of an mRNA transcript
    e. Activator proteins binding an enhancer
    Textbook Reference: *16.3 How Is Eukaryotic Gene Transcription Regulated? p. 352*
12. Which of the following statements about histone modifications is *false*?
    a. They cause some genes to be transcriptionally activated.
    b. They can result in the repression of gene transcription.
    c. They are inherited from parental cells in a Mendelian fashion.
    d. They cause Barr bodies to form.
    e. All of the above are false.
    Textbook Reference: *16.4 How Do Epigenetic Changes Regulate Gene Expression? pp. 357–358*
13. What would happen initially to cells that lack a functional ubiquitin?
    a. Nothing would happen.
    b. Transcriptional initiation would increase.
    c. Protein degradation would decrease.
    d. Histone modifications would increase.
    e. Translation of proteins would be more efficient.
    Textbook Reference: *16.5 How Is Eukaryotic Gene Expression Regulated After Transcription? p. 362*
14. Which of the following would *not* affect gene expression in a eukaryotic cell?
    a. Deletion of a promoter
    b. Deletion of an enhancer
    c. Lack of modification of the cap structure on mRNA.
    d. Inability of transcription factor to bind promoter
    e. Deletion of a DNA *ori* site
    Textbook Reference: *16.3 How Is Eukaryotic Gene Transcription Regulated? pp. 352–356*
15. Transcription factors
    a. have a particular motif that allows them to interact with mRNA.
    b. assemble at the promoter to help translation begin.
    c. mark particular proteins for degradation.
    d. interact with RNA polymerase to initiate transcription.
    e. help stabilize the mRNAs in the cytoplasm.
    Textbook Reference: *16.3 How Is Eukaryotic Gene Transcription Regulated? pp. 353–355*

Application Questions

1. Animal viruses are termed obligate parasites. What is the meaning of this term?
   Textbook Reference: *16.1 How Do Viruses Regulate Their Gene Expression? p. 343*
2. Why are antibiotics useless in combating animal viral infections?
   Textbook Reference: *16.1 How Do Viruses Regulate Their Gene Expression? pp. 346–348*
3. Suppose that a cell has a mutation that deletes the gene encoding the repressor for a certain operon, and a plasmid is introduced into the host cell that carries a wild-type copy of the gene for the repressor. Is normal regulation of this operon restored in the presence of this plasmid?
   Textbook Reference: *16.2 How Is Gene Expression Regulated in Prokaryotes? pp. 349–351*
4. Suppose that a cell has a mutation that deletes the gene encoding the operator for a certain operon, and a plasmid is introduced into the host cell that carries a wild-type copy of the operator. Is normal regulation of this operon restored in the presence of this plasmid?
   Textbook Reference: *16.2 How Is Gene Expression Regulated in Prokaryotes? pp. 349–351*
5. Suppose you are engineering gene *Y*, such that when it is inserted into a eukaryotic chromosome it will be expressed continuously. Which specific sequences must be part of this gene so that it will be expressed?
   Textbook Reference*: 16.3 How Is Eukaryotic Gene Transcription Regulated? pp. 353–354, Figures 16.14, 16.15*
6. Suppose that you are engineering a new plant in which gene *Y* will be activated under drought conditions. What kinds of DNA sequences need to be present to ensure activation of the gene under these conditions?
   Textbook Reference*: 16.3 How Is Eukaryotic Gene Transcription Regulated? pp. 355–356, Figure 16.17*

Answers

Diagram Exercise Answer

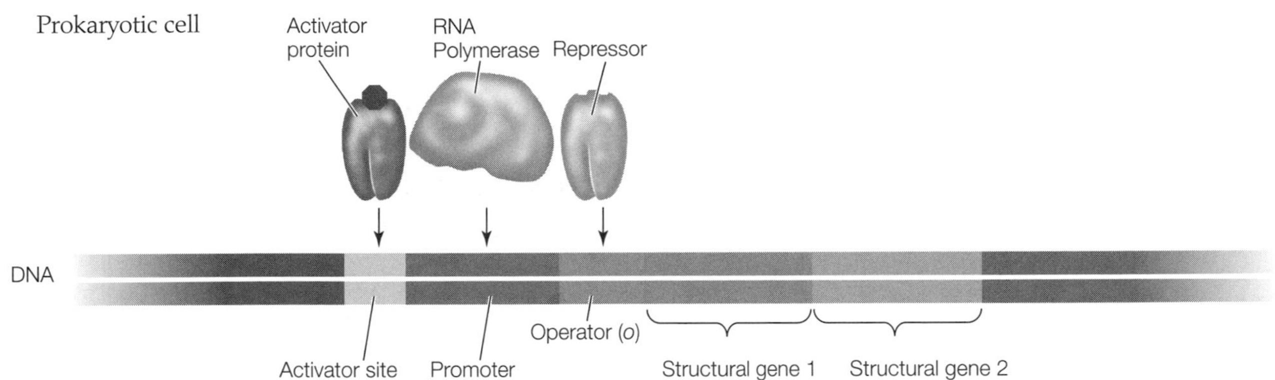

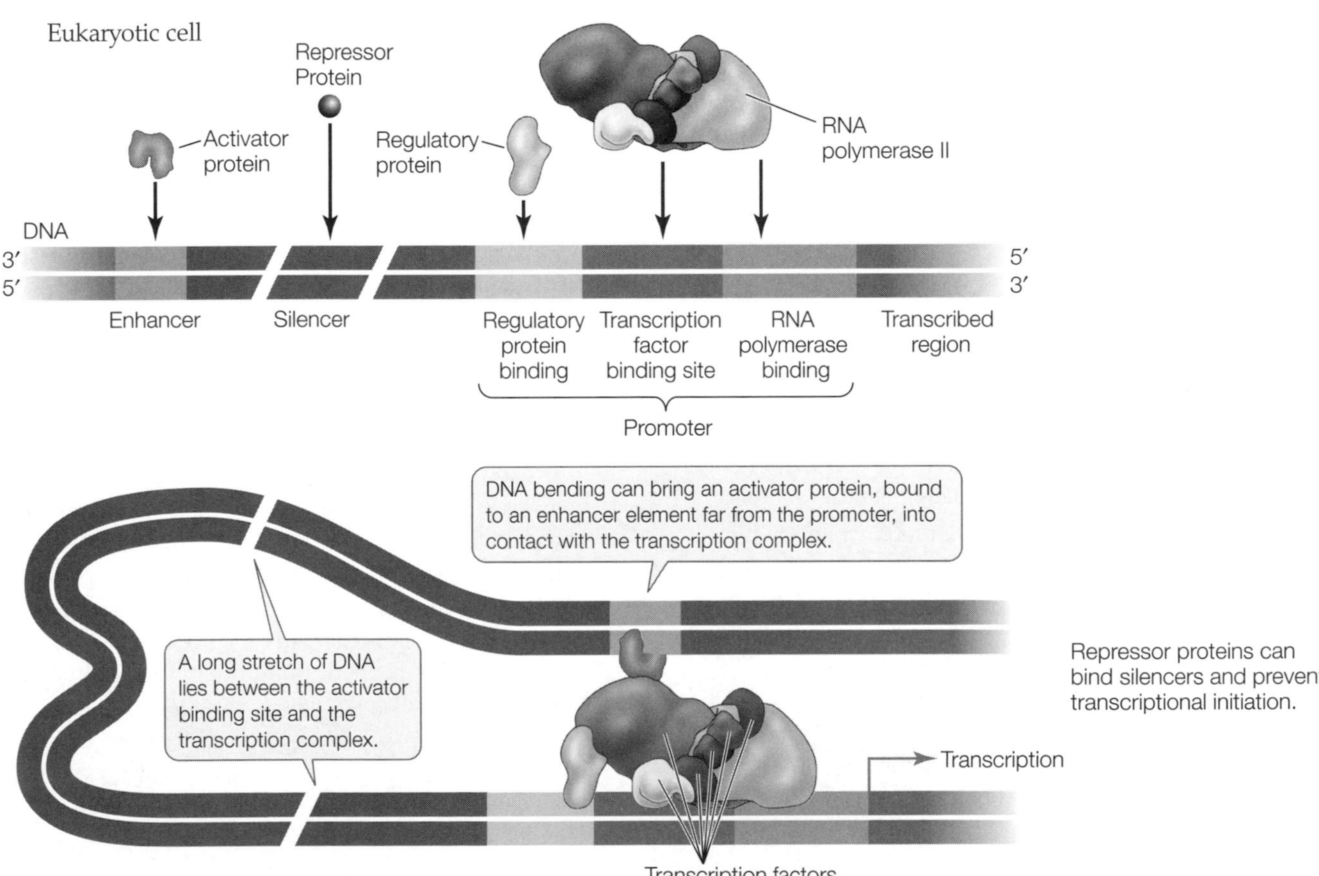

Knowledge and Synthesis Answers

1. **d.** Nucleic acids do not form capsids; cell walls are found in bacterial and plant cells, not viruses. Viruses are organized so that the nucleic acid is surrounded by protein (not membranes), and a membrane may surround the protein capsid.
2. **d.** This sequence includes the most complete details of the viral life cycle. Viral transcription and translation have to occur first so that viral gene products needed for viral replication will be synthesized.
3. **c.** Answer **b** describes a provirus, which is a bacterial virus that has inserted its genome into a host chromosome. Animal viruses replicate their RNA genomes using reverse transcriptase and do not lyse the cells they infect.

4. **b.** An operon is a set of genes that are all transcribed from the same promoter, which is the site where RNA polymerase binds. The repressor binds at the operator site, which overlaps the promoter. Answer **d** is not correct because the operon is regulated transcriptionally, not translationally.
5. **d.** If the *lac* repressor is nonfunctional, it cannot bind the operator site, and transcription of the *lac* operon will occur at all times, whether or not lactose is present.
6. **a.** If the repressor can no longer bind tryptophan, then it cannot bind the operator, and transcription of the *trp* operon will always be on, whether tryptophan levels in the cell are high or low.
7. **e.** Answer **a** refers to the *lac* and *trp* repressors, answer **b** to the CRP protein, and answer **c** to promoters that have different transcriptional efficiencies. Answer **d** refers to the *lac* operon.
8. **c.** The "decision" to become a prophage by bacteriophage λ is made when cI binds the promoter early in infection. This decision is made if the host is experiencing rich nutrient conditions and results in lysogeny, not cell lysis.
9. **d.** Gene *X* will not be transcribed or translated, since methylation sites on DNA are transcriptionally inactive.
10. **d.** Acetylation leads to loosening of the nucleosomes, resulting in more DNA sequence being accessible for transcription. If transcription factors receive the appropriate environmental signal, transcription will occur. Deacetylation leads to tighter packing of the nucleosomes.
11. **e.** Answer **a** refers to translational regulation; answer **b** refers to the regulation of protein longevity; answers **c** and **d** refer to posttranscriptional regulation.
12. **d.** Histone modifications are inherited in a non-Mendelian fashion. Depending on the parent in which the modification (genomic imprinting) occurs (i.e., Angelman syndrome versus Prader-Willi), the resulting phenotype can be different, even though the genotype is the same.
13. **c.** In the absence of ubiquitin, protein degradation would decrease, since ubiquitin targets proteins for degradation in the proteasome.
14. **e.** Deletion of an *ori* site on a eukaryotic chromosome would affect DNA replication, not transcription. All of the other alterations would affect transcription.
15. **d.** Transcription factors interact with RNA polymerase to help transcriptional (not translational) initiation. They do have particular protein motifs, but they bind DNA, not RNA. They do not stabilize mRNAs in the cytoplasm, or mark proteins for degradation.

Application Answers

1. Animal viruses cannot express their genes, replicate their genomes, or multiply unless they are in the cytoplasm of the host cell. They use the host cell's components (ribosomes, ATP) to grow and reproduce.
2. Most antibiotics attack bacterial cells by inhibiting prokaryotic translation. Animal viruses use the host translation machinery, which is different enough from the bacterial translation machinery to be unaffected by antibiotics.
3. Yes. The repressor gene can be transcribed and translated from the plasmid DNA, and normal regulation will be restored.
4. No. The operator site on the plasmid cannot restore regulation unless it recombines with the host operator site in such a way that it replaces the mutant operator on the host chromosome. The DNA site on the plasmid would bind repressor, but because that site is not adjacent to the promoter or the structural genes on the chromosome, normal regulation of those genes cannot occur.
5. You will need a promoter that binds transcription factors (such as a TATA box), an RNA polymerase, and regulatory binding sites that bind activator proteins. You will also need to put it into a chromosomal region that has not been silenced by condensed nucleosomes.
6. The plant will need stress response elements (SREs) in front of the promoter for gene *Y*. SREs are bound by transcription factors that are sensitive to drought, and genes with SRE sequences in front of their promoters can be coordinately regulated.

17 Genomes

The Big Picture

- Some diseases in humans can be directly related to a change in the genome, causing the production of an abnormal gene product. Other molecular diseases are affected by the individual's environment. Understanding the molecular alterations in a particular gene product and their effects can lead to therapeutic approaches, new diagnostic tools, and preventative measures in health care.
- The science of genomics is the study and comparison of genomes from different organisms and is used to identify genes, potential genes, and their regulatory sequences. Information from these studies has been used to design new medicines to combat pathogens and to understand the minimal requirements for a cell to sustain life.

Common Problem Areas

- It may be difficult to understand why so little of the eukaryotic genome codes for proteins. However, noncoding regions (transposons, introns, and functional regions of the chromosome) may have played a role in creating new functional genes over the course of evolution.
- There are many methods for analyzing genomes and proteomes, including those of pharmagenomics and metagenomics. Review what kinds of sequences are analyzed by each of these methods and give examples of how those analyses have furthered our understanding of the functions of genes, proteins, and cells in different organisms.

Study Strategies

- Describe the steps that must be followed in order to generate a genome sequence from a particular organism.
- Comparing genomes between different organisms has allowed us to determine which genes are essential for all cells and which genes are designated for specific adaptive functions. Compare prokaryotic and eukaryotic genomes and list genes that are essential for both types of organisms and genes that are specialized for each type of organism.
- Compare the information and applications of genome and proteome data.
- List the agricultural, medical, and environmental benefits of genome sequencing.
- Go to yourBioPortal.com to review the following tutorials and activity:

  Animated Tutorial 17.1 Sequencing the Genome

  Animated Tutorial 17.2 High-Throughput Sequencing

  Web Activity 17.1 Concept Matching

Important Concepts

How are genomes sequenced?

- The Human Genome Project, completed in 2003, sequenced all 3.3 billion base pairs in haploid cells.
- To sequence genomic DNA, it must first be cut into smaller fragments (500-base-pair fragments), separated, and sequenced.
- Bioinformatics was developed to analyze DNA sequences using complex mathematics and computer programs.
- Two methods can be used to align the different fragments: hierarchical sequencing and shotgun sequencing.
- Hierarchical DNA sequencing involves the careful ordering of DNA sequences using marker sequences along the chromosome (see Figure 17.1A).
  - Genomic libraries of human DNA were created using restriction enzymes that recognize 8–12 base-pair recognition sequences, resulting in much larger DNA fragments. These large DNA fragments were cloned into a bacterial artificial chromosome (BAC), which can accommodate 55,000–2 million base pairs of inserted DNA. These BAC clones were cut into smaller overlapping pieces and sequenced.
  - These fragments were then arranged in proper order using markers and by comparing those sequences to DNA libraries made with different restriction endonucleases (see Figure 17.1A).

- Shotgun sequencing involves cutting the DNA into random small fragments, and then cloning and sequencing those fragments. They are then analyzed for overlapping sequences by computer and aligned (see Figure 17.1B). This method was facilitated by the development of bioinformatics and increasingly powerful computers.
- The nucleotide sequence of DNA can be determined using dideoxyribonucleoside triphosphate sequencing (see Figure 17.2).
  - The DNA to be sequenced is mixed with DNA polymerase, artificially synthesized primers, a mixture of the four deoxyribonucleotides (dNTPs), and small amounts of four fluorescently labeled dideoxynucleotides (ddNTPs).
  - The DNA fragment is denatured, the primers bind to the DNA, and DNA synthesis proceeds until a ddNTP is incorporated, which terminates the DNA chain.
  - The fragments are separated by size using electrophoresis, and the fluorescent tag at the end of each fragment is read by a scanner. Those signals are then processed by computer, which aligns the fragments to provide the complete sequence (see Figure 17.2).
- High-throughput sequencing methods have increased the speed and output of DNA sequencing for large genomes (see Figure 17.3).
- Genome sequences yield several kinds of information: open reading frames, amino acid sequences of proteins, regulatory sequences, RNA genes, and other noncoding sequences.
- Sequence information is also used for comparative genomics, in which new sequences or entire genomes are compared with genome sequences from other organisms. This information can help establish evolutionary relationships.

The sequencing of prokaryotic genomes led to new genomics disciplines.

- Genomic sequencing of prokaryotes has revealed the genes that are used for different cellular functions and how those specialized functions are carried out.
- Functional genomic analysis has been used to assign functions to gene products. Genes can be identified that are involved in the prokaryote's metabolism, transport, and its infectious properties.
- Comparative genome analysis has been used to compare genomes from different organisms to understand their physiology.
- Genome sequencing has enabled scientists to understand more about transposable elements and transposons, mobile genetic elements that can move from one site in the genome to another.
- Genomic analysis for prokaryotes has revealed useful information, including newly identified virulence genes in *Rickettsia* and genes for ATP synthesis in *Chlamydia*; newly discovered antibiotic-producing genes in *Streptomyces*; unique genes in pathogenic strains of *E. coli*, *Salmonella*, and *Shigella*; methane-producing and methane-oxidizing genes in *Methanococcus* and *Methlyococcus*; and several novel proteins in the virus responsible for severe acute respiratory syndrome (SARS). Some of the genome information will be potentially useful in developing new vaccines against pathogens.

Metagenomics allows us to describe new organisms and ecosystems.

- Metagenomics analyzes genes without isolating the intact organism. The polymerase chain reaction (PCR) can be used to amplify specific sequences from an environmental sample without the need to culture those organisms. Using this method, thousands of new viruses and bacteria have been found in seawater, marine sediment, and mine water runoff (see Figure 17.7).
- Genome sequencing can be used to determine the minimal number of genes needed for life. The *Mycoplasma genitalium* genome has been mutated to determine the smallest number of genes needed for survival, which is 382 genes, and experiments are now under way to make a synthetic genomes based on that of *M. genitalium.*
- This new knowledge has the potential to help us create new microbes to degrade oil spills, reduce tooth decay, or convert cellulose to ethanol for use as fuel.

What have we learned from sequencing eukaryotic genomes?

- Eukaryotic genomes are larger than those of prokaryotic genomes, and have more protein-coding genes and regulatory sequences. Much of eukaryotic DNA is noncoding. Eukaryotes have multiple chromosomes.
- Model organisms include *Saccharomyces cerevisiae*, *Caenorhabditis elegans*, *Drosophila melanogaster*, *Arabidopsis thaliana*, and *Oryza sativa*.
  - *Saccharomyces cerevisiae* (budding yeast) is a simple eukaryote that has 12.5 million bp on 16 chromosomes (per haploid genome). *S. cerevisiae* has 5,770 genes, including many whose products are involved in protein targeting and organelle function. The genomes of *S. cerevisiae* and *E. coli* have the same number of genes for the basic functions of cell survival.
  - The nematode *Caenorhabditis elegans* is a simple multicellular organism used to study development. The worm's body is transparent, and its growth from a fertilized egg to a thousand-celled differentiated adult organism takes just three days. The genome of *C. elegans* is eight times larger than that of yeast (97 million bp) and has 3.5 times the number of protein-coding genes (19,427 proteins versus 6,000 proteins). Gene inactivation studies have revealed that the worm can survive with only 10 percent of those genes. Other gene products in *C. elegans*

include proteins used for holding cells together to form tissues, for cellular differentiation, and for intracellular communication.

- The fruit fly *Drosophila melanogaster* has ten times more cells than *C. elegans*. Its genome has 123 million bp but it contains fewer genes than *C. elegans*.
- The genome of the *Arabidopsis thaliana* plant has only 115 million bp with 28,000 protein-coding genes. Many of these genes are duplicates of each other. When the duplicates are removed, only 15,000 genes remain. *Arabidopsis* contains genes unique to plants, including genes involved in photosynthesis, water transport into the root, cell-wall synthesis, uptake and metabolism of inorganic substances, and defense against herbivores. Many of the genes in *Arabidopsis* can be found in rice, *Oryza sativa*.

- Eukaryotes have gene families—sets of duplicate or closely related genes (such as the β-globin genes). Gene families provide the organism with a functional gene while allowing mutations in other members of the gene family. Some of these mutations may create new genes that are advantageous to the organism.
- In humans, the globin gene family consists of three α-globin genes and five β-globin genes (see Figure 17.11).
  - Different globins are expressed at different times in development. γ-globin, which is expressed in the fetus, binds oxygen more tightly to ensure oxygen transfer from the mother across the placenta to the fetus.
- Many gene families include pseudogenes, which are inexact copies of genes. Pseudogenes are nonfunctional because they lack promoters and/or recognition sites for intron removal.

Eukaryotic genomes contain many repetitive sequences.

- Highly repetitive sequences are short (less than 100 bp) and are repeated thousands of times in tandem in the genome. They can be densely packed in heterochromatic regions or scattered around the chromosome (as seen for short tandem repeats, STRs).
- The number of repeats varies in individual organisms and provides unique molecular markers that can be used to identify individuals.
- Moderately repetitive DNA sequences include sequences that are repeated 10–1000 times and include tRNA and rRNA genes.
- Multiple copies of the tRNA and rRNA genes are needed to provide the cell with high concentrations of components needed for protein translation.
- In mammals there are four different rRNA molecules: 18S, 5.8S, 28S, and 5S. The first three are all transcribed from one promoter, producing a large precursor RNA, which is enzymatically trimmed to yield the mature rRNA molecules (see Figure 17.12).
- In humans there are 280 copies of these rRNA gene clusters, located on five different chromosomes.
- Transposons, or transposable elements, are moderately repetitive sequences that are not stably integrated into the DNA but can move from place to place in the genome. There are four kinds of transposable elements: SINEs, LINEs, retrotransposons, and DNA transposons.
  - SINEs are *s*hort *in*terspersed *e*lements of up to 500 bp; they are transcribed but not translated.
  - LINEs are *l*ong *in*terspersed *e*lements of up to 7,000 bp; some are transcribed and translated. LINEs make up about 17 percent of the human genome. They include the 300 bp *Alu I* element.
  - SINEs and LINEs (greater than 100,000 copies per genome) make an RNA copy of DNA; this RNA is then used as a template for DNA synthesis, which is inserted at a new site in the genome.
  - Retrotransposons are transposons that make an RNA copy when they move to a different site in the genome. These transposable elements make up about 8 percent of the human genome.
  - DNA transposons do not replicate when they move to a new site on the chromosome, nor do they use RNA as an intermediate. Transposons adversely affect the cell by inserting their DNA into different genes and inactivating them. Transposons can replicate themselves and adjacent chromosomal genes, resulting in gene duplication. Transposons can also pick up adjacent genes during transposition and move them to a new site on the chromosome, potentially creating a new gene that will be advantageous to the organism's survival. Transposons help explain why some mitochondrial and chloroplast genes are located in the nucleus, whereas other organelle genes remain inside the organelle.

What are the characteristics of the human genome?

- Less than 2 percent of the DNA in the human genome contains coding regions (a total of 54,000 genes).
- The average gene has 27,000 base pairs.
- Virtually all human genes have many introns.
- Over 50 percent of the genome contains highly repetitive sequences.
- Almost all genes (97 percent) are the same in all people. Scientists have matched over 7 million single nucleotide polymorphisms (SNPs) in humans.
- Genes are not evenly distributed over the genome.
- Comparisons of genes from different organisms have revealed evolutionary relationships (see Figure 17.4). 95 percent of the human genome is shared with the chimpanzee.

Human genomics has potential benefits in medicine.

- Complex phenotypes are determined by multiple genes interacting with the environment.

- Haplotype mapping uses SNPs that are inherited on individual chromosomes.
- Over 500,000 SNPs can be placed on a chip and used to analyze disease states.
- Statistical measures of association of SNP data can be used to determine increased risk for particular diseases (see Figure 17.15 and Table 17.5).
- SNP testing will eventually be replaced with DNA sequencing.
- Pharmagenomics is the study of how an individual's genome can affect his or her response to drugs or other outside agents (see Figure 17.16).

The proteome is more complex than the genome.

- Because of alternative splicing and posttranslational modifications, the sum total of proteins produced (the proteome) is more complex than the genome (see Figure 17.17A).
- Proteome analysis includes two-dimensional gel electrophoresis (see Figure 17.17B) and mass spectrophotometry.
- Proteome analysis has revealed a common set of proteins that provide the basic metabolic functions of a eukaryotic cell in humans, worms, flies, and yeast.
- The unique proteins in each organism may result from a reshuffling of the same domains that exist in other organisms.

Metabolomics is the study of chemical phenotype.

- Metabolomics is the quantitative description of all the small molecules in a cell or organism.
- Primary metabolites are involved in normal processes such as glycolysis pathways.
- Secondary metabolites unique to particular organisms are often involved in special responses to the environment.
- Patterns of metabolites may be helpful in diagnosing diseases or may provide insights into how plants cope with stress.

Test Yourself

Diagram Exercise

Diagram the steps required to sequence a DNA molecule using the dideoxy sequencing method. In your diagram be sure to list all of the necessary components and technologies that are utilized for DNA sequencing.
Textbook Reference: *17.1 How Are Genomes Sequenced? pp. 368–369*

Knowledge and Synthesis Questions

1. Functional genomics
   a. assigns functions to the products of genes.
   b. assigns functions to regulatory sequences.
   c. compares genes in different organisms to see how those organisms are related physiologically.
   d. Both a and c
   e. All of the above
   Textbook Reference: *17.2 What Have We Learned from Sequencing Prokaryotic Genomes? pp. 371–372*
2. Comparative genomics
   a. assigns functions to the products of genes.
   b. assigns functions to regulatory sequences.
   c. compares genes in different organisms to see how those organisms are related physiologically.
   d. Both a and c
   e. All of the above
   Textbook Reference: *17.2 What Have We Learned from Sequencing Prokaryotic Genomes? p. 372*
3. Which of the following statements is *false*?
   a. Cancer can be caused by the insertion of transposons in genes in somatic cells.
   b. Cancer can be caused by genes that are inherited from one or both parents.
   c. Cancer can be caused by environmental conditions but not by genetic factors.
   d. Cancer can be caused by multiple genes, epigenetic effets, and environmental conditions.
   e. Cancer can be linked to several key SNPs.
   Textbook Reference: *17.4 What Are the Characteristics of the Human Genome? p. 381*
4. Genes that cause cancer when mutated
   a. can be analyzed by means of linked markers such as SNPs.
   b. can be analyzed to determine which cancer treatment will work the best for an individual.
   c. do not have any homologs in other organisms.
   d. can help us predict which individuals are more likely to develop cancer.
   e. a, b, and d
   Textbook Reference: *17.4 What Are the Characteristics of the Human Genome? pp. 381–382*
5. The sequencing of the human genome has allowed scientists to
   a. understand regulatory sequences that are important for gene expression.
   b. locate genes that cause disease.
   c. understand evolutionary relationships by comparing human genes to genes in other organisms.
   d. investigate gene families and their origins.
   e. All of the above
   Textbook Reference: *17.4 What Are the Characteristics of the Human Genome? pp. 380–382*
6. Proteomics has been used to compare
   a. DNA sequences between closely related species.
   b. gene expression during embryonic development.
   c. protein sequences between closely related species.
   d. shotgun cloned sequences.
   e. prokaryotic genomes.
   Textbook Reference: *17.5 What Do the New Disciplines of Proteomics and Metabolomics Reveal? pp. 382–383*

7. Which of the following was *not* a component of the sequencing of the human genome?
   a. Large segments of DNA cloned into bacterial artificial chromosomes (BACs)
   b. cDNA cloning
   c. Computer alignment of overlapping pieces of chromosomes
   d. Sequencing of DNA with dideoxy nucleotides
   e. Bioinformatics

   Textbook Reference: *17.1 How Are Genomes Sequenced? pp. 366–369*

8. Which of the following statements about transposable elements is *false*?
   a. They can inactivate genes into which they are inserted.
   b. They can contain gene sequences.
   c. They are mobile genetic elements that move from RNA molecule to RNA molecule.
   d. They may be spliced out of one region of the genome and inserted into another.
   e. They replicate themselves before moving to another site on the genome.

   Textbook Reference: *17.2 What Have We Learned from Sequencing Prokaryotic Genomes? pp. 372–373*

9. Gene inactivation studies have allowed us to
   a. determine the minimal number of genes humans need to survive.
   b. demonstrate that *C. elegans* needs most of its genes.
   c. investigate the minimal number of genes needed to sustain life.
   d. create artificial life in a test tube.
   e. All of the above

   Textbook Reference: *17.2 What Have We Learned from Sequencing Prokaryotic Genomes? p. 374; 17.3 What Have We Learned from Sequencing Eukaryotic Genomes? p. 376*

10. Comparisons of yeast and bacterial cell genomes have revealed that
    a. yeast cells have more genes devoted to the basic functions of survival than bacteria do.
    b. eukaryotic cells are structurally similar to bacterial cells in terms of complexity.
    c. there are more genes for targeting proteins to organelles in yeast than in bacteria.
    d. the histones of bacteria are very similar to those of yeast.
    e. bacteria and yeast are both haploid.

    Textbook Reference: *17.3 What Have We Learned from Sequencing Eukaryotic Genomes? pp. 375–376*

11. Which one of the following does *not* represent information that we have gained from genome sequencing?
    a. Mycobacteria have many genes devoted to metabolizing fats.
    b. There is extensive genetic exchange between different kinds of bacteria.
    c. Genomes can be sequenced even when organisms cannot be cultured.
    d. Much of the eukaryotic genome contains coding sequences.
    e. The majority of genes in *C. elegans* are cell signaling genes.

    Textbook Reference: *17.2 What Have We Learned from Sequencing Prokaryotic Genomes? p. 373; 17.3 What Have We Learned from Sequencing Eukaryotic Genomes? pp. 375–377; 17.4 What Are the Characteristics of the Human Genome? p. 380*

12. Which of the following was *not* one of the discoveries that resulted from the genome sequencing of plants?
    a. There are more protein-coding genes in animals than plants.
    b. Many *Arabidopsis* genes are duplicated due to chromosomal rearrangements.
    c. There are more genes in plants that are similar to each other than there are genes that are unique.
    d. Plants have many genes whose products are used for defense against microbes and herbivores.
    e. All of the above are true.

    Textbook Reference: *17.3 What Have We Learned from Sequencing Eukaryotic Genomes? p. 377*

13. Which of the following statements about eukaryote retrotransposons is *false*?
    a. They are translated before they are transcribed.
    b. They use an RNA intermediate to move from one region of the genome to another.
    c. They are highly repetitive sequences found throughout the genome.
    d. They can encode gene products that are required for their own transposition.
    e. All of the above are true.

    Textbook Reference: *17.3 What Have We Learned from Sequencing Eukaryotic Genomes? p. 379*

14. Which of the following statements about the human genome is *false*?
    a. Over 50 percent of the genome contains transposons.
    b. Almost every gene has introns.
    c. Genes are evenly distributed over the genome.
    d. About 2 percent of the genome codes for genes.
    e. Humans have about the same number of genes as fruit flies have.

    Textbook Reference: *17.4 What Are the Characteristics of the Human Genome? p. 380*

15. Single nucleotide polymorphisms (SNPs)
    a. can be used to map unlinked genes in order to follow the inheritance of disease traits.
    b. can be used to predict if a patient is at risk for a particular disease.
    c. can be linked in order to generate gene sequences.
    d. have limited sequence variations.
    e. All of the above

    Textbook Reference: *17.4 What Are the Characteristics of the Human Genome? p. 381*

16. The study of proteomes allows scientists to compare
    a. the proteome with the genome to see if the gene sequences are correct.
    b. proteome sequences between species to see if similar proteins are expressed in all species.
    c. transcriptional patterns in different organisms.
    d. highly repetitive DNA sequences in different organisms.
    e. how noncoding regions of the genome differ in different organisms.

    Textbook Reference: *17.5 What Do the New Disciplines of Proteomics and Metabolomics Reveal? pp. 382–383*

Application Questions

1. Analyzing the human genome has revealed many genes that cause disease when mutated. Why have we not been able to eliminate those diseases now that we know the genes that are involved?

   Textbook Reference: *17.4 What Are the Characteristics of the Human Genome? p. 380*
2. Why do different organisms show more similarities in their proteomes than in their genomes?

   Textbook Reference: *17.5 What Do the New Disciplines of Proteomics and Metabolomics Reveal? pp. 382–383*
3. Genomics has revealed that the bacterium that causes tuberculosis has more 250 genes that metabolize lipids. What does this finding suggest about the bacterium and about medicinal approaches to this disease?

   Textbook Reference: *17.2 What Have We Learned from Sequencing Prokaryotic Genomes? p. 373*
4. Suppose you use a radioactively labeled DNA probe to identify a DNA fragment from an *Eco*RI digestion of human DNA. When you complete your hybridization, you notice that the probe has hybridized to several different fragments in the DNA sample. How would you explain this result?

   Textbook Reference: *17.4 What Are the Characteristics of the Human Genome? pp. 380–382*

Answers

Diagram Exercise Answer

1. Synthesize a single-stranded DNA primer complementary to the DNA strand you are sequencing.
2. Add the following to the test tube:
   a. DNA primer
   b. DNA template
   c. All four deoxynucleotides (at a high concentration)
   d. All four dideoxynucleotides, fluorescently tagged (at a low concentration)
   e. DNA polymerase
   f. Buffer/salts
3. DNA synthesis stops when a dideoxynucleotide is incorporated into the replicating strand.

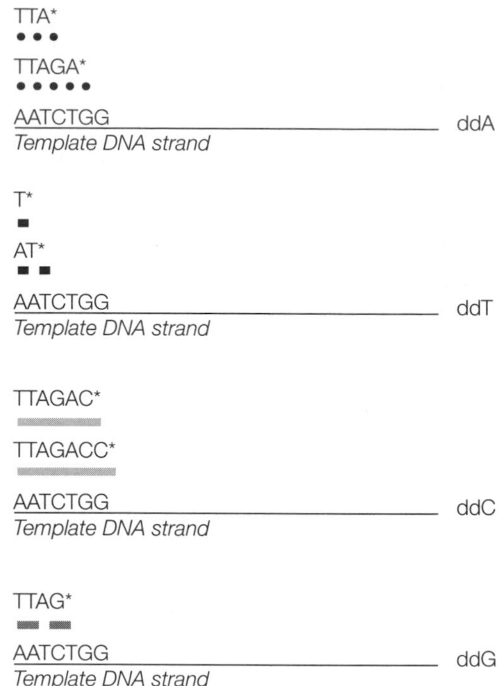

4. The resulting DNA bands are loaded onto gel, subjected to electrophoresis, and then are exposed to a laser beam, which causes them to fluoresce. A detector measures their fluorescence and the data is printed out as peaks that correspond to the individual nucleotides.

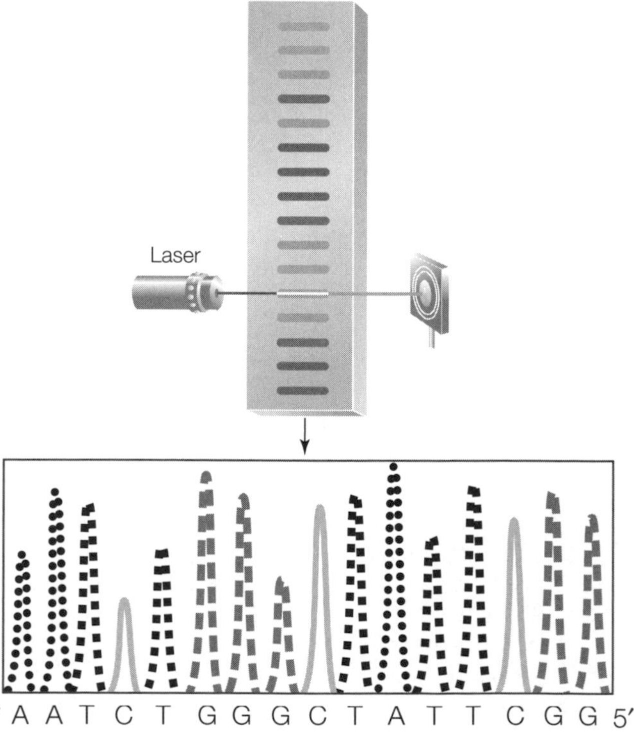

Knowledge and Synthesis Answers

1. **a.** Functional genomics assigns functions to gene products, not to the regulatory sequences. Comparing the genes of different organisms is the work of comparative genomics.
2. **c.** Comparative genomics compares genes between different organisms to see which genes are possessed by one organism but not another. These comparisons can be related to the physiology of the organisms being compared.
3. **c.** Multiple genes, environmental influences, and epigenetic effects all contribute to the development of cancer. Transposon insertion into a somatic gene may also result in cancer, and a small percentage of cancers are inherited. Cancer genes have been linked to many different SNPs.
4. **e.** Mutated genes that are linked to cancer can be linked to particular SNPs, can be used to predict if a person might be susceptible to developing cancer, and are being used to try to determine the best treatment. Those genes do have homologs in other organisms.
5. **e.** Human genome sequencing provides all these advantages.
6. **c.** Proteomics is used to compare protein sequences between different organisms. Answers a, b, and d all refer to techniques used in genomics.
7. **b.** cDNA cloning is used to determine what mRNAs are being expressed in cells. All of the other techniques were utilized to sequence the human genome and order those sequences.
8. **c.** Some transposons do contain gene sequences. They can be spliced out of one region and inserted into another and also replicate themselves and then move to another site. They do not move from one RNA molecule to the next.
9. **c.** Gene inactivation studies have been used in some organisms (not humans) to determine minimal number of genes needed by cells in order to survive. *C. elegans* needs only about 10 percent of its genes to survive. Artificial life has not yet been created in a test tube, but scientists are actively pursuing this possibility.
10. **c.** A comparison of yeast and bacterial cell genomes revealed that the number of genes needed for survival in these two organisms was roughly the same but that yeast cells had many genes for targeting proteins to organelles. Yeast cells are structurally more complex than bacterial cells. Bacterial cells do not have histones. Yeast cells can be haploid or diploid; bacteria are haploid (which had been known long before comparative genomics was developed).
11. **d.** Only 2 percent of the human genome codes for protein-coding genes.
12. **a.** The reverse is true: There are more protein-coding genes in plant cells than in animal cells.
13. **a.** Retrotransposons can move from one region of the genome to another using an RNA intermediate to make a DNA copy of themselves. They are highly repetitive sequences found throughout the genome. Some transposons encode gene products needed for their transposition. Retrotransposons are not translated before they are transcribed.
14. **c.** Genes are not evenly distributed over the genome; some chromosomes have many more genes than other chromosomes.
15. **b.** SNPs that are linked to a disease gene can be used to analyze patients' DNA samples to determine if they are at risk for that disease. SNPs are quite variable. They are small sequences and are not used to generate gene sequences.
16. **b.** Proteomics studies can be used to compare proteins in different organisms. The gene sequence will determine the protein sequence. Proteomic studies cannot be used to study transcriptional patterns, which are cellular processes involving RNA. Proteins are coding regions of the genome and highly repetitive sequences do not code for protein.

Application Answers

1. We have not been able to eliminate genetic diseases because, except in a few rare cases, we are not able to replace the mutated genes in the diseased individual. If those genes play a role in development, they need to be identified and replaced in the early embryo.
2. More sequence similarities are revealed when proteomes are compared than when genomes are compared because the genetic code is redundant. There can be more than one codon for a particular amino acid, so the DNA sequences can vary more and still generate the same amino acid sequences.
3. The tuberculosis bacterium must use lipids as a source of energy-rich compounds; inhibiting lipid synthesis in this bacterium may inhibit the growth of this bacterium.
4. The DNA sequence of the probe is not for a unique gene in humans and may be part of a gene family or a repetitive sequence.

18 Recombinant DNA and Biotechnology

The Big Picture

- The ability to isolate DNA from any organism, ligate it to vector DNA, introduce that DNA into host cells, and propagate those cells has had an enormous impact on our understanding of genetics, molecular biology, and cell function and development. These techniques have been used to elucidate evolutionary relationships among different organisms and to understand gene regulation and function in greater depth. Applications of these techniques have been used to develop new medicines and diagnostics, agricultural products, and powerful forensic tools.

Common Problem Areas

- There are a variety of ways to clone DNA fragments, and trying to remember all the different cloning methods can be challenging. Consider the following as you review the different cloning procedures: the size of the cloned DNA, the host cell in which the DNA should be cloned, and the expression of the cloned DNA in the host cell. Different vectors can be used in different host cells to address each of these considerations.
- Many different experimental questions can be answered using cloning. Ask yourself what cloning strategies could be used to answer the following questions, and review the textbook for answers: What is the sequence of a gene? What sequences are important for regulation of that gene? What sequences are important for targeting that gene to a particular site in a eukaryotic cell? What is the difference in function between a mutant gene product and a wild-type gene? What kinds of genes are expressed during the development of an organism? How can cloned genes be expressed in plants or in the milk of mammals?

Study Strategies

- Outline the specific steps needed to clone a gene in a bacterial cell. Include how the gene is initially isolated, what kind of vectors can be used, how to introduce the recombinant DNA into the cell, and how to confirm that the recombinant molecule is in the host cell. Then outline the steps needed to clone a gene in a eukaryotic cell.
- Complementary base pairing is important for many aspects of biotechnology. Describe each technique in gene cloning that uses complementary base pairing, detailing specifically how base pairing is involved.
- Go to yourBioPortal.com to review the following tutorial and activity:

  Animated Tutorial 18.1 DNA Chip Technology

  Web Activity 18.1 Expression Vectors

Important Concepts

Recombinant DNA

- DNA ligase (the enzyme that covalently joins the Okazaki fragments during DNA replication and mends broken DNA; see Chapter 13) is used to form a covalent bond on each DNA strand of the recombinant molecule (see Figure 18.2).
- Restriction endonucleases recognize palindromic sequence in DNA and cut at or near those sites.
- Some restriction endonucleases cut between the same bases on both DNA strands, producing blunt ends.
- When restriction endonucleases cut DNA, they often leave ends that have 5′ or 3′ overhangs of single-stranded DNA. These ends are called "sticky ends" and can form complementary base pairs with other DNA molecules that have the same sticky ends (see Figure 18.2).
- DNA fragments cut with the same restriction enzyme can be joined together by ligase, even if they are from different species.

How are genes inserted into cells?

- Recombinant DNA is inserted into a host cell by transformation (or transfection, if the host cell is an animal cell), creating a transgenic cell or organism.
- Selectable markers such as genes that confer resistance to antibiotics are often included on the recombinant DNA molecule.

Genes can be cloned into prokaryotes or eukaryotes.

- Prokaryotes have been used to clone many genes. Eukaryotic gene expression is studied by cloning genes in eukaryotic cells or multicellular organisms.

- Yeast cells have a rapid cell division cycle (2–8 hours), are easy to grow, have a small genome size (12 million base pairs), and have been used to clone many eukaryotic genes.
- Many plant cells are totipotent. They can be grown in culture, transformed with recombinant DNA, and manipulated to form an entire new transgenic plant containing the recombinant DNA molecule.

Recombinant DNA must be replicated to be maintained in the host cell.

- Recombinant DNA can be inserted into the host chromosome, where it is replicated when the chromosome is replicated. Such insertion might happen randomly after the DNA is introduced into the cell. Alternatively, recombinant DNA molecules can enter the host cell by being part of a vector that already has an origin of replication.
- Vectors can be plasmids or viruses. Plasmids must be able to replicate independently, have restriction sites, selectable markers, and a small size.
- Plasmid vectors in *E. coli* are small (from 2,000 to 6,000 base pairs) circular DNA molecules. They have a single set of unique restriction sites where the cut DNA is inserted; a drug resistance marker allows the investigator to confirm the presence of the plasmid in the *E. coli* cell. Plasmids synthesize their DNA independently from their own origins of replication, making many copies of plasmid DNA per cell.
- Both prokaryotic and eukaryotic virus vectors are used to clone larger DNA sequences (up to 20,000 base pairs of inserted DNA). Viruses infect cells naturally, allowing easy entry of cloned sequences into the cytoplasm of the cell.
- *Agrobacterium tumefaciens*, a bacterium that causes crown gall in plants, harbors a Ti plasmid. This plasmid contains a transposon, T DNA, which inserts copies of itself into the host plant cell's DNA when the bacteria infect the plant (see Figure 18.3).

Reporter genes identify host cells that contain recombinant DNA.

- Selectable markers such as markers for antibiotic resistance can be used to determine if the host cell contains the recombinant DNA molecule (see Figure 18.4).
- Reporter genes such as β-galactosidase and green fluorescent protein (GFP) are selectable markers that can be used for detection of expression of recombinant DNA molecules in host cells. Reporter genes can be attached to promoters of gene coding regions to follow transcription or protein localization in cells.

Sources of DNA for cloning

- Genomic libraries are a collection of DNA fragments from the entire genome of an organism.
- Genomic libraries of DNA are usually cloned into bacteriophage λ because each virus can accommodate 20,000 base pairs of DNA.
- To make cDNA (complementary DNA), the messenger RNA must first be isolated from the cell. A DNA molecule complementary to that RNA is synthesized using reverse transcriptase, and the new molecule can then be cloned. (see Figure 18.6).
- cDNA libraries consist of all the genes that are being transcribed in a particular tissue.
- cDNA clones are used to compare gene expression in different tissues at different stages of development. One-third of all genes in an animal are expressed only during prenatal development.
- Synthetic DNA can be made using PCR to amplify a specific sequence.
- Artificial genes can be made using the amino acid sequence of the gene. Sequences for transcriptional and translational initiation and termination can be added to the gene sequence.
- Mutant genes can be synthesized and compared to wild-type genes to analyze gene function.

Other techniques can be used to study DNA.

- Gene function can be eliminated by creating knockout genes, which are the result of a homologous recombination event in the cell in which a normal gene is replaced by an inactivated form of the gene. The inserted DNA carries a reporter gene, so its expression can be followed in the host's cells.
- In constructing transgenic mice, a plasmid containing the inactivated marker gene is transfected into a mouse stem cell (see Figure 18.7). If recombination occurs, the marker gene will be expressed. The transfected stem cell is then transplanted into an early mouse embryo, and the resulting phenotype is analyzed.

Antisense messenger RNA and RNAi can prevent the translation of specific genes.

- Antisense RNA will base pair with mRNA in the cytoplasm and form a double-stranded RNA molecule, which cannot be translated and will be degraded by the cell.
- Interference RNAs, known as small interfering RNAs (RNAi), are short (about 20 nucleotides) double-stranded RNA molecules that bind specific mRNAs and target them for degradation.
- Small inhibitory RNAs are more effective at inhibiting translation than antisense RNAs are.

DNA microarrays reveal RNA expression patterns.

- Microarrays are small glass chips containing thousands of copies of different DNA sequences per chip. These sequences (greater than 20 base pairs long) are attached to the chip in a precise order.
- For example, cDNA copies of cellular mRNA are synthesized and amplified using PCR (see Chapter 13). Those cDNA molecules are coupled with fluorescent dyes and allowed to hybridize with the DNA chips. The chip is then exposed to fluorescent light to determine

which sequences formed a hybrid with the cDNA (see Figure 18.9).

- Chip technology has been used to look at gene expression from different breast cancers to predict the prognosis for patients and determine treatments (see Figure 18.10).

Biotechnology

- Biotechnology is the use of living cells to produce useful materials for people, including food, medicines, and chemicals.
- For cells to produce a cloned gene product, the vector (an expression vector) must have DNA sequences that allow the cloned gene to be expressed.
- Prokaryotic expression vectors require a promoter, a termination site for transcription, and a ribosome-binding site.
- Eukaryotic expression vectors require the same elements as well as a poly A–addition site, transcription factor binding sites, and enhancers.
- Modifications of expression vectors include the addition of inducible promoters (which respond to a specific signal), tissue-specific promoters, and signal sequences.

How biotechnology is changing medicine, agriculture, and the environment

- Medically useful products that have been cloned in expression vectors include tissue plasminogen activator, human insulin, and vaccine proteins (see Table 18.1).
- Recombinant DNA offers breeders the opportunity to choose specific genes that will be incorporated into an organism, to introduce any gene into a plant or animal species, and to generate new organisms quickly.
- Transgenic plants have been created that express toxins for insect larva, have resistance to herbicides, produce extra nutrients, or can grow in high-salt conditions.
- Recombinant bacteria have been used to clean up the environment in composting programs, wastewater treatments, oil spills, and other such efforts.
- The creation of transgenic plants has raised some concerns that these crops could be unsafe for human consumption, that it is unnatural to interfere with nature, and that transgenes could escape into other noxious plants. Transgenic plants are extensively field tested, and scientists are examining these issues and proceeding cautiously.

Test Yourself

Diagram Exercise

1. a. You are working at a biotech company and your project is to clone a eukaryotic gene (*X*) so that you can isolate large amounts of protein X. Diagram the steps you would take to successfully complete this project.

   b. You have your clone but you realize that protein X, which is being made from this clone, is nonfunctional. What do you have to change in your procedure to clone gene *X* so the expressed protein will be functional?

   Textbook Reference: *18.2 How Are New Genes Inserted into Cells? pp. 389–391*

Knowledge and Synthesis Questions

1. Cloning a gene may involve
   a. restriction endonucleases and ligase.
   b. plasmids and bacteriophage λ.
   c. transformation or transfection.
   d. selectable markers and/or reporter genes.
   e. All of the above

   Textbook Reference: *18.1 What Is Recombinant DNA? p. 387*

2. Complementary base pairing is important for
   a. ligation reactions with blunt-end DNA molecules.
   b. hybridization between DNA and transcription factors.
   c. restriction endonucleases for cutting cell walls.
   d. synthesizing cDNA molecules from mRNA templates.
   e. the transcriptional activation of expression vectors.

   Textbook Reference: *18.3 What Sources of DNA Are Used in Cloning? pp. 392–393*

3. For a prokaryotic vector to be propagated in a host bacterial cell, the vector needs
   a. an origin of replication.
   b. telomeres.
   c. centromeres.
   d. drug-resistance genes.
   e. reporter genes.

   Textbook Reference: *18.2 How Are New Genes Inserted into Cells? p. 389*

4. For a Ti plasmid to be propagated in a host plant cell, the vector needs
   a. telomeres.
   b. centromeres.
   c. an origin of replication.
   d. a reporter gene.
   e. a, b, and c

   Textbook Reference: *18.2 How Are New Genes Inserted into Cells? p. 390*

5. Reporter genes include genes for
   a. drug resistance.
   b. bioluminescence.
   c. DNA origins.
   d. restriction endonucleases.
   e. Both a and b

   Textbook Reference: *18.2 How Are New Genes Inserted into Cells? pp. 390–391*

6. Vectors include
   a. bacterial plasmids.
   b. viruses.
   c. plant plasmids.

d. bacteriophage λ.
e. All of the above

Textbook Reference: *18.2 How Are New Genes Inserted into Cells? pp. 389–390*

7. A cDNA clone is
   a. mostly cytosine.
   b. a copy of the DNA identical to the nuclear gene.
   c. a copy of noncoding DNA.
   d. a DNA molecule complementary to an mRNA molecule.
   e. a fragment of DNA inserted into the host chromosome.

   Textbook Reference: *18.3 What Sources of DNA Are Used in Cloning? p. 392*

8. Gene expression can be inhibited by
   a. antisense RNA.
   b. knockout genes.
   c. DNA microarrays.
   d. microRNA.
   e. a, b, and d

   Textbook Reference: *18.4 What Other Tools Are Used to Study DNA Function? pp. 394–395*

9. Expression vectors are different from other vectors because they contain
   a. drug-resistance markers.
   b. telomeres.
   c. regulatory regions that permit the cloned DNA to produce a gene product.
   d. DNA origins.
   e. reporter genes that are expressed in the host.

   Textbook Reference: *18.5 What Is Biotechnology? pp. 397–398*

10. RNAi
    a. is more effective than antisense RNA in inhibiting translation.
    b. inhibits transcription in eukaryotes.
    c. is produced only by viruses.
    d. requires other proteins to modify the RNAi.
    e. Both a and d

    Textbook Reference: *18.4 What Other Tools Are Used to Study DNA Function? p. 395*

11. DNA chip technologies can be used to
    a. predict who will get cancer.
    b. show transcriptional patterns in an organism during different times of development.
    c. clone DNA.
    d. make transgenic plants.
    e. inhibit transcription of disease genes.

    Textbook Reference: *18.4 What Other Tools Are Used to Study DNA Function? pp. 395–397*

12. Which of the following is *not* an application of recombinant DNA technology?
    a. Generating large amounts of tissue plasmonigen factor to help dissolve blood clots
    b. Making plants more resistant to insect larva
    c. Creating vaccines for pathogens
    d. Reducing the salt in environmental soil so that plants can grow
    e. Creating bacteria that can accelerate the breakdown of wood chips and paper

    Textbook Reference: *18.6 How Is Biotechnology Changing Medicine, Agriculture, and the Environment? pp. 400–402*

13 A disadvantage of recombinant DNA technology is that it
    a. has the capacity to spread transgenes from crops to other species.
    b. makes weeds herbicide-resistant.
    c. creates transgenic plants that kill beneficial insects.
    d. creates transgenic plants that could be harmful for human consumption.
    e. All of the above

    Textbook Reference: *18.6 How Is Biotechnology Changing Medicine, Agriculture, and the Environment? pp. 402–403*

14. Expression vectors
    a. are useful for analyzing RNA transcription patterns.
    b. are useful for isolating large amounts of DNA.
    c. can be used to make large amounts of protein in *E. coli* cells.
    d. are useful only in prokaryotic cells, and cannot be expressed in eukaryotic cells.
    e. are used to create SNPs for DNA microarrays.

    Textbook Reference: *18.5 What Is Biotechnology? p. 397*

15. Which of the following is minimally required for a transgene to be expressed in a eukaryotic host?
    a. A promoter, a transcriptional termination site, and a ribosome binding site
    b. A DNA origin and an antibiotic resistance marker
    c. Transcription factor binding sites, enhancers, and a poly A–recognition sequence
    d. An inducible promoter and repressor proteins
    e. a, c, and d

    Textbook Reference: *18.5 What Is Biotechnology? pp. 397–398*

Application Questions

1. Describe three useful products that have been produced using biotechnology. Outline two specific dangers that could result from producing organisms that contain foreign genes.

   Textbook Reference: *18.6 How Is Biotechnology Changing Medicine, Agriculture, and the Environment? pp. 398–402*

2. Suppose that your lab assistant has cloned gene *X* into yeast and confirmed that the recombinant DNA molecule is present in the yeast cells. However, the yeast cell is unable to synthesize X protein. Suggest why this part of the experiment is not working and what modifications are needed in the cloning procedure.

   Textbook Reference: *18.5 What Is Biotechnology? pp. 397–398*

3. What kind of techniques can be used to study gene expression during development?
   Textbook Reference: *18.3 What Sources of DNA Are Used in Cloning? pp. 392–393; 18.4 What Other Tools Are Used to Study DNA Function? pp. 395–397*
4. Hybridization is a useful technique in biotechnology. Describe three ways hybridization is used experimentally in DNA recombinant technology.
   Textbook Reference: *18.1 What Is Recombinant DNA? p. 388; 18.3 What Sources of DNA Are Used in Cloning? pp. 392–393; 18.4 What Other Tools Are Used to Study DNA Function? pp. 395–397*

Answers

Diagram Exercise Answer

a. You would use an expression vector that has a promoter, a transcriptional termination site and a ribosome binding site that will be recognized by the host cell (*E. coli*), so that the protein will be expressed in that host cell.

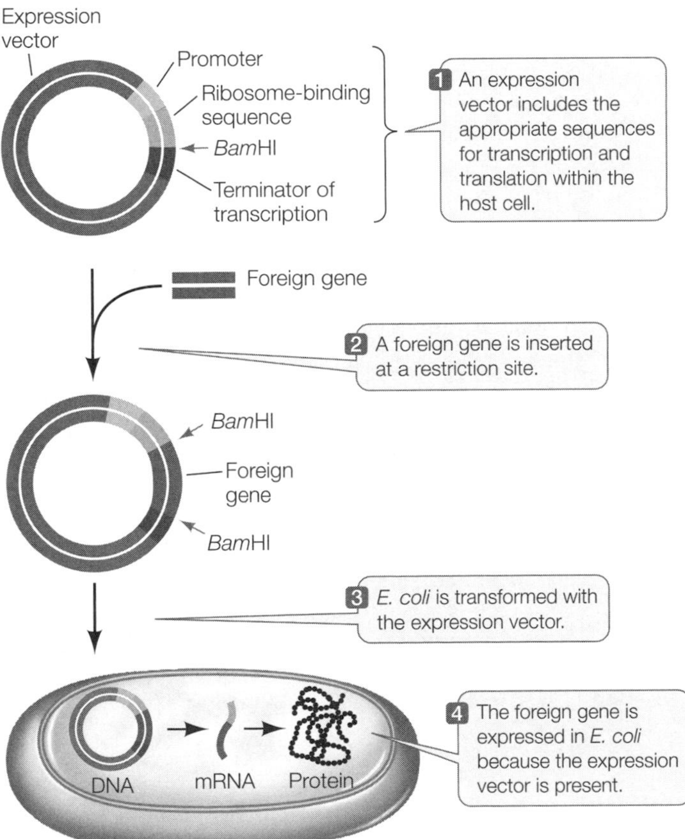

Steps:

1. Isolate DNA containing gene X from your cell sample.
2. Create a genomic library with DNA.
3. Using a probe for gene X, select a colony containing gene *X* from the genomic library.
4. Isolate DNA from the selected clone and cut with restriction endonuclease.

(or)

1. Isolate DNA containing gene *X* from your cell sample.
2. Utilizing the PCR technique and primers flanking gene *X*, amplify gene *X*.
3. Purify the PCR fragment and cut with restriction endonuclease.
4. Cut expression vector with same restriction endonuclease as gene *X*.
5. Ligate gene *X* into expression vector.
6. Transform *E. coli* with expression vector.
7. Transcriptionally activate gene *X* and isolate protein X from *E. coli* cells.

b. The reason that protein X is nonfunctional is that you cloned gene *X* from genomic DNA, which still had introns in the gene sequence.You should use a cDNA clone (which is made from the mRNA from gene *X* and thus lacks introns) as a source for gene *X*.

Knowledge and Synthesis Answers

1. **e.** Cloning a gene requires restriction endonucleases to cut the gene of interest, vectors, including plasmids and bacteriophage λ, and ligase to covalently join the DNA to the vector. Transformation or transfection is needed to introduce the vector into the host cells and reporter genes and selectable markers can be used to detect the presence of the gene in the host cells.
2. **d.** No complementary base pairing can occur between blunt-end cut DNA molecules. Transcription factors are proteins that bind DNA through interactions with their side chains (which are amino acids) and the nucleotides of the DNA. Restriction endonucleases cut double-stranded DNA, not cell walls. The activation of transcription vectors does not require complementary base pairing
3. **a.** A prokaryotic vector needs an origin of replication to be propagated in a prokaryotic cell.
4. **c.** The Ti plasmid requires an origin of replication but not telomeres or centromeres to propagate itself inside the plant cell.
5. **e.** Reporter genes include genes for drug resistance and bioluminescence.
6. **e.** All of these molecules can serve as vectors for cloned DNA.
7. **d.** cDNA clones are not clones that contain mostly cytosine, nor are they copies of noncoding genes. A cDNA clone is generated by making a DNA copy of a particular messenger RNA using reverse transcriptase. The cDNA clone is not identical to the nuclear gene because in the messenger RNA (which served as a template for the cDNA) the introns have been

removed, leaving coding sequence and 5′ and 3′ flanking sequences.

8. **e.** Gene expression can be inhibited by antisense RNA, which complementarily base pairs with the target messenger RNA, making it inaccessible to the translation machinery in the cell. MicroRNAs bind mRNAs and target them for degradation. Knockout genes are genes that have been inactivated by the insertion of DNA into their coding sequences. DNA microarrays are used in hybridization experiments and do not inhibit gene expression.
9. **c.** Expression vectors may contain drug-resistance markers and reporter genes, and they must contain DNA origins, but none of these features distinguishes them from other vectors. Expression vectors are unique because they contain regulatory sequences that allow the cloned gene to be expressed in the host cell.
10. **e.** RNAi is more effective than antisense RNA at inhibiting translation (not transcription). It is produced by viruses and by eukaryotic cells in small amounts and requires other modifying proteins.
11. **b.** DNA chips can be used to analyze gene expression at different times in development and to predict if an individual is at risk of developing cancer. Other factors (e.g., environmental) also determine if a person will get cancer. They are not used to make transgenic plants, to clone DNA, or to inhibit transcription of disease genes.
12. **d.** Soils have not been made less salty using biotechnology; plants have been made more salt-tolerant by means of recombinant DNA techniques.
13. **e.** All of the problems listed are disadvantages of recombinant DNA technology.
14. **c.** Expression vectors have the appropriate control regions that allow them to be expressed (make protein) in both prokaryotic and eukaryotic host cells. They are not used to create SNPs, nor are they used to isolate DNA or study transcription.
15. **e.** A promoter, a site for transcription factors, enhancers, a transcriptional termination sequence including a poly A–sequence, and a ribosome site are minimally required for an expression vector to produce its cloned gene product in a host eukaryotic cell. Inducible promoters and repressors will regulate gene expression but are not required for expression. A tissue-specific promoter will cause the gene to be expressed in particular cells (and would be required for the expression vector to work in a particular tissue), and signal sequences will target the protein to particular compartments in the cell, but they are not part of the minimal requirements for expression.

Application Answers

1. Useful products include rice grains that produce β-carotene, plants that are resistant to herbicides and insect larvae, the production of human growth hormone in cow's milk, and others (see Tables 18.1 and 18.2). Dangers include the creation of genetically engineered foods that could adversely affect human nutrition, the transfer of herbicide- and insect-resistant genes from crop plants to noxious weeds, and the introduction into the wild of new organisms that might have unforeseen ecological consequences.
2. The *X* gene can be cloned into a yeast cell, but unless the vector has the appropriate regulatory signals (promoters, poly A–addition sites, translational initiation, and termination signals), no expression of gene *X* will occur. Recloning gene *X* in a yeast expression vector will result in the expression of the *X* gene.
3. cDNA libraries can be made from different developing tissues in order to see what genes are being expressed. DNA microarrays can also be used to analyze transcriptional patterns during development.
4. Hybridization of complementary base pairs allows ligase to seal DNA fragments with the same sticky ends to create a recombinant molecule. Hybridization of newly made DNA to mRNA allows reverse transcriptase to generate cDNAs. Hybridization is also used in DNA microarray technology.

19 Differential Gene Expression in Development

The Big Picture

- The development of a mature differentiated organism from a fertilized egg involves the differential activation of genes in response to environmental signals. In many organisms, positional determinants may be present in different regions of the egg due to maternal factors. The asymmetric distribution of these factors will activate genes differentially, and as cell division proceeds, that asymmetric gene expression continues. As a result, different cells experience different environmental signals based on their position in the developing embryo, and they respond by activating different genes. This differential gene activation determines the fate of the cell.
- The zygote and early embryonic cells are totipotent; they can develop into any structure in the adult organism. Cells of the inner cell mass of the blastocyst are pluripotent, and adult stem cells are multipotent. As development proceeds, the developmental potential of embryonic cells narrows. In the adult organism, differentiated cells express particular genes that give those tissues a particular structure and function, even though all the genes are still present in the nucleus of those cells.
- The concentration of specific genes and their products are involved in the regulation of polarity during development. These genes are classified as maternal effect genes, and their products are supplied by the mother.

Common Problem Areas

- Sometimes it is difficult to visualize how a single cell can divide and grow to produce a mature functional organism with highly differentiated tissues. At some very early point in development (either before fertilization or in one of the first set of divisions), different genes begin to be expressed in different cells. This gene expression can be in response to cytoplasmic factors or signals from other cells. Early gene expression sets up positional determinants that activate another wave of genes that further divides regions of the embryo into different developmental areas. As cell division proceeds in the embryo, these developmental genes continue to be differentially expressed, resulting in differentiation and morphogenesis in the organism.

Study Strategies

- Transplantation experiments help explain totipotency. Describe the transplantation experiments in frogs using early embryonic tissues and those using adult somatic cells to address the totipotency of those tissues. Make sure that you understand the different potentials of totipotent, pluripotent, and multipotent stem cells.
- Review the experiments that demonstrated how tissue induction causes the lens in the vertebrate eye to form.
- Review some of the important genes whose products direct development in the organisms discussed in this chapter. Identify homologs from different organisms.
- Go to yourBioPortal.com to review the following tutorials and activity:

  Animated Tutorial 19.1 Embryonic Stem Cells

  Animated Tutorial 19.2 Early Asymmetry in the Embryo

  Animated Tutorial 19.3 Pattern Formation in the *Drosophila* Embryo

  Interactive Tutorial: Cell Fates: Genetic versus Environmental Influences

  Web Activity 19.1 Stages of Development

Important Concepts

Development includes four processes: determination, differentiation, morphogenesis, and growth.

- During development, an organism progresses through successive forms as it moves through its life cycle. The embryo is an early stage of development that is nourished directly, via a placenta, or indirectly, via nutrients stored in the egg or seed.
- The developmental fate of a cell is set during determination and is influenced by internal and external conditions. The fate of a cell is set by differential gene expression and morphogenesis.

- During differentiation, cells become specialized to contain specific structures and to perform particular functions.
- Morphogenesis is the creation of form as seen in body shape and organs. It is a result of pattern formation during development, which helps direct differentiated tissues to form specific structures. Cell movements are also important in morphogenesis.
- Growth is an increase in size due to cell division (an increase in the number of cells) or cell expansion (an increase in the size of existing cells). Cell division is important in the development of plants and animals; cell expansion is particularly important in the development of plants.
- Programmed cell death (apoptosis) is an important part of development in both plants and animals.
- Transplantation experiments have shown that the environment in which early embryonic cells exist can redirect them along different developmental paths.
- Embryonic cells eventually become committed to a particular developmental fate, even though they are not yet differentiated. This is because their fate has been determined. When these cells are transplanted to other locations in the embryo, they continue to develop into the original differentiated tissue, regardless of their environment (see Figure 19.2).
- Cell fate becomes apparent as cells differentiate.

Some cells, such as a zygote, are totipotent, but their cellular descendants lose this ability as they develop.

- Early embryonic cells are totipotent: they can develop into any of the different kinds of cells in the mature organism. Developmental possibilities narrow as cell determination and differentiation occur.
- In plants, differentiation is reversible in some cells. Under certain conditions, plant cells in culture can give rise to a new, genetically identical plant (a clone). Plant cells first dedifferentiate, and then produce a callus (mass of cells), which develops into a plant embryo when placed in the appropriate medium. Thus, the original differentiated cells contained all the genetic information needed to express genes in the correct sequence to generate a whole plant.
- Initially, the fertilized egg or zygote of an animal is totipotent and can give rise to every type of cell in the adult body. Cell descendants lose their totipotency as they become committed to a developmental path. A cell's fate first becomes determined, and then the cell differentiates to form specialized cells.
- The totipotency of early embryonic animal cells has been shown in experiments in which nuclei isolated from frog embryos were fused to enucleated eggs, resulting in the development of tadpoles and mature frogs. Such experiments show that cytoplasm has a large influence on the fate of a cell.
- The principle of genomic equivalence states that no information is lost from the nuclei of cells as they pass through the early stages of development.
- Mammals can be cloned by fusing somatic cells with enucleated eggs and implanting the resulting embryos in surrogate mothers. The cloning of mammals from somatic cells has been accomplished by manipulating the donor cells so that they are in G1 of the cell cycle. Upon fusing with the enucleated egg, the donor cells are stimulated by cytoplasmic factors in the egg to enter the S phase. Several mammals, including sheep, mice, and cattle, have been cloned by the technique of nuclear transfer.

Under the right environmental conditions, cells differentiate.

- The areas of undifferentiated cells in the growing tips of plant roots are known as meristems. These cells can become any cell type found in the root or stem.
- Undifferentiated dividing cells in animals are known as stem cells. Stem cells in adult animals are specific for the kinds of tissue they replace, typically skin, the lining of the intestines, and blood cells. These stem cells are multipotent, meaning that they can differentiate into a limited number of cell types. Growth factors and adjacent cells influence the differentiation of stem cells.
- Bone marrow contains two types of multipotent stem cells: hematopoietic and mesenchymal. Hematopoietic stem cells produce red and white blood cells, whereas mesenchymal stem cells produce bone and muscle cells. In hematopoietic stem cell transplantation, stem cells are removed from the blood, stored, and encouraged to increase in number while the patient receives cancer treatments, and then returned to the patient once the treatments have been completed.
- In embryonic mammals, cells from a part of the blastocyst (the inner cell mass) are pluripotent, which means they are capable of forming nearly every type of cell. These pluripotent cells are more restricted than totipotent cells because they cannot form cells of the placenta.
- Future medical applications include the use of pluripotent stem cells. These cells could be taken from human embryos made available through in vitro fertilization or by making induced pluripotent stem cells from skin cells. The latter technique does not destroy human embryos or provoke an immune response in recipients.

Differentiated cells retain all their original genetic content, even though they express only a very small subset of genes.

- Scientists use nucleic acid hybridization techniques to look for the presence of genes in cells. These experiments have shown that even though a specific gene is not expressed and not present as mRNA, it is still present in the nuclear DNA.

- Transcription factors are proteins that bind DNA and regulate the expression of genes.
- In response to the transcription factor MyoD (myoblast-determining gene), undifferentiated muscle precursor cells stop dividing; this step is necessary for the eventual differentiation of these cells into mature muscle cells.
- Sometimes a single transcription factor causes a cell to differentiate. In other cases, several different transcription factors promote differentiation.

Unequal distribution of cytoplasmic factors in eggs, zygotes, and early embryos establishes the initial pathways for precursor cells in developing embryos.

- Cytoplasmic segregation of factors in the egg results in an unequal distribution of maternal elements in each of the cells of the embryo.
- Unequal distribution of cytoplasmic determinants directs embryonic development and controls the polarity of the organism. For example, if a sea urchin embryo is cut in half at the eight-cell stage and split from left to right by that cut, the resulting larvae are normal, but small. If the sea urchin is cut so that its upper half is split from its lower half, the cells from the upper half do not develop at all, and the cells from the lower half develop into abnormal larvae (see Figure 19.8).

Cells or tissues in developing embryos can induce other cells or tissues to follow a particular developmental path.

- Cells can induce other cells to differentiate by secreting induction factors.
- The lens of the vertebrate eye forms when a portion of the forebrain bulges outward, forming the optic vesicle, which then contacts cells at the surface of the head. The optic vesicle induces the surface cells to form a lens placode, which eventually develops into the lens. If a barrier is placed between the optic vesicle and surface cells in a developing embryo, then no lens develops. If the optic vesicle is cut out before it can contact the surface cells, then no lens develops. Under normal conditions, the developing lens moves away from the surface tissue and induces it to form the cornea (see Figure 19.10).
- The nematode *Caenorhabditis elegans* is a model organism used in developmental studies. The egg of *C. elegans* develops into a larva in 8 hours and an adult in 3.5 days. The animal has a transparent body, making it easy to follow the developmental fate of its 959 somatic cells. The adult is hermaphroditic, having female and male reproductive organs. Eggs are laid through a pore called the vulva. The vulva is induced to form from a single cell called an anchor cell. If this cell is destroyed, no vulva develops.
- The anchor cell determines the fates of six cells on the ventral surface of *C. elegans* by producing a primary inducer (LIN-3 protein) that diffuses toward adjacent cells, establishing a concentration gradient. The closest cell is exposed to the highest concentration of LIN-3 and becomes the primary precursor cell. It produces a secondary inducer (known as the lateral signal) that causes the next closest cells to become secondary precursors. Descendants of primary and secondary precursor cells form the vulva. The final three cells (which are farthest from the anchor cell) become epidermal cells (see Figure 19.11).
- LIN-3, the primary inducer for the development of the vulva in *C. elegans*, is homologous to mammalian epidermal growth factor. When LIN-3 binds to a receptor on the surface of the closest cell, it causes a signal transduction cascade involving the Ras protein and MAP kinases. The end result of this cascade is the differentiation of vulval cells.
- Molecular switches, such as inducers, allow a developing cell to proceed down one of two alternative pathways.

Pattern formation is the spatial organization of tissue that results in the appearance of body form (morphogenesis).

- During morphogenesis, some cells are programmed to die by death genes through the process of apoptosis.
- As *C. elegans* develops from a fertilized egg into an adult, 1,090 cells are produced. However, 131 of these cells are programmed to die due to the sequential expression of *ced-4* and *ced-3* genes. *Ced-9* inhibits the actions of *ced-3* and *ced-4* during development.
- In human embryos, apoptosis occurs in the webs of skin that initially form between fingers and toes. Caspases, a group of enzymes homologous to *ced-3* in *C. elegans*, cause this apoptosis. The human homolog to *ced-9* is Bcl-2.
- The genes controlling apoptosis are crucial for proper development in many organisms. These genes have been conserved in organisms separated by 600 million years of evolution (nematodes and humans).
- During plant development, organs such as leaves, roots, and flowers are produced. Flowers have four types of organs: sepals, petals, stamens, and carpels. Stamens are male reproductive organs and carpels are female reproductive organs. Flower organs occur in whorls organized around a central axis and are derived from meristematic tissue on the plant (see Figure 19.14).
- Plant geneticists have studied the development of flower organs in *Arabidopsis*. There are four whorls of organs in *Arabidopsis*. Three genes expressed in the whorls act as organ identity genes to guide differentiation in each whorl. Gene A is expressed in whorls 1 and 2, which form the sepals and petals, respectively; gene B is expressed in whorls 2 and 3, which form petals and stamens, respectively; and gene C is

expressed in whorls 3 and 4, which form stamens and carpels, respectively.

- The three genes, A, B, and C, all encode transcription factors that are active as dimers (proteins with two polypeptide subunits). Gene regulation is combinatorial, and the combination of different dimers determines which genes are activated. A dimer of the transcription factor encoded by gene A will activate genes that make sepals. A dimer of transcription factor A with transcription factor B will result in petals.
- LEAFY is a transcription factor that regulates the transcription of the genes A, B, and C. Plants bearing a mutation in *LEAFY* can produce leaves but no flowers.
- In plants, cell fate is often determined by MADS box genes.
- The homeotic genes, A, B, and C, as well as *LEAFY*, may be able to be genetically modified so that the flower organs can produce more fruit and seeds (which come from the carpels).
- Positional information allows developing cells to determine where they are in the developing organism. Morphogens are signals that establish positional information. They act directly on the target cell, and differential concentrations within the embryo cause different effects.
- Concentration gradients of the morphogen Sonic hedgehog, secreted by the zone of polarizing activity in the limb bud, determine the anterior–posterior axis of the developing vertebrate limb. According to the "French flag" model, cells in the limb bud form different digits depending on the concentration of the morphogen.

A cascade of transcription factors controls development in *Drosophila*.

- The body of *Drosophila* is segmented, consisting of a head (which is formed from fused segments), three thoracic segments, and eight abdominal segments.
- The first step of development in *Drosophila* is to establish anterior–posterior and dorsal–ventral polarity. Polarity is based on the cytoplasmic distribution of mRNA and proteins produced by maternal effect genes.
- Mutations in maternal effect genes (*Bicoid* and *Nanos*) create larvae lacking anterior structures (*Bicoid*) or abdominal segments (*Nanos*). Bicoid encodes a transcription factor that positively affects some genes (*Hunchback*) and negatively affects others. Nanos protein inhibits the translation of *Hunchback*.
- Segmentation genes are expressed when the embryo is at the 6,000-nuclei stage and determine the number, boundaries, and polarity of the segments. Three types of segmentation genes are involved in regulation of segmental development: gap genes, pair rule genes, and segment polarity genes.
- Gap genes organize large areas along the anterior–posterior axis of the developing embryo. Gap mutants produce larvae that are missing consecutive larval segments. Pair rule genes divide the larva into units of two segments each. Mutations in the pair rule genes produce larvae that are missing every other segment. Segment polarity genes determine the boundary and the anterior–posterior organization of each segment. Mutations in segment polarity genes produce larvae in which the posterior structures in the segments have been replaced by reversed anterior structures.
- Gap genes control the expression of pair rule genes, some of which are transcription factors that control segment polarity.
- Differences between segments are encoded by Hox genes, which give each segment an identity. For instance, in response to Hox gene expression, cells in a thoracic segment produce legs and cells in the head segment produce eyes.
- Homeotic mutants include ***antennapedia*** (producing legs in place of antennae; see Figure 19.20) and *bithorax* (producing a fly with an extra set of wings). Homeotic genes are clustered on the chromosome in the order of the segments they determine. The first set of genes in the homeotic cluster specifies anterior segments (head and then thoracic segments). The second cluster begins with a gene specifying the last thoracic segment and then has genes for anterior and posterior abdominal segments.
- Nucleic acid hybridization experiments confirm that Hox genes have arisen from duplications of a single gene in an ancestral unsegmented organism.
- A DNA sequence, the homeobox, encodes a 60-amino acid sequence, the homeodomain. The homeodomain includes a DNA-binding motif and acts as a transcription factor.

Test Yourself

Diagram Exercise

Create a flow chart detailing the three different types of stem cells present in a developing human. Name and define each type of stem cell, and include the developmental stage (and location, when possible) at which each type of stem cell is present.
Textbook Reference: *19.2 Is Cell Differentiation Irreversible? pp. 408, 410–411*

Knowledge and Synthesis Questions

1. Cell differentiation
   a. results from the loss of particular genes from the nucleus of the differentiated cell.
   b. results from the differential expression of genes that are responsive to environmental signals.
   c. involves the persisting totipotency of early embryonic cells in the mature organism.

d. results from mutations in genes that control the synthesis of DNA.
e. precedes cell determination.

Textbook Reference: *19.1 What Are the Processes of Development? p. 406*

2. A totipotent cell is a cell
a. whose developmental fate has been decided.
b. that has differentiated into a specialized tissue.
c. that is fated to form a particular structure.
d. whose developmental potential is extremely broad.
e. whose developmental potential is narrower than that of a pluripotent cell.

Textbook Reference: *19.2 Is Cell Differentiation Irreversible? p. 408*

3. Apoptosis
a. occurs during embryonic development of invertebrates, but not vertebrates.
b. prompts cell differentiation.
c. is controlled by similar genes in humans and nematodes.
d. does not occur during morphogenesis.
e. involves caspase enzymes in the nematode *C. elegans*.

Textbook Reference: *19.5 How Does Gene Expression Determine Pattern Formation? pp. 417–418*

4. The fate of a cell
a. refers to the cell's original type.
b. describes its genes.
c. refers to the type of cell into which it will differentiate.
d. describes its death.
e. cannot by modified by its cytoplasmic environment.

Textbook Reference: *19.2 Is Cell Differentiation Irreversible? p. 409*

5. Mammals can be cloned using donor somatic cells fused to enucleated eggs if the
a. donor cell is in the S phase of the cell cycle.
b. donor cell is in the G1 phase of the cell cycle.
c. mammalian embryo can be propagated in culture.
d. donor cell is a mature red blood cell.
e. donor cell and enucleated egg are derived strictly from individuals of the same species.

Textbook Reference: *19.2 Is Cell Differentiation Irreversible? pp. 409–410*

6. The protein MyoD
a. is a transcription factor that controls the expression of genes involved in the differentiation of muscle cells.
b. controls segment identity in mice.
c. is a transcription factor that activates a gene causing the cell cycle to start in muscle precursor cells.
d. is absent from adult muscle.
e. is the only transcription factor involved in differentiation of mesoderm cells into mature muscle cells.

Textbook Reference: *19.3 What Is the Role of Gene Expression in Cell Differentiation? p. 413*

7. Induction occurs when
a. nuclear genes are lost in some tissues.
b. one cell contacts another and fails to alter its developmental fate.
c. a cell or tissue sends a chemical signal to another, causing differentiation of that cell or tissue.
d. two tissues come into contact with one another, releasing transcription factors.
e. factors are unequally distributed in the cytoplasm of an egg.

Textbook Reference: *19.4 How Is Cell Fate Determined? pp. 414–416*

8. Which of the following statements about plant development is *false*?
a. Both cell division and cell expansion contribute to growth in early plant embryos.
b. Differentiation in plant cells is irreversible.
c. Meristems at the tips of the roots and stems consist of undifferentiated cells.
d. Flower development in a plant involves organ identity genes.
e. In plants, the fertilized egg is called a zygote.

Textbook Reference: *19.1 What Are the Processes of Development? p. 407; 19.2 Is Cell Differentiation Irreversible? pp. 408–409; 19.5 How Does Gene Expression Determine Pattern Formation? pp. 418–419*

9. It may be possible to genetically engineer plants to produce more seeds and fruits by
a. providing more fertilizer in the spring.
b. manipulating their gap genes.
c. altering the homeotic genes that control carpel development.
d. inducing a mutation that eliminates *LEAFY* gene function.
e. altering the homeotic genes that control stamen development.

Textbook Reference: *19.5 How Does Gene Expression Determine Pattern Formation? p. 419*

10. Morphogens
a. diffuse within the embryo to set up a concentration gradient.
b. are expressed as positional signals in developing embryos.
c. direct differentiated cells to form organs.
d. act directly on target cells.
e. All of the above

Textbook Reference: *19.5 How Does Gene Expression Determine Pattern Formation? p. 419*

11. Maternal effect genes
a. begin setting up positional axes in the egg prior to fertilization.
b. are the same as segmentation genes.
c. are masked by paternal genes.
d. operate late in the gene cascade regulating development of *Drosophila* embryos.
e. determine cell fate within each segment.

Textbook Reference: *19.5 How Does Gene Expression Determine Pattern Formation? p. 421*

12. Hox genes
    a. encode protein domains that are important in development and that have been highly conserved over evolutionary time.
    b. are found in diverse organisms.
    c. can produce the wrong structure in the wrong place when mutated.
    d. help determine cell fate within each segment of a developing *Drosophila* embryo.
    e. All of the above

    Textbook Reference: *19.5 How Does Gene Expression Determine Pattern Formation? pp. 422–423*

13. What principle of development states that the nuclei of cells do not lose any genetic information during the early stages of development?
    a. Genomic equivalence
    b. Apoptosis
    c. Totipotentcy
    d. Transcription
    e. Fate mapping

    Textbook Reference: *19.2 Is Cell Differentiation Irreversible? p. 409*

14. What is the role of cytoplasmic segregation in determining the fate of a cell?
    a. It keeps cells totipotent.
    b. It determines polarity within the organism.
    c. It ensures that gradients do not develop in the developing organism.
    d. It stops cell division.
    e. It has no role in determining cell fate.

    Textbook Reference: *19.4 How Is Cell Fate Determined? pp. 413–414*

15. Organ identity genes are found in
    a. bacteria.
    b. animals.
    c. plants.
    d. fungi.
    e. None of the above

    Textbook Reference: *19.5 How Does Gene Expression Determine Pattern Formation? p. 419*

Application Questions

1. Experiments have shown that fully differentiated cells still contain all their genes, even though they express only a small set of those genes. Which type of hybridizations have demonstrated this? How do such hybridizations work?
   Textbook Reference: *19.3 What Is the Role of Gene Expression in Cell Differentiation? pp. 412–413*
2. Describe one application for human health of multipotent stem cells and one application for human health of cloned mammals.
   Textbook Reference: *19.2 Is Cell Differentiation Irreversible? pp. 410–411*
3. How would you define "model organism"? What traits do you think characterize model organisms?
   Textbook Reference: *19.4 How Is Cell Fate Determined? p. 415; 19.5 How Does Gene Expression Determine Pattern Formation? pp. 419–420*
4. In cloning whole organisms by fusing nuclei (as seen in frogs) or whole cells (as seen in sheep, cows, and goats) with enucleated eggs, why is the original nucleus removed from the egg?
   Textbook Reference: *19.2 Is Cell Differentiation Irreversible? pp. 409–410*
5. Describe three ways in which plant and animal development differ.
   Textbook Reference: *19.1 What Are the Processes of Development? p. 407; 19.5 How Does Gene Expression Determine Pattern Formation? p. 419*

Answers

Diagram Exercise Answer

Name: Totipotent stem cell
Definition: Can specialize to be any type of cell
Developmental stage: Zygote

↓

Name: Pluripotent stem cell
Definition: Can specialize to be nearly every type of cell (cannot form cells of the placenta)
Location and developmental stage: Inner cell mass of the blastocyst

↓

Name: Multipotent stem cell
Definition: Can specialize to be many types of cells
Location and developmental stage: Tissues that need frequent cell replacement such as bone marrow, brain, and skin of an adult

Knowledge and Synthesis Answers

1. **b.** Genes are not lost from differentiated cells. There are no embryonic cells in a mature organism. Mutations in cells that affect DNA replication would result in cells that were unable to replicate their DNA. Cell differentiation occurs after cell determination.
2. **d.** A totipotent cell is a cell that can develop in any number of ways. Its fate has not been determined, it has not differentiated, and it is not expressing morphogens.
3. **c.** Apoptosis is controlled by similar genes in humans and nematodes, suggesting the importance of this pathway (i.e., mutations are extremely harmful, so the pathway is conserved).
4. **c.** The fate of a cell is the type of cell it will eventually become once it has differentiated.

5. **b.** The donor somatic cell must be in G1. Mammalian embryos are grown in surrogate mothers, not in culture. Mature red blood cells in mammals have no nuclei and could not be used for cloning. Cloning can be used to preserve endangered species, such as when a banteng (a species closely related to domestic cattle) was cloned using a donor banteng cell and an enucleated cow egg. Thus, the donor cell and enucleated egg do not have to come from exactly the same species; closely related species can be used.
6. **a.** MyoD is a transcription factor that controls the expression of genes involved in the differentiation of mesoderm cells into mature muscle cells. It activates a gene that causes the cell cycle to stop, so that differentiation can begin. The transcription factor MyoD is one of several involved in the differentiation of muscle cells, and it is found in adult muscle cells where it may play a role in repair.
7. **c.** Induction occurs when one cell or tissue secretes factors that influence the developmental fate of other cells or tissues. Nuclear genes are not lost in developing cells (red blood cells are an exception). Transcription factors are found in the nucleus and are not released to adjacent tissues.
8. **b.** In general, it is easier to reverse differentiation in plant cells than it is in animal cells, as demonstrated by the cloning of a carrot from a differentiated storage cell in its root.
9. **c.** Fertilization is not genetic engineering. Gap genes are found in *Drosophila*, not plants. The LEAFY protein transcriptionally activates the homeotic genes that control organ development (and subsequent seed and fruit formation). In its absence, no fruit or seeds would develop. Altering the homeotic genes that control carpel (not stamen) development could lead to more fruit and seeds.
10. **e.** Morphogens diffuse in the embryo to form concentration gradients that act as positional signals to direct differentiated tissues to form organs. They act directly on target cells rather than triggering a secondary signal that acts on target cells.
11. **a.** Beginning before fertilization, maternal effect genes determine the anterior–posterior axis. These genes, which operate early in the gene cascade, also induce segmentation genes. The paternal genotype does not affect the expression of maternal gene products in the egg. Hox genes, and not maternal effect genes, determine cell fate within each segment.
12. **e.** Hox genes encode proteins that are highly conserved and important for directing development in many different organisms. Mutations in these genes can produce an abnormal developmental event in the organism. In *Drosophila* embryos, Hox genes determine segment identity.
13. **a.** Genomic equivalence is a fundamental principle of developmental biology. It states that none of the genetic information contained in the original zygote is lost during subsequent cell division and growth.
14. **b.** Cytoplasmic segregation refers to the unequal distribution of factors within the developing embryo. This asymmetry helps set up, for example, the polarity (distinct top and bottom) of the developing organism.
15. **c.** Organ identity genes are found in plants, where they help to direct the development of structures such as flowers.

Application Answers

1. Hybridizations were carried out using a specific DNA probe to nuclear DNA and mRNA. If the DNA probe was complementary to a gene that was not being expressed in the cell, it would not hybridize to the mRNA isolated from that cell, but it still would hybridize to the nuclear DNA. This indicated that the gene was still present in the nucleus of the cell, even though it was not being expressed in the cytoplasm.
2. Hematopoietic stem cells are multipotent stem cells that produce red and white blood cells. If left in the body, these dividing cells are killed during cancer treatments along with cancer cells. In hematopoietic stem cell transplantation, these stem cells are removed prior to cancer treatment, stored in a laboratory where they are provided with growth factors, and then replaced in greater numbers once treatment has been completed. Through cloning we can increase the number of particularly valuable animals, such as transgenic cows that produce human growth hormone, and thereby increase the supply of growth hormone needed for treating hormone deficiencies.
3. Model organisms are research organisms that are expected to yield insights into how other organisms work. Some characteristics of Model organisms generally share genetic, cellular, and/or molecular characteristics with these other organisms. They also have short generation times, are easy to obtain and maintain, are amenable to experimental manipulation, and are not of concern from a conservation standpoint.
4. The nucleus of the egg contains a haploid set of chromosomes. Fusing a diploid somatic cell with an egg cell that still contained a nucleus would result in a cell with three copies of each chromosome and little chance of developing normally.
5. The MADS box gene family plays a role in regulating plant development; these genes are not involved in animal development, where *homeobox* genes are important. With regard to morphogenesis and growth, cell division is important in plants and animals; cell expansion is particularly important in plants. Also, whereas cell movement is critical to animal development, plants cells exist within rigid cell walls, making cell movement during development impossible.

20 Development and Evolutionary Change

The Big Picture

- Evolutionary developmental biology is the study of how organisms evolved by examining the differences in developmental gene expression and regulation between species. Animals have highly conserved genes, such as homeobox genes, that control development.
- Genetic switches determine where and when genes will be expressed. Changes in these switches can result in the evolution of species differences.
- Heterochrony describes how differences in organisms have arisen because of differences in the timing of different developmental processes. The timing of gene expression can differ between organisms because of the modular makeup of organisms.
- Developmental plasticity occurs in response to environmental signals. These signals can either provide an accurate prediction of the future or be poorly correlated with future conditions.

Common Problem Areas

- The concept of heterochrony can be difficult for students to grasp. Try to remember that when we discuss heterochrony, we are talking about differences between species, not within species.

Study Strategies

- This chapter introduces many concepts. Learn the concepts involved in evolutionary developmental biology first. Then go back and see how the examples fit with the concepts.
- Go to yourBioPortal.com to review the following tutorial and activity:

  Animated Tutorial 20.1 Modularity

  Web Activity 20.1 Plant and Animal Development

Important Concepts

Evolutionary developmental biology (often shortened to evo-devo) is the contemporary study of evolution and development.

- A major focus of evo-devo studies is the comparison of genes that regulate development across different multicellular organisms.
- Evo-devo studies use the techniques of molecular genetics.

Organisms have a genetic toolkit that directs development.

- Many of the genes involved in controlling development are conserved across different organisms. Thus, these genes can be very similar in organisms that have very different appearances. For example, genes controlling eye development are similar in mice and fruit flies. The major differences in body form arise because genes are turned on and off at different times and locations during development.
- The organism's genes and transcription factors, along with the way they interact with extracellular signals, can be thought of as a genetic took kit used to assemble the organism.
- Genes involved in regulating development in plants and animals have been highly conserved. As an example, the development of an anterior–posterior axis in insects and mammals is due to the same types of homeobox-containing genes that are expressed at either the anterior or posterior end of the developing embryo (see Figure 20.2).

Changes in gene expression of homologous genes can be localized to single modules.

- The many parts of the developing embryo can change independently from one another because the organism consists of developmental modules. A module, such as the genes and signaling pathways that lead to development of the heart, can change independently of the other modules because some developmental genes affect only one module.
- Genetic switches, which consist of promoters and the transcription factors that bind them, control the way in which a gene is used during development. Each gene is controlled by multiple switches that influence where and when it is expressed.

- For example, genetic switches control spatial development of the embryo. The pattern of development depends on the Hox genes that are expressed within the module. This can be seen in the development of wings in *Drosophila*, where Hox genes are differentially expressed in the three thoracic segments of the embryo (see Figure 20.3).

The modularity of embryos allows different developmental processes to shift independently of one another.

- Heterochrony occurs when the relative timing of developmental processes independently shifts in different species; it is possible because of the modularity of the developing embryo.
- Salamander embryos have webbed feet, and in most species a particular gene is expressed during development that prompts apoptosis (programmed cell death) in the webbing. Lack of webbing between the toes makes walking on the ground much easier. In one arboreal species of salamander, however, the gene is not expressed, so apoptosis does not occur in the webbing. This leads to feet that are much better suited to an arboreal lifestyle because they work like suction cups on vertical surfaces. Thus, a simple change in gene expression can lead to different morphologies (free digits or webbed digits) and lifestyles (terrestrial versus arboreal).
- Like salamander embryos, bird embryos have webbed feet. The webbing is retained in adult ducks but not in chickens. The signaling protein bone morphogenetic protein 4 (BMP4) causes the cells of the webbing to undergo apoptosis. Although BMP4 is present in the webbing of embryonic ducks and chickens, these species differ in whether the gene *Gremlin* is expressed. *Gremlin* encodes a protein that inhibits BMP4. In ducks, *Gremlin* is expressed, encoding a protein that inhibits BMP4, thereby preventing apoptosis and producing webbed feet.
- Like almost all mammals, the giraffe has seven cervical vertebrae in its neck. Thus, its long neck cannot be explained by the presence of additional vertebrae. Instead, its long neck results from a delay in the signaling process that stops bone growth, and this delay allows the vertebrae of the neck to grow longer. This example of heterochrony involves changes in the timing of expression of genes that control bone formation.

Genetic switches can explain the evolution of major morphological differences among species.

- Evolutionary changes in morphology have occurred due to the change in the timing or spatial expression of regulatory genes during development. The insect *Ultrabithorax* (*Ubx*) homeotic gene has a mutation that results in the repression of the gene involved in limb formation, the *Distal-less* gene. The *Ubx* gene is expressed in the abdomen during development, leading to the absence of limbs on the abdomen in insects. Other arthropods, such as centipedes, do not have this mutation and have abdominal legs. Thus, a mutation in a Hox gene changed the number of legs in insects relative to other arthropods.
- In vertebrates, the spatial pattern of Hox gene expression regulates the number of each kind of vertebra (cervical, thoracic, lumbar, sacral, and caudal). In an embryonic mouse, the transition from cervical to thoracic vertebrae occurs at the anterior limit of *Hoxc6* expression, resulting in seven cervical vertebrae. In the embryonic chicken, the anterior limit of *Hoxc6* expression is further down the vertebral column, resulting in more than seven cervical vertebrae.

The environment can greatly influence the development of plants and animals.

- Organisms have the ability to exhibit different phenotypes based on a single genotype. The phenotype is dependent on the environment in which the organism develops and lives. This response is known as developmental (or phenotypic) plasticity.
- Sex determination in mammals is based on sex chromosomes: XX individuals are female and XY individuals are male. In contrast to the chromosomal sex determination of mammals, some reptiles exhibit temperature dependent sex determination, whereby the temperature at which an embryo is incubated while in the egg determines its gender. The specific effects of incubation temperature can vary among species, but the general phenomenon is an example of developmental plasticity.
- In vertebrates, the biosynthesis of sex steroid hormones begins with cholesterol, which first forms testosterone. Testosterone may be converted via the enzyme aromatase to estrogens (female sex steroids) or via the enzyme reductase to other androgens (male sex steroids). In animals with temperature-determined sex, incubation temperature influences gender by influencing the amount of aromatase expressed. When aromatase is expressed in large quantities, estrogens dominate and female sex organs develop. When aromatase is not expressed, testosterone dominates and male sex organs develop. Some experiments have demonstrated the fitness value of temperature-determined sex by showing that this form of sex determination is associated with sex-specific differences in fitness.
- An example of developmental plasticity in invertebrates concerns the spring and summer caterpillar forms of the moth *Nemoria arizonaria*. Whereas spring caterpillars resemble the oak tree flowers on which they feed, summer caterpillars resemble oak twigs and feed on oak leaves. Spring and summer caterpillars look alike at hatching; the different appearances are triggered by their different diets. Thus, developmental plasticity allows both forms of caterpillars to avoid predation by visually hunting predators.

- Environmental signals can either be accurate predictors of the future environment or poorly correlated with the future environment. Day length is an accurate predictor of future conditions. Many insects use day length to time periods of developmental arrest called diapause, and in mammals, such as deer and elk, antlers are lost in response to changes in day length. Day length is critical to timing reproduction in many animals.
- Light is an important environmental signal in plant development. In plants, low light levels result in elongation of cells in the stem. This allows a plant to grow tall and reach a sunny patch more quickly than a compact plant could. When light is abundant, plants invest their energy in growing leaves rather than in lengthening their stems.

Evolution is constrained by existing structures and highly conserved genes.

- Evolution works on existing genes and their expression and can occur through changes in Hox gene expression in different modules. Thus, most evolutionary innovations are modifications of previously existing structures. For example, wings evolved three times in vertebrates (birds, bats, and extinct pterosaurs) but rather than being new structures they are modified limbs. Evolutionary losses are also controlled by Hox genes, such as the loss of forelimbs in the ancestors of present-day snakes.
- Similar traits evolve repeatedly because of the highly conserved nature of the genetic code. This can lead to parallel phenotypic evolution of a trait, such as the evolutionary loss of body armor in numerous species of freshwater stickleback fish. Freshwater populations of sticklebacks are descended from marine populations. In marine sticklebacks, the gene *Pitx1* codes for a transcription factor that stimulates the production of protective plates and spines. The *Pitx1* gene has changed in various freshwater populations of sticklebacks, leading to the loss of body armor in these populations.

Test Yourself

Diagram Exercise

In this chapter you learned about some environmental signals that influence development. Write "Developing organism" on a piece of paper and then indicate, using arrows pointing toward the term, at least six enviro-nmental signals that could potentially influence development. This will require you to go beyond the number of environmenal signals covered in the textbook.

Textbook Reference: *20.4 How Does the Environment Modulate Development? p. 435*

Knowledge and Synthesis Questions

1. Genes that regulate development are highly conserved. This means that
   a. large differences have evolved among multicellular organisms.
   b. they have changed very little over the course of evolution.
   c. they are always turned on.
   d. they have undergone mutation.
   e. they have been lost in some lineages.
   Textbook Reference: *20.1 What Is Evo-Devo? p. 428*
2. _______ genes are involved in the development of the anterior–posterior axis in mammals and insects.
   a. Developmental plasticity
   b. Heterochrony
   c. Hox
   d. *Pax6*
   e. *Gremlin*
   Textbook Reference: *20.1 What Is Evo-Devo? p. 428*
3. Heterochrony
   a. can explain the long neck of the giraffe.
   b. involves shifts in the relative timing of developmental processes.
   c. can lead to major morphological changes and new lifestyles.
   d. can explain whether webbing remains between the digits of adult salamanders and birds.
   e. All of the above
   Textbook Reference: *20.2 How Can Mutations with Large Effects Change Only One Part of the Body? p. 430*
4. Module development has allowed for evolutionary changes to occur while still resulting in a viable organism because
   a. all modules must change together.
   b. modules can change independently of one another.
   c. modules are unimportant for development.
   d. modules prevent heterochrony.
   e. developmental genes always exert their effects on more than one module.
   Textbook Reference: *20.2 How Can Mutations with Large Effects Change Only One Part of the Body? p. 429*
5. Homologous genes
   a. are adjacent to one another on the same chromosome.
   b. never involve whole sets (clusters) of genes.
   c. are found exclusively in mammals.
   d. evolved from a gene present in a common ancestor.
   e. are expressed at the same time in an organism.
   Textbook Reference: *20.1 What Is Evo-Devo? p. 428*
6. Developmental plasticity
   a. is the ability of an organism to change its development in response to environmental conditions.
   b. is limited to plants.
   c. is the ability of an organism to express only one phenotype under different environmental conditions.

d. means that an organism can begin as one genotype and later change genotype in response to environmental conditions.
e. is also called genotypic plasticity.

Textbook Reference: *20.4 How Does the Environment Modulate Development? p. 433*

7. Which of the following statements is *false*?
a. Diet can initiate developmental changes.
b. Day length is a reliable environmental signal.
c. Temperature can initiate developmental changes.
d. Cells of plant stems elongate in response to low light levels.
e. The cycle by which deer form and lose their antlers occurs in response to temperature.

Textbook Reference: *20.4 How Does the Environment Modulate Development? p. 435*

8. Genetic switches
a. stop development.
b. help various modules develop differently.
c. ensure that all modules develop into the same type.
d. do not involve promoters and transcription factors.
e. play no role in integrating positional information in the embryo.

Textbook Reference: *20.2 How Can Mutations with Large Effects Change Only One Part of the Body? p. 429*

9. The difference in appearance of *Nemoria arizonaria* caterpillars in spring and summer is triggered by
a. temperature.
b. day length.
c. the relative availability of food.
d. different genotypes.
e. light intensity.

Textbook Reference: *20.4 How Does the Environment Modulate Development? pp. 434–435*

10. Which of the following statements is correct?
a. Wings have evolved independently four times in vertebrates.
b. Wings arose as a result of new "wing genes."
c. Wings arose as modifications to already existing structures.
d. Wings have different skeletal components in birds and bats.
e. Wings have similar structural components in pterosaurs and insects.

Textbook Reference: *20.5 How Do Developmental Genes Constrain Evolution? pp. 436–437*

11. Which of the following statements is *false*?
a. All bird embryos have webbed feet, although some adult birds do not.
b. Bone morphogenetic protein 4 (BMP4) directs cells in the foot webbing of bird embryos to undergo apoptosis.
c. The protein Gremlin inhibits BMP4 protein from signaling for apoptosis, causing webbed feet.
d. The gene *Gremlin* is expressed in the webbing of both chicken and duck embryos.
e. Adding the protein Gremlin to a developing chicken foot will cause the foot to resemble that of a duck.

Textbook Reference: *20. 2 How Can Mutations with Large Effects Change Only One Part of the Body? pp. 430–431*

12. Which of the following statements is *false*?
a. In arthropods, the gene *Distal-less* causes legs to form from segments.
b. In centipedes, the Hox gene *Ubx* activates expression of the gene *Distal-less*, causing legs to form from segments.
c. A mutation in the Hox gene *Ubx* in the ancestor of insects resulted in legs forming from thoracic but not abdominal segments in this lineage.
d. Changes in the spatial pattern of Hox gene expression can account for the different number of cervical vertebrae in mouse and chicken embryos.
e. Morphological differences among species do not evolve through mutations in Hox genes.

Textbook Reference: *20.3 How Can Differences among Species Evolve? p. 432*

13. Temperature-determined sex
a. occurs in mammals.
b. is an example of developmental plasticity.
c. is not associated with sex-specific fitness differences.
d. relies on a mechanism whereby high levels of the enzyme aromatase make testosterone dominant and lead to development of male sex organs.
e. relies on a mechanism that makes high temperatures produce males.

Textbook Reference: *20.4 How Does the Environment Modulate Development? p. 433*

14. Which of the following statements is true?
a. Development ceases when an organism reaches maturity.
b. Development continues until an organism dies.
c. Development is fixed and cannot change in response to environmental conditions.
d. In every organism, the genes governing development are unique to that organism.
e. Development cannot constrain evolution.

Textbook Reference: *20.1 What is Evo-Devo? p. 428; 20.4 How Does the Environment Modulate Development? p. 435; 20.5 How do Developmental Genes Constrain Evolution? p. 436*

15. Which of the following statements is *false*?
a. Body armor is absent from sticklebacks living in freshwater environments due to a genetic mutation.
b. The *Pitx1* gene promotes the production of protective plates and spines in marine populations of sticklebacks.
c. Similar traits are likely to evolve repeatedly because of the highly conserved nature of developmental genes.
d. The *Pitx1* gene is inactive in freshwater stickleback populations.

e. Environmental conditions induce the loss of body armor in freshwater sticklebacks.

Textbook Reference: *20.5 How Do Developmental Genes Constrain Evolution? pp. 436–437*

Application Questions

1. There are morphological differences between adult chickens and ducks that relate to the duck's aquatic lifestyle. Why were these adaptations able to occur without influencing the development of the rest of the duck's morphology?

   Textbook Reference: *20.2 How Can Mutations with Large Effects Change Only One Part of the Body? pp. 429–431*

2. Day length and temperature are both potential environmental signals that may influence development. What do changes in these two environmental factors tell an organism about future conditions?

   Textbook Reference: *20.4 How Does the Environment Modulate Development? pp. 435–436*

3. Why do insects have legs only on thoracic segments, whereas centipedes have legs on thoracic and abdominal segments?

   Textbook Reference: *20.3 How Can Differences among Species Evolve? p. 432*

4. Conservation efforts to restore sea turtle populations often rely on collecting eggs, incubating them in captivity, and then releasing hatchlings. Why would knowledge of temperature-determined sex be critical to such efforts?

   Textbook Reference: *20.4 How Does the Environment Modulate Development? pp. 433–434*

5. Why do scientists consider the gene *eyeless* in fruit flies and the gene *Pax6* in mice to be homologous?

   Textbook Reference: *20.1 What Is Evo-Devo? p. 428*

Answers

Diagram Exercise Answer

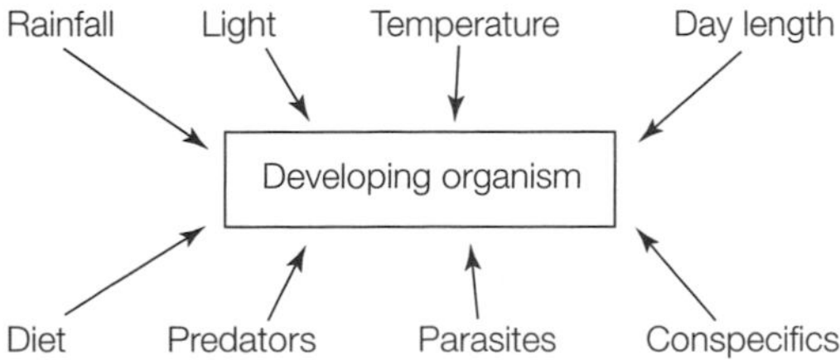

Knowledge and Synthesis Answers

1. **b.** Conserved genes are genes that are found in many organisms and have undergone very little change.
2. **c.** Hox genes regulate development along the anterior–posterior axis in animals.
3. **e.** Heterochrony is the process by which the timing of gene expression differs between two or more species. Heterochrony can explain the long neck of the giraffe (delays occur in the signaling process that stops bone growth, so the neck vertebrae grow longer) and whether webbing remains between the digits of adult salamanders and birds.
4. **b.** Embryos are composed of self-contained units called modules that can change independently of one another. This allows for the evolution of one module without disrupting the other modules of the animal.
5. **d.** Homologous genes evolved from a gene present in a common ancestor. They can be clusters of genes (such as Hox genes) and they are not restricted to mammals.
6. **a.** Developmental plasticity is found in both plants and animals. It allows a developing organism to express different phenotypes, depending on the environment.
7. **e.** Day length, not temperature, influences antler development in deer.
8. **b.** Genetic switches help each module develop into its appropriate form. They work by turning Hox genes on and off. They are involved in integrating positional information in the embryo, and they do involve promoters and transcription factors.
9. **c.** The difference in appearance of the *Nemoria arizonaria* spring caterpillar and summer caterpillar results from differences in diet.
10. **c.** Wings arose as modifications of preexisting structures. In vertebrates, for example, wings are modified forelimbs.
11. **d.** The *Gremlin* gene is expressed in the foot webbing cells of ducks, but not chickens; the Gremlin protein inhibits the BMP4 protein from signaling for apoptosis, so the feet of adult ducks are webbed.
12. **e.** Morphological differences between species can arise through mutations in Hox genes. This is exemplified by the mutation in the *Ubx* gene of insects, which causes leg formation in abdominal segments to be inhibited rather than prompted.
13. **b.** Temperature-determined sex is an example of developmental plasticity in which the sex of some reptiles is determined by the temperature at which eggs incubate rather than by sex chromosomes (as in mammals).
14. **b.** Development occurs throughout an organism's lifespan and ends when it dies.
15. **e.** Absence of body armor in freshwater sticklebacks results from a mutation in the *Pitx1* gene; the absence of armor is not induced by environmental conditions.

Application Answers

1. Organisms are made up of modules. Change can occur in a module independent of other modules. The webbing found in duck feet is due to expression of the *Gremlin* gene in the foot module. Expression of *Gremlin* in the webbing cells of the feet results in a

protein that inhibits the BMP4 protein responsible for prompting apoptosis.

2. Day length and temperature are accurate predictors of the future environment. Changes in day length act as an environmental signal that accurately predicts when future conditions will change. Shorter days signify that fall and winter are approaching. Similarly, changes in temperature occur with changes in the seasons and can signal to an organism that a change in the season soon will occur.
3. Leg development in arthropods involves the homeotic gene *Ultrabithorax* (*Ubx*) and the gene *Distal-less* (*dll*). Expression of *dll* in a segment results in the development of a pair of legs in that segment. In insects, the *Ubx* gene has a mutation that suppresses the expression of *dll* in the abdomen. As a result, no legs will develop in the abdomen. Centipedes do not have this mutation, therefore legs develop in thoracic and abdominal segments.
4. Knowing that sea turtles have temperature-determined sex would be critical to conservation efforts because if all eggs were incubated at the same temperature, then hatchlings would all be the same sex, a situation unlikely to be helpful to such efforts. Knowledge of which temperatures produce males and which produce females also would allow scientists to manipulate incubation temperature to produce the sex ratio best conducive to the restoration of sea turtle populations.
5. The gene *eyeless* in fruit flies and the gene *Pax6* in mice each control the developmental switch that turns on eye development. *Eyeless* and *Pax6* exhibit DNA sequence similarity even though fruit flies and mice have very different eyes. These sequence similarities suggest that *eyeless* and *Pax6* evolved from a gene present in a common ancestor of fruit flies and mice.

21 Evidence and Mechanisms of Evolution

The Big Picture

- Most biologists rank Darwin's *The Origin of Species* as the most significant book ever written about their subject. Its importance is twofold. First, it presents a vast amount of evidence for the fact of evolution. Second, it describes a mechanism—natural selection—to explain evolutionary change. Evolution through natural selection remains the most important unifying concept in biology.
- Population genetics unites the concepts of Darwin with the insights into the mechanisms of heredity provided by Mendel. It enables biologists to study evolution with quantitative rigor.
- The loss of genetic variation that occurs during a population bottleneck is a major concern for biologists working to preserve endangered species. As the discussion of the greater prairie chicken makes clear, genetic uniformity in a population renders it more vulnerable to extinction.

Common Problem Areas

- A common mistake made by students when attempting to solve Hardy–Weinberg problems is the use of the wrong equation. For example, if you are given phenotypic frequencies for a trait that is at genetic equilibrium and shows dominance, remember that the frequency of the recessive phenotype (and genotype) is equal to q^2, not q. By taking the square root of q^2, you can obtain the frequency of the recessive allele and can then determine the frequency of the dominant allele by subtraction ($p = 1 - q$).
- It is important to bear in mind that for any gene locus with two alleles, the allele frequencies must total unity (i.e., it is always the case that $p + q = 1$). The equation that specifies genotype frequencies ($p^2 + 2pq + q^2 = 1$) is valid only for a population at genetic equilibrium.

Study Strategies

- Population genetics is learned best by doing problems, so be sure you can solve problems similar to those presented in the "Test Yourself" section of this chapter of the *Study Guide*. Your instructor may provide a worksheet with additional examples.
- Go to yourBioPortal.com to review the following tutorials:

  Animated Tutorial 21.1 Natural Selection

  Animated Tutorial 21.2 Hardy–Weinberg Equilibrium

  Animated Tutorial 21.3 Assessing the Costs of Adaptations

  Interactive Tutorial: Genetic Drift Simulation

  Interactive Tutorial: Molecular Evolution: Selection Pressures

Important Concepts

Darwin's main contribution to biology was the theory of evolution by natural selection.

- Evolutionary theory encompasses our understanding of the mechanisms that lead to biological changes in populations over time.
- A population is a group of individuals of a single species that live and interbreed in a particular area at the same time. Populations evolve, whereas individuals do not.
- The theory of evolution by natural selection rests on two facts:
  - All populations have the capacity to increase in number exponentially. Because populations rarely increase rapidly, high death rates must balance the potential growth rates.
  - All populations show variations among individuals with regard to numerous inherited characteristics, some of which may affect their chances of surviving and reproducing.
- From these two facts, Darwin made the following inference: Those individuals whose heritable traits enable them to survive and reproduce more successfully than others will pass those traits on to greater numbers of offspring.
- In short, natural selection is the differential contribution of offspring to the next generation by various genetic types belonging to the same population.

- When selection of individuals with desirable traits is carried out by humans, such as plant and animal breeders, it is referred to as artificial selection.
- An adaptation is a characteristic that helps its bearer survive and reproduce; the term also refers to the evolutionary process that produces such characteristics.
- The rediscovery of Mendel's laws of inheritance led to the development of population genetics, which applies these laws to entire populations in order to understand the genetic basis of evolutionary change.

Evolution is any genetic change in a population from generation to generation.

- A genotype is the genetic constitution of a particular trait of an individual. Evolution occurs when individuals with different genotypes survive or reproduce at different rates.
- Agents of evolution actually act on the phenotype, which is the physical expression of the genotype.
- The features of a phenotype are its *characters*. A *trait* is a specific form of a character.
- The gene pool of a population is the sum of all the alleles (alternative versions of a gene) found within it. Although a single diploid individual can have at most two different alleles for any gene, a population may contain many alleles for that gene. The gene pool contains all the variations that result in individuals with differing phenotypes on which agents of evolution act.
- Studies of both domestic and wild species show that virtually all populations contain a certain amount of genetic variation.
- Genotypes do not uniquely determine phenotypes. For example, if one allele is dominant to another, two genotypes (e.g., *AA* and *Aa*) can determine the same phenotype. Conversely, one genotype can produce different phenotypes because of the interaction of genetic and environmental factors during development.
- A Mendelian population is a locally interbreeding group within a geographic population. Such groups are often the subject of evolutionary studies.
- Allele frequencies measure the amount of genetic variation in a population, whereas genotype frequencies show how this variation is distributed among its members. Populations that have the same allele frequencies may nevertheless differ in their genotype frequencies.
- The sum of all allele frequencies at a locus is equal to 1, as is the sum of all genotype frequencies. For a locus with two alleles, p and q are typically used to represent the frequencies of the dominant and recessive alleles, respectively, and thus $p + q = 1$. Biologists estimate allele frequencies by measuring numbers of alleles in a sample of individuals from the population.
- If there is only one allele at a locus, the population is *monomorphic*, and the allele is *fixed*. If two or more alleles exist, the population is *polymorphic* at that locus.
- The genetic structure of a population is described by the frequencies of different alleles at each locus and the frequencies of different genotypes.

A population at Hardy–Weinberg equilibrium is not changing genetically and hence is not evolving.

- To be at Hardy–Weinberg equilibrium, a population must meet five conditions:
  - Mating must be random.
  - The population must be infinite.
  - There must be no migration into or out of the population.
  - There must be no mutation.
  - Natural selection does not affect the survival of particular genotypes.
- If these conditions are met, allele frequencies at a locus remain the same from generation to generation. Moreover, after one generation of random mating, the genotype frequencies will remain in the proportions $p^2 + 2pq + q^2 = 1$, where p^2, $2pq$, and q^2 represent the frequencies of the homozygous dominant, heterozygous, and homozygous recessive genotypes, respectively.
- Though populations in nature can never fully meet the conditions of the Hardy–Weinberg equilibrium, this equation is often useful for predicting the approximate genotype frequencies in a population. It is also important because deviations from it may show that evolution is occurring. Moreover, the pattern of deviations is useful in identifying the agents of evolutionary change operating on the population.

Evolutionary agents are forces that change the allele and genotype frequencies in a population.

- The ultimate source of genetic variation in a population is mutation. Because mutations are random changes in genetic material, most are harmful or neutral, but it is the environment that determines whether a particular mutation is disadvantageous or adaptive.
- Though mutations cannot be totally prevented, their frequency is usually so low as to cause little deviation from Hardy–Weinberg expectations.
- Gene flow occurs when individuals migrate from one population to another and breed in their new location. Gene flow can add new alleles to a population's gene pool, change the frequencies of alleles already present, or both.
- Genetic drift is caused by chance events that alter allele frequencies in a population. It has its greatest impact on small populations, in which it may even cause harmful alleles to increase in frequency.
- During a population bottleneck, when a large population is severely reduced in size, allele frequencies may shift drastically, and genetic variation may be reduced as a result of genetic drift.
- The change in genetic variation that occurs when a few individuals originate a new population is called the

founder effect. As in the case of a population bottleneck, some alleles found in the source population will be missing from the founding population, and others will occur with altered frequencies.

- The preferential mating of individuals with others either of the same genotype or of a different genotype is called nonrandom mating. The effect of nonrandom mating is a population with either fewer or more heterozygous individuals than would be expected in a population in Hardy–Weinberg equilibrium. The effect of self-fertilization, another form of non-random mating, is a reduction in the frequency of heterozygotes. In most types of nonrandom mating (except for sexual selection), the allele frequencies remain the same despite the changes in genotype frequencies.
- Natural selection acts on the phenotype (the physical features expressed by an organism) rather than directly on the genotype. Fitness is the contribution of a phenotype to the composition of later generations, relative to the contribution of alternative phenotypes. The fitness of a phenotype is determined by the average rates of survival and reproduction of individuals with that phenotype, as compared to individuals with other phenotypes.
- Differences in fitness among phenotypes (and hence in the controlling genotypes) lead to changes in allele frequencies in the gene pool of later generations. Changes in allele frequencies due to differences in fitness result in adaptation.
- Natural selection, unlike other agents of evolution, adapts organisms to their environment. In cases in which the distribution of a phenotype approximates a bell-shaped curve because it is controlled by many gene loci, selection can produce any one of three results:
  - Stabilizing selection reduces variation in the population by favoring average individuals.
  - Directional selection changes the mean value for a character by favoring individuals that vary in one direction.
  - Disruptive selection, by favoring both extremes, leads to a population with two peaks in the distribution of the character.
- Sexual selection favors traits that benefit their bearers (generally males) in the competition for access to members of the other sex, or make their bearers more attractive to members of the other sex. Sexual selection often results in sexually dimorphic species, in which males and females differ in appearance.

Most populations maintain substantial genetic diversity.

- Genetic drift, stabilizing selection, and directional selection tend to lessen the genetic variation within a population.
- Neutral mutations—that is, mutations that do not affect the fitness of an individual—tend to accumulate and thereby increase the genetic variation in a population.
- By generating new combinations of alleles, sexual reproduction amplifies existing genetic variation.
- Sexual reproduction has short-term disadvantages, including disruption of adaptive combinations of genes and reduction both of the rate at which females pass genes on to their offspring and of the overall reproductive rate. Possible advantages of sexual reproduction that may account for its evolution include facilitation of DNA repair, elimination of harmful mutations, and defense against pathogens and parasites.
- Frequency-dependent selection, in which the less-common genotype or phenotype is favored by natural selection, preserves variation as a polymorphism.
- If heterozygous individuals have a selective advantage over homozygotes for either allele, both alleles will persist in a population.
- Geographically distinct subpopulations of a population often vary genetically because they are adapted to different environments. Clinal variation is gradual change in phenotype across a geographic gradient.

There are constraints on the evolutionary process.

- Evolutionary changes must be based on modifications of previously existing traits.
- The evolution of adaptations depends on the trade-off between costs and benefits.
- Long-term evolutionary changes are strongly influenced by events that occur so infrequently or so slowly that they are rarely observed during short-term evolutionary studies. To understand the long-term course of evolution, biologists seek evidence of rare events and search for trends in the fossil record.

Test Yourself

Diagram Exercise

Construct a concept map with the theme of "Evolutionary Agents." Include in your map the following terms: evolutionary agents, allele frequencies, directional, disruptive, founder effects, gene flow, genetic drift, genetic structure of a population, genotype frequencies, mutation, natural selection, nonrandom mating, phenotypic variation, population bottlenecks, and stabilizing. Connect these terms with verbs or short phrases to indicate the relationships among them.

Textbook Reference: *21.1 What Facts Form the Base of Our Understanding of Evolution? pp. 440–448; 21.2 What Are the Mechanisms of Evolutionary Change? pp. 448–451; 21.3 How Does Natural Selection Result in Evolution pp. 451–456*

Knowledge and Synthesis Questions

1. Evolution occurs at the level of
   a. the individual genotype.
   b. the individual phenotype.
   c. environmentally based phenotypic variation.
   d. the population.
   e. the species
   Textbook Reference: *21.1 What Facts Form the Base of Our Understanding of Evolution? p. 444*
2. Natural selection acts on
   a. the gene pool of the species.
   b. the genotype.
   c. the phenotype.
   d. multiple gene inheritance systems.
   e. the environment.
   Textbook Reference: *21.3 How Does Natural Selection Result in Evolution? p. 451*
3. Which of the following statements about Mendelian populations is *false*?
   a. It must consist of members of the same species.
   b. It must have members that are capable of interbreeding.
   c. It must show genetic variation.
   d. It must have a gene pool.
   e. All of the above
   Textbook Reference: *21.1 What Facts Form the Base of Our Understanding of Evolution? pp. 445–446*
4. In comparing several populations of the same species, the population with the greatest genetic variation would have the
   a. greatest number of genes.
   b. greatest number of alleles per gene.
   c. greatest number of population members.
   d. largest gene pool.
   e. None of the above
   Textbook Reference: *21.4 How Is Genetic Variation Maintained within Populations? pp. 456–458*
5. The ability to taste the chemical PTC (phenylthio-carbamide) is determined in humans by a dominant allele *T*, with tasters having the genotypes *Tt* or *TT* and nontasters having *tt*. If 36 percent of the members of a population cannot taste PTC, then according to the Hardy–Weinberg rule, the frequency of the *T* allele should be
   a. 0.36.
   b. 0.4.
   c. 0.6.
   d. 0.64.
   e. 0.8
   Textbook Reference: *21.1 What Facts Form the Base of Our Understanding of Evolution? pp. 446–448*
6. A gene in humans has two alleles, *M* and *N*, that code for different surface proteins on red blood cells. If you know that the frequency of allele *M* is 0.2, according to the Hardy–Weinberg rule, the frequency of the genotype *MN* in the population should be
   a. 0.16.
   b. 0.2.
   c. 0.32.
   d. 0.64.
   e. 0.8
   Textbook Reference: *21.1 What Facts Form the Base of Our Understanding of Evolution? pp. 446–448*
7. If the frequency of allele *b* in a gene pool is 0.2, and the population is in Hardy–Weinberg equilibrium, the expected frequency of the genotype *bbbb* in a tetraploid (4*n*) plant species would be
   a. 0.0016.
   b. 0.04.
   c. 0.08.
   d. 0.2.
   e. The answer cannot be determined from this information.
   Textbook Reference: *21.1 What Facts Form the Base of Our Understanding of Evolution? pp. 446–448*
8. Random genetic drift would probably have its greatest effect on a
   a. small, isolated population.
   b. large population in which mating is nonrandom.
   c. large population in which mating is random.
   d. large population with regular immigration from a neighboring population.
   e. large population with a high mutation rate.
   Textbook Reference: *21.2 What Are the Mechanisms of Evolutionary Change? pp. 449–450*
9. Allele frequencies for a gene locus are *least* likely to be significantly changed by
   a. mutation.
   b. the founder effect.
   c. self-fertilization.
   d. gene flow.
   e. natural selection.
   Textbook Reference: *21.2 What Are the Mechanisms of Evolutionary Change? pp. 448–451*
10. Which of the following evolutionary agents would produce nonrandom changes in the genetic structure of a population?
    a. Self-fertilization
    b. Population bottlenecks
    c. Mutation
    d. Natural selection
    e. Both a and d
    Textbook Reference: *21.3 How Does Natural Selection Result in Evolution? pp. 451–453*
11. Suppose that a particular species of flowering plant that lives only one year can produce red, white, or pink blossoms, depending on its genotype. Biologists studying a population of this species count 300 red-flowering, 500 white-flowering, and 800 pink-

flowering plants in a population. When a census of the population is taken the following year, 600 red-flowering, 900 white-flowering, and 1,000 pink-flowering plants are observed. Which color has the highest fitness?

a. Red
b. White
c. Pink
d. All colors are equally fit.
e. The answer cannot be determined from this information.

Textbook Reference: *21.3 How Does Natural Selection Result in Evolution? p. 451*

12. The graph below shows the range of variation among population members for a trait determined by multiple genes.

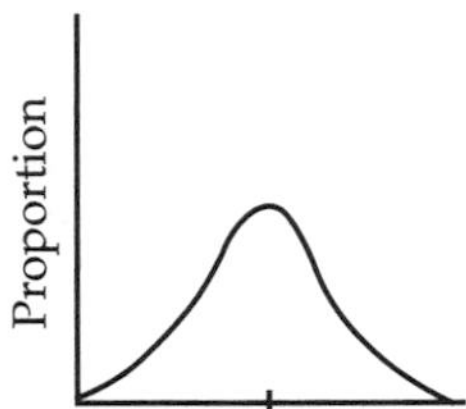

If this population is subject to *stabilizing selection* for several generations, which of the distributions would be most likely to result?

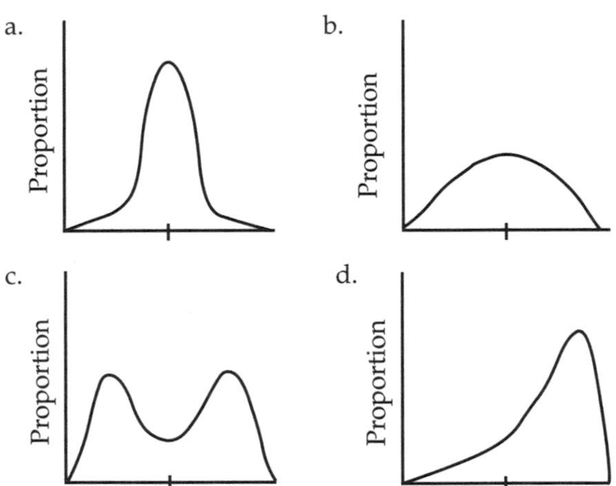

e. Both a and b

Textbook Reference: *21.3 How Does Natural Selection Result in Evolution? p. 452*

13. In areas of Africa in which malaria is prevalent, many human populations exist in which the allele that produces sickle-cell disease and the allele for normal red blood cells occur at constant frequencies, despite the fact that sickle-cell disease frequently causes death at an early age. This phenomenon is an example of

a. the founder effect.
b. a stable polymorphism.
c. mutation.
d. nonrandom mating.
e. Both b and c

Textbook Reference: *21.4 How Is Genetic Variation Maintained within Populations? p. 457*

14. Which of the following is *not* a disadvantage of sexual reproduction?

a. When it involves separate genders, it reduces the overall reproductive rate.
b. It breaks up adaptive combinations of genes.
c. It reduces the rate at which females pass genes on to their offspring.
d. It increases the difficulty of eliminating harmful mutations from the population.
e. All of the above are disadvantages of sexual reproduction.

Textbook Reference: *21.4 How Is Genetic Variation Maintained within Populations? p. 456*

15. Genetic variation within a population may be maintained by

a. frequency-dependent selection.
b. the accumulation of neutral alleles.
c. sexual recombination.
d. heterozygote advantage.
e. All of the above

Textbook Reference: *21.4 How Is Genetic Variation Maintained within Populations? pp. 456–457*

16. Which of the following can act as a constraint on the evolutionary process?

a. The trade-off between the cost and benefit of an adaptation
b. The occurrence of rare catastrophic events, such as meteorite impacts
c. The fact that all evolutionary innovations are modifications of previously existing structures
d. Both a and c
e. All of the above

Textbook Reference: *21.5 What Are the Constraints on Evolution? pp. 459–460*

Application Questions

1. Discuss the main application of the Hardy–Weinberg rule in evolutionary biology.

Textbook Reference: *21.1 What Facts Form the Base of Our Understanding of Evolution? p. 448*

2. The Hardy–Weinberg expression is an expansion of a binomial, or $(p + q)^2 = p^2 + 2pq + q^2$. In this equation, the left-hand side represents _______ frequencies, and the right-hand side represents _______ frequencies.

Textbook Reference: *21.1 What Facts Form the Base of Our Understanding of Evolution? pp. 446–447*

3. In a population with 600 members, the numbers of individuals of three different genotypes are *AA* = 350, *Aa* = 100, *aa* = 150.

a. What are the genotype frequencies in this population?
1. *AA* =
2. *Aa* =
3. *aa* =

b. What are the allele frequencies in this population?
   1. *A* =
   2. *a* =

c. What would be the expected genotype frequencies if this population were in genetic equilibrium?
   1. *AA* =
   2. *Aa* =
   3. *aa* =

d. Is this population in genetic equilibrium? Explain.

Textbook Reference: *22.1 What Facts Form the Base of Our Understanding of Evolution? pp. 446–447*

4–5. The following data were collected in a study in which dark and light moths of the peppered moth (*Biston betularia*) were released and recaptured a few days later in several different areas (*a*, *b*, *c*, and *d*).

4. Based on these data, in which area is the light phenotype most advantageous? (Assume that recapture rates reflect the relative survival of the moths, not the ability of the investigators to find and recapture the moths.) What does this tell us about the fitness of light moths in this area?

Area	Moth type	Released	Recaptured
a	dark	100	25
	light	200	50
b	dark	100	25
	light	200	100
c	dark	200	50
	light	400	100
d	dark	200	100
	light	300	75

Textbook Reference: *21.3 How Does Natural Selection Result in Evolution? pp. 451–452*

5. As peppered moths rest on bark surfaces during the day, they are subject to predation by birds. If one can assume that moths that are more camouflaged will be preyed on less, what would be the likely color of bark surfaces in area *d*?

Textbook Reference: *21.3 How Does Natural Selection Result in Evolution? pp. 451–452*

6. Describe what is meant by sexual selection and how the sexual selection of tail length in male widowbirds was studied by behavioral ecologists as described in the textbook.

Textbook Reference: *21.3 How Does Natural Selection Result in Evolution? pp. 453–455*

Answers

Diagram Exercise Answer

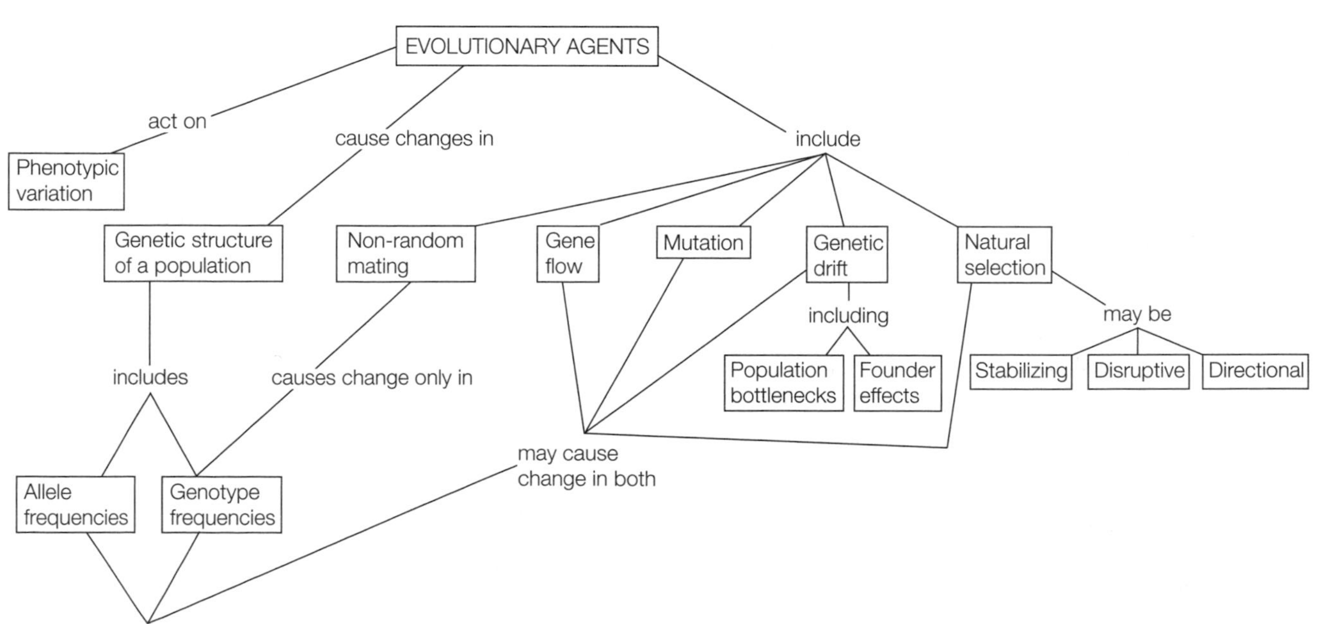

Knowledge and Synthesis Answers

1. **d.** Evolution is defined as changes in the genetic structure of a population over time, so evolution occurs at the level of the population.
2. **c.** Natural selection acts on phenotypes, not genotypes. For example, a harmful recessive allele is "invisible" to natural selection when it occurs in a heterozygote, where its harmful effect is masked by the dominant allele.
3. **c.** Although most populations show genetic variation, this condition is not part of the definition of a Mendelian population.
4. **b.** Genetic variation is related to the number of different alleles per gene. Regarding choice **a**, recall that all species members have the same number of genes. Population size per se has little to do with genetic variation, so choices **c** and **d** are also incorrect.
5. **b.** Recall that $q^2 = 0.36$ is the frequency of the *tt* genotype, so q (0.6, the square root of 0.36) is the frequency of the t allele. If there are only two alleles for this trait, then $T + t = 1$. The frequency of the T allele is therefore $1 - t = 1 - 0.6 = 0.4$.
6. **e.** Because $p = 0.2$ and therefore, $q = 1 - 0.2 = 0.8$, the frequency of the *MN* genotype is $2pq$, or $2 \times 0.2 \times 0.8 = 0.32$.
7. **a.** The probability of one allele b in a genotype is equal to its frequency, or 0.2, so the probability that all four of the alleles in a tetraploid organism will be b would be $(0.2)^4$, or 0.0016.
8. **a.** Genetic drift is most significant in small populations.
9. **c.** Unlike the founder effect, mutation, gene flow, and natural selection, all of which may change allele frequencies in a population, self-fertilization (like other types of nonrandom mating, with the exception of sexual selection) only causes a deviation from the frequency of heterozygotes predicted by Hardy–Weinberg equilibrium.
10. **e.** Self-fertilization (**a**) leads to increased numbers of homozygous individuals, and natural selection (**d**) is also nonrandom.
11. **a.** Fitness measures the relative contribution of a genotype or phenotype to subsequent generations. The red-flowering plants, which doubled in number, had the greatest percentage increase of any of the plants and thus had the highest fitness.
12. **a.** Stabilizing selection results when individuals that are intermediate in phenotype make a larger contribution to future generations than individuals of more extreme phenotype. This leads to reduced variation for the trait and causes the curve to be higher and narrower. Curve b shows greater variation, curve c would result from disruptive selection, and curve d would result from directional selection.
13. **b.** Polymorphism in a population is the existence of two or more alleles at a particular gene locus. If phenotypes are stable through time, then the underlying alleles will also be constant. In this instance, the polymorphism is stable because malaria is a significant cause of mortality in some parts of Africa, and heterozygotes have greater resistance to this disease than individuals with a "normal" phenotype. Thus the superior fitness of the heterozygotes maintains both alleles in the population.
14. **d.** Sexual recombination produces some individuals in a population who are less fit than others because they carry a greater than average number of deleterious mutations. Because these individuals are selected against, sexual reproduction is able to reduce the number of deleterious mutations in the population over time.
15. **e.** Frequency-dependent selection, the accumulation of neutral alleles, sexual recombination, and heterozygote advantage are the four major forces that maintain genetic variation in a population.
16. **e.** Cost–benefit trade-offs, developmental constraints, and major environmental disruptions can all act as constraints on natural selection and thus on the evolution of adaptive traits.

Application Answers

1. No population meets the Hardy–Weinberg conditions for genetic equilibrium. However, by determining how a population deviates from the expectations of Hardy–Weinberg, evolutionary biologists can identify which evolutionary agents are affecting the population.
2. allele; genotype
3. a
   1. $AA = 350/600 = 0.58$
   2. $Aa = 100/600 = 0.1$
   3. $aa = 150/600 = 0.25$

   b.
   1. $A = (700 + 100)/1200 = 0.67$
   2. $a = (300 + 100)/1200 = 0.33$

   c.
   1. $AA = p^2 = (0.67)2 = 0.45$
   2. $Aa = 2pq = 2 \times 0.67 \times 0.33 = 0.44$
   3. $aa = q^2 = (0.33)2 = 0.11$

   d. No. The observed genotypic frequencies differ from those predicted by the Hardy–Weinberg rule using the observed allele frequencies for p and q.
4. The light phenotype is most advantageous in area b. In areas a and c, recaptures represented 25 percent of releases for all moths, whereas in area b, 50 percent of the light moths were recaptured, indicating that moths with this phenotype survived best in this area. This suggests that fitness is higher for light moths in area b. However, it would not be possible to determine this for certain until the moths reproduced, because fitness, by definition, depends on the relative contribution of individuals to subsequent generations. (It is possible, for example, that light moths will have a harder time finding mates in this area).

5. Dark moths seem to survive best in area *d*. If the increased survival of the moths is due to cryptic coloration, then the bark surfaces in that area should also be dark.
6. Sexual selection is the spread of a trait that improves the reproductive success of an individual. The trait may improve the ability of its bearer to compete with other members of its sex for access to mates, or it may make its bearer more attractive to members of the opposite sex. By artificially lengthening or shortening the tails of male widowbirds, behavioral ecologists were able to show that increased tail length attracted more females but did not confer an advantage in interactions with other males.

22 Reconstructing and Using Phylogenies

The Big Picture

- It is fair to say that several decades ago systematics was widely regarded as one of the less dynamic arenas of research within the biological sciences. Several factors have combined in recent years to bring new excitement to the field. First was the development of a new, rigorous approach to systematics, known as cladistics. More recently, new methods of sequencing DNA and RNA, coupled with enormous increases in computer power, have enabled biologists to apply sophisticated mathematical techniques to phylogenetic studies, such as maximum likelihood analyses. As a result, systematists are now contributing to a wide array of biological studies, including subjects of medical importance such as the origins and types of human immunodeficiency virus (HIV) and other infectious organisms.
- The parsimony principle, which holds that one should prefer the simplest hypothesis capable of explaining the known facts, is fundamental not only to the reconstruction of phylogenies but also to every other field of scientific research.
- The concept of evolution revolutionized taxonomy. Before Darwin's time, Linnaeus and others had developed "natural" systems of classification based primarily on similarity of morphology. Modern biologists realize that similarities of organisms (other than homoplasies) are the result of descent from a common ancestor and believe that a truly natural system of classification should reflect evolutionary relationships.
- Knowing that organisms are evolutionarily related enables biologists to make predictions about their characteristics. This knowledge can provide important hints in the search for organisms with valuable properties, such as the ability to produce medically useful drugs.

Common Problem Areas

- The concept of homology can be confusing. Figure 22.4 provides a helpful image of the difference between homologous and homoplastic traits.
- Distinguishing monophyletic, paraphyletic, and polyphyletic groups is difficult for many students. Bear in mind that a monophyletic group is analogous to a branch (or twig) of a tree: a single "cut" can remove it from a phylogenetic tree. This analogy should help you work out which kinds of "cuts" would result in paraphyletic and polyphyletic groups.

Study Strategies

- Here is a mnemonic for the hierarchy of taxa (kingdom, phylum, class, order, family, genus, species) in the Linnaean system of classification: "Kindly Professors Cannot Often Fail Good Students." Note that the plural of genus is genera; the words *general* and *generic* come from the same root; this can help you remember that the genus is the more general taxon in the genus/species binomial. *Species* and *specific* also come from the same root, and the species is, of course, the most specific taxon.
- Go to yourBioPortal.com to review the following tutorials and activities:

  Animated Tutorial 22.1 Using Phylogenetic Analysis to Reconstruct Evolutionary History

  Interactive Tutorial: Phylogeny: Evolution of Traits

  Web Activity 22.1 Constructing a Phylogenetic Tree

  Web Activity 22.2 Types of Taxa

Important Concepts

Phylogenetic trees show the evolutionary patterns of life on Earth.

- A phylogeny is a description of the evolutionary history of relationships among organisms or their genes. Phylogenetic trees display the order in which lineages are hypothesized to have split. Each split (or node) in a phylogenetic tree represents a point at which lineages diverged in the past. The common ancestor of all the organisms in the tree forms the root of the tree. The timing of separations between lineages of organisms is shown by the positions of nodes on a time or divergence axis.

- A taxon is any named group of species. A taxon consisting of all the evolutionary descendants of a common ancestor is called a clade. Just as species that are each other's closest relatives are called sister species, so clades that are each other's closest relatives are called sister clades.
- Systematics is the study and the classification of the diversity of life.
- All of life is connected through its evolutionary history, known as the tree of life. The evolutionary relationships among species, as shown by the tree of life, form the basis for biological classification.
- Homologous traits are keys to reconstructing phylogenetic trees because they are features that are shared by members of a lineage owing to their descent from a common ancestral trait. A trait that differs from its ancestral form is called a derived trait. Conversely, a trait that was present in the ancestor of a group is known as an ancestral trait for that group.
- Synapomorphies are derived traits that are shared among a group of organisms and are viewed as evidence of the common ancestry of the group.
- Homoplasies (homoplastic traits) create confusion in reconstructing the evolutionary history of a lineage because they are features that are similar for some reason other than descent from a common ancestral trait.
- Two processes generate homoplasies: convergent evolution and evolutionary reversals. In convergent evolution, features that evolved independently become superficially similar. In an evolutionary reversal, a character reverts from a derived state to an ancestral one.

Reconstructing accurate phylogenies involves several steps.

- The first step in reconstructing a phylogeny is the choice of the ingroup and the appropriate outgroup. The ingroup is an assemblage of organisms whose phylogeny is to be determined. One method of distinguishing ancestral and derived traits is to compare the ingroup to an outgroup, which can be any species or group of species outside the group of interest. Ancestral traits should be present in both the ingroup and the outgroup, whereas derived traits should occur only in the ingroup. The root of the tree is determined by the relationship of the ingroup to the outgroup.
- In reconstructing phylogenies, systematists are guided by the parsimony principle, which states that the preferred explanation of the observed data is the simplest explanation. In practice, this means that the best phylogenetic reconstruction is the one that minimizes the number of evolutionary changes that need to be assumed over all characters in all groups in the tree. In other words, the best hypothesis is one that requires the fewest homoplasies.
- Phylogenies are constructed from many sources of data.
  - Morphology—the presence, size, shape, and other attributes of body parts—is an important source of traits for phylogenetic analysis.
  - Similarities in developmental pattern may reveal evolutionary relationships. Structures in early developmental stages sometimes reveal evolutionary relationships that are not evident in adults.
  - The morphology of fossils is particularly useful in helping to distinguish ancestral and derived traits. The fossil record also reveals when lineages diverged. In groups with few living representatives, information on extinct species may be critical to an understanding of the large divergences between the surviving species.
  - Behavior, if genetically determined, is a useful source of information about evolutionary relationships.
  - Like morphological characters, molecules are heritable characteristics that may diverge among lineages. The complete genome of an organism contains an enormous set of traits (the individual nucleotide bases of DNA) that can be used to analyze phylogenies. Both nuclear and organelle DNA sequences are used in phylogenetic studies, as are sequences in gene products (such as the amino acid sequences of proteins).
  - Because the chloroplast genome has changed slowly over evolutionary time, it is often used for the study of relatively ancient phylogenetic relationships among plants. Animal mitochondrial DNA has changed more rapidly, making it useful for studies of evolutionary relationships among closely related animal species.
- Biologists have conducted experiments both in living organisms and with computer simulations that have demonstrated the effectiveness and accuracy of phylogenetic methods.
  - The maximum likelihood method uses computer analysis for the reconstruction of phylogenies. A likelihood score of a tree is based on the probability that the observed data evolved on the specified tree, given an explicit mathematical model of evolution for the characters. An advantage of maximum likelihood analyses is that they incorporate more information about evolutionary change than parsimony methods do.

Phylogenetic trees provide important information to biologists trying to answer many kinds of questions.

- Phylogenetic trees are used to determine how many times a particular trait has evolved within a lineage. They can also be used to determine when, where, and how zoonotic diseases (diseases caused by infectious organisms that have been transferred to humans from another animal host) first entered human populations.

- Phylogenetic analysis is used by biologists to help them make relevant biological comparisons among genes, populations, and species.
- Phylogenetic methods are used not only to discover the evolutionary relationships among lineages of living organisms, but also to reconstruct morphological, behavioral, and molecular characteristics of ancestral species.
- In a molecular clock analysis, biologists assume that particular DNA sequences evolve at a reasonably constant rate and can be used as a metric to gauge the time of divergence for a particular split in a phylogeny. Molecular clocks must be calibrated with independent data such as the fossil record, known times of divergence, or biogeographic dates.

Biological classification systems are designed to express evolutionary relationships among organisms.

- The system of binomial nomenclature developed by Linnaeus in 1758 assigns each species two names, one identifying the species itself and the other the genus to which it belongs. The generic name (always capitalized) is followed by the species name (lower case), and both are italicized.
- In the Linnaean system of classification, species are grouped into higher-order taxa. The hierarchy of taxa, ranked from most to least inclusive, is kingdom, phylum, class, order, family, genus, species. Thus, a genus includes one or more species, a family includes one or more genera, and so forth.
- Biologists today recognize the tree of life as the basis for classification and often name clades without placing them into any Linnaean rank.
- Taxa in biological classifications are expected to be monophyletic. A monophyletic taxonomic group (a clade) contains an ancestor and all descendants of that ancestor and no other organisms.
- A polyphyletic group does not include its common ancestor, whereas a paraphyletic group includes some, but not all, descendants of a particular ancestor. Virtually all taxonomists agree that polyphyletic and paraphyletic groups are inappropriate as taxonomic units.
- Several sets of rules govern the use of scientific names, with the goal of providing unique and universal names for biological taxa. One such rule is that if a species is named more than once, the valid name is the first name proposed.
- In the past, different sets of taxonomic rules were developed by zoologists, botanists, and microbiologists, with the result that there are many duplicated names. Today, taxonomists are developing rules to ensure that every taxon has a unique name.

Test Yourself

Diagram Exercises

1. All but one of the trees shown in the diagram below portray the same phylogenetic relationships among taxa A, B, C, D, E, F, and G. Which one depicts a different phylogeny?

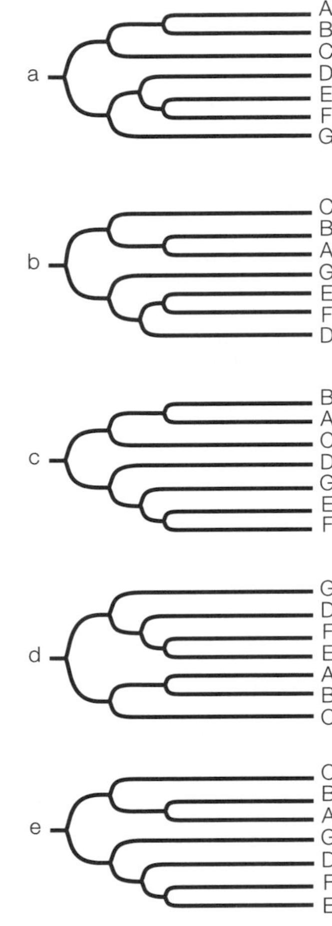

Textbook Reference: *22.1 What Is Phylogeny? p. 466*

2. In the phylogeny shown below, which group of taxa would *not* constitute a clade?

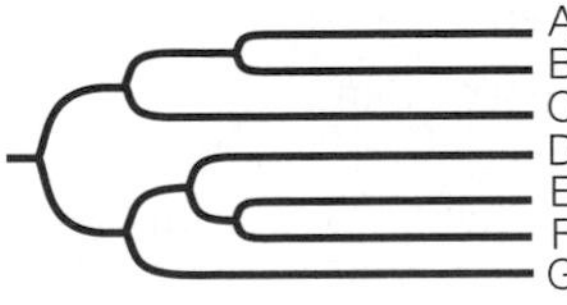

a. A, B, C, D, E, F, and G and their common ancestor
b. E, F, and G and their common ancestor
c. A, B, and C and their common ancestor
d. C, D, E, and F
e. Both b and d

Textbook Reference: *22.4 How Does Phylogeny Relate to Classification? pp. 477–478*

Knowledge and Synthesis Questions

1. A group that consists of all the evolutionary descendants of a common ancestor is called
   a. a grade.
   b. a taxon.
   c. a homology.
   d. an ingroup.
   e. a clade.
   Textbook Reference: *22.1 What Is Phylogeny? p. 466*
2. A synapomorphy is
   a. the product of convergent evolution.
   b. the result of an evolutionary reversal.
   c. a shared derived characteristic.
   d. a trait that was present in the ancestor of a group.
   e. a phylogenetic tree.
   Textbook Reference: *22.1 What Is Phylogeny? p. 467*
3. Members of genus X, a hypothetical taxon of invertebrates, have antennae with a variable number of segments. Species A and B have 10 segments; species C and D have 9 segments; species E has 8 segments. In all other genera in this family (including genus Y), all species have antennae with 10 segments. Which of the following character states is a synapomorphy that would be useful for determining evolutionary relationships within genus X?
   a. 10-segment antennae in species A and B
   b. 10-segment antennae in genus Y and in two species of genus X
   c. Antennae with fewer than 10 segments in species C, D, and E
   d. 8-segment antennae in species E
   e. Both c and d
   Textbook Reference: *22.1 What Is Phylogeny? p. 467*
4. Which of the following would *not* be expected to result in homoplasy?
   a. Convergent evolution
   b. The independent evolution of similar structures in different lineages
   c. Selection for traits that perform similar functions
   d. The inheritance of ancestral traits
   e. An evolutionary reversal
   Textbook Reference: *22.1 What Is Phylogeny? p. 467*
5. A derived trait is one that
   a. differs from its ancestral form.
   b. is homologous with another trait found in a related species.
   c. is the product of an evolutionary reversal.
   d. has the same function, but not the same evolutionary origin, as a trait found in another species.
   e. is found only in members of the outgroup.
   Textbook Reference: *22.1 What Is Phylogeny? p. 467*
6. Which of the following statements about reconstructing phylogenies is *false*?
   a. Traits found in the outgroup as well as in the ingroup are likely to be ancestral traits.
   b. Shared traits are generally assumed to be homoplastic until they can be proven to be homologous.
   c. Phylogenetic trees do not always provide an explicit time scale by which to date the splits between lineages.
   d. In a phylogenetic tree, branches can be rotated around any node without changing the meaning of the tree.
   e. A particular trait may be either ancestral or derived depending on the point of reference of the phylogeny.
   Textbook Reference: *22.2 How Are Phylogenetic Trees Constructed? pp. 468–469*
7. Which of the following is the most significant limitation of fossils as a source of information about evolutionary history?
   a. It is sometimes impossible to determine when a fossil organism lived.
   b. The fossil record for many groups is fragmentary or even nonexistent.
   c. Most fossils contain no nucleic acids or proteins and therefore are worthless for studies of molecular evolution.
   d. It is impossible to determine if morphologically similar fossils belong to the same species, because one cannot know if the fossil species could have interbred.
   e. Most fossils provide no information about the morphology of soft anatomical structures or about external characteristics such as color.
   Textbook Reference: *22.2 How Are Phylogenetic Trees Constructed? p. 470*
8. Which of the following sources of molecular data would be most helpful for a study of the evolutionary relationships of closely related animal species?
   a. Chloroplast DNA
   b. Mitochondrial DNA
   c. The amino acid sequences of a protein found in all animals, such as cytochrome *c*
   d. Ribosomal RNA sequences
   e Both b and d
   Textbook Reference: *22.2 How Are Phylogenetic Trees Constructed? p. 471*
9. Which of the following statements about the use of molecular clocks in phylogenetic analyses is true?
   a. A given gene usually evolves at the same rate in two different species regardless of differences in generation time of the species.
   b. Because changes in DNA sequences occur very slowly, molecular clocks can be used only to date evolutionary divergences that occurred millions of years ago.
   c. Molecular clocks must be calibrated with independent data, such as the fossil record.
   d. Even in a group of closely related species, different genes have been found to evolve at different rates.

e. Both c and d

Textbook Reference: *22.3 How Do Biologists Use Phylogenetic Trees? pp. 475–476*

10. Which of the following statements describes a purpose for which biologists use phylogenetic trees?
    a. For human diseases once found only in other animals, phylogenetic trees are helpful in determining when and where the infectious organisms first entered human populations.
    b. Phylogenetic trees are useful for determining how many times a particular trait may have evolved independently within a lineage.
    c. Phylogenetic trees can be used to reconstruct ancestral traits.
    d. Phylogenetic trees can be used in conjunction with molecular clocks to estimate the timing of evolutionary events.
    e. All of the above

    Textbook Reference: *22.3 How Do Biologists Use Phylogenetic Trees? pp. 473–476*

11. The *most* important attribute of a biological classification scheme is that it
    a. avoids the ambiguity created by the use of common names.
    b. reflects the evolutionary relationships among organisms.
    c. helps us remember organisms and their traits.
    d. improves our ability to make predictions about the morphology and behavior of organisms.
    e. groups together organisms with similar traits.

    Textbook Reference: *22.4 How Does Phylogeny Relate to Classification? p. 477*

12. Suppose you are writing a scientific paper about a unicellular green alga called *Chlamydomonas reinhardtii.* What is the proper way to refer to this species after the full binomial has been used once?
    a. *Chlamydomonas reinhardtii*
    b. *Chlamydomonas* spp.
    c. *Chlamydomonas* sp.
    d. *C. reinhardtii*
    e. *Chlamydomonas r.*

    Textbook Reference: *22.4 How Does Phylogeny Relate to Classification? p. 476*

13. The organisms that make up a class are _______ diverse and _______ numerous than those in a family within that class. The organisms that make up a phylum all diverged from a common ancestor _______ recently than did the organisms in an order within that phylum.
    a. more; more; less
    b. more; more; more
    c. more; less; less
    d. less; less; more
    e. less; less; less

    Textbook Reference: *22.4 How Does Phylogeny Relate to Classification? pp. 476–477*

14. Which of the following lists of taxonomic categories ranks them properly (from most inclusive to least inclusive)?
    a. Phylum, order, family, genus
    b. Class, phylum, order, species
    c. Order, class, family, genus
    d. Family, order, class, kingdom
    e. Kingdom, class, species, genus

    Textbook Reference: *22.4 How Does Phylogeny Relate to Classification? pp. 476–477*

15. The ratites are a group of flightless birds comprising the ostrich, emu, cassowaries, rheas, and kiwis. All share certain morphological similarities (such as a breastbone without a keel) not found in other birds, but they live on different continents. In the past, some ornithologists regarded their similarities as homoplasies, but they are now thought to be synapomorphies. Based on this information, you would conclude that the ratites were once regarded as a _______ group but are now believed to be _______.
    a. polyphyletic; paraphyletic
    b. paraphyletic; monophyletic
    c. polyphyletic; monophyletic
    d. monophyletic; polyphyletic
    e. monophyletic; paraphyletic

    Textbook Reference: *22.4 How Does Phylogeny Relate to Classification? pp. 477–478*

Application Questions

1. The phylogenetic tree below shows the evolutionary relationships of five species (A–E) relative to five traits (1–5). Based on this tree, fill in the table below it, using 1 to indicate the presence of a derived trait and 0 to indicate the presence of an ancestral trait.

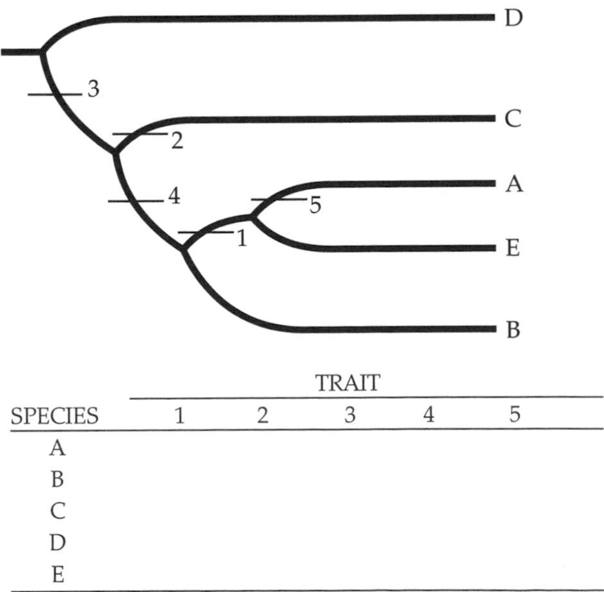

SPECIES	TRAIT 1	2	3	4	5
A					
B					
C					
D					
E					

Textbook Reference: *22.2 How Are Phylogenetic Trees Constructed? pp. 468–469*

2. In the phylogenetic tree shown in Question 1, which species is considered the outgroup?
Textbook Reference: *22.2 How Are Phylogenetic Trees Constructed? pp. 468–469*

3. The following table shows the ancestral and derived traits of five species (A–E). Based on the table, and following conventions presented in the textbook, construct a phylogenetic tree that represents the evolutionary relationships of this group. In this table, the ancestral state of each trait is indicated by 0 and the derived state is indicated by 1.

	TRAIT				
SPECIES	1	2	3	4	5
A	1	1	1	0	0
B	0	0	0	0	0
C	1	1	0	1	0
D	0	1	0	0	0
E	1	1	0	1	1

Textbook Reference: *22.2 How Are Phylogenetic Trees Constructed? pp. 468–469*

4. Discuss the application of the parsimony principle in the construction of phylogenetic trees.
Textbook Reference: *22.2 How Are Phylogenetic Trees Constructed? p. 470*

5. Discuss the implications of the following statement for the field of systematics: "DNA is the genetic material for all prokaryotes and eukaryotes."
Textbook Reference: *22.1 What Is Phylogeny? pp. 466–467*

Answers

Diagram Exercise Answers

1. **c.** Recall that branches of a phylogenetic tree can be rotated around any node without changing the meaning of the tree. Tree "c" is different from all the others because it shows G (rather than D) as the sister taxon to taxa E and F.
2. **e.** E, F, and G and their common ancestor constitute a paraphyletic group; to be a clade, D would have to be included in the group. C, D, E, and F represent a polyphyletic group because the common ancestor of these taxa is not included in the group. (If the common ancestor of these four taxa were included, they would make up a paraphyletic group.)

Knowledge and Synthesis Answers

1. **e.** A clade can be thought of as a complete branch on the tree of life. It includes the ancestor of a group, all the ancestor's descendants, and no other organisms.
2. **c.** Synapomorphies are traits that are not found in the ancestor of a group (hence they are derived) and that are found in more than one member of a group (hence they are shared).
3. **c.** Because antennae with 10 segments are found in all genera in this family except for genus X, the trait of 10-segment antennae is best regarded as ancestral; hence, having fewer than 10 segments in the antennae is a derived trait. The presence of 8-segment antennae in species E is not a synapomorphy because it is a trait found only in that species.
4. **d.** Homoplasy is the appearance of similar structures in different lineages that were not present in the common ancestor.
5. **a.** Derived traits are those that have undergone a change during evolution from the ancestral (original) character state.
6. **b.** Most shared traits, especially in species with a recent common ancestor, are likely to be homologous, not homoplastic. Therefore, the assumption that traits are homologous until proven homoplastic is more consistent with the parsimony principle than the reverse assumption is.
7. **b.** The incompleteness of the fossil record is by far the greatest limitation of its usefulness in determining phylogenies.
8. **b.** Mitochondrial DNA in animals changes rapidly over evolutionary time and hence would be most useful in determining evolutionary relationships among species that have diverged from one another only recently.
9. **e.** It is true that molecular clocks must be calibrated with independent data, and it is true that different genes and other DNA sequences evolve at different rates (see Chapter 24). But it is not true that the rate of gene evolution is independent of generation time or several other biological factors, or that molecular clocks cannot be used to date comparatively recent events (e.g., the evolution of HIV-1).
10. **e.** The textbook describes specific examples of all four of the purposes listed as answers.
11. **b.** Although all of the statements listed are important attributes of biological classification schemes, the most important attribute is that biological classification reflects evolutionary relationships.
12. **d.** After a scientific name is referenced once in a text, the genus typically is abbreviated but the species name is given in full.
13. **a.** Organisms in a higher taxon are *less* similar, have diverged from a common ancestor *less* recently, and include *more* species than organisms in a lower included taxon.
14. **a.** The complete hierarchy of taxonomic categories, from most to least inclusive, is: kingdom, phylum, class, order, family, genus, species.
15. **c.** If the shared characteristics of the ratites are homoplasies (meaning that they evolved independently by convergent evolution), then the group did not have a single common ancestor and is polyphyletic. If (as is now accepted) the shared traits

are synapomorphies shared among all ratites and not found in other birds, then the group is monophyletic.

Application Answers

1.

SPECIES	TRAIT 1	2	3	4	5
A	1	0	1	1	1
B	0	0	1	1	0
C	0	1	1	0	0
D	0	0	0	0	0
E	1	0	1	1	0

2. Species D. The outgroup has the ancestral form for all traits.

3.

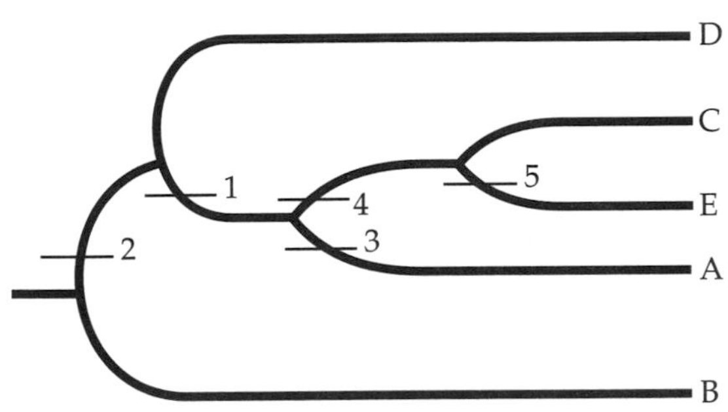

4. In the construction of a phylogenetic tree, the initial assumption is that derived traits appear only once and never disappear. Given a set of traits for a group of species, these restrictions sometimes must be relaxed to produce a phylogenetic tree for the group. Parsimony involves arranging the species so that the number of required reversals and multiple origins is minimized. Generally, the simplest explanation is most likely to be the most accurate.

5. Some of the implications of this statement are that DNA evolved as the genetic material before eukaryotes had diverged from prokaryotes, that DNA is an ancestral and general homologous trait, and that all surviving eukaryotes have DNA as their genetic material.

23 Species and Their Formation

The Big Picture

- Speciation is the process that has produced the millions of life forms—each adapted to a particular environment and way of life—that constitute life on Earth.
- Archipelagos such as the Galápagos and Hawaiian Islands have been called natural laboratories of evolution. Studies of the evolutionary radiations of the Galápagos finches and of several Hawaiian groups, such as *Drosophila* and the silverswords, have provided crucial insights to evolutionary biologists since the time of Darwin.
- Chapter 59, "Conservation Biology," discusses the reasons for the rapid pace of human-caused species extinctions, including the disappearance of many members of groups that have undergone evolutionary radiations, such as Australian marsupials and Hawaiian honeycreepers. It also describes some strategies for the preservation of Earth's precious biological diversity.

Common Problem Areas

- Some students have difficulty in understanding why hybrids between species that have different numbers of chromosomes are inevitably sterile unless the hybrid is an allopolyploid. Recall that in meiosis I, homologous chromosomes undergo synapsis. This process cannot occur properly if the haploid sets of chromosomes inherited from the parents contain different numbers of chromosomes, because it is then impossible for every chromosome to have a homolog. As a consequence, meiosis does not proceed normally and few if any normal gametes will be produced. Because allopolyploids possess four sets of chromosomes (two from each parent), their chromosomes can synapse normally and they can produce viable gametes.

Study Strategies

- Because most instructors consider the heart of this chapter to be the discussions of allopatric (geographic) speciation and reproductive isolating mechanisms, you should pay particular attention to these topics.
- Go to yourBioPortal.com to review the following tutorials and activity:

  Animated Tutorial 23.1 Founder Events and Allopatric Speciation

  Animated Tutorial 23.2. Speciation Mechanisms

  Interactive Tutorial: Speciation: Trends

  Web Activity 23.1 Concept Matching

Important Concepts

Species are distinct lineages on the tree of life.

- The morphological concept of species used by early biologists grouped organisms into species on the basis of their appearance. This concept has limitations, because in some instances not all members of a species look alike and in other instances two or more cryptic species are morphologically indistinguishable but do not interbreed.
- Speciation is the process by which one species splits into two species. Determining whether two populations constitute different species may be difficult because speciation is frequently a gradual process.
- The lineage species concept regards species as the smallest branches on the tree of life. The lineage splitting may be sudden or gradual, but in either case the lineages are thereafter independent of each other, allowing biologists to consider species over evolutionary time.
- The biological species concept defines a species as a group of actually or potentially interbreeding natural populations that are reproductively isolated from other such groups. It emphasizes that reproductive isolation is essential to keep sexual lineages on the tree of life separated from one another. This definition cannot be applied to organisms that reproduce asexually, and it is limited to a single point in evolutionary time.

The two principal modes of speciation are allopatric speciation and sympatric speciation.

- Evolutionary change can occur without speciation. A single lineage may change through time without diverging into two species.

- Speciation requires that the gene pool of the original species divide into two isolated gene pools. After separation, according to the Dobzhansky-Muller model, the isolated populations will accumulate allelic differences at gene loci (or chromosomal differences) that eventually will make it impossible for members of the two populations to interbreed successfully if they come together again.
- In some cases, complete reproductive isolation may take millions of years to develop, whereas in other cases it may take only a few generations.
- In allopatric speciation, the population is initially divided by a geographic barrier.
    - If the barrier results from a geological or climatic change, the two isolated populations are often large and genetically similar. These populations diverge not only because of genetic drift but especially because the environments in which they live are, or become, different.
    - Alternatively, separation may occur when some members of a population cross a barrier and found a new, isolated population.
    - Evidence suggests that allopatric speciation is the most common mechanism of speciation in most groups of organisms.
- Sympatric speciation occurs without geographic subdivision of the gene pool of the original species.
    - Disruptive selection, in which different genotypes have high fitness on one of two different food resources, may be a widespread mechanism of sympatric speciation among insects.
- Sympatric speciation by polyploidy, the production of duplicate sets of chromosomes within an individual, is common in plants. Polyploidy produces new species because the polyploid organisms cannot interbreed with members of the parent species. Polyploid species that have a single ancestor are called autopolyploids, whereas those that have resulted from the hybridization of two species are referred to as allopolyploids. New species may arise by polyploidy much more easily among plants than among animals because plants of many species can reproduce by self-fertilization.

Speciation requires that reproductive barriers evolve to prevent previously allopatric populations from exchanging genes if and when they become sympatric.

- Prezygotic barriers prevent members of different species from mating.
    - Species may simply mate in different areas or different parts of a habitat (habitat isolation).
    - Species may not be able to interbreed because they are fertile at different times (temporal isolation).
    - Differences in reproductive organs may prevent interbreeding (mechanical isolation).
    - The sperm and egg may be chemically incompatible (gametic isolation).
    - The two species may not recognize or respond to each other's mating behaviors, or in flowering plants, the behavioral preferences of the pollinating animals may prevent interbreeding (behavioral isolation).
- Postzygotic barriers can prevent effective gene flow between species, even if mating occurs.
    - Hybrid zygotes may not mature normally (low hybrid zygote viability).
    - Hybrids may survive less well than either parent species (low hybrid adult viability).
    - Hybrids may be infertile (hybrid infertility).
- The evolution of more effective prezygotic reproductive barriers is known as reinforcement. It may occur if the hybrid offspring of two species survive poorly.

If two populations reestablish contact before reproductive isolation is complete, several results are possible.

- If hybrid offspring are not at a selective disadvantage, they may spread through both populations with the result that the gene pools of the populations combine. Thus no new species would result from the period of isolation.
- If hybrid offspring are less successful, reinforcement may strengthen prezygotic reproductive barriers.
- If hybrid offspring are at a disadvantage but reinforcement fails to occur, a stable, narrow hybrid zone may form.

Several factors affect the rate of speciation of different evolutionary lineages.

- Characteristics of a species that make it prone to speciation include membership in a large evolutionary lineage and poor dispersal ability.
- Among plants, high rates of speciation are found in groups with specialized animal pollinators.
- Among animals, dietary specialization and sexual selection typically stimulate speciation.
- In an evolutionary radiation, many daughter species arise from a single ancestor. An adaptive radiation results in an array of species that differ significantly in their habitats and resource utilization.
- The native biota of the Hawaiian Islands illustrates that populations colonizing environments that have underutilized resources are particularly likely to produce adaptive radiations.

Test Yourself

Diagram Exercise

Construct a concept map whose theme is "Species." Include in your map the following terms: species, allopatric, allopolyploidy, autopolyploidy, biological, concepts, founder events, independent evolution,

interruption of gene flow, lineage, morphological, physical similarity, postzygotic barriers, prezygotic barriers, reproductive isolation, speciation, and sympatric. Connect these terms by verbs or short phrases to indicate the relationships among them.
Textbook Reference: *23.1 What Are Species? pp. 482–484; 23.2 How Do New Species Arise? pp. 484–489; 23.3 What Happens When Newly Formed Species Come Together? pp. 489–493*

Knowledge and Synthesis Questions

1. It is difficult to apply the biological species concept to groups of organisms that
   a. are asexual.
   b. produce hybrids only in captivity.
   c. show little morphological diversity.
   d. exist only in the fossil record.
   e. Both a and d
   Textbook Reference: *23.1 What Are Species? p. 483*
2. Which of the following statements about allopatric speciation is *false*?
   a. It can sometimes involve small populations.
   b. It occurs only in species that are widely distributed.
   c. It always involves a physical barrier that interrupts gene flow.
   d. It sometimes can involve chance events.
   e. It is the dominant mode of speciation in most groups of organisms.
   Textbook Reference: *23.2 How Do New Species Arise? p. 485*
3. A long, narrow hybrid zone exists in Europe between the ranges of the fire-bellied toad and the yellow-bellied toad. The persistence of this zone can be attributed to which of the following factors?
   a. Reinforcement strengthens the prezygotic barriers between the two species.
   b. Hybrid offspring have the same fitness as nonhybrid offspring.
   c. Both species travel long distances over the course of their lives.
   d. Individuals from outside the hybrid zone regularly move into the hybrid zone.
   e. None of the above
   Textbook Reference: *23.3 What Happens When Newly Formed Species Come Together? pp. 492–493*
4. Which type of speciation is most common among flowering plants?
   a. Geographic
   b. Sympatric
   c. Allopatric
   d. Disruptive
   e. None of the above
   Textbook Reference: *23.2 How Do New Species Arise? p. 488*
5. Which of the following would *not* be considered an example of a prezygotic reproductive isolating mechanism?
   a. One bird species forages in the tops of trees for flying insects, whereas another forages on the ground for worms and grubs.
   b. The males of one species of moth cannot detect and respond to the sex attractant chemicals produced by the females of another species.
   c. Sperm of one species of sea urchin are unable to penetrate the egg plasma membrane of another species.
   d. Mosquitoes of one species are active in foraging and searching for mates at dusk, whereas those of another species are active at dawn.
   e. Flowers of one orchid species mimic female bees of species A, whereas flowers of another orchid species mimic female bees of species B.
   Textbook Reference: *23.3 What Happens When Newly Formed Species Come Together? pp. 489–490*
6. Which of the following factors would *not* be expected to increase the rate of speciation in a group of organisms?
   a. Fragmentation of populations
   b. Poor dispersal ability
   c. High birthrates
   d. Dietary specialization
   e. Pollination by animals
   Textbook Reference: *23.4 Why Do Rates of Speciation Vary? pp. 493–494*
7. Which of the following is *not* a suggested reason for the adaptive radiation of silverswords on the Hawaiian archipelago?
   a. Water is an effective barrier for many organisms.
   b. Because islands are small compared with mainland areas, more species would be expected to develop there.
   c. Competition is frequently reduced on islands.
   d. More ecological opportunities exist on islands that have not been colonized by many species.
   e. Neither b nor d is a suggested reason.
   Textbook Reference: *23.4 Why Do Rates of Speciation Vary? pp. 494–495*
8. More than 800 species of *Drosophila* occur in the Hawaiian Islands, representing 30 to 40 percent of all the species in this genus. The occurrence of so many *Drosophila* species in this island chain is
   a. the result of many founder events followed by genetic divergence.
   b. an example of an evolutionary radiation.
   c. largely the result of sympatric speciation.
   d. evidence that the genus *Drosophila* first evolved in the Hawaiian Islands.
   e. Both a and b
   Textbook Reference: *23.2 How Do New Species Arise? p. 486*
9. Which of the following statements about speciation is *false*?
   a. A small founding population can be involved in speciation.

b. Speciation always involves interruption of gene flow between different groups of organisms.
c. The rate of speciation can vary for different groups of organisms.
d. Speciation always requires many generations.
e. Speciation may occur because certain genotypes within a population prefer distinct microhabitats where mating takes place.

Textbook Reference: *23.2 How Do New Species Arise? p. 488*

10. Which of the following observations would constitute conclusive evidence that two overlapping populations that had been geographically separated have *not* diverged into distinct species?
a. Matings between members of the two populations produce viable hybrids.
b. A stable hybrid zone exists where their ranges overlap.
c. Interbreeding is common between members of the two populations.
d. All of the above
e. None of the above

Textbook Reference: *23.3 What Happens When Newly Formed Species Come Together? pp. 492–493*

11. In allopatric speciation, which process is likely to be *least* important?
a. A founder event
b. Allopolyploidy
c. Behavioral isolation
d. Genetic drift
e. Both b and d

Textbook Reference: *23.2 How Do New Species Arise? pp. 485–488*

12. A field contains two related species of flowering plants. Species A has a diploid chromosome number of 16, and species B has a diploid number of 18. If a third species arises as a result of hybridization between A and B, how many chromosomes will it have?
a. 17
b. 32
c. 34
d. 36
e. 68

Textbook Reference: *23.2 How Do New Species Arise? p. 488*

13. Speciation by polyploidy occurs far more often in plants than in animals because
a. plants are more likely to be capable of self-fertilization than animals.
b. plant cells can tolerate extra sets of chromosomes, whereas animal cells cannot.
c. plants as a rule have higher reproductive rates than animals.
d. many plants are specialized with respect to their pollinating agent.
e. All of the above

Textbook Reference: *23.2 How Do New Species Arise? p. 488*

14. Two species of narrowmouth frogs in the United States have mating calls that differ more in their region of sympatry than in those parts of their ranges that do not overlap. If this difference in their vocalizations has the function of preventing hybridization between the two species, it is an example of
a. a hybrid zone.
b. reinforcement.
c. sympatric speciation.
d. a postzygotic reproductive barrier.
e. allopatric speciation.

Textbook Reference: *23.3 What Happens When Newly Formed Species Come Together? p. 491*

15. Four hypothetical families of birds (1) are endemic either to a single large island or to an island group (archipelago) far from the nearest continental land mass, and (2) have either monogamous or promiscuous mating systems in their species. In which of the four would you predict the highest rate of speciation?
a. The single-island family whose species are promiscuous
b. The archipelago family whose species are monogamous
c. The single-island family whose species are monogamous
d. The archipelago family whose species are promiscuous
e. The rate of speciation cannot be predicted from the factors mentioned above.

Textbook Reference: *23.2 How Do New Species Arise? p. 486; 23.4 Why Do Rates of Speciation Vary? p. 494*

Application Questions

1. Discuss the conditions on the Galápagos Islands that led to the evolution of the birds known as Darwin's finches.

Textbook Reference: *23.2 How Do New Species Arise? pp. 486–487*

2. In autopolyploidy, a new species of plant can arise by the doubling of chromosome numbers in a single individual of one species (provided that the individual is capable of self-fertilization). Why is it virtually impossible for such a tetraploid plant to interbreed successfully with diploid individuals of the "same" species?

Textbook Reference: *23.2 How Do New Species Arise? p. 488*

3. Discuss what is known about the evolutionary radiations on islands, based on studies of Hawaiian silverswords and tarweeds.

Textbook Reference: *23.4 Why Do Rates of Speciation Vary? pp. 494–495*

4. Suppose that members of two populations are separated by a geographic barrier and begin to diverge genetically. Many generations later, when the barrier is removed, the two populations can interbreed, but the hybrid offspring do not survive and reproduce well. Explain how natural selection might lead to the

evolution of more effective prezygotic barriers in these species.
Textbook Reference: *23.3 What Happens When Newly Formed Species Come Together? pp. 491–493*

5. The yellow-rumped warbler was formerly split into two species (myrtle and Audubon's warblers), but in 1973 it was reclassified as a single species. "Myrtle" and "Audubon's" warblers have largely allopatric ranges but hybridize where they are sympatric in the Canadian Rockies. They are similar in appearance but are readily distinguished by experienced birders. What further data about these two forms should ornithologists collect and analyze in order to decide whether they should continue to be classified as a single species?
Textbook Reference: *23.3 What Happens When Newly Formed Species Come Together? pp. 492–493*

Answers

Diagram Exercise Answer

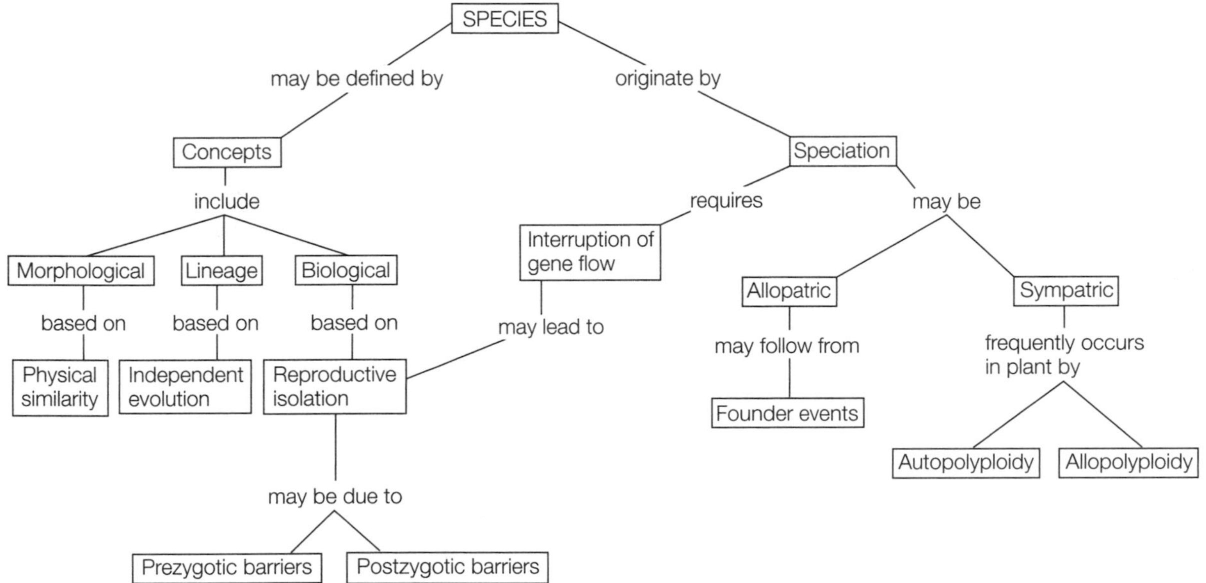

Knowledge and Synthesis Answers

1. **e.** The key criterion of a biological species is that its members are reproductively isolated from other such groups. This criterion is impossible to evaluate in asexual and fossil species.
2. **b.** A wide distribution is not a requisite for allopatric speciation.
3. **d.** These toads are an example of related species in which reinforcement does not strengthen prezygotic barriers, even though hybrid offspring are only half as fit as nonhybrid offspring. The reason for this is that toads from outside the hybrid zone (not subject to the selective pressure against hybridizing) regularly move into the hybrid zone and mate with members of the other species. The hybrid zone remains narrow because toads do not travel long distances; natural selection removes hybrids from the population before they can disperse very far.
4. **b.** Sympatric speciation is most common among flowering plants. It has been estimated that about 70 percent of all flowering plant species are polyploid.
5. **a.** Provided that the two species are active in the same locality at the same time, a difference in the habitat in which they forage would not in itself be a barrier to interbreeding (though seeking mates in different habitats might well be a barrier to interbreeding). All the other choices describe reproductive barriers that would act prior to fertilization.
6. **c.** Birthrates per se do not seem to affect the rate of speciation in organisms. All other factors have been shown to increase speciation rates in the lineages of some organisms.
7. **b.** Actually, biogeographers have found that larger land masses tend to have more species than smaller land masses, so you might expect the reverse effect.
8. **e.** Island groups are frequently sites of evolutionary radiations through repeated allopatric speciation events that are initiated by individuals (or groups) dispersing from one island to another. The numerous species of *Drosophila* in the Hawaiian Islands are believed to have originated in this way.
9. **d.** New species formed by polyploidy can arise in only two generations.
10. **e.** Interbreeding, production of viable hybrids, and establishment of a hybrid zone do not necessarily mean that speciation is incomplete. If, however, the hybrids were successful, fertile, and bred freely with

members of both original populations, their gene pools would merge, and you would conclude that speciation had not taken place.

11. **b.** Genetic drift, founder events, and behavioral isolation may all play a role in allopatric speciation. Allopolyploidy, however, can occur only as the result of hybridization between individuals of different species; hence the two parent species cannot be allopatric.
12. **c.** If haploid gametes of species A and B joined, the result would be a zygote with 17 chromosomes. A mature plant with 17 chromosomes would be sterile, because chromosomes would be unable to pair properly during prophase and metaphase of meiosis I. A fertile allopolyploid would therefore have to have 34 chromosomes so that each chromosome would have a homolog with which to pair. Autopolyploids of species A and B would have 32 and 36 chromosomes, respectively.
13. **a.** If a polyploid plant or animal is capable of self-fertilization, then a new species can arise from a single individual. The ability to self-fertilize is far more common among plants than animals.
14. **b.** Reinforcement is defined as the evolutionary strengthening of prezygotic barriers to interbreeding within the zone of sympatry of two closely related species.
15. **d.** Speciation occurs more readily in archipelagos than on isolated islands because the establishment of geographical isolation through founder events involving dispersing individuals occurs much more readily within archipelagos than on a single island. Speciation would be more rapid in a family with promiscuous mating systems because species with this system are frequently characterized by a high degree of sexual dimorphism and the capacity of individuals to make subtle discriminations both between members of their own species and between members of their own species and other species. As a consequence, even slight differences in the appearance or behavior of members of different populations may lead to the evolution of reproductive barriers based on the mating preferences of individuals in the different populations.

Application Answers

1. The relatively great distance between the Galápagos Islands and the South American mainland and also between each of the islands in the archipelago ensured that once immigrants had arrived on an island, they would be genetically isolated for a substantial period of time. Also, because the islands differ greatly in climate and vegetation, the resident birds were subject to different selection pressures. This, in combination with reduced gene flow between the islands, led to a rapid evolutionary radiation of finches.
2. Recall from Chapter 11 that pairs of homologous chromosomes synapse during prophase and metaphase of the first division of meiosis. The homologs then separate, so that each cell resulting from meiosis I is haploid, as are the products of meiosis II. In tetraploids as in diploids, meiosis is normal because pairing of homologs can occur. Any offspring of a cross between tetraploid and diploid individuals would be triploid, however, and therefore sterile because correct synapsis of homologs could not occur. Because the tetraploid product of autopolyploidy is reproductively isolated from its diploid relatives, it is a new species.
3. Studies of the silverswords of the Hawaiian archipelago show that taxa that evolve on islands frequently show great morphological diversity because of the reduced competition that immigrants encounter on islands. Thus Hawaiian silverswords, unlike their mainland tarweed relatives, have evolved tree- and shrublike species because there were few resident tree and shrub species with which they had to compete.
4. Recall that natural selection tends to remove from a population traits that reduce survival or reproductive success. Individuals that interbreed between populations will have lower fitness (they will contribute fewer offspring to future generations) than those who breed within their own population. If the tendency to avoid interbreeding is heritable (and not just the result of chance), the frequency of alleles that prevent interbreeding will increase in each population. How might such a trait be heritable? Any of the prezygotic barriers to interbreeding might be heritable traits. For example, if the species in question are frogs, and if their mating calls started to diverge while the populations were separated, the following traits might be heritable: a tendency to make a call that is more distinct from that of the other population, or the ability to distinguish between the existing calls of the two populations (coupled with a preference for the call of one's own population). As alleles for these traits increased in frequency, they would contribute to the behavioral isolation of the two populations and perhaps eventually to more complete speciation.
5. During allopatric speciation, divergence of two populations often occurs gradually. In such cases it is inevitable that intermediate stages of speciation occur, and it may be a matter of opinion whether two forms have diverged sufficiently to be considered separate species. With respect to these two warbler populations, biologists would seek answers to these questions: Are hybrid offspring as fit as those resulting from mating of individuals of the same population? Is there evidence that the zone of hybridization is expanding, indicating that the gene pools of the populations are combining? Is there evidence of reinforcement of prezygotic barriers to interbreeding (e.g., a greater difference in the songs of the two forms in the area of hybridization than in allo-patric parts of their ranges)? Lesser fitness of hybrid offspring, a stable, narrow zone of hybridization, and evidence of reinforcement would favor the conclusion that the populations are best regarded as separate species.

24 Evolution of Genes and Genomes

The Big Picture

- The growing importance of the study of macromolecules in the reconstruction of phylogenies is an example of the far-reaching impact that advances in molecular biology have had in the decades since the discovery of the structure of DNA by Watson and Crick in 1953.
- The principles of molecular evolution are critical for our understanding of how new diseases—especially those caused by viruses—originate and evolve. This understanding in turn may lead to more effective strategies for combating these diseases.

Common Problem Areas

- The concept of concerted evolution and the two mechanisms by which it can occur may seem confusing. Study Figure 24.11 to understand the difference between unequal crossing over and biased gene conversion.
- The difference between orthologs and paralogs may also be confusing. Study Figure 24.12 to understand the distinction.

Study Strategies

- Study Figure 24.1 carefully to understand how a similarity matrix is constructed. Similarity matrices are essential to reconstructing phylogenies based on molecular data.
- Go to yourBioPortal.com to review the following tutorial and activities:

  Animated Tutorial 24.1 Concerted Evolution

  Web Activity 24.1 Amino Acid Sequence Alignment

  Web Activity 24.2 Similarity Matrix Construction

  Web Activity 24.3 Gene Tree Construction

Important Concepts

The genome of an organism is its entire set of genes as well as any noncoding regions of the DNA (or, in some viruses, RNA).

- The genome of eukaryotes includes both the nuclear genes located on chromosomes and genes present in mitochondria and chloroplasts.
- A successful gene must be able to function in a wide variety of genetic backgrounds. Among the genes of an individual, there are both divisions of labor and strong interdependencies.

The field of molecular evolution is the study of the evolution of genomes and of particular nucleic acids and proteins.

- Scientists study genomes and particular nucleic acids and proteins to determine how rapidly and why they have changed. The answers to these questions are crucial to an understanding of the evolutionary history of genes and of the organisms that carry them.
- The evolution of nucleic acids and proteins depends on variation introduced by mutation. Nucleotide substitution mutations may result in amino acid replacements on the encoded proteins.
- Evolutionary changes in genes and proteins are detected by comparing the nucleotide and amino acid sequences among different organisms.

The structures of macromolecules can be determined and compared.

- By the use of the sequence alignment technique, biologists identify homologous sequences within nucleic acids or proteins. The concept of homology (similarity that results from common ancestry) extends down to particular positions in nucleotide or amino acid sequences.
- Having identified homologous regions of a nucleic acid or protein, biologists construct a similarity matrix to measure the minimum number of changes that have occurred during the divergence between pairs of organisms. The assumption is that the longer the molecules have been evolving separately, the more differences they will have.
- For several reasons, the observed number of differences in homologous DNA sequences in two species almost certainly underestimates the number of substitutions that actually have occurred since the sequences diverged from a common ancestor. Molecular

evolutionists use mathematical models to correct for this undercounting.

- Because viruses, bacteria, and unicellular eukaryotes have short generation times, biologists use them to study molecular evolution in the laboratory.

Molecular evolution occurs by diverse mechanisms.

- A synonymous (silent) mutation replaces a nucleotide base in a codon but does not change the amino acid specified by the codon. Because synonymous mutations do not affect the functioning of a protein, they are unlikely to be affected by natural selection.
- A nonsynonymous mutation changes the amino acid specified by the codon. Though such mutations are likely to be harmful, they are sometimes selectively neutral, or nearly so, and are occasionally advantageous.
- Within functional genes, nucleotide substitution rates are highest at nucleotide positions that do not change the amino acid being expressed. The rate of substitution is higher in pseudogenes—duplicate copies of genes that are never expressed—than in functional genes.
- The neutral theory of molecular evolution postulates that most evolutionary change in macromolecules, as well as much of the genetic variation within species, is the result of random genetic drift, rather than natural selection. The rate of fixation of neutral mutations is theoretically constant and equal to the mutation rate.
- According to the neutral theory of molecular evolution, it is possible to distinguish among evolutionary processes by comparing the rates of synonymous and nonsynonymous substitutions in a protein-coding gene.
  - If an amino acid substitution is neutral in its effect on fitness, then the rates of synonymous and nonsynonymous substitutions in the corresponding DNA sequences are expected to be very similar.
  - If an amino acid position is under strong selection for change, then the rate of nonsynonymous substitutions in the corresponding DNA is expected to exceed the rate of synonymous substitutions.
  - If an amino acid position is under purifying selection, then the rate of synonymous substitutions in the corresponding DNA is expected to be much higher than the rate of nonsynonymous substitutions.
- The enzyme lysozyme, while serving in almost all animals as an important first line of defense against invading bacteria, also has evolved to take on an essential role in the digestive process of several groups of foregut fermenters. By comparing the lysozyme-coding sequences in foregut fermenters with several of their non-fermenting relatives, molecular evolutionists have discovered that neutral evolution, purifying selection, and selection for change have all occurred as lysozyme evolved to take on its new function. The independent evolution in several groups of foregut fermenters of a type of lysozyme adapted to its new environment and function shows that convergent evolution occurs at the molecular level.
- Genome size and organization also evolve. The size of the coding portion of the genome is larger in more complex organisms. Thus, eukaryotes have many times more genes than prokaryotes, and multicellular eukaryotes with tissue organization have more genes than single-celled eukaryotes.
- Most of the variation in genome size of various organisms is due not to differences in the number of functional genes, but in the amount of noncoding DNA.
- Although much of the noncoding DNA appears to be nonfunctional, it may alter the expression of surrounding genes. Important categories of noncoding DNA include pseudogenes and parasitic transposable elements.
- Studies of the rate of retrotransposon loss show that species differ greatly in the rate at which they gain or lose apparently functionless DNA. The reason for the differences between species is unclear. It may be related to the rate at which the organism develops or to its population size.

Gene numbers within a genome can increase over evolutionary time by lateral transfer or by gene duplication.

- Lateral gene transfer occurs when a species picks up fragments of foreign DNA directly from the environment, or via a virus, or through hybridization with another species. This process increases the genetic variability of the species and thus provides additional raw material on which natural selection can act. It can also result in the spread of genetic functions between distantly related species.
- The endosymbiotic events that gave rise to mitochondria and chloroplasts in the eukaryotic lineage can be viewed as lateral transfers of entire bacterial genomes. Lateral transfer appears to occur much more commonly among species of bacteria than among most eukaryotic lineages, a notable exception being the high level of hybridization among closely related plant species. Species boundaries among bacteria are frequently blurred because of lateral transfer of genes.
- When a gene is duplicated, four evolutionary outcomes are possible:
  - Both copies retain the gene's original function.
  - Both copies retain the ability to produce the original gene product, but the expression of the genes diverges in different tissues or at different times of development.
  - One copy becomes a functionless pseudogene.
  - One copy retains its original function, while the second mutates so extensively that it can perform a different function.

- When an entire genome is duplicated (as in polyploid organisms), there are massive opportunities for new gene functions to evolve. Genome duplication events that occurred in the ancestor of the jawed vertebrates have permitted many individual vertebrate genes to become highly tissue-specific in their expression.
- Successive rounds of gene duplication and mutation can result in a gene family—a group of homologous genes with related functions.
- The globin gene family illustrates that gene diversification can produce molecules with different functions (hemoglobin and myoglobin) as well as functionless pseudogenes.
- Concerted evolution results in similar DNA sequence changes in all copies of highly repeated genes, such as those that code for ribosomal RNA. Two separate mechanisms can cause concerted evolution:
  - If homologous chromosomes align imprecisely during meiosis, unequal crossing over may occur, with the result that one chromosome will gain extra copies of a highly repeated gene and the other chromosome will be left with fewer copies.
  - If one favored copy of a repeated gene on one homologous chromosome is used as the template for the repair of damage to the copies of the gene on the other homolog, the result is biased gene conversion: the rapid spread of the favored sequence across all the copies of the gene.

Molecular evolution has many applications.

- A gene tree depicts the evolutionary history of a particular gene or of the members of a gene family. Orthologs are genes found in different organisms that arose from a single gene in their common ancestor. Paralogs are related genes that have resulted from gene duplication in a single lineage.
- The principles of molecular evolution help us understand function and diversification of function in many proteins. For example, detection of strong selection for change in a nucleotide sequence can help us identify molecular changes that have resulted in functional changes.
- Molecular evolutionary principles underlie the field of in vitro evolution, in which new molecules are produced in the laboratory to perform particular desired functions. The basis of in vitro evolution is the creation of random molecular variation followed by selection by the experimenter.
- Biomedical scientists are using principles of molecular evolution to identify and combat human diseases.

Test Yourself

Diagram Exercise

Construct a concept map whose theme is "Genome." Include in your map the following terms: genome, duplication, endosymbiosis, eukaryotes, genes, gene families, hybridization, lateral gene transfer, noncoding DNA, organelle DNA, paralogs, prokaryotes, pseudogenes, and size. Connect these terms by verbs or short phrases to indicate the relationships among them.
Textbook Reference: *24.1 How Are Genomes Used to Study Evolution pp. 499–505; 24.2 What Do Genomes Reveal About Evolutionary Processes? pp. 505–509; 24.3 How Do Genomes Gain and Maintain Functions? pp. 509–512; 24.4 What Are Some Applications of Molecular Evolution? pp. 512–515*

Knowledge and Synthesis Questions

1. The genome of a eukaryotic organism is best defined as
   a. all of the organism's protein-coding genes.
   b. all of the organism's DNA contained in its nucleus.
   c. all of the organism's genetic material.
   d. a haploid set of all the organism's chromosomes.
   e. all of the organism's DNA that is transcribed.
   Textbook Reference: *24.1 How Are Genomes Used to Study Evolution? p. 499*
2. The sequence alignment technique
   a. permits comparison of sequences of amino acids in proteins or sequences of nucleotides in DNA.
   b. enables the detection of deletions and insertions in sequences that are being compared.
   c. enables the detection of back substitutions and parallel substitutions in sequences that are being compared.
   d. Both a and b
   e. All of the above
   Textbook Reference: *24.1 How Are Genomes Used to Study Evolution? pp. 500–501*
3. Experimental molecular evolutionary studies have shown that
   a. a heterogeneous environment favors adaptive radiation.
   b. a heterogeneous environment induces an increase in the mutation rate.
   c. even in bacteria, substantial molecular evolution cannot be observed in less than a year.
   d. Both a and b
   e. None of the above
   Textbook Reference: *24.1 How Are Genomes Used to Study Evolution? pp. 502–504*
4. In a eukaryote, one would expect to find the lowest rate of nonsynonymous nucleotide substitutions in an
   a. intron of a protein-coding gene.
   b. exon of a protein-coding gene.
   c. intron of a pseudogene.
   d. exon of a pseudogene.
   e. intron or exon of a protein-coding gene.
   Textbook Reference: *24.2 What Do Genomes Reveal About Evolutionary Processes? pp. 505–506*
5. Which of the following statements about mutations is *false*?
   a. A silent mutation results in no change in the amino acid sequence of a protein.

b. According to the neutral theory of molecular evolution, most substitution mutations are selectively neutral and accumulate through genetic drift.
c. A base substitution mutation in the third codon position is more likely to be neutral than a substitution at the first or second codon position.
d. Nonsynonymous mutations are virtually always deleterious to the organism.
e. The rate of fixation of neutral mutations is equal to the mutation rate and is independent of population size.

Textbook Reference: *24.2 What Do Genomes Reveal About Evolutionary Processes? pp. 505–506*

6. Studies of the structure of proteins such as cytochrome *c* and lysozyme in different species have shown that
a. the rate of evolution of particular proteins is often relatively constant over time.
b. many nucleotide substitutions result either in no change in the amino acid sequence of the protein or in a change to a functionally equivalent amino acid.
c. there are fewer differences in the amino acid sequences of proteins whose organismal sources were closely related than in those whose sources were distantly related.
d. functionally important regions of a protein can be discovered by identifying the regions with the most amino acid substitutions.
e. All of the above

Textbook Reference: *24.1 How Are Genomes Used to Study Evolution? pp. 502–504; 24.2 What Do Genomes Reveal About Evolutionary Processes? pp. 506–508*

7. Which of the following statements about the enzyme lysozyme is true?
a. A small group of closely related mammals has evolved a special form of lysozyme that functions in digestion.
b. The lysozymes found in the foregut fermenters resulted from convergent evolution.
c. Lysozyme could not have evolved a secondary function if it had been an enzyme with a vital primary function.
d. A higher mutation rate in the foregut fermenters allows their lysozymes to evolve rapidly.
e. Lysozyme first evolved as a defense against bacteria in the common ancestor of mammals.

Textbook Reference: *24.2 What Do Genomes Reveal About Evolutionary Processes? pp. 506–507*

8. Which of the following ranks the organisms correctly in terms of the expected total amount of coding DNA in their genomes (from least coding DNA to most coding DNA)?
a. Bacterium, single-celled eukaryote, *Drosophila*, bird
b. Bacterium, *Drosophila*, bird, single-celled eukaryote
c. Single-celled eukaryote, bacterium, *Drosophila*, bird
d. *Drosophila*, single-celled eukaryote, bird, bacterium
e. *Drosophila*, bacterium, single-celled eukaryote, bird

Textbook Reference: *24.2 What Do Genomes Reveal About Evolutionary Processes? p. 508*

9. Which of the following ranks the organisms correctly in terms of the proportion of coding DNA to noncoding DNA in their genomes (from smallest proportion to largest proportion)?
a. *E. coli*, yeast, *Drosophila*, human
b. Human, yeast, *E. coli*, *Drosophila*
c. Human, *Drosophila*, yeast, *E. coli*
d. *Drosophila*, human, yeast, *E. coli*
e. Yeast, *E. coli*, *Drosphila*, human

Textbook Reference: *24.2 What Do Genomes Reveal About Evolutionary Processes? p. 509*

10. Which of the following would be the *least* likely result of gene duplication?
a. The gene produces less of its product than it did before duplication.
b. The two copies of the gene are expressed at different stages in the development of the organism.
c. As a result of evolutionary divergence, one copy retains its original function, and the other copy acquires a different function.
d. One copy remains functional, and the other copy evolves into a functionless pseudogene.
e. All of the above are about equally likely results.

Textbook Reference: *24.3 How Do Genomes Gain and Maintain Functions? p. 510*

11. Which of the following statements about gene families is *false*?
a. Gene families evolve via gene duplication.
b. Pseudogenes are quickly removed from gene families by deletion.
c. Members of a gene family can include several functional genes.
d. Examples of gene families include the *engrailed* and globin gene families in vertebrates.
e. In some gene families, the members do not evolve independently of one another.

Textbook Reference: *24.3 How Do Genomes Gain and Maintain Functions? pp. 510–512*

12. Gene duplication via the mechanism of polyploidy
a. results in the duplication of the entire genome, apart from extranuclear DNA.
b. has occurred in the evolutionary history of many plants.
c. is believed not to have occurred in the evolutionary history of any animal groups.
d. Both a and b
e. All of the above

Textbook Reference: *24.3 How Do Genomes Gain and Maintain Functions? p. 510*

13. Nonindependent evolution of some repeated genes within a species
a. is called concerted evolution.
b. can be caused by biased gene conversion.
c. can be caused by unequal crossing over.
d. Both a and c
e. All of the above

Textbook Reference: *24.3 How Do Genomes Gain and Maintain Functions? pp. 511–512*

14. Orthologous genes are genes that can be traced back to a common _______ event.
    a. duplication
    b. substitution
    c. speciation
    d. deletion
    e. duplication and speciation

    Textbook Reference: *24.4 What Are Some Applications of Molecular Evolution? p. 512*

15. In vitro evolution
    a. can produce both nucleic acid and protein molecules unknown in living organisms.
    b. requires many rounds of production of many variant molecules and the selection of those having (or beginning to have) the desired properties.
    c. often involves techniques and molecules employed in recombinant DNA technology, such as PCR and cDNA.
    d. Both b and c
    e. All of the above

    Textbook Reference: *24.4 What Are Some Applications of Molecular Evolution? p. 514*

Application Questions

1. The table below shows the amino acid sequences for an eight-residue section of a small protein for five different species (1–5).

Position	1	2	3	4	5	6	7	8
Species 1:	Arg	Cys	Leu	Leu	Ser	Thr	Asn	Met
Species 2:	Arg	Cys	Phe	Leu	Leu	Ser	Thr	Asn
Species 3:	Arg	His	Leu	Leu	Ser	Thr	Asn	Met
Species 4:	Arg	Cys	Leu	Ser	Ser	Thr	Asn	Met
Species 5:	Arg	His	Leu	Leu	Ser	Gln	Asn	Met

Complete the following similarity matrix using these sequences:

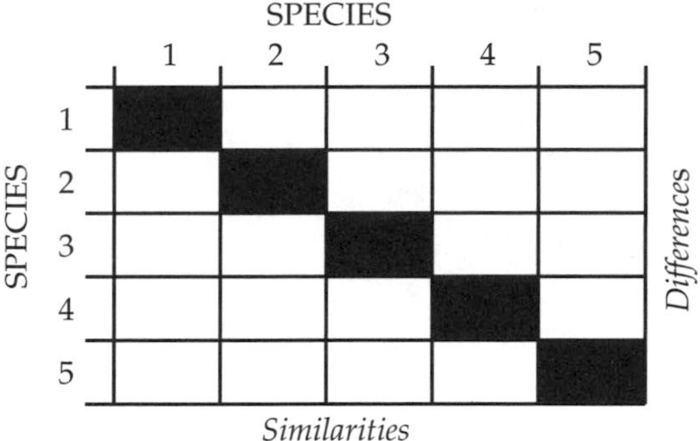

Which species differ from Species 1 because of an amino acid insertion?

Which species differ from Species 1 because of amino acid substitutions?

Textbook Reference: *24.1 How Are Genomes Used to Study Evolution? pp. 500–501*

2. How might the proportion of coding to noncoding DNA in the genome of a species be related to the relative importance of natural selection and genetic drift in the evolution of the species?

   Textbook Reference: *24.2 What Do Genomes Reveal About Evolutionary Processes? p. 509*

3. How is one of the mechanisms of concerted evolution related to the pairing of homologous chromosomes that occurs during meiosis?

   Textbook Reference: *24.3 How Do Genomes Gain and Maintain Function? pp. 511–512*

4. Compare in vitro evolution with molecular evolution in organisms.

   Textbook Reference: *24.4 What Are Some Applications of Molecular Evolution? p. 514*

5. How is the study of molecular evolution important in efforts to combat HIV and other viral pathogens that have recently emerged?

   Textbook Reference: *24.4 What Are Some Applications of Molecular Evolution? pp. 514–515*

Answers

Diagram Exercise Answer

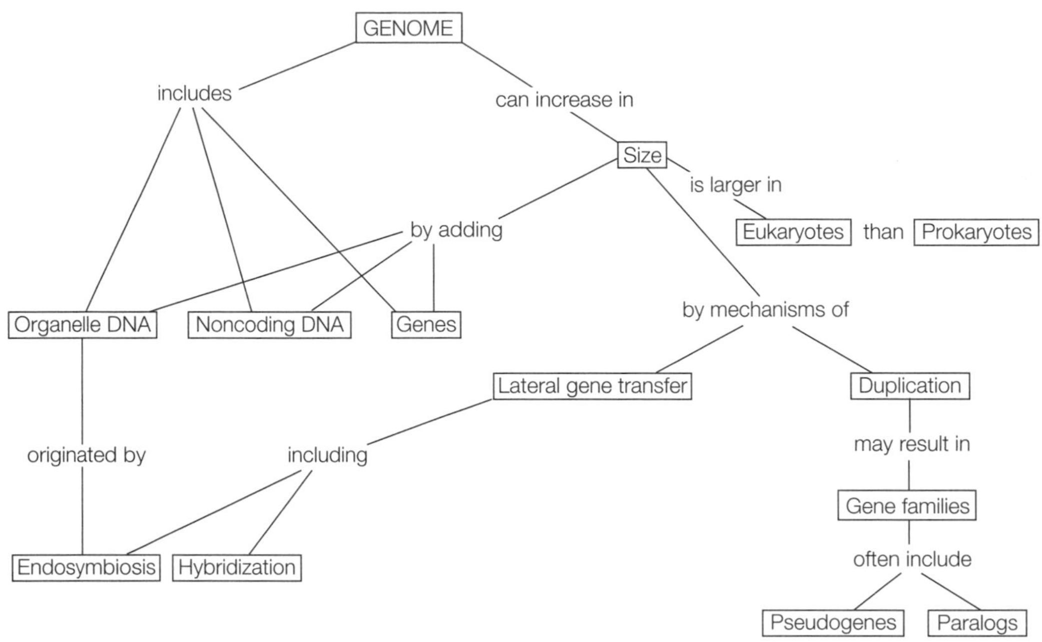

Knowledge and Synthesis Answers

1. **c.** The genome includes not only protein-coding genes but all other DNA in the organism, including noncoding DNA and genes in organelles.
2. **d.** The sequence alignment technique can be employed to compare both amino acid sequences and nucleotide sequences. Although it enables the locations of deletions and insertions to be pinpointed, it is not capable of detecting such phenomena as back substitutions and parallel substitutions.
3. **a.** In the experiments of Rainey and Travisano, heterogeneous environments favored diversification in their bacterial cultures because of natural selection, not because of a difference in either the rates or kinds of mutations that occurred in homogeneous and heterogeneous environments. Experiments of this kind can produce observable evolutionary changes in a matter of months, at most.
4. **b.** Recall from Chapter 14 that eukaryotic genes usually contain both protein-coding regions (exons) and noncoding regions (introns). A substantial proportion of nonsynonymous substitutions occurring in an exon of a gene would most likely be deleterious and therefore be eliminated by selection. Because introns and pseudogenes are not expressed, nonsynonymous substitutions occurring in them cannot be selected against.
5. **d.** A nonsynonymous substitution mutation can be selectively neutral if, as sometimes happens, it results in an amino acid change that has no significant effect on the shape (and hence the functional properties) of the protein.
6. **d.** Functionally important regions of a protein can be discovered by identifying the regions with the fewest amino acid substitutions. Because the sequences in these regions have been optimized by natural selection to accomplish the function of the molecule, random changes in these areas are not tolerated.
7. **b.** Because the animals that evolved a similar lysozyme do not share a recent common ancestor, the mechanism involved is convergent evolution. All other statements are false.
8. **a.** Recall that there is a rough relationship between the amount of coding DNA and organismal complexity.
9. **c.** As shown in Figure 24.9, multicellular eukaryotes with relatively small populations and slow rates of development have the lowest proportion of coding to noncoding DNA, whereas quickly reproducing prokaryotes such as *E. coli* have the highest proportion. It is unclear whether developmental rate or population size (or both) is chiefly responsible for this trend.
10. **a.** If there are two copies of the gene, it is more likely to produce more of its product than less of it.
11. **b.** Pseudogenes (nonfunctional DNA) may be removed via deletion, but numerous pseudogenes persist in many gene families.
12. **d.** Polyploidy is a common event in plant evolution (as discussed in Chapter 23), and it does result in the duplication of the entire nuclear genome. But it is also

believed to have occurred at least twice in the lineage that gave rise to the jawed vertebrates.

13. **e.** Concerted evolution is the phenomenon in which all the copies of a highly repeated gene (such as a gene coding for ribosomal RNA) remain highly similar despite experiencing nucleotide substitutions or other mutations that generally cause genes to diverge. Unequal crossing over during meiosis and biased gene conversion during DNA repair are two mechanisms that can produce concerted evolution.
14. **c.** Orthologs are homologous genes whose divergence can be traced to the speciation events that gave rise to the species in which the genes occur.
15. **e.** In vitro evolution can create novel molecules not known to occur in living organisms. The process starts with a very large pool of variant molecules (nucleic acid or protein) and involves selection on the part of the experimenters of those molecules that show some inkling of the desired property. Many rounds of production of variant molecules and selection are typically needed to produce the final product.

Application Answers

1.

SPECIES

SPECIES	1	2	3	4	5
1	■	*1*	*1*	*1*	*2*
2	*7*	■	*2*	*2*	*3*
3	*7*	*6*	■	*2*	*1*
4	*7*	*6*	*6*	■	*3*
5	*6*	*5*	*7*	*5*	■

Similarities (below diagonal); *Differences* (above diagonal)

Species 2 shows an amino acid insertion (Phe) at position 3. There are no deletions.

Species 3, 4, and 5 all show amino acid substitutions.

2. According to one hypothesis, the proportion of coding to noncoding DNA in the genome of a species is related to the sizes of the populations typical of the species. As discussed in Chapter 21, genetic drift has more influence on the evolution of small populations than on the evolution of large ones. If individuals of a species differ in their fitness only slightly as a function of the proportion of noncoding DNA in their genomes, and if the populations of the species are typically small, then genetic drift could lead to retention of the noncoding DNA (or even an increase in its proportion), despite its selective disadvantage. However, the effect of genetic drift is minimal in very large populations. Therefore, in species characterized by large populations, even a slight disadvantage in fitness for individuals with a greater proportion of noncoding DNA should lead to a decrease in its proportion through the action of natural selection.
3. As discussed in Chapter 11, during the synapsis of homologous pairs of chromosomes in meiosis I, the alignment of the homologs is normally extremely precise. This ensures that the result of crossing over will be the exchange of equivalent sections of genetic material in the two chromatids involved in the crossover event. In the case of highly repeated genes, however, it sometimes happens that the alignment of the homologs is imprecise, as shown in Figure 24.11. The result is that crossing over yields one chromatid with extra copies of the gene, whereas the other has fewer copies. Subsequently, during meiosis II, the chromatids will become independent chromosomes that now possess the altered number of copies of the repeated gene.
4. Variation and selection are involved in both in vitro evolution and in molecular evolution in organisms. With in vitro evolution, a huge number of variant molecules are created in the laboratory, and the human experimenters select those that show any sign of having the desired property. Many repetitions of these two steps—production of variants and selection—eventually produce the targeted molecule. In molecular evolution in organisms, naturally occurring mutations in previously existing DNA sequences are the ultimate source of variation, and the environment determines through natural selection which mutations are the favorable variants. But also recall that according to the neutral theory of molecular evolution, much evolutionary change in DNA (and in the encoded proteins) is not adaptive but rather the result of genetic drift.
5. The principles of molecular evolution are critical for understanding the origin and evolutionary development of such pathogenic viruses as hantaviruses, the SARS virus, and HIV. By studying the evolutionary changes that occur in such viruses, medical researchers gain valuable insights into such questions as how some viruses are able to switch from an animal to a human host and how vaccines can remain effective as viruses evolve. In the future, as more extensive genomic databases and evolutionary trees for viruses are developed, it will be possible to identify and treat a much wider array of human diseases.

25 The History of Life on Earth

The Big Picture

- The evidence provided by nineteenth-century geologists that Earth is very ancient was crucial to Darwin's theory of evolution by natural selection.
- Radiometric dating is an excellent example of how a discovery in one field of science (nuclear physics, in this instance) can lead to advances in other fields (geology and paleontology). Prior to the development of this technique, no absolute dates could be assigned accurately to any of the important geological and evolutionary events described in this chapter.
- The concept of continental drift, first widely accepted in the late 1960s, has revolutionized our understanding of geological processes and of the history of life on Earth.
- The accumulating evidence of the sudden and catastrophic effect of the impact of large meteorites and of other rapid changes (e.g., widespread volcanic eruptions) on the environment of Earth has given paleontologists a new perspective on the causes of mass extinctions and on the dynamics of the evolutionary process in general.

Common Problem Areas

- The concept of the half-life of radioactive isotopes may seem confusing. Bear in mind that if half of a radioisotope decays in a given time period (its half-life), then at the end of the first half-life only one-half of the original quantity of the isotope is still present, one-half of which will decay in the second half-life. Thus after two half-lives, one-quarter of the original quantity of radioisotope is still present. The same line of reasoning applies to all further half-lives. Study Figure 25.1, which explains this point.

Study Strategies

- This chapter contains many names of geological eras and periods and many dates. Instructors vary with respect to the importance they place on memorization of this information. In general, it is best to focus on broad evolutionary patterns and trends. Most instructors will also place more importance on the phylogenetic sequences (e.g., amphibians gave rise to amniotes, which gave rise to reptiles and to mammals) than on the dates at which these events occurred.
- Making a table, chart, or timeline to summarize and organize the events that occurred in each geological era and period is an excellent study strategy.
- Go to yourBioPortal.com to review the following tutorial and activity:

  Animated Tutorial 25.1 Evolution of the Continents

  Web Activity 25.1 Concept Matching

Important Concepts

Evolutionary changes may occur over both short and long time frames.

- The development of the science of biology, particularly Darwin's theory of evolution by natural selection, depended on evidence supplied by geologists that Earth is very old.
- Short-term evolutionary changes occur rapidly enough to be studied directly.
- Long-term evolutionary changes involve the appearance of new species and evolutionary lineages. The fossil record provides evidence of such changes.

Several types of evidence have been used to estimate the age of rocks, of fossils, and of Earth itself. All the evidence indicates that both Earth and life on Earth are very old.

- Sedimentary rocks (formed by the accumulation of grains at the bottom of bodies of water) are deposited in strata (layers). The oldest layers are found at the bottom, and successively higher strata are progressively younger. Fossils, which are the preserved remains of ancient organisms, are useful for establishing the relative ages of the sedimentary rocks in which they occur.
- The regular pattern of decay of radioactive isotopes provides a means of estimating the absolute ages of fossils and rocks. The half-life of a radioisotope is the time period during which half of the remaining

radioactive material decays to become a different, stable isotope. The ratio of radioactive carbon-14 to its stable isotope, carbon-14, is used to date fossils less than 50,000 years old. The decay of potassium-40 to argon-40 is widely used to date ancient evolutionary events.

- Paleomagnetic dating is based on the fact that the age of sedimentary and igneous rocks can be determined because they preserve a record of Earth's magnetic field at the time they were formed.
- Earth's geological history is divided into eras, which are subdivided into periods. The boundaries between these divisions are marked by changes in the types of fossils found in successive layers of sedimentary rock.
- The first forms of life evolved early in the Precambrian era, which lasted for more than three billion years and was marked by enormous physical changes on Earth.

The history of Earth has been marked by oscillatory changes, as well as by occasional collisions with extraterrestrial objects.

- The theory of plate tectonics proposes that Earth's crust consists of solid lithospheric plates floating on a fluid layer of magma. Convection currents in the magma result from heat emanating from Earth's core. The currents cause continental drift, which is the gradual shift in the position of the plates and the continents they contain. The drifting of continents has had profound effects on climate, sea level, oceanic circulation, and the distributions of organisms.
- No free oxygen was present in the atmosphere until certain bacteria evolved the ability to use water as a source of hydrogen ions for photosynthesis. The gradual increase in oxygen concentration in the atmosphere resulted in the dominance of organisms using aerobic metabolism and in the evolution of larger eukaryotic cells and of multicellular organisms. The exceptionally high oxygen concentrations that occurred during the Carboniferous and Permian periods are associated with the evolution of giant amphibians and flying insects.
- The climate of Earth has alternated between hot/humid and cold/dry conditions. Though most major climatic changes have been gradual, some have occurred within periods of 5,000 to 10,000 years or less. The rapid climate change occurring today is thought to be caused by a buildup of atmospheric CO_2, primarily from the burning of fossil fuels.
- The climatic shifts caused by massive volcanic eruptions associated with continental drift are implicated in several mass extinctions. The late Permian mass extinction, which occurred at the end of the Paleozoic era, was caused by volcanism triggered by the collision of continents that formed the supercontinent Pangaea.
- Several mass extinctions have probably been caused by collisions of the Earth with meteorites or comets, such as the event 65 million years ago that is thought to have resulted in the extinction of dinosaurs.

The fossil record reveals patterns in life's history.

- All of the organisms found in a particular place or time constitute its biota. The plant component of a biota is its flora, whereas the animal component is its fauna.
- The earliest life on Earth appeared about 3.8 billion years ago, but the fossil record of organisms that lived in the Precambrian is fragmentary.
- The first organisms were unicellular prokaryotes. The first eukaryotes evolved about 1.5 billion years ago.
- Because an oxygen-rich environment favors rapid decomposition, organisms that become fossils are likely either to have lived in a poorly oxygenated environment or to have been transported to such a site soon after death.
- The 300,000 known fossil species represent only a tiny fraction of the species that have ever lived. Some groups, such as hard-shelled marine animals, are much better represented in the fossil record than others.
- The fossil record shows that an organism of any specific type can predictably be found in rocks of a particular age.
- Living organisms resemble recent fossils more closely than they resemble fossils from more ancient periods.
- For most of the Precambrian, which lasted for over 3 billion years, life consisted of microscopic prokaryotes. The first unicellular eukaryotes evolved about 1.5 billion years ago.
- The best Precambrian fossil deposits, dating from about 600 million years ago, contain diverse soft-bodied invertebrates, some of which may represent lineages with no living descendents.

Beginning with the Cambrian period, the Paleozoic era (542–251 mya) was marked by a dramatic increase in the diversity of life.

- During the Cambrian period, the oxygen concentration in the atmosphere approached its current level, and several large continents formed. The rapid increase in the diversity of multicellular life forms that occurred at this time is known as the "Cambrian explosion."
- The Ordovician period was marked by a proliferation of marine filter feeders living on the sea floor. At the end of this period, massive glaciers formed over the southern continents, the sea level and ocean temperatures dropped, and the majority of animal species became extinct.
- The Silurian period witnessed the evolution of swimming marine animals and the diversification of jawless fish. It also marked the appearance of the first terrestrial arthropods and vascular plants.
- During the Devonian period, all major groups of fishes evolved. On land, the first insects and amphibians evolved, and forests of club mosses, horsetails, and tree

ferns appeared. A mass extinction of approximately three-quarters of all marine species occurred at the end of this period, possibly caused by the collision of two large meteorites with Earth.

- In the Carboniferous period, swamp forests consisting largely of giant tree ferns and horsetails were widespread. The fossilized remains of these plants formed coal. The first winged insects evolved during this period, while amphibians became better adapted to life on land and gave rise to the lineage leading to the amniotes (whose eggs can be laid in dry places).
- The Permian period was marked by the formation of a single supercontinent called Pangaea. As the climate cooled drastically, amniotes split into two lineages, the reptiles and one leading to the mammals. The occurrence of the most extensive of all mass extinctions brought the Permian period to a close.

During the Mesozoic era (251–65 mya), the continents that formed Pangaea slowly separated to form Laurasia and Gondwana, and distinct assemblages of plants and animals evolved on each continent.

- During the Triassic period, seed ferns and conifers were the dominant forms of terrestrial vegetation, and reptiles were the dominant vertebrates. A mass extinction occurred at the end of the Triassic.
- The Jurassic period was marked by the complete separation of Pangaea to form Laurasia in the north and Gondwana in the south. It also witnessed the radiation of the dinosaurs, the evolution of flying reptiles, and the first appearance of mammals. The earliest fossils of flowering plants date from late in this period. In the sea, ray-finned fishes also began a great radiation.
- By the early Cretaceous period, Laurasia and Gondwana had begun to break apart into the present-day continents. The flowering plants began the radiation that led to their current dominance. The dinosaurs continued as the dominant land vertebrates, despite the presence of many groups of mammals. At the end of this period, the collision of a large meteorite with Earth caused the extinction of the dinosaurs and of many other animal and plant lineages.

Modern groups of plants and animals evolved during the Cenozoic era (65 mya–present).

- During the Tertiary period, the continents drifted toward their present positions. As the climate became cooler and drier, extensive grasslands appeared. Many groups of land vertebrates—especially frogs, snakes, lizards, birds, and mammals—radiated extensively.
- The current geological period, the Quaternary, is divided into the Pleistocene and Holocene (Recent) epochs. Modern humans evolved during the Pleistocene, which was a time of severe climatic fluctuations, including four major periods of extensive glaciation. As humans spread geographically, many species of large birds and mammals became extinct.
- Phylogenetic trees help reconstruct the timing of evolutionary events and clarify relationships among modern species.

Test Yourself

Diagram Exercise

Below is a chronologically scrambled list of important events in the history of life on Earth. Draw four lines on a sheet of paper to represent the time lines of events occurring in the Precambrian, the Paleozoic era, the Mesozoic era, and the Cenozoic era. Place the listed events on the correct line and in proper sequence.

Conifers become dominant
Cambrian "explosion"
Evolution of *Homo*
Fifth mass extinction
First eukaryotes
First flowering plant fossils
First forests ("fern" forests); first jawed fish
First fossils of multicellular animals
First mammals, dinosaurs diversify
First mass extinction
First photosynthetic eukaryotes
First vascular plants and terrestrial arthropods
Fourth mass extinction
Grasslands spread
Origin of amniotes
Origin of life
Origin of photosynthesis
Second mass extinction
Third mass extinction
Rapid radiation of mammals

Textbook Reference: *25.3 What Are the Major Events in Life's History? pp. 526–534*

Knowledge and Synthesis Questions

1. The half-life of an isotope is the
   a. time a fixed fraction of isotope material will take to change from one form to another.
   b. age over which the isotope is useful for dating rocks.
   c. ratio of one isotope species to another in a sample of organic matter.
   d. Both a and b
   e. None of the above
   Textbook Reference: *25.1 How Do Scientists Date Ancient Events? p. 520*
2. Mountain ranges are ultimately the result of
   a. plates in Earth's crust that move against one another on top of a fluid layer of molten rock.

b. climate changes and the movement of glacial ice sheets.
c. leftover debris from ancient collisions with an asteroid or meteor.
d. the breakup of Laurasia and Gondwana.
e. Both a and b

Textbook Reference: *25.2 How Have Earth's Continents and Climates Changed over Time? p. 522*

3. Which of the following would likely be most conducive to the occurance of fossilization?
a. The surf zone along a sandy beach
b. A shallow, cool swamp with good deposition rates of mud sediments
c. The bottom of a hot, dry cave with no running water
d. A fast-running mountain stream
e. It is not possible to decide based on the information provided.

Textbook Reference: *25.3 What Are the Major Events in Life's History? p. 526*

4. Despite being incomplete as a whole, the fossil record is rather detailed for
a. soft-bodied insects.
b. cnidarians and sponges.
c. most terrestrial animals.
d. hard-shelled mollusks.
e. Both c and d

Textbook Reference: *25.3 What Are the Major Events in Life's History? p. 526*

5. During which geological period did Pangaea became fully divided and distinctive assemblages of plants and animals begin to arise on different continents?
a. Cambrian
b. Tertiary
c. Devonian
d. Permian
e. Jurassic

Textbook Reference: *25.3 What Are the Major Events in Life's History? p. 532*

6. Which of the following statements about the Ordovician period is *false*?
a. Marine filter feeders flourished.
b. The number of classes and orders increased.
c. Modern mammals appeared.
d. Many groups became extinct at the end of the period.
e. The continents were located primarily in the Southern Hemisphere.

Textbook Reference: *25.3 What Are the Major Events in Life's History? p. 528*

7. Most of human evolution has occurred during the
a. Paleozoic era.
b. Devonian period.
c. Quaternary period.
d. Carboniferous period.
e. Cretaceous period.

Textbook Reference: *25.3 What Are the Major Events in Life's History? p. 533*

8. For terrestrial animals and plants, the most recent mass extinction event that occurred prior to the evolution of humans took place approximately _______ million years ago.
a. 10
b. 65
c. 200
d. 250
e. 400

Textbook Reference: *25.3 What Are the Major Events in Life's History? p. 533*

9. One of the main factors that distinguishes the Cambrian explosion from all others is that
a. evolutionarily, it was the most recent explosion.
b. many new major groups of animals appeared at this time in contrast to other explosions.
c. it was the time when the dinosaurs became extinct.
d. there was a dramatic drop in species diversity, especially among marine organisms.
e. it was a time of massive volcanic eruptions.

Textbook Reference: *25.3 What Are the Major Events in Life's History? pp. 527–528*

10. During which of the following geological times did the most new kinds of body plans appear?
a. Carboniferous
b. Triassic
c. Jurassic
d. Devonian
e. Cambrian

Textbook Reference: *25.3 What Are the Major Events in Life's History? p. 527*

11. The sudden disappearance of the dinosaurs some 65 mya may have been the result of
a. Earth's collision with a large meteorite.
b. slow climate changes due to planetary cooling.
c. competition from better-adapted organisms.
d. the rise of birds and mammals.
e. the formation of Pangaea.

Textbook Reference: *25.3 What Are the Major Events in Life's History? p. 533*

12. Fossil insects many times larger than any insects alive today have been dated to the Carboniferous and Permian periods. These fossils are evidence that during these periods
a. the climate was much warmer than it is today.
b. there were fewer predators on insects than there are today.
c. the carbon dioxide concentration of the atmosphere was lower than it is today.
d. insects had fewer competitors for food than modern insects.
e. the oxygen concentration of the atmosphere was significantly higher than it is today.

Textbook Reference: 25.2 How Have Earth's Continents and Climates Changed Over Time? pp. 523–524

13. Which of the following statements about patterns or processes in the evolution of life is *false*?
    a. ^{14}C can be used to date the age of dinosaur bones.
    b. The supercontinent Pangaea formed during the Permian period.
    c. Mass extinctions of marine organisms have coincided with periods of low sea levels.
    d. The ends of five geological periods have been marked by mass extinctions.
    e. All of the above are true.

    Textbook Reference: 25.1 How Do Scientists Date Ancient Events? p. 520

14. Which of the following pairs of organisms were *not* present on Earth in their living forms at the same time?
    a. Tree ferns and ray-finned fish
    b. Amphibians and birds
    c. Gymnosperms and insects
    d. Humans and dinosaurs
    e. Jawless fish and vascular plants

    Textbook Reference: 25.3 What Are the Major Events in Life's History? p. 533

15. The most severe mass extinction event, linked to the formation of Pangaea and massive volcanic eruptions, occurred at the end of the _______ period.
    a. Ordovician
    b. Devonian
    c. Permian
    d. Triassic
    e. Cretaceous

    Textbook Reference: 25.3 What Are the Major Events in Life's History? p. 532

Application Questions

1. ^{14}C decays to ^{12}C with a half-life of about 5,700 years. Suppose you find a fossil in which the amount of ^{14}C is only $1/16$ of what one would find in a living organism with the same carbon content. Approximately how old is the fossil?

   Textbook Reference: 25.1 How Do Scientists Date Ancient Events? p. 520

2. For each geological period, make a chart summarizing its major geological and evolutionary events.

   Textbook Reference: 25.1 How Do Scientists Date Ancient Events? pp. 520–521

3. In what way was the evolution of eukaryotic cells linked to the increase in the oxygen concentration in the atmosphere that occurred during the Precambrian?

   Textbook Reference: 25.2 How Have Earth's Continents and Climates Changed over Time? p. 523

4. Scientists believe that if there are no controls on the emission of CO_2 from the burning of fossil fuels, the concentration of this gas could double by the end of the current century, leading to a significant rise in the average temperature of Earth. What would be some of the likely evolutionary effects of this climatic change?

   Textbook Reference: 25.2 How Have Earth's Continents and Climates Changed over Time? pp. 524–525

5. In 2006 paleontologists announced that fossils of *Tiktaalik roseae*, an important link between fish and tetrapods, had been discovered in freshwater sediments in the Canadian Arctic that were 375 million years old. How were geologists able to determine the age of these sediments?

   Textbook Reference: 25.1 How Do Scientists Date Ancient Events? pp. 519–521

Answers

Diagram Exercise Answer

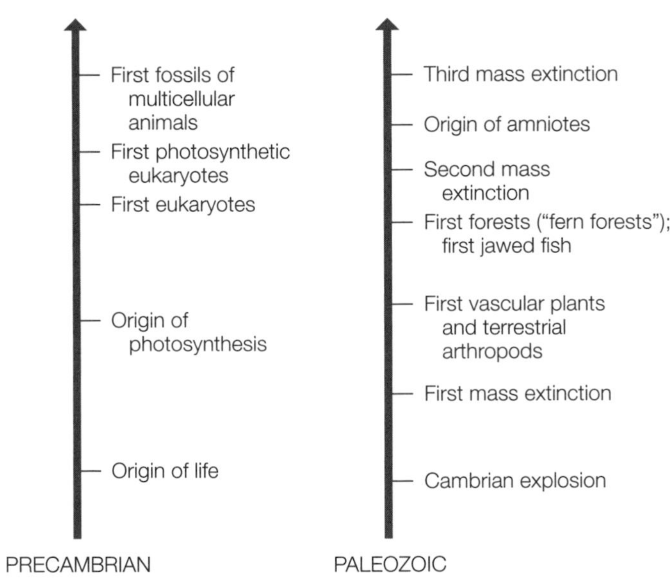

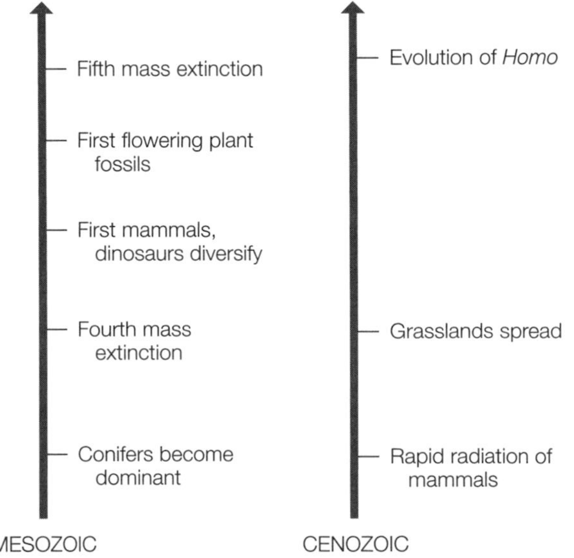

See also Figures 25.10 and 25.12 in the textbook.

Knowledge and Synthesis Answers

1. **a.** Radioactive decay is measured as the time it takes for one-half the amount of a substance to spontaneously convert into another substance.
2. **a.** As the crustal plates are pushed together, one may move underneath the other, pushing up mountain ranges.
3. **b.** Fossilization occurs best in areas of low oxygen concentration and rapid sedimentation, where scavengers cannot destroy the body.
4. **d.** Hard-shelled animals such as mollusks are good candidates for fossilization because their shells can withstand decay long enough for them to be buried. Many also tend to live in quiet, shallow waters.
5. **e.** The division of Pangaea began during the Triassic period, but Laurasia and Gondwana did not become completely separated until the Jurassic.
6. **c.** The modern mammals did not appear until the Tertiary, more than 350 million years after the Ordovician.
7. **c.** The Pleistocene epoch, which is part of the Quaternary, was the time of most hominid evolution.
8. **b.** The mass extinction at the end of the Cretaceous, 65 mya, was the most recent mass extinction affecting terrestrial life before the evolution of humans.
9. **b.** The Cambrian explosion produced many novel groups of animals characterized by distinctive body plans. Later explosions caused an increase in diversity only within already existing major lineages.
10. **e.** For the reasons outlined in the previous answer, the Cambrian period is the correct answer.
11. **a.** Paleontologists believe that a large meteorite's collision with Earth so altered conditions that the dinosaurs rapidly succumbed during what we know as the great Cretaceous mass extinction.
12. **e.** There is experimental evidence that insects can grow larger when raised in hyperbaric conditions (see Figure 25.6). The current level of atmospheric O_2 appears to limit the evolution in body size of flying insects.
13. **a.** Carbon dating is generally reliable only for fossils less than 50,000 years old.
14. **d.** Dinosaurs disappeared at the end of the Cretaceous, over 60 million years before humans evolved.
15. **c.** Though mass extinctions occurred at the end of all the periods listed, the highest percentage of species became extinct during the Permian period.

Application Answers

1. If only $\frac{1}{16}$ of the ^{14}C remains, then four ^{14}C half-lives have passed since the fossil was formed ($\times \frac{1}{2} \times \frac{1}{2} \times \frac{1}{2} = \frac{1}{16}$). Because the half-life of ^{14}C is roughly 5,700 years, the fossil must be about 22,800 years old ($4 \times 5{,}700 = 22{,}800$).
2. See Table 25.1.
3. Small prokaryotic cells can obtain enough oxygen by diffusion even when oxygen concentrations are very low. Since eukaryotic cells are larger, they have a lower surface area-to-volume ratio and hence need a higher concentration of oxygen in their environment for diffusion to meet their requirement for this gas.
4. The fossil record shows that major environmental changes occurring over a short time interval sometimes lead to large-scale rapid extinctions that appear "instantaneous" in the fossil record. The possible effects of global warming on biodiversity are discussed in more detail in Chapter 59.
5. Recall that radioisotope dating is used to determine the absolute age of rocks. Igneous rocks, formed from volcanic ash or lava flows, are required for this purpose. Such rocks may have been found near the discovery site in Canada, or they may have been found at one or more other sites that have sediments of the same relative age. The relative age of the sediments can be determined by the types of fossils found in them.

26 Bacteria and Archaea: The Prokaryotic Domains

The Big Picture

- All organisms fall within one of three domains. This chapter focuses on the domains Archaea and Bacteria. Subsequent chapters focus on the Eukarya. Archaea and Bacteria are prokaryotic organisms and retain many features of the earliest forms of life on Earth. They have no membrane-enclosed organelles, circular chromosomes, or cell walls of peptidoglycans or proteins. Their structures are simple, and they exist as single cells. Some prokaryotes are stationary, while others move by means of flagella. Their metabolic processes are highly varied, as are their modes for procuring energy. Many prokaryotic organisms make valuable contributions to their environments through nitrogen fixation, photosynthesis, and other metabolic processes. The majority of prokaryotes are free-living, but some live in mutualistic relationships and others are pathogenic to other organisms.
- The evolutionary relationships among prokaryotes are still highly disputed and best understood at the gene level. The most ubiquitous extant prokaryotes are the bacteria. They are grouped according to physical and biochemical characteristics, but these groupings do not necessarily reflect their evolutionary relationships. The archaea are more closely related to the eukarya, and they exist in some of the harshest known environments.
- As defined in this text, viruses do not meet the criteria for life. While there is no single definition of life used by all scientists, many do not consider viruses living organisms even though they arose from cells. They require a host cell to carry out their basic functions. Viral genomes are varied in form and structure. These include positive- and negative-stranded RNA, single-stranded and double-stranded DNA, and they can be linear or circular. Currently, the genome structure is the basic classification system since the evolutionary history of viruses is unknown and unlikely to be determined.

Common Problem Areas

- Because this chapter covers a large number of diverse microscopic organisms included in two of the three domains of life, the details and exceptions can seem overwhelming. Many of the names of these groups are long, complex, and unfamiliar. Because many of the distinguishing features of the different groups of bacteria and archaea are biochemical, you may need to review information from Part III of the textbook.

Study Strategies

- This chapter is an overview of two domains of organisms and of viruses. A single chapter can cover only a few points of prokaryotic diversity. For the two domains, focus your study on what makes each one distinct and on the unique features that indicate a closer evolutionary relationship between archaea and eukaryotes than between archaea and bacteria. Move on to learning about sample organisms from each domain and groups within each domain.
- The temptation is to focus on the pathogenic organisms, but remember that only a very small percentage of prokaryotes are pathogenic.
- Go to yourBioPortal.com to review the following tutorial and activity:

  Animated Tutorial 26.1 The Evolution of the Three Domains

  Web Activity 26.1 Gram Stain and Bacteria

Important Concepts

Domains attempt to reflect evolutionary relationships among organisms.

- All living things can be placed into one the following three domains: Archaea, Bacteria, and Eukarya. This system reflects evolutionary relationships among these monophyletic groups.
- All organisms have similarities; only derived similarities indicate evolutionary relationships. Eukarya have nuclei,

membrane-enclosed organelles, and cytoskeletons. All organisms that possess these are descended from the most recent common ancestor of the Eukarya. The absence of these features in Archaea and Bacteria is a retained characteristic inherited from the common ancestor of all life on Earth (a prokaryote). Unique and distinguishing features of Archaea and Bacteria have been discovered through biochemistry and genetic studies.

- All three domains arose from a common ancestor possessing a circular chromosome and machinery for transcription and translation of the DNA code to protein. Divergence from this ancestor occurred billions of years ago. The domain Archaea is more closely related to the Eukarya than to the Bacteria.
- ***For review, go to Diagram Exercise 1.***

Prokaryotes have many features in common.

- Though they are members of two different domains, Archaea and Bacteria have several structural similarities that are indicative of evolutionary adaptation to similar environments.
- Prokaryotes are prolific and are the most numerous of all organisms. Because they are single-celled (there are a few exceptions) and very small, they generally escape human notice. They play extremely important roles in the biosphere: for instance, in nutrient cycling. They can also metabolize many substances that eukaryotes cannot.
- Prokaryotes generally exist in one of three shapes: spherical (cocci), rod-shaped (bacilli), and spiral-shaped. All of the different shapes may aggregate to form chains or plates of cells. Advanced associations may result in filaments that are encased in a protective sheath.
- Prokaryotes do not have membrane-enclosed organelles; therefore, the organization and replication of DNA proceeds differently in prokaryotes than it does in eukaryotes. Also, the processes of cellular respiration and photosynthesis are carried out differently and rely on infoldings of the cell membrane. Finally, prokaryotes lack a cytoskeleton; therefore, true mitosis does not occur. Prokaryotes go through an asexual process called fission after replicating their DNA.
- Many prokaryotes are mobile and move by gliding, rolling, regulating gas vesicles, or flagella. Bacterial flagella are whiplike and are composed of a single protein called flagellin. In contrast, the flagellum of eukaryotes is complex.
- The cell wall of prokaryotes is thick and rigid and composed of peptidoglycans (bacteria) or protein (archaea). This cell wall is very different from that of plant and fungal cell walls. Cell wall composition is one of many ways of differentiating bacteria and archaea.
- Cell wall structure within the bacteria also serves as a diagnostic characteristic among different bacteria. This difference is used in the Gram staining technique. Whether a cell is Gram-negative or Gram-positive relates to the amount of peptidoglycan in the cell wall.
  - Gram-positive cells stain purple and have a thick layer of peptidoglycan.
  - Gram-negative cells do not retain stain and appear red due to a thin layer of peptidoglycan.
- Some antibiotics work by interfering with the synthesis of peptidoglycan-containing cell walls. Eukaryotic cells are not harmed because their cell walls lack this material.
- Prokaryotes reproduce asexually but may exchange genetic information through transformation, conjugation, or transduction.
- Metabolic processes in prokaryotes are varied and diverse.
- Obligate anaerobes are oxygen-sensitive and can survive only in oxygen-poor conditions. Facultative anaerobes can shift between aerobic and anaerobic modes of respiration. Aerotolerant anaerobes do not respire aerobically but are not harmed by an oxygen-rich atmosphere as obligate anaerobes are. Obligate aerobes require oxygen and cannot survive in the absence of it.
- Photoautotrophs are photosynthetic and convert light energy to chemical energy. Some photoautotrophs use chlorophyll *a* and give off oxygen as a by-product. Others use bacteriochlorophyll and produce sulfur as a by-product.
- Photoheterotrophs use light for energy but also require outside carbon sources.
- Chemolithotrophs oxidize inorganic substances for energy. Some chemolithotrophs fix carbon dioxide into carbohydrate. Others fix nitrite or ammonia, hydrogen gas, hydrogen sulfide, and other materials.
- Chemoheterotrophs obtain energy and carbon from consuming complex organic molecules.
- Some bacteria carry out respiratory electron transport using nitrogen or sulfur compounds as electron acceptors. These organisms are extremely important in nutrient cycling.
- Prokaryotes make diverse contributions to their environments.
- The nitrogen cycle depends on nitrification by bacteria. Nitrifying bacteria use nitrogen to make ATP, which is used to fix carbon dioxide into glucose and other food molecules. Through this process, unusable nitrogen sources are converted into biologically available nitrogen sources.
- Cyanobacteria, using photosynthesis, enriched the atmosphere with oxygen. Oxygen levels became too high for many existing organisms; many others evolved to fill those niches, and new organisms arose.
- Prokaryotes live in many mutualistic associations. Many of your digestive processes rely on the bacteria that inhabit your intestine.
- A small minority of bacteria are pathogenic. Bacteria cause disease because they produce toxins that are harmful to host tissues. Endotoxins are released when

bacterial cells burst, but are rarely fatal. Exotoxins are secreted by bacteria and may be fatal.

- The German physician Robert Koch came up with criteria for determining if a bacterium was a disease-causing agent:
  - The organism must always be found in the individual with the disease.
  - The organism must be taken from the host and grown in pure culture.
  - A sample of the culture must produce disease if injected into a healthy individual.
  - The newly infected individual must yield a new pure culture of the same organism.
- In order to be pathogenic, an organism must be able to reach a new host, invade the host, evade the host's defenses, multiply, and infect a new host. Relatively few organisms can do this.
- Evolutionary relationships among prokaryotes are determined at the genetic level.
- Physical characteristics are used in classification for identification purposes, but to understand the true evolutionary relationships among very small organisms, such as prokaryotes, it is necessary to look at nucleic acid sequences and biochemistry.
- Ribosomal RNA gives us information about relationships among organisms. It is useful for this purpose because it is evolutionarily ancient, is found in all organisms, has the same function in all organisms, and is highly conserved (it evolved slowly). Scientists compare signature sequences to make evolutionary inferences.
- rRNA sequences have not provided clear data. Lateral gene transfer—genetic transfer between species—makes these data hard to understand at the species level. At higher taxonomic levels, rRNA sequence data is not as problematic since stable portions of multiple genomes can be compared.
- Mutations are important in prokaryotic variation. Because prokaryotes have single circular DNA pieces, mutations cannot be masked and are passed on through fission. This is in contrast to diploid organisms, in which mutation plays a small role in genetic variation.
- Prokaryotes evolve comparatively rapidly because of rapid reproduction, persistent mutation, and selection. Mutations leading to antibiotic resistance are of extreme concern and can occur in a matter of just a few years.
- ***For review, go to Diagram Exercise 2.***

The majority of known prokaryotes are bacteria.

- Bacteria can be subdivided into a series of monophyletic groups. Not all scientists agree about the composition (membership) of these groups, and this is an active area of research.
- Spirochetes are corkscrew-shaped prokaryotes characterized by the presence of axial filaments that run through their periplasmic space. Many spirochetes are parasites of humans.
- Chlamydias are the smallest bacteria and exist only as parasites. They have a two-stage life cycle and are responsible for several human diseases.
- Actinobacteria, or high-GC Gram-positive bacteria, develop an elaborately branched system of filaments, some with chains of spores at the tips. This group includes the pathogen *Mycobacterium tuberculosis* (tuberculosis). However, one of our most potent antibiotics comes from this group. Streptomycin is produced by the actinobacteria *Streptomyces*. Most currently used antibiotics come from the actinobacteria.
- Cyanobacteria undergo photosynthesis similar to eukaryotic photosynthesizers and contribute to the current oxygen-rich environment. These organisms frequently exist in colonies with distinct divisions of labor. Some cells are vegetative and photosynthesize. Others form resting stages called spores that can withstand harsh environmental conditions. Still others, called heterocysts, fix nitrogen.
- The diversity of low-GC Gram-positive bacteria belies the DNA sequence evidence supporting the grouping of these organisms into a monophyletic clade. These bacteria include a few that are Gram-negative or lack a cell wall entirely. Some produce endospores that are capable of surviving in extremely harsh conditions. Many medically important bacteria fall within this group, including *Clostridium botulinum* (botulism), *Staphylococcus* (many skin infections), and *Bacillis anthracis* (the causative agent of anthrax). Another subgroup, the mycoplasmas, may represent the smallest life form.
- The five groups of proteobacteria (Delta, Epsilon, Alpha, Beta, and Gamma) represent the largest number of known species. This group is believed to be the origin of the eukaryotic mitochondria. The early ancestor of this group was probably a photoautotroph, but today the group encompasses multiple metabolic schemes. Included among the proteobacteria are *E. coli*, the nitrogen-fixing *Rhizobium*, and many human pathogens.
- ***For review go to Diagram Exercise 3.***

Archaea live in some of the harshest known environments.

- Very little is known about the phylogeny of the Archaea, but some of the shared characteristics distinguishing them from the Bacteria are well-established. They lack peptidoglycan cell walls, and some have cell walls composed of proteins. Archea also have very distinctive lipids linked by ether linkages in their cell membranes. No other organisms possess this type of lipid.
- Currently, Archea has been divided into two main groups (Euryarchaeota and Crenarchaeota) and two

recently discovered groups (Korarchaeota and Nanoarchaeota).

- Crenarchaeota live in hot, acidic environments and are thermophilic and acidophilic. They require temperatures above 55°C and can survive in an environment with a pH as low as 0.9 while maintaining an internal pH of 7.
- Some members of Euryarchaeota produce methane gas and are obligate anaerobes. These organisms are prevalent in the guts of herbivores. Others live near volcanic vents. Another group of Euryarchaeota live in extremely salty conditions and are referred to as halophiles.
- Korarchaeota are archaea known only from DNA isolated from hot springs. Scientists have been unable to grow any species of Korarchaeota in a pure culture.
- Nanoarchaeota, obtained from a deep-sea hydrothermal vent, is minute and lives attached to cells of *Ignicoccus*, a crenarchaeote.

Viruses are not true organisms, but are derived from cells of other organisms.

- Viruses are classified based on their genomic structure. Currently, viral groups include negative-sense single-stranded RNA, positive-sense single-stranded RNA, double-stranded RNA viruses, RNA retroviruses, single-stranded DNA, and double-stranded DNA.
- Viruses are obligate parasites, relying on their host cell to carry out basic functions such as replication and metabolism.
- Many diseases important to humans are caused by viruses. These include HIV, the cause of AIDS; influenza, which causes the flu; and herpes viruses, the cause of chicken pox, shingles, and genital herpes. Other viruses cause diseases in important crops and can have a devastating impact on agriculture.

Test Yourself

Diagram Exercises

1. Label the three basic shapes of bacteria shown in the micrograph below.

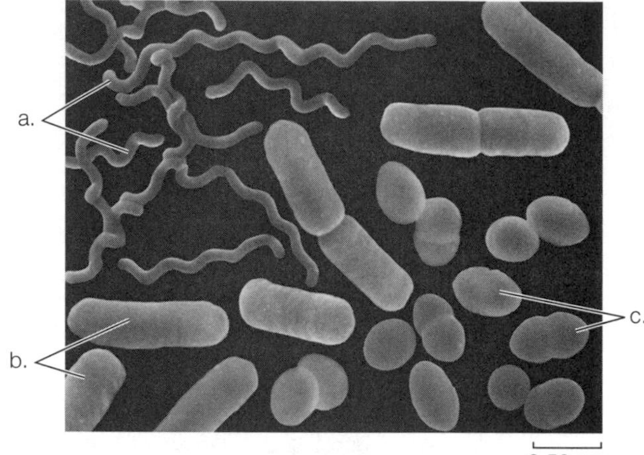

Textbook Reference: *26.1 How Did the Living World Begin to Diversify? p. 539, Figure 26.2*

2. In the diagram below, label the cells as Gram-postive or -negative, and identify each portion of the structure.

This is a Gram-__________ bacteria.

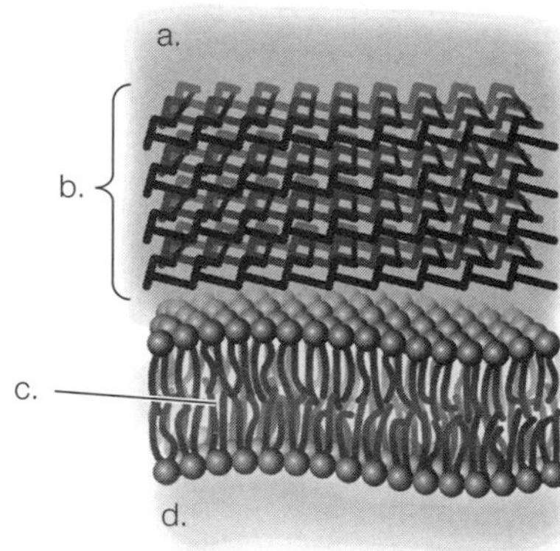

This is a Gram-__________ bacteria.

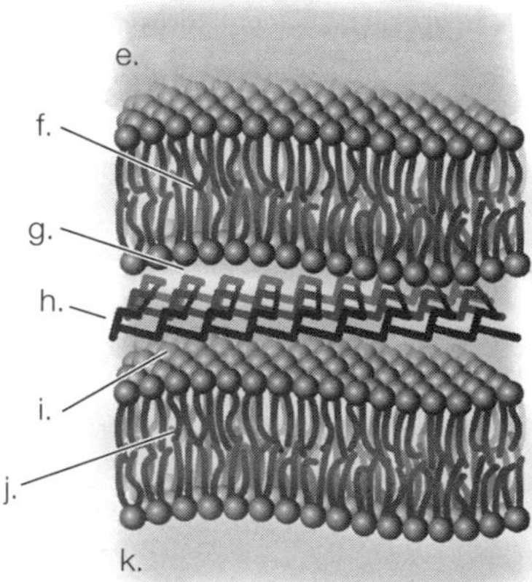

Textbook Reference: *26.2 What Are Some of the Keys to Success of Prokaryotes? p. 541, Figure 26.5*

3. Complete the table below.

Domain	Group Name	Key Features	Example Genera or Associated Diseases
	Chlamydias		
	Crenarchaeota		
	Cyanobacteria		
	Euyarchaeota		
	High-GC Gram-positives		
	Korarchaeota		
	Low-GC Gram-positives		
	Nanoarchaeota		
	Proteobacteria		
	Spirochetes		

Textbook Reference: *26.4 What Are the Major Known Groups of Prokaryotes? pp. 547–553*

Knowledge and Synthesis Questions

1. Which of the following characteristics is unique to prokaryotes?
   a. Lack of membrane-enclosed organelles
   b Presence of cell walls
   c. Presence of plasma membranes
   d. Presence of a cytoskeleton
   e. Absence of ribosomes
   Textbook Reference: *26.1 How Did the Living World Begin to Diversify? pp. 537–538*
2. Rod-shaped bacteria are referred to as
   a. bacilli.
   b. cocci.
   c. spiral.
   d. helici.
   e. roddi.
   Textbook Reference: *26.1 How Did the Living World Begin to Diversify? p. 539*
3. Which of the following statements concerning prokaryotes is *true*?
   a. Because prokaryotes do not contain organelles, they cannot photosynthesize or carry out cellular respiration.
   b. Prokaryotes have no chromosomes and therefore lack DNA.
   c. Prokaryote flagella are similar in structure to eukaryote flagella.
   d. Prokaryotes undergo mitosis.
   e. None of the above
   Textbook Reference: *26.1 How Did the Living World Begin to Diversify? p. 537*
4. Gram-negative bacteria stain pink because
   a. they have specialized lipids in their cell walls.
   b. their peptidoglycan layer is thin.
   c. their peptidoglycan layer is thick.
   d. they are receptive to antibiotics.
   e. their cell walls are composed largely of proteins.
   Textbook Reference: *26.2 What Are Some Keys to the Success of Prokaryotes? p. 541*
5. Archaea are more closely related to _______ than to any other monophyletic group.
   a. bacteria
   b. eukaryotes
   c. bacteria and eukaryotes
   d. fungi
   e. protists
   Textbook Reference: *26.1 How Did the Living World Begin to Diversify? p. 538*
6. The dense films laid down by many prokaryotes are
   a. endotoxins.
   b. denitrifiers.
   c. biofilms.
   d. pathogens.
   e. endospores.
   Textbook Reference: *26.2 What Are Some Keys to the Success of Prokaryotes? p. 539*
7. Which of the following groups of bacteria has the highest proportion of pathogens?
   a. Proteobacteria
   b. Cyanobacteria
   c. Chlamydias
   d. High-GC Gram-positives
   e. Low-GC Gram-positives
   Textbook Reference: *26.4 What Are the Major Known Groups of Prokaryotes? p. 547*
8. The mitochondria of eukaryotes were derived by endosymbiosis from
   a. proteobacteria.
   b. chemoheterotrophs.
   c. eukaryotes.
   d. archaea.
   e. viruses.
   Textbook Reference: *26.4 What Are the Major Known Groups of Prokaryotes? p. 550*

9. Autoclaves, which sterilize medical and laboratory equipment by means of pressurized heat, must pass a "spore test" in many states to demonstrate that they are able to work correctly. The spore-producing bacteria used for this test are most likely taken from which of the following groups?
   a. Low-GC Gram-positive bacteria
   b. Proteobacteria
   c. Cyanobacteria
   d. Chlamydias
   e. High-GC Gram-positive bacteria
   Textbook Reference: *26.4 What Are the Major Known Groups of Prokaryotes? pp. 548–549*
10. Archaea often live in harsh environments. Those that live in extremely salty conditions are referred to as
   a. thermophiles.
   b. halophiles.
   c. salinophiles.
   d. brinophiles.
   e. sodiophiles.
   Textbook Reference: *26.4 What Are the Major Known Groups of Prokaryotes? p. 552*
11. Methane gas contributes to the greenhouse effect that is raising atmospheric temperatures. A large portion of all methane emission is from grazing cattle because cows harbor methane-producing archaea from the _______ group.
   a. Crenarchaeota
   b. Euryarchaeota
   c. Anarchaeota
   d. Proteobacteria
   e. Nanoarchaeota
   Textbook Reference: *26.4 What Are the Major Known Groups of Prokaryotes? p. 552*
12. One of the major diagnostic characteristics that distinguishes archaea from bacteria is the absence of
   a. peptidoglycan cell walls in archaea.
   b. peptidoglycan cell walls in bacteria.
   c. ribosomes in archaea.
   d. chemoautotrophy in bacteria.
   e. pseudopeptidoglycan cell walls in archaea.
   Textbook Reference: *26.4 What Are the Major Known Groups of Prokaryotes? pp. 551–552*
13. A bacterium that requires a carbon source other than carbon dioxide, yet can convert light energy to chemical energy, is called a
   a. photoautotroph.
   b. photoheterotroph.
   c. chemoautotroph.
   d. chemoheterotroph.
   e. chemolithotroph.
   Textbook Reference: *26.2 What Are Some Keys to the Success of Prokaryotes? pp. 543–544, Table 26.2*
14. A bacterium that *cannot* live in the presence of oxygen is called a(n)
   a. obligate aerobe.
   b. facultative aerobe.
   c. obligate anaerobe.
   d. facultative anaerobe.
   e. aerotolerant anaerobe.
   Textbook Reference: *26.2 What Are Some Keys to the Success of Prokaryotes? p. 543*
15. Which of the following is *not* a form or method of prokaryotic locomotion?
   a. Rolling
   b. Corkscrew-like twisting
   c. Adjustment of gas vesicles
   d. Flagella-driven movement
   e. Paddling with cilia
   Textbook Reference: *26.2 What Are Some Keys to the Success of Prokaryotes? p. 541*
16. Which of the following is *not* caused by a negative-sense single-stranded RNA virus?
   a. Measles
   b. Shingles
   c. Rabies
   d. Influenza
   e. Mumps
   Textbook Reference: *26.6 Where Do Viruses Fit into the Tree of Life? pp. 555–556*
17. Which of the following virus types inserts a double-stranded DNA copy of its genome into the host cell's genome?
   a. Double-stranded RNA viruses
   b. Double-stranded DNA viruses
   c. Retroviruses
   d. Negative-sense single-stranded RNA viruses
   e. Positive-sense single-stranded RNA viruses
   Textbook Reference: *26.6 Where Do Viruses Fit into the Tree of Life? pp. 555–557*
18. Which of the following has *not* made the study of prokaryotic phylogeny difficult?
   a. The rapid mutation rates of rRNA genes
   b. The process of lateral gene transfer
   c. The very small size of prokaryotic cells
   d. Our inability to grow many prokaryotic cells in a pure culture (containing only one species)
   e. The harsh and often inaccessible habitats in which many bacteria live
   Textbook Reference: *26.3 How Can We Resolve Prokaryote Phylogeny? pp. 545–546*
19. Which of the following is *not* one of Koch's postulates?
   a. The introduction of the microorganism to a new, healthy host causes the same disease that existed in the original host.
   b The microorganism is always found in the person with the disease.
   c. after the microorganism is introduced into a healthy host who gets the disease, the same microorganism can be isolated from this patient.
   d. The microorganism is susceptible to antibiotic treatment.

e. The microorganism can be isolated from an infected individual and grown in culture.
Textbook Reference: *26.5 How Do Prokaryotes Affect Their Environments? pp. 553–554*

Application Questions

1. You work for a wastewater treatment plant. An organism is getting past your treatment process, and fish are dying in a stream that the runoff enters. You are able to isolate the organism. Using a microscope, how can you tell if the organism is a prokaryote or a eukaryote?
Textbook Reference: *26.1 How Did the Living World Begin to Diversify? p. 538*
2. You work with the Centers for Disease Control in Atlanta. You have been called to a remote area of Uganda to study a mysterious disease that is causing respiratory ailments in a small village. You isolate a bacterium from several patients that seems like a good candidate for the pathogen. How can you determine if this bacterium is causing the illness?
Textbook Reference: *26.5. How Do Prokaryotes Affect Their Environments? p. 554*
3. As a U.S. Department of Agriculture field representative, you counsel a young farmer to plant alfalfa in fields with soils that have low nitrogen levels. You know alfalfa roots are hosts for nitrogen-fixing bacteria. How will the alfalfa and its associated bacteria help "fertilize" this farmer's soil?
Textbook Reference: *26.2 What Are Some Keys to the Success of Prokaryotes? p. 544*
4. A patient comes to your medical practice with a bacterial infection. You are terribly concerned because the bacteria responsible for the patient's infection produce an exotoxin. Why are exotoxins often more dangerous than endotoxins?
Textbook Reference: *26.5 How do Prokaryotes Affect Their Environments? p. 555*
5. Describe how Gram staining depends on bacterial cell wall structure. Would Gram staining work for archaea?
Textbook Reference: *26.2 What Are Some Keys to the Success of Prokaryotes? p. 541*
6. Differentiate between the following terms:
obligate anaerobe/obligate aerobe
photoautotroph/photoheterotroph
chemolithotroph/chemoheterotroph
Textbook Reference: *26.2 What Are Some Keys to the Success of Prokaryotes? p. 543, Table 26.2*
7. A friend tells you she "hates" bacteria and declares her intention to eradicate all bacteria from her home. What do you tell her?
Textbook Reference: *26.5 How Do Prokaryotes Affect Their Environments? pp. 553–554*

Answers

Diagram Exercise Answers

1.
   a. Helical bacteria
   b. Bacilli
   c. Cocci
2. positive; negative
   a. Outside of cell
   b. Cell wall (peptidoglycan)
   c. Plasma membrane
   d. Inside of cell
   e. Outside of cell
   f. Outer membrane of cell wall
   g. Periplasmic space
   h. Peptidoglycan layer
   i. Periplasmic space
   j. Plasma membrane
   k. Inside of cell

3.

Domain	Group Name	Key Features	Example Genera or Associated Diseases
Bacteria	Chlamydias	Gram-negative cocci. Very small, obligate parasites. Life cycle involves two forms of cells called elementary bodies and reticulate bodies.	Cause human eye disease, the STD chlamydia, and some types of pneumonia
Archaea	Crenarchaeota	Most live in hot and/or acidic environments such as sulfur hot springs. Internal cellular environment is close to pH 7.	*Sulfolobus*
Bacteria	Cyanobacteria	Photoautotrophs. Use chlorophyll *a* for photosynthesis. May form colonies with differentiated cells (see Figure 26.13). Some also fix nitrogen. Thought to be the source of eukaryotic chloroplasts.	*Anabaena* (see Figure 26.15)
Archaea	Euyarchaeota	Many are methanogens living in the guts of ruminant herbivores or termites and cockroaches. Others are extreme halophiles. *Thermoplasma* lacks a cell wall and lives in coal deposits.	*Methanopyrus thermoplasma*
Bacteria	High-GC Gram-positives	Gram-positive, GC rich genomes, may form spores at tips of filaments. Source of most of our antibiotics.	*Mycobacterium tuberculosis*, the cause of tuberculosis and *Streptomyces*, the source of streptomycin.
Archaea	Korarchaeota	Poorly characterized, only DNA has been isolated directly from hot springs.	None available
Bacteria	Low-GC Gram positives	Gram positive, although some are Gram negative and lack a cell wall. Genome is not GC rich. May produce endospores capable of surviving extreme conditions. Mycoplasmas are the smallest cellular organisms known.	*Bacillus anthracis:* causes anthrax, *Clostridium:* a cause of food poisoning, *Staphylococcus aureus*: can colonize healthy individuals or cause infection.
Archaea	Nanoarchaeota	Poorly characterized. Lives attached to cells of *Ignicoccus,* a crenarchaeote. Discovered at a deep-sea thermal vent near Iceland's coast.	*Nanoarchaeum equitans*
Bacteria	Proteobacteria	Largest group of identified species. Very diverse group classified into five groups based on their metabolic pathways.	*Rhizobium:* important for nitrogen fixation, *Yersinia pestis:* cause of bubonic plague, *Vibrio cholerae:* cause of cholera, *Salmonella typhinurium*: cause of gastrointestinal disease, and *Escherichia coli:* well characterized.
Bacteria	Spirochetes	Gram-negative, helical structure (See Figure 26.12) hemoheterotrophic. Move via axial filaments.	Some species cause syphilis and lyme disease.

Knowledge and Synthesis Answers

1. **a.** Prokaryotes are differentiated from eukaryotes by their lack of membrane-enclosed organelles. They also have a single circular piece of genomic DNA and lack a cytoskeleton.
2. **a.** A rod-shaped bacterium is known as a bacillus (plural, bacilli). Cocci are roughly spherical.
3. **e.** Bacteria do carry out both photosynthesis and multiple forms of cellular respiration. All life forms have DNA. The flagella of bacteria consist of a single protein called flagellin rather than multiple proteins as found in eukaryotes. Bacteria replicate by binary fission.
4. **b.** Gram-negative bacteria have thin peptidoglycan layers that do not retain crystal violet stain.
5. **b.** The Archaea and the Eukarya share a more recent common ancestor with each other than either does with the Bacteria. One piece of evidence for this is that a signature sequence from rRNA has been found in all archaea and eukaryotes tested so far, but in none of the bacteria.
6. **c.** Biofilms are gel-like polysaccharide matrices that are laid down by prokaryotes and trap other bacteria.
7. **c.** Chlamydias are all parasitic and cause several human diseases.
8. **a.** Endosymbiosis of proteobacteria gave rise to the mitochondria of eukaryotes.
9. **a.** Among the groups listed, only low-GC Gram-positive bacteria, high-GC Gram-positive bacteria, and

cyanobacteria produce spores. Low-GC Gram-positive spores can withstand extreme environmental condition, so they would make good spores for testing an autoclave's ability to sterilize equipment.

10. **b.** The term *halophile* means "salt loving."
11. **b.** Methanogenic bacteria that reside in the guts of cows belong to the Euryarchaeota.
12. **a.** Archaea lack peptidoglycans. Instead they have a unique lipid in their cell walls; other species have pseudopeptidoglycans in their cell walls.
13. **b.** Photoheterotrophs require an outside carbon source other than carbon dioxide, yet are able to harvest light energy.
14. **c.** Oxygen gas is toxic to obligate anaerobes.
15. **e.** True cilia are found in eukaryotes.
16. **b.** Shingles is caused by a double-stranded DNA herpes viruses.
17. **c.** Retrovirues, such as HIV, insert a form of their genome into the host's DNA. This provirus is replicated every time the cell divides by mitosis.
18. **a.** The mutation rates of rRNA genes are quite slow and thus rRNA has evolved slowly. This has made study of rRNA sequences quite valuable in the taxonomy of prokaryotes.
19. **d.** Koch's postulates were developed in the 1880s, when even the role of microorganisms in disease was not understood. Antibiotic therapies to target these microorganisms came later.

Application Answers

1. The most immediately apparent difference between a prokaryote and a eukaryote would be the much smaller size of the prokaryote and the absence of membrane-enclosed organelles in prokaryotes. With proper staining techniques, the nuclei of any eukaryote would likely be visible with the light microscope.
2. Koch's postulates stipulate that in order for an organism to be identified as a disease-causing agent (1) it must always be found in individuals with the disease, (2) it must be taken from the host and grown in pure culture, (3) a sample of the culture must produce disease if injected into a healthy individual, and (4) the newly infected individual must yield a new pure culture of the same organism. That said, it is unethical to knowingly infect a person with a pathogen just for the purposes of identification! Therefore, a diagnosis can be made based on the first three postulates but not on the fourth one.
3. Nitrogen-fixing bacteria can convert atmospheric nitrogen into ammonia. Other bacteria can then convert the ammonia into nitrates that plants can use. (This topic is covered in greater detail in Chapter 36.)
4. Exotoxins are continuously produced by the infecting bacteria, whereas endotoxins are produced only at cell death. Only a limited number of endotoxin molecules are released, but exotoxins can be released until the host dies.
5. Gram staining results depend on the arrangement and amounts of peptidoglycans in the bacterial cell wall. This technique does not work with archaea because they do not have peptidoglycan in their cell walls.
6. Obligate anaerobes die in the presence of oxygen gas, whereas obligate aerobes die without oxygen gas for cellular respiration. Photoautotrophs convert light energy to chemical energy using carbon dioxide as their carbon source; photoheterotrophs rely on an outside carbon source other than carbon dioxide but still can convert light energy to chemical energy. Chemolithotrophs can produce what they need from inorganic molecules, whereas chemoheterotrophs depend on nutrients from external sources.
7. This effort to eradicate all bacteria will not be successful, since humans are hosts to innumerable bacteria both on and in our bodies. Furthermore, most bacteria are not harmful. In fact, many are beneficial. For example, bacteria are essential in nutrient cycling in the ecosystem. In some cases the attempt to eradicate bacteria has created dangers of its own. The abuse and overuse of antibiotics, for instance, creates selective pressures that lead to an increase in antibiotic-resistant pathogenic bacteria, rendering previously effective treatments useless. Your friend would be better-served by declaring a truce with the bacteria and recognizing them as part of the world we inhabit.

27 The Origin and Diversification of Eukaryotes

The Big Picture

The many lineages of protists, as well as all fungi, animals, and plants, are eukaryotic. Eukaryotic cells are characterized by membrane-enclosed organelles, the presence of a cytoskeleton, and a nuclear envelope. The evolution of the eukaryotic cell was a major evolutionary milestone. The eukaryotic cell allows for compartmentalization of processes and increased adaptability. Though the evolutionary lineage is still debated, it is thought that eukaryotes evolved via multiple symbiotic events.

- Protists as a group are not monophyletic. The most recent common ancestor of protists gave rise to the plants, animals, and fungi, groups that are not included in the protists. Protists are diverse, ranging from microscopic single-celled organisms to complex multicellular organisms. Several subgroups of protists are covered in this chapter, some of which are monophyletic while others are not. The five major eukaryotic clades are chromalveolates, Plantae, excavates, rhizaria, and unikonts.
- Certain body plans and modes of nutritional acquisition are repeated throughout the microbial eukaryote groups and are good examples of form following function. Protists as a group play a vast role in the environment and medicine and are economically valuable.

Common Problem Areas

- There is a tendency for students to become overwhelmed with organism names and characteristics. Focus on the trends outlined in the chapter and use lab opportunities to understand the organism groupings. For a start, see Figure 27.1 and Table 27.1.

Study Strategies

- This chapter contains a great deal of information. It is best not to try to learn it all at once, but to break it up into smaller pieces and study it over several sessions.
- To learn the organism groups, create a chart with key characteristics. This will help you compare and contrast the groups.
- On one side of a 4 × 6 index card, write down a particular clade or subgroup; on the other side, note the defining characteristics. Do the same with sample organisms, with the genus specified on one side of the card and the clade and subgroup on the other side. Then shuffle the deck and sort them into different types of piles. For example, you can sort all of the subgroups into their proper clades. You can match sample organisms with their subgroups or clades. Or you can sort the groups according to their reproductive strategies or other criteria of your choosing.
- Go to yourBioPortal.com to review the following tutorials and activities:

  Animated Tutorial 27.1 Family Tree of Chloroplasts

  Animated Tutorial 27.2 Food Vacuoles Handle Digestion and Excretion

  Web Activity 27.1 An Isomorphic Life Cycle

  Web Activity 27.2 A Haplontic Life Cycle

  Web Activity 27.3 Anatomy of *Paramecium*

Important Concepts

Protists are a paraphyletic group of very diverse organisms.

- The term "protist" lacks any real taxonomic meaning. It is really shorthand for all eukaryotes that are neither land plants, fungi, nor animals. You may also see the term "microbial eukaryotes" used in place of protists.
- Because of the "catch-all" nature of this grouping, it is extremely diverse, containing many different organisms that are not necessarily closely related to one another. Most are unicellular, but many are colonial and some are multicellular.
- Most eukaryotes can be classified in one of five major clades, and "protists" can be found in each of these clades. The five clades are chromalveolates, Plantae, excavates, rhizaria, and unikonts.

The evolution of the eukaryotic cell consisted of a series of pivotal events.

- The origin of the eukaryotic cell is still being debated. The evolution of the eukaryotic cell was revolutionary and occurred at a time of great environmental change.

- The modern eukaryotic cell contains a flexible cell surface, a cytoskeleton, a nuclear envelope, digestive vesicles, and organelles.
- The first step in the path toward a eukaryotic cell may have involved loss of the rigid cell wall, which allowed infoldings of the cell surfaces to increase the surface area-to-volume ratio. This allows cells to become larger. It also allows vesicles to form from bits of infolded membrane that pinch off. Such infolding may have been the origin of the nuclear envelope (see Figure 27.2).
- New structures evolved. Ribosomes became associated with internal membranes to form the endoplasmic reticulum, a cytoskeleton of actin and microtubules formed, and digestive vesicles developed. The cytoskeleton opened up modes of locomotion with actin/myosin-based flagella. How these structures formed is currently conjecture.
- It is thought that these early motile cells may have been phagocytes that engulfed prokaryotic cells.
- Subsequent endosymbiotic events may have led to the evolution of mitochondria and chloroplasts.
- See Figure 27.2 for a diagram of events in the evolution of the eukaryotic cell.
- An understanding of the evolution of eukaryotes is complicated by lateral gene transfer. The prokaryotic genes in many eukaryotes are so numerous that lateral transfer cannot explain them all, nor can endosymbiosis. Additional hypotheses are being tested.

Endosymbiosis is the origin of chloroplasts.

- Some chloroplasts are enclosed in double membranes, and others are enclosed in triple membranes. This phenomenon can be explained by endosymbiosis.
- All chloroplasts can be traced to the engulfment of an ancestral cyanobacterium with a single membrane. Its engulfment in a vesicle of the phagocyte's membrane resulted in a double membrane. This initial event is called primary endosymbiosis. Primary endosymbiosis gave rise to the green and red algae.
- Photosynthetic euglenoids, with their triple membranes, arose from secondary endosymbiosis. In this case, a chlorophyte was ingested with its housed chloroplast, resulting in three membranes. Ultimately all of the constituents of the chlorophyte except the chloroplast were lost. Another line is thought to have arisen by means of a red algal endosymbiont.

Protists have diverse ways to accomplish the same functions.

- Most protists are aquatic. Even those that seem terrestrial live in soil water or other extremely moist environments.
- Mechanisms of metabolism are relatively uniform among protists, but nutritional acquisition varies. Some protists are autotrophic, some are heterotrophic, and some have the ability to switch between the two modes. Many protists are involved in symbioses from mutualism to parasitism.
- Most protists are motile and move via pseudopodia, flagella, or cilia. Cilia and eukaryotic flagella are identical in cross section; they differ only in length. The amoeboid body plan (using pseudopods) is highly effective in nutrient-rich environments. Some organisms have an amoeboid body plan at different stages of their life cycle.
- Many protists are microscopic, though some algae literally form aquatic forests of very large, multicellular organisms. Single-celled protists are able to increase their cell size with the presence of vesicles, which effectively increase surface area. Vesicles assist with water regulation (contractile vacuoles) and nutrient procurement (food vacuoles).
- Many protists have cell walls or "shells" to protect their cell membranes. The chemical composition of these structures varies among organisms.
- Many protists contain endosymbionts. Some harbor other protists that photosynthesize or carry out other metabolic processes.
- Most protists undergo both sexual and asexual reproduction, but some lack the ability to reproduce sexually. Sexual and asexual reproductive practices are as diverse as the protists themselves.
- ***For review, go to Diagram Exercise 1.***

Clade: Chromalveolates, subgroup: Alveolates

- The alveolates are a monophyletic group consisting of dinoflagellates, apicomplexans, and ciliates.
- The synapomorphy that defines the Alveolata is the possession of cavities called alveoli just below their cell surfaces.
- Dinoflagellates are mostly marine organisms with two flagella. They are yellowish in color due to photosynthetic and accessory pigments and are primary producers in marine environments. Many live in symbiosis with other organisms. They are common endosymbionts in corals, for example. Dinoflagellates such as *Pfiesteria piscicida* and *Gonyaulax* are responsible for "red tides" in warm marine waters and can damage fish. Toxins produced by *Gonyaulax* species can accumulate in shellfish and be fatal to people eating the shellfish.
- Apicomplexans are parasitic protists. Their name comes from the cluster of organelles at the apex of their cells, which assist with invasion of their host organism. The life cycles of these parasites are complex. Apicomplexans in the genus *Plasmodium* cause malaria, a disease that claims more than a million lives annually (see Figure 27.9).
- Ciliates move via cilia and are characterized by multiple macronuclei and micronuclei (see Figure 27.12). Most ciliates are heterotrophic. *Paramecium* is a common and well-studied ciliate genus.

- Paramecia move by means of cilia, their membranes are protected by a pellicle, and they protect themselves with trichocysts, which act as sharp darts.
- Paramecia reproduce asexually by binary fission. Genetic recombination is accomplished through conjugation, which involves the exchange of equal amounts of genetic material (see Figure 27.12). The resulting recombined paramecia then go through binary fission.

Clade: Chromalveolates; subgroup: Stramenopiles

- Stramenopiles consist of diatoms, brown algae, oomycetes, and a few smaller groups, and are characterized by their unequal haired flagella. Those that are not flagellated have lost their flagella over the course of evolution.
- Diatoms are yellowish brown and store carbohydrates and oils as photosynthetic products. They are most noted for their silica-containing cell walls, which occur in two halves and fit together like a petri plate (see Figure 27.19) and have been used as filters, insulation, and insecticides. All diatoms are symmetrical and unicellular. They reproduce both asexually and sexually.
- Brown algae are multicellular and are the largest of the protists.
  - Brown algae are almost exclusively marine. They produce leaflike growths or branched filaments. The giant kelp that grow in large oceanic "forests" are brown algae.
  - Brown algae are brown because of chlorophylls *a* and *c* and the carotenoid fucoxanthin. They have specialized regions called holdfasts, which anchor them to a substrate and allow them to resist water movement.
  - Brown algae are commercially important for the presence of alginic acid, used as a binder in many food and cosmetic products.
  - Like many other protists and all plants, brown algae go through alternation of generations, in which a diploid sporophyte gives rise to haploid spores through meiosis to form the haploid gametophyte, which in turn produces haploid gametes (see Figure 27.13). Heteromorphic alternation of generations occurs when the two generations differ morphologically, and isomorphic alternation of generations occurs when the two generations are morphologically similar.
- Oomycetes are a nonphotosynthetic group of stramenopiles. These are commonly known as water molds and downy mildews, but they are not fungi. Oomycetes are coenocytic, and have multiple nuclei per cell. Many are saprobic and feed on dead material. Others are infectious to plants. The Irish potato famine was caused by an oomycete.

Clade: Plantae; subgroup: Red algae

- The red algae are mostly unicellular.
- The red algae are characterized by the pigment phycoerythrin, which gives them their characteristic red color. Depending on light intensity, red algae alter the ratio of pigments in the chloroplast. In high light conditions, red algae may appear very green.
- Most red algae are large marine organisms and have specialized structures.
- Red algae store the products of photosynthesis as floridean starch and have no motile cells at any point in their life cycles.
- Red algae are the source of agar, which is important in food products as well as in the study of biology.

Clade: Plantae; subgroup: Chlorophyta

- Chlorophyta is the largest lineage of green algae in the Plantae clade and is a monophyletic group.
- The green algae are not a monophyletic group. This is because plants are also descended from the most recent common ancestor of all green algae, but plants are not included among the green algae.
- Chlorophytes contain chlorophylls *a* and *c*, store photosynthetic products as starch in plastids, and are mainly aquatic.
- Morphology of the chlorophytes is diverse and ranges from simple single-celled organisms to very complex multicellular organisms. Colonial and aggregate forms exist. Colonial forms show the first indications of cell specialization without multicellularity. Individual cells range from single-nucleated to multi-nucleated.
- Life cycles within the chlorophytes are diverse. (see Figures 27.14 and 27.15 on the life cycle of *Ulva* and *Ulothrix*, respectively.) Some chlorophytes are isogamous, meaning that their gametes are identical. Others are anisogamous, with female gametes much larger than the male gametes. Some chlorophytes have isomorphic life cycles and others are heteromorphic. In one variation of the heteromorphic life cycle—the haplontic life cycle—a multicellular haploid individual (gametophyte) produces gametes that fuse to form a zygote. Some chlorophytes exhibit diplontic life cycles much like animals, in which meiosis of sporophytes produces gametes directly.

Clade: Excavates

- The excavates are a monophyletic group consisting of five major subgroups: the euglenoids, the kinetoplastids, the heteroloboseans, the diplomonads, and the parabasalids. They are single-celled flagellates that reproduce asexually by binary fission.
- Euglenids have anterior flagella. Many are photosynthetic and are characterized by a complex cell structure (see Figure 27.25). They are flexible in nutritional requirements and may switch between autotrophism and heterotrophism. Euglenid chloroplasts have triple membranes.

- Kinetoplastids are parasitic and are characterized by their single large mitochondrion. The mitochondrion is unique in having a kinetoplast housing multiple circular DNA molecules and associated proteins. Many tropical human diseases are caused by kinetoplastids, including African sleeping sickness (see Table 27.2).
- Heteroloboseans have a life cycle that alternates between an amoeboid stage and a flagellated stage. One species of Naegleria (*Naegleria fowleri*) can cause a fatal brain infection in humans. The amoeba enters the body through the nose and moves through the olfactory nerve to the brain where it destroys tissues.
- Diplomonads and parabasalids lack mitochondria. *Giardia lamblia* is a well-known diplomonad causing the human intestinal disorder giardiasis. *Trichomonas vaginalis* is a parabasalid responsible for a sexually transmitted disease in humans.

Clade: Rhizaria

- The rhizaria comprise three related groups of unicellular aquatic eukaryotes including the cercozoans, foraminiferans, and radiolarians.
- Cercozoans are very diverse, with either amoeboid or flagellated forms and aquatic or soil habitats.
- Limestone deposits are often the result of discarded foraminiferan shells made of calcium carbonate. Foraminiferans live either as plankton or on the sea bottom. Their pseudopods are used to trap food.
- Radiolarians have stiff, microtubule-reinforced pseudopods that help the cells float in their marine environment and provide additional surface area.

Clade: Unikonts

- Unikonts are thought to be close to the root of the Eukaryote tree. They have two major groups: the Opsithokonts and the Amoebozoans.
- Opsithokonts include fungi, animals, and choanoflagellates.
  - Choanoflagellates are structurally similar to sponges, the most basal animal lineage, and are thought to be the closest relatives of the animals. They are flagellated, and the majority are colonial.
- Amoebozonans include the loboseans, the cellular slime molds, and the plasmodial slime molds.
  - Loboseans live as independent, single cells. They feed by phagocytosis. A few have shells (see Figure 27.7A).
  - Slime molds are motile, feed by endocytosis, form spores on erect fruiting bodies, and undergo drastic changes during their life cycle.
  - Plasmodial slime molds form multinucleate masses (see Figure 27.28). During the vegetative stage, an acellular slime mold is a wall-less mass of cytoplasm with multiple diploid nuclei. It oozes over a network of strands called plasmodium. This phenomenon is called cytoplasmic streaming and is used to move and engulf food. If exposed to adverse environmental conditions, the slime mold will form a dormant sclerotium from which it can turn back into a plasmodium when environmental conditions again become favorable. It may also transform into a fruiting structure called a sporangiophore.
  - Cellular slime molds are made of large numbers of cells called myxamoebas with single haploid nuclei. They reproduce by mitosis and fission, and this life cycle can continue as long as conditions are favorable. Once conditions become unfavorable, myxamoebas secrete cAMP, aggregate, and form a motile slug that eventually produces fruiting bodies. Spores from the fruiting bodies germinate to form more myxamoebas. Sexual reproduction occurs with the fusion of two myxamoebas, which then undergo meiosis to release haploid myxamoebas (see Figure 27.29).
- ***For review, go to Diagram Exercises 2–5.***

Test Yourself

Diagram Exercises

1. Organize the following list of clades and subgroups according to the evolutionary relationships described in the textbook: Alveolates, Amoebozoans, Chromalveoates, Excavates, Opisthokonts, Plantae, Rhizaria, Stramenopiles, Unikonts.

   1. ____________
      a.____________
      b.____________
   2. ____________
   3. ____________
   4. ____________
   5. ____________
      a.____________
      b.____________

   Textbook Reference*: 27.1 How Did the Eukaryotic Cell Arise? pp. 562–563, Figure 27.1, Table 27.1*

2. Identify a.–k. using the following terms (some may be used more than once): cyst,male gametocyte, female gamete, human, male gamete, merozoites, female gametocyte, mosquito, sporozoites, and zygote. What kind of life cycle is this?

   ***Textbook Reference**: 27.4 How Do Protists Reproduce? p. 573, Figures 27.9 and 27.13*

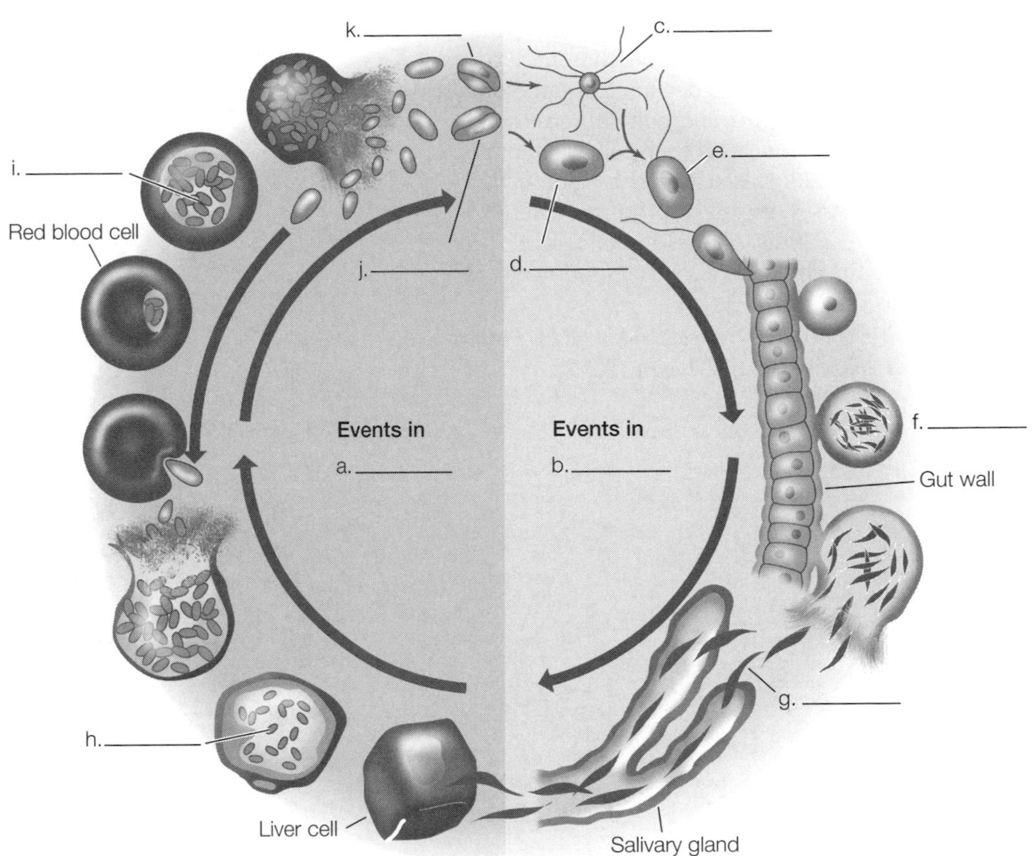

3. Identify a.–o. using the following terms (some may be used more than once or not used at all): diploid, fusing gametes, haploid gametes (n), gametes, gametophyte, haploid, spores, sporophyte, zoospores, zygote. Indicate if each cell is haploid (n) or diploid ($2n$). Also label the reproductive processes (fertilization, meiosis, and mitosis) in the closed boxes. What kind of life cycle is this?

   ***Textbook Reference**: 27.4 How Do Protists Reproduce? p. 574, Figure 27.14*

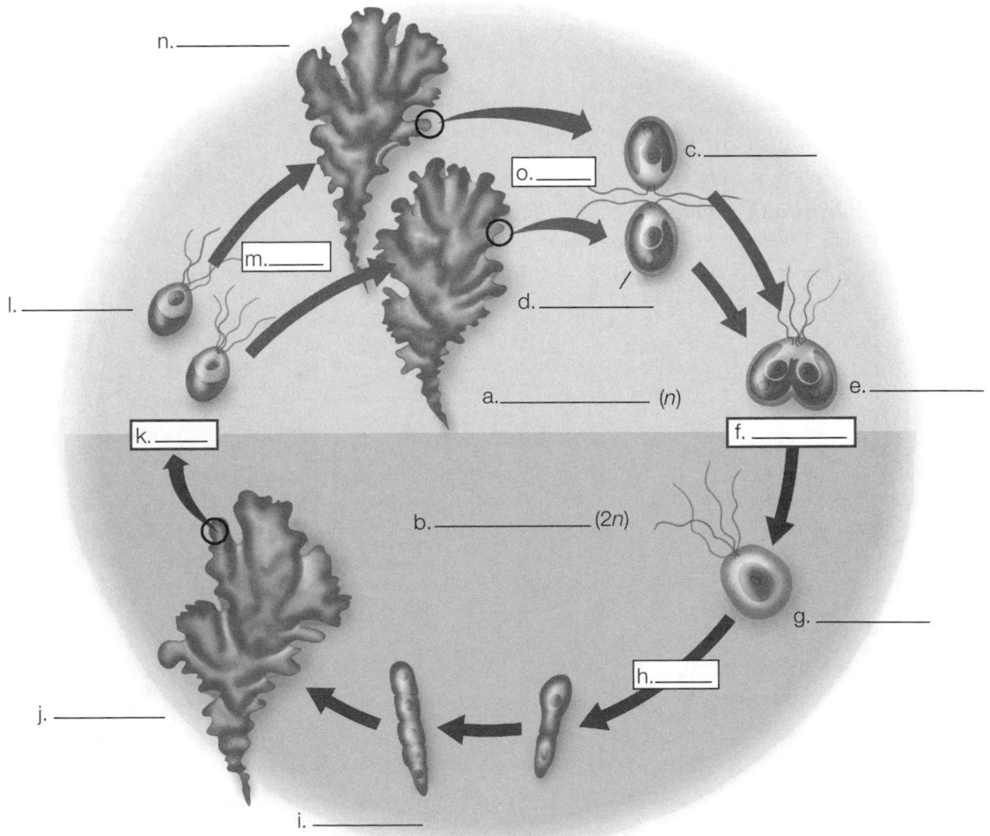

4. Identify a.–k. using the following terms (some may be used more than once or not used at all): diploid, gametes, gametophyte, haploid, spores, sporophyte, zoospores, zygote. Indicate if each cell is haploid (n) or diploid ($2n$). Also label the reproductive processes (asexual reproduction, fertilization, meiosis, mitosis, sexual reproduction) in the closed boxes. What kind of life cycle is this?
   ***Textbook Reference**: 27.4 How Do Protists Reproduce? p. 574, Figure 27.15*

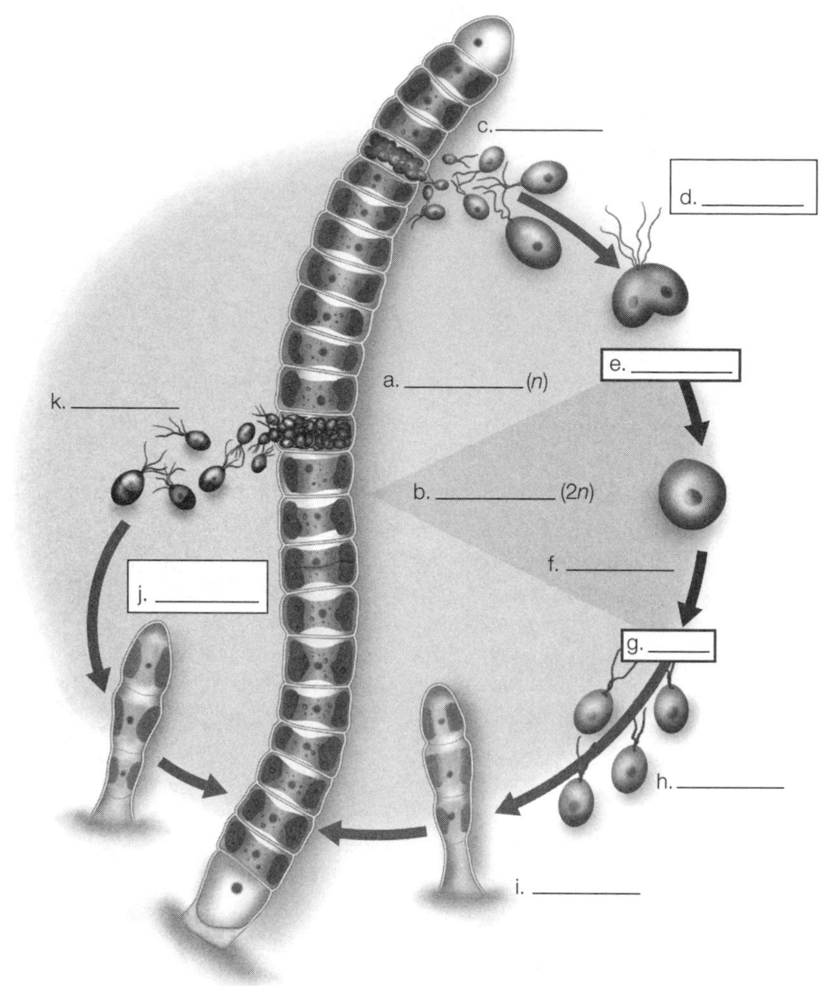

5. Identify a.–m. in the *Paramecium* shown at right.
   ***Textbook Reference**: 27.5 What Are the Evolutionary Relationships among Eukaryotes? p. 577, Figure 27.18*

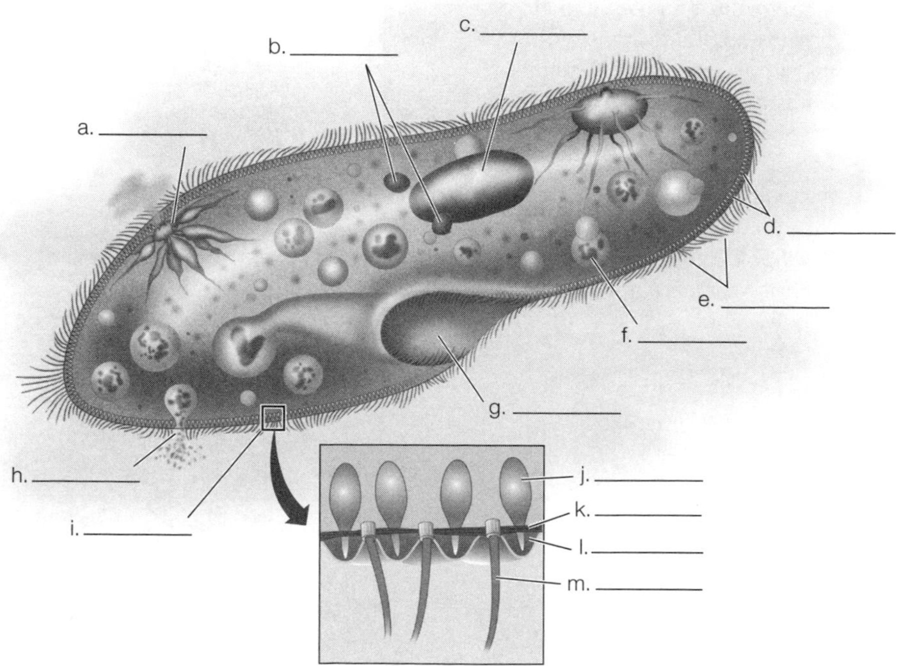

Knowledge and Synthesis Questions

1. Which of the following modes of reproduction can be found in at least some protists?
   a. Binary fission
   b. Sexual reproduction
   c. Spore formation
   d. Multiple fission
   e. All of the above

   Textbook Reference: *27.4 How Do Protists Reproduce? p. 572*

2. During the evolution of eukaryotes from prokaryotes, which of the following did *not* occur?
   a. Infolding of the flexible cell membrane
   b. Loss of the cell wall
   c. A switch from aerobic to anaerobic metabolism
   d. Endosymbiosis of once free-living prokaryotes
   e. Development of a cytoskeleton

   Textbook Reference: *27.1 How Did the Eukaryotic Cell Arise? pp. 564–565*

3. Which of the following statements about protists is *false*?
   a. Apicomplexans are the only microbial eukaryote group without parasitic representatives.
   b. Foraminiferans and radiolarians are shelled protists.
   c. Ciliates have great control over the direction their cilia beat.
   d. Although they appear structurally simple, amoebas are not primitive organisms.
   e. All diatoms are unicellular, although a few individual species associate in filaments.

   Textbook Reference: *27.3 How Do Protists Affect the World Around Them? p. 569–571*

4. Ciliates, as represented by *Paramecium*, have defensive organelles called _______ in their pellicles.
   a. trichonympha
   b. tridents
   c. trichomes
   d. trichocysts
   e. trochlea

   Textbook Reference: *27.5 What Are the Evolutionary Relationships among Eukaryotes? pp. 584–585*

5. A major difference between the vegetative states of cellular and plasmodial slime molds is that plasmodial slime molds _______ and cellular slime molds _______.
   a. have haploid nuclei; have diploid nuclei
   b. produce fruiting bodies; do not produce fruiting bodies
   c. undergo aggregation when conditions become adverse; do not undergo aggregation under adverse conditions
   d. exist as a coenocytic mass; exist as individual myxamoebas
   e. grow almost indefinitely in favorable conditions; must aggregate into myxamoebas every seven generations

   Textbook Reference: *27.5 What Are the Evolutionary Relationships among Eukaryotes? pp. 583–585*

6. Organisms that exhibit isogamy have
   a. similar male and female gametes.
   b. female gametes that are significantly larger than male gametes.
   c. male gametes that are significantly larger than female gametes.
   d. a diplontic mode of reproduction.
   e. a primarily asexual mode of reproduction.

   Textbook Reference: *27.4 How Do Protists Reproduce? p. 574*

7. Red tides often cause massive fish kills and human illness in those eating shellfish. Which group of protists is responsible for red tides?
   a. Parabasalids
   b. Red algae
   c. Euglenozoans
   d. Dinoflagellates
   e. Radiolarians

   Textbook Reference: *27.3 How Do Protists Affect the World Around Them? p. 571*

8. Holdfasts and alginic acid are characteristic of which group of protists?
   a. Parabasalids
   b. Red algae
   c. Brown algae
   d. Stramenopiles
   e. Unikonts

   Textbook Reference: *27.5 What Are the Evolutionary Relationships among Eukaryotes? p. 578*

9. A water mold is discovered growing on the surface of a freshwater green alga. Which two groups of protists are involved?
   a. Choanoflagellates and red algae
   b. Oomycetes and chlorophytes
   c. Apicomplexans and chlorophytes
   d. Apicomplexans and ciliophora
   e. Stramenopiles and oomycetes

   Textbook Reference: *27.5 What Are the Evolutionary Relationships among Eukaryotes? pp. 578–579*

10. Which of the following statements regarding conjugation in *Paramecium* is *false*?
    a. Conjugation results in genetic recombination.
    b. Conjugation results in clones.
    c. Conjugation results in offspring.
    d. Conjugation is a sexual process.
    e. Conjugation results in the production of no new cells.

    Textbook Reference: *27.4 How Do Protists Reproduce? p. 573*

11. Which of the following statements regarding protists is true in general?
    a. Protists are always parasitic.
    b. Protists are all single celled.
    c. Protists are all heterotrophic.
    d. Protists are always photosynthetic.
    e. Protists are always aquatic, at least on a small scale.

    Textbook Reference: *27.3 How Do Protists Affect the World Around Them? p. 566–569*

12. According to Figure 27.1, which of the following statements about the evolution of eukaryotes is true?
    a. The alveolates are more closely related to the stramenopiles than to any other group.
    b. The most recent common ancestor of the parabasilids and euglenoids also gave rise to the animals.
    c. The red algae are more closely related to the brown algae than they are to any other taxon.
    d. The opisthokonts are polyphyletic.
    e. Euglenids and Cercozoans are subgroups of the clade Rhizaria.

    Textbook Reference: *27.1 How Did the Eukaryotic Cells Arise? Figure 27.1, p. 562*
13. Contractile vacuoles
    a. are important for acquiring food.
    b. are mechanism for defense against predators.
    c. are lined with cilia alveoli to maximize surface area.
    d. are a mechanism for removing excess water from the organism.
    e. are more important in marine protists than they are in freshwater protists.

    Textbook Reference: *27.2 What Features Account for Protist Diversity? p. 567*
14. Which of the following statements about protists is *false*?
    a. Protists are photosynthetic and are one of the foundations of the marine food web.
    b. Protists are synthesizers of materials that form many sandy beaches
    c. Protists neutralize the sulfuric acid in deep sea vents.
    d. Protists aid in ruminant herbivore digestion.
    e. Protists aid in the formation of what we call crude oil.

    Textbook Reference: *27.3 How Do Protists Affect the World Around Them? p. 569*
15. Which of the following statements about the protist cytoskeleton is *false*?
    a. It allows for the formation of pseudopods.
    b. It is the main structural component of cilia.
    c. It is the primary component of the diatoms' outer shell.
    d. It is the main structural component of flagella.
    e. It helps in removing excess water via the contractile vacuoles.

    Textbook Reference: *27.2 What Features Account for Protist Diversity? p. 567*

Application Questions

1. Explain alternation of generations. Why are the protists the first group in which this process could be seen?
   Textbook Reference: *27.4 How Do Protists Reproduce? pp. 573–574*
2. Differentiate between asexual and sexual reproduction. Select one mode of microbial eukaryote sexual reproduction and explain it fully.
   Textbook Reference: *27.4 How Do Protists Reproduce? pp. 572–575*
3. What would be the criteria for placing a newly discovered organism within the protists?
   Textbook Reference: *27.1 How Did the Eukaryotic Cell Arise? p. 561*
4. Explain one line of thought regarding the origin of the eukaryotic cell. Why are scientists unsure of the evolutionary relationships among the protists?
   Textbook Reference: *27.1 How Did the Eukaryotic Cell Arise? pp. 563–565*
5. Most protists are motile. Describe some of the means of motility.
   Textbook Reference: *27.2 What Features Account for Protist Diversity? p. 567*
6. Describe the origin of double-membrane-enclosed and triple-membrane-enclosed chloroplasts.
   Textbook Reference: *27.1 How Did the Eukaryotic Cell Arise? pp. 565–566*
7. Many protists are significant human pathogens. Describe the pathogenic protist that causes malaria, describe its life cycle, and identify the microbial eukaryote group to which it belongs.
   Textbook Reference: *27.3 How Do Protists Affect the World Around Them? p. 570*
8. Draw the phylogeny of the three domains of life and add the various lineages of protists to it. Then add the plants, animals, and fungi.
   Textbook Reference: *27.1 How Did the Eukaryotic Cell Arise? p. 562*

Answers

Diagram Exercise Answers

1.
   1. Chromalveolates
      a. Alveolates
      b. Stramenopiles
   2. Plantae
   3. Excavates
   4. Rhizaria
   5. Unikonts
      a. Opisthokonts
      b. Amoebozoans
2. This is a protist life cycle.
   a. Human
   b. Mosquito
   c. Male gamete
   d. Female gamete
   e. Zygote
   f. Cyst
   g. Sporozoites
   h. Merozoites
   i. Merozoites
   j. Female gametocyte
   k. Male gametocyte

3. This type of life cycle is called isomorphic.
   a. Haploid
   b. Diploid
   c. Haploid (n) gametes
   d. Haploid (n) gametes
   e. Fusing gametes
   f. Fertilization
   g. Zygote ($2n$)
   h. Mitosis
   i. Sporophyte ($2n$)
   j. Sporophyte ($2n$)
   k. Meiosis
   l. Spores (n)
   m. Mitosis
   n. Gametophyte
   o. Mitosis
4. This type of life cycle is called haplotic.
   a. Haploid
   b. Diploid
   c. Gametes (n)
   d. Sexual reproduction
   e. Fertilization
   f. Zygote ($2n$)
   g. Meiosis
   h. Zoospores (n)
   i. Gametophyte (n)
   j. Asexual reproduction
   k. Zoospores (n)
5.
   a. Contractile vacuole
   b. Micronuclei
   c. Macronucleus
   d. Alveoli
   e. Cilia
   f. Food vacuole
   g. Oral groove
   h. Anal pore
   i. Pellicle
   j. Trichocyst
   k. Fibrils
   l. Alveolus
   m. Cilium

Knowledge and Synthesis Answers

1. **e.** Methods of reproduction are quite varied among the protists.
2. **c.** At the time the first eukaryotes evolved, the environment was becoming oxygen-rich. There was a switch from anaerobic to aerobic metabolism, not vice versa.
3. **a.** Malaria is caused by *Plasmodium*, which is an apicomplexan. In fact, all apicomplexans are parasitic.
4. **d.** Trichocysts are defensive barbs ejected from ciliates when they are disturbed.
5. **d.** The vegetative (feeding) state of an acellular slime mold is called a plasmodium; it consists of multiple diploid nuclei enclosed in a single membrane. The vegetative state of a cellular slime mold is a myxamoeba with a single haploid nucleus. Although the myxamoebas of cellular slime molds do aggregate to form fruiting structures; individual myxamoebas never fuse into multinucleated structures.
6. **a.** "Isogamy" means "same gametes." The gametes appear identical.
7. **d.** Dinoflagellates are the cause of red tides.
8. **c.** The brown algae are multicellular protists notable for their organ and tissue differentiation and the presence of alginic acid in their cell walls.
9. **b.** Water molds are oomycetes, and green algae are chlorophytes.
10. **d.** Conjugation does result in genetic recombination but does not result in the production of clones or offspring.
11. **e.** Protists do not have a unifying characteristic other than being eukaryotic organisms that do not fit into the kingdoms Plantae, Animalia, or Fungi. If an aquatic environment is considered to be a single droplet of water, however, protists would meet this criteria. For a single-celled organism, a droplet of water is a very large environment.
12. **a.** The only choice that is supported by the cladogram in Figure 27.1 is that the alveolates are more closely related to the stramenopiles than to any other group.
13. **d.** Contractile vacuoles remove excess water from several freshwater protists.
14. **c.** Protists are a very diverse group, but they do not release bases or neutralize environmental acids. Members of Archaea are likely to be found in an acidic environment.
15. **c.** The cytoskeleton, consisting of actin filaments, microtubules, and intermediate filaments, is a feature of eukaryotic cells. In protists, the cytoskeleton is important in anchoring organelles, movement of materials in the cell, movement of the cell in the environment, and controlling the shape of the cell. (The cytoskeleton was covered in detail in Chapter 4.)

Application Answers

1. An organism that exhibits alternation of generations exists in a haploid gamete-producing form and a diploid spore-producing form. Prokaryotes have a single chromosome, so only in eukaryotes, which have multiple copies of chromosomes, is a diploid stage of a life cycle possible.
2. Asexual reproduction results in clones of the original organism, and there is no genetic recombination or variation associated with the creation of offspring. Sexual reproduction allows for new genetic combinations. Protists undergo varied types of sexual reproduction, from fusing haploid myxamoebas in cellular slime molds, to haplontic alternation of generations, to a diplontic type of reproduction. Conjugation between paramecia is an example of sexual recombination without reproduction.
3. The organism must be a eukaryote and not fit the criteria for land plants, animals, or fungi.
4. See Figure 27.2. The evolutionary relationships are difficult to understand due to limited fossilization, lateral gene transfers, and the contributions of symbiotic prokaryotes.
5. Protists move by flagella, cilia, and pseudopodia. Because many are aquatic, additional modifications allow for floating and movement along currents.
6. Double-membrane-enclosed chloroplasts most likely arose from the endosymbiosis of a cyanobacteria in a process called primary endosymbiosis; triple-membrane-enclosed chloroplasts most likely arose from the endosymbiosis of a chlorophyte or a rhodophyte (see Figure 27.3).

7. You may select from a variety of pathogenic protists: malaria is caused by one of the apicomplexans, African sleeping sickness caused by a kinetoplastid, or others. See Figure 27.9 for the life cycle of *Plasmodium*, the protist that causes malaria.

8. This takes practice! Begin by drawing just the Bacteria, Archaea, and Eukarya. Now "add" the branches representing groups of protists, animals, plants, and fungi. Check your final tree against Figure 27.17.

28 Plants without Seeds: From Water to Land

The Big Picture

- Land plants are photosynthetic eukaryotes that utilize chlorophylls *a* and *b*, undergo alternation of generations, and develop from multicellular embryos that are protected by the parent plant. The ten extant (surviving) groups can be classified as nonvascular plants (those without highly developed vascular tissue) and vascular plants (those with highly developed vascular tissue). This chapter focuses on the nonvascular plants and the nonseed vascular plants.
- In order to colonize land, plants had to evolve strategies for coping with desiccation and gravity. This involved mechanisms for extracting water from soil, means of transporting water throughout the plant, methods of ensuring fertilization, and modes of protecting developing embryos.
- This chapter introduces and describes the liverworts, hornworts, and mosses, all of which are considered nonvascular plants, and the club mosses, horsetails, whisk ferns, and ferns, which are the nonseed vascular plants.

Common Problem Areas

- It is very tempting to memorize land plant types and characteristics, but it is more important to focus on the trends and relationships among them. Make it a goal to understand the evolutionary relationships among them.

Study Strategies

- Focus on relationships among the land plants and how to differentiate one group from another. Many of the important distinctions between groups are related to adaptations for terrestrial life.
- When studying life cycles, be sure to note whether the sporophyte or the gametophyte is dominant and make note of where mitosis and meiosis take place.
- Go to yourBioPortal.com to review the following tutorial and activities:

  Animated Tutorial 28.1 Life Cycle of a Moss

  Web Activity 28.1 Homospory

  Web Activity 28.2 Heterospory

  Web Activity 28.3 The Fern Life Cycle

Important Concepts

The plant kingdom is monophyletic.

- All land plants descend from a single common ancestor and form a branch of the evolutionary tree of life.
- Plants are photosynthetic eukaryotes that utilize chlorophylls *a* and *b* to produce and store carbohydrates as starches.
- All land plants (embryophytes) develop from an embryo that is protected by tissues of the parent plant. This is a key shared trait or *synapomorphy* of land plants.
- Land plants undergo alternation of generations (see Figures 28.3 and 28.4).
  - Haploid gametophytes produce haploid gametes through mitosis, and diploid sporophytes produce haploid spores through meiosis.
  - The diploid sporophyte arises from the fusion of gametes to produce a diploid zygote that develops into the mature sporophyte.
- The extent of the development of the two life stages is characteristic of the different phyla.
  - Nonvascular plants typically have a dominant gametophyte. The small sporophyte is nutritionally dependent on the gametophyte, even if the sporophyte is photosynthetic.
  - In the nonseed vascular plants, the sporophyte is dominant, but the gametophyte, though small, is nutritionally independent of the sporophyte.
  - In the most recently evolved vascular plants, the sporophyte is dominant, while the gametophyte is greatly reduced.
- There are ten extant groups of land plants (see Table 28.1). Land plants as a whole are monophyletic, but the nonvascular plants as a group are paraphyletic.
- Molecular and fossil evidence indicates that land plants arose from a green algae ancestor. Most likely arose from a group of green algae called the Charales (see Figure 28.2) based on the following synapomorpies:

(1) plasmodesmata that join the cytoplasm of adjacent cells, (2) branching and apical growth, and (3) similar peroxisome contents, chloroplast structure, and mechanics of cellular division.

Survival on land poses a series of challenges.

- To survive on land, aquatic organisms must have ways to reduce their dependence on water and develop mechanisms to avoid lethal desiccation (drying). They require a mechanism to move water to all their cells, to prevent loss of water to evaporation, and to prevent desiccation of the developing embryo.
- Whereas aquatic organisms can float in their environment, terrestrial plants have to support themselves against gravity in order to grow upward.
- Whereas the sperm of aquatic organisms can swim in search of an egg, terrestrial plants need some other mechanism to move sperm from one plant to another.
- The first land plants that met at least some of these challenges are the nonvascular land plants.

Many characteristics that distinguish land plants from green algae are adaptations to a terrestrial environment.

- The earliest land plants developed the following modifications for terrestrial life: a waxy cuticle, gametangia, embryos, protective pigments, thick spore walls, mutualistic associations with fungi, and (with the exception of the liverworts) the presence of stomata.
  - Waxy cuticles, though minimal in early land plants, prevented water loss from tissues as they were exposed to dry air. The development of a cuticle was likely one of the earliest and most important adaptations to land. The extent of the cuticle varies among plant groups and is most well developed in the more recently evolved gymnosperms and angiosperms.
  - Multicellular gametangia mark the transition from algae to plant. Gametangia enclose the gamete-producing tissue and protect the developing gametes.
  - Plant embryos are significant in that they contain the developing sporophyte and are housed within protective tissues. Recently evolved angiosperms have a complex protective mechanism for the embryo.
  - Water protects aquatic organisms from damaging ultraviolet radiation. To compensate for the loss of this protection, land plants have evolved pigments that screen out UV radiation.
  - Spore walls protect spores from desiccation during dispersal. This is particularly important in the nonseed land plants.
  - Mutualistic associates with fungi help with the absorption of nutrients from the soil that might otherwise be inaccessible to land plants.
  - With the exception of the liverworts, today's land plants have stomata, which are small openings in leaves used to regulate gas exchange. The stomata of mosses and hornworts cannot close.

The development of vascular tissue was key to the further adaptation to a terrestrial existence.

- Vascular tissue provides for the transport of water and food through a plant. Xylem is a vascular tissue responsible for the transport water and minerals from the soil to the aerial parts of the plant. Vascular plants also contain phloem, which conducts the products of photosynthesis from the location that they are produced to where they are stored. Some mosses have hydroid cells that die and leave a tiny channel that allows for the movement of water and nutrients. These are not true tracheids because they lack lignin (a stiffening substance) and a cell-wall structure.
- In all vascular plants except the angiosperms (the flowering plants), the tracheid is the principal conducting element of the xylem. The tracheid is a strawlike cell that conducts water in the plant. Angiosperms also have xylem and have retained tracheids as part of their vascular system.
- This vascular system also provides a rigid structural support, which allows vascular plants to grow upward and complete for sunlight. Increased height also allows for increased dispersal of spores.
- Not all land plants have vascular tissue; therefore, it is not an absolute requirement for terrestrial life. Land plants that lack vascular tissue, however, cannot grow very large or live far from a water source. The evolution of vascular tissue has allowed land plants to occupy many niches in the biosphere that would otherwise be inaccessible to them.

Nonvascular plants were the first plant colonizers of land.

- Land plants without vascular tissue depend on abundant water sources, thin tissues that can absorb water easily, mechanisms for capturing water vapor in the air, and capillary action to move water through the plant. They lack the support of lignin and hug the ground closely.
- The life cycle of nonvascular land plants is dominated by the gametophyte generation. The sporophyte is very tiny and is completely dependent on the gametophyte. Figure 28.4 illustrates the life cycle of a moss as an example of the nonvascular plant life cycle. You should be able to follow this life cycle and understand which stages are haploid, which are diploid, and where in the cycle meiosis takes place.
- Liverworts (*Hepatophyta*) may be the oldest surviving plant clade.
  - The liverwort gametophyte can be either leafy or a flat plate of cells (see Figure 28.11). Gametophytes produce antheridia and archegonia on their upper surfaces (sometimes on stalks) and rhizoids (rootlike structures for anchoring and water absorption) on their lower surfaces.

  - Liverwort sporophytes are shorter than those of mosses and hornworts (often only millimeters long). In most species, the cells of the sporophyte expand, allowing the stalk to elongate. This aids in spore dispersal.
  - In some species of liverworts, spores are not released until the surrounding sporangium rots. Other species have structures that forcefully eject spores.
  - Liverworts can reproduce asexually by shedding small clumps of cells called gemmae. Some liverworts have gemmae cups that promote dispersal by raindrops (see Figure 28.11C).
- Hornworts (*Anthocerophyta*) evolved simple stomata, pores that help with the exchange of gas and water vapor. This trait is shared with the mosses. Liverworts do not have stomata.
  - Hornworts are distinguished from other nonvascular plants by two characteristics: the presence of a single, platelike chloroplast in each cell, and sporophytes that are capable of indeterminate growth. In very moist environments, the sporophyte can be as tall as 20 centimeters. Growth is limited by the lack of a true vascular system. The hornwort sporophyte resembles a long, slender horn (see Figure 28.14).
  - Hornworts often have symbiotic relationships with cyanobacteria that are able to fix atmospheric nitrogen and make it available to the hornwort.
- Mosses (*Bryophyta*) are the most familiar nonvascular plants.
  - Mosses have specialized cells called hydroids. Hydroids are thought to be precursors to the tracheid seen in vascular plants, but they lack lignin for structural support (see the section on vascular plant terrestrial adaptation above). Like tracheids, the hydroid cell dies and leaves a tiny channel that can transport water through the plant.
  - Some mosses also have cells called leptoids that allow for the limited transport of sugars.
  - Mosses also differ from other nonvascular plants in that they exhibit apical cell division. This characteristic is seen in "higher" land plants and allows for regular, upright growth.
  - The moss life cycle is presented in Figure 28.4.

• ***For review, go to Diagram Exercise 1.***

The vascular plants have specialized vascular tissues.

- All vascular plants are characterized by the presence of water-conducting tracheids. Tracheids were a significant advance in the evolution of land plants, providing a means of water transport and support for growth. (See the section above on terrestrial adaptation for further details).
- Vascular plants fossilize much more readily than nonvascular plants. Thus, our morphological knowledge of early vascular plants is much more extensive than our knowledge of nonvascular plants.
- Vascular plants are also characterized by branching, independent sporophytes.
- Figure 28.5 illustrates the relationships among the different groups of vascular plants.
- The first vascular plants were members of a now-extinct group of land plants called the rhinophyta. *Rhynia* had characteristics of extant vascular plants and showed the earliest features of this group, including an independent branched sporophyte and a simple vascular system of xylem and phloem, but they lacked more advanced characteristics such as true leaves and roots (see Figure 28.7). Not all had the tracheids found in modern vascular plants.

Significant new features arose in early vascular plants.

- Roots are believed to have evolved over time from branches that grew underground. Because branches above-ground and below-ground would have been exposed to different selective pressures, underground branches might have given rise to what we know today as roots.
- True leaves are flattened photosynthetic structures with vascular tissue. Vascular plants have evolved two different kinds of leaves: microphylls and megaphylls (see Figure 28.8).
  - Microphylls are present in the club mosses. They probably evolved from sterile sporangia.
  - Most familiar leaves are megaphylls. These are thought to have evolved when photosynthetic tissue developed between the ends of small lateral branches.
- Heterospory appears to have evolved several times in the vascular plants (see Figure 28.10).
  - Most early vascular plants exhibit homospory. In homospory, spores (and the gametophytes that grow from them) are all of the same type; gametophytes pro-duce both archegonia and antheridia. In heterospory, one spore type called a megaspore gives rise to the female, egg-producing megagametophyte; another spore type called a microspore develops into a male, sperm-producing microgametophyte.
  - Land plants typically produce many more microspores than megaspores.

Extant nonseed vascular plants are varied and abundant.

- Club mosses (*Lycophyta*) diverged from the tracheid lineage relatively early and are the most "primitive" of the extant nonseed vascular plants. They have simple leaves arranged spirally on the stem. The sporangia of many club mosses are held in apical strobili, which are simple branching structures of fertile sporangia and sterile leaves (see Figure 28.15). Other club mosses have their sporangia on the upper surfaces of leaves called sporophylls. Club mosses may either be homosporous or heterosporous.

- The horsetails, whisk ferns, and ferns were once thought to be only distantly related but are now in their own clade, the monilophytes. As with seed plants, there is differentiation between the main stem and side branches.
  - Horsetails are represented by few extant species. They have true roots, their sporophytes are large and independent, and their gametophytes are highly reduced but also independent. The leaves of horsetails are simple, forming whorls around the stem (see Figure 28.16). This group is one of the few plant groups that exhibits basal growth.
  - Whisk ferns once were thought to be the "missing link" to *Rhynia,* but molecular evidence indicates that they evolved much later from more highly complex land plants. Their relatively simple body plan demonstrates that evolution does not necessarily move toward greater complexity. Wisk fern gametophytes live below the surface, lack chlorophyll, and depend on fungal partners for nutrition.
  - Ferns are the largest surviving group of nonseed vascular plants. They are characterized by relatively large, complex leaves with branching vascular strands. Like all other nonseed vascular plants and nonvascular plants, ferns continue to be dependent on water to carry motile sperm. They have advanced vascular structures and can reach great heights, but they do not produce true wood, and their root systems are poorly developed. The ferns consist of more than 12,000 species. Their life cycle is illustrated in Figure 28.19.

The fern life cycle has independent sporophytes and gametophytes.

- The fern life cycle is dominated by the sporophyte, but the gametophyte is an independent photosynthetic structure (see Figure 28.19).
- In most species of ferns, the sporangia are found in clusters called sori.
- Most ferns are homosporous, but some fern groups are heterosporous. A few genera of ferns produce gametophytes that depend on a mutualistic fungus for nutrition.
- ***For Review, go to Diagram Exercise 2.***

Test Yourself

Diagram Exercises

1. Label the structures in the moss life cycle in the diagram below. Indicate the sporophyte generation, gametophyte generation, and diploid or haploid status. Also indicate, in the open boxes, if the process occuring is meiosis, mitosis, or fertilization.

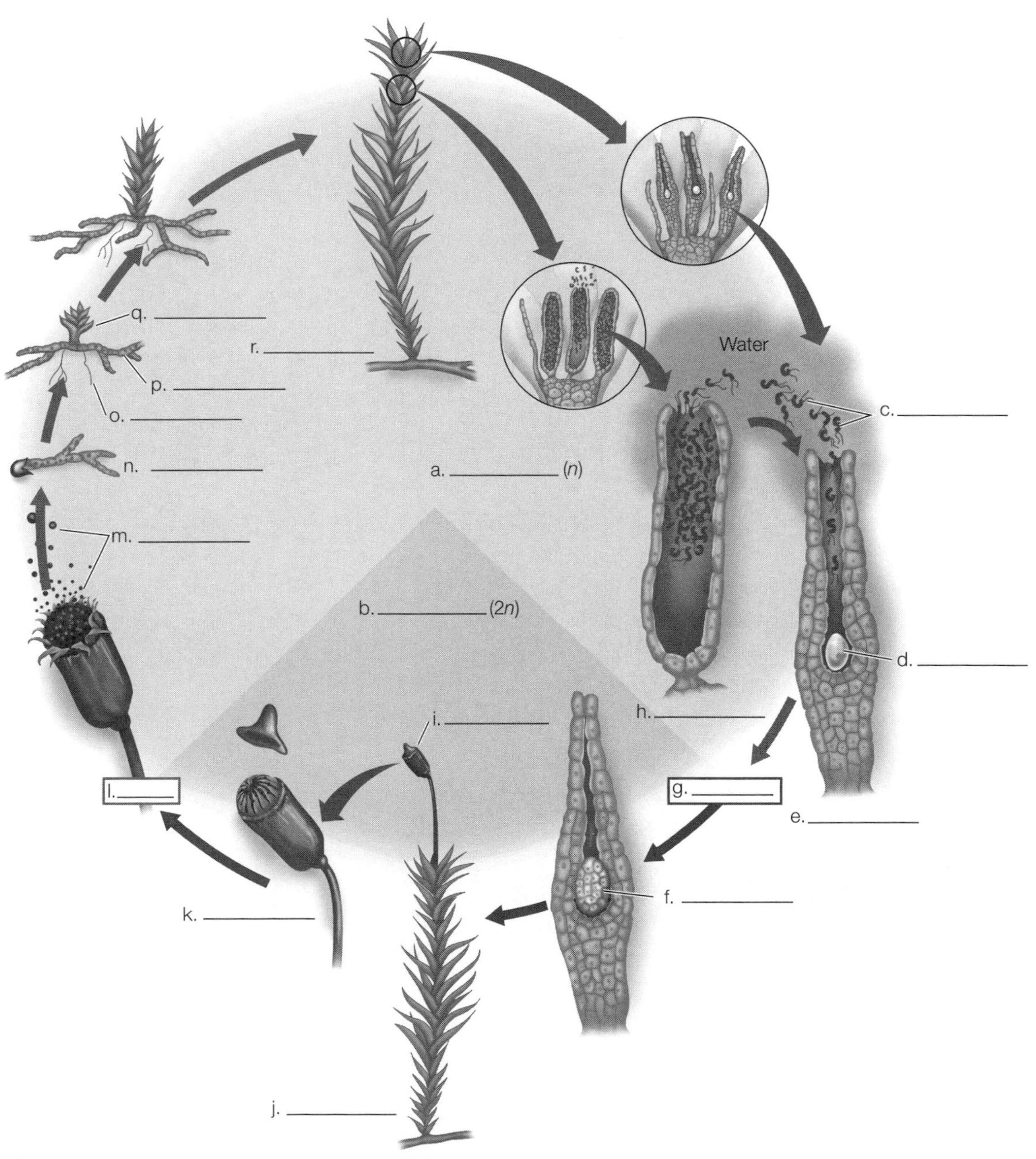

***Textbook Reference**: 28.2 How Did Plants Colonize and Thrive on Land? p. 593, Figure 28.4*

2. Label the structures in the homosporous fern life cycle in the diagram below. Indicate the sporophyte generation, gametophyte generation, and diploid or haploid status. Also indicate, in the open boxes, if the process occuring is meiosis, mitosis, or fertilization.

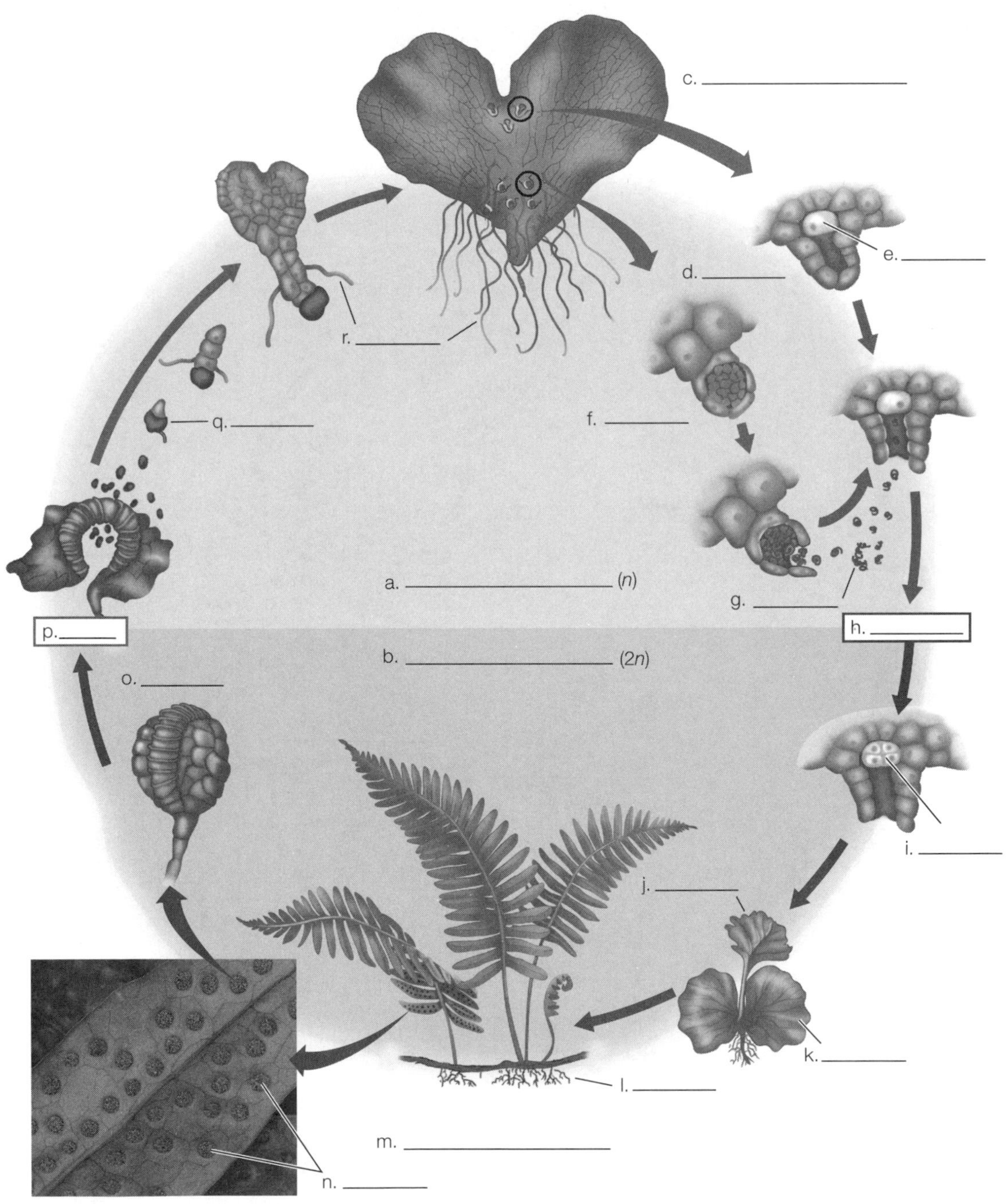

Textbook Reference: *28.4 What Are the Major Clades of Seedless Plants? p. 604, Figure 28.19*

Knowledge and Synthesis Questions

1. A land plant may be reliably distinguished from green algae by which of the following characteristics?
   a. Chlorophyll type
   b. The presence of an embryo protected by parent tissue
   c. The presence of roots
   d. Swimming sperm
   e. All of the above
   Textbook Reference: *28.1 How Did the Land Plants Arise? p. 589; 28.2 How Did Plants Colonize and Thrive on Land? p. 591*
2. Which of the following characteristics was *not* necessary for plants to colonize land?
   a. Vascular tissue for moving water throughout the plant
   b. A waxy cuticle to reduce water loss
   c. The ability to screen ultraviolet radiation
   d. The development of thick spore walls to protect the spores from dehydration
   e. Development of embryos protected inside other tissues
   Textbook Reference: *28.2 How Did Plants Colonize and Thrive on Land? p. 591*
3. The main difference between nonvascular plants and vascular plants is that nonvascular plants
   a. lack gametophytes.
   b. produce spores.
   c. have tracheids.
   d. reproduce sexually.
   e. All of the above
   Textbook Reference: *28.2 How Did Plants Colonize and Thrive on Land? p. 592*
4. In alternation of generations, the sporophyte generation is _______ and the gametophyte generation is _______.
   a. haploid; diploid
   b. diploid; haploid
   c. haploid; haploid
   d. diploid; diploid
   e. either haploid or diploid; either haploid or diploid
   Textbook Reference: *28.2 How Did Plants Colonize and Thrive on Land? pp. 592–593*
5. Which of the following characteristics helps distinguish the liverworts from the mosses?
   a. The presence of hydroids
   b. Gametophyte dominance
   c. Sporophyte dominance
   d. Swimming sperm
   e. Chloroplast structure
   Textbook Reference: *28.4 What Are the Major Clades of Seedless Plants? pp. 600–601*
6. During a plant's life cycle, meiosis takes place in the _______ to produce _______.
   a. gametophyte; haploid gametes
   b. sporophyte; haploid gametes
   c. sporophyte; haploid spores
   d. gametophyte; diploid spores
   e. gametophyte; haploid spores
   Textbook Reference: *28.2 How Did Plants Colonize and Thrive on Land? pp. 592–593, Figure 28.3*
7. Asexual reproduction in liverworts is accomplished by
   a. gametophytes.
   b. spores.
   c. gemmae.
   d. physical separation of gametophyte parts.
   e. Both c and d
   Textbook Reference: *28.4 What Are the Major Clades of Seedless Plants? p. 600*
8. Which of the following limits the size of a hornwort's sporophyte?
   a. The sporophyte's ability to distribute water to all its cells
   b. The gametophyte's ability to produce enough nutrients for the sporophyte
   c. Developmental genes that prevent growth of the sporophyte beyond a certain size
   d. A tendency to collapse under its own weight once it grows beyond a certain size
   e. All of the above
   Textbook Reference: *28.4 What Are the Major Clades of Seedless Plants? p. 601*
9. You are walking along a roadside and find a plant with the following characteristics: a very thin, waxy cuticle, stomata, simple leaves in whorls around a central stem, independent sporophytes and gametophytes, and sporangia in strobili. This plant is most likely a member of which of the following groups?
   a. Bryophyta
   b. Monilophyta
   c. Anthocerophyta
   d. Lycopodiophyta
   e. Cycadophyta
   Textbook Reference: *28.4 What Are the Major Clades of Seedless Plants? p. 602*
10. Which of the following groups has a unique chloroplast?
    a. Mosses
    b. Liverworts
    c. Hornworts
    d. Club mosses
    e. Horsetails
    Textbook Reference: *28.4 What Are the Major Clades of Seedless Plants? p. 601*
11. Which of the following groups has hydroids?
    a. Mosses
    b. Liverworts
    c. Hornworts
    d. Club mosses
    e. Horsetails
    Textbook Reference: *28.4 What Are the Major Clades of Seedless Plants? p. 600*

12. Which of the following groups lacks stomata?
    a. Mosses
    b. Liverworts
    c. Hornworts
    d. Club mosses
    e. Horsetails
    Textbook Reference: *28.4 What Are the Major Clades of Seedless Plants? p. 600*
13. Which of the following groups has basal growth?
    a. Ferns
    b. Whisk ferns
    c. Liverworts
    d. Club mosses
    e. Horsetails
    Textbook Reference: *28.4 What Are the Major Clades of Seedless Plants? p. 602*
14. In which of the following groups are sporangia arranged in strobili?
    a. Horsetails
    b. Whisk ferns
    c. Hornworts
    d. Club mosses
    e. Ferns
    Textbook Reference: *28.4 What Are the Major Clades of Seedless Plants? p. 602*
15. Which of the following groups has large leaves with branching vascular strands?
    a. Horsetails
    b. Whisk ferns
    c. Hornworts
    d. Club mosses
    e. Ferns
    Textbook Reference: *28.4 What Are the Major Clades of Seedless Plants? p. 603*

Application Questions

1. Explain why the largest mosses are less than a meter tall.
   Textbook Reference: *28.2 How Did Plants Colonize and Thrive on Land? p. 592*
2. Diagram the evolutionary relationships among the nonvascular plants and nonseed vascular plants. Label the major differentiating characteristics at each branch of your tree.
   Textbook Reference: *28.3 What Features Distinguish the Vascular Plants? p. 595*
3. Discuss how the relationship between sporophytes and gametophytes changes as one moves from the first nonvascular plants to later nonseed vascular plants.
   Textbook Reference: *28.2 How Did Plants Colonize and Thrive on Land? pp. 592–594*
4. Compare and contrast the moss life cycle and the fern life cycle.
   Textbook Reference: *28.2 How Did Plants Colonize and Thrive on Land? p. 593; 28.4 What Are the Major Clades of Seedless Plants? p. 604*
5. Discuss the challenges of terrestrial life that land plants have addressed in order to survive.
   Textbook Reference: *28.2 How Did Plants Colonize and Thrive on Land? p. 591*
6. Compare and contrast homospory and heterospory. Which reproductive structures result from meiosis in each type of life cycle, and which structures result from mitosis?
   Textbook Reference: *28.3 What Features Distinguish the Vascular Plants? pp. 597–599*

Answers

Diagram Exercise Answers

1.
   a. Haploid; Gametophyte generation
   b. Diploid ; Sporophyte generation
   c. Sperm (n)
   d. Egg (n)
   e. Archegonium (n)
   f. Embryo ($2n$)
   g. Fertilization
   h. Antheridium (n)
   i. Sporophyte ($2n$)
   j. Gametophyte (n)
   k. Sporangium
   l. Meiosis
   m. Ungerminated spores
   n. Germinating spore
   o. Rhizoid
   p. Protonema
   q. Bud
   r. Gametophytes (n)
2.
   a. Haploid
   b. Diploid
   c. Mature gametophyte
   d. Archegonium
   e. Egg
   f. Antheridium
   g. Sperm
   h. Fertilization
   i. Embryo
   j. Sporophyte
   k. Gametophyte
   l. Roots
   m. Mature sporophyte
   n. Sori
   o. Sporangium
   p. Meiosis
   q. Germinating spore
   r. Rhizoids

Knowledge and Synthesis Answers

1. **b**. According to the definition presented in the textbook, all land plants produce embryos that are protected by tissue of the parent plant. Green algae and land plants make use of the same types of chlorophyll. Not all land plants have roots, so a plantlike organism lacking roots will not necessarily be a green algae.
2. **a.** Several successful groups of terrestrial plants lack vascular tissue.
3. **c.** Tracheids are found only in the vascular plants.

4. **b.** Meiosis occurs in the diploid sporophyte to produce haploid spores that develop into the haploid gametophyte.
5. **a.** Mosses are the only group with hydroids, the precursors to true vascular tissue. Hornworts have chloroplasts that are different than those of mosses and liverworts.
6. **c.** The outcome of meiosis is four cells, each of which has half the genetic material of the parent cell. A haploid cell already has only half the normal number of chromosomes of most eukaryotes, so a cell must be diploid (or have higher ploidy) to undergo meiosis. In all plant life cycles, the sporophyte is diploid and the gametophyte is haploid; therefore, only the sporophyte can undergo meiosis. Spores are the products of sporophyte meiosis.
7. **e.** Liverworts have specialized asexual reproductive structures called gemmae; they can also reproduce by fragmentation of the gametophyte.
8. **a.** The sporophyte of a hornwort is nutritionally dependent on the gametophyte, but its growth is indeterminate—the sporophyte will continue to grow as long as water is able to reach all its cells.
9. **b.** The plant is a horsetail, which is very common along roadsides in damp ditches, particularly in the Midwest.
10. **c.** The hornworts have a single, large, platelike chloroplast.
11. **a.** Mosses have hydroids, which are cells similar to the tracheid.
12. **b.** Unlike the mosses, hornworts, and vascular plants, liverworts do not have any stomata for gas exchange.
13. **e.** This growth at the base of the stem, though uncommon in plants, is seen in both hornworts and horsetails.
14. **d.** The strobili are conelike structures of aggregated sporangia.
15. **e.** Some ferns uncurl as their leaves grow.

Application Answers

1. Mosses have very rudimentary water transport cells called hydroids. Hydroids lack the waterproofing and support molecule lignin. Because of this, they can carry water only short distances and cannot support tall growth.
2. See Figure 28.5 and Table 28.1.
3. Early nonvascular plants have reduced sporophytes that are highly dependent on the gametophyte for nutrition. Nonseed vascular plants, which have evolved more recently, have independent sporophytes and gametophytes, and the gametophyte is highly reduced.
4. See Figures 28.4 and 28.19. Pay particular attention to the relative dominance of the sporophyte versus the gametophyte.
5. Terrestrial life poses problems of support, water conduction, UV radiation, water loss, and embryo protection. The tracheid is an important evolutionary innovation that helped with both water conduction and support. Special pigments protect against UV radiation and a waxy cuticle prevents water loss from cells. Plant embryos are protected in early development by parental tissue.
6. See Figure 28.10. In homospory, meiosis results in a single type of spore, but in heterospory, meiosis results in megaspores and microspores. Mitosis occurs throughout both types of life cycle as cells divide and plants grow, but the reproductive cells that result from mitosis in both homospory and heterospory are sperm and eggs. In homospory, the antheridium produces sperm by mitosis, and the archegonium produces eggs; in heterospory, the microgametophyte produces sperm and the megagametophyte produces eggs.

29 The Evolution of Seed Plants

The Big Picture

- Seed plants can be divided into two main groups: the gymnosperms and the angiosperms. Seed plants have developed complex ways of protecting embryos, dispersing gametes, and moving water, nutrients, and food throughout the plant. Both gymnosperms and angiosperms are characterized by the presence of seeds for protecting the embryo until conditions favor germination. The main difference between gymnosperms and angiosperms is the presence of flowers and ovaries in angiosperms. Both groups have highly developed vascular tissue, may exhibit secondary woody growth, and show significant diversity.
- There is great variation among flower types, but the basic flower consists of the structures shown in Figure 29.5. Many plants and animals have coevolved in such a way that the flower structure and the animal are integrally linked for nutrition of the animal and pollination of the plant. The angiosperm life cycle is different from that of all other plants in that double fertilization occurs.
- Both the angiosperm and gymnosperm life cycles are dominated by the sporophyte. Both groups are heterosporous. The roles of megasporangia and microsporangia are similar in both groups.

Common Problem Areas

- Not all gymnosperms are pines. The examples in this chapter focus on the pine, but gymnosperms are a varied group, and cones are not the defining characteristic of this group. Cones define the conifers only.
- Angiosperms are extremely varied as well. Not all flower parts are found on all plants, and some are highly modified. Also note that cultivated flowers are often sterile, have aberrant parts, and may be altered through selective breeding and treatment.

Study Strategies

- Much of the terminology in this chapter was introduced in Chapter 28, and it is worth reviewing that chapter in light of the material presented here. The details of heterospory can be particularly confusing; reviewing Figure 28.10 can help you sort them out.
- Be sure that you truly understand what sets angiosperms and gymnosperms apart. Use Figures 29.8 and 29.15 to compare their life cycles.
- Make sure you know how flowers can vary from the generalized structure of a perfect flower presented in Figure 29.5 (see Figures 29.10 and 29.11).
- Go to yourBioPortal.com to review the following tutorials and activities:

  Animated Tutorial 29.1 Life Cycle of a Conifer

  Animated Tutorial 29.2 Life Cycle of an Angiosperm

  Web Activity 29.1 Flower Morphology

  Web Activity 29.2 Life Cycle of a Pine Tree

Important Concepts

Seed plants have characteristics that set them apart from nonseed plants.

- Seed plants consist of two groups of vascular plants (tracheophytes): the gymnosperms and the angiosperms. Gymnosperms consist of four major phyla: Cycadophyta (cycads), Ginkgophyta (ginkgos), Coniferophyta (conifers), and Gnetophyta (gnetophytes). The angiosperms are the most diverse group consisting of a number of different phyla.
- Seed plants have highly reduced gametophyte generations that are nutritionally dependent on the sporophyte for survival. This nutritional dependence of the gametophyte sets the seed plants in contrast to the seedless vascular plants discussed in Chapter 28. When you look at a seed plant, what you see is the sporophyte. (See Figure 29.3 for a general overview of the relationship between the sporophyte and gametophyte.)
- Very few seed plants (e.g., the cycads and ginkos) have retained swimming sperm. Most have evolved other means for dispersing male gametes.
- All seed plants are heterosporous in that they produce two types of spores. One becomes the female gametophyte and the other becomes the male gametophyte.

Microspores and megaspores develop in specialized cones or in flowers.

- In the megasporangium, meiosis produces four megaspores, but in most seed plants only one of the four megaspores is retained. This megaspore divides by mitosis to form the multicellular (yet tiny) megagametophyte (female gametophyte), which produces the eggs.
- When the eggs are ready to be fertilized, they are surrounded by megagametophyte cells that are still within the megasporangium. The megasporangium is surrounded by sterile sporophyte tissues (the integument). Together, the megasporangium and the integument constitute the ovule.
- Microspores develop into pollen grains, the male gametophytes. Pollen grains consist of sperm and supporting cells; they are dispersed by wind and animals. The wall of the pollen grain contains sporopollenin. Sporopollenin is one of the most chemically resistant biological compounds known, and represents a major advantage for land colonization by plants.
- Because the female gametophyte is retained within sporophyte tissue, pollen grains do not have direct access to gametophytes and their eggs. The sporophyte housing a gametophyte creates tissue for receiving pollen grains; upon reaching this tissue, pollen produces pollen tubes that deliver the sperm through the sporophyte tissue to the eggs for fertilization. This process is known as pollination (see Figure 29.5).
- The embryo resulting from fertilization grows to a certain size and then becomes dormant within the surrounding tissues. This dormant protected embryo and its surrounding tissues constitute the seed.

Seeds are complex structures that may contain tissue from three plant generations.

- Recall that alternation of generations involves a multicellular sporophyte generation that alternates with a multicellular gametophyte generation. The embryo within a seed represents the beginning of a new sporophyte generation. It is still surrounded by the female gametophyte tissue, which will provide the embryo with nutrients to begin growth (particularly if it is a gymnosperm). The tough coat, or seed coat, that surrounds the seed consists of tissue provided by the embryo's sporophyte parent. This tissue is derived from the integument of the diploid parent.
- Seeds protect the embryo until conditions are right for germination. Seeds can remain viable or dormant for many years, waiting for favorable conditions to germinate. Many seeds have adaptations that allow dispersal by wind or another vector.

Wood provides support to many seed plants, allowing them to grow taller to capture more light for photosynthesis.

- The younger portions of wood consists of vascular tissue allowing for water transport.
- As wood becomes older, it becomes clogged with materials and provides the plant with support. New layers of vascular tissue are produced underneath the wood.
    - This is an example of secondary growth; and is one reason why seed plants became today's dominant vegetation.
- Not all seed plants have wood (e.g., grasses). During the course of evolution, some seed plants lost woody structures but gained other advantages that allow them to adapt to their environments.

Gymnosperms are "naked-seed" plants.

- Gymnosperms are seed plants that do not form flowers or true fruits. They are diverse and live in a variety of habitats, from sparse deserts to vast forests. There are four groups within the gymnosperms (see Figure 29.6). The Coniferophyta group is the most abundant.
- Gymnosperms have significant vascular tissue and are capable of woody secondary growth. With the exception of the gnetophytes, their vascular tissue is less complex than that of the angiosperms. Tracheids are the only support and water-conducting cells within the xylem. Despite this simplicity, some of the largest trees known are gymnosperms (see Figure 29.6D).
- The gymnosperm life cycle can be illustrated by that of the pine, a conifer (see Figure 29.8). Conifers differ from other gymnosperms in that they produce cones (megastrobili) and strobili (microstrobili), which are specialized structures for reproduction (see Figure 29.7). Cones house the megasporangia and produce megaspores, megagametophytes, and eggs; strobili house the microsporangia and produce microspores, microgametophytes (pollen), and sperm.
- Pines do not have swimming sperm, and their pollen is modified for wind dispersal. Pollen lands on the female cone and lodges in the pollen chamber. The pollen tube grows through the maternal sporophyte tissue to the female gametophyte, where it releases two sperm. Only one of the two sperm will fertilize an egg. The resulting zygote develops into an embryo that remains encased in the tissues of the megasporangium and gametophyte. The seed is also protected by the scale of the cone (sporophyte tissue) until it is mature and ready for dispersal. Dispersal is aided by modifications of the seed coat.
- Some conifer species have soft, fleshy modifications that surround the seed. Examples include the "berries" found on yew and juniper plants. These are not true fruits, however. The fruits of angiosperms are ripened ovaries, and gymnosperms do not have ovaries.
- ***For review, go to Diagram Exercise 1.***

Angiosperms are characterized by flower formation.

- Angiosperms are a diverse group of plants that produce flowers and thus are called flowering plants. They are currently the dominant plant form on Earth.

- Angiosperms have the most reduced gametophyte generation and the most highly developed sporophyte generation of all plant lineages.
- Angiosperms differ from all other plants in that they produce triploid endosperm (the nutritive tissue), they have double fertilization, their ovules and seeds are enclosed in a carpel, they produce flowers and fruits, and their xylem and phloem have multiple modified cell types, including vessel elements, fibers, and companion phloem cells.

Flower structure is varied among plants but consists of similar structural elements.

- All flower parts are modified leaf structures (see Figure 29.12). Although flowers are very diverse, they are constructed from the same small set of structures (see Figure 29.5) and have evolved over time.
- The male structures in a flower are stamens. Each stamen consists of a filament and an anther; the anther contains microsporangia, which produce pollen. Flowers often have several stamens.
- The female structure in a flower is the pistil, which consists of a stigma, a style, and an ovary. The stigma is modified to receive pollen. The style is a stalk that separates the stigma from the ovary. The ovary contains one or more ovules, each of which houses a megasporangium.
- Nonreproductive floral structures include petals, sepals, and the receptacle. Petals and sepals are often modified to attract animal pollinators. Sepals also serve to protect the developing flower bud. The receptacle is the attachment site for the sepals, petals, stamens, and carpels.
- Flowers that have both megasporangia and microsporangia are called "perfect." If either structure is missing, the flower is called "imperfect." Species that have both megasporangiate flowers and microsporangiate flowers on the same plant are called "monoecious." If they are on separate plants, they are called "dioecious."
- Flowers can be single or grouped together to form an inflorescence (see Figure 29.10).
- Many plants and animals have coevolved in such a way that the nutrition of the animal and pollination of the plant are interdependent. Some plants have evolved methods of limiting pollination to a single species of insect, but most can be pollinated by a range of species. Many flowers entice animals by providing food rewards such as nectar. Pollen grains themselves can be a food reward, and this pollen can be carried from one plant to another. Bee-pollinated flowers often have nectar guides that can be seen only in the ultraviolet spectrum visible to bees.
- *For review, go to Diagram Exercise 2.*

The angiosperm life cycle is dominated by the sporophyte.

- The angiosperm life cycle differs from that of all other plants in that double fertilization occurs (see Figure 29.15). In double fertilization, one sperm unites with the egg to produce a diploid zygote; the other sperm unites with two other haploid cells of the female gametophyte to produce a triploid cell. This triploid cell divides mitotically to create the (triploid) endosperm tissue. The endosperm provides nutrition for the developing embryo.
- Angiosperm embryos have one or two seed leaves called cotyledons. The number of cotyledons distinguishes the two major clades of flowering plants. Monocots have one cotyledon; eudicots have two. A few other relatively small groups of angiosperms do not fit into these two lineages (see Figures 29.17 and 29.18).
- Fruits develop from the ovary and its supporting tissues. Fruit types (see Figure 29.16) depend on the number of carpels associated with the fruit and the extent of support tissue incorporated into the fruit structure.
- *For review, go to Diagram Exercise 3.*

The evolutionary relationships among angiosperms have not yet been resolved.

- Different phylogenetic methods have led to different conclusions regarding angiosperm phylogeny. Investigators are still working on the question of how angiosperms first arose.
- There are other angiosperms that are not monocots or eudicots (see Figure 29.18). The overall relationship among the angiosperm clades is shown in Figure 29.17.
- Molecular and morphological evidence now points to *Amborella* as the living species most similar to the first angiosperms. *Amborella* has a variable number of carpels and stamens, and it lacks vessel elements.

Test Yourself

Diagram Exercises

1. Illustrated below is the pine tree life cycle. Label the processes in the boxes, indicate the haploid, diploid, sporophyte and gametophyte generations, and label all of the structures indicated.

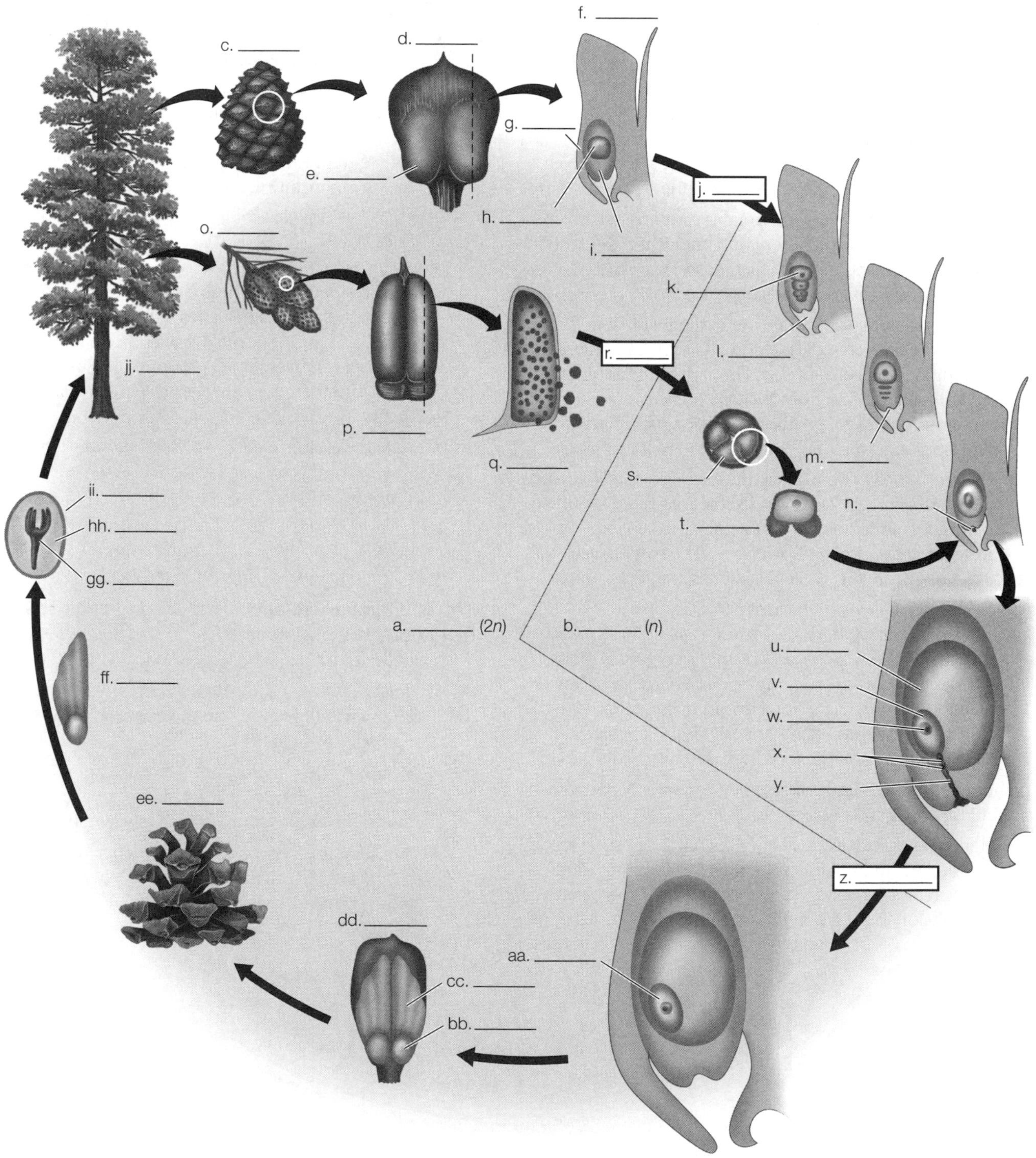

Textbook Reference: *29.2 What Are the Major Groups of Gymnosperms? p. 614, Figure 29.8*

2. Label the flower in the diagram below.

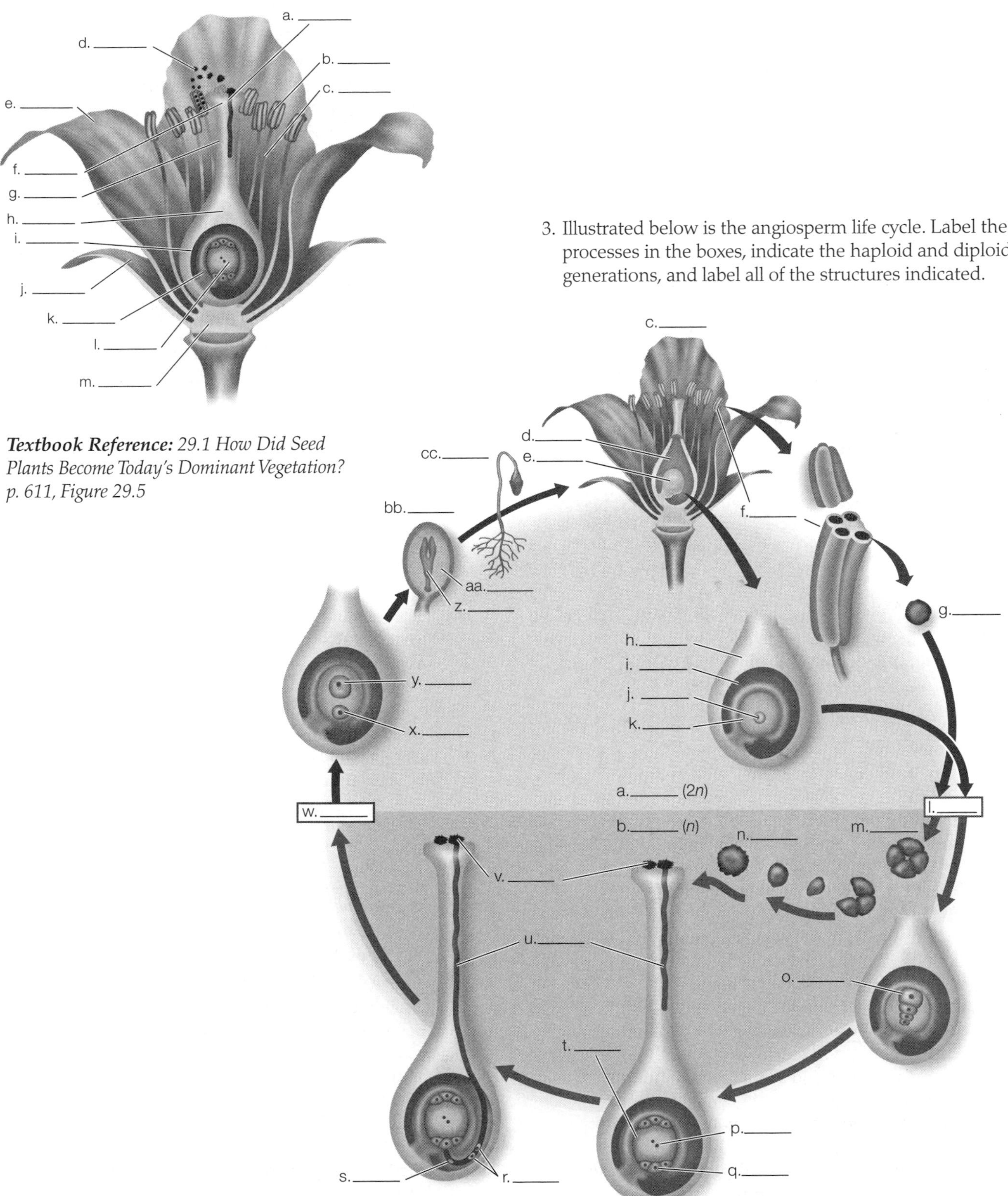

Textbook Reference: *29.1 How Did Seed Plants Become Today's Dominant Vegetation? p. 611, Figure 29.5*

3. Illustrated below is the angiosperm life cycle. Label the processes in the boxes, indicate the haploid and diploid generations, and label all of the structures indicated.

Textbook Reference: *29.3 What Features Contributed to the Success of the Angiosperms? p. 619, Figure 29.15*

Knowledge and Synthesis Questions

1. You are enjoying a stroll in the botanical gardens. You notice a plant with a beautiful flower. Upon closer inspection you find that the flower has a pistil but no stamens. You look at several more flowers on the same plant, but you are unable to find any that have stamens. This plant is _______ and its flowers are _______.
   a. monoecious; perfect
   b. monoecious; imperfect
   c. dioecious; perfect
   d. dioecious; imperfect
   e. hermaphroditic; perfect
   Textbook Reference: *29.3 What Features Contributed to the Success of the Angiosperms? p. 616*
2. Gymnosperms are referred to as naked-seed plants because they
   a. lack ovules.
   b. lack ovaries.
   c. do not protect their embryos.
   d. do not have seed coats.
   e. make only very small fruits.
   Textbook Reference: *29.2 What Are the Major Groups of Gymnosperms? p. 612*
3. Which of the following statements about seeds is true?
   a. Seeds provide a mechanism for a plant's dispersal.
   b. Seeds protect the plant embryo.
   c. Seeds allow an embryo to remain dormant until optimum growth conditions are available.
   d. Seeds provide nutrients for the developing embryo.
   e. All of the above
   Textbook Reference: *29.1 How Did Seed Plants Become Today's Dominant Vegetation? pp. 610–611*
4. Which of the following roles do animals play in the life cycle of plants?
   a. They act as pollinators.
   b. They assist in dispersal of seeds.
   c. They promote fertilization.
   d. They contribute to genetic diversity of plants.
   e. All of the above
   Textbook Reference: *29.3 What Features Contributed to the Success of the Angiosperms? p. 618*
5. In gymnosperms, fertilization results in a _______. In angiosperms, fertilization results in a(n) _______.
   a. diploid zygote; haploid zygote
   b. diploid zygote; endosperm nucleus
   c. diploid zygote; diploid zygote and an endosperm nucleus
   d. haploid zygote and an endosperm nucleus; diploid zygote
   e. haploid zygote; diploid zygote and an endosperm nucleus
   Textbook Reference: *29.3 What Features Contributed to the Success of the Angiosperms? p. 615*
6. The identifying characteristic of an angiosperm is the presence of
   a. multiple carpels.
   b. woody growth.
   c. a flower.
   d. secondary growth.
   e. All of the above
   Textbook Reference: *29.3 What Features Contributed to the Success of the Angiosperms? p. 616*
7. Which of the following statements regarding gymnosperms is true?
   a. All gymnosperms produce cones.
   b. Gymnosperms are heterosporous.
   c. Gymnosperm seeds have no protection.
   d. Only some living gymnosperms are woody.
   e. Gymnosperms all have swimming sperm.
   Textbook Reference: *29.1 How Did Seed Plants Become Today's Dominant Vegetation? p. 608*
8. Which of the following *incorrectly* characterizes the calyx?
   a. The calyx is a collection of modified leaves.
   b. The calyx functions to protect immature flower parts within the bud.
   c. The calyx is a source of gametes.
   d. The calyx is a term for all of the sepals together.
   e. The calyx, together with the corolla, forms the perianth.
   Textbook Reference: *29.3 What Features Contributed to the Success of the Angiosperms? p. 616*
9. Which of the following statements about the function of a fruit is true?
   a. It aids in dispersal of seeds.
   b. It protects seeds until they are mature.
   c. It attracts pollinators.
   d. It provides nutrients to the embryo.
   e. Both a and b
   Textbook Reference: *29.3 What Features Contributed to the Success of the Angiosperms? pp. 618–619*
10. Vascular tissue in angiosperms is highly developed. The purpose of this vascular tissue is to move
    a. water.
    b. food.
    c. nutrients.
    d. water and anything dissolved in the water.
    e. All of the above
    Textbook Reference: *29.3 What Features Contributed to the Success of the Angiosperms? p. 616*
11. More than half of the world's population relies on the seeds of the _______ plant as food.
    a. corn
    b. rice
    c. soybean
    d. common bean
    e. wheat
    Textbook Reference: *29.4 How Do Plants Support Our World? p. 623*
12. Which of the following drugs is derived from foxglove and used to strengthen contractions of the heart?

a. Atropine
b. Ephedrine
c. Morphine
d. Tubocurarine
e. Digitalin
Textbook Reference: *29.4 How Do Plants Support Our World? p. 623, Table 29.1*

13. Which of the following represents the correct path followed by the sperm from the pollen grain to the female gametophyte in an angiosperm?
a. Stigma, style, ovary, ovule, egg
b. Sepal integument, style, ovary, ovule
c. Anther, filament, pollen tube, style, ovary
d. Ovule, ovary, style, stigma, egg
e. Receptacle, pistil, sepal, integument, egg
Textbook Reference: *29.1 How Did Seed Plants Become Today's Dominant Vegetation? pp. 610–611, Figure 29.5; 29.3 What Features Contributed to the Success of the Angiosperms? p. 619, Figure 29.15*

14. How many generations of material are present in a seed?
a. One: the haploid female gametophyte tissue
b. Two: the haploid female gametophyte tissue and the embryo of the diploid sporophyte
c. Two: the diploid female gametophyte tissue and the embryo of the diploid sporophyte
d. Three: integument tissue from the haploid sporophyte parent, the haploid female gametophyte tissue, and the embryo of the diploid sporophyte
e. Three: integument tissue from the diploid sporophyte parent, the diploid female gametophyte tissue, and the embryo of the diploid sporophyte
Textbook Reference: *29.1 How Did Seed Plants Become Today's Dominant Vegetation p. 610*

15. Which of the following is *not* an ecological service performed by plants?
a. Removing CO_2 from the atmosphere
b. Reducing erosion
c. Increasing atmospheric humidity
d. Increasing soil humidity
e. Aiding in soil formation
Textbook Reference: *29.4 How Do Plants Support Our World? p. 622*

Application Questions

1. Discuss the benefits of double fertilization in angiosperms.
Textbook Reference: *29.3 What Features Contributed to the Success of the Angiosperms? p. 620*

2. How have gymnosperms surmounted the obstacles that all plants faced when they began to colonize terrestrial environments?
Textbook Reference: *29.1 How Did Seed Plants Become Today's Dominant Vegetation? pp. 608–611*

3. Pollen is found only in seed-producing vascular plants. What role did pollen play in vascular plant evolution? Is pollen a sporophyte or gametophyte?
Textbook Reference: *29.1 How Did Seed Plants Become Today's Dominant Vegetation? pp. 609–610*

4. Where would you find a female gametophyte in an angiosperm? Explain your answer.
Textbook Reference: *29.3 What Features Contributed to the Success of the Angiosperms? p. 615*

5. What is a cotyledon? What role does it play in the classification of angiosperms into two monophyletic groups?
Textbook Reference: *29.3 What Features Contributed to the Success of the Angiosperms? p. 620*

6. Discuss the origin and development of the angiosperm fruit? From which types of tissues does the fruit arise?
Textbook Reference: *29.3 What Features Contributed to the Success of the Angiosperms? p. 620*

7. Match each of the following flower structures with its function.

_____ Ovule	a. Assists in attracting pollinators
_____ Anther	b. Secretes sticky material to help pollen adhere
_____ Stigma	c. Site of ovule development
_____ Style	d. Protects the immature flower bud
_____ Ovary	e. Houses the megasporangium
_____ Petal	f. Holds the stigma in position
_____ Sepal	g. Houses microsporangia

Textbook Reference: *29.3 What Features Contributed to the Success of the Angiosperms? p. 616*

Answers

Diagram Exercise Answers

1. a. Diploid
b. Haploid
c. Immature megastrobilus
d. Scale of megastrobilius
e. Ovule
f. Section through scale
g. Integument
h. Megasporocyte
i. Megasporangium
j. Meiosis
k. Functional megaspore
l. Pollen chamber
m. Micropyle
n. Pollen grain
o. Microstrobili
p. Scale of microstrobilus
q. Section through scale
r. Meiosis
s. Microspores
t. Pollen grain
u. Femal gametophyte
v. Archegonium
w. Egg
x. Sperm
y. Male gametophyte (germinating pollen grain)
z. Fertilization
aa. Zygote
bb. Seed
cc. Wing
dd. Scale of megastobilus
ee. Mature megastrobilus
ff. Winged seed
gg. Embryo
hh. Female gametophyte (provides nutrition for developing embryo)
ii. Seed coat
jj. Sporophyte (about 10–110 m)

2. a. Pollen tube
   b. Anther
   c. Filament
   d. Pollen grains
   e. Petal
   f. Stigma
   g. Style
   h. Ovary
   i. Ovule
   j. Sepal
   k. Integument
   l. Megagametophyte
   m. Receptacle

3. a. Diploid
   b. Haploid
   c. Flower of mature sporophyte
   d. Ovary
   e. Ovule
   f. Anther
   g. Microsporocyte
   h. Ovary
   i. Ovule
   j. Megasporocyte ($2n$)
   k. Megasporangium
   l. Meiosis
   m. Microspores (4)
   n. Pollen grain
   o. Surviving megaspore (n)
   p. Polar nuclei (2)
   q. Egg
   r. Sperm (2)
   s. Tube cell nucleus
   t. Megagametophyte (n)
   u. Pollen tube
   v. Pollen grains (microgametophyte, n)
   w. Double fertilization
   x. Zygote ($2n$)
   y. Endosperm nucleus ($3n$)
   z. Embryo
   aa. Endosperm
   bb. Seed
   cc. Seedling

Knowledge and Synthesis Answers

1. **d.** This is most likely a dioecious plant in which male and female flowers appear on separate specimens. The absence of stamens makes these flowers imperfect.
2. **b.** Gymnosperms lack ovaries and thus lack the ability to produce fruit. Their embryos are protected, and they have ovules and seed coats.
3. **e.** Seeds protect the embryo, provide a mechanism for dispersal, provide nutrients to the embryo, and allow an embryo to remain dormant.
4. **e.** By acting as pollination vectors, animals promote fertilization of the plant, thus increasing genetic diversity. Animals also assist with dispersal of seeds.
5. **c.** Angiosperms differ from gymnosperms by having double fertilization that results in a triploid endosperm nucleus in addition to the zygote.
6. **c.** Only angiosperms have flowers. It is true that only angiosperms can have multiple carpels, but multiple carpels are not a necessary characteristic of an angiosperm. Not all angiosperms undergo secondary (woody) growth.
7. **b.** All gymnosperms are heterosporous. Only conifers are cone producers. Only the earliest groups of angiosperms had swimming sperm.
8. **c.** The calyx is the collective term for the sepals, which are specialized leaves. The sepals can be showy and play a role in attracting pollinators, but are not a source of gametes.
9. **e.** By the time fruit forms, a flower has already been pollinated, and there is no need to attract additional pollinators. The seed provides nutrients to the embryo, but the surrounding fruit does not participate in this process.
10. **e.** Vascular tissues in angiosperms move food, water, and nutrients dissolved in the water (such as sugars and minerals) throughout the plant.
11. **b.** Rice is the seed consumed by over one-half of the world's population. There are 12 vital crops listed in your textbook, but the part humans consume is not always limited to the seed tissue, as specified by the question.
12. **e.** Digitalin is derived from *Digitalis purpurea*, commonly called foxglove.
13. **a.** The pollen grain lands on the top of the pistil and two sperm move through a pollen tube from the stigma to the style, then to the ovary and into the ovule, where a sperm meets with the egg. (Chapter 38 explains how the other sperm fuses with polar nuclei to form triploid endosperm.)
14. **d.** The plant's seed coat is the first generation, the female gametophyte is the second generation, and the embryo is the third generation.
15. **d.** Plants increase atmospheric humidity by reducing soil humidity. They also reduce erosion by blocking wind and holding soil in place. Plants also consume CO_2 and use it to build organic compounds.

Application Answers

1. Double fertilization allows for a diploid embryo and a triploid endosperm. The endosperm provides nutrition to the developing embryo at the time of seed germination.
2. Gymnosperms have a sophisticated vascular system that allows for movement of nutrients and also provides support for large plants competing for sunlight. Wind-dispersed pollen eliminates the dependence on water for fertilization and increases the range of gymnosperms. Seeds protect embryos from desiccation and provide a mechanism for dispersal.
3. Pollen is the male gametophyte of seed vascular plants; in other words, it is the male multicellular haploid structure of the seed-producing vascular plant's life cycle (see Figure 29.15). As a gametophyte, pollen's most fundamental function is to create gametes (sperm, in this case), but it also plays a role in helping those gametes reach and fertilize female gametes. Pollen is significant in that it provides a means of dispersal that eliminates the need for water in fertilization.
4. The female gametophyte is highly reduced and is part of the ovule of the angiosperm.
5. A cotyledon is the first "seed leaf" produced in an angiosperm embryo. Monocots have only one seed leaf, whereas eudicots have two.
6. Angiosperm fruit arises from tissues of the sporophyte and the gametophyte. How that fruit develops depends on fruit type. Different fruits encompass more or less of the parent sporophyte tissues.
7. e. Ovule
   g. Anther
   b. Stigma
   f. Style
   c. Ovary
   a. Petal
   d. Sepal

30 Fungi: Recyclers, Pathogens, Parasites, and Plant Partners

The Big Picture

- Fungi are heterotrophic organisms that absorb nutrients from the environment. Fungi can be unicellular or multicellular. The mycelium is the body of a multicellular fungus and is composed of many tubular filaments known as hyphae. Many fungi form symbiotic relationships with photosynthetic organisms, creating mycorrhizae and lichens.
- Reproduction in the fungi occurs both sexually and asexually. Asexual reproduction involves the production of spores, breakage, fission, or budding. Sexual reproduction in multicellular fungi requires the union of hyphae with two different mating types.
- Fungi are classified into six major groups: microsporidia, chytrids, Zygomycota, Glomeromycota, Ascomycota, and Basidiomycota.
  - Chytrids are aquatic fungi with flagellated gametes. Many chytrids display alternation of generations.
  - Zygomycota have coenocytic hyphae. Sexual reproduction occurs when hyphae of two mating types join to form a zygosporangium from which a sporangium will eventually grow.
  - Glomeromycota are terrestrial, coenocytic, and asexual. They are the predominant arbuscular fungi and as such are essential to plant life.
  - Ascomycota produce an ascus as their reproductive structure. They reproduce by budding, fission, or the production of an ascus. Many have a dikaryotic ($n + n$) stage during reproduction.
  - Basidiomycota have a reproductive structure called the basidium. The fusion of haploid hyphae forms a dikaryotic mycelium that produces the fruiting body, a basidiocarp.

Common Problem Areas

- Students are often confused by the numerous ways that fungi reproduce. Create flowcharts to help you understand the ploidy level at each stage of the life cycle in the different phyla of fungi. Make sure to include both the sexual and asexual stages in your charts. Comparisons of your flowcharts for the different phyla will help you learn the differences among the groups.

Study Strategies

- Organisms are placed in particular systematic groupings because they share unique characteristics. Look for patterns that distinguish the six major fungal groups from one another. Most fungi are grouped according to the reproductive structures that they produce.
- Go to yourBioPortal.com to review the following tutorial and activities:

  Animated Tutorial 30.1 Life Cycle of a Zygomycete

  Web Activity 30.1 Fungal Phylogeny

  Web Activity 30.2 Life Cycle of a Dikaryotic Fungus

Important Concepts

Fungi are absorptive heterotrophs characterized by hyphae.

- Most fungi belong to one of six major groups: microsporidia, chytrids, Zygomycota, Glomeromycota, Ascomycota, and Basidiomycota. Based on evidence from DNA analysis, the chytrids and zygomycota appear to be paraphyletic, while the other three groups are clades (i.e., monophyletic; see Figure 30.2 and Table 30.1). The chytrids are aquatic; the other four groups are terrestrial.
- Unicellular fungi are commonly referred to as yeasts. Yeasts may reproduce by budding (see Figure 30.3), by fission, or sexually.
- The body of a multicellular fungus is called a mycelium. A mycelium is made up of individual tubular filaments called hyphae. Hyphae grow rapidly into a substrate; the hyphae in a single mycelium may collectively grow as much as 1 kilometer per day. Hyphal cell walls are strengthened by the polysaccharide chitin.
- Hyphae may be septate (divided into compartments by chitinous walls) or coenocytic (continuous and multinucleate) (see Figure 30.4).
- Most fungi are multicellular, but multicellularity in fungi is significantly different from the familiar multi-

cellularity of plants and animals. In plants and animals, each cell is usually enclosed in its own membrane and has only one nucleus. In multicellular fungi, cells are separated by porous septa that do not completely block the movement of organelles (see Figure 30.4). In many species, each cell has two nuclei; in some, even nuclei can pass through the septa.

- Hyphae provide a large surface area-to-volume ratio, thus a large surface area for absorption of nutrients from the substrate.
- Fungi are tolerant of hypertonic environments and temperature extremes.
- ***For review, go to Diagram Exercise 1.***

Fungi are saprobes, pathogens, predators, or symbionts.

- Fungi absorb the nutrition needed for their survival from dead matter (saprobes), from living hosts (parasites), and from mutually beneficial symbiotic relationships with other organisms (mutualists).
- Adaptations of hyphae allow fungi to exploit different sources of nutrition. Among fungi that are plant parasites, for example, hyphae enter a leaf from a spore on the surface and form a mycelium within the leaf (see Figure 30.7). Rhizoids are hyphae that anchor fungi to their substrate.
- Parasitic fungi are either facultative (possessing the ability to grow independently) or obligate (dependent on their living host for growth).
- Some parasitic fungi not only derive nutrition from their hosts, but also sicken or kill the host. Such fungi are called pathogens.
- Some fungi are active predators that trap microscopic protists or animals in a sticky substance; others form a constricting ring around their prey (see Figure 30.8).

Many fungi are beneficial to other organisms.

- Saprobic fungi secrete enzymes into the environment that help in absorption of dead matter and the decomposition and recycling of elements—especially carbon—used by living organisms. As decomposers, saprobic fungi are essential life on Earth.
- Fungi form two crucial types of symbiotic, mutualistic relationships: lichens and mycorrhizae.
  - Lichens are associations of fungi with a unicellular photosynthetic alga or cyanobacteria. The fungus provides the photosynthetic partner with minerals and water, and the photosynthesizing organism provides the fungus with organic compounds. Lichens are among Earth's hardiest organisms, and can thrive in barren and extreme environments such as Antarctica. Lichens are characterized by their appearance as either crustose (crusty), foliose (leafy), or fruticose (shrubby) (see Figure 30.9).
  - Mycorrhizae are associations between fungi and the roots of plants in which a fungus obtains the products of photosynthesis from the plant and provides minerals and water to the plant. The mycorrhizal symbiosis is essential to the survival of most plants, and the evolution of this relationship may have been the most important step in allowing plants to colonize land.
- Ectomycorrhizal fungi wrap their hyphae around a plant root. The hyphae of arbuscular mycorrhizal fungi penetrate the root cell walls and form treelike (arbuscular) structures that provide the plant with nutrients (see Figure 30.11).
- Endophytic fungi are symbionts living within the aboveground plant parts. They help certain plants (especially grasses) resist pathogens, herbivores, and stresses such as drought and salty soil; their role in other plants is not well understood.

Fungi have many different life cycles and can reproduce both sexually and asexually.

- The different fungal groups display different life cycles. Most display alternation of generations. Glomeromycota reproduce only asexually.
- Fungi have several means of asexual reproduction: They can produce spores enclosed within sporangia or naked spores at the tips of hyphae known as conidia. Unicellular fungi can reproduce by fission or budding. Just about any part of a mycelium is capable of living independently of the rest; therefore, simply the division of a mycelium into two or more parts is a method of reproduction.
- Members of the different fungal groups can be distinguished from one another by their mechanisms of sexual reproduction, which involves two or more mating types. Self-fertilization is prevented by the incapacity of individuals of the same mating type to mate with each other (see Figure 30.12).
  - The first step in sexual reproduction is fusion of two hyphae of different mating types. Eventually, two nuclei (one from each mating type) will fuse to create a zygote nucleus. The zygote nucleus may be the only diploid nucleus in the life cycle of a fungus.
  - Zygospores have a unique multinucleate zygospore within the zygosporangium. In ascomycota and basidiomycota, a unique dikaryon condition is the first stage of sexual reproduction (see Figure 30.12). The cytoplasms of two individuals fuse, but the nuclei do not fuse immediately, and the resulting hyphae contain genetically different haploid nuclei. A fruiting structure is eventually formed, and the two nuclei fuse to form a zygote. This form of reproduction has no gamete cells (only gamete nuclei). The hyphae are not truly diploid ($2n$), but dikaryotic ($n + n$).
- Many parasitic fungi have complex life cycles that require a number of different hosts.
- Sexual stages have not yet been identified in many fungi, so these species have been classified according to data from DNA sequence analysis.
- ***For review, go to Diagram Exercise 2.***

Chytrids are aquatic fungi with flagellated gametes.

- Chytrids likely resemble the common ancestor of all fungi more closely than any other fungal group.
- Chytrids are aquatic, have chitinous cell walls, and are the only fungi with flagellated gametes. They can be either parasitic or saprobic.
- The genus *Allomyces* displays alternation of generations. A haploid zoospore becomes a small organism (the gametophyte) with both female and male gametangia that produce gametes (see Figure 30.12A). When the gametes fuse, they form a diploid zygote and a diploid organism (the sporophyte). The sporophyte produces diploid zoospores that grow into other diploid sporophytes. Eventually, a sporophyte produces sporangia that give rise to haploid zoospores via meiosis.

Zygomycota have coenocytic hyphae and (usually) no fleshy fruiting body.

- Over 1,000 species of zygomycota have been described. The hyphae of a zygomycota spread randomly over a substrate, periodically producing stalked sporangiophores bearing sporangia (see Figure 30.15).
- The zygomycota include black bread mold (*Rhizopus stolonifer*) and many other saprobic and parasitic species.
- Sexual reproduction in zygomycota occurs between adjoining individuals with different mating types (see Figure 30.12B). Branches from each individual grow toward each other until they fuse to produce gametangia. A thick-walled zygosporangium with a zygospore forms at the fusion site. A sporangium, containing haploid nuclei incorporated into new spores, sprouts from the zygospore.

Glomeromycota form arbuscular mycorrhizae.

- Glomeromycota are entirely asexual and coenocytic. They are also entirely terrestrial; about half the fungi found in soils are glomeromycota.
- The 200 species of glomeromycota form arbuscular mycorrhizae with the roots of 80 to 90 percent of all plants (see Figure 30.11B).

Ascomycota (sac fungi) include baker's yeast, truffles, and *Penicillium*.

- The ascomycota are distinguished by their ascus, a sexual reproductive structure (see Figure 30.12C). There are approximately 64,000 species of ascomycota, which were historically divided into two groups. Based on new DNA sequence analysis, these groupings have recently been abandoned.
- Sexual reproduction of filamentous sac fungi involves two different mating types moving into a dikaryotic stage before forming a diploid cell population. A dikaryon is produced during sexual reproduction by the fusing of two mating structures (see Figure 30.12C). From the dikaryon hyphae ($n + n$), dikaryotic asci form where the nuclei fuse. The zygote ($2n$) undergoes meiosis and then produces an ascus with haploid ascospores (n). The ascospores germinate and grow into new mycelia.
- Asexual reproduction in ascomycota involves the production of conidia at the tips of hyphae (see Figure 30.17).
- Ascomycota include molds and the cup fungi and have a number of uses for humans (see Figure 30.16). Mold from the genus *Penicillium* produces the antibiotic penicillin; other molds are important in making cheese. Brown molds are used in brewing sake (a Japanese alcoholic beverage) and soy sauce.
- The species of yeast used to make bread and alcoholic beverages, *Saccharomyces cerevisiae,* is a unicellular sac fungus. Reproduction in these yeasts is asexual and accomplished by budding.

The fruiting structures of basidiomycota are familiar as mushrooms.

- Included in the 30,000 species of basidiomycota fungi are puffballs and mushrooms, which produce the group's most spectacular fruiting structures.
- In basidiomycota, meiosis and nuclear fusion occur in the basidium at the tips of hyphae (see Figure 30.12D). After meiosis, four basidiospores are formed on tiny stalks. The basidiospores are released into the environment and germinate to form haploid hyphae. The hyphae grow and fuse with different mating types to form the dikaryotic mycelium. The fruiting structure, or basidiocarp, grows from the dikaryotic mycelium. The cap of a mushroom is the fruiting structure of a basidiomycota (see Figure 30.18). Basidia form on the underside of the cap along gills and discharge their spores, which will form the new developing basidium.

Test Yourself

Diagram Exercises

1. Label the diagram below using the following terms: hypha, nuclei, cell wall, septa.

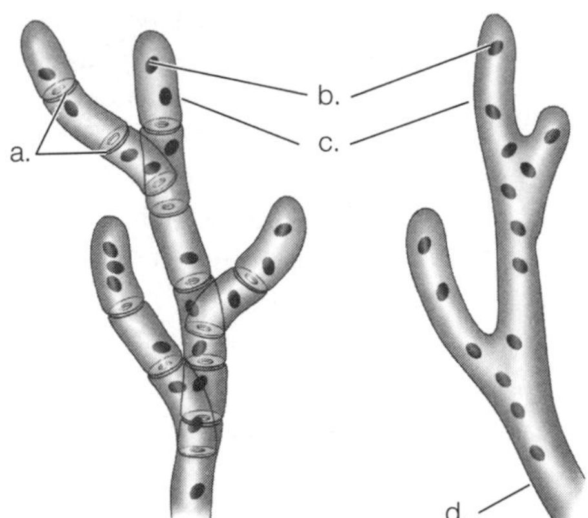

Textbook Reference: *30.1 What Is a Fungus? p. 629, Figure 30.4*

2. Four major fungal life cycles are represented below. Label each according to the major group it represents. Indicate the haploid, diploid, and dikaryotic stages. In the boxes, label the process as mitosis, meiosis, or fertilization. Then label each structure using the appropriate terms.

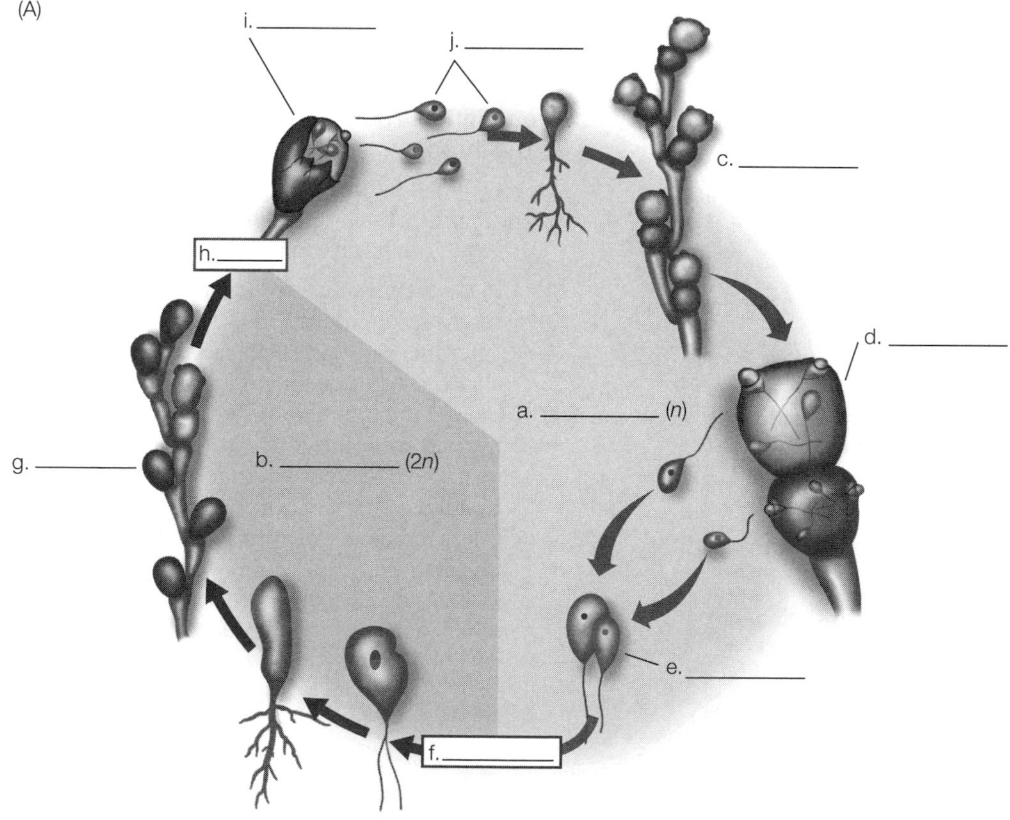

Textbook Reference: *30.3 What Variations Exist Among Fungal Life Cycles? pp. 636–637, Figure 30.12*

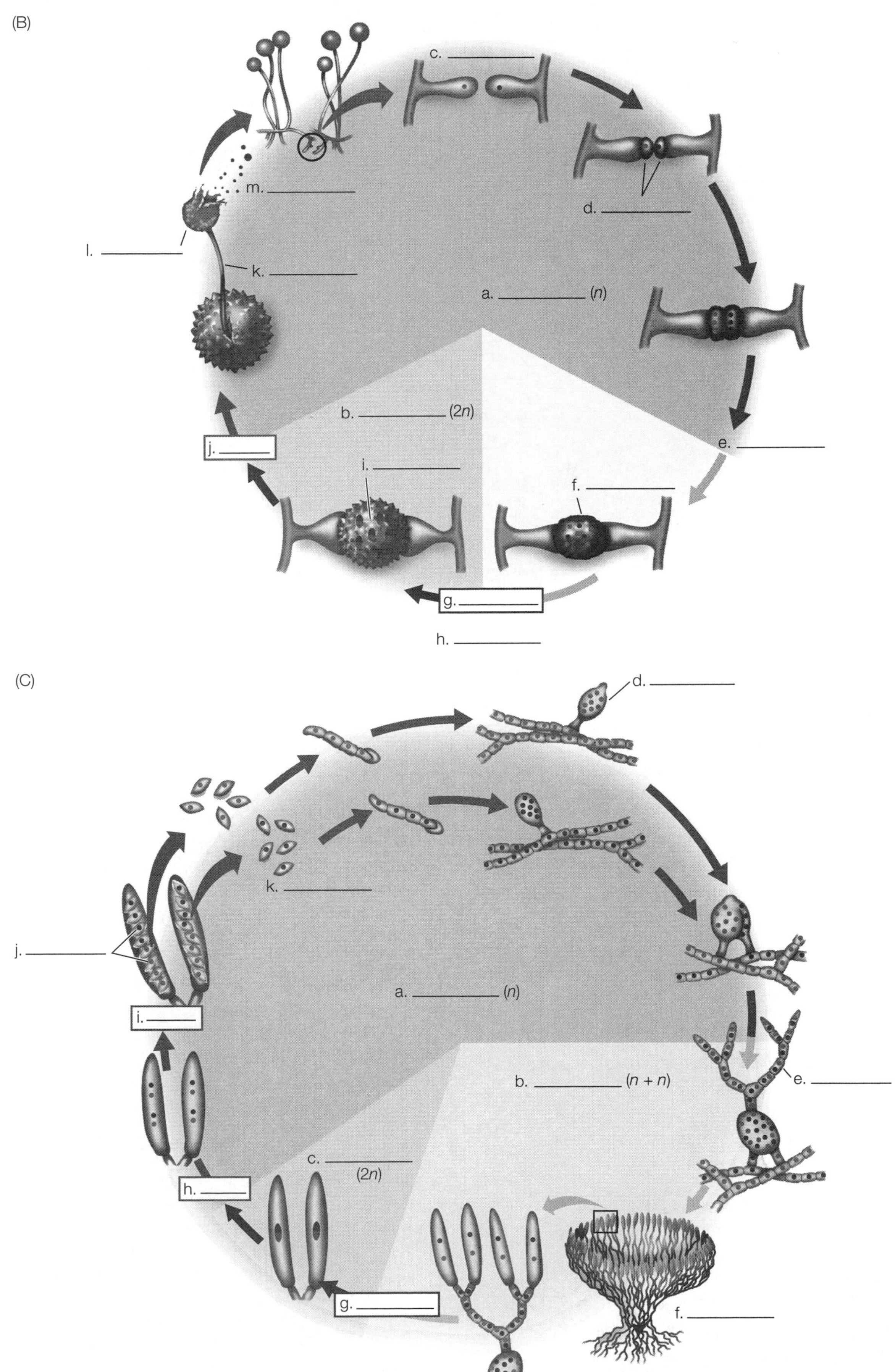
(B)
c.
d.
m.
l.
k.
a. (n)
b. (2n)
j.
e.
i.
f.
g.
h.
(C)
d.
k.
j.
a. (n)
i.
b. (n + n)
e.
c.
(2n)
h.
g.
f.

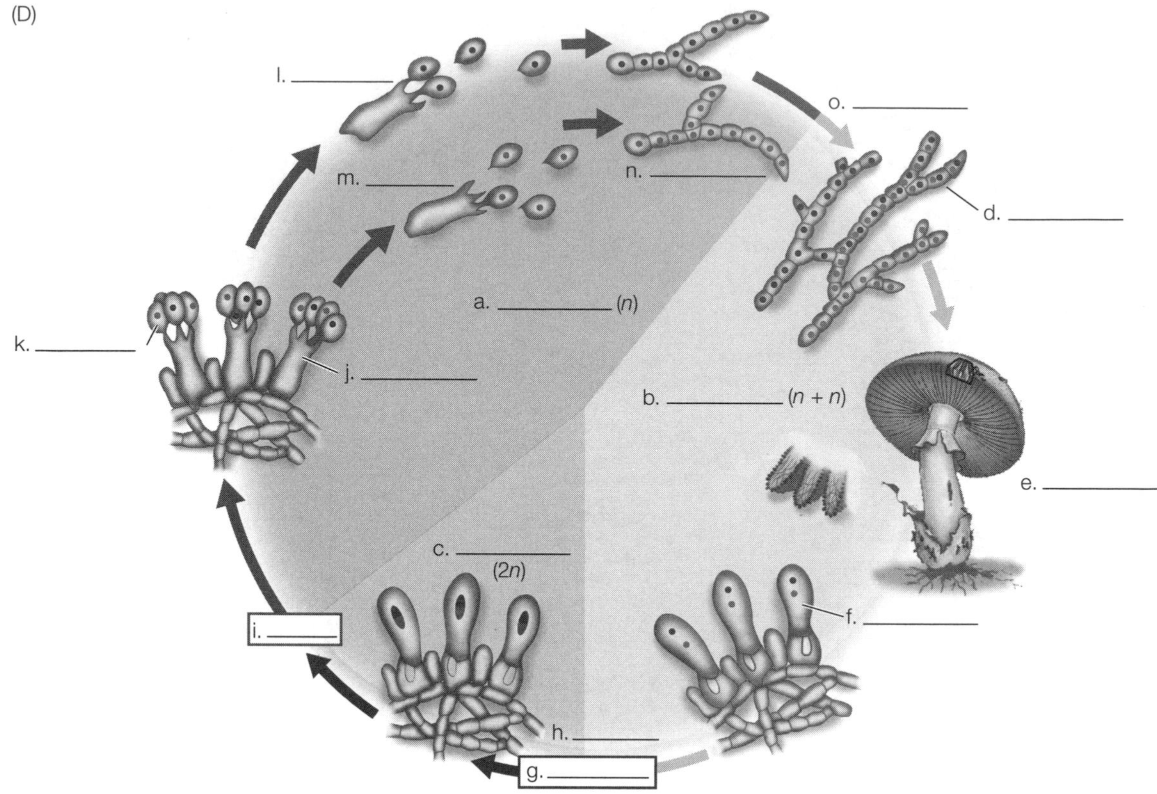

Knowledge and Synthesis Questions

1. Many species of fungi can be placed into one of six main groups based on
   a. methods of sexual reproduction.
   b. whether or not gametes have a flagella.
   c. the presence or absence of septa in the hyphae.
   d. DNA sequence analysis.
   e. All of the above
   Textbook Reference: *30.1 What Is a Fungus? p. 627*
2. Which of the following would *not* be found in any of the typical fungal life cycles?
   a. Haploid nuclei
   b. Diploid nuclei
   c. Spores
   d. Chloroplasts
   e. A dikaryotic stage
   Textbook Reference: *30.3 What Variations Exist among Fungal Life Cycles? pp. 635–638*
3. Fungi are absorptive heterotrophs. Which of the following is an adaptation that greatly aids this mode of nutrient procurement?
   a. Dikaryosis
   b. A large surface area-to-volume ratio
   c. Conjugation
   d. A complex life cycle
   e. a small surface area-to-volume ratio
   Textbook Reference: *30.1 What Is a Fungus? p. 630*
4. Assume that two normal hyphae of different fungal mating types meet. After a period of time, the cell walls between these hyphae will dissolve, producing a
   a. mycelium.
   b. fruiting body.
   c. zygote.
   d. spore.
   e. dikaryotic cell.
   Textbook Reference: *30.3 What Variations Exist among Fungal Life Cycles? pp. 635–638*
5. Which of the following is the best way to represent the ploidy of dikaryotic hyphae?
   a. $1/2\ n$
   b. n/n
   c. n
   d. $n + n$
   e. $2n$
   Textbook Reference: *30.3 What Variations Exist among Fungal Life Cycles? pp. 635–638*
6. Which of the following statements about sexual reproduction in fungi is *false*?
   a. Motile gametes are present in all fungal species.
   b. An aquatic environment is not required for fertilization to occur in most fungi.
   c. There is no true diploid tissue in the life cycle of most sexually reproducing fungi.
   d. Sexual reproduction often begins with contact between hyphae of different mating types.
   e. Alternation of generations and sexual reproduction can occur in the same species.
   Textbook Reference: *30.4 How Have Fungi Evolved and Diversified? pp. 638–642*

7. A mycorrhiza is
   a. a specialized type of lichen.
   b. the fruiting structure of a basidiomycota.
   c. a symbiotic association between a fungus and cyanobacterium or green algae.
   d. a reproductive stage of sac fungi.
   e. a symbiotic association between a fungus and a plant.

   Textbook Reference: *30.2 How Do Fungi Interact with Other Organisms? pp. 633*

8. Suppose a scientist investigating the classification of a fungus that has never been observed to reproduce sexually discovers that it has DNA sequences characteristic of basidiomycota. If this scientist could coax fungi of this species to reproduce sexually, which of the following would most likely be observed?
   a. Dikaryotic hyphae segmented by septa
   b. Dikaryotic hyphae without septa
   c. Asymmetrical cell division
   d. A dikaryotic ascus
   e. Flagellated gametes

   Textbook Reference: *30.4 How Have Fungi Evolved and Diversified? pp. 642–643*

9. Which of the following fungi have coenocytic hyphae and stalked sporangiophores?
   a. Chytrids
   b. Zygomycota
   c. Ascomycetes
   d. Basidiomycota
   e. Glomeromycota

   Textbook Reference: *30.3 What Variations Exist among Fungal Life Cycles? pp. 635–636*

10. Which of the following fungi display alternation of generations?
    a. Chytrids
    b. Glomeromycota
    c. Ascomycota
    d. Basidiomycota
    e. Zygomycota

    Textbook Reference: *30.3 What Variations Exist among Fungal Life Cycles? pp. 635–636*

11. Which of the following is thought to be the most ancient group?
    a. *Penicillium*
    b. *Fusarium*
    c. *Saccharomyces*
    d. *Aspergillus*
    e. *Allomyces*

    Textbook Reference: *30.4 How Have Fungi Evolved and Diversified? pp. 640–641*

12. A saprobe is an organism that
    a. absorbs nutrients from the sap of a host plant.
    b. reproduces in the sap of a plant.
    c. undergoes asexual reproduction.
    d. is mutualistic.
    e. absorbs nutrients from dead organic matter.

    Textbook Reference: *30.1 What Is a Fungus? p. 627*

13. Which of the following is *not* a form of asexual reproduction in fungi?
    a. Budding
    b. Formation of haploid spores in sporangia
    c. Formation of dikaryotic mycelia
    d. Fission
    e. Formation of haploid spores in conidia

    Textbook Reference: *30.1 What Is a Fungus? p. 630, Figure 30.5*

14. Which of the following statements about fungi is *false* (i.e., a common misconception)?
    a. Fungi grow only in warm, wet environments.
    b. Fungi lose water rapidly in a dry environment.
    c. Fungi can grow in environments too hypertonic to sustain bacteria.
    d Fungi are eukaryotes and have multiple mating types.
    e. Male and female fungi have no distinctive morphology.

    Textbook Reference: *30.1 What Is a Fungus? p. 630*

15. Which of the following scientific groups is correctly matched with its correct common name?
    a. Chytrids–microspore fungi
    b. Zygomycota–sac fungi
    c. Glomeromycota–mycorrhizial fungi
    d Basidiomycota–sac fungi
    e. Ascomycota–club fungi

    Textbook Reference: *30.1 What Is a Fungus? p. 628, Table 30.1*

Application Questions

1. Early taxonomists considered fungi to be members of the plant kingdom. What evidence indicates that they are in fact more closely related to animals?

   Textbook Reference: *30.1 What Is a Fungus? p. 627*

2. Explain the different ways in which fungi are important to plants.

   Textbook Reference: *30.2 How Do Fungi Interact with Other Organisms? p. 635*

3. How is basic research on fungi relevant to research on the prevention and cure of HIV/AIDS?

   Textbook Reference: *30.2 How Do Fungi Interact with Other Organisms? p. 632*

4. Review the material in Chapter 7 (Cell Signaling and Communication). List and explain three examples of cell signaling in fungi.

   Textbook Reference: *30.2 How Do Fungi Interact with Other Organisms? pp. 631–635*

5. Why have scientists abandoned the classification of fungi based on reproductive cycles? Explain and evaluate the justifications for this decision.

   Textbook Reference: *30.1 What Is a Fungus? p. 627*

Answers

Diagram Exercise Answers

1.
   a. Septa
   b. Nuclei
   c. Cell wall
   d. Hypha
2. (A) Chytrid life cycle
   a. Haploid
   b. Diploid
   c. Multicellular haploid chytrid (n)
   d. Female gametangium
   e. Gametes
   f. Fertilization
   g. Multicellular diploid chytrid ($2n$)
   h. Meiosis
   i. Sporangium
   j. Haploid zoospores (n)

   (B) Zygospore fungi (Zygomycota) life cycle
   a. Haploid
   b. Diploid
   c. Hyphae
   d. Gametangia (n)
   e. Plasmogamy
   f. Zygosporangium
   g. Fertilization
   h. Karyogamy
   i. Multinucleate zygospore within zygosporangium
   j. Meiosis
   k. Sporangiophore
   l. Sporangium
   m. Spores

   (C) Sac Fungi (Ascomycota)
   a. Haploid
   b. Dikaryotic
   c. Diploid
   d. Mating structure
   e. Dikaryotic mycelium ($n + n$)
   f. Ascoma (fruiting structure)
   g. Fertilization
   h. Meoisis
   i. Mitosis
   j. Ascospores
   k. Ascospores (n)

   (D) Club fungi (Basidiomycota)
   a. Haploid
   b. Dikaryotic
   c. Diploid
   d. Dikaryotic mycelium ($n + n$)
   e. Basidioma (fruiting structure)
   f. Developing basidium ($n + n$)
   g. Fertilization
   h. Karyogamy
   i. Meiosis
   j. Basidium
   k. Basidiospores
   l. + Mating type
   m. –Mating type
   n. Mycelial hyphae
   o. Plasmogamy

Knowledge and Synthesis Answers

1. **d.** Traditionally, a fungus's method of sexual reproduction was the primary criterion in its classification, but all the traits listed are useful in classifying fungi.
2. **d.** Fungi have haploid and diploid nuclei at different stages of their life cycle, and many have a dikaryotic stage. They also produce spores. The chloroplasts are not found in the fungi, but in plants.
3. **b.** The large surface area-to-volume ratio of the hyphae increases the ability of a fungus to absorb nutrients.
4. **e.** When two hyphae of different mating types fuse, they form a dikaryotic hyphae.
5. **d.** Dikaryotic hyphae are neither truly diploid ($2n$) nor haploid (n). Because dikaryotic hyphae include genetic material from two haploid nuclei that remain separate, the best way to represent their ploidy is as $n + n$.
6. **a.** Not all fungi have motile gametes; only the gametes of chytrids are motile.
7. **d.** Mycorrhizae are associations between fungi and the roots of plants. Lichens are symbiotic relationships between fungi and cyanobacteria or green algae.
8. **a.** If this fungus is indeed a basidiomycota, the fusing of hyphae of different mating types will most likely result in dikaryotic hyphae that are segmented by septa.
9. **b.** Coenocytic hyphae are characteristic of both the chytrids and the zygomycota, but only the zygomycota regularly produce sporangiophores.
10. **a.** Among the fungi, only the chytrids have a multicellular haploid stage and a multicellular true diploid stage.
11. **e.** Chytrids are thought to be the most ancient group, and *Allomyces* is a well-characterized genus of chytrids.
12. **e.** Saprobes are organisms that absorb nutrients from dead matter. Some bacteria are saprobes.
13. **c.** The dikaryotic mycelium is a structure formed in the sexual reproduction life cycle.
14. **a.** Fungi are plentiful in warm, wet environments, but many also survive extreme temperatures and very dry conditions.
15. **c.** The correct associations are as follows: Microsporidia/microsporidia; Chytrids/chytrids; Zygomycota/zygospore; Glomeromycota/mycorrhizae; Ascomycota/sac fungi; Basidiomycota/club fungi.

Application Answers

1. At the cellular level, fungi bear very little resemblance to plants or even to the surviving green algae that are the most likely common ancestor of all plants. They do not produce chlorophyll, nor do they have plastids for

storing reserves of photosynthetic products. Fungi and plants both have cell walls, but fungal cell walls contain chitin, a polysaccharide molecule (see Chapter 3) that plants do not produce. Chitin is found in some animals (particularly among the ecdysozoans) and in choanoflagellates, the protist group most closely related to the animals. It is unlikely that this complex molecule evolved more than once, so it is likely that fungi and animals shared a chitin-producing ancestor that is more recent than any ancestor shared by fungi and plants or by animals and plants.

2. Plants rely on mycorrhizae for adequate absorption of water and nutrients from the soil. Scientists hypothesize that this is the relationship that allowed plants to colonize land in the first place. Other fungi can provide their hosts with some resistance to herbivores, increased resistance to drought, and protection from some pathogens. The mechanisms of these protective actions are not all understood, although they have been adopted in some agricultural practices.
3. Basic research in mycology (the study of fungi) may directly improve the lives of many HIV-positive individuals who suffer from fungal infections, many of which can be fatal. Better understanding of how fungal-based pneumonia, diarrhea, and esophagitis develop in the human body will help us treat these patients and alleviate suffering. Funding for basic mycology research may also have very direct impact on treating humans in the latter stages of AIDS.
4. Many examples of cell signaling can be found among the fungi. For example, chemical signaling is the mechanism by which plants attract fungi. The introduction of Chapter 30 of the textbook has a discussion of how the parasite *Striga* can detect these chemical signals to seek out a host. Predatory fungi make use of a detection mechanism in order to contract around a nematode, and cell signaling is necessary in the coordination of many different cells to form structures such as basidiomas, ascomas, diploid chytrids, and sporangia. Even proper mating types are identified by means of cell signaling.
5. Scientists have always used the tools available to them. In a time when molecular data was not available, a classification system based on morphological features made sense, since many of these structures reflect the reproductive strategies and thus the common ancestries of different organisms. This approach is limited, however, since it does not account for changes in reproductive strategies over time. As organisms have evolved and moved into new environments, their relationships to other species may not be reflected in physical features and current reproductive strategies. DNA sequence analysis allows scientists to examine the entire genome and compare conserved sequences that may not be expressed as visible changes. This new technology thus gives us new insights into the relationships among organisms. As we develop new tools in the future, we will understand even more about the evolution of the fungi.

31 Animal Origins and the Evolution of Body Plans

The Big Picture

- The animals are a monophyletic group sharing a number of morphological and genetic traits. Animals are motile, multicellular organisms that must ingest nutrients.
- Animals are distinguished by the number of cell layers found in their embryos, their type of body cavity, and their body symmetry.
- The importance of acquiring nutrition (food) has led to the evolution of a variety of feeding strategies that maximize the available sources of nutrition. Life cycles also vary considerably among the animals, and involve a considerable number of trade-offs.
- The eumetazoans encompass all animals except the three groups of sponges. Ctenophores and cnidarians are diploblastic eumetazoans that are not bilaterally symmetrical.
- Sponges can be classified in three different groups based on molecular evidence and the morphology of their spicules.
- Both ctenophores and cnidarians are characterized by the presence of a largely inert layer of gelatinous mesoglea. Ctenophores have a complete gut with a mouth and an anus, whereas cnidarians have a blind gastrovascular cavity with a single opening that ingests food and expels wastes. Cnidarians have simple nerve nets and muscle fibers, which allow them a level of control over their movements that the ctenophores do not have.

Common Problem Areas

- Many of the same evolutionary "themes" can be found in widely divergent species. Venomous or venom-producing structures, for example, are found in many different animal groups, from the cnidarians to spiders to snakes. As the different animal groups are described, create a table of the various characteristics or mechanisms that consistently appear. Use this table to help you organize your thinking about evolution and the creation of diversity.

Study Strategies

- The phylogenetic trees in the textbook are good frameworks on which you can add more information. Doing so will provide you with a point of reference when you study the different groups. Remember that animals have a set of traits shared with and inherited from their ancestors, as well as distinctive derived traits.
- It is important that you understand what features of an animal group may give it an advantage in its environment and the compromises (trade-offs) that are made in other areas.
- Don't forget that even though particular animals may be placed in a group or clade, not all species have all of the features that characterize the group. For example, think about the reasons that sponges are considered as animals even though they lack many of the features of other animals.
- Go to yourBioPortal.com to review the following tutorial and activities:

  Animated Tutorial 31.1 Life Cycle of a Cnidarian

  Web Activity 31.1 Sponge and Diploblast Classification

  Web Activity 31.2 Animal Body Cavities

Important Concepts

What Characteristics Distinguish the Animals?

- Although there are exceptions, the following are characteristics that are commonly associated with animals:
  - Multicellularity: Most animal life cycles feature a complex pattern of development from a single-celled zygote.
  - Heterotrophic metabolism: All animals are heterotrophs and must have some means to take nutrients from their environment.
  - Internal digestion: Most animals (but not all) have an internal gut where digestion takes place.
  - Movement: Most animals either move to their food or have some means to bring it to them. However, some animals have stages in their life cycle in which movement does not take place.

- Phylogenetic analyses of animal gene sequences support the conclusion that animals are monophyletic (see Figure 31.1).
- Animals generally share the following morphological and genetic synapomorphies: tight junctions, desmosomes, and gap junctions between their cells; extracellular matrix molecules, including collagen and proteoglycans; and Hox genes that specify body pattern and axis formation. These traits were probably possessed by the common ancestor of all animals but have been lost in some groups.
- The common ancestor of modern animals may have been a colonial flagellated protist such as a choanoflagellate (see Figure 27.27).
- Differences in patterns of embryonic development, including cleavage patterns, gastrulation patterns, and the number of cell layers present, show the evolutionary relationships among animals. On these bases, animals can be grouped as follows:
  - Diploblastic animals have two cell layers: the endoderm and the ectoderm.
  - Triploblastic animals have three layers: the endoderm, mesoderm, and ectoderm.
  - Triploblastic animals are divided further into protostomes ("mouth first"; the blastopore becomes the mouth and the anus forms later) and deuterostomes ("mouth second"; the blastopore becomes the anus and the mouth forms later).
- Sequencing data indicate that the protostomes and deuterostomes are two different animal clades. Together, they are known as the bilaterians (see Figure 3.1) and account for most of the animal species.

Every animal has a body plan.

- The overall organization of an animal's body is known as its body plan. The features of an animal's body plan include symmetry or asymmetry, body cavity structure, support (skeletal) structure, segmentation, and the presence or absence of appendages.
- An animal that can be divided along at least one plane into similar halves is said to be symmetrical. Asymmetrical animals such as Placozoans and most sponges have no plane of symmetry. The sea anemone is radially symmetrical; any plane running along a sea anemone's main axis will divide it into roughly equal halves (see Figure 31.3A).
- Animals exhibiting bilateral symmetry can be divided into two mirror images by only one plane (see Figure 31.3B). Bilateral symmetry is common among animals that are able to move quickly.
- Bilaterally symmetrical animals often have sense organs and nervous tissue concentrated at the anterior end; this type of organization is known as cephalization (from the Greek word for "head").
- Animals have three different types of body cavities (see Figure 31.4).
  - Acoelomates have no enclosed body cavity.
  - Pseudocoelomates have a liquid-filled space known as the pseudocoel in which many of the internal organs are located.
  - Coelomates have a true body cavity, a coelom, developing within the mesoderm. In coelomates, the internal organs are in pouches of the peritoneum.
  - ***For review, go to Diagram Exercise 1.***

The structure of the body plan influences movement.

- Fluid-filled body cavities act as hydrostatic skeletons for many animals. Other animals evolved rigid supportive skeletons that can be internal (bones or cartilage) or external (a shell or cuticle). Muscles attached to hard skeletons allow the animal to move.
- Segmentation allows specialization of the different body regions and can improve control of movement. In some animals, segments are not apparent (e.g., vertebrae column). In other animals, similar body segments are repeated many times, and in yet others the body segments differ (see Figure 31.5).
- Appendages, especially jointed limbs, enhance locomotion. Jointed limbs in the arthropods and vertebrates are a major factor in their evolutionary success. Other appendages are specialized and can be used to sense the environment (e.g., antennae) or can be used to capture prey.

All animals are ingestive heterotrophs.

- To obtain food, animals either move through the environment to the food or move the environment and the food to them.
- Most animals are motile, but some are sessile (nonmoving).
- As heterotrophs, animals have feeding strategies that fall into a few broad categories. Animals may be filter feeders (see Figure 31.6), herbivores, predators (see Figure 31.8), parasites, or detritivores. Some animals change their feeding strategies at different developmental stages. Other animals are omnivores and eat both plants and other animals.

There are a variety of animal life cycles.

- The life cycle of an animal encompasses embryonic development, birth, growth to maturity, reproduction, and death.
- In direct development, newborns look very similar to adults. In many species, however, newborns differ strikingly from adults and pass through distinct larva, pupa, and adult stages. Metamorphosis refers to the dramatic changes that can occur between the larval and adult stages, when the individual is a pupa (see Figure 31.10). In species that exhibit these stages, one stage is often specialized for feeding while the other is primarily for reproduction. If both stages feed, what is eaten can change with the stage.
- All life cycles have a least one dispersal stage. In general terms, dispersal refers to the movement of an

organism from the parent or from the population. As a specific example, the larva is the dispersal stage for most sessile organisms. For animals that live on the sea floor, the common larval types are trochophore and nauplius (see Figure 31.11). Both types feed on plankton before settling down on the ocean floor, where they develop into adults.

- Every life cycle necessarily involves evolutionary trade-offs. Often, an adaptation that improves performance in one activity comes at a cost at the reduced performance of another activity. Common trade-offs involve reproduction. For example, there is a choice between producing many eggs with a small amount of nutrients or a small number of large eggs that store more nutrients (see Figure 31.12). The trade-off is the amount of offspring produced versus the amount of nutrients the offspring receives.
- In some bird and mammal species, offspring are altricial, meaning that the young are not fully developed and require a significant amount of parental care such as feeding. With others, the young are precocial and can care for and feed themselves soon after birth.
- For parasites, the dispersal stage has evolved to overcome host defenses. The life cycle of parasites can be fairly complex and involve multiple hosts and larval stages (see Figure 31.14). This complexity can facilitate parasite dispersal.
- In some groups of animals, asexual reproduction occurs without fission, resulting in colonies of individuals. To the eye, a colony looks like one, single, integrated organism. In some species, colonies are composed of individuals that function alike while in some other species, the colonies are specialized for different functions. Examples of how the lines between an individual and a population are blurred are the bryozoan (see Figure 31.15) and the Portuguese man-of-war (see Figure 31.9).

Animals are classified into broad groups.

- A summary of the living members of the major animal groups is presented in Table 31.1.
- The simplest animals are the sponges, which have no body symmetry and no distinct cell layers. Placozoans have only four cell types and weakly differentiated layers of tissue. All animals other than the sponges and placozoans make up the eumetazoans. Eumetazoans have body symmetry, a defined gut, a nervous system, and distinct organs.
- The bilaterians are a monophyletic group that includes all of the eumetazoans except the ctenophores and the cnidarians. Bilaterian synapomorphies include bilateral symmetry, three cell layers, and the presence of at least seven Hox genes. The Bilateria has two major subgroups, the protostomes (see Chapter 32) and the deuterostomes (see Chapter 33).
- ***For review, go to Diagram Exercise 2.***

Sponges have differentiated cells but no true organs.

- Although the sponge body plan is relatively simple, the three sponge groups all have cells that are differentiated for specific functions.
- Most of the 8,000 species of sponges are marine filter feeders that remove small organisms and nutrient particles from seawater as it flows through pores in the walls of their inner cavity. Choanocyte cells use flagella to divert water through the pores and filter out food particles. The water exits the sponge through large openings called oscula (see Figure 31.7). A few sponges are carnivores, and can trap prey on hook-shaped spicules that are on the outside of the body surface.
- Sponges have a supporting skeleton composed of branching spines (spicules). The spicules of glass sponges and demosponges (the largest sponge group) are made of silicon. The calcareous sponges take their name from their calcium carbonate spicules and are the sponge group most closely related to the eumetazoans.
- In addition to spicules, sponges also have an extracellular matrix composed of collagen, adhesive glycoproteins, and other molecules that hold the cells together.
- The huge variety of sponge body sizes and shapes is a response to the different movement patterns of water, specifically tides and currents. Sponges that live in environments that have strong wave action tend to be firmly attached to substratum, while those that live in slowly moving water are generally flat and oriented at right angles to the current flow.
- Sponges reproduce sexually by producing both egg and sperm, and asexually by budding and fragmentation.

Placozoans are structurally very simple, with a diploblastic body plan.

- Placozoans do not have a mouth, gut, or true nervous system, and they consist of only a few different cell types. They have upper and lower epithelial cell layers with contractile fiber cells in between the layers.
- Based on phylogenetic analysis, their structural simplicity may have been secondarily derived (i.e., some of their common features were probably lost sometime during evolution from a common ancestor).
- The life cycle of placozoans remains relatively unknown because their transparency makes it difficult to observe them in nature. It is known that they have a pelagic (free-swimming) stage in the ocean and that they are capable of sexual and asexual reproduction.

Ctenophores have two cell layers separated by mesoglea.

- The ctenophores (comb jellies) are marine diploblastic animals that are radially symmetrical (see Figure 31.18). There are 150 known species, and they are found primarily in the open ocean, where they feed on planktonic organisms that are filtered by sticky filaments on the tentacles. Although ctenophores are classified as eumetazoans, they lack most of the Hox genes.

However, they have a complete gut (i.e., a gut with an entrance and exit, or mouth and anus).

- Ctenophores get their name from the eight rows of comblike plates of cilia, known as ctenes. The cilia are used to propel the animal through the water. Prey is caught on sticky filaments on the tentacles or body (see Figure 31.18).
- The ectoderm and endoderm of ctenophores are separated by a thick, gelatinous layer called the mesoglea.
- They can overpopulate protected bodies of water and can cause significant damage to local ecosystems.
- Ctenophores have a simple life cycle and reproduce sexually. In most species, the externally fertilized egg hatches into a miniature ctenophore (i.e., direct development).

The cnidarian life cycle has two stages: the polyp and the medusa.

- All cnidarians are diploblastic and have radial symmetry.
- The cnidarian gastrovascular cavity is a blind sac, so cnidarians do not have complete gut. There is only one opening, which serves as both mouth and anus. The gastrovascular cavity functions in food digestion, respiratory gas exchange, and circulation. It also lends support as a hydrostatic skeleton.
- As in the ctenophores, a large amount of mesoglea is found between the two cell layers of cnidarians. Because of the inert nature of the mesoglea, cnidarians have low metabolic rates. Many species, however, are able to capture large prey; this combination of characteristics makes it possible for them to live in environments with little prey.
- Cnidarian tentacles have specialized cells, called cnidocytes, which inject toxins into their prey with the help of stingers called nematocysts (see Figure 31.9).
- The life cycle of most cnidarians includes a sessile polyp stage and a motile medusa stage (see Figure 31.19). The polyp stage usually reproduces asexually. The medusa stage reproduces sexually, with the fertilized egg becoming a planula larva that eventually develops into a polyp.
- Cnidarians possess muscle fibers that enable them to move and simple nerve nets that integrate their activities. They also possess the structural molecules collagen, actin, and myosin, which, along with their Hox genes, link them to the bilaterians.
- All but a few of the 11,000 or so species of cnidarians are marine. The smallest individuals are almost microscopic, while individual jellyfish can be quite large. Many cnidarians are colonial.
    - Anthozoans comprise about 6,000 species of sea anemones, sea pens, and corals (see Figures 31.20A and 31.20B). They lack a medusa stage altogether. Their polyps can reproduce both sexually and asexually. Sexual reproduction produces a planula that will develop into a polyp. Sea anemones are solitary and occasionally motile; corals are sessile and colonial. The coral skeleton is composed of calcium carbonate and varies greatly between species. Corals live in symbiosis with photosynthetic protists that provide nutrients for the coral colony through photosynthesis, contributing to the success of corals in clear, nutrient-poor tropical waters.
    - Scyphozoans include the marine jellyfish, which have medusae with thick mesoglea. These species spend most of their lives in the medusa stage. Polyps are produced sexually by adult medusae. The polyps produce young medusae by budding (see Figures 31.19 and 31.20C).
    - Hydrozoans have a life cycle that typically is dominated by the polyp stage, though some have only medusae and others have only polyps. The hydrozoans tend to be colonial, with many polyps sharing a common gastrovascular cavity (see Figure 31.20D). Within the colony, some polyps have tentacles with numerous nematocysts that capture prey. Not all hydrozoans have tentacles; others have finger-like projections that defend the colony with their nematocysts.
    - ***For review, go to Diagram Exercises 3–4.***

Test Yourself

Diagram Exercises

1. For each of the geometric shapes or characters shown below, determine if it has radial symmetry, bilateral symmetry, or no symmetry. Hint: draw some lines through the images to help you decide.

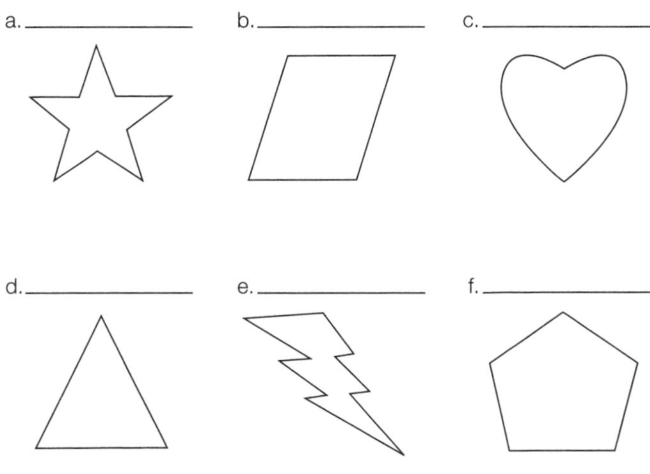

Textbook Reference: *31.2 What Are the Features of Animal Body Plans? p. 649*

2. In the diagram below, identify the following groups: Protostomes, Deuterostomes, Arrow worms, Calcareous sponges, Chordates, Cnidarians, Ctenophores, Demosponges, Ecdysozoans, Echinoderms, Glass sponges, Hemichordates, Lophotrochozoans, and Placozoans. Additionally, identify the distinguishing traits marked with dots on the phylogenetic tree: Bilateral symmetry along an anterior–posterior axis; three embryonic cell layers, Blasotopore develops into mouth, Blastopore develops into anus, Choanocytes; spicules, Distinct organ systems, Exoskeleton molting, Notochord, Radial symmetry (used twice in tree), Silicaceous spicules, Two embryonic cell layers, Unique cell junctions; collagen and proteoglycans in extracellular matrix.

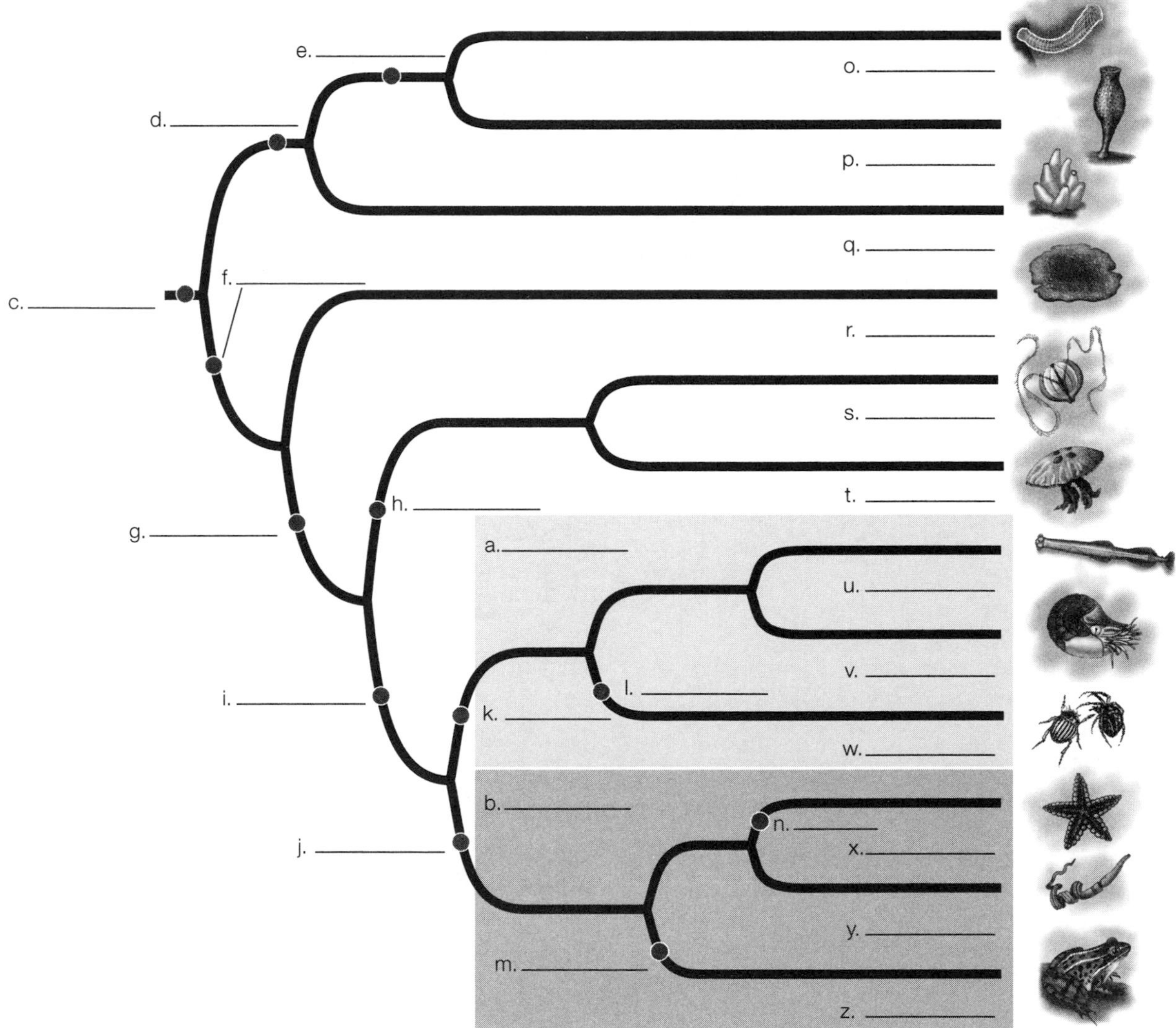

Textbook Reference: *31.1 What Characteristics Distinguish the Animals? p. 647, Figure 31.1*

3. The diagram at right shows a typical jellyfish life cycle. Identify the diploid and haploid stages, the process occurring in the boxes (mitosis, meiosis, fertilization, fusion), and label the identified structures.
   Textbook Reference: *31.5 What Are the Major Groups of Animals? p. 661, Figure 31.19*

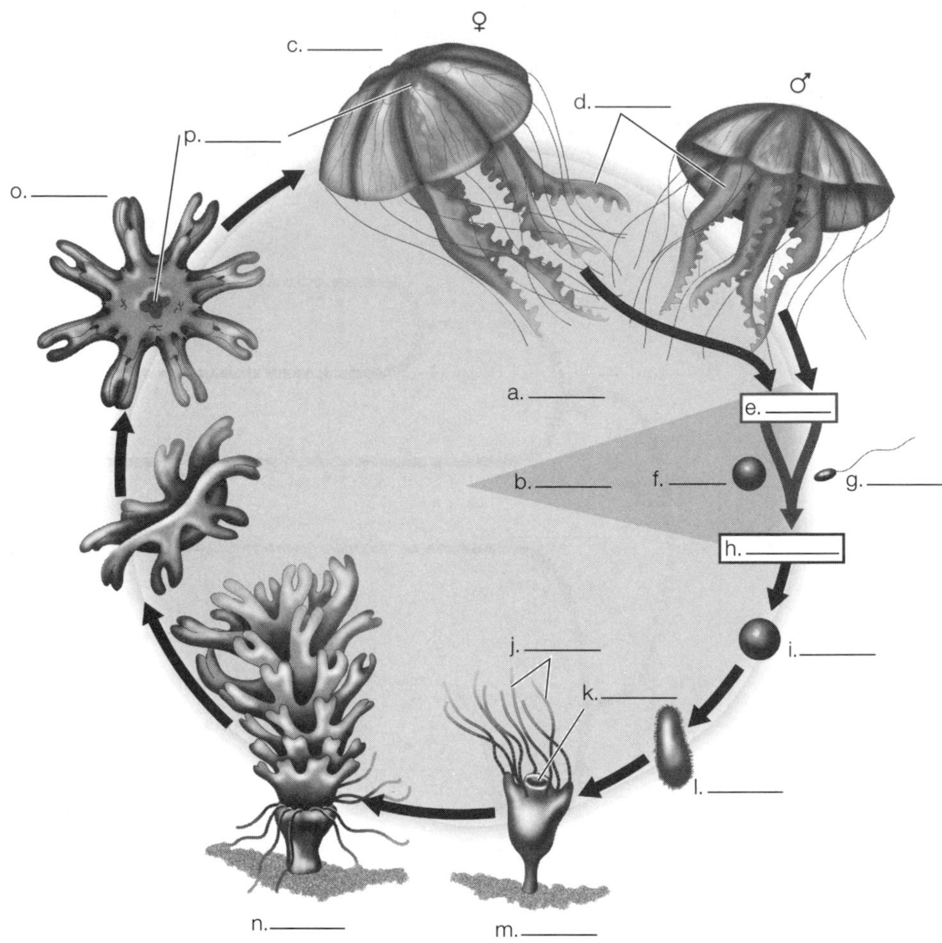

4. The diagram at right shows a typical hydrozoan life cycle. Identify the diploid and haploid stages, the process occurring in the boxes (mitosis, meiosis, fertilization, fusion), and label the identified structures.
   Textbook Reference: *31.5 What Are the Major Groups of Animals? p. 663, Figure 31.22*

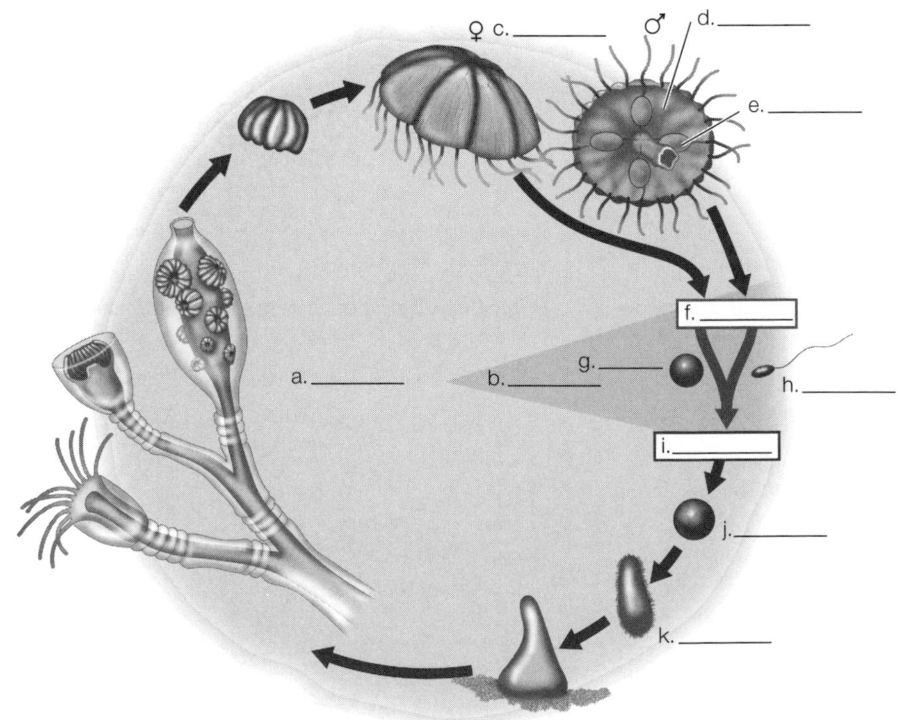

Knowledge and Synthesis Questions

1. Which of the following is *not* a derived trait that is shared by all animals?
   a. Hox genes
   b. The extracellular matrix molecule collagen
   c. Tight junctions, desmosomes, and gap junctions
   d. Bilateral symmetry
   e. Absence of a cell wall
   Textbook Reference: *31.1 What Characteristics Distinguish the Animals? p. 647*
2. Which of the following statements about a deuterostome is *false*?
   a. Three distinct layers of tissue were present during development.
   b. If a coelom is present, it formed within the embryonic mesoderm.
   c. Its early embryonic cleavage pattern was radial.
   d. It is diploblastic.
   e. Gastrulation occurred during development.
   Textbook Reference: *31.1 What Characteristics Distinguish the Animals? p. 648*
3. An important factor contributing to the evolution of diversity among animals is the considerable variation in their
   a. multicellularity.
   b. cell junctions.
   c. methods of food acquisition.
   d. symmetry.
   e. segmentation patterns.
   Textbook Reference: *31.3 How Do Animals Get Their Food? p. 651*
4. Which of the following statements about the body cavity of animals is true?
   a. The body cavity of coelomates develops from the embryonic ectoderm.
   b. The body cavity of acoelomates is filled with liquid.
   c. The pseudocoel of the pseudocoelomates has a peritoneum.
   d. The acoelomates do not have an enclosed body cavity.
   e. The coelomates have a body cavity surrounded by peritoneum.
   Textbook Reference: *31.2 What Are the Features of Animal Body Plans? p. 650*
5. Which of the following is *not* one of the mechanisms of movement in animals?
   a. Hydrostatic skeleton
   b. Jointed appendages
   c. Being sessile
   d. Segmentation
   e. Body cavities
   Textbook Reference: *31.2 What Are the Features of Animal Body Plans? pp. 649–650*
6. Which one of the following feeding strategies requires a life cycle in which much time and energy are devoted to dispersal?
   a. Herbivory
   b. Parasitism
   c. Predation
   d. Filter feeding
   e. Detritivory
   Textbook Reference: *31.4 How Do Life Cycles Differ among Animals? p. 656*
7. Which of the following would likely be true of a species of bird with precocial young?
   a. Their eggs would have a small amount of yolk.
   b. They would have to care for the young for a long period of time.
   c. They would have a long incubation period.
   d. They would produce many small eggs.
   e. They would have a short incubation period followed by a long period of feeding the young.
   Textbook Reference: *31.4 How Do Life Cycles Differ among Animals? p. 656*
8. Which of the following statements about sponge structure or function is *false*?
   a. Choanocytes are flagellated cells that play a role in feeding.
   b. Large species are found in areas of heavy wave action, where food is most abundant.
   c. Individual sponges are both male and female.
   d. Water enters a sponge through pores and exits via one or more oscula.
   e. Sponges have an extensive extracellular matrix holding the cells together.
   Textbook Reference: *31.5 What Are the Major Groups of Animals? p. 659*
9. Which of the following statements about ctenophores and cnidarians is *false*?
   a. Most of them are marine organisms.
   b. Both have radial symmetry.
   c. Both have complete guts.
   d. Both have feeding tentacles.
   e. Fertilization generally takes place in open water.
   Textbook Reference: *31.5 What Are the Major Groups of Animals? pp. 661–663*
10. In what way does the mesogleal layer found in both ctenophores and cnidarians help these organisms survive, even when prey is scarce?
    a. Prey adheres to the sticky surface of the mesoglea.
    b. The mesoglea is biologically inert.
    c. The mesoglea allows the animal to capture large prey items.
    d. Mesogleal molecules provide nutrients to the host animal.
    e. The mesoglea houses photosynthetic bacteria that supplement the nutrition obtained from prey.
    Textbook Reference: *31.5 What Are the Major Groups of Animals? pp. 660–661*
11. Which of the following traits is *not* shared by sea anemones and jellyfishes?
    a. A medusa as the dominant stage in the life cycle
    b. Possession of a gastrovascular cavity

c. Sexual reproduction
d. Nematocysts on the tentacles
e. A primarily carnivorous diet
Textbook Reference: *31.5 What Are the Major Groups of Animals? pp. 661–663*

12. Which cnidarian group is dominated by species that live symbiotically with photosynthetic protists?
a. Hydrozoans
b. Ctenophores
c. Scyphozoans
d. Anthozoans
e. Placozoans
Textbook Reference: *31.5 What Are the Major Groups of Animals? p. 662*

13. Which of the following groups is matched *incorrectly* with a descriptor?
a. Mollusks: trochophore
b. Polychaete worms: trochophore
c. Crustaceans: nauplius
d. Flatworms: acoelomate
e. Earthworms: pesudocoelomate
Textbook Reference: *31.2 What Are the Features of Animal Body Plans? p. 650, Figure 31.4; 31.4 How Do Life Cycles Differ among Animals? p. 655*

14. Which of the following organisms is matched *incorrectly* with its feeding strategy?
a. Rabbit: herbivore
b. Earthworm: detritivore
c. Tiger: predator
d. Flamingo: carnivore
e. Tapeworm: parasite
Textbook Reference: *31.3 How Do Animals Get Their Food? pp. 651–654*

15. Which of the following is the group of animals about which we know the least?
a. Sponges
b. Ctenophores
c. Placozoans
d. Bilateria
e. Cnidaria
Textbook Reference: *31.5 What Are the Major Groups of Animals? p. 660*

Application Questions

1. Discuss two evolutionary trade-offs involving reproduction that animals are confronted with.
Textbook Reference: *31.4 How Do Life Cycles Differ among Animals? pp. 655–656*

2. Discuss the two major strategies animals use to get food, and indicate which of these would most likely be used by sessile organisms.
Textbook Reference: *31.3 How Do Animals Get Their Food? p. 651*

3. What are some advantages to segmentation?
Textbook Reference: *31.2 What Are the Features of Animal Body Plans? p. 649*

4. Explain the fundamental difference between protostomes and deuterostomes. What evidence supports dividing organisms into these two clades?
Textbook Reference: *31.1 What Characteristics Distinguish the Animals? p. 648*

5. What are some of the limitations imposed by a hydrostatic skeleton? Are the limitations the same for terrestrial animals?
Textbook Reference: *31.2 What Are the Features of Animal Body Plans? p. 650*

Answers

Diagram Exercise Answers

1. a. Radial symmetry
b. Bilateral symmetry
c. Bilateral symmetry
d. Radial symmetry
e. No symmetry
f. Radial symmetry

2. a. Protostomes
b. Deuterostomes
c. Unique cell junctions; collagen and proteoglycans in extracellular matrix
d. Choanocytes; spicules
e. Silicaceous spicules
f. Two embryonic cell layers
g. Distinct organ systems
h. Radial symmetry
i. Bilateral symmetry along an anterior–posterior axis; three embryonic cell layers
j. Blastopore develops into anus
k. Blastopore develops into mouth
l. Exoskeleton molting
m. Notochord
n. Radial symmetry
o. Glass sponges
p. Demosponges
q. Calcareous sponges
r. Placozoans
s. Ctenophores
t. Cnidarians
u. Arrow worms
v. Lophotrochozoans
w. Ecdysozoans
x. Echinoderms
y. Hemichordates
z. Chordates

3. a. Diploid
b. Haploid
c. Medusa (“jellyfish”)
d. Tentacles
e. Meiosis
f. Egg
g. Sperm
h. Fertilization
i. Fertilized egg
j. Tentacles
k. Mouth/anus
l. Planula larva
m. Polyp
n. Mature polyp
o. Young medusa
p. Mouth/anus

4. a. Diploid
b. Haploid
c. Medusa
d. Oral surface
e. Gonad
f. Meiosis
g. Egg
h. Sperm
i. Fertilization
j. Fertilized egg
k. Planula larva

Knowledge and Synthesis Answers

1. **d.** Not all animals have bilateral symmetry (e.g., sponges, cnidarians, ctenophores).
2. **d.** Deuterostome embryos have three layers, the ectoderm, the mesoderm, and the endoderm, making deuterostomes triploblastic, not diploblastic.
3. **c.** Animals eat foods from all kingdoms of life. Thus, they have evolved a wide variety of methods for acquiring diverse types of food. Segmentation patterns are a product of evolution, not the cause of evolution.
4. **d.** The body cavity of coelomates develops from the mesoderm and contains a peritoneum. The acoelomates lack a body cavity.
5. **c.** An animal that is sessile is stationary and does not move. All the other structures or characteristics are involved with movement in some way.
6. **b.** Because many parasites die when the host dies, parasites must have a way to disperse their progeny to new hosts.
7. **c.** Because precocial young hatch as well-developed individuals that can forage for themselves, they undergo more of their development in the egg and tend to be incubated for longer periods of time.
8. **b.** Because they are not structurally robust, large, upright sponges would be destroyed by heavy wave action.
9. **c.** Ctenophores and cnidarians are mostly marine, have radial symmetry, and are composed largely of mesoglea. Fertilization occurs in open water. Ctenophores have a complete gut with a mouth and two anal pores; cnidarians have a blind gut with only one opening.
10. **b.** The biologically inert mesoglea does not require energy; thus a large portion of the body needs no nutritional support, allowing ctenophores and cnidarians to maintain a very low metabolic rate and survive for an extended period of time with minimal amounts of food.
11. **a.** Although the jellyfish have both a medusa and polyp stage in their life cycle, the sea anemones have lost the medusa stage and spend their lives as polyps.
12. **d.** Anthozoans include the corals and sea anemones, both of which contain many species that live symbiotically with photosynthetic protists. Placozoans and Cnetophores are not cnidarians.
13. **e.** Earthworms are coelomates and have a body cavity within the mesoderm.
14. **d.** Flamingos are filter feeders that strain small organisms out of the mud using their beaks.
15. **c.** Placozoans are not easily observed in nature and were unknown until 1883. They are structurally simple, but very abundant in warm marine environments.

Application Answers

1. One trade-off is the production of many small eggs or several large eggs. The trade-off is the amount of offspring produced versus the amount of energy resources the offspring receives. Another trade-off is the production of altricial young that hatch early but require much parental care versus the production of precocial young that require a longer incubation period but little parental care after they hatch.
2. Animals can either move through the environment to where the food is located or they can move the environment and the food to themselves. Sessile organisms would most likely have adaptations to move the environment and food to themselves.
3. Segmentation provides several advantages, including facilitating specialization of body regions and improved muscle coordination, as each segment can be manipulated independently.
4. Protostomes and deuterostomes differ in the structure that becomes the anus and the structure that forms the mouth. The sequence in which these structures develop is also different. In protostomes, the mouth develops from the blastopore; in the deuterostomes, the anus develops from the blastopore. In both, the opposite structure forms later. These groups were first described according to observations of their developmental patterns, but sequence data supports this classification.
5. Hydrostatic skeletons provide for controlled movement of body parts. In terrestrial organisms, the surrounding environment does not support the body, so most organisms with hydrostatic skeletons are very small and soft bodied. Larger bodies are possible with exoskeletons or hard interior skeletons that provide structural support and protection to soft tissues.

32 Protostome Animals

The Big Picture

- The protostomes include two major groups: the lophotrochozoans and the ecdysozoans. The arrow worms are not placed in either group; they may be sister to the protostomes as a whole, or they may be more closely related to the lophotrochozoans. Protostomes have an anterior brain and a ventral nervous system. Many species in both groups have a wormlike appearance.
- The lophotrochozoans get their name from two structures: the lophophore (a feeding and gas-exchange structure found in a number of groups in this clade) and the trochophore larvae.
- A number of lophotrochozoan groups, including the annelids (segmented worms) and mollusks, undergo spiral cleavage during early development.
- The segmentation found among the annelid worms helps improve their locomotion. The unique molluscan body plan is based on a muscular foot structure, a visceral mass containing the organs, and a mantle that covers and protects the organs. Diverse adaptations of this plan are seen in the major molluscan groups.
- The ecdysozoans include more species than all other lineages combined. They are characterized by a rigid external covering (cuticle or exoskeleton) that must be molted periodically in order to allow the animal to grow.
- The wormlike ecdysozoans, which include the priapulids and the kinorhynchs, are unsegmented and have a thin cuticle. Horsehair worms and nematodes are also unsegmented, and have a tougher cuticle.
- The arthropods are characterized by a segmented body with a hard exoskeleton and jointed appendages. They are Earth's dominant animals in both number of species and number of individuals. The trilobites are extinct arthropods. Today four arthropod groups—crustaceans, hexapods, myriapods, and chelicerates—are found in all environments. The hexapods (which include the numerous and diverse insects), chelicerates, and myriapods are found in terrestrial and aquatic environments. The crustaceans are the dominant arthropods in marine environments.

Common Problem Areas

- With the advent of molecular phylogenetic studies (see Chapter 24), we have had to rethink many of our previous ideas about the classification of organisms. The arrow worms, for example, used to be considered deuterostomes, but have recently been placed among the protostomes, based primarily on molecular (i.e., gene sequence) studies. Compared to our understanding of deuterostome phylogeny (which you will study in Chapter 33), our conceptions of protostome phylogeny are in a state of flux and there is a great deal of disagreement among systematists about the ancestry and monophyly—and thus the classification—of many protostome groups.

Study Strategies

- Try to use the phylogeny presented in Figure 32.1 as a basis for learning about different groups. Understanding that certain characteristics are shared among related groups can help you remember which characteristics define each group.
- Sometimes learning the Latin root of a group can help you organize your thinking. Although the names are unfamiliar at first, they all make sense if you understand what they mean.
- The end of the chapter provides an overview of protostome evolution. Use the overview as a guide in studying the body forms and life cycles of the different groups.
- Go to yourBioPortal.com to review the following activities:

  Web Activity 32.1 Protostome Classification

  Web Activity 32.2 Features of the Protostome

Important Concepts

Protostomes are diverse and exhibit many body plans.

- After the origin of diploblastic animals, a third embryological germ layer (between the ectoderm and endoderm) evolved, called the mesoderm. Triploblastic animals that possess a mesoderm fall into one of two major clades: the protostomes and the deuterostomes.

- Although their body patterns vary extensively from group to group, protostomes have the following general characteristics:
  - The blastopore of the embryo develops into the mouth in almost all protostomes.
  - Protostomes have an anterior brain that surrounds the entrance to the digestive tract.
  - Protostomes have a ventral nervous system consisting of paired or fused longitudinal nerve cords.
- With the exception of the arrow worms, the protostomes are divided into two major clades based on DNA sequence analysis: the lophotrochozoans and the ecdysozoans (see Figure 32.1). A number of representatives in both groups have a wormlike body plan.
- The common ancestor of the protostomes had a coelom (a fluid-filled cavity within the mesoderm). However, the protostomes include some groups that are coelomate and some that are pseudocoelomate. One important group, the flatworms, is acoelomate (lacks a coelom). Two prominent protostome groups, the arthropods and the mollusks, have had secondary evolutionary modifications of the coelom. In the arthropods, the coelom has become a hemocoel; the mollusks have returned secondarily to a virtually open circulatory system.
- The lophotrochozoans share several characteristics.
  - Several (but not all) lophotrochozoans groups are characterized by a complex U-shaped structure, the lophophore, which is used both as a feeding apparatus and in gas exchange (see Figure 32.2). These groups are distantly related, and it appears that the lophophore has evolved independently several times. Nearly all animals that have a lophophore are sessile as adults.
  - Many groups of lophotrochozoans also have a type of free-living ciliated larva known as a trochophore (hence, "lophotrochozoans").
  - Several different lophotrochozoan lineages (e.g., flatworms, ribbon worms, annelids, and mollusks) exhibit a derived form of early development known as spiral cleavage. These are sometimes grouped into the spiralians, but gene sequence analysis indicates that the different spiralian groups are not monophyletic.
  - Many lophotrochozoans have a wormlike body plan; one group, the mollusks, are an exception.
- The ecdysozoans share several characteristics.
  - Ecdysozoans have a cuticle that provides both protection and support. These animals grow by molting their cuticles (*ecdysis* is the Greek word for "shedding"). Increasing molecular evidence, including a set of Hox genes shared by all ecdysozoans, supports the monophyly of these animals and suggests that cuticle molting is a lifestyle that may have evolved only once.
  - Ecdysozoans with a wormlike body plan often have a thin and flexible cuticle that allows for gas, water, and mineral exchange. Most of these ecdysozoans are confined to moist habitats.
  - Other ecdysozoans, most notably the arthropods, have a hard cuticle known as an exoskeleton containing many layers of protein and a strong, waterproof polysaccharide called chitin. An exoskeleton limits locomotion and gas exchange.
  - Jointed appendages controlled by muscles allow for rapid locomotion in the arthropod ecdysozoans. Over evolutionary time, different arthropod groups have experienced extensive modification of the appendages to suit life in a multitude of different environments. Arthropods are the dominant animals on Earth today, in both number of species and number of individuals.
- The arrow worms are an enigmatic group. Because of their early developmental morphology, they were once classed with the deuterostomes, but molecular evidence now clearly identifies them as protostomes. It is unclear whether the arrow worms are a sister group of the entire protostome clade, or whether they are most closely related to the lophotrochozoans. Arrow worms have no circulatory system and no larval stage. They are major predators of small organisms in the open ocean (see Figure 32.5).
- ***For review, go to Diagram Exercise 1.***

Many lophotrochozoans have unsegmented bodies.

- The 4,500 species of bryozoans are colonial; strands of tissue connect individuals in each colony, and in some species individuals are specialized for feeding, reproduction, defense, and support (see Figure 32.6). Individuals have a great deal of control in manipulating their lophophores to increase contact with prey (see Figure 32.2). Colonies grow via asexual reproduction of the founding members. Sexual reproduction also occurs, and larvae emerge to seek suitable sites to form new colonies. Bryozoans can cover large areas of coastal rock and can even form small reefs in shallow seas.
- Recent genomic studies indicate that the flatworms and rotifers are related, despite their structural differences.
  - Flatworms lack respiratory organs and have only simple cells for waste removal. The flat shape of the animal helps in oxygen transport and waste removal (see Figure 32.7). The digestive tract consists of a mouth that opens into a blind sac, which has many branches that aid in nutrient absorption. Although there are some free-living flatworm species, most are parasites, including about 25,000 species of tapeworms or flukes. Parasitic flatworms feed on the nutrient-rich body tissues of their host animal and disperse their eggs in the host's feces.
  - Most of the 1,800 species of rotifers are very small (some smaller than single-celled protists), but they have specialized internal organs, including a complete gut and a pseudocoel that serves as a hydrostatic skeleton (see Figure 32.8). Cilia are used to propel the rotifers through the water. Most species live in freshwater habitats and feed using a ciliated organ called a corona. Some species have both males

and females; some have only one sex and reproduce asexually. This is the only group of animals known to have existed for millions of years without the benefits of sexual reproduction.

- Nemerteans, or ribbon worms, have a complete digestive tract with two openings. Small ribbon worms use their cilia for movement, and large ribbon worms move by means of muscle contractions. The feeding organ of the ribbon worms is a proboscis that lies within a rhynchocoel, or fluid-filled cavity (see Figure 32.9). The proboscis has a sharp stylet and can be forcefully ejected from the body to catch prey. Most of the 1,000 or so species are marine, although a few species are found in fresh water or on land. Most are small, but some species can be up to 20 meters long. DNA sequence analyses indicate that they are closely related to the phoronids and brachiopods.
- The phoronids include 20 species of tiny sessile worms that live in chitinous tubes and extract food from the water with their lophophores (see Figure 32.10). While the lophophore is similar in function to that of the bryozoans, these groups are not closely related based on genetic analyses. This indicates that the lophophore structure has involved more than once.
- Brachiopods are solitary marine animals that live attached to the substratum. Their divided shell gives them a superficial resemblance to bivalve mollusks (e.g., clams), but the two halves are dorsal and ventral instead of lateral (see Figure 32.11). The lophophore is located in the shell, and cilia help draw water and food into the shell. More than 26,000 fossil brachiopod species have been described, but only about 335 species are known to exist today.

Annelids undergo spiral cleavage and have segmented bodies.

- The annelids consist of approximately 16,500 species of segmented worms living in marine, freshwater, and moist terrestrial environments. Their thin body wall serves as a surface for gas exchange, and they are restricted to these environments because the thin covering causes them to lose moisture rapidly when exposed to air. A segmented body plan gives these worms extremely good control of their movement. A separate nerve center called a ganglion controls the movement of each segment, and in most cases each segment also contains an isolated coelom (see Figure 32.12).
- More than half of all annelids are polychaetes, meaning "many hairs"; this term is descriptive, however, rather than the name of a clade. Annelids generally are found in marine environments in the sediment. Polychaetes can have more than one pair of eyes and tentacles (see Figure 32.13A). Outgrowths called parapodia used in gas exchange extend from segments laterally over much of the body. Setae extending from the parapodia help attach the animal to the substrate and aid in movement. A relatively recently discovered polycheate group, the pogonophorans, have secondarily lost their digestive tract and secrete tubes made of chitin and other substances that come from their surroundings (see Figure 32.13B). Pogonophorans are found in the deep ocean near hydrothermal vents. They harbor a number of endosymbiont bacteria in a specialized organ known as a trophosome, and the bacteria provide much of their nutrition.
- There are two major clades of clitellate annelids. The evolutionary relationship of the clitellates to the different polychaete groups is unclear, and the latter may be a paraphyletic group. They lack parapodia, eyes, or tentacles.
  - The oligochaetes ("few hairs") include the most familiar annelids, called the earthworms. Oligochaetes live mainly in freshwater and terrestrial environments and are hermaphroditic, containing both male and female reproductive organs in the same individual. Sperm is exchanged between two individuals (see Figure 32.13C).
  - The second clitellate group, the leeches, are also hermaphroditic species that live either in fresh water or on land (see Figure 32.13D). The coelom of these parasitic annelids is not segmented but is composed of undifferentiated tissue. Clusters of segments at their anterior and posterior ends are modified into suckers, which the leech attaches to the substratum for movement, or to a host mammal from which it sucks blood (its nutritional source). To keep the blood from clotting, the leech secretes an anticoagulant.
  - ***For review go to Diagram Exercise 2.***

The three-part molluscan body plan has undergone dramatic evolutionary radiation.

- Mollusks are another group that undergoes spiral cleavage. The mollusks have evolved into a morphologically diverse group based on a distinctive three-part body plan (see Figure 32.14).
  - All species have a large muscular foot. In some groups, such as the clams, this foot is a burrowing organ, while in squids and octopuses, the foot has been modified and consists of arms and tentacles; the tentacles bear complex sensory organs.
  - Organs such as the heart, the digestive tract, and the reproductive system are concentrated centrally in a visceral mass.
  - A tissue fold, known as the mantle, covers a visceral mass of internal organs. In many species, the mantle secretes a hard, calcareous shell. In most species, the mantle is extended to create a mantle cavity holding the gills used in respiration. Mollusks have a secondarily reduced coelom (see Figure 32.14).
- The blood vessels of the mollusks do not form a closed circulatory system. Instead, blood and other fluids empty into a hemocoel through which fluid moves around the animal to deliver oxygen to the internal organs. Mollusks have a heart that moves the blood back into the blood vessels.

- The molluscan body plan has resulted in a diverse array of some 95,000 species that fall into four major modern groups: the chitons, bivalves, gastropods, and cephalopods (see Figure 32.15).
  - Chitons are marine mollusks that feed on algae, bryozoans, and other organisms that they scrape off rocks using a razorlike body structure called the radula. A chiton's shell consists of eight overlapping plates that are surrounded by a girdle (see Figure 32.15A). They have simple internal organs, multiple gills, and bilateral symmetry.
  - The bivalve mollusks include the familiar clams, oysters, scallops, and mussels. They are found in both salt and fresh water, but they are all aquatic. Bivalves use an opening called an incurrent siphon to bring water into their two-part hinged shells; they are filter feeders that extract foodstuffs from these water currents. Large gills inside the shell extract the food and also function as respiratory organs. Water and gametes exit from an excurrent siphon. Among the clams, the molluscan foot has been modified into a digging device that allows the animal to burrow into the mud or sand.
  - Gastropods are the most species-rich and widely distributed of the mollusks, and are found in all environments. There are shelled and unshelled gastropod species, including the snails and slugs (the only terrestrial mollusks), as well as the marine nudibrachs (sea slugs), whelks, limpets, and abalones. Gastropods use their foot either to crawl or swim. Land snails and slugs are able to survive on land, as the mantle tissue is modified into a highly vascularized lung.
  - Cephalopods include the octopuses, squids, and nautiluses. The excurrent siphon is modified to give cephalopods the ability to control water movement into the mantle, allowing them to use ejected water as a jet for propulsion. They capture prey with their tentacles, and their greatly enhanced mobility makes them dominant ocean predators. They are also able to control gas movement in the mantle, which helps in buoyancy control. As is typical of active, rapidly moving predators, cephalopods have a head with complex sensory organs, most notably the eyes, which are in many ways comparable to those of vertebrates.
  - A fifth group, the monoplacophorans, was abundant 500 million years ago, but only a few species survive today. Unlike in other mollusks, the gas exchange organs, muscles, and excretory pores are present throughout the body.
  - ***For review, go to Diagram Exercise 3.***

Ecdysozoans are characterized by an external skeleton that does not grow.

- Ecdysozoans have an exoskeleton that ranges from thin and flexible to hard and rigid. Exoskeletons provide protection and support, but they present obstacles to growth, locomotion, and gas exchange.
- Ecdysozoans have overcome the obstacles of the exoskeleton to become the most numerous and diverse of all animal groups.
- Some marine ecdysozoans have a thin exoskeleton called a cuticle that is molted periodically. The cuticle allows for gas exchange with the environment, but it offers little protection. The priapulids, kinorhynchs, and loriciferans are groups of tiny, wormlike marine animals that live burrowed in ocean sediments (see Figure 32.16). Most priapulids have a larval stage; the kinorhynchs do not. The loriciferans were only recently discovered in 1983, and they are still being described.

  Nematodes, or roundworms, range in size from microscopic to up to 9 meters in length. About 25,000 species of this diverse group have been described, and these may represent less than a quarter of extant nematode species. They live in soil, on the bottoms of lakes and streams, and in marine sediments. Nematodes use their gut for both gas exchange and nutrient uptake (see Figure 32.17). Many species prey upon protists and other microscopic organisms, but of most significance to humans is the large number of parasitic species; several of these, including *Trichinella spiralis*, are dangerous to humans. The free-living nematode *Caenorhabditis elegans* is a widely used model organism for geneticists and developmental biologists.
- The 320 species of horsehair worms are long and very thin, as their name suggests. They live in fresh water or very damp soil near the edges of ponds and streams and feed in the larval stage as parasites of terrestrial and aquatic insects and crabs (see Figure 32.18). Adults have no mouth and have reduced guts. In some species the adults may not feed. The adults of other species continue to grow, so it is also possible that adult worms may absorb nutrients from the environment during the time between the molting of the old cuticle and the hardening of the new one.

Arthropods are Earth's dominant animals.

- Arthropods have a hard exoskeleton of chitin that provides protection, waterproofing, and muscle attachment sites. Their bodies are segmented, with individual muscles attached to the exoskeleton that operate each segment. The jointed appendages that give the group its name (*arthros* = joint, *poda* = limb) allow for a greater range of movement and the specialization of different appendages for different purposes. Chitin is waterproof and keeps the animal from dehydrating.
- The relationships among the arthropod groups are currently being revised based on gene sequence data. Current data supports the idea that arthropods may be monophyletic.
- Two groups are closely related to the arthropods, and one extinct group left an important fossil record.
  - The onychophorans (velvet worms) were once thought to be more closely related to the annelids, but recent genetic analysis links them to the arthropods. They have unjointed legs and thin cuticles composed

of chitin (see Figure 32.19A). Onychophorans have probably changed relatively little from their common ancestor with the arthropods.

- Like the onychophorans, the tardigrades have fleshy, unjointed legs and use their fluid-filled body cavities as hydrostatic skeletons (see Figure 32.19B). Tardigrades lack a circulatory system and do not have gas exchange organs. When their environment dries out, they shrink in size drastically and can survive in a dormant state for over a decade.
- The jointed appendages that characterize the arthropods appeared during the Cambrian, among a once-widespread group known as the trilobites ("three sections"). The trilobites died out during the great Permian mass extinction, but their heavy exoskeletons were readily fossilized and left an abundant record of their existence (see Figure 32.20). At least 10,000 species have been described based on the fossil record.

- Four major arthropod groups survive: the myriapods (centipedes and millipedes), the chelicerates (including the arachnids—spiders, mites, ticks, etc.), the crustaceans (crabs, lobsters, scallops, barnacles, etc.), and the hexapods (insects and their relatives).

Myriapods and chelicerates have only two body regions.

- The myriapods include 3,000 described species of centipedes and 11,000 described species of millipedes. Members of both groups have similar body plans, consisting of a well-formed head and a long, segmented trunk. Centipedes have one pair of legs on each trunk segment; millipedes have two pairs on each segment (see Figures 32.21A and 32.21B).
- The chelicerates include the pycnogonids, the horseshoe crabs, and the arachnids. All chelicerates have two body parts, and most have eight legs.
  - The pycnogonids, or sea spiders, are a group of about 1,000 exclusively marine species. Most are very small and are rarely seen (see Figure 32.22A). There are only four living species of horseshoe crabs. These animals have changed so little in morphology over evolutionary time that they are often referred to as "living fossils" (see Figure 32.22B).
  - The arachnids are the most prominent chelicerates and include the spiders, scorpions, mites, and ticks (see Figure 32.23). They have a simple life cycle, with young that resemble small adults. Spiders, the most familiar arachnids, build webs of protein threads that they use to capture prey. Spider webs are strikingly varied, often species-specific, and increase the predatory ability of spiders in many different environments. Mites and ticks are vectors for a variety of organisms that cause diseases in plants and animals.

Most marine arthropods are crustaceans.

- Crustaceans include many familiar animals (shrimp, lobsters, crabs, barnacles) as well as the less-familiar isopods, copepods, krill, and others. About 50,000 species have been described so far (see Figure 32.24).
- The crustacean body is divided into the head, thorax, and abdomen (see Figure 32.25). In many species, a carapace extends dorsally from the head to protect and cover the body. The thorax and abdomen have one pair of appendages each; crustacean appendages are specialized for walking, swimming, feeding, sensation, and gas exchange. In some species, complex branched appendages have evolved.
- Krill are oceanic and very abundant. They are an important food source for a variety of large marine vertebrates. Copepods are another important food source in the open oceans.
- Crustaceans may not be monophyletic, according to recent DNA sequencing. One group, the branchiopods, appears to be more closely related to the hexapods.
- Fertilized eggs remain attached to the body of the female during the early stages of development. Upon hatching, the young are released as larvae in some species. The typical crustacean larva, called a nauplius, has three pairs of appendages and one eye. With other species, the juveniles are similar in form to the adults. With some species, the fertilized eggs are released into the water or attached to an object.

Insects dominate the terrestrial environment.

- More than a million species of insects have been described so far, and many biologists believe this is only a small fraction of the species that actually exist. The major insect groups are listed in Table 32.2. Based on studies carried out in the tropical rainforest, Terry Erwin estimated that there may be up to 50 million different insect species on Earth. Even if this estimate is inaccurate, it is likely that many insect species have yet to be discovered.
- Like the crustaceans, the insect body has three parts: the head with a pair of antennae, the thorax with three pairs of legs, and the abdomen. Unlike in the crustaceans, the abdominal segments do not bear appendages (see Figure 32.26). Gas exchange occurs in a system composed of a series of air sacs and tubular channels called tracheae that extend from external openings called spiracles. Other distinguishing characteristics of insects include paired antennae with a sensory receptor known as Johnston's organs, three pairs of legs on the thorax, and external mouthparts.
- Three groups of wingless hexapods—the springtails, two-pronged bristletails, and proturans—are related to the insects and probably resemble the insect ancestral form most closely (see Figure 32.27). Members of these three groups differ from insects in having internal rather than external mouthparts. Springtails have a very simple lifestyle in which the juveniles resemble the adults and are probably the most abundant hexapod.
- There are two classes of insects: the wingless apterygotes and the winged pterygotes (some species of

which have secondarily become wingless). The apterygotes include the jumping bristletails and silverfish. Pterygotes typically have two pairs of wings attached to the thorax, but in some groups (e.g., parasitic lice and fleas, some beetles, and worker ants), one or both pairs of wings have been secondarily lost.

- Hatching insects do not look like the adults, and they undergo changes at each molt. Each stage between molts is called an instar. If the changes are gradual, the insect is said to undergo incomplete metamorphosis. If a drastic change occurs between some instars, then the insect is said to undergo complete metamorphosis. The most dramatic example is the change that occurs when a caterpillar transforms into a pupa in which the adult form develops. In insects that undergo complete metamorphosis, each stage is often specialized with regard to the environment and food source.
- Insects generally can be divided into three groups: those that cannot fold their wings over their bodies, those that can fold their wings and undergo incomplete metamorphosis, and those that can fold their wings and undergo complete metamorphosis (see Table 32.2 and Figure 32.29). Those that can fold their wings over their bodies are collectively known as the neopterans.
  - Insects that cannot fold their wings over their bodies include dragonflies and mayflies. The aquatic larvae can be predatory or herbivorous. This is an ancestral form for pterygotes. Dragonflies are active predators, while adult mayflies do not have a functional digestive tract.
  - Insects that undergo incomplete metamorphosis include grasshoppers and crickets, termites, stone flies, earwigs, thrips, true bugs, aphids, cicadas, and many others. In these groups, the hatchlings resemble small, usually wingless adults; as they molt from one stage to the next, they gradually acquire more adult characteristics.
  - Complete metamorphosis occurs in lacewings, beetles, caddisflies, butterflies, moths, flies, wasps, bees, and ants, among others. In these groups, the larvae and adult forms are substantially different; the young pass through at least two stages (larva and pupa) before becoming adults. They form a subgroup of the neopterans called the holometabolous insects and account for about 80 percent of the known insect groups.
- The evolution of wings probably occurred only once during insect evolution. Studies of homologous genes in insects and crustaceans suggest that insect wings are modified from a respiratory structure on the leg of an ancestral crustacean (see Figure 32.30).

Test Yourself

Diagram Exercises

1. In the diagram below, label the following structures: tentacles, mouth, anus, outer covering (chitinous tube), and gut. Is the organism a phoronid or an ecdysozoan?

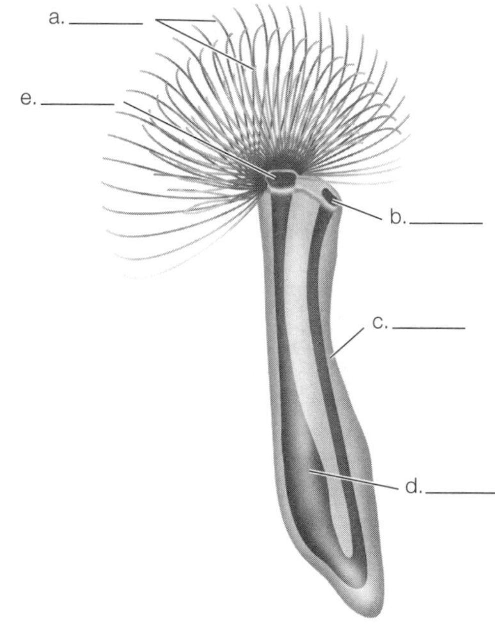

Textbook Reference: *32.2 What Features Distinguish the Major Groups of Lophotrochozoans? p. 674, Figure 32.10*

2. In the diagram below of an annelid segment cross section, label the following: blood vessel, intestine, circular muscle, longitudinal muscle, setae, septum, nerve cord, coelom, and excretory organ.

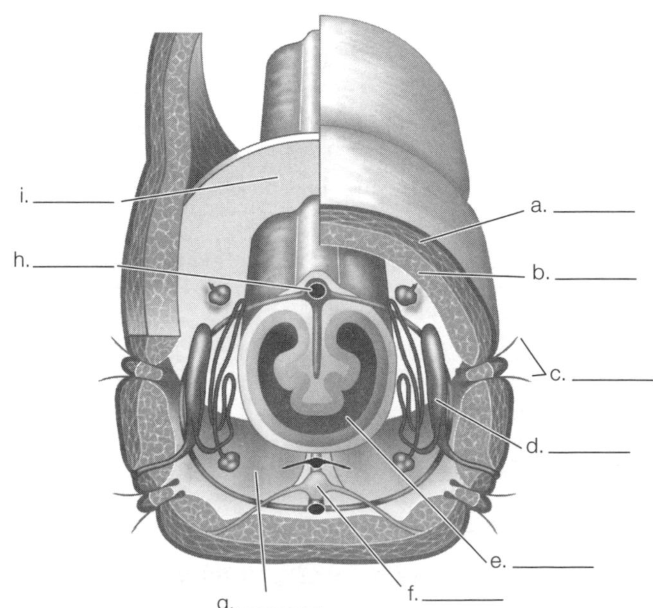

Textbook Reference: *32.2 What Features Distinguish the Major Groups of Lophotrochozoans? p. 675, Figure 32.12*

3. In the diagram below, first identify the cephalopod, chiton, bivalve, and gastropod. Then identify the following structures for each organism that has them: stomach, salivary gland, intestine, foot, anus, radula, siphon, gill(s), mantle, shell, shell plates, heart, mouth, digestive gland.

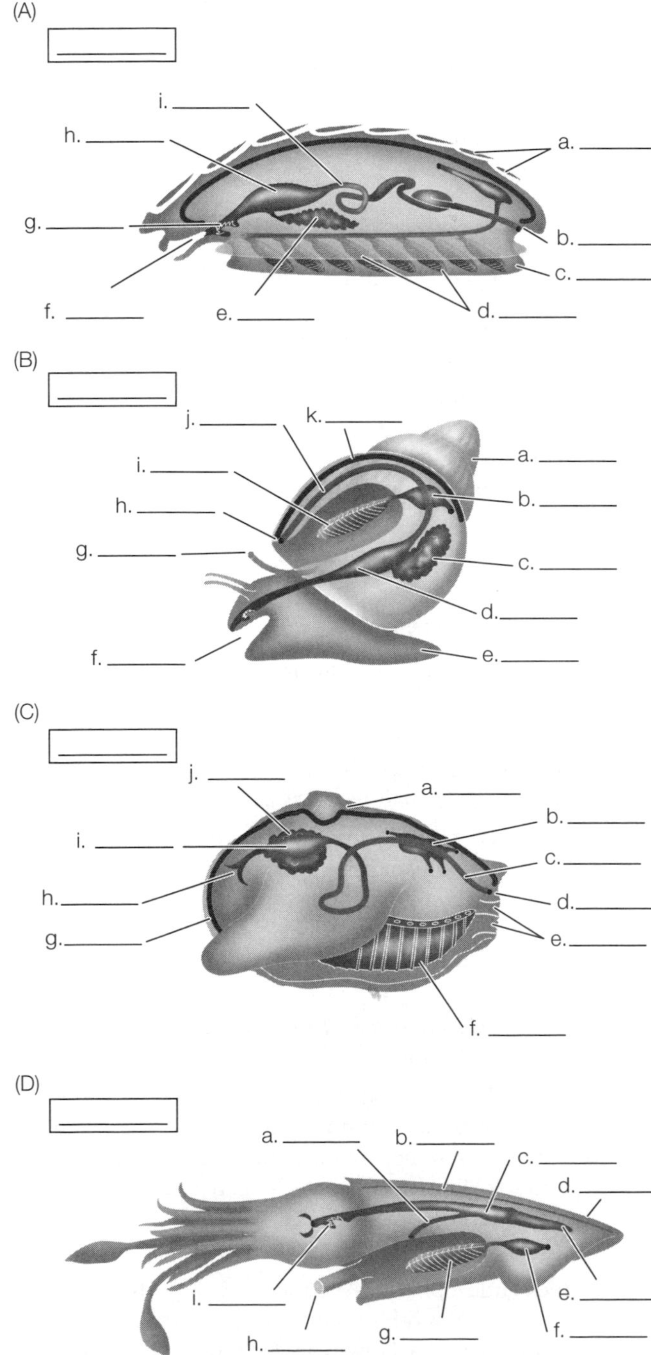

Textbook Reference: *32.2 What Features Distinguish the Major Groups of Lophotrochozoans? p. 677, Figure 32.14*

Knowledge and Synthesis Questions

1. The rhynchocoel is a body plan feature found only in
   a. bryozoans.
   b. flatworms.
   c. rotifers.
   d. ribbon worms.
   e. brachiopods.
   Textbook Reference: *32.2 What Features Distinguish the Major Groups of Lophotrochozoans? p. 673*
2. Which of the following does *not* possess lophophores?
   a. Bryozoans
   b. Phoronids
   c. Annelids
   d. Brachiopods
   e. All of the above have lophophores
   Textbook Reference: *32.1 What Is a Protostome? p. 668*
3. Which of the following statements about rotifers is *false*?
   a. They have a complete gut with an anterior mouth and posterior anus.
   b. They are coelomates.
   c. The corona is a ciliated organ used in acquiring food.
   d. They have a hydrostatic skeleton.
   e. They are very small, between 50 and 500 μm long.
   Textbook Reference: *32.2 What Features Distinguish the Major Groups of Lophotrochozoans? pp. 672–673*
4. A restaurant appetizer of escargot (snails), clams on the half shell, and calamari (octopus) would contain which types of mollusks?
   a. Chitons, bivalves, and gastropods
   b. Bivalves, gastropods, and cephalopods
   c. Chitons, gastropods, and cephalopods
   d. Bivalves and gastropods
   e. Chitons and cephalopods
   Textbook Reference: *32.2 What Features Distinguish the Major Groups of Lophotrochozoans? pp. 677–679*
5. The combination of a true coelom and repeating body segmentation allowed the annelids to
   a. evolve complex body shapes and control movement more precisely.
   b. move through loose marine sediments.
   c. become hermaphroditic.
   d. inject paralytic poisons into their prey.
   e. become larger than their ancestors.
   Textbook Reference: *32.2 What Features Distinguish the Major Groups of Lophotrochozoans? p. 675*
6. Lobsters, millipedes, and butterflies share which of the following traits?
   a. Parapodia
   b. Setae
   c. Gas exchange across the skin
   d. Jointed appendages
   e. Spiracles
   Textbook Reference: *32.4 Why Are Arthropods So Diverse? p. 681*

7. Which of the following have a specialized internal gas exchange system?
   a. Priapulids
   b. Arrow worms
   c. Hexapods
   d. Horsehair worms
   e. Annelids
   Textbook Reference: *32.4 Why Are Arthropods So Diverse? p. 686*
8. Which of the following insect groups is made up of species that undergo complete metamorphosis and can fold their wings back over their bodies?
   a. Coleoptera (beetles)
   b. Apterygota (silverfish and springtails)
   c. Orthoptera (grasshoppers and their kin)
   d. Odonata (dragonflies and damselflies)
   e. Ephemeroptera (mayflies)
   Textbook Reference: *32.4 Why Are Arthropods So Diverse? pp. 686–690*
9. Which of the following insect groups consists of winged species that undergo incomplete metamorphosis and can fold their wings back over their bodies?
   a. Coleoptera (beetles)
   b. Apterygota (silverfish and springtails)
   c. Orthoptera (grasshoppers and their kin)
   d. Odonata (dragonflies and damselflies)
   e. Ephemeroptera (mayflies)
   Textbook Reference: *32.4 Why Are Arthropods So Diverse? pp. 686–690*
10. Which of the following insect groups consists of winged species that cannot fold their wings back over their bodies?
    a. Coleoptera (beetles)
    b. Apterygota (silverfish and springtails)
    c. Orthoptera (grasshoppers and their kin)
    d. Odonata (dragonflies and damselflies)
    e. None of the above
    Textbook Reference: *32.4 Why Are Arthropods So Diverse? p. 689*
11. The majority of terrestrial arthropod species are
    a. myriapods.
    b. hexapods.
    c. cirripeds.
    d. tardigrades.
    e. onychophorans.
    Textbook Reference: *32.4 Why Are Arthropods So Diverse? p. 687*
12. The typical crustacean larva is called a(n)
    a. nauplius.
    b. pupa.
    c. instar.
    d. cyst.
    e. juvenile.
    Textbook Reference: *32.4 Why Are Arthropods So Diverse? p. 686*
13. Which of the following living arthropod groups has a long fossil history showing little morphological change?
    a. Pycnogonids (sea spiders)
    b. Arachnids (mites, ticks, and spiders)
    c. Horseshoe crabs
    d. Trilobites
    e. Onychophorans
    Textbook Reference: *32.4 Why Are Arthropods So Diverse? p. 683*
14. Which of the following ecdysozoan groups is extinct?
    a. Chelicerates
    b. Crustaceans
    c. Tardigrades
    d. Trilobites
    e. Onychophorans
    Textbook Reference: *32.4 Why Are Athropods So Diverse? p. 682*
15. Which of the following attributes is *not* seen in the arthropod body plan?
    a. Segmentation
    b. Jointed appendages
    c. A closed circulatory system
    d. A hard exoskeleton
    e. Complete digestive tract
    Textbook Reference: *32.4 Why Are Athropods So Diverse? p. 681*

Application Questions

1. The mollusks are a very diverse group morphologically, but they share three common morphological traits. Describe these traits and their functions.
   Textbook Reference: *32.2 What Features Distinguish the Major Groups of Lophotrochozoans? pp. 676–677*
2. The following are all characteristics of earthworms: segmented bodies, a spiral cleavage pattern, a terrestrial lifestyle, a complete gut, a blastopore that develops into a mouth, tight junctions between cells, and bilateral symmetry. In what order did these characteristics evolve?
   Textbook Reference: *32.1 What Is a Protostome? pp. 667–668; Figure 32.1*
3. Scientists have not found a fossilized body of the species that is the common ancestor of all bilaterally symmetrical animals. What is the likelihood that they will do so? How have scientists been able to learn about this common ancestor?
   Textbook Reference: *32.1 What Is a Protostome? p. 669*
4. A thick cuticle provides an animal with protection, but it also can seal it off from the world, obstructing the exchange of gases. How would you design a simple, hypothetical animal with a covering that both provided protection and allowed gas exchange?
   Textbook Reference: *32.4 Why Are Arthropods So Diverse? p. 686*

5. The letters *pter* occur in the names of many insect groups. Can you discover the Greek word that is the root of this nomenclature? (Hint: the same word is the root name for a well-known dinosaur group, the pterodactyls.) Most good dictionaries give the meanings of word prefixes and roots; using such a dictionary, determine the meaning of the following names of insect groups:
   Apterygota
   Ephemeroptera
   Orthoptera
   Isoptera
   Hemiptera
   Homoptera
   Coleoptera
   Trichoptera
   Lepidoptera
   Diptera
   Hymenoptera
   Siphonaptera

Answers

Diagram Exercise Answers

1. The organism shown is a phoronid.
   a. Tentacles
   b. Anus
   c. Outer covering (chitinous tube)
   d. Gut
   e. Mouth
2. a. Circular muscle
   b. Longitudinal muscle
   c. Setae (bristles)
   d. Excretory organ
   e. Intestine
   f. Nerve cord
   g. Coelom
   h. Blood vessel
   i. Septum between segments
3. (A) Chiton
   a. Shell plates
   b. Anus
   c. Foot
   d. Gills in mantle cavity
   e. Digestive gland
   f. Mouth
   g. Radula
   h. Stomach
   i. Intestine

   (B) Gastropod
   a. Shell
   b. Heart
   c. Salivary gland
   d. Stomach
   e. Foot
   f. Mouth
   g. Siphon
   h. Anus
   i. Gill
   j. Intestine
   k. Mantle

   (C) Bivalve
   a. Shell
   b. Heart
   c. Intestine
   d. Anus
   e. Siphons
   f. Gill
   g. Mantle
   h. Mouth
   i. Stomach
   j. Digestive gland

   (D) Cephalopod
   a. Intestine
   b. Shell
   c. Stomach
   d. Mantle
   e. Digestive gland
   f. Heart
   g. Gill
   h. Siphon
   i. Radula

Knowledge and Synthesis Answers

1. **d.** The rhynchocoel is a fluid-filled cavity containing a proboscis, a feeding organ found in the nemerteans.
2. **c.** Annelids belong to the spiralian lineage, not the lophophorates.
3. **b.** Rotifers are pseudocoelomates and have a pseudocoel.
4. **d.** Snails are in the Gastropoda, clams are in the Bivalvia, and octopuses are in the Cephalopoda.
5. **a.** The segmentation of the annelids allows for more complex, coordinated movement.
6. **d.** All of these animals have jointed appendages.
7. **c.** The Hexapoda have an internal system of tracheae for gas exchange.
8. **a.** The Coleoptera, or beetles, undergo complete metamorphosis and are able to fold their wings over their bodies.
9. **c.** The Orthoptera (grasshoppers, crickets, roaches, etc.) undergo incomplete metamorphosis and are able to fold their wings over their bodies.
10. **d.** The Odonata (dragonflies and damselflies), along with the Ephemeroptera (mayflies), have wings that they cannot fold over their bodies. Apterygotes (silverfish and springtails) are wingless.
11. **b.** The 1 million species of insects that have been described so far are only a small fraction of all insect species.
12. **a.** The larvae of crustaceans are known as nauplius larvae.
13. **c.** Horseshoe crabs have changed little during their fossil history; they are sometimes referred to as living fossils.
14. **d.** The Chelicerata include spiders, sea spiders, and horseshoe crabs; the Crustacea include decapods (crabs, shrimp, and others), barnacles, isopods, and copepods; the Tardigrada are a small (but not extinct) group also known as water bears. Onychophorans (velvet worms) are not extinct. Of the groups listed, only the Trilobita are extinct.
15. **c.** Arthropods have a hard exoskeleton, a segmented body, jointed appendages, a complete digestive tract, and an open circulatory system (see Table 32.1).

Application Answers

1. The mollusk body plan has three morphological components shared by all molluscan groups. The first trait is a muscular foot used for locomotion. In

cephalopods the foot has been modified into arms and tentacles. The second trait is a mantle, which covers the third trait, the visceral mass. The mantle secretes a calcareous shell in many species and often houses the gills. The visceral mass consists of the internal organs.

2. The first of these characteristics to evolve was the tight junction. This type of cell junction occurs in all animals and was probably present in the common ancestor of all animals. The evolution of a complete gut came next, followed by bilateral symmetry (see Figure 32.1). After the evolution of bilateral symmetry, the phylogenetic tree splits into the deuterostomes and protostomes; as protostomes, the annelids have a mouth that develops from their blastopore. The next characteristic to evolve was a segmented body. Since the annelids are overwhelmingly aquatic, and even terrestrial annelids are confined to moist habitats, it seems likely that the terrestrial lifestyle was the most recent of the listed characteristics to evolve.
3. The vast majority of species that have lived on Earth have gone extinct without leaving fossils, so it is very unlikely that scientists will ever discover a fossilized body of the common ancestor of all bilaterally symmetrical animals. Some fossilized animal tracks from the Precambrian appear to have been made by a bilaterally symmetrical animal, but these offer limited information. The best source of information about the common ancestor of any group of animals is the set of characteristics that are common to all members of that group. All bilaterally symmetrical animals share a triploblastic organization and certain Hox genes, and it is probable that these were also present in the common ancestor of this group.
4. The animal would need special mechanisms for gas exchange with the environment. The insects have overcome this problem with their system of tracheae, a network of tubes found throughout the body that is in close contact with all the cells of the animal and opens to the environment. In an aquatic environment, the animal could have an internal chamber with gills used for respiration.
5. As a root or prefix, *pter* comes from the Greek word *pteris* and refers to wings.

   Apterygota: The prefix *a-* in this case means *without*. Apterygote insects have no wings.

   Ephemeroptera: *Ephemeral* means *short-lived*. Adult mayflies typically live only a few hours or days.

   Orthoptera: *Ortho-* means *straight*. Orthopterans tend to have very straight wings (see Figure 32.29A).

   Isoptera: *Iso-* means *equal* or *uniform*. All four wings on termites tend to be very similar.

   Hemiptera: *Hemi-* means *half*. In many true bugs, the front half of the forewing is hardened, and they appear to have half a shell (like a beetle's) and half wings.

   Homoptera: *Homo-* means *same*. Unlike their close relatives the hemipterans, homopterans have forewings and hind wings that are similar to each other.

   Coleoptera: *Coleo-* means *sheath*. The forewings of beetles are modified into a protective sheath that usually covers the hind wings and abdomen.

   Trichoptera: *Tricho-* means *hair*. The bodies and wings of adult caddisflies are covered in small hairs.

   Lepidoptera: *Lepido-* means *scale*. Butterfly and moth wings are covered with tiny scales.

   Diptera: *Di-* means *two*. Unlike most winged insects, flies have only two wings.

   Hymenoptera: *Hymen-* means *membrane*. Bees, wasps, and the reproductive castes of ants have membranous wings.

   Siphonaptera: *Siphon-* means *tube* or *pipe*. This is a reference to the flea's feeding method. Note that this name ends in *-aptera*, not *-optera*: Fleas do not have wings (*a-* means *without*), though as pterygotes they are descended from winged ancestors, having secondarily lost their wings over the course of evolution.

33 Deuterostome Animals

The Big Picture

- The deuterostomes are not as diverse or numerous as the protostomes. Most fit into one of two groups, the echinoderms and the chordates.
- Echinoderms have an internal skeleton of calcified plates and a water vascular system used in feeding and gas exchange. They include the sea stars (starfish), sea lilies, and sea cucumbers, all of which exhibit pentaradial symmetry as adults.
- The chordates have several shared traits: pharyngeal slits, a nerve cord, a ventral heart, a tail beyond the anus, and a notochord. The chordates consist of the ascidians, lancelets, and vertebrates. Vertebrates have a dorsal vertebral column that supports a rigid endoskeleton.
- The most numerous vertebrates, both in terms of number of species and number of individuals, are the fishes. The amphibians adapted to life on land, but their life cycle requires them to return to water to reproduce.
- The evolution of the amniote egg freed reptiles from water-based reproduction. During the Carboniferous, amniotes diverged into two major clades, the reptiles and the mammals. Reptiles were the dominant vertebrates for many millennia. One reptilian group gave rise to the now-extinct dinosaurs and to the birds. Among the birds, the reptilian scales evolved into feathers, allowing for flight.
- The extinction of the dinosaurs made the radiation of the mammals possible. Most mammals belong to one of 20 major groups of eutherians. The largest eutherian group is the rodents; the primate eutherians include the lemurs, monkeys, apes, and humans.
- Humans (genus *Homo*) are primates that evolved from bipedal australopithecine ancestors on the African continent. Several species of *Homo* arose and went extinct over evolutionary time. In the lineage leading to *Homo sapiens*, the only surviving species, brain size expanded greatly relative to body size, allowing humans to develop culture and learn language.

Common Problem Areas

- As you work your way through the phylogenies presented in the textbook, try to recall the representative organisms that you already know from each group and what their special traits are. If you base your learning on information with which you are already familiar, you will have an easier time learning the facts.

Study Strategies

- Many students are confused by the evolution of the chordates and vertebrates and the traits that separate the vertebrates from other chordates. Create a timeline with all the major chordate classes and insert the specific traits that distinguish one class from another. This will help you organize the evolution of the chordates so that memorizing it becomes a manageable task.
- Go to yourBioPortal.com to review the following tutorial and activities:

  Animated Tutorial 33.1 Life Cycle of a Frog

  Web Activity 33.1 Deuterostome Phylogeny

  Web Activity 33.2 Amniote Egg

  Web Activity 33.3 Deuterostome Classification

Important Concepts

The deuterostomes include the echinoderms, hemichordates, and chordates.

- There are many fewer species of deuterostomes than there are of protostomes, but the deuterostomes are of special interest because mammals—a group that includes the largest living animals as well as the human lineage—are deuterostomes.
- Historically, deuterostomes were classified into one group based on early developmental patterns. The development of the blastopore into the anus is a shared trait, although the development may be ancestral for bilaterians in general. The strongest support for a shared common ancestor comes from phylogenetic analysis of DNA gene sequences.

- Deuterostomes fall into three major clades: the echinoderms, the hemichordates, and the chordates (see Figure 33.1). All are triploblastic, coelomate animals. Skeletal support features, when present, are internal. A few species have segmented bodies, but the segments are not visible. The chordates are characterized by a structure called the notochord and a dorsal nerve cord, and are further divided into three groups: the cephalochordates, the urochordates, and the vertebrates.
- Fossil beds, most of them from a 520-year-old site in China, have revealed insights into deuterostomate ancestry. Fossils and phylogenetic analysis of living species indicate that the earliest deuterostomes were bilaterally symmetrical, segmented animals with a slitted pharynx and external gills (see Figure 33.2). The unique pentaradial symmetry of the echinoderms likely evolved later, while other deuterostome groups retained bilateral symmetry.

The echinoderms and hemichordates are collectively known as the ambulacrarians.

- While 23 major groups consisting of about 13,000 species of echinoderms have been described from the fossil records, only six groups containing 7,000 species live today. All known species live in marine environments. Only 100 species of hemichordates are known to exist.
- Although echinoderm larvae have bilateral symmetry, the echinoderm adults display a unique pentaradial symmetry (radial symmetry in fives or multiples of five). They have no heads. Echinoderms move slowly but can move equally well in any direction. They do not have anterior–posterior (head–tail) body organization, but rather an oral-aboral orientation: the bottom-facing side contains the mouth while the top-facing surface contains the anus (see Figure 33.3).
- In addition to their pentaradial symmetry, echinoderms are unique in possessing a system of water-filled canals called a water vascular system. The canals end in external structures called tube feet. The water vascular system functions in feeding, gas exchange, and locomotion. Echinoderms also have an internal skeleton made up of calcified plates.
- Echinoderms are divided into two major clades. The first of these, the crinoids, includes about 80 species of sea lilies and feather stars (see Figure 33.4A). The body form is a cup-shaped structure and has anywhere from five to several hundred arms. The arms have tube feet that catch food particles. Cilia on the tube feet move the food to a groove that leads to the mouth. The tube feet also provide a surface for respiratory gas exchange and nitrogenous waste excretion.
- The remaining echinoderms are generally mobile and are divided into two main groups: the echinozoans (sea urchins and sea cucumbers; see Figures 33.4B and 33.4C) and the asterozoans (sea stars and brittle stars; see Figures 33.4D and 33.4E).
  - Sand dollars and sea urchins feed on algae and organic debris. Neither have arms, but the sea urchins have spines of varying shapes and sizes that are attached to the skeleton with ball-and-socket joints. Spines provide effective protection. Sand dollars are flat, disc-shaped relatives.
  - Sea cucumbers have tube feet that are used for anchoring the animal to the substrate rather than for moving. The anterior tube feet are large, feathery, sticky tentacles that are used to capture prey. Sea cucumbers are unique among the echinoderms in that they have an anterior–posterior body organization.
  - The sea stars, or starfish, are major predators, feeding on marine animals including polychaetes, mollusks, crustaceans, and fish. Their tube feet are used for locomotion, gas exchange, and attachment. Each tube foot has an internal ampulla that is connected to an external suction cup by a muscular tube. With so many tube feet, sea stars can capture large prey. Some species use their tube feet to pry open the shells of bivalve mollusks. Once the bivalve is open, the sea star inserts its stomach into the shell and digests the mollusk's internal organs.
  - Brittle stars have five flexible arms that are made of hard, jointed plates. Most feed on surface sediments and ingest the accompanying organic material through their single opening to the digestive tract. Some species can filter food particles from the water, and others are able to capture small animals.
- The hemichordates include the acorn worms and the pterobranchs. Both groups have a body plan composed of a trunk, a collar, and a sticky proboscis used for catching prey and digging (see Figure 33.5A). In the acorn worm, cilia move food from the proboscis to the mouth. Behind the mouth are pharynx and an intestine. The pharynx can open to the outside through pharyngeal slits that contain vascularized tissue for gas exchange. Acorn worms breath by pumping water through the mouth and out the pharyngeal slits. They burrow in sand or mud and extract food items from the substrate.
- Pterobranchs are sedentary marine animals that are either solitary or colonial (see Figure 33.5B). Behind the proboscis are one to nine pairs of arms that are used to capture prey and also function in gas exchange.

All chordates have a notochord at some point in their development.

- The features that reveal the evolutionary relationship between the echinoderms and chordates, as well among the chordates, are primarily observed in the larval stage or during the early states of development.
  - All three chordate clades share the following at some point during development: (1) A hollow dorsal nerve

cord; (2) a tail that extends beyond the anus; (3) a dorsal supporting rod called the notochord; and (4) pharyngeal slits.

- The notochord is the main distinguishing characteristic of the chordates. In urochordates, the notochord is lost during the metamorphosis from larval to adult forms. In vertebrates, the notochord is replaced by skeletal structures called the vertebrae (spinal column).
- Ancestral pharyngeal slits are generally lost in adults, but are retained in some urochordates and cephalochordates (see Figure 33.6). A pharynx develops around the slits and becomes large in some chordates, forming a pharyngeal basket.

- Most chordates also have some form of a ventral heart.
- The 30 species of cephalochordates (lancelets) are small, fishlike animals that retain the notochord for their entire life and use their pharyngeal baskets to catch prey (see Figure 33.6). Fertilization of eggs takes place in the water.
- The three major urochordate groups are the ascidians (sea squirts, also known as tunicates), thaliaceans, and larvaceans. All members of these three groups are marine, and more than 90 percent of urochordate species are ascidians. The body form for an adult ascidian is baglike in shape and surrounded by a tough tunic (hence the name "tunicate") (see Figure 33.7A).
- Ascidian larvae have pharyngeal slits, a nerve cord, and a notochord; it is the larva, not the adult form, that suggests their chordate ancestry. The adults lose the notochord and nerve cord but have an enlarged pharynx—the baglike pharyngeal basket—for catching food. Ascidians sometimes reproduce asexually by budding, thus forming colonies.
- Thaliaceans can live as a single individual or in colonies (see Figure 33.7B). They can live in tropical waters at depths up to 1,500 meters.
- Larvaceans retain a notochord and dorsal hollow nerve cord throughout adult life. They are solitary and small, but some species that live near the bottom of the ocean surround themselves with an extensive casing of mucus more than a meter wide. They capture organic particles with filters built into the mucus.

Vertebrates evolved an internal skeleton.

- With the exception of some of the jawless fishes, the vertebrate body plan can be categorized by the following traits: (1) a jointed, dorsal vertebral column that replaces the notochord and supports an internal skeleton with two pairs of appendages; (2) an anterior head with a large brain and protective skull; (3) a closed circulatory system; and (4) a large coelom in which the internal organs are suspended. This body plan is capable of supporting large-sized animals (see Figure 33.10).
- The nonvertebrate deuterostomes are primarily marine. The group that led to the vertebrates is also thought to have evolved in a marine environment, possibly in estuarine waters (places where fresh and salt water mix).
- The vertebrates are divided into two groups: the jawless fishes and the gnathostomes (see Figure 33.8 for the phylogenetic tree that shows the evolutionary relationships among the vertebrates).
- The elongate, eel-like hagfishes are the sister group of the remaining vertebrates (see Figure 33.9A). These jawless fishes look superficially similar to the lampreys, which are also jawless, but hagfishes lack true vertebrae and have only a partial cranium (skull). Lampreys have true vertebrae (see Figure 33.9B). The hagfishes also have a weak circulatory system with three small accessory hearts; they lack a stomach, and their skeleton is composed of cartilage. Although they lack a jaw, they have a tonguelike structure with toothlike rasps that they use to capture prey. There is some debate as to the accuracy of placing of the hagfishes with the other vertebrates in the same phylogenetic tree. The analyses of gene sequences suggest that the hagfishes and lampreys are related, and it is possible that during evolution the ancestor of the hagfishes lost some of the structural characteristics that define vertebrates.
- Lampreys resemble hagfishes, but they differ biologically. They have a complete braincase and distinct vertebrae made of cartilage. Unlike hagfishes, which undergo direct development, lampreys undergo complete metamorphosis. The larvae, called ammocoetes, are filter feeders, while the adult forms are often parasitic. The mouth is a rasping, sucking organ used by the lampreys to attach themselves to prey. The adults of a few lamprey species do not feed as adults, surviving only long enough to breed.
- Jawless fishes were dominant during the Devonian, but during that time jaws evolved via modifications in the skeletal arches supporting the gills (see Figure 33.11A). This led to greatly enhanced feeding efficiency. Today, the 100 or so species of hagfishes and lampreys are the only surviving jawless vertebrates.
- The remaining vertebrates fall into the category of gnathostomes, or "jaw mouths." The evolution of the jaw—and then of teeth—led to greatly enhanced feeding efficiency (see Figure 33.11B). Jaws provided access to new food sources, and teeth made for extremely effective predation. Jawed fishes were the major predators of the Devonian. The evolution of unjointed appendages called fins helped control movement through the water.
- The chondrichthyan fishes, including sharks, skates, rays, and chimaeras, have flexible and leathery skin and a skeleton composed of cartilage (see Figure 33.12). Sharks move forward by laterally undulating their body and tail (caudal) fin; skates and rays move vertically by undulating their pectoral fins. Almost all chondrichthyans are marine, but a few live in estuarine waters or migrate to lakes and rivers.

- Lunglike, gas-filled sacs called swim bladders evolved in the ancestral fishes of the bony fishes. With the aid of the swim bladder and fins, many fish can control their position in the water column with minimal energy expenditure.
- The ray-finned fishes and virtually all other vertebrates have a bony skeleton. The outer body is covered by thin, lightweight scales, which aid in movement and provide protection. A hard flap called the operculum, which covers the gills, allows for the movement of water over the gills, and thus for gas exchange. The approximately 30,000 species of ray-fins encompass a remarkable variety of shapes, sizes, and lifestyles, and they exploit every kind of food source in the aquatic environment (see Figure 13.13). Some ray-finned fishes are solitary and some form aggregate groups in open waters called schools. The eggs of many fishes tend to sink, which is why the shallow coastal waters and estuary environments are important to their life cycles. Fishes such as salmon are anadromous, meaning that they live in salt water but must move to fresh water to spawn.
- ***For review, go to Diagram Exercise 1.***

The evolution of jointed fins aided the colonization of land by vertebrates.

- The evolution of lunglike swim bladders in some ray-finned fishes set the stage for the move to the land. Some fishes may have used these sacs to supplement their oxygen supply in low-oxygen freshwater environments, and the sacs may have allowed some fishes to survive out of water. The fins of these fishes were unjointed, so they could only flop around on land.
- Jointed fins evolved in the ancestors of the coelocanths and the lungfishes, which along with the tetrapods are known as sarcopterygians.
  - Coelacanths were thought to have died out 65 million years ago, but two species were discovered in the twentieth century. One genus, *Latimeria,* is a predator and can weigh up to 80 kilograms (see Figure 33.14A). Unlike other fishes, *Latimeria* has a skeleton composed primarily of cartilage; this is thought to be a derived feature, since the ancestors to this group had bony skeletons.
  - Only six species of lungfishes remain, all of them in the southern hemisphere (see Figure 33.14B). The jointed fins of lungfishes are connected to the body by a single large bone, and they have lungs as well as gills. During dry periods when the water from ponds evaporates, they burrow deep in the mud and can survive in an inactive state for many months, breathing air.
- The change in fin structure allowed these fish to support themselves in shallow water and, eventually, to move onto land and exploit food sources there. These early land-dwelling fishes gave rise to the tetrapods—the four-legged vertebrates. In 2006, an important fossil was discovered that is believed to be an intermediate between the fishes and the tetrapods (see Figure 33.14C). The pectoral fins have skeletal structures found in tetrapod limbs that may have allowed for front-to-rear movement needed for walking and making brief trips out of the water.
- Amphibians evolved from a common ancestor shared with sarcopterygian fishes. Their walking legs allowed them to exploit the new dry territory. However, most spend at least part of their life cycle in the water (see Figure 33.15). Many species return to water to lay their eggs, producing aquatic larvae. The skin surface is a common site of respiratory gas exchange in amphibians, but many also have lungs.
- There are approximately 6,500 species of extant amphibians, falling into three groups:
  - The caecilians are wormlike, legless, burrowing or aquatic amphibians found only in moist tropical regions (see Figure 33.16A).
  - Anurans include the tail-less frogs and toads, and account for the vast majority of amphibian species (see Figure 33.16B). Anurans undergo metamorphosis from an aquatic to a terrestrial life form. The adults have a very short vertebral column and a pelvic region modified for hopping on the hind legs. Some species have adapted to life in very dry, even desert environments, while others have returned to an entirely aquatic lifestyle.
  - Salamanders are tailed amphibians that exchange respiratory gases through both lungs and gills (see Figures 33.16C and 33.16D). One group relies on gas exchange through its skin and mouth as all amphibians do, but it does not have lungs. Paedomorphic evolution (retention of the juvenile form in the adult) has led to several entirely aquatic salamander species. Most species have internal fertilization. Sperm is transferred in a jellylike capsule called a spermatophore.
- The social behaviors of many amphibians are quite complex; males make species-specific calls to females and can defend their own breeding territories. The amount of care given to the eggs varies. Some amphibians lay large numbers of eggs that are abandoned, while others guard a few fertilized eggs. A few species are viviparous, meaning that the adult gives birth to well-developed juveniles.

The amniote egg allowed tetrapods to invade dry environments.

- Amphibians are largely confined to moist environments by their need to reproduce in water. In one ancestral lineage, the amniote egg evolved (see Figure 33.17). The egg has a protective calcium-based shell that inhibits dehydration while allowing oxygen and carbon dioxide to pass through. Within the shell, extraembryonic membranes further protect the embryo, and large quantities of yolk provide nutrition. The amniote animals—reptiles (including birds) and mammals—were thus freed from reliance on water to reproduce.

- In several groups of amniotes the egg became modified to allow the embryo to grow inside the mother. In mammals, the egg has lost the shell and the embryonic membranes have been retained and modified (see Figure 33.18).
- Other evolutionary adaptations to life on land appeared among the amniotes, including a tough, impermeable skin covered with scales or modified scales (i.e., hair or feathers). The excretory systems of amniotes also evolved adaptations that allowed these animals to excrete nitrogenous wastes in the form of concentrated urine with a minimal loss of valuable water.
- During the Carboniferous period, about 250 mya, the amniote animals diverged into two groups, the reptiles and the mammals. Birds, the only living members of the now-extinct dinosaurs, would evolve from one of the reptilian groups (see Figure 33.18).
- ***For review, go to Diagram Exercise 2.***

Reptiles and birds belong to a monophyletic group.

- One relatively small reptilian group, the turtles and tortoises, has a unique body plan that has changed very little over the millennia (see Figure 33.19A). These animals are characterized by dorsal and ventral bony plates that evolved from the ribs to form a protective shell. The relationship of turtles and tortoises to other reptiles is not clear. For example, it is not known how the pectoral girdles, unlike in other vertebrates, evolved to be inside the ribs.
- The lepidosaurs include the squamates (lizards, snakes, and amphisbaenians—another group of legless, wormlike burrowers) and the tuataras (See Figures 33.19B–33.19D). Their body is covered in scales that greatly reduce the loss of water but make their skin surface unavailable for gas exchange. Gases are exchanged through the lung, which is larger in surface area compared to that of the amphibians. The lepidosaurs have a three-chambered heart that partially separates oxygenated from deoxygenated blood. The limbless condition of snakes is a product of secondary evolution. Most lizards and all snakes are carnivores, and adaptations to the jaws of snakes allow them to swallow prey much larger than themselves. Tuataras are represented by only two living species.

Birds and crocodilians share a common ancestry with the dinosaurs.

- The archosaurs include the extant crocodilians (crocodiles and alligators), the extinct dinosaurs, and the "living dinosaurs" we know as birds (see Figure 33.18). The crocodilians—crocodiles, alligators, caimans, and ghalials—live only in tropical and warm temperate regions. They spend most of their lives in water, but build nests on land. They are all carnivores, feeding on other vertebrates, including large mammals.
- Dinosaurs arose among the reptiles around 215 mya and survived for about 150 million years. Both the fossil record and molecular evidence support the position of birds as a sister group of the saurischian dinosaurs.
- Birds probably evolved from an ancestral theropod, a bipedal, predatory dinosaur that appears to have had hollow bones, a furcula (wishbone), limbs with three digits, and a backward-thrusting pelvis.
- Living birds diverged from the common flying ancestor and fall into one of two groups. The palaeognaths are flightless and include rheas, emu, kiwis, cassowaries, and the largest bird, the ostrich (see Figure 33.20B). The neognaths, most of which have retained the ability to fly, represent the remaining 9,600 species of birds.
- Recent fossil discoveries in China have demonstrated that some dinosaurs had scales that were highly modified to form feathers. In one species, *Microraptor gui*, the feathers were structurally similar to those of modern birds (see Figure 33.21A).
- *Archaeopteryx* lived 150 mya and represents the oldest known fossil of a bird. It was covered in feathers and had well-developed wings and a wishbone (see Figure 33.21B). It also had teeth, which were lost in later members of this lineage.
- Feathers are complex and strong, but also lightweight (see Figure 33.22). Their evolution was a major factor in diversification. The bones of dinosaurs and birds are hollow, and along with the development of feathers, they assisted in the evolution of flight.
- Along with the ability to fly came a high metabolic rate needed to fuel flight. The metabolic rate of birds means that they generate a great amount of heat, and their feathers are adapted to allow for heat loss. The lungs of birds also function differently from those of other vertebrates, another adaptation to the needs of flight. Different bird groups feed on many different types of animal and plant material, including carrion. Birds that feed on fruits and seeds are major agents of plant dispersal.
- ***For review, go to Diagram Exercise 3.***

Mammals radiated after the dinosaurs went extinct.

- The earliest mammals lived side by side with reptiles from the first split of the two lineages early in the Mesozoic era. However, only after the large dinosaurs became extinct did mammals begin to flourish in size and number.
- A number of unique traits distinguish the mammals. Sweat glands in the skin produce secretions that are crucial to cooling the body. Adaptations to the sweat glands may have led to the mammary glands that provide nutrient fluid for their young and that give the group its name. Modifications of scales became the external body hair that protects and insulates most mammals. Mammals such as cetaceans and humans have a greatly reduced amount of hair. In cetaceans, thick layers of insulating blubber protect against heat loss. A four-chambered heart that completely separates oxygenated from deoxygenated blood evolved in the

mammals. It also evolved convergently (i.e., separately) in the crocodilians and birds.

- In most mammals, the amniote egg became modified for growth of the embryo within the mother's uterus. There are five species of egg-laying mammals. The prototherians exist today only in Australia and New Guinea (see Figure 33.24). They supply milk to their young as do other mammals, but they do not have nipples on their mammary glands
- The remaining mammals are classed as therians. Marsupial therians give birth to tiny, underdeveloped young that they nurture externally, usually in a pouch on the mother's belly (see Figure 33.25). Marsupials were once widespread, but today the approximately 330 marsupial species are largely found in Australia and South America, with minor representatives in North America.
- The 4,500 species of eutherian mammals develop in the mother's uterus. This group is widely known by the name placental mammals, from the placenta that provides nourishment for the growing embryo. However, some marsupials also have placentas.
- Grazing and browsing by many herbivorous eutherian groups have transformed the terrestrial environment. In plants, this lead to the evolution of spines and tough leaves. Despite such defenses, herbivores have developed adaptations to their teeth and digestive systems in order to consume these plants. This is an example of coevolution.
- Eutherians exist in virtually all of Earth's environments and vary greatly in form (see Figure 33.26). Several groups, most notably the cetaceans (whales and dolphins) returned to a marine lifestyle. Flight evolved in the bats (which represent the second largest number of eutherian species, after the rodents) (see Table 33.1). The 235 species of primates are the best-studied eutherians because human beings belong to this group.

Primate evolution is well understood.

- The primate ancestor was a small, arboreal, insectivorous mammal. Primates are identifiable by the presence of opposable digits (i.e., thumbs). Early in their evolutionary history—about 65 mya—the primates split into two clades, the prosimians (lemurs, bush babies, and lorises) and the anthropoids (tarsiers, monkeys, apes, and humans) (see Figure 33.27). Prosimian species were once found on all continents, but today they are restricted to Africa, especially the island of Madagascar (see Figure 33.28).
- Soon after the prosimian–anthropoid split, the anthropoids split further into the New World and Old World monkeys. The breakup of the African and South American continents meant that the two groups evolved in isolation. About 35 mya, an Old World lineage broke off that would lead to modern apes and humans. Another split occurred around 22 mya. The Asian apes (gibbons and orangutans) descended from two groups in this lineage (see Figures 33.27 and 33.30).
- New World monkeys are tree-dwellers (arboreal) (see Figure 33.29A). A long prehensile tail, which is unique to the New World monkeys, allows them to grasp branches.
- The ancestor of modern African apes is believed to belong to the extinct genus *Dryopithecus*. The African apes include the modern gorillas, chimpanzees, and humans (see Figures 33.30C and 33.30D).

Early human ancestors evolved bipedalism.

- Ardipithecines (the earliest protohominids) and their descendants, the australopithecines, were adapted for bipedalism, which freed up their hands for other tasks and elevated the eyes to enable seeing over tall vegetation. Bipedal movement also requires less energy. *Australopithecus afarensis* is currently regarded as the ancestor of the modern humans (genus *Homo*; see Figure 33.31). There is disagreement as to how many species are represented by australopithecine fossils. It is clear that these different groups of hominids coexisted in Africa several million years ago. The genus *Homo* arose from one of the smaller lineages of australopithecines. In parallel, the genus *Parathropus* arose from larger australopithecines
- The oldest known member of the genus *Homo, Homo habilis,* lived about 2 mya, and tools have been found with their bones. *Homo erectus* appeared about 1.6 mya. They were as large as modern humans, used stone tools, and cooked with fire. However, their brains were smaller than those of modern humans and they had a relatively thick skull.
- A number of species of *Homo* existed simultaneously over evolutionary time between 1.5 mya and 250,000 years ago. One species, *Homo neanderthalensis,* appeared about 500,000 years ago and became widespread in Europe and Asia. Although they were short, they were powerfully built. Their skulls housed a brain that is somewhat larger than that of modern humans. They manufactured tools and were able to hunt large animals.
- By 200,000 years ago *H. sapiens* was predominant in Africa, and they began to extend their range around 60,000–70,000 years ago. Around 35,000 years ago, they coexisted alongside *H. neanderthalensis.* It is likely that the two species interacted, but Neanderthals abruptly became extinct 28,000 years ago, possibly exterminated by *H. sapiens*. Genetic evidence suggests that there was little or no interbreeding between the two species.
- By about 20,000 years ago, all other *Homo* species had been supplanted by *Homo sapiens* (modern humans), the only currently existing human species. It was about this time that *H. sapiens* reached North America.
- Brains of *H. sapiens* had reached modern size by about 160,000 years ago. Larger brain size appears to have evolved concurrently with a smaller, less muscular jaw

structure, suggesting that the two traits may be functionally correlated. The rapid increase in brain size may have been favored by an increasingly complex social life that thrived on ever more sophisticated communication. Any trait that allowed more effective communication would have been favored in a society based on cooperative hunting and other complex social interactions.

- The expansion of language and other behavioral abilities allowed humans to develop culture and pass knowledge and traditions from generation to generation. Human societies were transformed from communities of hunters and gatherers to those of farmers and eventually urban dwellers.

Test Yourself

Diagram Exercises

1. Label the different groups of organisms in the phylogenetic tree shown below and identify the key features of each branch as indicated by the dot. What are the names of the groups identified by brackets?

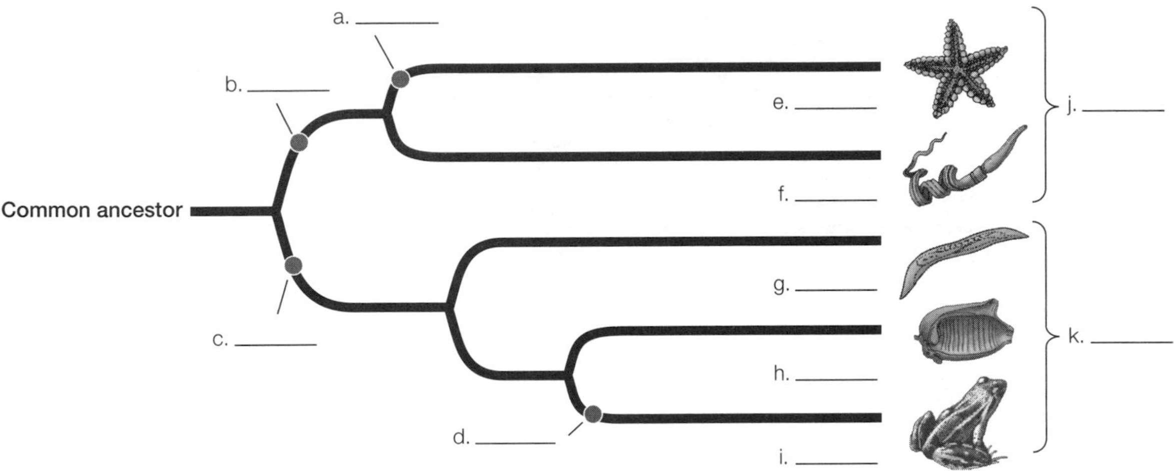

Textbook Reference: *33.1 What Is a Deuterostome? p. 694, Figure 33.1*

2. Label the different groups of organisms in the phylogenetic tree shown below and identify the key features of each branch as indicated by the dot. What is the name of the groups identified by brackets?

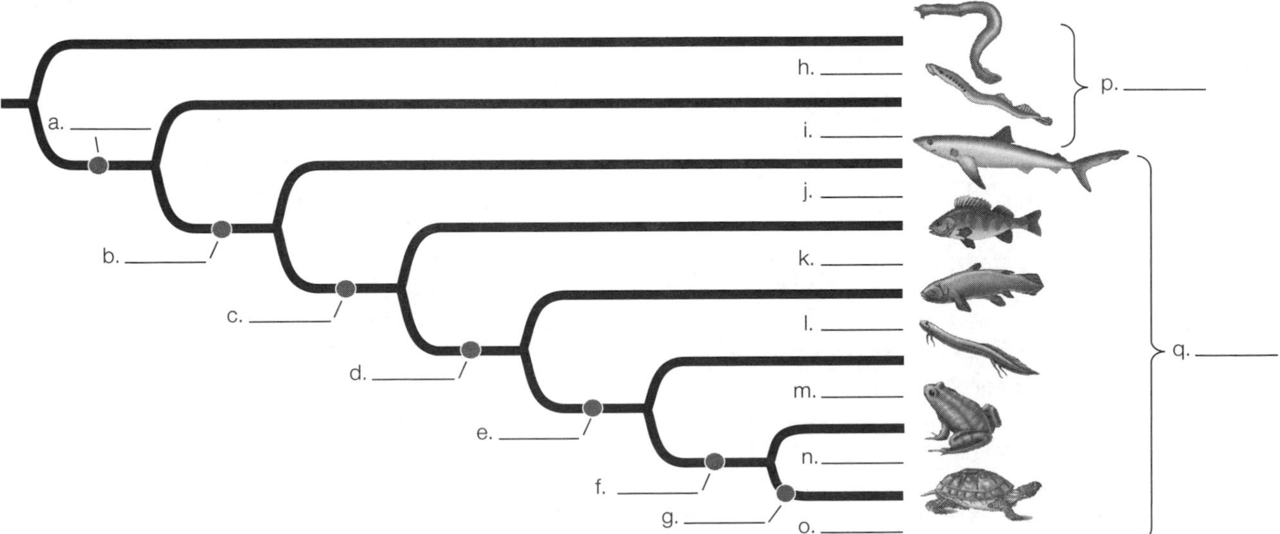

Textbook Reference: *33.3 What New Features Evolved in the Chordates? p. 699, Figure 33.8*

3. Label the different groups of organisms in the phylogenetic tree shown below. What are the names of the groups identified by brackets? Which groups are extinct?

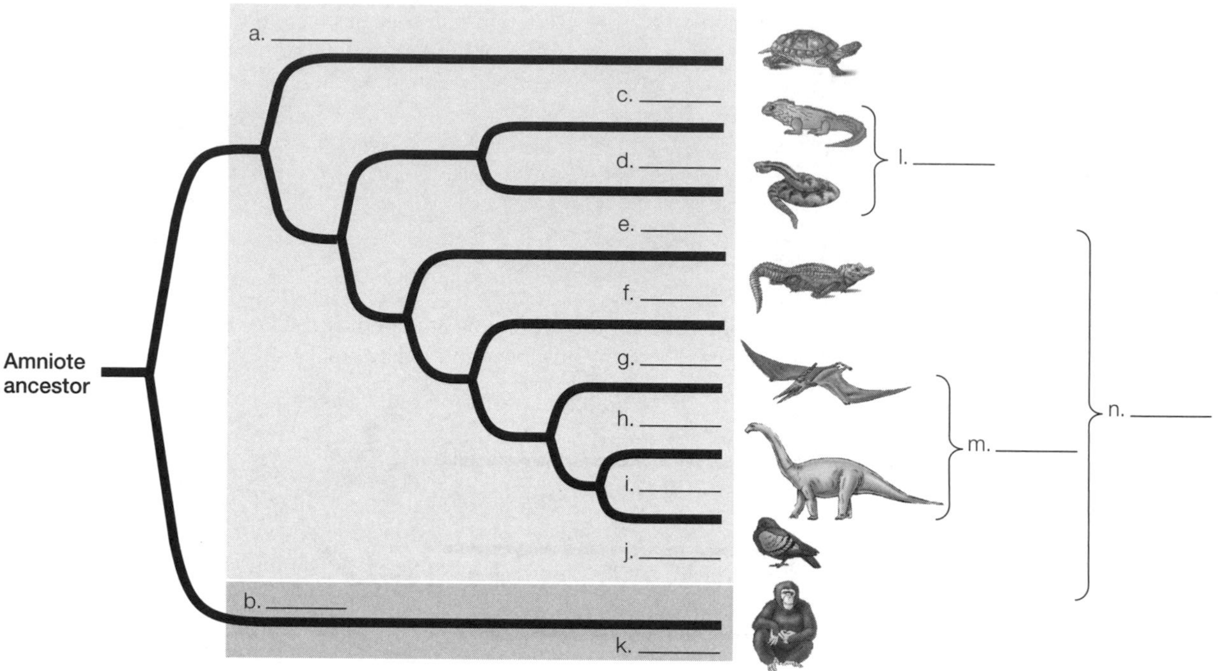

Textbook Reference: *33.4 How Did Vertebrates Colonize the Land? p. 706, Figure 33.18*

Knowledge and Synthesis Questions

1. Which of the following statements about echinoderms is *false?*
   a. They have a water vascular system.
   b. They have an internal skeleton.
   c. They are protostomes.
   d. The larvae have bilateral symmetry.
   e. The adult form lacks a head.
   Textbook Reference: *33.2 What Are the Major Groups of Echinoderms and Hemichordates? p. 695*
2. In which of the following echinoderm groups are the extinct members more numerous than the living members?
   a. Sea stars and brittle stars
   b. Sea lilies and feather stars
   c. Sea urchins and sea cucumbers
   d. Sea daisies
   e. Asterozoans
   Textbook Reference: *33.2 What Are the Major Groups of Echinoderms and Hemichordates? p. 695*
3. Which of the following chordate groups evolved before the appearance of cartilaginous fishes?
   a. Ray-finned fishes and sea squirts
   b. Ascidians, lancelets, and hagfishes
   c. Ascidians, lancelets, and ray-finned fishes
   d. Lampreys and ray-finned fishes
   e. Coelacanths and ray-finned fishes
   Textbook Reference: *33.3 What New Features Evolved in the Chordates? pp. 698–699 and Figure 33.8*
4. Which of the following is *not* a feature of the vertebrate body plan?
   a. A ventral spinal cord
   b. An internal skeleton
   c. A well-developed circulatory system
   d. Organs suspended in the coelom
   e. An anterior skull encasing a proportionally large brain.
   Textbook Reference: *33.3 What New Features Evolved in the Chordates? p. 700*
5. The swim bladder of many fishes that evolved from a lunglike sac has the important function of
   a. aiding in prey capture.
   b. controlling swimming speed.
   c. controlling buoyancy.
   d. aiding in reproduction.
   e. providing balance.
   Textbook Reference: *33.3 What New Features Evolved in the Chordates? p. 701*
6. Which of the following traits is shared by the chondrichthyans and the ray-finned fishes?
   a. Gills as the major site of gas exchange
   b. A skeleton composed of cartilage
   c. An outer surface covered with bony plates
   d. A swim bladder
   e. Propulsion by means of dorsal/ventral movements of their tails
   Textbook Reference: *33.3 What New Features Evolved in the Chordates? pp. 701–703*

7. The transition from aquatic to terrestrial lifestyles required many adaptations in the vertebrate lineage. Which of the following is *not* one of those adaptations?
   a. A shift from gills to air-breathing lungs
   b. Improvements in the water resistance of skin
   c. An alteration in the mode of locomotion
   d. Development of feathers for insulation
   e. Modifications to the nitrogen elimination system
   Textbook Reference: *33.4 How Did Vertebrates Colonize the Land? pp. 703–706*
8. The amniotes evolved the ability to reproduce by laying eggs that have shells. The major advantage of shelled eggs is that
   a. the embryo needs only a small amount of yolk for development.
   b. they do not have to be laid in a moist environment.
   c. the shells increase evaporation from the egg.
   d. nitrogenous wastes can be excreted across the shell.
   e. gas exchange with the environment is more efficient.
   Textbook Reference: *33.4 How Did Vertebrates Colonize the Land? p. 706*
9. Which of the following statements about birds and reptiles is true?
   a. Birds have a lower metabolic rate than reptiles.
   b. Reptiles give birth to live young, whereas birds lay eggs.
   c. Birds can breathe and run at the same time, whereas reptiles cannot.
   d. Birds are amniotes, reptiles are not.
   e. Reptiles supplement their lung action with gas exchange through their skin, whereas birds depend entirely on their lungs for gas exchange.
   Textbook Reference: *33.4 How Did Vertebrates Colonize the Land? pp. 707–708*
10. Which of the following is *not* a trait that could be used to identify an animal as a mammal rather than an amphibian?
    a. Mammary glands
    b. Hair
    c. Sweat glands
    d. Kidneys
    e. Four-chambered heart
    Textbook Reference: *33.4 How Did Vertebrates Colonize the Land? pp. 709–710*
11. Which of the following vertebrate groups is a living representative of the dinosaur lineage?
    a. Crocodilians
    b. Birds
    c. Lobe-finned fishes
    d. Snakes
    e. Lepidosaurs
    Textbook Reference: *33.4 How Did Vertebrates Colonize the Land? p. 706*
12. Which of the following statements about human evolution is *false?*
    a. Bipedalism was a hominid adaptation for life on land.
    b. Increases in the size of hominid brains preceded the appearance of language and culture.
    c. The extinction of the Neanderthals was caused by the emergence of *Homo habilis*.
    d. Humans are not the direct descendants of modern-day chimpanzees.
    e. *Homo sapiens* coexisted with *H. neanderthalensis* in portions of Europe and Asia.
    Textbook Reference: *33.5 What Traits Characterize the Primates? pp. 715–716*
13. A key difference between Old World monkeys and New World monkeys is that the latter
    a. have a prehensile tail.
    b. are aboreal.
    c. have a placenta.
    d. are less closely related to tarsiers.
    e. All of the above
    Textbook Reference: *33.5 What Traits Characterize the Primates? p. 713*
14. Which of the following is *not* part of the evidence that birds are closely related to dinosaurs?
    a. The presence of feathers in representatives of both groups
    b. The presence of hollow bones in representatives of both groups
    c. DNA sequence data comparing birds to other living reptiles
    d. A bipedal stance in fossilized therapods
    e. The ability to fly
    Textbook Reference: *33.4 How Did Vertebrates Colonize the Land? pp. 707–708*
15. Which of the following are not deuterostomes?
    a. Echinoderms
    b. Hemichordates
    c. Cephalochordates
    d. Ecdysozoans
    e. Chordates
    Textbook Reference: *33.1 What Is a Deuterostome? p. 694*

Application Questions

1. Ascidians, or sea squirts, are armless and legless organisms that spend their adult lives attached to a substrate under water. They feed by pulling water into one tube, filtering out planktonic organisms, and pushing the water out another tube. Grasshoppers have legs, move around freely on land, and feed by ingesting food through a mouth. What evidence has led biologists to believe that humans are more closely related to sea squirts than to grasshoppers?
   Textbook Reference: *33.3 What New Features Evolved in the Chordates? pp. 697–698*
2. The evolution of a hinged jaw resulted in an extensive radiation of the fishes into the many modern jawed forms. What was the main advantage and significance of hinged jaws in this radiation?
   Textbook Reference: *33.3 What New Features Evolved in the Chordates? pp. 700–701*

3. How do mammals and reptiles address some of the challenges of a terrestrial habitat?
   Textbook Reference: *33.4 How Did Vertebrates Colonize the Land? pp. 707–710*
4. Suppose that you discover the fossil of a new organism that has radial symmetry and was in a marine environment (as shown by examining the surrounding rock and adjacent fossils in the rock layer). What other features would you look for in order to determine if this fossil is an echinoderm?
   Textbook Reference: *33.2 What Are the Major Groups of Echinoderms and Hemichordates? p. 695*
5. Amphibians are dependant on water and moist environments for their life cycle. Which stages require an aquatic environment and why is such an environment necessary? Do any amphibians have strategies that allow them to avoid the aquatic stage? If so, describe them.
   Textbook Reference: *33.4 How Did Vertebrates Colonize the Land? pp. 704–705*

Answers

Diagram Exercise Answers

1.
   a. Radical symmetry as adults, calcified internal plates, loss of pharyngeal slits
   b. Ciliated larvae
   c. Notochord, dorsal hollow nerve cord, post-anal tail
   d. Vertebral column, anterior skull, large brain, ventral heart
   e. Echinoderms
   f. Hemichordates
   g. Cephalochordates
   h. Urochordates
   i. Vertebrates
   j. Ambulacrarians
   k. Chordates

2.
   a. Vertebrae
   b. Jaws, teeth, paired fins
   c. Bony skeleton, swim bladder/lung
   d. Lobe fins
   e. Internal nares
   f. Terrestrial limbs and digits
   g. Anmiote egg
   h. Hagfishes
   i. Lampreys
   j. Chondrichthyans
   k. Ray-finned fishes
   l. Coelacanths
   m. Lungfishes
   n. Amphibians
   o. Amniotes
   p. Jawless fishes
   q. Gnathostomes

3.
   a. Reptiles
   b. Mammals
   c. Turtles
   d. Tuataras
   e. Squamates
   f. Crocodilians
   g. Pterosaurs (extinct)
   h. Ornithischians (extinct)
   i. Sauropods (extinct)
   j. Birds
   k. Mammals
   l. Lepidosaurs
   m. Dinosaurs
   n. Archosaurs

Knowledge and Synthesis Answers

1. **c.** Species from the echinoderm clade are deuterostomes.
2. **b.** The crinoids (sea lilies and feather stars) are a relict group with many more extinct than extant species.
3. **b.** The ascidians, lancelets, and hagfishes all evolved before the cartilaginous fishes.
4. **a.** Along with an internal skeleton, well-developed circulatory system, and organs suspended in a coelom, the vertebrates have a dorsal spinal cord.
5. **c.** The swim bladder of modern-day fishes is involved in controlling buoyancy.
6. **a.** In both the ray-finned fishes and cartilaginous fishes, the major site of gas exchange is the gills. The ray-finned fishes have a skeleton of bone and a swim bladder. These traits are not shared with the cartilaginous fishes. Neither group has bony plates in its outer surface. They all move their tails laterally, in contrast to cetaceans, which move their tails vertically.
7. **d.** The move onto land did not require the development of feathers for insulation. Amphibians and reptiles do not have an insulation layer, and many mammals have hair for insulation.
8. **b.** The shelled egg of the birds and reptiles allowed them to occupy dry terrestrial habitats because the shell decreases water loss from the egg.
9. **c.** Birds and reptiles differ in the morphology of the muscles that control movement and breathing. Birds can run and breathe at the same time, whereas reptiles must stop running to take a breath.
10. **d.** Kidneys are present in reptiles, birds, fishes, and mammals. The presence of kidneys is not a trait that could be used to identify a mammal.
11. **b.** Birds are now thought to be direct descendants of a group of dinosaurs.
12. **c.** The extinction of Neanderthals is thought to have been due in part to the presence of Cro-Magnon. *Homo habilis* was extinct for more than a million years by the time the Neanderthals emerged.

13. **a.** Only the New World monkeys have a prehensile tail.
14 **e.** Not all birds fly and not all dinosaurs were able to fly.
15. **d.** Ecdysozoans are a type of protostome (see Figure 31.1).

Application Answers

1. Although adult ascidians and adult humans are very different animals, their embryonic stages share certain important characteristics, including the presence of a notochord (an in all deuterostomes) and a blastopore that develops into the anus (see Chapter 31). A look at embryonic grasshoppers, however, shows them to be fundamentally different. The embryonic grasshopper's blastopore develops into the mouth, placing grasshoppers squarely among the protostomes. Grasshoppers have an external rather than an internal skeleton.
2. The hinged jaw of the fishes evolved during the Devonian period. The evolution of the jaw opened up a new food source for these animals. With a jaw, fish could now grasp and kill larger living prey and chew and tear body parts.
3. Mammals and reptiles are both amniotes. Their eggs minimize water loss and still permit gas exchange. The skin of both groups of organisms is modified to reduce water loss, and the kidneys eliminate nitrogen while minimizing water loss. Both groups of organisms have lungs designed to exchange gases with the atmosphere.
4. You would look for evidence of an internal skeleton and a vascular system composed of water-filled canals. These two traits are defining characteristics of echinoderms.
5. Amphibians generally require an aquatic environment for fertilization, larval development, and metamorphosis into a terrestrial adult. Some amphibians develop from eggs laid on land and move directly into adultlike forms, skipping the aquatic form entirely.

34 The Plant Body

The Big Picture

- A plant can be thought of as having vegetative structures that carry out the major functions of day-to-day life and reproductive structures that are responsible for reproducing the plant. You will see that there are cases in which the vegetative portions of the plant are quite adept at reproducing asexually. The vegetative plant body consists of the root system, which anchors the plant and absorbs water and nutrients, and the shoot system, which carries out photosynthesis and supports the plant against gravity. Modifications of the root and shoot systems lead to specialization of the plant.
- Plant cells are uniquely suited for support, transport, and the carrying out of cellular functions. Groups of cells form tissues that have specific roles within a plant. Patterns of development are under the control of hormones and regulatory genes. Indeterminate growth and the modular organization of plants allow for regeneration of parts lost to damage and disease.
- Plant growth occurs from meristems. Apical meristems allow for elongation of the plant, whereas lateral meristems allow for secondary or woody growth. Not all plants exhibit secondary growth. All tissue types arise from the meristems and go through a process of elongation, then differentiation. Review stem, root, and leaf anatomy using the figures in the textbook.

Common Problem Areas

- This chapter covers plant anatomy. Basic plant anatomy is not difficult, but it is new to most students. The more time you spend looking at diagrams and live specimens, the easier it will be to understand the anatomy. As you study, think about the function of each structure. You will find that "form follows function."
- This chapter exposes you to many new vocabulary words. Do not merely memorize these. By looking at the words and understanding their roots, you will be better able to understand the terms.

Study Strategies

- The best study strategy for this material is to study the figures, pictures, and live material. Plant anatomy is the study of structure. Structures are three-dimensional and best understood visually.
- Make a vocabulary list. Though you should not try to memorize all the terms, a list will help you organize your study.
- Go to yourBioPortal.com to review the following tutorial and activities:

  Animated Tutorial 34.1 Secondary Growth: The Vascular Cambium

  Web Activity 34.1 Eudicot Root

  Web Activity 34.2 Monocot Root

  Web Activity 34.3 Eudicot Stem

  Web Activity 34.4 Monocot Stem

  Web Activity 34.5 Eudicot Leaf

Important Concepts

Vegetative structures serve a variety of roles, including nutrient procurement, support, water uptake, and transport.

- Plants are composed of root systems and shoot systems (see Figure 34.1). Root systems are responsible for mineral and water uptake and support. Shoot systems consist of leaves (and leaf derivatives) involved in photosynthesis and stems for support.
- Root systems develop from the radical, an embryonic root. In most eudicots, a primary taproot develops as a single large root growing deep within the ground. In monocots, numerous thin, adventitious roots make up a fibrous root system that absorbs water efficiently and helps the plant cling to the soil.
- The shoots of plants are laid out in modules, or units, known as phytomers. A phytomer consists of a node and its attached leaf, the internode (a section of stem) below the node, and the axillary buds at the base of the internode. Buds are embryonic shoots. At each node where leaves meet stem, an axillary bud is produced

that may generate a new branch. At each stem tip a terminal bud is responsible for elongation of that stem.

- Stems may be modified into tubers (such as potatoes) or runners. Stems may remain photosynthetic or become nonphotosynthetic woody material.
- Leaves are the site of photosynthesis but may have modifications to allow for specific jobs. The blade of a leaf is attached to the stem by a petiole. The leaf blade, with help from the petiole, is able to maintain a constant orientation toward sunlight for maximum photosynthesis.
- Some leaves are modified for storage of nutrients or water. Some are modified into structures like tendrils to help the plant hold onto a support.

Plants have rigid cell walls.

- Plant cells differ from other eukaryotic cells in that every plant cell is bounded by cellulose-containing cell walls, and some have plastids such as chloroplasts and a central vacuole that provides turgor pressure on the cell wall. The vacuole contains enzymes, amino acids, and sugars that have been pumped across the vacuolar membrane, the tonoplast.
- Cell walls play important roles by helping to regulate the volume and shape of the plant cell. Cell walls are formed after cytokinesis. The daughter cells are separated by a cell plate and a middle lamella, which is a thin layer between the two daughter cells. The primary cell wall is produced as the cells secrete cellulose, hemicellulose, and pectin. Once the cell is full size, it may begin secreting other substances to form a secondary wall.
- Plant growth occurs through cell expansion. As the living content of the cell (protoplast) increases in size, the cellulose microfibril linkages in the cell wall loosen with the help of expansins. The cell expands, and new polysaccharides are added to the cell wall. After expansion, some cells produce a thick secondary cell wall that provides mechanical support for the plant.
- Cell walls do not block communication between cells. Water and mineral ions are able to cross the cell wall. Small cytoplasmic strands called plasmodesmata extend through the walls, allowing movement of substances from cell to cell.

Plant cells are organized into tissues.

- The body plan of a plant is established in the embryo along a basal–apical axis and a radial axis. Meristems are undifferentiated cells found at the tips of the embryonic shoot and root that will become the organs of the plant as it grows.
- Two unequal daughter cells are produced as the zygote goes through a mitotic division. This results in the development of a thin suspensor and a globular embryo. The cotyledons (seed leaves) begin to form as the embryo enters the heart stage. As the cotyledons elongate, the embryo enters the torpedo stage and the internal tissues begin to differentiate. The root and shoot apical meristems develop between the cotyledons.
- Tissues are composed of cells that function together. Different tissue types are grouped into the dermal, vascular, and ground tissue systems.
- The dermal tissue system makes up the outer covering of the plant and includes the epidermis and the layer of cuticle it secretes. Special epidermal cells include stomatal guard cells, trichomes, and root hairs.
- The ground tissue system is found between the dermal and vascular tissue and is involved in storage, support, and photosynthesis. Thin-walled parenchyma cells have large central vacuoles and are frequently photosynthetic or used for storage. They may also continue dividing and proliferate in a wounded area. Collenchyma cells are support cells with special thickenings at the cell wall corners. They are elongated and allow for support without rigidity, which is necessary for plants in windy areas. Sclerenchyma cells have highly thickened cell walls and are either elongated fibers or variously shaped sclereids. Many undergo apoptosis and provide rigid support after they die. Fibers strengthen bark and woody stems. Densely-packed sclereids are found in the shells of nuts and in some seed coats and produce the gritty texture of pears and other fruit.
- The vascular tissue system is made of xylem and phloem, and is the conductive tissue of the plant.
  - Xylem transports water and mineral ions from the roots to the rest of the plant. The tracheary elements involved in the water movement are functional after they die. Tracheids make up the xylem transport in the gymnosperms, while vessel elements are found in angiosperms.
  - Living phloem moves carbohydrates and nutrients. Individual phloem cells are called sieve tube elements. Plasmodesmata enlarge where sieve tube elements join, making sieve plates. Sieve tube elements may lose nuclei and other organelles to prevent clogging of the sieve plates. Adjacent companion cells regulate the function of the sieve tube elements.

Plant meristems grow continually.

- Plants differ from animals in that their growth is indeterminate—they continue to grow throughout their life span. This is possible because plants have regions of continual cell division called meristems.
- Primary growth occurs as the shoots and roots lengthen and proliferate to produce the nonwoody primary plant body. Woody plants increase their girth during secondary growth and produce a secondary plant body composed of wood and bark.
- Meristems are undifferentiated cells that are able to produce new cells indefinitely. These initial cells are similar to human stem cells. Meristems are classified as either apical or lateral, depending on whether they contribute to primary or secondary growth.

- Two types of apical meristems, shoot and root, give rise to the primary plant body and are located in buds and at the tips of stems and roots. They are responsible for primary growth—the elongation of the plant body. Apical meristems give rise to primary meristems that produce the primary plant body. These primary meristems are called the protoderm (dermal tissue system), the ground meristem (ground tissue system), and the procambium (vascular tissue system). All plant parts arise from division of the apical meristems.
- Root apical meristems (protoderm, ground meristem, and procambium) produce root tissues. At the tip of a root, the root apical meristem forms a root cap and quiescent center. The zone of cell division includes the apical and primary meristems. The zone of elongation is found above this and is the site of new cell formation. The upper layer is the zone of maturation where the cells differentiate and take on special functions.

- The root has three primary tissue systems. The protoderm gives rise to the epidermis and root hairs, both involved in water and mineral ion uptake. The ground meristem gives rise to the cortex and endodermis. The endodermis is specialized with a waxy coating of suberin to assist with water movement. The procambium gives rise to the stele, which houses three tissues: the pericycle, the xylem, and the phloem. In eudicots, the very center of the root is xylem, but in monocots the center is pith tissue (see Figures 34.13).
- Shoot growth occurs as the plant adds repeating units of phytomers from the terminal and axillary buds. The shoot primary meristem also gives rise to three primary meristems that produce the three tissue systems. Leaves arise from leaf primordia with bud primordia forming at each leaf base. Shoot vascular tissues are arranged in vascular bundles containing both xylem and phloem. In eudicots, the vascular bundles are arranged in a cylinder, allowing for woody growth. In monocots the vascular bundles are scattered.
- Vegetative meristems produce the leaves of the plant. Most eudicot leaves have two zones of photosynthetic cells called mesophyll. The upper level of cylindrical mesophyll is called palisade mesophyll, and the lower level is called spongy mesophyll. Air space around mesophyll cells is necessary for carbon dioxide to reach the photosynthesizing cells. Vascular tissue extends throughout leaves as a network of veins. Veins carry water to cells and transport carbohydrates to sink tissues. The entire leaf is covered by a protective epidermis and is waterproofed by a waxy cuticle. Gas exchange occurs through guarded stomata. Guard cells open and close stomata to limit water loss.
- Secondary growth in eudicots involves the laying down of wood and bark by the two lateral meristems, vascular cambium and cork cambium. Vascular cambium arises from the lateral meristem and forms new secondary xylem (wood) and secondary phloem (bark). The cork cambium produces new dermal tissues to accommodate increasing diameter and inhibits water loss with the production of periderm cells. The action of the vascular and cork cambiums is called secondary growth. Wood results from secondary xylem, and bark is made from the cork cambium, cork, phelloderm, and secondary phloem.
    - Stretching, breaking, and flaking off of epidermis and cortex produces bark. This leaves the secondary phloem at risk. Cells at the surface of the phloem produce a protective layer of cork that is thickened and reinforced with waterproof suberin. New cork is produced as secondary growth proceeds. The areas that allow gas exchange through the bark are known as lenticels.
    - The annual rings seen in wood are a result of climate shifts in temperate zones, particularly water availability. Tropical trees do not undergo seasonal growth and do not lay down such visible rings.

Humans have domesticated plants.

- Humans have artificially selected plants to improve crop yield. This is possible because of the morphological variation found within wild plant species.

Test Yourself

Diagram Exercise

Draw a typical eudicot plant. Label the root system and shoot system. Indicate on your drawing where you would find an axillary bud and where you would find a terminal bud. Label the following structures: leaf blade, internode, petiole, taproot, and lateral roots.
Textbook Reference: *34.1 What Is the Basic Body Plan of Plants? p. 721*

Knowledge and Synthesis Questions

1. Which of the following is a not a component part of a phytomer?
   a. Leaf
   b. Root hair
   c. Axillary buds
   d. Internode
   e. All of the above are components of a phytomer.
   Textbook Reference: *34.1 What Is the Basic Body Plan of Plants? p. 722*
2. Suppose you are studying tropical plants in a Costa Rican cloud forest and find a tree that is a eudicot in the forest ecosystem. This plant most likely has a(n) _______ system.
   a. fibrous root
   b. taproot
   c. adventitious root
   d. rhizoid root
   e. terminal root
   Textbook Reference: *34.1 What Is the Basic Body Plan of Plants? p. 721*

3. Some plants, such as sweet peas, will attach themselves to a fence by means of tendrils, which are modifications of
   a. stems.
   b. roots.
   c. branches.
   d. leaves.
   e. seeds.
   Textbook Reference: *34.1 What Is the Basic Body Plan of Plants? p. 722*
4. Plant cells are easily distinguished from animal cells by their
   a. rigid cell walls.
   b. plastids.
   c. large vacuoles.
   d. chloroplasts.
   e. All of the above
   Textbook Reference: *34.2 How Does the Cell Wall Support Plant Growth and Form? p. 723*
5. Plant cells that are photosynthetically active are found in the _______ layer of the leaf and are _______ cells.
   a. mesophyll; parenchyma
   b epidermis; parenchyma
   c. mesophyll; sclerenchyma
   d. epidermis; sclerenchyma
   e. xylem; mesophyll
   Textbook Reference: *34.4 How Do Meristems Build a Continuously Growing Plant? p. 732*
6. Water is conducted in _______ tissue, and carbohydrates and nutrients are transported in _______ tissue.
   a. xylem; phloem
   b. phloem; xylem
   c. parenchyma; phloem
   d. parenchyma; xylem
   e. mesophyll; xylem
   Textbook Reference: *34.3 How Do Plant Tissues and Organs Originate? p. 728*
7. Plants are capable of indeterminate growth because of
   a. regions of nondividing cells.
   b. meristem tissues.
   c. the epidermis.
   d. their xylem.
   e. All of the above
   Textbook Reference: *34.4 How Do Meristems Build a Continuously Growing Plant? p. 730*
8. Which of the following best describes the origin of wood?
   a. Xylem cells enlarge and deposit large amounts of lignin.
   b. Primary meristems increase the amount of xylem deposited.
   c. Lateral meristems contribute to continuous increases in vascular tissue.
   d. Spongy mesophyll cells become cork cambium.
   e. None of the above
   Textbook Reference: *34.4 How Do Meristems Build a Continuously Growing Plant? p. 733*
9. Which of the following is *not* a component of the primary cell wall?
   a. Cellulose
   b. Hemicelluloses
   c. Pectins
   d. Collenchymas
   e. All of the above are components of the primary cell wall.
   Textbook Reference: *34.2 How Does the Cell Wall Support Plant Growth and Form? p. 723*
10. Which of the following best describes the function of the cork cambium?
    a. It lays down a protective cork covering over exposed phloem tissue.
    b. It inhibits the sloughing off of epidermal tissue.
    c. It allows for diameter shrinking in stems and roots.
    d. It supplies the secondary xylem.
    e. All of the above
    Textbook Reference: *34.4 How Do Meristems Build a Continuously Growing Plant? p. 733*
11. Sieve tube elements have sieve plates where they join other sieve tube elements. Which of the following statements about the sieve plates is true?
    a. Sieve plate pores are enlargements of meristems.
    b. They allow conduction between sieve tube cells through plasmodesmata.
    c. They allow for the joining of cytoplasm between adjacent stomata.
    d. They contain the organelles of the cell.
    e. None of the above
    Textbook Reference: *34.3 How Do Plant Tissues and Organs Originate? p. 728*
12. Plants regulate gas exchange and water loss via
    a. the cuticle.
    b. xylem.
    c. coated pits.
    d. sieve plates.
    e. guarded stomata.
    Textbook Reference: *34.4 How Do Meristems Build a Continuously Growing Plant? p. 733*
13. The protoderm becomes the _______ tissue system.
    a. dermal
    b. ground
    c. vascular
    d. All of the above
    e. None of the above
    Textbook Reference: *34.4 How Do Meristems Build a Continuously Growing Plant? p. 730*
14. Primary growth occurs at the
    a. lateral meristems.
    b. fruit.
    c. quiescent center.
    d. apical meristems.
    e. wood.
    Textbook Reference: *34.4 How Do Meristems Build a Continuously Growing Plant? p. 730*

15. Vascular bundles are composed of _______ and _______.
    a. root hairs; xylem
    b. cork; phloem
    c. xylem; phloem
    d. wood; cork
    e. mesophyll; xylem
    Textbook Reference: *34.4 How Do Meristems Build a Continuously Growing Plant? p. 732*

Application Questions

1. How have humans domesticated plants? What are the advantages domestication has provided?
   Textbook Reference: *34.5 How Has Domestication Altered Plant Form? p. 736*
2. Differentiate between apical and lateral meristems. Do all plants have apical meristems? Do all plants have lateral meristems?
   Textbook Reference: *34.4 How Do Meristems Build a Continuously Growing Plant? pp. 729–730*
3. Describe the development of a xylem cell in a monocot from its origin in the apical meristem.
   Textbook Reference: *34.3 How Do Plant Tissues and Organs Originate? p. 728; 34.4 How Do Meristems Build a Continuously Growing Plant? p. 730*
4. Draw a growing root. Label the primary meristems, root cap, cortex, stele, and epidermis. Discuss the function of each of these structures.
   Textbook Reference: *34.4 How Do Meristems Build a Continuously Growing Plant? p. 730*
5. Look at the leaf in Figure 34.15 in the textbook. Which surface of that leaf faces the sun? How do you know? Why are the stomata opposite the palisade layer?
   Textbook Reference: *34.4 How Do Meristems Build a Continuously Growing Plant? p. 733*

Answers

Diagram Exercise Answer

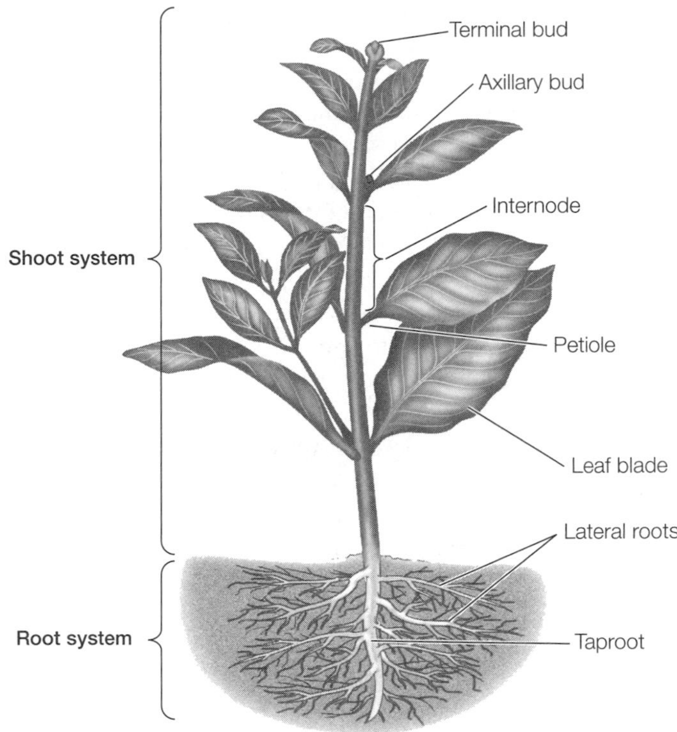

Knowledge and Synthesis Answers

1. **b.** The phytomer is the repeating unit of a plant from node to node that includes the leaves, an internode, and one or more axillary buds.
2. **b.** A taproot would be necessary to anchor a plant of that size. Most eudicots have taproots.
3. **d.** Tendrils are modified leaves. Some climbing plants produce tendrils, and others produce suckers.
4. **e.** Plant cell shape is maintained by their cell walls. No animal cells have cell walls. Plants have vacuoles and plastids, including chloroplasts. Though these characteristics are seen in some protists, no animal cells have them.
5. **a.** Palisade and spongy mesophyll cells are photosynthetically active and derived from parenchyma cells.
6. **a.** Xylem tissue transports water from the roots throughout the plant. Phloem tissue transports carbohydrates and nutrients from source tissue to sink tissue.
7. **b.** The apical and (when present) lateral meristems are regions of continually dividing cells that can contribute to the growth of a plant throughout its life. A modular body plan also contributes to the ability of plants to grow continually.
8. **c.** Lateral meristems are responsible for the growth of new xylem and phloem. The secondary xylem gives rise to wood.

9. **d.** The primary cell wall is composed of three polysaccharides: cellulose, hemicelluloses, and pectins.
10. **a.** The cork cambium is a meristematic region outside the secondary phloem. As girth increases and splits the epidermis, causing loss of those protective layers, the cork cambium produces new cells to cover the expanding vascular tissue. It can also produce cells inside the plant known as phelloderm.
11. **b.** The sieve plates allow for the conduction of sap from one sieve tube cell to another sieve tube cell. The pores are due to enlargements of the plasmodesmata. These cells may also lose nuclei and organelles so that carbohydrates can easily pass through the sieve tubes.
12. **e.** Guard cells at the edges of stomata respond to changes in osmotic pressure. These changes lead to the opening and closing of the stomata to regulate gas exchange and water loss.
13. **a.** The protoderm becomes the dermal tissue system of the growing plant.
14. **d.** The apical meristems are the site of primary growth. Apical meristems are found at the tips of the roots, in stems, and in buds.
15. **c.** Vascular bundles are composed of the vascular tissue of the plant, which includes the xylem and phloem.

Application Answers

1. Humans have taken advantage of the large variation in plant shape and size within a species. They have selectively used the seeds from plants with desirable characteristics, ensuring these characteristics remain in the crop. This has allowed humans to produce more productive crops.
2. Apical meristems are responsible for elongation of the plant body. Lateral meristems are responsible for an increase in girth. Lateral meristems are found only in woody eudicots and are responsible for creating wood. All plants have apical meristems.
3. The apical meristem gives rise to the protoderm, the ground meristem, and the procambium. The procambium gives rise to the vascular tissue system, including xylem. As the vascular tissue develops, it undergoes apoptosis and becomes the tracheary elements of the xylem. Monocots do not have secondary xylem, so only primary xylem exists in this plant.
4. See Figure 34.11 for a diagram of root growth. The protoderm gives rise to the epidermis for protection. The ground meristem gives rise to the cortex for storage. The procambium gives rise to the stele for transport. The root cap protects the meristem as it pushes through the soil.
5. The "top" of the diagram in the textbook would be the surface that faces the sun. This is the surface where photosynthetic cells, which need maximum sun exposure, are located. The stomata are on the opposite side to reduce water loss due to evaporation during photosynthesis.

35 Transport in Plants

The Big Picture

- In terrestrial plants, water must be acquired from the soil, transported through the plant, and used in the leaves for photosynthesis. At the same time, the nutritional products of photosynthesis must be transported throughout the plant to nonphotosynthetic tissues. This two-way transport is achieved through specialized cells that make up the vascular tissue of the plant. Water with dissolved mineral nutrients is absorbed through the roots and transported to cells in the xylem of the plant, where it is pulled up the stem to the leaves via the transpiration–cohesion–tension mechanism. Sugars and solutes are moved out of the leaves and to the rest of the plant through cells in the phloem.
- Water uptake is regulated by osmotic and water potentials in the root cells, and the rate of transport is controlled by the rate of evaporation at the leaf surface. Guard cell activity in the leaves regulates the opening and closing of stomata to match water availability, light, and drying conditions. Sucrose movement is regulated by active transport and facilitated diffusion in the phloem tissue. The rate of sucrose transport depends on the rates of loading and unloading at source and sink tissues.

Common Problem Areas

- Students often fail to look at the anatomy of the structures that are used in transport. Each structure is uniquely suited to its function. The process is much easier to understand if you think in terms of form coupled with function.

Study Strategies

- Water and sucrose transport are pathways that can be understood by visually tracing a molecule of water or sucrose through the plant. Use the figures in your textbook to follow these pathways.
- In order to understand water transport, you should understand osmosis and the properties of water molecules. If you are struggling with this chapter, review these concepts from earlier chapters.
- The basis of nutritional transport in plants is cell-to-cell transport. Review the sections in the textbook on diffusion, osmosis, and active transport.
- Go to yourBioPortal.com to review the following tutorials and activity:

  Animated Tutorial 35.1 Xylem Transport

  Animated Tutorial 35.2 The Pressure Flow Model

  Web Activity 35.1 Apoplast and Symplast of the Root

Important Concepts

Water and nutrients are taken up in the roots of plants.

- The movement of water across a semipermeable membrane is a special type of diffusion known as osmosis (see Figure 35.2).
  - For osmosis to occur across a semipermeable membrane, there must be a solute potential (or difference in solute concentrations) great enough to initiate movement and a pressure potential (turgor pressure in plants) small enough to allow for movement.
  - The overall tendency of a solution to take up water across a membrane is called water potential and is the sum of the negative solute potential and the positive pressure potential. All three parameters can be measured in megapascals (MPa). Water always moves to a region of more negative water potential.
- The structure of plants is maintained by osmotic phenomena. If a plant loses turgor pressure by a decrease in pressure potential, it wilts. Movement of water from cell to cell depends on the gradient of water potential. Long-distance movement depends on pressure potential and is referred to as bulk flow.
- Specialized membrane channel proteins in plant cells called aquaporins can increase the rate of water movement by allowing water to cross the plasma membrane without interacting with the hydrophobic bilayer that slows water flow. Though aquaporins can increase the rate of osmosis, they cannot influence the direction of flow.
- Mineral ion uptake from the soil solution requires active transport via proteins. When mineral concentra-

tions are greater in the soil solution than in the plant, they are taken up by facilitated diffusion. If minerals are in smaller concentrations outside the plant than inside the plant, or if they must be moved against an electrochemical gradient, then the plant must rely on active transport.

- Plants rely on a proton pump for active transport of minerals into cells. Plants actively pump protons out of cells, causing the area outside the cell to become more positive. This assists facilitated diffusion of positive ions through protein channels. It also drives the movement of negatively charged ions into the cell by active transport (see Figure 35.5). The result of this pumping action is that the internal environment of the plant cell becomes highly negative compared to its environment. This difference in charge is called membrane potential. The proton gradient that develops across the membrane can also facilitate secondary active transport of ions such as Cl^-.
- Water moves into a root because the root has a more negative water potential than the soil solution it is bathed in. Movement inside the root takes place because the stele (vascular tissue) has a more negative water potential than the cortex. Minerals dissolved in water are moved via bulk flow of water once it is in the vascular system.
- Minerals follow two paths for reaching the vascular tissue: the rapid apoplast or the slower symplast.
    - The apoplast is formed by cell walls and intercellular spaces. Water and minerals may move unregulated through this space without ever having to cross a membrane.
    - The symplast is the living portion of the plant and is enclosed in plasma membranes. Movement of water and minerals in the symplast is highly regulated.
    - Water and minerals can travel through the apoplast as far as the endodermis. At the endodermis, water and minerals are stopped by the Casparian strips, which are waxy structures surrounding the endodermal cells. Because of this, water can reach the stele only via the symplast. The transport proteins in the endodermal cells determine which minerals enter the stele.
    - Once past the endodermal barrier, water and minerals can again leave the symplast and move back to the apoplast with the aid of parenchyma cells.
- Ultimately, water and minerals from the soil solution end up in xylem cells and are referred to as xylem sap.

Water and nutrients are moved through the plant in the xylem.

- It was originally thought that a pumping mechanism for the movement of fluids might be active in plants. This was shown to be false by experiments in 1893 by Eduard Strasburger. Tree trunks immersed in poison showed progressive death of all living cells as the poison progressed through the plant. This experiment led to three important conclusions:
    - Because movement continued even as cells were killed, no "pumping" cells were active.
    - Leaves were critical to transport because transport continued until the leaves died.
    - Transport did not depend on the roots because it occurred in the absence of roots.
- Root pressure, as shown by guttation (the forcing of water out of openings in leaves), was another theory for water movement. It was thought that the pressure exerted by root tissue might be sufficient to force water up the xylem. In actuality, xylem sap is under negative pressure as it is ascending.
- Pulling forces known as tension due to transpiration at the leaf surface are responsible for the movement of water through the xylem. Water evaporates from mesophyll cells during transpiration, creating tension on the water associated with the mesophyll cell wall. Transpiration generates tension on the water molecules in the xylem water column, which pulls them up from the roots and through the apoplast of the leaves. Water molecules are cohesive, meaning that they stick together enough to resist the tension, and they pull other water molecules along because of their hydrogen-bonding. This results in bulk flow (see Figure 35.8).
- The entire mechanism that pulls water up from the roots through the plant is known as the transpiration–cohesion–tension mechanism. This is a passive process requiring no energy input by the plant. Minerals are drawn passively along with the water column. Transpiration also assists with temperature regulation through evaporative cooling of the leaves.
- Per Scholander measured the tension in the xylem sap with a pressure chamber (see Figure 35.8). To conduct the experiments, Scholander cut the stem of the plant and placed the stem and leaves in a pressure chamber, leaving the cut portion of the stem out of the chamber. By placing the leaves under pressure and measuring the pressure needed to push the sap back to the surface of the cut end, the tension in the xylem was measured.

Water loss in a plant must be controlled.

- Leaf surfaces are covered with a waxy cuticle to prevent excessive water loss. However, the leaf must take up CO_2 for photosynthesis. Any time a plant surface is open enough to allow gas exchange, water is lost to the environment.
- Stomata with guard cells are pores that regulate gas exchange and water loss from a leaf. Guard cells, in response to osmotic differences, shrink and swell to open and close the stomata. Guard cells open when light is sufficient to maintain photosynthesis and carbon dioxide levels are low. Guard cells are also regulated by water potential. If the water potential in mesophyll cells is low, mesophyll cells release the hormone abscisic acid, which causes the guard cells to close.

- Blue light and low CO_2 levels stimulates a proton pump that helps regulate guard cell activity. Guard cells open when potassium ions diffuse into the cell as a result of the electrical gradient set up by the proton pump. High potassium levels cause water to move in by osmosis. Pressure potential builds in the guard cells, and they are pulled apart to reveal the stoma. For guard cells to shut, the proton pumps stop, and potassium ions move back across the membrane. Water follows, and the cells go limp and seal off the stoma (see Figure 35.9).

Phloem moves materials from sources to sinks by translocation.

- Sources are organs that produce more sugars than are used by metabolism, storage, and growth. Sinks are organs that do not make enough sugar for their own growth or storage needs. Sugars, amino acids, minerals, and other substances are translocated between sources and sinks in the phloem.
- Translocation proceeds in both directions along the stem. Translocation stops if phloem tissue is killed, and it is inhibited whenever respiration and the availability of ATP are limited.
- Sieve tube elements are the cells of the phloem. They are connected end-to-end by sieve plates containing plasmodesmata, allowing for movement between cells. Because the sieve tube elements do not contain organelles, they rely on companion cells to provide them with all they need to survive.
- Scientists can sample the sieve tube sap of a single sieve tube member using aphids. An aphid drills into a single cell, and sap is forced out. Once the aphid begins eating, it is frozen and its feeding organ is used as a tap to collect the sap from a single sieve tube.
- The pressure flow model explains how materials move through the phloem. Once in the sieve tubes, sieve tube sap moves via bulk flow, which requires no energy input by the plant. Energy is required for the loading of the sieve tubes at the sources and the unloading of solutes when the sink is reached. Sucrose is actively transported into sieve tubes at the sources. This causes water to move into sieve tubes by osmosis, thereby increasing the pressure potential at the source end and pushing the contents toward the sink end. The result is bulk flow.
- For the pressure flow model to work, the sieve plates must be open to allow uninterrupted flow of sieve tube sap from one sieve tube member to another. Microscopic analysis indicates that the sieve tube plates are open in undamaged cells.
- There must also be a mechanism for loading and unloading sucrose and amino acids. Neighboring cells assist with the loading and unloading of sucrose at sources and sinks.

Test Yourself

Diagram Exercise

Create a flow chart of the path taken by a water molecule as it moves from the soil solution to the stele of a plant. Identify where the molecule is traveling through the apoplast and where it is traveling through the symplast.
Textbook Reference: *35.1 How Do Plants Take Up Water and Solutes? p. 744*

Knowledge and Synthesis Questions

1. The function of the Casparian strips is to
   a. divert water and minerals through the membranes of endodermal cells.
   b. prevent water and minerals from entering the stele through the apoplast.
   c. provide regulation for water and mineral movement in the plant.
   d. All of the above
   e. None of the above
   Textbook Reference: *35.1 How Do Plants Take Up Water and Solutes? p. 744*
2. The primary difference between the apoplast and the symplast is that
   a. the apoplast consists of nonliving spaces and cell walls, whereas the symplast consists of living cells.
   b. apoplast movement is tightly regulated and symplast movement is not.
   c. the symplast consists of nonliving spaces and cell walls, whereas the apoplast consists of living cells.
   d. apoplast movement is slow and symplast movement is fast.
   e. the apoplast transports only ions and the symplast transports only water.
   Textbook Reference: *35.1 How Do Plants Take Up Water and Solutes? p. 743*
3. Which of the following statements about water transport is true?
   a. Root pressure is sufficient to drive xylem sap movement.
   b. Bulk flow is not a mechanism by which water and minerals are transported.
   c. The cohesive nature of water is central to water movement in a plant.
   d. Water transport is an active process.
   e. None of the above
   Textbook Reference: *35.2 How Are Water and Minerals Transported in the Xylem? p. 746*
4. Tension in the xylem is a result of
   a. transpiration at the leaf surface.
   b. the cohesive nature of water.
   c. the narrowness of the xylem tube.
   d. the surface area of the phloem.
   e. All of the above
   Textbook Reference: *35.2 How Are Water and Minerals Transported in the Xylem? p. 746*

5. The fact that water transport continues as long as leaves are alive and active indicates that
   a. leaves pump water.
   b. leaves are necessary for transport of water.
   c. roots are active.
   d. water is not needed for leaves to remain alive.
   e. sieve tube elements are inactive.
   Textbook Reference: *35.2 How Are Water and Minerals Transported in the Xylem? p. 745*
6. Which of the following statements regarding transport in phloem is true?
   a. It always moves in the direction of leaves to roots.
   b. It proceeds from source tissue to sink tissue.
   c. It requires no energy inputs from the plant.
   d. It is the same process as transport in xylem.
   e. None of the above
   Textbook Reference: *35.4 How Are Substances Translocated in the Phloem? p. 750*
7. If the pressure potential of a plant's cells is 0.16 megapascals (MPa) and the solute potential is –0.24 MPa, then the water potential would be
   a. 0.04 MPa.
   b. 0.08 MPa.
   c. –0.08 MPa.
   d. –0.24 MPa.
   e. –0.04 MPa.
   Textbook Reference: *35.1 How Do Plants Take Up Water and Solutes? p. 740*
8. Which of the following represents the correct ordering of the water potential of these root cells or regions, from least negative to most negative?
   a. Xylem, cortex apoplast, stele apoplast, soil next to root
   b. Soil next to root, xylem, stele apoplast, cortex apoplast
   c. Xylem, stele apoplast, cortex apoplast, soil next to root
   d. Stele apoplast, cortex apoplast, xylem, soil next to root
   e. Soil next to root, cortex apoplast, stele apoplast, xylem
   Textbook Reference: *35.1 How Do Plants Take Up Water and Solutes? pp. 740–744*
9. The movement of water up the stems of tall plants is *least* dependent on which of the following factors?
   a. Guttation
   b. Transpiration
   c. Cohesion of water molecules
   d. Tension within water columns
   e. All of the above are of equal importance.
   Textbook Reference: *35.2 How Are Water and Minerals Transported in the Xylem? pp. 745–746*
10. Which of the following statements about xylem transport and phloem transport is true?
   a. Both are passive processes that do not require energy from the plant.
   b. Both rely on only living cells.
   c. Both rely on a water potential gradient.
   d. The direction of flow can reverse in both.
   e. The driving force for both is in the leaves.
   Textbook Reference: *35.2 How Are Water and Minerals Transported in the Xylem? pp. 746–747; 35.4 How Are Substances Translocated in the Phloem? p. 750*
11. Stomatal opening and closing are regulated by
   a. abscisic acid levels.
   b. light levels.
   c. carbon dioxide concentrations.
   d. All of the above
   e. None of the above
   Textbook Reference: *35.3 How Do Stomata Control the Loss of Water and the Uptake of CO_2? pp. 748–749*
12. The opening and closing of the stomata are accomplished by the
   a. sieve tube.
   b. guard cells.
   c. process of translocation.
   d. aquaporins.
   e. xylem.
   Textbook Reference: *35.3 How Do Stomata Control the Loss of Water and the Uptake of CO_2? p. 748*
13. Regulators of stomatal opening and closing work by activating the
   a. proton pump in guard cells.
   b. proton pump in stomata.
   c. sodium–potassium pump in guard cells.
   d. sodium–potassium pump in stomata.
   e. All of the above
   Textbook Reference: *35.3 How Do Stomata Control the Loss of Water and the Uptake of CO_2? p. 748*
14. Mineral ions enter the cell due to the force of an electrochemical gradient set up by the pumping of _______ out of the cells.
   a. K^+
   b. Ca^{2+}
   c. Na^+
   d. H^+
   e. Cl^-
   Textbook Reference: *35.1 How Do Plants Take Up Water and Solutes? p. 743*
15. In the pressure flow model of translocation, the movement of water by osmosis occurs
   a. from the xylem to the phloem at the sink.
   b. from the phloem to the xylem at the source.
   c. from the xylem to the phloem at the source.
   d. in both directions at the sink.
   e. None of the above
   Textbook Reference: *35.4 How Are Substances Translocated in the Phloem? p. 752*

Application Questions

1. Differentiate between source and sink tissues. What happens relative to phloem in each?
   Textbook Reference: *35.4 How Are Substances Translocated in the Phloem? p. 750*

2. Under what conditions does transpiration occur most rapidly? What effect does increased transpiration have on water flow in a plant? What happens if adequate water for the plant is not available?
   Textbook Reference: *35.2 How Are Water and Minerals Transported in the Xylem? pp. 746–747*
3. Describe the role of the proton pump in moving minerals into the root.
   Textbook Reference: *35.1 How Do Plants Take Up Water and Solutes? p. 743*
4. Explain how transpiration, cohesion, and tension work together to move water in a large plant.
   Textbook Reference: *35.2 How Are Water and Minerals Transported in the Xylem? p. 746*
5. Describe the pressure flow model of phloem transport.
   Textbook Reference: *35.4 How Are Substances Translocated in the Phloem? pp. 751–752*

Answers

Diagram Exercise Answer

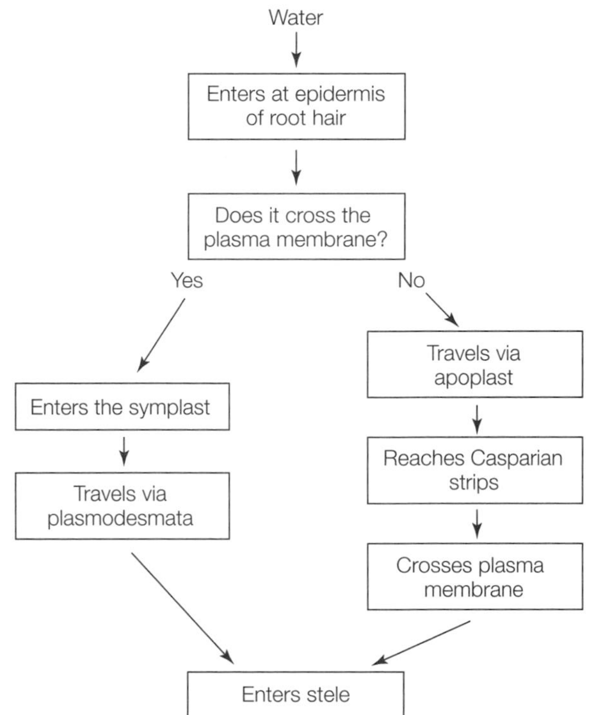

Knowledge and Synthesis Answers

1. **d.** Not all minerals that enter the apoplast of a plant's root are beneficial to the plant. The Casparian strips prevent water and minerals from reaching the stele through the apoplast, diverting them instead through the plasma membranes of the endodermal cells. Channel proteins in these plasma membranes determine which minerals can enter the symplast, and from there, the stele. The Casparian strips thus contribute to the regulation of water and mineral movement in the plant.
2. **a.** The intercellular spaces and cell walls of the plant constitute the apoplast.
3. **c.** Water movement depends on the cohesive nature of water and its capacity to withstand the tension placed on the water column by transpiration.
4. **a.** Transpiration causes tension.
5. **b.** Leaves are necessary for transpiration to take place.
6. **b.** Transport in phloem does not always go from leaf to root, but it always proceeds from source tissue to sink tissue. The plant must contribute energy to create the water pressure gradient by pumping solutes into the phloem at the source and out of the phloem at the sink.
7. **c.** Water potential is equal to pressure potential plus solute potential.
8. **c.** The xylem would be the most negative, followed by the stele, then the cortex, then the area outside the root.
9. **a.** Guttation occurs under extremely humid conditions when water is plentiful.
10. **c.** Both xylem transport and phloem transport depend on water potential.
11. **d.** Abscisic acid, light, and carbon dioxide levels all regulate stomatal opening and closing.
12. **b.** Guard cells are specialized epidermal cells that regulate the opening and closing of the stomata by covering the stomata opening.
13. **a.** Stomatal regulators work by activating and deactivating the proton pump in guard cells.
14. **d.** Cells pump H^+ ions out into the soil with the help of proton pumps.
15. **c.** In the pressure flow model, water moves by osmosis from the xylem into the phloem at the source and from the phloem to the xylem at the sink. This movement is all driven by the solute concentrations at each location.

Application Answers

1. Source tissues produce sugars in excess of what can be used and stored. Phloem loading occurs in source tissues and creates a pressure potential that results in bulk flow of sieve tube sap toward sink tissues. Sink tissues produce fewer sugars than can be stored or used and unload phloem through active transport.
2. Transpiration occurs most rapidly in high light conditions when stomata are open, along with high wind conditions and low humidity when evaporation is greatest. This results in faster bulk flow through the xylem and increased water demands by the plant. If water is not available, plant cells lose turgor and the plant wilts.
3. Plants rely on a proton pump for active transport of minerals into cells. Plants pump protons out of cells, which causes the area outside the cell to be more positive. This assists facilitated diffusion of positive ions

through protein channels. It also drives the movement of negatively charged ions into the cell by active transport (see Figure 35.5).

4. The transpiration–cohesion–tension mechanism pulls water from the roots up through the plant. Water evaporates from mesophyll cells during transpiration. This puts tension on the film of water associated with the mesophyll cell wall. The tension at the mesophyll cell draws water from the xylem of the nearest vein. This creates tension in the entire xylem column, and the column is drawn upward from the roots.
5. The difference in solute concentration between sources and sinks creates a pressure potential along sieve tubes, resulting in bulk flow. For this to occur, sugars must be loaded at the source tissue and unloaded at the sink tissue through active transport, and the sieve plates must remain open and unclogged along the phloem column.

36 Plant Nutrition

The Big Picture

- Plants require specific macro- and micronutrients. Deficiencies in any of these challenge the health of the plant. Essential nutrients must be available, and no substitutions will sustain the plant. Nutrients are procured from the soil solution that bathes the roots of a plant. The availability of nutrients depends on the quantity, solubility, and structure of soil. Many agricultural practices deplete soils of nutrients, and these must be replenished through fertilization.
- Plants interact with fungi and bacteria that help them obtain needed nutrients. Nitrogen availability is essential to plant growth. Bacteria and plants are intrinsically linked in the nitrogen cycle (see Figure. 36.11). Agriculture makes use of biological nitrogen fixation in the practice of crop rotation, in which farmers plant nonharvested crops that harbor nitrogen-fixing bacteria in their root nodules. Commercial nitrogen fixation by means of chemical fertilizers is extremely energy demanding.
- A small number of plants do not photosynthesize. These heterotrophic plants are often parasites of other plants and acquire nutrients solely through their hosts.

Common Problem Areas

- The interactions between nitrogen-fixing bacteria and plants can become confusing. Remember that it is a mutualistic relationship and that the bacteria have the enzymes necessary to fix atmospheric nitrogen.
- The most difficult concept of this chapter is the nitrogen cycle. The figures in the textbook should help you visualize this process and understand how organisms interact with the environment.

Study Strategies

- Be sure you understand the consequences of nutrient shortages in plants.
- Go to yourBioPortal.com to review the following tutorial and activity:

  Animated Tutorial 36.1 Nitrogen and Iron Deficiencies

  Web Activity 36.1 The Nitrogen Cycle

Important Concepts

All organisms must acquire nutrients through the cycling of other compounds or by uptake from their environment.

- The basic nutrient requirements for all living things are carbon, hydrogen, oxygen, and nitrogen. These elements are the fundamental building blocks of all macromolecules.
- Plants are autotrophs and carbon is incorporated through photosynthesis. Oxygen and hydrogen enter plants as water. Nitrogen's entry into plants is dependent on nitrogen-fixing bacteria in the soil.
- Mineral nutrients are essential to life. Sulfur, phosphorus, magnesium, and iron are all essential components of many macromolecules.
- Mineral nutrients enter biological organisms through soil solutions, which plants take up through their roots.
- Plants are sessile, meaning that they cannot move around and therefore cannot "search" for nutrients. They overcome this problem by growing toward new resources. A plant grows taller to procure more sunlight for itself and to outcompete nearby plants, and its roots spread to acquire mineral nutrients in the soil.

Mineral nutrients are vital to proper plant growth.

- Every plant requires specific essential nutrients for growth and development. These nutrients cannot be replaced by another element and must be obtained directly for the day-to-day functioning of the plant. A deficiency in any essential nutrient leads to an unhealthy plant.
- Macronutrients are essential elements required at a rate of 1 g per 1 kg of dry plant matter, and micronutrients are essential elements required at a rate of 100 mg per 1 kg of dry plant matter.
- See Table 36.1 for a list of macro- and micronutrients, their sources, and their functions. Though many nutrient deficiencies ultimately lead to plant death, specific symptoms of deficiency are evident in a plant before it dies. Deficiency can be corrected by the addition of fertilizers to supplement nitrogen in soils.

- Scientists determine which elements are essential by growing plants hydroponically, which allows for manipulation of the nutrients that are available to the plant.

Soils are nature's nutrient sink.

- Plants and soils interact in a complex fashion. Plants change soils, and soils influence the growth of plants.
- Soils are composed of living and nonliving matter. The living portion of soil contains roots, protists, bacteria, fungi, and many small animals. The nonliving portion consists of rock fragments, clay, water, air spaces, and dead organic matter. The air spaces provide the O_2 needed for a plant to survive.
- All soils have a soil profile consisting of two or more horizons (layers). Water-soluble nutrients are leached to deeper horizons through rainfall. Three major horizons (also called zones), termed A, B, and C, can be identified in soils.
    - The A horizon is topsoil. It is organically rich, very biologically active, and the most agriculturally important layer. A loam is a topsoil with an optimal mixture of sand, silt, and clay; it has high nutrient content, plentiful water, and adequate air spaces. Soils with too much sand typically do not hold nutrients or water well, and clays are too dense for the trapping of air.
    - The B horizon is the subsoil, which holds many leached nutrients.
    - The C horizon is parent rock, which roots cannot penetrate.
- Soils form from mechanical and chemical weathering that breaks down rocks. Mechanical weathering is cause by freeze and thaw cycles, rain, and drying. Chemical weathering is caused by oxidation, hydrolysis, or the breakdown by acids, and it can change the composition of rock. Chemical weathering is very important in clay formation.
- Mineral nutrients are tied to clay particles in the soil. Because many nutrients are positively charged cations, clays with a negative charge can hold these nutrients and make them available to plants. Roots release protons into the soil that bind to clay particles, and the cation minerals are released. The CO_2 released from the roots can also form bicarbonate and free protons, which bind with clay. These processes are called ion exchange. Nutrients that are negatively charged and therefore do not participate in ion exchange are rapidly leached from soil. Thus, nitrate and sulfate are often not available to plants.
- Agricultural fertilizers add nitrogen, phosphorus, and potassium to soils and are rated by their "N-P-K" percentages. A 10-10-10 fertilizer contains 10 percent nitrogen, 10 percent phosphate, and 10 percent potash (potassium source). These nutrients need to be replenished in soils because they are negatively charged and leach from soils rapidly.
- Fertilizers may be organic in nature (manures, compost) or inorganic (chemical fertilizer). Organics release nutrients much more slowly and do not leach as quickly as inorganics. Inorganic fertilizers provide a much more rapid release of nutrients.
- The pH of soils affects nutrient availability. Specific plants have specific pH needs. Soils tend to become slightly acidic from leaching and rain. The practice of liming raises soil pH. This has the secondary effect of making calcium available to plants. Adding sulfur reduces soil pH.
- Copper, iron, and manganese can be added by foliar spraying of nutrients.
- Plants affect the pH of soil, adding decaying matter in the form of humus and altering soil temperature. The plant roots also can alter the soil pH by excreting H^+ or OH^- ions.

Bacteria and fungi help plants obtain nutrients.

- Fungi form associations with roots to form mycorrhizae. The fungi help expand the surface area of the root to increase nutrient uptake. The fungi help plants take up phosphorus, while the plants provide the fungi with sugars for energy.
- Formation of mycorrhizae is stimulated by the release of strigolactones by the root. A prepenetration apparatus helps guide the fungi as they grow into the root. Arbuscules in the cortical cells of the roots are the site of nutrient exchange.
- Nitrogen gas is readily available in the atmosphere, but plants are unable to break the triple bonds between the two nitrogens. Only a few bacteria species can fix nitrogen gas into biologically usable ammonia through nitrogen fixation.
- The fixation of nitrogen gas requires the enzyme nitrogenase to catalyze the reaction, lots of energy (ATP), and a strong reducing agent. Nitrogenase is inhibited by oxygen and therefore is active only under anaerobic conditions. The nitrogen-fixing bacteria are typically found in root nodules that maintain very low oxygen levels. The protein leghemoglobin binds with oxygen to keep oxygen levels low.
- The most biologically important nitrogen fixers are associated with plant roots and release up to 90 percent of the nitrogen they fix to the soil.
    - Rhizobium bacteria have a close association, or mutualistic symbiosis, with the roots of plants in the legume family; both species benefit from the association. These bacteria and the associated legumes are important agriculturally and are the basis of crop rotation.
    - The anaerobic rhizobium bacterium resides in a root nodule of the legume, where it is protected from oxygen. The bacteria are provided with a low-oxygen

growing environment, and the plant benefits from the released ammonia. To establish the symbiosis, the plant releases flavonoids to attract the bacteria. In response to the flavonoids, the bacteria turn on the production of Nod factors, which cause formation of the nodule by the plant. Once housed in the nodule, the bacteria form swollen bacteroids capable of nitrogen fixation. The plant surrounds the bacteroids with leghemoglobin to support respiration (see Figure 36.9).

- Bacterial nitrogen fixation is not adequate to supply all the nitrogen needed for agriculture. Currently, nitrogen fertilizers are produced via industrial fixation through the Haber process, an energy-expensive process.
- In high levels, ammonia is toxic to plants, but at low levels it is used to produce amino acids. Therefore, soil bacteria that convert ammonia to nitrate are necessary. These bacteria are called nitrifiers. Plants, through their metabolism, reduce nitrate to ammonia. This is accomplished by the plant's own enzymes.
- The entire nitrogen cycle is completed by denitrifiers, which convert the nitrogen from waste and dead organic matter back to nitrogen gas. (See Figure 36.11 for a review of the nitrogen cycle.)

Some plants are heterotrophic.

- Some plants that live in nitrogen- or phosphorus-deficient soils are carnivorous. Carnivorous plants acquire nitrogen from the proteins of trapped decaying animals. Examples of carnivorous plants are Venus flytraps, pitcher plants, and sundews. Although these plants can survive and grow without consuming insects, they thrive when they have a continuous supply of them.
- Some plants have lost the ability to photosynthesize and must acquire their nutrients from other sources. Some are parasitic and acquire some or all of their nutrients from a host plant at the host plant's expense.

Test Yourself

Diagram Exercise

Diagram the nitrogen cycle.
Textbook Reference: *36.4 How Do Fungi and Bacteria Increase Nutrient Uptake by Plant Roots? p. 766*

Knowledge and Synthesis Questions

1. Which of the following nutrients is *not* considered essential for plant growth?
   a. Cadmium
   b. Nitrogen
   c. Manganese
   d. Potassium
   e. All are essential.
   Textbook Reference: *36.2 What Mineral Nutrients Do Plants Require? p. 757*

2. Compared to micronutrients, macronutrients are
   a. larger.
   b. needed in greater quantities.
   c. more essential.
   d. of equal importance.
   e. less essential.
   Textbook Reference: *36.2 What Mineral Nutrients Do Plants Require? p. 757*

3. Nitrogen and potassium are acquired from
   a. the soil solution.
   b. heterotrophs.
   c. air.
   d. micronutrients.
   e. All of the above
   Textbook Reference: *36.1 How Do Plants Acquire Nutrients? p. 756*

4. A tomato plants whose young leaves are very yellow, but whose older leaves are still green, most likely has a(n)_______ deficiency.
   a. nitrogen
   b. carbon
   c. water
   d. iron
   e. phosphorus
   Textbook Reference: *36.2 What Mineral Nutrients Do Plants Require? p. 757*

5. Years of cotton farming in the South have stripped away much of the A horizon of the soils. Subsequent agriculture
   a. has been problematic, because the A horizon contains the most available nutrients.
   b. has not been affected, because the B horizon contains significantly more available nutrients.
   c. has been affected slightly, because the C horizon is most conducive to root growth.
   d. has been problematic, because stripping of the A horizon leaches the C horizon of its nutrients.
   e. All of the above
   Textbook Reference: *36.3 How Does Soil Structure Affect Plants? p. 760*

6. Clay particles in soils are important for
   a. holding the soil together.
   b. ion exchange.
   c. holding water.
   d. All of the above
   e. None of the above
   Textbook Reference: *36.3 How Does Soil Structure Affect Plants? p. 760*

7. Most clays form from the _______ of rock.
   a. mechanical weathering
   b. chemical weathering
   c. heaving
   d. grinding
   e. All of the above
   Textbook Reference: *36.3 How Does Soil Structure Affect Plants? p. 760*

8. The label "10-20-10" on a package of commercial fertilizer refers to the _______ the fertilizer.
   a. percentages of nitrogen, phosphate, and potassium in
   b. percentages of nitrogen, carbon, and oxygen in
   c. percentages of phosphate, iron, and potassium in
   d. rate at which nitrogen is released from
   e. ratio of organic to inorganic matter in

   Textbook Reference: *36.3 How Does Soil Structure Affect Plants? p. 761*
9. The relationship between rhizobium bacteria and the roots of legumes can best be described as
   a. parasitic.
   b. one-sided.
   c. mutualistic.
   d. carnivorous.
   e. detrimental.

   Textbook Reference: *36.4 How Do Fungi and Bacteria Increase Nutrient Uptake by Plant Roots? p. 762*
10. Nitrogen gas is reduced to ammonia by which of the following enzymes or processes?
   a. Rhizobium
   b. Nitrogenase
   c. Nitrification
   d. Denitrification
   e. Rhizobenase

   Textbook Reference: *36.4 How Do Fungi and Bacteria Increase Nutrient Uptake by Plant Roots? p. 763*
11. Plants are able to take up and use nitrogen in the form of _______ and _______.
   a. ammonia; nitrate
   b. ammonia; nitrogen gas
   c. nitrogen gas; nitrate
   d. nitrogen gas; nitrous oxide
   e. All of the above

   Textbook Reference: *36.4 How Do Fungi and Bacteria Increase Nutrient Uptake by Plant Roots? p. 766*
12. Carnivorous plants are often found in acidic and nutrient-poor environments. The main selective pressure for carnivory is
   a. lack of nitrogen and phosphorus sources.
   b. lack of iron and calcium sources.
   c. incomplete ion exchange.
   d. lack of water sources.
   e. All of the above

   Textbook Reference: *36.5 How Do Carnivorous and Parasitic Plants Obtain a Balanced Diet? p. 767*
13. Nitrate and sulfate tend to leach from the soil because
   a. they bind with ions such as K^+ and Mg^{2+}.
   b. the H^+ ions released by the roots push them out.
   c. they are unable to bind with the negatively charged clay particles.
   d. they bind with the positively charged clay particles.
   e. All of the above

   Textbook Reference: *36.3 How Does Soil Structure Affect Plants? pp. 760*
14. A plant lowers the pH of soil by means of
   a. ion exchange.
   b. leaching.
   c. Na^+ pumping.
   d. proton pumping.
   e. All of the above

   Textbook Reference: *36.3 How Does Soil Structure Affect Plants? pp. 760–761*
15. The role of leghemoglobin is to maintain _______ levels in the root nodule.
   a. high O_2
   b. high CO_2
   c. low O_2
   d. low CO_2
   e. high N_2

   Textbook Reference: *36.5 How Do Fungi and Bacteria Increase Nutrient Uptake by Plant Roots? p. 763*

Application Questions

1. Explain how scientists determined which plant nutrients are the essential nutrients.

   Textbook Reference: *36.2 What Mineral Nutrients Do Plants Require? p. 758*
2. Describe how plants and bacteria interact to form nitrogen-fixing root nodules. Why do many farmers plant crops such as alfalfa and soybeans without harvesting them?

   Textbook Reference: *36.4 How Do Fungi and Bacteria Increase Nutrient Uptake by Plant Roots? pp. 764–765*
3. Most carnivorous plants are found in boggy, wet, acidic environments. What is the effect of such an environment on nutrient availability?

   Textbook Reference: *36.5 How Do Carnivorous and Parasitic Plants Obtain a Balanced Diet? p. 767*
4. Explain how plants "grow into their nutrients."

   Textbook Reference: *36.1 How Do Plants Acquire Nutrients? p. 756*
5. Differentiate between micro- and macronutrients. Where are most of these nutrients acquired?

   Textbook Reference: *36.2 What Mineral Nutrients Do Plants Require? p. 757*

Answers

Diagram Exercise Answer

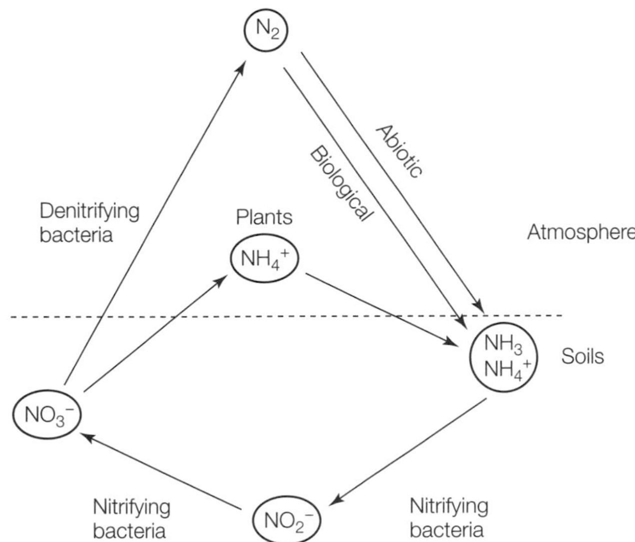

Knowledge and Synthesis Answers

1. **a**. Cadmium is not one of the 14 essential micro- and macronutrients.
2. **b.** The main difference between micronutrients and macronutrients is in the quantity of each needed by a plant for survival.
3. **a.** Nitrogen and mineral nutrients are acquired in soil solution (dissolved in water).
4. **d.** If the older leaves look normal, but the younger leaves are yellow, an iron deficiency should be suspected.
5. **a.** Topsoil, or the A horizon, is most conducive to root growth. Ample nutrients are available, as are air spaces and water for ease of root growth.
6. **d.** Clay particles are critical for ion exchange. They are also important for retaining water and for the integrity of the soil.
7. **b.** Chemical weathering leads to clay formation.
8. **a.** The fertilizer is 10 percent nitrogen, 20 percent phosphorus, and 10 percent potash (a potassium source).
9. **c.** The relationship is mutualistic, in that both the plant and the bacteria benefit from the association.
10. **b.** Nitrogenase catalyzes the reduction of nitrogen gas to ammonia. This is an energy-expensive process.
11. **a.** Plants can take up and use nitrogen that is in the form of ammonia or nitrate.
12. **a.** Carnivory supplements insufficient nitrogen and phosphorus availability.
13. **c.** Nitrate and sulfate are negatively charged anions. Without any positively charged molecules in the soil with which they can interact, they leach from the soil.
14. **d.** Plants decrease the pH in the surrounding soil by pumping protons out of the roots.
15. **c.** Leghemoglobin is a protein produced by the root nodule of plants to help maintain low levels of O_2 so that the nitrogen-fixing bacteria can have an anaerobic environment.

Application Answers

1. Plants were grown in cultures that lacked specific nutrients. If a plant could not complete its life cycle, the missing nutrient was said to be essential. These experiments were well-controlled hydroponic studies because nutrient content can easily be manipulated in water.
2. See Figure 36.9. By rotating crops (and especially by rotating and plowing under legume crops) farmers can organically add nitrogen to depleted soils. This results in significantly larger yields of the harvested crops.
3. Acidic environments limit decomposition and thus nitrogen sources. They also limit ion exchange. Both conditions result in reduced nutrient availability that can be offset by carnivory.
4. Plants cannot move physically to areas with greater nutrients. They are limited to extending their roots into soils that contain a larger nutrient reserve. When nutrients are limited, plant growth matches the availability of resources.
5. Macronutrients are needed at a rate of 1 g/1 kg of dry plant tissue. Micronutrients are needed at a rate of 100 mg/1 kg of dry plant tissue. The majority of these nutrients are in soil solution and are taken up as water is drawn into roots.

37 Regulation of Plant Growth

The Big Picture

- Plant growth is controlled by interactions between a plant's environment, hormones, and genetic makeup. Changes in plant growth are dependent on the information transmitted by hormones to hormone receptors, the subsequent turning on of signal transduction pathways, and the eventual alteration of gene expression. Receptors are often highly specific and respond to select hormones. Hormones are produced in a specific region of the plant body and translocated throughout the plant; therefore, their effects are often concentration dependent. All growth in a plant is a result of changes in cell division, cell expansion, and cell differentiation.
- Seed germination, growth of the vegetative structures, reproduction, and senescence all proceed in defined patterns. Seed dormancy is broken, and development begins when the seed coat is abraded, inhibitory chemicals are diluted, and the seed imbibes water. This begins a series of events that mobilize nutrients and induce growth. Once a seedling emerges from the soil, light begins to influence subsequent development under the control of hormones.
- Hormones exist in several classes, each with its own effects on growth and development. Some of these effects are antagonistic, and therefore control is maintained via relative concentration of several hormones. Hormones influence every step of development, from the breaking of seed dormancy to senescence.
- Light regulates plant processes through photoreceptors. Photoreceptors respond to very specific wavelengths and induce cascades that lead to changes in a plant.

Common Problem Areas

- Understanding how phytochrome shifts from the red to far-red forms frequently gives students difficulty. This is understandable, because phytochromes have puzzled researchers for many years.
- There is tendency when studying hormones simply to memorize functions. You need to understand the effects in a plant of different ratios of hormones, not what one individual hormone does.
- The amount of information in this chapter may seem overwhelming. Take your time in learning how growth is regulated and the effects of the various hormones.

Study Strategies

- Avoid focusing too much on details. Try to assimilate the big picture of how the environment, receptors, hormones, and genome interact. From there, begin to work toward the details. A big mistake is to jump in and memorize functions of hormones or sequences of development without understanding the broad picture.
- Go to yourBioPortal.com to review the following tutorials and activities:

  Animated Tutorial 37.1 Tropisms

  Animated Tutorial 37.2 Went's Experiment

  Animated Tutorial 37.3 Auxin Affects Cell Walls

  Web Activity 37.1 Monocot Shoot Development

  Web Activity 37.2 Eudicot Shoot Development

  Web Activity 37.3 Events of Seed Germination

Important Concepts

Plant development is influenced by multiple regulatory factors.

- Plant seeds are dormant and remain so until seed germination. The germination stage is complete once the embryonic root, known as the radicle, emerges from the seed. A plant is considered a seedling when its radicle breaks through the seed coat to end germination.
- Dormancy, which lasts for different periods of time depending on the plant, involves exclusion of water or oxygen from the embryo by the seed coat, mechanical restraint of the embryo by the seed coat, and chemical inhibition of the embryo. Dormancy ensures survival through adverse conditions. Some seeds rely on other environmental cues before germination, such as cold temperatures, light, or the passage of a specific amount of time. These cues help ensure that the seed will germinate in the correct location and at the correct time.

- Germination is triggered by one or more mechanical or environmental cues. To germinate, a plant must imbibe water and draw polysaccharides, fats, and protein nutrients from the endosperm or cotyledons. Dormancy can be broken by mechanical abrasion or fire. Prolonged exposure to water may leach chemical inhibitors away from the seed and induce germination.
- Plant development depends on the interplay of environmental cues, receptors that sense these cues, and hormones that mediate the effects of the environment. Enzymes influence development at all stages.
- Hormones are regulatory compounds that are produced in one area of a plant and translocated throughout it. The effects of the hormones are determined by their relative concentrations, and they play multiple regulatory roles in plants.
- Photoreceptors are pigment proteins that sense light and are altered by light quality to induce changes within a plant.
- Signal transduction mediates hormone and photoreceptor action. A receptor receives an environmental cue, a biochemical signal transduction pathway is initiated, and ultimately a cellular response is generated through a protein kinase.
- A plant's genome ultimately regulates its growth and development, under the influence of hormones and photoreceptors.

Gibberellins regulate growth from germination to fruiting.

- Gibberellins are produced in both plants and fungi. The first gibberellin was isolated from a fungus that infects rice plants and causes them to grow tall and spindly. Initial experiments indicated that the medium that the fungus was grown in was sufficient to cause rice plants to be spindly; therefore, the agent that caused the phenomenon was a chemical produced by the fungus.
- Gibberellins are involved in stimulating elongation of the plant stem.
  - Experiments with dwarf and normal corn and tomato plants showed that plants have innate gibberellins. Normal corn plants exposed to gibberellins showed no alteration in appearance, but dwarf plants exposed to gibberellins showed shoot elongation to near normal lengths. These experiments proved that the chemical is present in plants.
  - In plants such as cabbage, gibberellins cause the rapid production of a tall stem of flowers. Inhibition of gibberellins produces a shorter plant.
- Gibberellins play a role in fruit development. Developing seeds produce gibberellins that enhance development of fruit tissue. It is common agricultural practice to spray seedless fruits (especially grapes) with gibberellins to enhance fruit growth. In the developing seed, gibberellins stimulate the aleurone layer, a tissue layer under the seed coat, to secrete enzymes that break down the seed coat.
- Scientists have studied mutant plants to work out the mechanisms by which gibberellins work. The two main mutant types are excessively tall plants and dwarf plants. In the tall plants, the pathway by which gibberellins act is always turned on. In the dwarf plants, the pathway by which gibberellins act is always off. These differences are due to a mutation in a repressor gene of a transcription factor for a growth-promoting gene. In normal plants, gibberellin removes this repressor, allowing the growth-promoting gene to be transcribed.

Auxin affects plant growth and form.

- The discovery of auxin (indole-3-acetic acid) was the result of work of Charles and Francis Darwin in the 1880s. They were interested in how plants grew toward light by phototropism and which part of an emerging plant coleoptile was responsible for sensing light. They found that the tip was the light-receptive portion, but that the growing region (responsible for the bending to or away from light) was some distance below the receptor region. From this they reasoned that some chemical must be transmitted from the tip to the growing region. Additional experiments with removed tips and tips on gelatin blocks indicated that a chemical was indeed moving from the tip to the growing region. Subsequent experiments by other researchers showed that gelatin exposed to the tips was sufficient to cause altered growth. The chemical was later isolated and determined to be auxin.
- Auxin movement in plant tissues is unidirectional and polar, from apex to base. Auxin enters the cell in its nonpolar acid form by passive diffusion, and proton pumps transport H^+ out of the cell, causing the nonpolar auxin to become an anion. Auxin anion efflux carriers are carrier proteins at the basal end of the cell responsible for export of auxin anions from cells, and they contribute to the unidirectional movement of auxin.
- Redistribution of auxin laterally is responsible for phototropism and gravitropism (see Figure 37.11). Phototropin in the membrane of plants is responsible for the phototropic response by stimulating the transport of auxin to the cells on the shaded side of the plant. The gravitational settling of starch-rich plastids may be the trigger for the release of auxin from the bottom of the root or shoot. Under both conditions, higher auxin concentrations on one side of the plant cause increased rates of growth along that side and lead to bending.
- Auxin affects vegetative growth by initiating root growth in cuttings and promoting and maintaining the growth of a single main stem (apical dominance). Auxin inhibits abscisson (the dropping of leaves) and can stimulate unfertilized fruit to form (parthenocarpy).
- The effects of auxin on growth are mediated by the cell walls, which determine the rate and direction of cell growth. Cells grow by taking up water. The amount of

water that can be taken up is restricted by a rigid cell wall. Cell walls must loosen, stretch, and add polysaccharides and cellulose to maintain structure as the cell expands. According to the acid growth hypothesis of cell expansion, auxin stimulates the production and insertion of proton pumps into the plasma membrane. The subsequent decrease in pH stimulates proteins called expansins to alter polysaccharide bonding so that they slide past one another during expansion.

- There is a receptor that binds with auxin and promotes gene expression. A similar pathway has been observed for the action of gibberellin.

Cytokinins have multiple roles within plants.

- Cytokinins are powerful stimulators of cell division and bud formation, aid in seed germination, inhibit stem elongation, and delay leaf senescence. They are synthesized primarily in the roots and are translocated throughout the plant.
- Auxins and cytokinins regulate organ development based on the relative concentrations of the two. High auxin levels favor root formation, and high cytokinin levels favor bud formation.
- Cytokinins act through a two-component system similar to that found in bacteria. A receptor (AHK) phosphorylates proteins, and a target transcription factor (ARR) acts as an effector. An intermediate protein (AHP) transfers the phosphate from the receptor to the transcription factor.

Ethylene promotes senescence.

- Ethylene is a gaseous hormone that promotes leaf senescence and fruit ripening. In many instances it is given off by rotting fruit. The use of ethylene spray to promote fruit ripening is a common commercial practice. Ethylene scrubbers are used in fruit storage to prevent ethylene from accumulating and causing fruit to spoil. Other chemicals are used in the flower industry to inhibit ethylene's effects on flower senescence.
- Ethylene plays a role in maintaining the apical hook on emerging eudicots by inhibiting cells on the inner portion of the hook. Ethylene inhibits stem elongation, promotes lateral swelling of stems, and inhibits sensitivity to gravitropic stimulation. These three responses are known as the triple response.
- The signal transduction pathway by which ethylene exerts its effects is initiated by the binding of ethylene to a receptor on the endoplasmic reticulum (see Figure 37.17). This binding initiates a cascade, starting with the activation of endoplasmic reticulum channels. Ultimately the activation of a transcription factor promotes the expression of genes, resulting in physiological changes.

Brassinosteroids are involved in the response to light.

- Brassinosteroids were originally isolated from a member of the Brassicaceae. They have been shown to stimulate cell elongation, pollen tube elongation, and vascular tissue differentiation and to inhibit root elongation, all of which are similar to the effects of auxin. The brassinosteriod receptor is found on the cell wall and initiates a transduction pathway that influences gene expression.

Light and photoreceptors interact to stimulate a variety of plant events.

- Photoreceptors in plants interpret intensity, duration, and wavelength of light. Light regulates a wide variety of plant processes, including germination, flower production, and shoot elongation.
- Phototropin, a blue-light receptor, is a protein kinase that stimulates cell elongation by auxin. Zeaxanthin and phototropin act together to regulate light-stimulated opening of the stomata. Cryptochromes absorb blue and ultraviolet light and influence seedling development and flowering.
- Red light stimulates photomorphogenesis in plants. These developmental and physiological events can include germination, flowering, and production of chlorophyll in seedlings. Exposure to far-red light reverses the effects of exposure to red light, and vice versa. This "switching" occurs because phytochrome can be shifted from one form (red P_r) to the other (far-red P_{fr}) upon absorption of light. When exposed to red light, P_r is converted to P_{fr}. When exposed to far-red light, P_{fr} is converted to P_r.
- Phytochromes are composed of a protein chain that interacts with transcription factors and the pigment chromophore. Red light changes the conformation of the protein from the P_r to P_{fr} form and exposes a nuclear localization sequence. The P_{fr} form then moves into the nucleus, where it stimulates gene expression through a transcription factor. It can also act as a kinase and phosphorylate other proteins.
- Biological organisms exhibit a daily cycle in their functioning known as a circadian rhythm. The phytochromes are probably involved in the circadian rhythms of plants.

Test Yourself

Diagram Exercise

Both gibberellin and auxin act in a similar fashion at the molecular level. Create a flow chart describing the signal transduction pathways for gibberellin and auxin. Note their similarities and their differences. (Remember that the pathways work by the same mechanisms; the differences are in the actual receptors, repressors, and transcription factors involved in each one.)

Textbook Reference: *37.2 What Do Gibberellins Do? pp. 779; 37.3 What Does Auxin Do? p. 784*

Knowledge and Synthesis Questions

1. Plant growth is regulated by
   a. environmental cues.
   b. hormones.
   c. signal transduction pathways.
   d. the expression of the plant's genome.
   e. All of the above
   Textbook Reference: *37.1 How Does Plant Development Proceed? p. 773*
2. The mechanism by which gibberellins act involves
   a. the adding of proton pumps to the plasma membrane.
   b. the phosphorylation of proteins.
   c. binding with a transcription factor.
   d. the removal of a repressor from a transcription factor.
   e. None of the above
   Textbook Reference: *37.2 What Do Gibberellins Do? p. 779*
3. Which of the following triggers germination of a seed?
   a. The imbibing of water
   b. Its release from the fruit
   c. Chemical changes
   d. The exclusion of water
   e. All of the above
   Textbook Reference: *37.1 How Does Plant Development Proceed? p. 774*
4. Which of the following may have the effect of breaking dormancy in seeds?
   a. Penetration of the seed coat
   b. Leaching of inhibitory compounds by water
   c. Exposure to fire
   d. Passing through an animal's digestive tract
   e. All of the above
   Textbook Reference: *37.1 How Does Plant Development Proceed? p. 773*
5. Which of the following hormones is responsible for bud break in the spring in deciduous trees?
   a. Auxins
   b. Cytokinins
   c. Gibberellins
   d. Ethylene
   e. Brassinosteroids
   Textbook Reference: *37.1 How Does Plant Development Proceed? p. 775*
6. Which of the following is *not* involved in the acid growth hypothesis for the regulation of cell expansion by auxin?
   a. The pumping of protons into the cell wall
   b. The pumping of protons into the cytosol
   c. Increased gene expression of the proton pump gene
   d. Increased insertion of proton pumps into the plasma membrane
   e. All of the above are involved.
   Textbook Reference: *37.3 What Does Auxin Do? p. 783*
7. Which of the following is *not* involved in the polar transport of auxin?
   a. Diffusion across a plasma membrane
   b. Membrane protein asymmetry of auxin transport carriers
   c. Proton pumping from the cytosol
   d. Ionization of auxin as a weak acid
   e. All of the above are involved.
   Textbook Reference: *37.3 What Does Auxin Do? p. 781*
8. A homeowner has installed an outdoor gas-burning grill on her back patio next to her favorite camellia bush. After the first few nights of using the grill, she notices that the camellia is beginning to lose its leaves. Which of the following is the best explanation for what is happening?
   a. The bush is getting too warm next to the grill.
   b. Ethylene is a by-product of the burning gas and is causing senescence in the plant.
   c. Abscisic acid is a by-product of the burning gas and is causing senescence in the plant.
   d. The plant is a biennial and is bolting.
   e. Auxin production is being inhibited.
   Textbook Reference: *37.4 What Are the Effects of Cytokinins, Ethylene, and Brassinosteroids? p. 786*
9. Cytokinins interact with which other hormone?
   a. Ethylene
   b. Abscisic acid
   c. Gibberellins
   d. Auxins
   e. Brassinosteroids
   Textbook Reference: *37.4 What Are the Effects of Cytokinins, Ethylene, and Brassinosteroids? p. 785*
10. Which of the following light receptors is responsible for absorbing blue and ultraviolet light?
    a. Phytochrome P_r
    b. Phytochrome P_{fr}
    c. Cryptochrome
    d. Phototropin
    e. Etiolatin
    Textbook Reference: *37.5 How Do Photoreceptors Participate in Plant Growth Regulation? p. 788*
11. Etiolated seedlings are produced by germinating seeds that are kept in total darkness. Plants that are kept in the dark will begin to synthesize chlorophyll after they are given a pulse of
    a. blue light.
    b. red light.
    c. red light followed by a pulse of far-red light.
    d. far-red light followed by a pulse of red light.
    e. ultraviolet light.
    Textbook Reference: *37.5 How Do Photoreceptors Participate in Plant Growth Regulation? p. 789*
12. Ethylene is produced by what part of a plant?
    a. The seedling
    b. The leaves
    c. The fruit

d. All of the above
e. None of the above
Textbook Reference: *37.4 What Are the Effects of Cytokinins, Ethylene, and Brassinosteroids? p. 786*

13. Auxin transport within a plant is said to be _______ and it is dependent on the action of _______ pumps.
a. polar; proton
b. nonpolar; potassium
c. polar; potassium
d. bidirectional; proton
e. polar; sodium
Textbook Reference: *37.3 What Does Auxin Do? p. 781*

14. Which of the following is *not* initiated by auxin?
a. Stimulation of root initiation
b. Inhibition of leaf abscission
c. Stimulation of leaf abscission
d. Maintenance of apical dominance
e. Auxin is involved in all of the above.
Textbook Reference: *37.3 What Does Auxin Do? pp. 782–783*

15. Red light activation of the phytochorome into the P_{fr} state leads to which of the following events?
a. Inhibition of chlorophyll
b. Leaf expansion
c. Hook folding
d. Hook unfolding
e. Both b and d
Textbook Reference: *37.5 How Do Photoreceptors Participate in Plant Growth Regulation? p. 789*

Application Questions

1. Trace the steps that occur between the planting of a pea seed and the emergence of the pea plant.
Textbook Reference: *37.1 How Does Plant Development Proceed? pp. 773–774*

2. What hormonal influences are affecting a pea plant from the moment it is planted until its emergence?
Textbook Reference: *37.1 How Does Plant Development Proceed? pp. 773–774; 37.2 What Do Gibberellins Do? p. 778; 37.4 What Are the Effects of Cytokinins, Ethylene, and Brassinosteroids? pp. 784–787*

3. Discuss how gibberellins were discovered. How did researchers determine that they are chemical in nature?
Textbook Reference: *37.2 What Do Gibberellins Do? pp. 776–777*

4. Explain how auxin distribution regulates phototropism and gravitropism.
Textbook Reference: *37.3 What Does Auxin Do? p. 781*

5. Auxins and cytokinins appear to cancel out the effects of each other. Why is a hormone that affects bud growth produced in the roots while the one that affects root growth is produced in the shoots? How does this relate to the polar distribution of hormones?
Textbook Reference: *37.3 What Does Auxin Do? p. 781; 37.4 What Are the Effects of Cytokinins, Ethylene, and Brassinosteroids? p. 785*

6. Genetic engineering has produced fruits that are deficient in the ability to produce ethylene. How is this deficiency useful in the storage and marketing of fruits?
Textbook Reference: *37.4 What Are the Effects of Cytokinins, Ethylene, and Brassinosteroids? p. 786*

7. Suppose that a scientist is experimenting in the laboratory with tissue culture methods. Pith tissue that was isolated has grown an undifferentiated mass of cells (called a callus), and that tissue has been divided up and placed in the following culture media: (1) necessary nutrients plus indoleacetic acid; (2) necessary nutrients plus zeatin (a cytokinin); (3) necessary nutrients plus equivalent concentrations of indoleacetic acid and zeatin; (4) necessary nutrients plus an excess of zeatin and minimal indoleacetic acid. Explain what happens to the mass of tissue under each condition.
Textbook Reference: *37.4 What Are the Effects of Cytokinins, Ethylene, and Brassinosteroids? p. 785*

Answers

Diagram Exercise Answer

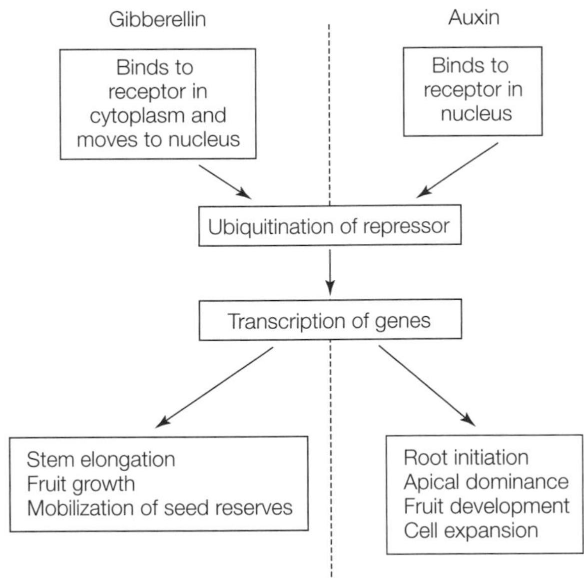

Knowledge and Synthesis Answers

1. **e.** The growth of a plant is regulated by all of these factors.
2. **d.** Gibberellin brings about a response in the plant cell by removing the repressor from a transcription factor, thus allowing transcription to occur.
3. **a.** The uptake of water by a seed begins the processes that lead to seed germination.
4. **e.** Mechanical abrasion, leaching of inhibitors by water, exposure to fire, and passing through a digestive tract may all trigger germination. Actual germination cannot begin until a seed imbibes water.
5. **c.** Gibberellins are responsible for bud break in deciduous trees.

6. **b.** The acid growth hypothesis involves an increase in the number of proton pumps in the plasma membrane and the pumping of protons into the cell wall.
7. **e.** All of the characteristics ensure that the transport of auxin is polar.
8. **b.** Ethylene gas promotes senescence and is one of the by-products of burning gas.
9. **d.** Cytokinins interact with auxins. The auxin-to-cytokinin ratio controls the bushiness of plants.
10. **c.** Cryptochromes respond to blue and ultraviolet light wavelengths.
11. **c.** The pulse of red light converts P_r to P_{fr}. Subsequent pulses of far-red light stimulate changes that lead to chlorophyll synthesis.
12. **d.** Ethylene can be produced by the fruit, seed, or leaves in most plants.
13. **a.** Auxin transport is a polar process, meaning that it moves in only one direction with the help of proton pumps and auxin anion efflux carriers (see Figure 37.10).
14. **c.** Auxin inhibits leaf abscission rather than stimulating it. Auxin is involved in all of the other processes.
15. **e.** The P_{fr} phytochrome stimulates chlorophyll synthesis, hook unfolding, and leaf expansion.

Application Answers

1. The watering of the pea seed after planting promotes the leaching of germination inhibitors. At this point the seed also begins to imbibe water, resulting in metabolic changes. DNA synthesis is halted until the radicle emerges from the seed coat. Breakdown of starch and protein reserves in the cotyledons and the endosperm begins to provide nutrients for the developing embryo. The apical hook begins to push through the soil. Upon its emergence and exposure to light, chlorophyll synthesis begins.
2. Dormancy of the seed is maintained by abscisic acid until it is leached from the seed by water. Once water is imbibed, cytokinins begin to influence germination. Gibberellins assist with the mobilization of storage products to the growing embryo. Ratios of auxins and cytokinins balance root and bud formation. The apical hook is maintained by ethylene.
3. The first gibberellin was isolated from a fungus that infects rice plants and causes them to grow tall and spindly. Initial experiments indicated that the medium in which the fungus was grown was sufficient to cause this effect; therefore, the agent that caused the phenomenon had to be a chemical produced by the fungus.
4. Lateral distribution of auxin controls phototropism and gravitropism. Auxin accumulates in the shaded portions of a stem and stimulates cell growth. This uneven cell growth results in a bending of the stem toward the light. The same mechanism works in response to gravity. An accumulation of auxin occurs where the gravitational pull is the strongest. Cells grow in response to auxin, and stems bend upward, away from the gravitational force.
5. Roots need shoots, and vice versa. Increases in roots require increases in shoots for photosynthesis. Regulation of the development of one by the other keeps growth in tandem. Because the distribution of the molecules is polar, a concentration gradient can be established. It is this relative gradient that controls development.
6. Fruits can be kept from ripening until they reach their destination. Once at market, they can be sprayed with ethylene to stimulate ripening. The result is fruit that can be shipped more easily yet can be ripe at any time in the market.
7. The results of the different treatments are the following: (1) roots will develop from the callus; (2) buds will develop from the callus; (3) both roots and buds will develop from the callus; (4) buds will develop from the callus.

38 Reproduction in Flowering Plants

The Big Picture

- Though plants can reproduce both asexually and sexually, maintenance of genetic variability depends on sexual reproduction. The flower is the basis of sexual reproduction in plants. The flower not only produces the necessary gametes, but it is also integrally involved in ensuring pollination. Eggs are produced through megasporogenesis, and pollen is produced through microsporogenesis. Angiosperms exhibit double fertilization, which results in a diploid embryo and a triploid endosperm.
- Embryo development takes place in the seeds of angiosperms and progresses until cotyledon(s) and the hypocotyl with its apical meristems are developed. At that point, the embryo halts development, the seed desiccates, and dormancy begins until conditions are right for germination. The seed protects the embryo, and in many cases a fruit is formed for protection until maturation and to aid in dispersal of the mature seed.
- Flowering and seed set is triggered by the length of continuous dark exposure. Some plants flower in response to short days, others to long days, and yet others to complex combinations of the two. Day-to-day functions of plants are triggered by photoperiod. Plants may flower, set seed, and die in one growing season or two growing seasons, or they may continue to do so for even longer periods of time.

Common Problem Areas

- Megasporogenesis is probably the most difficult concept in this chapter, but it is relatively simple if you look at the source of each cell.
- Recall that flowering, like all other plant processes, is a result of environmental signals, receptors, enzyme cascades mediated by hormones, and alterations in gene expression.

Study Strategies

- Refer to the figures in the textbook to understand flower structure, megasporogenesis, microsporogenesis, and fertilization.
- Spend some time thinking about the advantages and disadvantages of sexual and asexual reproduction, as related questions are often given in exams.
- Be sure you understand the particular features of double fertilization.
- Go to yourBioPortal.com to review the following tutorials and activity:

  Animated Tutorial 38.1 Double Fertilization

  Animated Tutorial 38.2 The Effect of Interrupted Days and Nights

  Web Activity 38.1 Early Development of a Eudicot

Important Concepts

Plants may reproduce asexually or sexually.

- Asexual reproduction results in plants with the same genetic makeup as the parent plant. In agriculture, this may be desirable.
- Sexual reproduction is necessary for genetic recombination. This promotes genetic variability, which provides the plants with the ability to adapt to their environment.

Sexual reproduction of angiosperms involves the flower.

- Plant reproduction involves the concept of alternation of diploid and haploid generations. The flower is the basis of sexual reproduction in angiosperms and is produced in the diploid sporophyte generation. The basic flower structures include carpels, stamens, petals, and sepals, all of which are modified leaves. The carpel and stamen are the male and female parts of the flower. Flowers with both a carpel and stamen are known as "perfect" and are found in monoecious species. Dioecious species have "imperfect" flowers that contain either the carpel or stamen, but not both.
- The flower produces haploid spores that develop into gametophytes. The female gametophytes, called embryo sacs, develop in the megasporangium. Male gametophytes, called pollen grains, develop in microsporangia.
- Within the ovule, a megasporocyte produces four haploid megaspores through meiosis. Only one of

these megaspores survives, and it divides mitotically to produce eight nuclei within a single large cell. The nuclei migrate to either end of the cell, but two remain in the middle. Cell walls form, isolating the three nuclei at either end into individual cells; the two nuclei in the middle remain together in one cell. At one end, the three cells become two synergid cells and one egg cell; at the other are three antipodal cells, which degenerate. In the large central cell are the two polar nuclei. This seven-celled embryo sac is the megagametophyte (see Figure 38.2).

- Pollen grains develop when a microsporocyte undergoes meiosis. All products of meiosis are retained and undergo mitosis to form a two-celled pollen grain composed of the tube cell and the generative cell. Further development is halted until after the pollen is transferred from the anther to the stigma during pollination.
- Pollination provides a means for fertilization to occur without water, allowing for the evolution of plants on land. Pollen grains are carried to female flowers via wind, animals, or other vectors. Some plants are able to self-fertilize within the same flower. However, most plants prevent self-fertilization by physical separation in either space or time of male and female gametophytes. Plants can also prevent self-fertilization by means of genetic self-incompatibility. A single gene, the *S* gene, regulates self-incompatibility and prevents self-fertilization in many plants.
- When a pollen grain germinates, a pollen tube grows down the style toward the embryo sac. Germination begins when the pollen grain begins to take up water from the stigma. Chemical signals in the form of small proteins from synergids within the ovule direct the growth of the pollen tube.
- The pollen grain consists of two cells at the time of pollination—the tube cell and the generative cell. The tube cell controls the growth of the pollen tube. As the tube is growing, the generative cell undergoes meiosis and produces two haploid sperm cells. Once the pollen tube enters the embryo sac, the two sperm cells are released into a synergid that disintegrates and releases the sperm nuclei. One nucleus fuses with the egg cell, the other with the polar nuclei. This results in the zygote ($2n$) and a triploid ($3n$) nucleus, the endosperm. All other cells disintegrate.
- Double fertilization resulting in a zygote and the nutritive endosperm is a characteristic feature of angiosperms (see Figure 38.6).

Embryos develop within seeds.

- The success of the embryo depends on its own development, the development of the endosperm, the integuments, and the carpel. Ultimately, a seed coat develops from the integuments and protects the dormant embryo.
- The zygote divides mitotically to produce the embryo itself and the suspensor. The suspensor pushes the embryo to the endosperm and becomes a route for nutrients. The suspensor continues to divide and becomes a long, thin structure, whereas the embryo produces a globular structure. This establishes polarity and symmetry in the embryo.
- In eudicots, the embryo progresses through the heart stage and torpedo stage as cotyledons and other organs are developed. The hypocotyl, a shoot apex, and a root apex are also established (see Figure 38.7).
- The endosperm develops and holds nutrient reserves. The cotyledons may absorb the nutrients from the endosperm and increase in size.
- Once development is nearly complete, the seed loses water, and the embryo ceases development. It remains in this state until germination.
- Abscisic acid regulates the development of the seed and the production of proteins that will protect against desiccation. Abscisic acid also inhibits germination of the seed on the plant. The condition of premature germination, in which a seed germinates while still on the plant, is called vivipary.
- As the embryo and seed are developing, the ovary begins to form the fruit. Other parts of the flower and plant may be included in the fruit, but to be considered a fruit, only the ovary wall and the seed need be involved. The fruit disperses by various means, including by traveling on the coats of animals or by being consumed and later deposited. The fruit also protects the seed from animals and microbial diseases.

Specific environmental factors signal the onset of reproduction through flower development.

- The beginning of flowering involves reallocation of the energy in the plant away from leaves and stem and toward flowers and gametes. Plants fall within three different life cycle patterns. Annual plants go from seed to seed set and die within one growing season. Biennial plants require one vegetative growing season before reproducing. Perennial plants repeatedly flower and live for many years.
- The first transition from vegetative growth to floral production is the transition of the apical meristem into an inflorescence meristem. The inflorescence meristem can produce bracts and floral meristems. The floral meristem differs from the apical meristem in that growth is determined. The floral meristem is programmed to produce four consecutive whorls of flower organs.
- A gene cascade leads to flower formation, which begins with the activation of a set of meristem identity genes. Two genes important in switching from vegetative growth to reproductive growth are *LEAFY* and *APETALA1*. The expressions of floral organ identity genes specify successive whorls.
- The gene cascade leading to flower development is controlled by environmental cues. Seasonal flowering

is a result of changes in photoperiod, specifically the duration of continuous darkness. Experiments have shown that plants actually respond to the length of the night, rather than amount of daylight. In experiments in which daylight was interrupted, there was little effect on flowering, but in experiments in which dark was interrupted, there were significant effects on flowering. The mechanism controlling this phenomenon is far from understood, however. Each plant type has a critical day length corresponding to light availability that induces flowering.

- Short-day plants flower when the amount of light available is shorter than the critical maximum. Long-day plants flower only when the day is longer (more light available) than the critical minimum. Some plants require complex combinations of day lengths.
- Plants, like all other organisms, have an internal mechanism for measuring the length of continuous dark periods. The duration of dark periods appears to be detected by special phytochromes that detect red light.
- The stimulus for flowering comes from the leaf of the plant. Florigen (FT) is thought to be the hormone responsible for flowering. The gene *FLOWERING LOCUS T* codes for the FT protein. High levels of FT induce the plant to flower. *CONSTANS* codes for a transcription factor (CO) that stimulates FT synthesis in the phloem cells. *FLOWERING LOCUS D* is a gene that codes for a transcription factor (FD) in the apical meristem that increases transcription of *APETALA1*.
- Vernalization is the induction of flowering by low temperatures. A transcription factor (FLC) coded by the *FLOWERING LOCUS C* inhibits the FT pathway described above. Cold temperatures decrease the production of the FLC protein, leading to a functional FT pathway. Gibberellin can also stimulate a plant to flower.
- Those plants that do not require an environmental cue to flower rely on an "internal clock" to trigger flowering. This most likely works through changes in FLC concentrations in the plant, stimulating the FT–FD pathway described above.

In certain conditions, asexual reproduction is advantageous to plants' success.

- Asexual reproduction occurs without genetic recombination. During the process of vegetative reproduction, asexual reproduction occurs through the modification of a vegetative organ. Stolons, tip layers, tubers, rhizomes, bulbs, corms, and suckers are all modifications of stems or roots that allow vegetative reproduction.
- Some plants such as dandelions reproduce asexually through seeds in a process called apomixis. Apomictic plants skip over meiosis and fertilization and produce diploid seeds with the identical genetic makeup of the maternal plant.
- Asexual reproduction is used in agriculture. Cuttings are commonly used in horticulture. Grafting is widely used in fruit crops in which the root-containing stock is grafted to a scion (shoot system). This allows for a hardy root stock to be combined with a good but often-fragile fruit producer. Tissue culture (production of entire new plants from a small tissue sample) is leading to daily advances in agriculture.

Test Yourself

Diagram Exercise

Florigen is a plant hormone actively involved in the initiation of flower formation. Draw a flow chart showing its mechanism of action including the interaction of the three genes involved in initiating flower formation.
Textbook Reference: *38.2 What Determines the Transition from the Vegetative to the Flowering State? p. 807*

Knowledge and Synthesis Questions

1. You manage a greenhouse that produces roses for Valentine's Day. Roses normally bloom in June. Which of the following will most likely be the best lighting schedule for your roses?
   a. 16 hours of light, followed by 8 hours of interrupted dark
   b. 16 hours of light, followed by 8 hours of uninterrupted dark
   c. 10 hours of light, followed by 14 hours of uninterrupted dark
   d. 10 hours of light, followed by 14 hours of interrupted dark
   e. None of the above
   Textbook Reference: *38.2 What Determines the Transition from the Vegetative to the Flowering State? p. 805*
2. After setting the correct photoperiod, the managers of a greenhouse still do not have blooming roses. Which of the following possibilities would most likely have contributed to the problem?
   a. The heating system allowed for fluctuations in temperature between 20°C and 25°C.
   b. The furnace mechanic accidentally turned off the lights for an hour two days in a row.
   c. The cleaning crew turned the lights on for an hour three nights in a row.
   d. All of the above
   e. None of the above
   Textbook Reference: *38.2 What Determines the Transition from the Vegetative to the Flowering State? p. 805*
3. You have moved into a new house. During the first summer you notice many of the plants do not bloom. During the second summer your yard is a sea of blooms. It is now spring of the third year, and there are no plants. This can best be explained by which of the following?
   a. The plants are annuals.
   b. The plants are biennials.

c. The plants are perennials.
d. The plants are being affected by drought.
e. The nights are too long for the plants.
Textbook Reference: *38.2 What Determines the Transition from the Vegetative to the Flowering State? p. 802*

4. You notice that a new houseplant sends out long stems with what look like "little plants" attached. You allow one of these to rest in a cup of water and note that roots form. This is an example of
a. asexual reproduction.
b. apomixis.
c. heterospory.
d. parthenogenesis.
e. vivipary.
Textbook Reference: *38.3 How Do Angiosperms Reproduce Asexually? pp. 809–810*

5. Self-pollination in plants that produce both pollen and eggs is prevented by
a. self-incompatibility genes.
b. physical barriers.
c. production of pollen and eggs at different times.
d. All of the above
e. None of the above
Textbook Reference: *38.1 How Do Angiosperms Reproduce Sexually? p. 798*

6. Most wine grape vines are grafted onto rootstock of another species. How does this practice increase grape yields?
a. A hardy rootstock can replace a weak rootstock.
b. A high-producing vine stock can replace a low-producing vine stock.
c. It allows vintners to select for pest resistance without losing grape quality.
d. All of the above
e. None of the above
Textbook Reference: *38.3 How Do Angiosperms Reproduce Asexually? p. 811*

7. The induction of flowering by means of exposure to low temperature is called
a. vernalization.
b. frigidation.
c. apomixis.
d. viviparity.
e. None of the above
Textbook Reference: *38.2 What Determines the Transition from the Vegetative to the Flowering State? p. 808*

8. The *LEAFY* and *APETALA1* genes are examples of _______ genes.
a. floral organ identity
b. meristem identity
c. viviparity
d. photoperiod
e. inflorescence
Textbook Reference: *38.2 What Determines the Transition from the Vegetative to the Flowering State? p. 803*

9. The production of seeds without fertilization is called
a. apomixis.
b. parthenogenesis.
c. conception.
d. circadian rhythm.
e. vernalization.
Textbook Reference: *38.3 How Do Angiosperms Reproduce Asexually? p. 810*

10. In the transition from vegetative growth to floral growth the _______ must be transformed into the _______. This involves a shift from _______ growth to _______ growth.
a. apical meristem; floral meristem; indeterminate; determinate
b. lateral meristem; floral meristem; indeterminate; determinate
c. apical meristem; floral meristem; determinate; indeterminate
d. apical cambium; floral cambium; determinate; indeterminate
e. floral meristem; apical cambium; determinate; indeterminate
Textbook Reference: *38.2 What Determines the Transition from the Vegetative to the Flowering State? p. 803*

11. Which of the following best describes the fate of the generative cell of the pollen grain?
a. It coordinates growth of the pollen tube.
b. It divides by meiosis to produce two sperm nuclei.
c. It divides by mitosis to produce two sperm nuclei.
d. It forms the pollen tube.
e. None of the above
Textbook Reference: *38.1 How Do Angiosperms Reproduce Sexually? pp. 796–797*

12. Which of the following is *not* part of a megagametophyte?
a. Pollen grain
b. Synergids
c. Antipodal cells
d. Polar nuclei
e. All of the above are not part of the megagametophyte.
Textbook Reference: *38.1 How Do Angiosperms Reproduce Sexually? pp. 796–797*

13. During double fertilization two sperm cells fuse (one each) with
a. an egg cell.
b. the two polar nuclei.
c. a pollen grain.
d. Both a and b
e. Both a and c
Textbook Reference: *38.1 How Do Angiosperms Reproduce Sexually? p. 800*

14. The shoot apex and the root apex appear during the _______ stage of embryo development.
    a. heart
    b. egg
    c. inflorescence meristem
    d. torpedo
    e. zygote
    Textbook Reference: *38.1 How Do Angiosperms Reproduce Sexually? p. 800*
15. Which of the following is *not* a part of an angiosperm seed?
    a. Seed coat
    b. Cotyledon
    c. Shoot apex
    d. Style
    e. Endosperm
    Textbook Reference: *38.1 How Do Angiosperms Reproduce Sexually? p. 801*

Application Questions

1. Compare and contrast asexual and sexual reproduction in plants. In which category does self-fertilization belong?
   Textbook Reference: *38.1 How Do Angiosperms Reproduce Sexually? pp. 795–802, 809*
2. Describe egg formation in angiosperms, beginning with the sporophyte.
   Textbook Reference: *38.1 How Do Angiosperms Reproduce Sexually? p. 797*
3. Much effort is spent detasseling corn (removing male flowers). Why doesn't it work simply to spray the corn plants with a meiosis inhibitor to halt pollen production? (Hint: Male and female flowers occur on the same corn plant.)
   Textbook Reference: *38.1 How Do Angiosperms Reproduce Sexually? pp. 796–798*
4. Flowering is stimulated when light sets off a gene cascade. Explain how this may be hormonally controlled.
   Textbook Reference: *38.2 What Determines the Transition from the Vegetative to the Flowering State? pp. 805–807*
5. Describe double fertilization. What is the ploidy level of the products of double fertilization?
   Textbook Reference: *38.1 How Do Angiosperms Reproduce Sexually? pp. 799–800*
6. When is asexual reproduction beneficial to plants? Under what conditions is sexual reproduction beneficial?
   Textbook Reference: *38.3 How Do Angiosperms Reproduce Asexually? p. 810*
7. Define a fruit. Which of the following are fruits: tomato, pear, potato, banana, cucumber, snow pea, peanut, sunflower seed?
   Textbook Reference: *38.1 How Do Angiosperms Reproduce Sexually? pp. 801–802*
8. Identify the following structures as occurring in the sporophyte or gametophyte generation.
   a. Embryo sac __________
   b. Antipodal cells __________
   c. Polar nuclei __________
   d. Integument __________
   e. Receptacle __________
   f. Ovary __________
   g. Anther __________
   h. Pollen grain __________
   Textbook Reference: *38.1 How Do Angiosperms Reproduce Sexually? pp. 795–797*

Answers

Diagram Exercise Answer

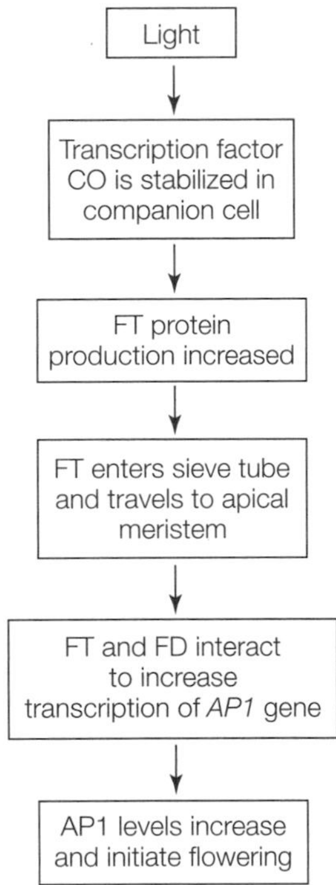

Knowledge and Synthesis Answers

1. **b.** Flowering is regulated by darkness. Interruptions in darkness can prevent flowering.
2. **c.** The interruptions in the dark cycle by the cleaning crew could have affected the signals that tell the plant to begin flowering.
3. **b.** The plants are most likely biennials that spend one season growing vegetatively and one season producing flowers before dying.
4. **a.** The runner (stolon) is propagating this plant asexually.
5. **d.** Some plants have self-incompatibility genes that prevent the growth of the pollen tube through the style. Others have mechanical barriers to their own pollen.

6. **d.** Grafting allows selection of hardy rootstock, produces well-producing vine stock, and contributes to disease resistance.
7. **a.** Vernalization, which is noted in winter wheat and is used agriculturally to grow high-yielding wheat.
8. **b.** These two genes are meristem identity genes that code for proteins involved in the initiation of flower formation.
9. **a.** Dandelions and other plants produce seeds by apomixis, without meiosis and fertilization. These seeds are genetically identical to the parent plant.
10. **a.** Apical meristems and floral meristems differ in that growth from the apical meristem is indeterminate and growth from the floral meristem is determinate, leading to four whorls of floral structures.
11. **b.** The two sperm nuclei involved in double fertilization are derived from the generative cell of the pollen grain.
12. **a.** The megagametophyte is the female gametophyte. It is initially called the embryo sac and contains three antipodal cells, two synergid cells, and two polar nuclei at the seven-cell stage.
13. **d.** Double fertilization involves the fertilization of the egg cell and the two polar nuclei by two sperm cells.
14. **d.** The torpedo-stage embryo has the shoot apex and the root apex on either end of the hypocotyl.
15. **d.** The seed coat, cotyledon, shoot apex, and endosperm are all parts of the seed. The style is a part of the ovary and is involved in the formation of a pollen tube.

Application Answers

1. Asexual reproduction does not involve meiosis, fertilization, or genetic recombination. The offspring from asexual reproduction are genetically identical to the parent plant. Sexual reproduction requires a meiotic event and fertilization. Self-fertilization is sexual reproduction, even though genetic recombination is limited. This is because a meiotic event occurs, and fertilization is necessary for reproduction to take place.
2. See Figure 38.2 in the textbook.
3. Meiosis occurs in the megasporocyte as well as the microsporocyte. Inhibition of pollen formation via a meiosis inhibitor will also inhibit egg formation.
4. Light induction begins with the leaves. A single isolated leaf may stimulate floral production throughout the plant. In experiments in which leaves from one plant were grafted onto other plants, an "induced" leaf caused a plant to flower even though the plant had not been exposed to the required amount of darkness. Therefore, the leaf must send a signal (the protein florigen) that begins the gene cascade leading to flower formation.
5. Double fertilization occurs when the two sperm nuclei produced from the generative cell of the pollen grain unite with the egg nucleus and two polar nuclei, respectively. The resulting zygote is diploid and the endosperm is triploid.
6. Asexual reproduction is beneficial in stable environments in which many genetically identical plants are sustainable. This process can help colonize a habitat or help a population of plants to spread. The disadvantage is lack of genetic diversity, which can be detrimental if the environment changes rapidly and adaptability of the population is required.
7. A fruit is the seed and the ovary wall, and it may include other structures of a flowering plant. With the exception of the potato, all of the listed structures are fruits.
8. a. Embryo sac — gametophyte
   b. Antipodal cells — gametophyte
   c. Polar nuclei — gametophyte
   d. Integument — sporophyte
   e. Receptacle — sporophyte
   f. Ovary — sporophyte
   g. Anther — sporophyte
   h. Pollen grain — gametophyte

39 Plant Responses to Environmental Challenges

The Big Picture

- Plants must respond to invasion by pathogens, physical damage by natural events and herbivory, variable water supplies, temperature fluctuations, and variable soil conditions. Natural selection has made plants uniquely suited to the areas they occupy. Some thrive in very harsh environments, but others cannot.
- Interactions between plants and pathogens stimulate a series of chemical changes that ward off further infection. Plant strategies to isolate pathogens include sealing plasmodesmata, releasing proteins that interact with pathogens, and signaling other parts of the plant.
- Herbivory can be advantageous for plants adapted to grazing or detrimental to those that are not. In the latter case, a plant produces chemicals to prevent herbivory and mounts defenses around the damaged tissues.
- Specific adaptations allow plants to withstand adverse conditions. Saline environments, water loss, and high temperatures can be survived if water uptake is maximized and water loss is minimized. Sunken stomata, reduction of surface area, and increased water potential of cells are adaptations to these conditions. Extremes in temperature are survived by means of heat shock proteins that stabilize metabolic proteins during temperature extremes.

Common Problem Areas

- Gene-for-gene resistance can be difficult to understand, but remember that in any pathogen situation, the pathogen and the host are continually interacting; therefore, over time, interplay among the genomes has been selected for.
- There are many examples of the ways plants deal with pathogens and herbivory. This chapter provides many examples; these are not exhaustive.

Study Strategies

- The best strategy is to study the various components of this chapter one at a time so that all of the defenses/modifications do not seem to run together. Focus first on pathogen interaction, then herbivory resistance, and so on.
- Think about how form follows function as you look at how plants are adapted to specific harsh environments.
- Go to yourBioPortal.com to review the following tutorials and activity:

  Animated Tutorial 39.1 Signaling Between Plants and Pathogens

  Interactive Tutorial: Water Uptake in Plants

  Web Activity 39.1 Concept Matching

Important Concepts

Plants and their pathogens interact, each giving clues to the other.

- The action of a pathogen signals a plant to mount a defense. In turn, the plant's defense signals the pathogen. These signals go back and forth until one or the other "wins." The defense of a plant can either be always turned on (i.e., constitutive) or turned on in response to damage or stress (i.e., induced).
- The first plant defense system is an attempt to prevent entry by pathogens. Cutin, suberin, and waxes cover the cork and epidermis of the plant to exclude pathogens. Plants cannot repair damaged and invaded tissues, so they seal them off to prevent systemic infection.
- Upon recognition of a pathogen, polysaccharides are deposited in the cell wall and seal off the plasmodesmata to form a barrier between cells. Subsequently, lignin is deposited, which reinforces the barrier and acts as a pathogen toxin.
- Gene-for-gene resistance is a highly specific type of resistance and depends on a plant's having an allele for a gene that matches an allele in the pathogen. Dominant *R* genes code for resistance in plants and dominant *Avr* genes code for elicitors. If the receptor from a plant's *R* gene and the elicitor from a pathogen's *Avr* gene match, a defensive response is elicited in the plant. Though this mechanism is not completely understood, it is thought to be mediated by nitric oxide and peroxide.

- Invaded plants begin a three-pronged response to the pathogen called the hypersensitive response. Once invaded, plants produce chemical defenses against pathogens. Phytoalexins are nonspecific antifungal and antibacterial compounds produced by the plant. Pathogenesis-related proteins (PR proteins) are enzymes that function to break down the walls of pathogens or serve as signaling molecules to pathogen-free cells in the plant.
- In the hypersensitive response, pathogen-containing tissue and surrounding tissues go through apoptosis upon infection and become necrotic lesions. This serves to contain and isolate the pathogen to the infected tissues. As these cells are dying, they release phytoalexins.
- Systemic acquired resistance protects a plant against further infection. Salicylic acid produced at the site of an infection stimulates PR protein production. Salicylic acid is also transported to other parts of the plant to signal the presence of an invader. Infected plants release methyl salicylate, which can become airborne and serve as a signal to other plant parts or to neighboring plants to mount a defense via PR proteins.
- Plants can use interference RNA (RNAi, also known as posttranscriptional gene silencing) to respond to attack by RNA viruses. A plant produces small pieces of small interfering RNA (siRNA) from interactions with the viral RNA. These siRNAs degrade the viral mRNA.

Herbivory confers advantages and disadvantages to plants.

- For some plants, coevolution of the plant and the herbivore has led to the plant's dependency on the grazing of herbivores to increase productivity. Removal of some leaves triggers increased photosynthesis in the remaining leaves and a better partitioning of resources such as nitrogen. Grazing also allows light to reach young leaves in plants such as grasses. Grazing can also stimulate some plants to produce more leaves and flower more abundantly.
- For the majority of plants, however, grazing is detrimental; therefore, defenses against grazing exist. Some plants use mechanical defenses to protect themselves from herbivores, including morphological features such as thorns, spines, and hairs. Some plants produce latex in response to injury from an herbivore.
- Plants also produce secondary metabolites to use as a defense from grazers. The effects of these chemicals are diverse, ranging from neurotoxins to hormone mimics (see Table 39.1).
- Canavanine is a defensive secondary metabolite that plays multiple roles in a plant, including assisting with nitrogen storage in the seed. In the herbivore, because it is similar to arginine, it is incorporated into the proteins made after consumption. Canavanine alters the tertiary structure of proteins and can be lethal to the organism consuming the plant.
- Plants perceive the damage caused by herbivores through membrane signaling and chemical signaling. When an herbivore begins to eat the plant, changes in the plasma membrane electrical potential are passed along to every cell within the plant. Chemical signaling occurs when the herbivore's saliva combines with plant fatty acids to elicit a local and systemic response in the plant.
- The response of a plant to perceived damage stimulates a signal transduction pathway involving jasmonic acid (jasmonate). Wounding causes the release of an elicitor, which travels to other parts of the plant and induces membrane breakdown. A by-product of this breakdown is jasmonate, which enters the nucleus and activates production of a protease inhibitor that inhibits digestion in the predator (see Figure 39.8).

Plants use a variety of strategies to prevent harming themselves with their own defensive mechanisms.

- Plants protect themselves from their own defenses by isolating toxic secondary metabolites into compartments. Water-soluble toxins are stored in vacuoles; hydrophobic poisons are stored in laticifers along with latex; other toxins are dissolved in the waxes covering the epidermis.
- Some toxic substances are produced only in already-damaged or dying tissue. The precursors of toxic material are stored separately from the enzymes that convert them into active poison, and these combine only once the cell is damaged.
- In plants that use the canavanine amino acid in defense, the plant's enzymes are able to distinguish between canavanine and arginine.
- Insects sometimes find ways to get around the mechanisms used by plants to deter them. One example is a beetle that feeds on the leaves of milkweed.

Some plants must be able to cope with limited or excess water availability.

- Some desert plants are drought avoiders. They survive in these areas by completing their entire life cycle in the short period that water is available.
- Xerophytes are plants specifically adapted to dry areas. They may have special modifications such as thickened cuticles, epidermal hairs, and stomatal crypts to prevent water loss. Succulents have water-storing leaves. Other plants drop leaves during drought, and yet others have water-storing stems and highly reduced leaves.
- Roots can also be adapted for drought conditions. Long taproots can reach water supplies far underground. Shallow, fibrous root systems that grow only during rainy seasons are also adapted to desert growth. Xerophytic plants are able to extract more water when it is available by changing the osmotic and water potential of their root cells. They do this by storing proline in their vacuoles.

- Too much water can be as dangerous to plant survival as too little. Roots submerged in water do not get adequate oxygen to sustain respiration. Some plants that grow in standing water overcome this by producing pneumatophores, which are root extensions that grow above water and provide oxygen to the entire root system. Others rely on fermentation for ATP production and thus grow very slowly. Yet others have leaf modifications called aerenchyma for buoyancy and oxygen storage.
- Inadequate water supply results in changes in membrane integrity and in the three-dimensional structure of proteins. Draught conditions stimulate the roots to release abscisic acid, which travels to the leaves. Abscisic acid closes the stomata and starts gene expression of late embryogenesis abundant (LEA) proteins. The LEA proteins stabilize membranes and proteins.

Temperature extremes pose yet another stress to plant survival.

- High temperatures denature proteins and destabilize membranes. Cold temperatures decrease membrane fluidity, and freezing causes rupturing of membranes if ice crystals form.
- Plants have evolved a number of adaptations to deal with heat, including hairs and spines to dissipate heat, changes in leaf profile to catch less sun, and CAM photosynthesis. Plants produce heat shock proteins that act as chaperonins that help stabilize protein structure against denaturation.
- Plants can adjust to cold through a process of cold-hardening, which involves repeated exposure to cool, nondamaging temperatures. This alters the saturated and unsaturated fatty acid composition of membranes, allowing them to remain fluid at cooler temperatures and making them more resistant to rupture. Heat shock proteins are also produced in response to cold temperatures. Some plants have antifreeze compounds that prevent ice crystal formation.

Some plants are adapted to survive in saline environments.

- Saline environments restrict the growth of angiosperms. Plants living in saline environments are exposed to an osmotic challenge.
- Halophytes (salt-loving plants) are adapted to saline environments. They are the only plants that accumulate sodium and chloride ions, which they store in leaf vacuoles. The increased salt concentration of the halophyte means it has a more negative water potential.
- Some plants are able to excrete salt so that it does not reach toxic proportions. Salt glands move salt to the leaf surface, where it can be lost to wind or rain. Salt glands on the leaf assist with water procurement from the roots and with reduction of water loss to evaporation.
- Many modifications, such as succulence, thick cuticles, and crassulacean acid metabolism, allow for survival in both drought and saline conditions.

Heavy metal accumulation in soils is detrimental to most plants.

- Chromium, mercury, lead, and cadmium are poisonous to most plants. Some geographic areas are rich in these metals as a result of normal geological processes, and others have been contaminated by human activity.
- Plants living in these areas have adapted to allow them to accumulate large quantities of heavy metals. These plants, known as hyperaccumulators, increase ion transport into the roots, increase translocation, accumulate ions in shoot vacuoles, and resist the toxin. These plants are important in bioremediation cleanup efforts.

Test Yourself

Diagram Exercise

It has long been a practice in rural areas to nail fencing to trees. Diagram how the act of nailing a fence to a tree trunk can introduce pathogens into the tree. Be sure to show what measures the tree takes to ward off disease.
Textbook Reference: *39.1 How Do Plants Deal with Pathogens? pp. 817–818*

Knowledge and Synthesis Questions

1. Defensive strategies in plants that are always turned on are called
   a. induced.
   b. heat shock proteins.
   c. alternating.
   d. constitutive.
   e. All of the above
   Textbook Reference: *39.1 How Do Plants Deal with Pathogens? p. 815*
2. Plants acquire systemic resistance in much the same way that people acquire resistance to pathogens; however, the mechanism of systemic acquired resistance is quite different in plants. Which of the following does *not* have a role in acquired resistance in plants?
   a. Salicylic acid
   b. *R* genes
   c. Methyl salicylate
   d. PR proteins
   e. All of the above have a role in acquired resistance in plants.
   Textbook Reference: *39.1 How Do Plants Deal with Pathogens? p. 818*
3. Gene-for-gene resistances depend on which of the following?
   a. Compatible alleles in plant and pathogen
   b. Incompatible alleles in plant and pathogen
   c. Recessive *Avr* genes
   d. Recessive *R* genes
   e. PR proteins
   Textbook Reference: *39.1 How Do Plants Deal with Pathogens? p. 817*

4. Upon infection by a pathogen, plant cells increase synthesis of polysaccharides. The function of these polysaccharides is to
   a. destabilizes the cell walls.
   b. synthesize antibodies to the pathogen.
   c. isolate the pathogen in the invaded tissue.
   d. break down the wax barrier.
   e. communicate about the pathogen to the leaves.

   Textbook Reference: *39.1 How Do Plants Deal with Pathogens? pp. 816–817*
5. Canavanine is toxic to many herbivores but not to plants. Which of the following statements regarding this differential toxicity is true?
   a. Canavanine is confused with arginine in plants but not in animals.
   b. Canavanine is incorporated into plant cells and causes them to fold properly.
   c. Plants are able to differentiate between arginine and canavanine, whereas animals cannot.
   d. Plants store canavanine in vacuoles, whereas animals metabolize it.
   e. None of the above

   Textbook Reference: *39.2 How Do Plants Deal with Herbivores? p. 821*
6. Which of the following best describes how plants produce their own insecticide?
   a. Wounded cells release an elicitor that directly induces synthesis of protease inhibitors, and protease inhibitors act as insecticides.
   b. Wounded cells release jasmonates, jasmonates stimulate elicitor synthesis, elicitor causes the production of protease inhibitors, and protease inhibitors act as insecticides.
   c. Wounded cells release an elicitor that causes membrane breakdown, membrane breakdown releases jasmonates, and jasmonates act as insecticides.
   d. Wounded cells release an elicitor that causes membrane breakdown, membrane breakdown releases jasmonates, jasmonates induce synthesis of protease inhibitors, and protease inhibitors act as insecticides.
   e. None of the above

   Textbook Reference: *39.2 How Do Plants Deal with Herbivores? p. 822*
7. Which of the following is *not* an adaptation to drought conditions?
   a. Water-storing tissues
   b. Leaf loss
   c. Sunken stomata
   d. Increased stomata number
   e. CAM photosynthesis

   Textbook Reference: *39.3 How do Plants Deal with Climatic Extremes? p. 824*
8. Halophytes are different from all other types of plants in that they
   a. can accumulate sodium and chloride ions.
   b. have a positive water potential.
   c. contain no sodium or chloride ions.
   d. contain no stomata.
   e. All of the above

   Textbook Reference: *39.4 How Do Plants Deal with Salt and Heavy Metals? p. 828*
9. Which of the following conditions stimulates the production of heat shock proteins?
   a. Abnormally high temperatures only
   b. Abnormally low temperatures only
   c. Both abnormally high and abnormally low temperatures
   d. Heat shock proteins are continually available in a plant, no matter what the temperature.
   e. None of the above

   Textbook Reference: *39.3 How Do Plants Deal with Climatic Extremes? p. 827*
10. The main function of heat shock proteins is to
    a. stabilize proteins necessary to a cell's survival.
    b. reinforce membranes that lose fluidity.
    c. cause the plant to enter dormancy.
    d. act as an antifreeze compound.
    e. stimulate bolting.

    Textbook Reference: *39.3 How Do Plants Deal with Climatic Extremes? p. 827*
11. Cold-hardening involves all of the following *except*
    a. production of antifreeze proteins.
    b. increased production of saturated fatty acids.
    c. production of phytoalexins.
    d. production of heat shock proteins.
    e. All of the above are involved in cold-hardening.

    Textbook Reference: *39.3 How Do Plants Deal with Climatic Extremes? p. 827*
12. What keeps plants that produce toxins for defense from poisoning themselves?
    a. The toxic substances are kept throughout all plant tissues.
    b. The toxic substances are stored in laticifers.
    c. The toxic substances are produced by every cell in the plant all the time.
    d. The toxic substances are stored in the same place as enzymes that convert them to the active form.
    e. Nothing; toxic substances do slowly poison the plants.

    Textbook Reference: *39.2 How Do Plants Deal with Herbivores? pp. 822–823*
13. Which of the following is *not* a secondary plant metabolite?
    a. Phenolics
    b. Alkaloids
    c. Terpenes
    d. Glucosinolates
    e. PR proteins

    Textbook Reference: *39.2 How Do Plants Deal with Herbivores? p. 820, Table 39.2*

14. The hypersensitive response involves the release of
    a. lignin.
    b. PR proteins.
    c. *R* genes.
    d. phytoalexins.
    e. secondary metabolites.

    Textbook Reference: *39.1 How Do Plants Deal with Pathogens? pp. 817–818*
15. A plant's immune response to RNA viruses involves
    a. heat shock proteins.
    b. secondary metabolites.
    c. small interfering RNA.
    d. PR proteins.
    e. All of the above

    Textbook Reference: *39.1 How Do Plants Deal with Pathogens? p. 818*

Application Questions

1. Describe strategies used by plants to prevent herbivory.
   Textbook Reference: *39.2 How Do Plants Deal with Herbivores? pp. 819–822*
2. Explain what is meant by coevolution of plants and herbivores. How can grazing increase the productivity of some plants?
   Textbook Reference: *39.2 How Do Plants Deal with Herbivores? p. 819*
3. At a T-intersection in a small town, a local business plants several ornamental shrubs. During the winter, snowplows push snow from the intersection around the shrubs. In an attempt to save the shrubs, the business covers the shrubs with burlap. Will this save the shrubs? What conditions in the intersection are likely to cause the greatest damage to the shrubs?
   Textbook Reference: *39.4 How Do Plants Deal with Salt and Heavy Metals? p. 828*
4. Bioremediation of mine and ore refinery sites is an area of intense study. How can plants help with the cleanup of sites contaminated with heavy metals?
   Textbook Reference: *39.4 How Do Plants Deal with Salt and Heavy Metals? pp. 828–829*
5. Many plants produce toxins that damage eukaryotic predators. Considering that the cell structure of both the predator and the plant is similar, why is the plant not affected by its own toxins?
   Textbook Reference: *39.2 How Do Plants Deal with Herbivores? pp. 822–823*
6. Pansies are planted in the winter in the southeastern United States. To ensure that the plants are ready to enter cool ground, nurseries cold-harden the plants for several days. Describe this process and explain why it is done.
   Textbook Reference: *39.3 How do Plants Deal with Climatic Extremes? p. 827*

Answers

Diagram Exercise Answer

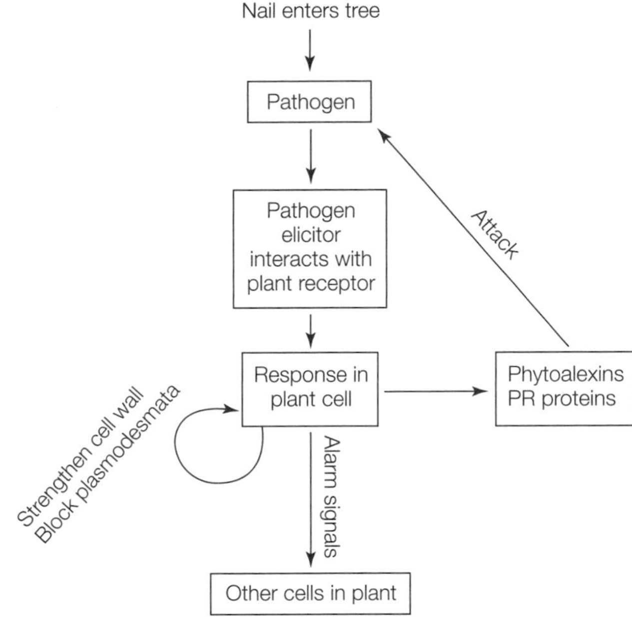

Knowledge and Synthesis Answers

1. **d.** If something is always turned on, it is said to be constitutive.
2. **b.** Plants that have been exposed to pathogens wall off invaded tissue. This tissue releases salicylic acid to the rest of the plant, which stimulates PR proteins. The tissue also releases airborne methyl salicylate, which causes release of PR proteins in distant regions of the plant and even in neighboring plants.
3. **a.** The plant must have an *R* allele, and the pathogen must have an *Avr* allele. Compatibility between these alleles leads to resistance.
4. **c.** Polysaccharides seal plasmodesmata and reinforce cell walls. The purpose of this is to contain the pathogen in a minimal number of cells.
5. **c.** Canavanine's mechanism of action is that it is taken up by the herbivore and incorporated into its protein structures. Because the structure of canavanine is different from that of the arginine it replaces, the herbivore's proteins take up a structure that is lethal to the herbivore. The plant itself and some herbivores have enzyme mechanisms to distinguish canavanine from arginine and prevent its incorporation into proteins.
6. **d.** See Figure 39.8 in the textbook.
7. **d.** Water storage tissues allow the plant to take advantage of any water that is available and hold on to it for lean times. Leaf loss, when photosynthesis cannot be sustained, prevents loss of resources. Sunken stomata

prevent excessive water loss to evaporation. CAM photosynthesis conserves water.

8. **a.** They can accumulate sodium and chloride ions. This makes their water potential more negative, and they are able to extract more water from their surroundings than plants that do not store the ions.
9. **c.** Heat shock proteins are produced in response to extreme high and low temperatures.
10. **a.** Heat shock proteins are chaperonin proteins. They function to stabilize and prevent denaturation (unfolding) of essential proteins.
11. **c.** Cold-hardening involves production of antifreeze proteins, heat shock proteins, and an increase in saturated fatty acids in the membrane. The production of phytoalexins is involved in chemical defenses against pathogens.
12. **b.** Plants that produce toxins as a defense typically store them in special locations such as laticifers where they will not damage tissue.
13. **e.** Alkaloids, phenolics, terpenes, and glucosinolates are all examples of secondary plant metabolites. PR proteins are proteins that function in defense.
14. **d.** The hypersensitive response involves the release of antibiotic phytoalexins.
15. **c.** Plants that are invaded by RNA viruses use the RNA of the invading virus to interfere with and block viral replication.

Application Answers

1. Plants prevent herbivory by producing secondary compounds that affect potential herbivores. These compounds may act as neurotoxins, inhibit digestion, or have a number of other consequences (see Table 39.1).
2. Coevolution of herbivores and plants occurs when the herbivore has resistance to toxins that the plant may produce, and the plant develops strategies to cope with herbivory. In some cases, regular grazing increases photosynthetic productivity of the plant by reducing shading and allowing better partitioning of resources.
3. The burlap will most likely do little for the shrubs. The roots are in most danger. Because streets are heavily salted in the winter, the snow around them creates a saline environment. If the shrubs are not adapted to a saline environment, little will help them survive the salt accumulation around their roots.
4. Only plants that have evolved a strategy for taking up and coping with heavy metals can grow in these contaminated sites. Deliberately planting them at these sites initiates the slow process of removing the metals from the soil.
5. Plants may isolate the toxins into vacuoles within their cells so that toxins do not come into contact with the active machinery of the cells. Plants may restrict toxin production to already-damaged tissue. Plants may also develop enzymes that differentiate toxins and render them harmless.
6. Cold-hardening is gradual exposure to adverse temperatures. This allows for structural changes in cell membranes to help them cope with extreme temperatures. It also allows for the production of heat shock proteins and other compounds necessary for survival at temperature extremes.

40 Physiology, Homeostasis, and Temperature Regulation

The Big Picture

- One of the most important roles of tissues, organs, and organ systems is to help maintain homeostasis within the body's cells. Only by maintaining relatively constant intracellular conditions can basic metabolic reactions continue.
- Animals either generate their own body heat (endotherms) or rely on the environment to determine body temperature (ectotherms). Endotherms typically have much greater insulation and higher metabolic rates than ectotherms. The metabolic rates of endotherms and ectotherms respond differently to changes in environmental temperature. Endotherms increase metabolism as environmental temperature decreases; they also increase metabolic rate at high temperatures when energy is needed for sweating and panting. The metabolic rate of ectotherms is temperature-dependent and falls with decreases in environmental temperature.
- Control of body temperature can occur by behavioral and physiological mechanisms. In both endothermic and ectothermic vertebrates, the brain's hypothalamus is the thermostat for temperature regulation.

Common Problem Areas

- Homeostasis does not necessarily mean holding every body variable absolutely constant, but rather making sure that body variables are held at the correct level for the circumstances. Even body temperature in humans, which we think of as being a constant 37°C, drops slightly in the early morning hours and rises considerably during extended vigorous exercise.
- Understanding the difference between ectotherms and endotherms can be difficult. These terms refer to the *source* of heat that determines an animal's temperature. Whereas body temperatures of ectotherms are determined primarily by external sources of heat, those of endotherms are determined by heat generated metabolically.

Study Strategies

- Refer to Figure 40.2 to understand how homeostasis is maintained through regulatory systems.
- Review the organization of physiological systems from cells to tissues to organs to organ systems. This will help you when you study each of the organ systems in later chapters.
- Go to yourBioPortal.com to review the following tutorial and activity:

  Animated Tutorial 40.1 The Hypothalamus

  Web Activity 40.1 Thermoregulation in an Endotherm

Important Concepts

Control of the internal environment is necessary for proper cell function.

- Due to their small size, single-celled organisms can receive all their required nutrients from the surrounding media. Small multicellular organisms, in which all the cells are only a few layers from the environment, can receive nutrients from the environment without the help of specialized cells. Larger multicellular organisms can maintain an internal environment that differs from the external environment. Specialized cells and groups of cells help control the makeup of the internal environment and maintain the differences between the internal and external environments (see Figure 40.1).
- Homeostasis is the state of having a constant, regulated internal environment, and homeostatic mechanisms help maintain this state.

Control systems help regulate physiology and maintain homeostasis.

- Physiological regulatory systems require both a set point at which the parameter is to be held, and feedback, which provides information about the state of the system (see Figure 40.2). Differences between the set point and the feedback information indicate to the regulatory mechanism which corrective measures are required to restore a parameter to its set point.

- Physiological control systems must take information from the regulator systems and effect changes. Negative feedback results in a slowdown or reversal of a process, returning the variable controlled by that process back toward its set point or regulated level.
- In contrast to negative feedback controls, positive feedback results in amplification of a response. Regulation by positive feedback is less common than regulation by negative feedback.
- The set point maintained by an organism for a given physiological process can also be changed by feedforward information.

Groups of cells are organized as tissues, organs, and organ systems.

- A tissue consists of a group of similar cells with the same form and function. There are four types of tissues: epithelial, connective, muscle, and nervous.
- Epithelial tissues make up the lining of the inner and outer body surfaces, such as the lining of the various organs and the skin. Some epithelial cells have secretory functions, whereas others have cilia to help move substances over surfaces. Still other epithelial cells function in protection, absorption, or the provision of information to the nervous system.
- Connective tissue consists of an extracellular matrix that holds together a dispersed group of cells. The protein collagen makes up the bulk of the extracellular matrix. Cartilage, bone, adipose tissue, and blood are all connective tissues.
- Muscle tissue is either skeletal, smooth, or cardiac. Cells of muscle tissue contract to generate force and movement.
- Two types of cells are found in nervous tissue—neurons and glial cells. Neurons generate and conduct electrochemical signals. Glial cells support and protect neurons.
- Organs are groups of tissues that carry out a specific function in the body. Most organs are composed of all four of the major tissue types. The integration of a number of organs to carry out a specific function is known as an organ system (e.g., the digestive system).
- ***For review, go to Diagram Exercise 1.***

Temperature plays an important role in life.

- Most living cells function only between a temperature of 0°C and 40°C, with most organisms having much narrower limits in their range of survival temperatures. For most cells, the upper limit for survival is 45°C.
- All physiological processes in organisms are sensitive to temperature. Increases in temperature usually increase the rate of a given process. The Q_{10} is a measure of change in a physiological process as the temperature increases or decreases by 10°C (see Figure 40.8). Most biological Q_{10} values range from 2 to 3. A Q_{10} of 2 means that the reaction rate doubles when temperature increases by 10°C.
- Acclimatization is the change in a physiological process that occurs over time in response to a change in the environment. For example, a fish experimentally exposed in summer to decreasing water temperatures will show slowed physiological functions. However, the physiological functions of the same fish in winter will not be as slow as they were when temperatures were experimentally reduced in summer. The physiology of the fish has acclimatized to seasonal changes in water temperature. Fish may catalyze reactions with one set of enzymes in the summer and a different set of enzymes in the winter. Acclimatization typically results in an enhanced probability of thriving in the new environmental conditions.

Animals can be classified as ectotherms or endotherms.

- Animals can be classified into two main groups based on their source of heat. Endotherms regulate their body temperature with internal heat production by means of a high metabolic rate and effective insulation. Ectotherms have a low metabolism and little insulation and must rely on the environment to provide body heat.
- Ectotherms have body temperatures that track environmental temperatures. In contrast, endotherms maintain a constant body temperature even as the environmental temperature changes (see Figure 40.9A). A hibernating mammal is an example of a heterotherm, an animal that sometimes behaves as an endotherm (in this case, in summer) and sometimes as an ectotherm (in this case, in winter).
- Both endotherms and ectotherms use behavioral mechanisms to regulate body temperature. Ectotherms shuttle between warm and cool environments during the day to maintain body temperature within a given range (see Figure 40.10A). Endotherms use behavioral thermoregulation as a first line of thermoregulation (see Figures 40.10B and 40.10C).
- Heat can be gained or lost to the environment through a number of routes (see Figure 40.11). Evaporation from the skin or respiratory tract cools the body. Solar radiation from the sun heats the body. Thermal radiation, convection, and conduction can either warm or cool the body, depending on the thermal environment.
- The total balance of heat production and heat exchange for an animal is its energy budget. To maintain a constant body temperature, the heat entering an animal must equal the heat leaving.
- Both ectotherms and endotherms use changes in blood flow to control body temperature. Increases in blood flow result in increased heat exchange with the environment, whereas decreases in blood flow result in decreased heat exchange (see Figure 40.12).
- Some large fishes have the ability to maintain a body temperature higher than the surrounding water. They possess special vascular countercurrent heat exchangers that retain metabolically-produced heat in blood that remains in the core (see Figure 40.13).

- A number of ectothermic insects are able to generate heat to raise their body temperature. Some do so by contracting their flight muscles. Others, such as honeybees, engage in group thermoregulation whereby individuals form clusters in winter and regulate temperature by adjusting their own metabolic heat production and the density of the cluster.
- *For review, go to Diagram Exercises 2 and 3.*

Endotherms have special adaptations to help in controlling body temperature.

- The metabolic rate of an organism is the total energy used by the animal and is usually measured as oxygen consumption.
- For endotherms, the basal metabolic rate per gram of tissue decreases as animals get larger (see Figure 40.15). The reason for this pattern is unclear.
- The range of environmental temperatures at which the metabolic rate of endotherms does not change is known as the thermoneutral zone (see Figure 40.16). The upper and lower critical temperatures are the environmental temperatures at which metabolic rate begins to increase. Increases in heat production occur by nonshivering and shivering thermogenesis at times when the environmental temperature is below the lower critical temperature. Nonshivering thermogenesis involves a specialized adipose tissue, brown fat (see Figure 40.17). Metabolism increases when environmental temperature is above the upper critical temperature because the animal must expend energy to cool down through panting or sweating.
- Animals that live in hot areas display increases in surface area to promote heat exchange with the environment (see Figure 40.18A). Decreases in surface area are seen in animals that live in cold climates (see Figure 40.18B). This decrease in surface area decreases the area over which heat can be lost.
- Fur and feathers provide insulation for mammals and birds in cold environments. These structures retain heat produced in the body and maintain a constant body temperature.
- Evaporative cooling by sweating or panting provides an avenue for heat loss to the environment.
- Countercurrent exchange in appendages is an important mechanism for maintaining core temperature in many endotherms.

The hypothalamus is the vertebrate thermostat.

- The hypothalamus, a structure at the bottom of the brain, is involved in the control of homeostasis for many regulatory systems, including thermoregulation. The hypothalamus has a set point temperature; changes in temperature of the hypothalamus result in thermoregulatory responses to offset any temperature change (see Figure 40.19). The hypothalamus also integrates thermal information received from the skin.
- In mammals, hypothalamic set points are different at different environmental temperatures. Other factors, including the sleep–wake cycle and level of activity, also influence set points.
- The body has the ability to increase the set point temperature in response to infections, resulting in fever. A fever is a rise in body temperature in response to compounds called pyrogens. Some pyrogens come from bacteria or viruses that invade the body; others are produced by cells of the immune system when it is challenged.
- Hypothermia is the general state of below-normal body temperature. Some animals are able to decrease body temperature either on a daily scale (daily torpor) or over longer time periods (hibernation) (see Figure 40.20). Daily torpor and hibernation are examples of regulated hypothermia that allow animals to survive during periods of low temperature and scarce food.

Test Yourself

Diagram Exercises

1. Label the organs and four main tissue types in the diagram below.

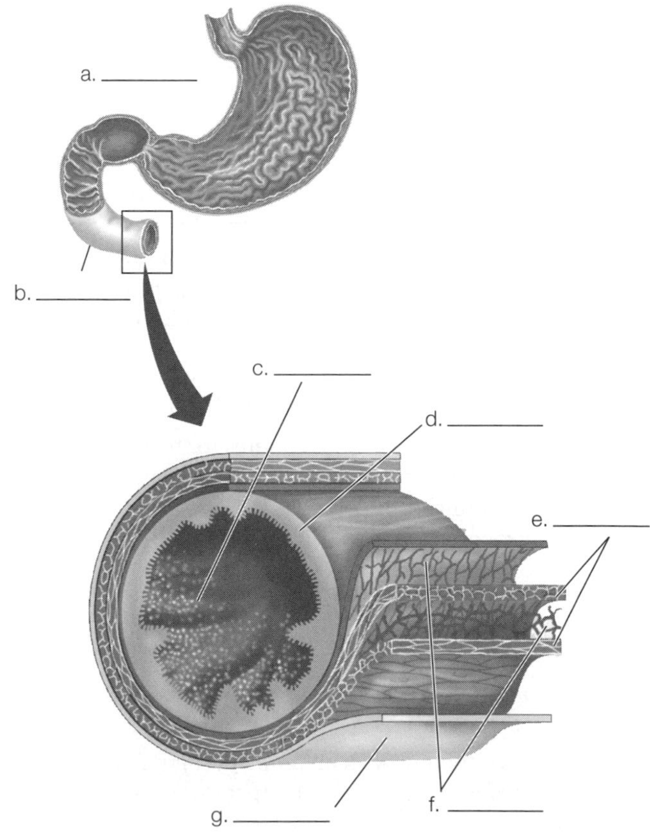

Textbook Reference: *40.1 How Do Multicellular Animals Supply the Needs of Their Cells? pp. 835–837*

2. Graph the lizard's predicted body temperature based on your knowledge of how ectotherms regulate their temperature with behavior.

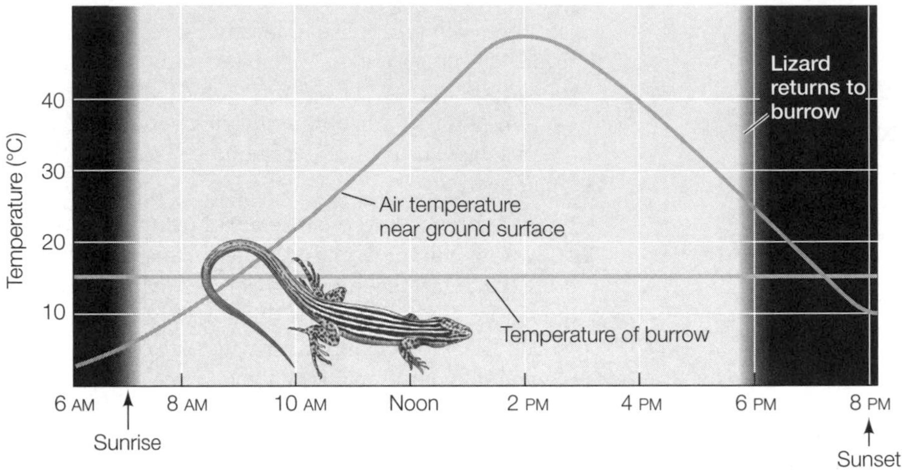

Textbook Reference: *40.3 How Do Animals Alter Their Heat Exchange with the Environment? p. 841*

3. Complete the table below, which compares some general traits of endotherms and ectotherms.

	Endotherm	Ectotherm
Heat source		
Efficiency of energy usage		
Resting metabolic rate		
Temperature control center		
Insulation		

Textbook Reference: *40.3 How Do Animals Alter Their Heat Exchange with the Environment? pp. 839–840*

Knowledge and Synthesis Questions

1. Which of the following contributes to the proper maintenance of homeostasis in the bodies of animals?
   a. pH buffering of the blood
   b. Control of blood flow to the skin
   c. Production of hormones
   d. Detection of external environmental conditions by the sensory system
   e. All of the above

   Textbook Reference: *40.1 How Do Multicellular Animals Supply the Needs of Their Cells? pp. 832–833*

2. In an environment with an ambient temperature lower than an animal's core body temperature, which of the following would be an *inappropriate* physiological or behavioral response if the animal needed to eliminate excess body heat?
   a. Wallowing in a pool of water
   b. Inhibiting blood flow to peripheral vessels
   c. Sweating
   d. Decreasing physical activity
   e. All of the above

   Textbook Reference: *40.3 How Do Animals Alter Their Heat Exchange with the Environment? p. 842*

3. A lizard lives in a desert environment where the temperature is low at night and high during the day. What might this animal do to maintain the most stable body temperature?
   a. Stay in a burrow during the night and shuttle between the sun and shade on the surface during the day
   b. Stay on the surface during the night and move to a burrow during the day
   c. Increase metabolism and heat production during the night and decrease it during the day
   d. Decrease metabolism and heat production during the night and increase it during the day.
   e. Move into a state of hypothermia during the night and a state of torpor during the day.

   Textbook Reference: *40.3 How Do Animals Alter Their Heat Exchange with the Environment? p. 840*

4. On a hot day, an active dog will pant heavily and visit his water bowl frequently. Which of the following statements about the dog's condition or behavior is *false*?
   a. He is using evaporative cooling to lower core body temperature.
   b. He is operating at a basal metabolic rate.
   c. He is at the upper end of the thermoneutral zone.
   d. He is sending more blood to his skin.
   e. He is adjusting behavior in order to thermoregulate.

   Textbook Reference: *40.4 How Do Mammals Regulate Their Body Temperatures? pp. 845–846*

5. Evaporative cooling is an effective way to increase heat loss, but it carries the physiological drawback of
   a. an increased use of ATP and substantial water loss.
   b. the lowering of the hypothalamic thermal set point.
   c. the decreased use of ATP and substantial water gain.
   d. the exhaustion of the supply of brown fat.
   e. a lowered basal metabolic rate.

   Textbook Reference: *40.4 How Do Mammals Regulate Their Body Temperatures? p. 846*

6. Which of the following statements about fever is true?
   a. It is a higher-than-normal body temperature that is always dangerous.
   b. It decreases the metabolic rate of the body to conserve energy.
   c. It is regulated by chemicals that reset the body's thermostat to a higher set point.
   d. It causes the liver to release large amounts of calcium, which seems to inhibit bacterial growth.
   e. It should always be controlled with fever-reducing drugs.

   Textbook Reference: *40.4 How Do Mammals Regulate Their Body Temperatures? p. 847*

7. Which of the following is a characteristic unique to connective tissue?
   a. Highly modified cells that show the special property of contractibility
   b. Diverse anatomy with distinct specializations for information transfer
   c. The composition of the lining of the inner surfaces of the intestines and lungs in mammals
   d. A loose array of cells embedded in an extracellular matrix
   e. The ability to synthesize large amounts of the protein collagen.

   Textbook Reference: *40.1 How Do Multicellular Animals Supply the Needs of Their Cells? pp. 836–837*

8. Elephants use their ears to release heat to the environment. What mechanisms might they employ to increase heat loss from the ears?
   a. Increased convection by means of ear flapping
   b. Moving into the sun
   c. Increased blood flow to the ears
   d. Covering their ears with dust
   e. Both a and c

   Textbook Reference: *40.3 How Do Animals Alter Their Heat Exchange with the Environment? pp. 840–841*

9. Which of the following organ systems is *not* involved in the maintenance of homeostasis in humans?
   a. Endocrine
   b. Digestive
   c. Muscle
   d. Reproductive
   e. All of the above organ systems are involved in maintaining homeostasis.

   Textbook Reference: *40.1 How Do Multicellular Animals Supply the Needs of Their Cells? pp. 833–834*

10. What is the difference between negative feedback mechanisms and positive feedback mechanisms?
    a. Negative feedback mechanisms exist only in the circulatory system.
    b. Negative feedback mechanisms return a system to a set point, and positive feedback mechanisms amplify a response.
    c. Negative feedback mechanisms move a system away from a set point, and positive feedback mechanisms stabilize a system toward a set point.
    d. Negative feedback mechanisms stabilize a system toward a set point, and positive feedback mechanisms reset the set point.
    e. There is essentially no difference between the two feedback systems.

    Textbook Reference: *40.1 How Do Multicellular Animals Supply the Needs of Their Cells? pp. 834–835*

11. A number of physiological processes can undergo acclimatization. Which of the following statements about acclimatization is *false*?
    a. It occurs in response to seasonal temperature changes.
    b. Acclimatization of metabolic rate occurs because enzyme expression changes.
    c. It involves the changing of a set point.
    d. All multicellular animals are capable of acclimating to all the environmental changes they face.
    e. Acclimatization is primarily seen in organisms with body temperatures tightly coupled to the environmental temperatures.

    Textbook Reference: *40.2 How Does Temperature Affect Living Systems? p. 839*

12. In fast-swimming cool-water fishes such as sharks and tunas, which of the following contribute(s) to the generation and maintenance of core body temperatures in excess of ambient water temperatures?
    a. Specializations in the size and arrangement of blood vessels
    b. Low rates of activity in the swimming muscles
    c. An ability to acclimatize rapidly to the surrounding water
    d. Metabolic rates that are insensitive to temperature change
    e. An insulating layer of fat located beneath the scaled skin.

    Textbook Reference: *40.3 How Do Animals Alter Their Heat Exchange with the Environment? p. 843*

13. Which of the following statements about hypothalamic function is *false*?
    a. Circulating blood temperature is monitored by the hypothalamus.
    b. Hypothalamic thermal set points never change.
    c. The hypothalamus is a part of the central nervous system.
    d. Different thermoregulatory responses have different hypothalamic set points.

e. The hypothalamus is important in fish, birds, amphibians, reptiles, and mammals.
Textbook Reference: 40.4 How Do Mammals Regulate Their Body Temperatures? pp. 846–847

14. Some animals use brown fat as a source of heat generation. Which of the following statements about brown fat is *false*?
   a. Brown fat is involved in nonshivering thermogenesis.
   b. The protein thermogenin in brown fat is involved in uncoupling proton movement from ATP production.
   c. Brown fat is found in endotherms.
   d. Brown fat is highly vascularized.
   e. Brown fat cells contain few mitochondria
   Textbook Reference: 40.4 How Do Mammals Regulate Their Body Temperatures? p. 845
15. During childbirth, the pressure exerted on the mother's cervix by the emerging infant leads to increased contraction of the uterus. This interaction can be explained by
   a. negative feedback.
   b. feedforward feedback.
   c. positive feedback.
   d. acclimatization.
   e. homeostasis.
   Textbook Reference: 40.1 How Do Multicellular Animals Supply the Needs of Their Cells? pp. 834–835
16. Which of the following statements about nervous tissue is true?
   a. Neurons outnumber glial cells in our nervous system.
   b. Glial cells generate and conduct electrochemical signals.
   c. Cell bodies are long extensions of neurons over which impulses travel to reach other neurons.
   d. Glial cells support and protect neurons.
   e. Glial cells are a type of connective tissue holding neurons in place.
   Textbook Reference: 40.1 How Do Multicellular Animals Supply the Needs of Their Cells? p. 837

Application Questions

1. Having measured the metabolism of two animals of similar size at 15°C and 25°C, you find that the metabolic rate of Animal 1 is 115 ml of oxygen per hour at 15°C and 55 ml of oxygen per hour at 25°C. The metabolic rate of Animal 2 is 5.5 ml of oxygen per hour at 15°C and 11.5 ml of oxygen per hour at 25°C. Calculate the Q_{10} for the metabolic rate in both animals and determine if they are endotherms or ectotherms.
   Textbook Reference: 40.2 How Does Temperature Affect Living Systems? p. 840
2. You have measured the metabolism of the first animal in Question 1 at both low and high environmental temperatures. You find that the metabolic rate of the first animal in the question above is higher at 40°C than at 30°C. What physiological phenomenon would explain why the metabolic rate of this animal rises when the environmental temperature is above the upper critical limit?
   Textbook Reference: 40.4 How Do Mammals Regulate Their Body Temperatures? p. 845
3. What types of insulation do mammals have, and what physiological controls exist to increase insulation?
   Textbook Reference: 40.3 How Do Animals Alter Their Heat Exchange with the Environment? p. 842
4. Animals have the ability to undergo acclimatization. Describe this process in relation to temperature regulation.
   Textbook Reference: 40.2 How Does Temperature Affect Living Systems? p. 839
5. Why is it important for an organism to maintain homeostasis?
   Textbook Reference: 40.1 How Do Multicellular Animals Supply the Needs of Their Cells? pp. 833–834

Answers

Diagram Exercise Answers

1. a. Stomach
   b. Small intestine
   c. Epthelial cells
   d. Mucosa
   e. Smooth muscle
   f. Nervous tissue
   g. Epithelial cells and connective tissue

2.

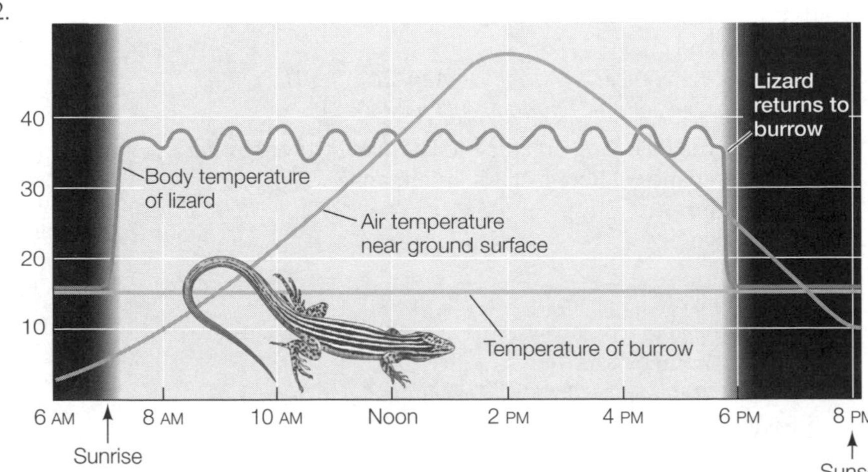

3.

	Endotherm	**Ectotherm**
Heat source	Behavioral and internal metabolism	Behavioral and environmental
Efficiency of energy usage	Leaky ion channels; inefficient	Non-leaky ion channels; more efficient
Resting metabolic rate	Higher	Lower
Temperature control center	Hypothalamus	Hypothalamus
Insulation	Fur, feathers, fat layer beneath skin	Little to none

Knowledge and Synthesis Answers

1. **e.** All of the answer choices are mechanisms that help maintain the internal environment of an animal.
2. **b.** If the environment is cooler than the animal, the response would be to lose heat to the environment across the skin. An increase in peripheral blood flow moves heat from the core of the animal to the surface, where it would be lost to the cooler environment.
3. **a.** For an ectothermic lizard in the desert to maintain the most stable body temperature, it should remain in a burrow during the night, where temperatures do not drop very much, and then shuttle between the sun and shade during the day.
4. **b.** After exercise, a dog will have a high metabolic rate and will pant in an attempt to cool down. Its body temperature will most likely be above the upper thermoneutral zone.
5. **a.** Evaporative cooling is a costly means to lower body temperature. ATP must be used either during sweating or panting. Water loss can also be high during evaporative cooling.
6. **c.** Fevers are brought about by a resetting of the thermoregulatory set point in the hypothalamus in response to pyrogens.
7. **d.** Connective tissues are blood, cartilage, and bone, all of which contain a loose population of cells embedded in an extracellular matrix.
8. **e.** Elephants use both physiological mechanisms (increased blood flow to the ears) and behavioral mechanisms (flapping of ears) to release heat into the environment.
9. **e.** All of these organ systems are involved in maintaining homeostasis. The endocrine system is a regulatory system, the digestive system acquires nutrients and raw materials for the entire body, the muscle system moves materials through the body and generates heat, and the reproductive system produces hormones that maintain secondary sex characteristics.
10. **b.** Negative feedback loops use information about the state of a system to bring the internal environment back toward the set point values. Positive feedback loops amplify the response of a system, moving the system away from the set point.
11. **d.** Physiological processes acclimate in response to a seasonal change in the environment. Acclimatization may involve changes in enzyme expression and a resetting of a physiological set point. Every organism has limits to its ability to acclimate.
12. **a.** Large fishes such as tuna are able to maintain core body temperature above ambient temperature by means of a number of mechanisms; they have a special countercurrent heat exchanger built into their blood vessels and high metabolic rates, thus they produce heat in their swimming muscles.
13. **b.** The hypothalamic thermal set points change daily, monthly, and seasonally.
14. **e.** Brown fat is found in endotherms. Thermogenin protein in brown fat uncouples proton movement from ATP production. This process of heat production is termed nonshivering thermogenesis. Brown fat is highly vascularized and rich in mitochondria.
15. **c.** Increased uterine contraction in response to pressure on the cervix is an example of a positive feedback.
16. **d.** Glial cells are a type of nervous tissue cell. They outnumber neurons in the nervous system and provide support and protection to neurons.

Application Answers

1. Metabolic rate can be calculated using the equation $Q_{10} = (R_T/R_{T-10})$, the Q_{10}. The second animal shows a direct correlation between temperature and metabolism, and compared to the first animal it has a lower overall metabolic rate at both temperatures. This information can be organized into a table such as the following:

	Metabolic rate at 15°C	Metabolic rate at 25°C	Q_{10}	Relationship of metabolism to increasing temperature	Endotherm or Ectotherm?
Animal 1	115 ml O_2/hour	55 ml O_2/hour	0.48 (=55/115)	Metabolism decreases (inverse relationship)	Endotherm
Animal 2	5.5 ml O_2/hour	11.5 ml O_2/hour	2.1 (=11.5/5.5)	Metabolism increases (direct correlation)	Ectotherm

2. An endotherm in a hot environment will attempt to keep core body temperature from increasing. The heat-loss mechanisms available to an endotherm—sweating and panting—require the use of ATP and result in an increased metabolic rate. Endotherms will expend energy to protect body temperature and lose heat to the environment.
3. Mammals use fur and fat as effective insulators. Fur is an effective insulator because it traps still, warm air from the body. Increases in the length of fur increases its effectiveness as an insulator. Fat works in a similar manner because it has a low thermal conductance. Humans do not have sufficient hair on their bodies for it to provide good insulation; therefore, clothes are used as insulation. Physiologically, mammals can change their blood flow patterns to increase or decrease the amount of blood that is received by the skin. When active, to get rid of excess heat, mammals transport heat to hairless skin surfaces.
4. Animals undergo acclimatization in response to changes in seasonal conditions such as temperature. Often these changes are brought about by the production of enzymes that function better at the new temperature. Because of acclimatization, metabolic functions are less sensitive to long-term changes in temperature than to short-term changes.
5. Homeostasis is the maintenance of a constant internal environment. The body functions with the help of proteins and enzymes. Changes in, for example, pH, temperature, glucose level, and oxygen and carbon dioxide levels of the internal environment can affect cellular function. Loss of homeostasis can lead to improper function of proteins and cell membranes and ultimately cell death.

41 Animal Hormones

The Big Picture

- Hormones regulate functions ranging from growth to sexual maturity. In numerous ways, the endocrine system complements the nervous system, both of which are the major regulators of body function and homeostasis.
- Circulating hormones travel via the bloodstream to distant sites within the body to affect target tissues. Other hormones, called paracrine hormones, act locally. Autocrine hormones act on the cells that secrete them.
- Hormone action is controlled by several different mechanisms. In many instances there is a "hormone cascade," in which a hormone controls the release of another hormone, which controls the release of another hormone, and so on. In a situation of negative feedback, a released hormone inhibits (negatively feeds back on) the tissue that may have stimulated its release in the first place. Tropic hormones influence other endocrine glands.

Common Problem Areas

- The pituitary, which is appropriately called the "master gland," produces numerous regulatory hormones. The functions of the anterior and posterior pituitary are quite distinct, and keeping them separate is important for understanding the endocrine system.
- Remember that most hormones are distributed broadly throughout the body and that all cells are exposed to the same level of hormone concentration. Hormones have specific actions because different cells have different levels of sensitivity, not because hormone concentrations differ throughout the body.

Study Strategies

- The many hormones may seem to constitute a long list of random compounds. Try to learn the function along with the name of each hormone, as this will help you remember the target tissues.
- Some hormones have extremely specific actions (e.g., follicle-stimulating hormone), whereas others have broad effects on many target tissues (e.g., epinephrine, thyroid hormones). Learning which hormones are "generalists" and which are "specialists" will help you better understand the endocrine system.
- Biologists studying the endocrine system frequently use abbreviations for the names of hormones. Learn these abbreviations along with the full names.
- Go to yourBioPortal.com to review the following tutorials and activities:

  Animated Tutorial 41.1 Complete Metamorphosis

  Animated Tutorial 41.2 The Hypothalamic–Pituitary–Endocrine Axis

  Animated Tutorial 41.3 Hormonal Regulation of Calcium

  Web Activity 41.1 The Human Endocrine Glands

  Web Activity 41.2 Concept Matching: Vertebrate Hormones

Important Concepts

Hormones are chemical signals used in physiological regulation.

- The genome, the endocrine system, the immune system, and the nervous system are the four major sources of information that animals use to develop, grow, and function. There are some differences in how these systems operate, but all these systems rely on the same basic function: one cell releases a signal that travels to cells with an appropriate receptor to trigger a response.
- Hormones are chemical signals secreted by endocrine cells, which exist either as individual cells or as groups of cells known as endocrine glands. Endocrine glands may secrete several different hormones.
- Hormones are secreted into the extracellular fluid, and then most diffuse into the blood. Once in the blood, they are carried to target cells, where they bind with a receptor to trigger the function of the hormone.
- Some hormones activate target cells that are far from their release site (e.g., testosterone, Figure 41.1B). But not all hormones act on a global scale; some are

secreted into the extracellular fluid and then taken up by local cells, without diffusing into the bloodstream. These local hormones are known as paracrine hormones (or simply paracrines) (see Figure 41.1B). Typically, these hormones are released in small amounts and are either taken up rapidly or inactivated by enzymes before they can travel far. An example is histamine, which acts locally in response to tissue damage by causing dilation of blood vessels. Hormones that act on the cells that secrete them are known as autocrine hormones.

- Unlike endocrine glands, exocrine glands secrete their products into ducts that empty onto the surface of the skin or into a body cavity. Examples of exocrine glands are sweat glands and salivary glands.
- Hormones should not be confused with neurotransmitters or pheromones. Neurotransmitters are released by neurons and act locally, often with the neuron itself. Pheromones are secreted by animals to the outside environment and transmit information to other animals.

Hormonal communication has existed since the evolution of multicellular organisms.

- Slime molds use a chemical signal to coordinate the coming together of individual protist cells into a multicellular fruiting structure.
- Even sponges have forms of extracellular communication using chemicals, and hormones control plant growth.
- Hormone structure is quite conserved over very broad groups of multicellular organisms. The same groups of compounds can be used by different organisms for extracellular signaling, but a hormone can function differently in the context of a given organism (e.g., prolactin; see Figure 41.2).
- Hormones control the processes of molting (shedding of the exoskeleton) and metamorphosis (transformation to the adult stage) in insects. (Recall from Chapter 32 that each growth stage is called an instar.) The hormone-receptor system involved in controlling arthropod metamorphosis is genetically similar to the anabolic steroid system.
- In an elegant set of experiments, Sir Vincent Wigglesworth used the blood-sucking bug *Rhodnius* to examine the control of molting (see Figure 41.3). Decapitation of these bugs at various times after feeding showed that a substance secreted in the heads of the bugs stimulated molting. If the head was removed after a blood meal, molting did not take place. However, if the head was removed a week after a meal, molting did take place. Wigglesworth hypothesized that a substance produced from the insect's head shortly after feeding diffused slowly, thus resulting in a time lag.
- The brains of *Rhodnius* and other insects produce prothoracicotropic hormone (PTTH; previously called brain hormone) that is stored in the corpora cardiaca attached to the brain. After a meal, PTTH diffuses in extracellular fluid to the prothoracic gland. The prothoracic gland releases a hormone called ecdysone, and ecdysone diffuses to the target tissues to stimulate molting.
- Ecdysone enters target cells via diffusion, and the receptor is expressed inside cells involved in secreting enzymes that digest the old cuticle. The mechanism of testosterone, which also plays an important role in development, is similar.
- Both the nervous and endocrine systems often act in concert to control growth and development, a link that is also common in many vertebrates.
- As many insects undergo several instar stages prior to becoming an adult, a hormone called juvenile hormone that prevents premature maturation is secreted by the corpora attata, which is attached to the corpora cardiaca. In insects with incomplete metamorphosis such as *Rhodnius*, the hormone is not secreted during the fifth instar, so the insect molts into an adult. With insects that undergo complete metamorphosis, such as butterflies or moths, the levels of juvenile hormone decrease as the larva molts a fixed number of times. When the juvenile hormone falls below a certain level, the larva spins into a cocoon and molts into a pupa (see Figure 41.4). No juvenile hormone is secreted during this stage, so the pupa molts into the adult.

Hormones act on the receptors of the target cells.

- Hormones can be divided into three chemical groups:
  - Most hormones are peptides or proteins. These hormones are water-soluble, but they do not pass easily through the cell lipid membrane. Before release, peptide and protein hormones are packaged in cellular vesicles and are released by exocytosis.
  - Steroid hormones are derived from cholesterol and are lipid-soluble. Steroid hormones are released by diffusion from the cells that make them.
  - Most of the amine hormones are derived from the amino acid tyrosine. Some are water-soluble and some are lipid-soluble, and their release differs accordingly.
- Most hormones are released in very small quantities. A hormone works only on cells that have specific receptors for it.
- Water-soluble hormones cannot pass through the plasma membrane and thus act on receptors at the surface of the cell. These surface receptors are large glycoprotein complexes with binding, transmembrane, and cytoplasmic domains. Binding to surface receptors usually triggers the target cell's response by activating protein kinases or phosphatases within the cell (see Figures 7.7 and 7.8 for a review). Lipid-soluble hormones are able to cross the plasma membrane and act on receptors inside the cell, ultimately altering gene expression. The steroid hormone-receptor complex often acts by changing gene expression in the cell's nucleus (see Figure 7.9).

- A single hormone can act on many different types of cells to produce different effects, and the response depends on the nature of the cell and its receptors. Epinephrine, secreted by the adrenal glands, activates the fight-or-flight response and causes changes in heart rate, blood flow, and the immune system. Epinephrine also stimulates the breakdown of glycogen in the liver and of fats in adipose tissue (see Figure 41.5).
- The list of known hormones is quite long (see figure 41.6).

The pituitary gland illustrates the close connection between the nervous and endocrine systems.

- The pituitary gland, which is attached to the hypothalamus, provides an example of how nervous and endocrine systems interact. The hypothalamus produces two hormones (antidiuretic hormone and oxytocin) that are stored and then secreted by the pituitary gland. Other hormones released by the hypothalamus influence release of pituitary hormones. Hormones released by the hypothalamus are usually referred to as neurohormones.
- The pituitary gland produces hormones that control many of the other endocrine glands in the body. The pituitary is made up of the anterior pituitary, which originates from the embryonic mouth cavity, and the posterior pituitary, which derives from the developing brain. Each utilizes different control mechanisms and releases different hormones.
- Long axons from the hypothalamus extend into the posterior pituitary. The posterior pituitary releases antidiuretic hormone and oxytocin, both of which are produced by neurons in the hypothalamus. They are stored in vesicles that move down the neurons into the posterior pituitary until an action potential causes their release. When blood pressure falls or when the blood becomes too salty, antidiuretic hormone (ADH, also called vasopressin) acts on the kidneys to stimulate reabsorption of water to help maintain blood pressure. Oxytocin stimulates uterine contractions during childbirth and the ejection of milk from mammary glands. It can be secreted simply in response to the sight and sound of a baby, a good example of how an external stimulus received by the nervous system controls a hormonal process. Oxytocin can also promote pair bonding and maternal bonding.
- The anterior pituitary gland releases four tropic hormones (thyrotropin, corticotropin, luteinizing hormone, and follicle-stimulating hormone) that control the function of the thyroid, adrenal cortex, testes, and ovaries, respectively. A different cell type produces each of the tropic hormones. The anterior pituitary also produces growth hormone, prolactin, endorphins, and enkephalins. Growth hormone (GH) stimulates the liver to release chemical signals called somatomedins, or insulin-like growth factors (IGFs), which stimulate bone and cartilage growth. GH also stimulates amino acid uptake by other cells. Over- or underproduction of GH can cause gigantism or pituitary dwarfism, respectively. In female mammals, prolactin stimulates milk production. Endorphins and enkephalins influence pain pathways in the brain.
- The anterior pituitary is composed of endocrine cells that also respond to neurohormones secreted by the hypothalamus and transported to the anterior pituitary by portal blood vessels (see Figure 41.8). Two of the controlling hormones are thyrotropin-releasing hormone (TRH), which causes the release of thyrotropin that stimulates the thyroid gland, and gonadotropin-releasing hormone (GnRH), which stimulates the release of hormones that control the gonads. Other neurohormones include prolactin-releasing and release-inhibiting hormones, and growth hormone-releasing hormone.
- The anterior pituitary gland is also controlled by direct and indirect negative feedback (see Figure 41.9). Hormones released from target glands may inhibit further release of the particular tropic hormone from the anterior pituitary. For example, cortisol is released in response to the production and secretion of corticotropin by the anterior pituitary. As cortisol levels rise, release of corticotropin is inhibited. Cortisol also inhibits the release of corticotropin-releasing hormone by the hypothalamus.

The thyroid gland regulates metabolic rate.

- The thyroid gland wraps around the front of the trachea, expanding into a lobe on each side. It helps regulate metabolic rate through the actions of thyroxine (T_4) and triiodothyronine (T_3), both of which increase metabolic rate in mammals. The precursor to T_4, thyroglobulin, is made by epithelial cells that surround a round hollow structure known as a follicle (see Figure 41.10A), where the thyroglobulin is stored. The tyrosine residues are iodinated with one or two atoms of iodine. When the thyroid gland is stimulated by thyrotropin produced by the anterior pituitary gland, the cells surrounding the follicle take up thyroglobulin. Thyroglobulin is then cleaved into a dipeptide consisting of two tyrosine residues (see Figure 41.10B). In T_4, all four possible iodination sites have an ion atom, while in T_3, only three out of the four possible sites are bound to an ion atom. T_4 is more abundant, but T_3 is the active form of thyroid hormone. Target cells have enzymes that can convert T_4 to T_3, thereby setting their own sensitivity to the hormone.
- Thyroxine plays a critical role in cellular metabolism. It is lipid soluble and can enter cells directly. Inside the cell, it enters the nucleus and binds to a receptor. The complex acts as transcriptional activator of genes that encode for proteins and enzymes such as those involved in metabolic pathways, transport of other substances, and structural proteins.
- Thyrotropin, or thyroid-stimulating hormone (TSH), from the anterior pituitary gland stimulates the thyroid

gland to produce thyroxine. TSH production and release is controlled by thyrotropin-releasing hormone (TRH) produced by the hypothalamus. The hypothalamus responds to environmental stimuli by increasing or decreasing TRH—another example of the links between the nervous system and the endocrine system. As with cortisol, these steps are controlled by a negative feedback loop; thyroxine inhibits the response of cells in the anterior pituitary to TRH, and it inhibits the production and release of TRH by the hypothalamus.

- Poor regulation of thyroxine production—either hyperthyroidism (thyroxine excess) or hypothyroidism (thyroxine deficiency)—can result in an enlarged thyroid gland, a condition known as goiter (see Figure 41.10C). Hyperthyroidism is often the result of an autoimmune response to the TSH receptor, which causes uncontrolled production and release of thyroxine. Negative feedback from high levels of thyroxine cause the TSH level to be low, but the thyroid gland is under a constant state of stimulation and enlarges. In hypothyroidism there is insufficient circulating thyroxine to turn off TSH production and the thyroid gland responds to high TSH levels by increased production of thyroglobulin. In cases of insufficient iodine intake, thyroglobulin is not converted efficiently to thyroxine and triiodothyronine. More and more thyroglobulin is made, the follicles enlarge, and the result is an enlarged thyroid gland.

The thyroid gland lowers blood calcium levels, while the parathyroid glands raise blood calcium levels.

- The careful regulation of calcium levels is critically important. Shifts below a narrow range can overstimulate the nervous system and result in spasms and seizures. Shifts above this narrow range cause muscles to weaken.
- Most of the calcium in the body (nearly 99 percent) is in bone tissue. The body has three sources for changing blood calcium levels: absorption from or addition to bone, retention or excretion by the kidneys, and absorption by the digestive system. These activities are controlled by calcitonin, parathyroid hormone, and vitamin D.
- Calcitonin, which is secreted by the thyroid gland, lowers blood calcium levels by stimulating osteoblasts (which deposit new bone) and inhibiting osteoclasts (which break down bone). In adults, bone turnover is not very high, so the other mechanisms of calcium regulation have increased importance.
- The parathyroid glands consist of four small glands that are attached to the posterior surface of the thyroid gland (see Figure 41.11). They produce parathyroid hormone (PTH), which increases blood concentrations of calcium by stimulating osteoclasts to dissolve bone. Circulating calcium binds to a surface receptor on the parathyroid cells, which inhibits PTH synthesis and release. Calcitonin (from the thyroid gland) and parathyroid hormone act antagonistically, countering the effects of each other. PTH also raises blood calcium by stimulating the kidneys to reabsorb it from the urine.
  - When PTH stimulates the release of calcium from bone, it also stimulates the release of phosphates. Because calcium and phosphate levels in the body are just below the point where they precipitate out of solution, an excess of both can lead to the formation of kidney stones and deposits in the arteries. To reduce this risk PTH also stimulates the kidneys to release phosphate in urine.
- Strictly speaking, vitamin D is not a vitamin but a hormone, since the body synthesizes it. It is produced by skin cells (where cholesterol is converted by ultraviolet light) and is activated when it passes through the liver and kidneys. Vitamin D helps raise blood calcium levels in a number of ways. In the kidneys, it works with parathyroid hormone to decrease loss of calcium in urine. In the bones, it stimulates bone turnover and thus the movement of calcium from bone to blood.

The pancreas regulates blood sugar levels.

- Endocrine cells in the pancreas called the islets of Langerhans produce the hormones insulin and glucagon, which regulate blood glucose levels. Glucose enters cells by diffusion, but it does not pass the lipid membrane easily. Proteins in the cells called glucose transporters assist in the entry of glucose into the cell. When insulin binds to receptors on cells, the glucose transporters move from the cytoplasm to the cell membrane, allowing glucose to enter the cell from the bloodstream.
- When insulin is absent or the receptors become insensitive to it, glucose accumulates in the blood, resulting in the disease diabetes mellitus. With type I diabetes (generally, juvenile onset), there is a lack of insulin, whereas in type II diabetes (adult onset) the insulin receptors become unresponsive to insulin. In either case, glucose accumulates in the blood. This causes water to move out of the cells, and the rise in blood volume increases urine production. Glucose in the tubules of the kidneys also pulls more water into the urine. Individuals with diabetes can become dehydrated, and they do not have adequate metabolic fuel.
- There are three different cell types in the islets of Langerhans that produce hormones. The beta (β) cells make and secrete insulin. Alpha (α) cells synthesize and secrete glucagon, and the delta (δ) cells produce somatostatin.
- When blood glucose levels fall well below normal, glucagon stimulates the liver to convert glycogen back into glucose and levels of glucose rise in the blood.
- A rapid rise in levels of glucose and amino acids in the blood stimulates the pancreas to release somatostatin, which inhibits the release of insulin and glucagon (giving somatostatin a paracrine function). Cells of the hypothalamus also produce somatostatin. Somatostatin from this source inhibits the release of growth

hormone and thyrotropin from the anterior pituitary gland.

The adrenal glands produce hormones with numerous regulatory functions.

- The adrenal glands have two regions: the adrenal medulla makes up the core of the glands, and the adrenal cortex surrounds the medulla (see Figure 41.12).
- Epinephrine and norepinephrine are produced in the adrenal medulla and are involved in the fight-or-flight response. The medulla develops from and is controlled by the nervous system. Both hormones are water soluble and interact with the alpha (α) and beta (β) adrenergic receptors. Each receptor stimulates different activities (see Figure 41.13).
- The adrenal cortex produces glucocorticoids, mineralocorticoids, and sex steroids from cholesterol (see Figure 41.14).
- The main glucocorticoid is cortisol, which mediates our response to stress by stimulating cells that are not necessary for the fight-or-flight response to decrease their use of glucose and to use fats and proteins as energy sources. The release of cortisol is controlled by corticotropin from the anterior pituitary, which in turn is controlled by corticotropin-releasing hormone that is secreted by the hypothalamus.
- The main mineralocorticoid is aldosterone, which is involved in the reabsorption of sodium and the excretion of potassium at the kidney. Compared to the amounts of sex steroids secreted by the gonads, the amounts produced by the adrenal cortex are negligible.

The gonads produce sex steroid hormones.

- In males, testes produce androgens (testosterone); in females, the ovaries produce estrogens and progesterone. Early in development, the sex organs of human embryos are similar. At about week 7, the presence of the Y chromosome in male embryos causes the undifferentiated gonads to begin producing androgens. In response to androgens, the reproductive system develops into a male system. In the absence of androgens, the reproductive system develops into a female system (see Figure 41.15).
- Sex steroids increase in concentration at the time of puberty. Both luteinizing hormone (LH) and follicle-stimulating hormone (FSH) from the anterior pituitary gland (collectively called gonadotropins) control the production of sex steroids. The production of these tropic hormones by the anterior pituitary is controlled by gonadotropin-releasing hormones (GnRH). At the onset of puberty, the GnRH-producing cells are no longer under negative feedback, so the level of GnRH increases, which stimulates the production of LH and FSH. Increased levels of LH and FSH in females and LH in males stimulate the gonads to produce more sex hormones. These hormones in turn have profound physiological effects on the development of secondary sexual characteristics.

The pineal gland produces melatonin.

- Melatonin is involved in biological rhythms, including photoperiodicity, the phenomenon whereby changing day length across the seasons prompts changes in physiology.
- The release of melatonin by the pineal gland occurs in the dark; light inhibits melatonin release.

Scientists have ways to detect and measure hormones and to characterize their receptors and signal transduction pathways.

- Binding of a hormone to its receptor activates signal transduction pathways within cells. Amplification of the signal occurs in many of these pathways, thus explaining how tiny quantities of a hormone can produce enormous physiological effects.
- Since many hormones are produced in small quantities, scientists have developed assays to measure their effects and their interactions with their receptors. Hormone concentrations are measured by immunoassays (see Figure 41.17). The development of immunoassays allows scientists to measure dose–response relationships (see Figure 41.18), which are important when a drug or compound is being considered for therapeutic use.
- A single hormone can have different receptors, and efforts to characterize receptors often rely on biochemical separation techniques such as affinity chromatography.
- The abundance of hormone receptors is sometimes regulated through negative feedback mechanisms. Downregulation involves a decrease in the number of receptors in response to high levels of a hormone. Upregulation can occur when levels of a hormone are suppressed; in response, a target cell may increase the number of receptors for that hormone.

Test Yourself

Diagram Exercise

In the diagram below, in (A), match the gland with its location in the body. In the case of gonads, you may use the same location as long as you identify the sex of the individual. In (B), label the structures indicated, and in (C) list the hormones secreted by each gland pictured.

Textbook Reference: *41.2 How Do the Nervous and Endocrine Systems Interact? pp. 858–861; 41.3 What Are the Major Mammalian Endocrine Glands and Hormones? pp. 859–868, Figures 41.6, 41.7, 41.11, 41.12*

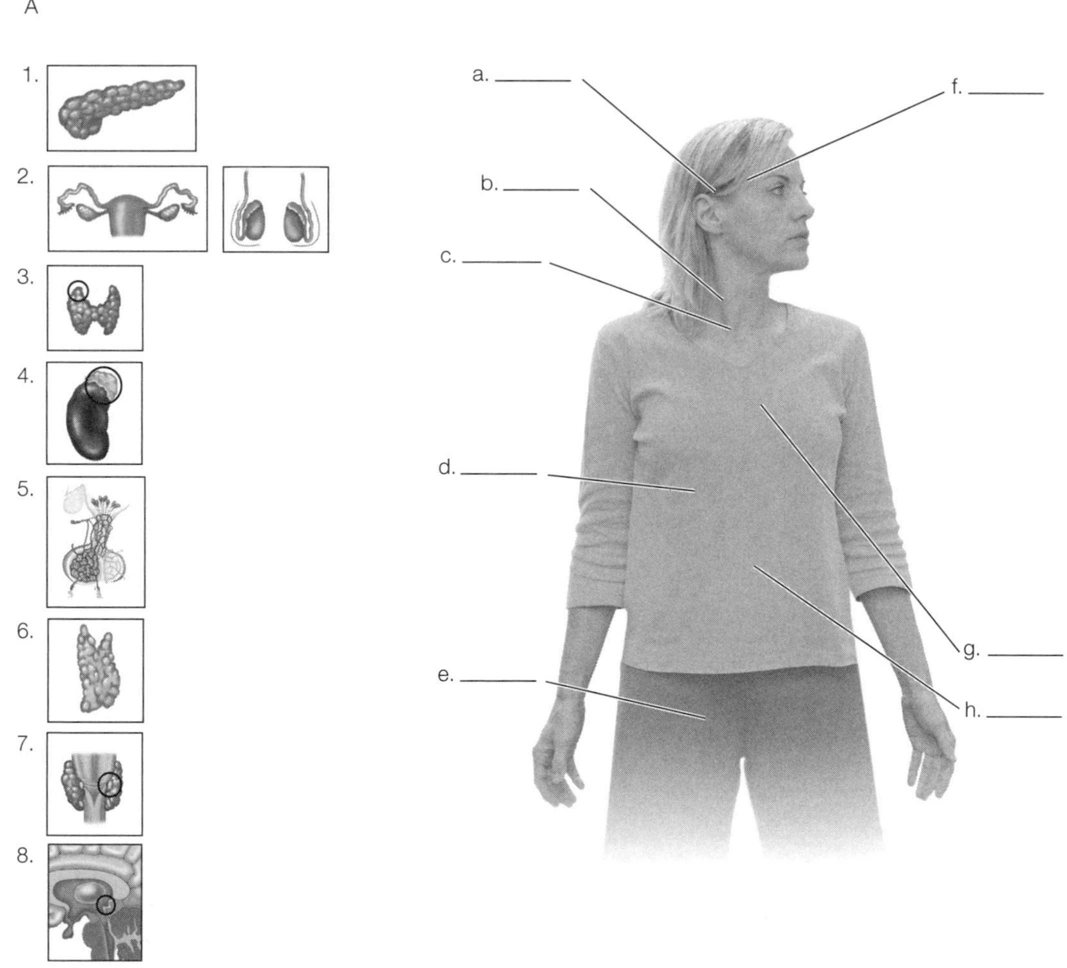

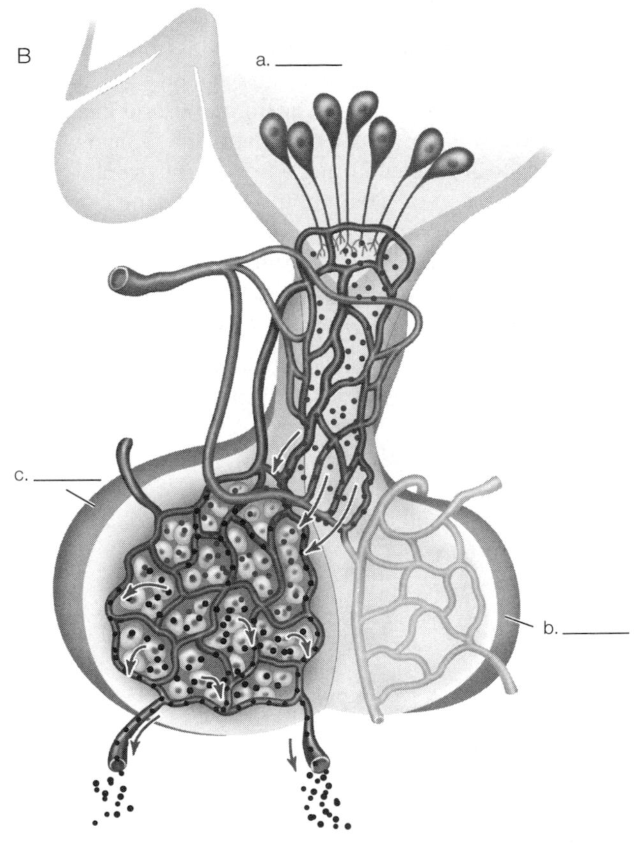

C

1.	Pancreas	______
2.	Gonads	______
3.	Thyroid	______
4.	Adrenal gland	______
5.	Hypothalamus Anterior Pituitary Posterior Pituitary	______
6.	Thymus	______
7.	Parathyroid	______
8.	Pineal gland	______

Knowledge and Synthesis Questions

1. A paracrine hormone
   a. circulates in the bloodstream and affects distant cells.
   b. always acts on a wide variety of target tissues.
   c. acts on nearby cells.
   d. acts on neuronal cells only.
   e. acts on glands only.
   Textbook Reference: *41.1 What Are Hormones and How Do They Work? p. 853*
2. A target cell's response to a hormone depends on
   a. the amount of hormone released.
   b. the number of receptors in or on that target cell.
   c. how well the hormone binds with the receptor.
   d. the presence of the proper receptor in or on the target cell.
   e. All of the above
   Textbook Reference: *41.1 What Are Hormones and How Do They Work? pp. 856–857*
3. Which of the following is the definition of "upregulation"?
   a. An increase in hormone receptors in response to low hormone levels
   b. An increase in hormone receptors in response to high neurotransmitter levels
   c. The increase in hormone levels produced by an increase in hormone receptors
   d. The decrease in hormone levels produced by a decrease in hormone receptors
   e. A change in hormone levels in response to a change in receptor levels
   Textbook Reference: *41.4 How Do We Study Mechanisms of Hormone Action? p. 870*
4. Which of the following statements does *not* describe hormonal functioning?
   a. Hormones act in very low concentrations.
   b. Many hormones act at sites distant from where they are produced.
   c. Hormones are transported in the blood.
   d. Hormones can alter the expression of genes
   e. All of the above describe hormonal functioning.
   Textbook Reference: *41.1 What Are Hormones and How Do They Work? pp. 856–857*

5. In insects, juvenile hormone
   a. is produced by the corpora cardiaca.
   b. is also known as brain hormone.
   c. prevents maturation.
   d. is produced in greatest concentration at the start of each instar stage.
   e. All of the above

   Textbook Reference: *41.1 What Are Hormones and How Do They Work? p. 855*
6. Which of the following statements about vitamin D is *false?*
   a. Vitamin D is a hormone secreted by skin cells.
   b. Vitamin D is activated as it passes through the liver and kidneys.
   c. Vitamin D decreases loss of calcium in the urine.
   d. Vitamin D can be acquired only from food.
   e. All of the above are true.

   Textbook Reference: *41.3 What Are the Major Mammalian Endocrine Glands and Hormones? p. 864*
7. Which of the following glands has one part that develops from the embryonic mouth cavity and a second part that develops from the floor of the developing brain?
   a. Adrenal gland
   b. Thyroid gland
   c. Hypothalamus
   d. Pituitary gland
   e. Parathyroid gland

   Textbook Reference: *41.2 How Do the Nervous and Endocrine Systems Interact? pp. 858–859*
8. Which of the following glands is involved in biological rhythms such as photoperiodicity?
   a. Pineal gland
   b. Thyroid gland
   c. Adrenal glands
   d. Ovaries
   e. Pituitary gland

   Textbook Reference: *41.3 What Are the Major Mammalian Endocrine Glands and Hormones? p. 868*
9. The half-life of hormones in the plasma ranges from _______ to _______.
   a. seconds; minutes
   b. minutes; hours
   c. minutes; days or weeks
   d. days; weeks
   e. hours; months

   Textbook Reference: *41.4 How Do We Study Mechanisms of Hormone Action? p. 869*
10. The target tissues of hormones are those tissues that
    a. can be penetrated by the particular hormones.
    b. have specific enzymes with which the hormones interact directly.
    c. have high concentrations of the"second messenger."
    d. have receptors for the particular hormones.
    e. have the right genes for the hormones to express.

    Textbook Reference: *41.1 What Are Hormones and How Do They Work? p. 857*
11. Which of the following vertebrate hormones are produced in the anterior pituitary gland?
    a. Somatostatin, antidiuretic hormone, and insulin
    b. Prolactin, growth hormone, and enkephalins
    c. Oxytocin, prolactin, and adrenocorticotropin
    d. Estrogen, progesterone, and testosterone
    e. Growth hormone, gonadotropin-releasing hormone, and thyroid-releasing hormone

    Textbook Reference: *41.2 How Do the Nervous and Endocrine Systems Interact? pp. 859–861*
12. Hormones that are secreted by one endocrine gland and control the activities of another endocrine gland are called _______ hormones.
    a. growth
    b. obstructive
    c. tropic
    d. selective
    e. paracrine

    Textbook Reference: *41.2 How Do the Nervous and Endocrine Systems Interact? p. 859*
13. Which of the following has both endocrine and exocrine functions?
    a. Pancreas
    b. Heart
    c. Testes
    d. Adrenal glands
    e. Pineal gland

    Textbook Reference: *41.3 What Are the Major Mammalian Endocrine Glands and Hormones? pp. 864–865*
14. Which of the following hormones have antagonistic effects?
    a. Insulin and glucagon
    b. Growth hormone and oxytocin
    c. Oxytocin and prolactin
    d. Cortisol and testosterone
    e. Thyroid and parathyroid

    Textbook Reference: *41.3 What Are the Major Mammalian Endocrine Glands and Hormones? pp. 864–865*
15. Steroid hormones are
    a. water-soluble.
    b. produced by the thyroid gland.
    c. lipid-soluble.
    d. derived from the amino acid tyrosine.
    e. hydrophilic.

    Textbook Reference: *41.1 What Are Hormones and How Do They Work? p. 856*

Application Questions

1. Describe the modes of action of lipid-soluble and water-soluble hormones.

   Textbook Reference: *41.1 What Are Hormones and How Do They Work? pp. 856–857*

2. The pituitary gland is made up of the anterior and posterior pituitary. What is the relationship between each of these parts and the hypothalamus?
   Textbook Reference: *41.2 How Do the Nervous and Endocrine Systems Interact? pp. 859–861*
3. Why would breast-feeding an infant soon after birth help a mother's uterus return to its prepregnancy size?
   Textbook Reference: *41.2 How Do the Nervous and Endocrine Systems Interact? p. 859*
4. Sir Vincent Wigglesworth conducted a number of studies on the insect *Rhodnius* to examine the control of molting. Describe what would happen to a fourth-instar *Rhodnius* if it was either partially or fully decapitated one week after a blood meal.
   Textbook Reference: *41.1 What Are Hormones and How Do They Work? pp. 854–856, Figure 41.4*
5. Would castration (removal of the testes) lead to an immediate and complete absence of testosterone in the person's bloodstream? Why or why not?
   Textbook Reference: *41.3 What Are the Major Mammalian Endocrine Glands and Hormones? p. 865*

Answers

Diagram Exercise Answer

A

a. 8. Pineal gland
b. 3. Thyroid gland
c. 7. Parathyroid gland
d. 4. Adrenal gland
e. 2. Gonads (ovaries; testes)
f. 5. Hypothalamus/Posterior pituitary/Anterior pituitary
g. 6. Thymus
h. 1. Pancreas

B

a. Hypothalamus
b. Posterior pituitary
c. Anterior pituitary

C

Pancreas: Insulin, glucagon, somatostatin

Gonads:

Ovaries: Estrogens, progesterone

Testes: Testosterone

Thyroid: Thyroxine (T_3 and T_4), calcitonin

Adrenal gland:

Cortex: cortisol, aldosterone, testosterone (in both sexes), estrogen (in both sexes)

Medulla: epinephrine, norepinephrine

Hypothalamus: ADH (antidiuretic hormone), oxytocin

Anterior pituitary: Thyrotropin (TSH), follicle stimulating hormone (FSH; produced in both sexes), Lutenizing hormone (LH, produced in both sexes), corticotropin (ACTH), growth hormone (GH), prolactin, melanocyte-stimulating hormone (MSH), endorphins, enkephalins.

Posterior pituitary: Releases (but does not make) ADH and oxytocin.

Thymus: Thymosin

Parathyroid: Parathyroid hormone (PTH)

Pineal gland: Melatonin

Knowledge and Synthesis Answers

1. **c.** Paracrine hormones act on nearby cells. Autocrine hormones act on the very same cells that secrete them. Circulating hormones enter the bloodstream and affect distant cells.
2. **e.** The response of target cells to a hormone depends on how much of the hormone is present and acting on the target, how many receptors are present on the surface of the cell that the hormone can act on, and how well the hormone binds with the receptor.
3. **a.** "Upregulation" of hormone receptors on a cell is the production of more receptors when a hormone is at low levels over time in the blood or in other fluids surrounding the cell.
4. **e.** All of the statements are correct. Hormones are found and act at low concentrations in the body. They are transported by the blood or diffuse through interstitial space to reach target cells that may be some distance from the secreting gland. Hormones can alter the expression of genes as well as cell metabolism.
5. **c.** Juvenile hormone is secreted continuously by the corpora allata of insects and prevents maturation. Prothoracicotropic hormone (brain hormone) is secreted by the corpora cardiaca and is involved with ecdysone in molting.
6. **d.** Vitamin D is a hormone (not a true vitamin) secreted by skin cells, and it becomes more active once it passes through the liver and kidneys. Vitamin D raises levels of calcium in the blood by reducing loss of calcium in the urine.
7. **d.** The pituitary gland has an anterior lobe that develops from an outpocketing of the embryonic mouth cavity and a posterior lobe that develops from an outpocketing of the floor of the developing brain.
8. **a.** The pineal gland secretes melatonin, a hormone that controls biological rhythms. One such rhythm is photoperiodicity, the phenomenon whereby seasonal changes in day length influence physiological processes in animals.

9. **c.** Hormones tend to act over time periods of minutes to days or weeks. Epinephrine has a half-life of up to three minutes, whereas cortisol has a half-life of many days.
10. **d.** For a hormone to act on a target cell, the target cell must have the receptors for the specific hormone to bind and trigger the hormonal action.
11. **b.** Prolactin, growth hormone, and enkephalins are produced in the anterior pituitary.
12. **c.** Hormones that control endocrine gland function are known as tropic hormones.
13. **a.** The pancreas has both endocrine and exocrine functions. It releases the hormones glucagon, insulin, and somatostatin to extracellular fluid and produces digestive enzymes that travel through ducts to the small intestine.
14. **a.** Insulin lowers blood glucose and glucagon raises blood glucose; thus these hormones have antagonistic effects.
15. **c.** Steroid hormones are lipid-soluble and are produced by the gonads and adrenal glands.

Application Answers

1. The plasma membrane of cells is hydrophobic. Lipid-soluble hormones, which (as their name suggests) dissolve in lipids (fats), cross the plasma membrane and act on receptors inside the target cells. Water-soluble hormones cannot cross the lipid-based plasma membrane and act on receptors on the outside of target cells. Such binding by water-soluble hormones initiates changes in the cell by activating enzymes. Because lipid-soluble hormones cross the plasma membrane and enter the cell, they tend to take longer to act, but they also remain active for a longer time period than water-soluble hormones do.
2. The hormones secreted by the posterior pituitary gland are produced by neurons in the hypothalamus. The neurohormones move down neurons into the posterior pituitary gland, where they are stored and secreted. The anterior pituitary produces the hormones it secretes. However, hormones from the hypothalamus control the release of the anterior pituitary's hormones.
3. Suckling by a baby stimulates the release of oxytocin from the posterior pituitary, and this leads to milk ejection from the mammary glands. Oxytocin also causes uterine contractions, and after birth this helps the mother's uterus return toward its prepregnancy size.
4. The insect *Rhodnius* can live for long periods of time after it has been decapitated. If the insect is partially decapitated, leaving the corpora allata, it will molt into a fifth-instar juvenile because the corpora allata produces juvenile hormone, which prevents maturation into an adult. An insect that is fully decapitated will molt into an adult rather than another juvenile instar. The fully decapitated animal has enough prothoracicotropic hormone (brain hormone) circulating one week after a meal to allow for molting; however, in the absence of juvenile hormone, it molts into an adult.
5. Castration would not lead to an immediate absence of testosterone in the bloodstream. Like many hormones, testosterone has a half-life of days to weeks. In addition, small amounts of the sex steroids are produced by the adrenal glands in both males and females.

42 Immunology: Animal Defense Systems

The Big Picture

- Animals have special systems to distinguish self from nonself. Some of these systems are nonspecific defenses (e.g., complement protein, phagocytotic cells). B and T cells can recognize specific antigens on the surface of pathogens and mount an immune response to destroy those pathogens. When the immune system is compromised, as seen in HIV-infected individuals, the individual becomes susceptible to a variety of pathogens.

Common Problem Areas

- The specificity of the immune response may seem complex initially, but there are really only two major components: B cells (which bind antigen) present antigen and secrete antibody, and T cells, which bind antigen on antigen-presenting cells and help direct the immune response.
- MHC proteins may seem complicated at first, but they have only a few essential functions in the immune system. These proteins are specific for each individual and are essential for presenting antigen on antigen-presenting cells (MHC II) to T_H cells and recognizing one's own cells (MHC I).
- The specificity of the immune response may seem complex, but it can be understood by reviewing cellular communication. Antibodies of B cells bind with antigen on the surface of pathogens; T cell receptors bind antigen presented on MHC I and MHC II proteins on the surface of cells. Once these membrane proteins interact, cytokines are secreted, and cells begin to divide and proliferate. Antibodies are secreted by B cells.
- The rearrangement of DNA in the nucleus of the B cell may seem foreign to students who think the integrity of the genetic information must be preserved. But this change in the genetic makeup of activated B and T cells is essential to provide the organism with a diverse and specific set of antibodies and cell receptors.

Study Strategies

- List the nonspecific responses that a human body can make in response to an invading pathogen. Next list cells that are part of the specific responses. Include proteins that are crucial to generating that specificity. Review how genetic rearrangements generate a diverse set of antibodies for B cells.
- Diagram an antibody and label the important components. Highlight those areas where mutations and alterations occur that change the amino acid sequence of the heavy and light chains.
- Review the expression and function of MHC I and MHC II molecules.
- Make a list of the functions of T_H cells and T_C cells.
- Draw a diagram of a chromosome and label it with 4 *V*s (*V*1–*V*4), 2 *D*s (*D*1 and *D*2), 3 *J*s (*J*1–*J*3) and 8 *C*s (*C*1–*C*8). Imagine that a B cell is maturing to produce antibody on its surface. Select one *V*, one *D*, one *J*, and one *C* segment to make an antibody. Repeat the process using different segments. Predict how many different antibodies can be generated from this set of gene segments.
- List some of the disorders of the immune system and their probable causes.
- Go to yourBioPortal.com to review the following tutorials and activities:

  Animated Tutorial 42.1 Cells of the Immune System
  Animated Tutorial 42.2 Pregnancy Test
  Animated Tutorial 42.3 Humoral Immune Response
  Animated Tutorial 42.4 Cellular Immune Response
  Animated Tutorial 42.5 A B Cell Builds an Antibody
  Web Activity 42.1 The Human Defense System
  Web Activity 42.2 Inflammation Response
  Web Activity 42.3 Immunoglobulin Structure

Important Concepts

Animal defense systems are based on distinguishing self from nonself and include nonspecific defenses and specific defenses.

- Pathogens are the harmful organisms and viruses that cause disease. The defensive response to a pathogen involves three phases: recognition, activation, and effector.
- Nonspecific defenses, or "innate defenses," are inherited mechanisms that act rapidly to protect the body from pathogens. Nonspecific defenses include physical barriers as well as cellular and chemical defenses (see Table 42.1).
- Specific defenses are adaptive mechanisms aimed at specific targets. The specific defenses involve antibodies and are found in the vertebrates.
- Nonspecific and specific defenses are both present in mammals. Nonspecific defenses respond quickly (often within minutes or hours) and are the first line of defense. Specific defenses take days or even weeks to develop.
- Lymphoid tissues, including thymus, bone marrow, spleen, and lymph nodes, are essential to a mammal's defense system (see Figure 42.1). The lymphatic system moves lymph fluid slowly from intercellular spaces, through tiny lymph capillaries and lager ducts, and finally to one large vessel called the thoracic duct that joins a major vein near the heart. Both lymph and blood contain white blood cells and platelets. Lymph nodes contain white blood cells and filter the lymph fluid, looking for nonself material.
- Lymphocytes (see Figure 42.2) are the most abundant class of white blood cells and include B and T cells. Immature T cells migrate via the blood to the thymus, where they mature. T cells participate in specific defenses against foreign or altered cells, including virus-infected cells and tumor cells. B cells circulate through the blood and lymph and make antibodies, which are proteins that bind to nonself or altered self.
- Granular cells are white blood cells that include the phagocytes involved in engulfing and digesting nonself materials. Macrophages are important phagocytes that are involved in the interaction of nonself materials and T cells.
- Cells and other components of the immune system interact with one another to fight off pathogens. There are four main players involved in the immune system response.
  - Antibodies, which are produced by B cells, are proteins that bind to substances that are nonself or altered self. Binding can inactivate pathogens and toxins and act as a tag to mark them for attack by immune system cells.
  - Major histocompatibility complex (MHC) proteins stick out from surfaces of most cells in the mammalian body and help in self-identity. MHC I proteins are on the surface of most cells, and MHC II proteins are on the surface of immune system cells.
  - T cell receptors are proteins on the surface of T cells that recognize and bind to nonself substances on the surfaces of other cells.
  - Cytokines are soluble signal proteins released by T cells, macrophages, and other cells. They can activate or inactivate B cells, macrophages, and T cells.

Barriers, local agents, chemicals, and cellular processes act as nonspecific defenses against invaders.

- Nonspecific defenses (see Table 42.1) include skin that acts as a physical barrier; lysozyme in tears, which attacks the cell wall of many bacteria; mucus in the nose, which traps airborne pathogens; and digestive enzymes, bile salts, and hydrochloric acid in the stomach. The bacteria and fungi that compose the normal flora living on the body also compete with pathogens and act as a means of defense.
- Mucous membranes produce peptides called defensins that are toxic to many bacteria and enveloped viruses. They are able to insert themselves into the membrane of the invader, making it permeable and thus killing it. Phagocytes also produce defensins.
- About 20 complement proteins act as antimicrobial proteins in the vertebrate blood. They function by destroying microbes, activating other immune responses, or by lysing microbes. They are an important part of the nonspecific and specific defense response.
- Small proteins, known as interferons, are produced in response to a viral infection. Interferons are glycoproteins that stimulate signaling pathways that inhibit viral reproduction inside infected cells and stimulate lysosome activity. They also send a signal cells that causes them to digest bacterial and viral proteins into smaller peptides. This is an important first step in specific immunity.
- Phagocyte cells ingest cells and viruses, which are then killed by defensins inside the phagocyte.
- Natural killer cells are white blood cells that lyse target cells that have been infected by a virus. They can also lyse target cells that have been tagged by an antibody and are important in recognizing and lysing abnormal "self" cells such as tumor cells.
- Inflammation defends the body against infectious agents through the release of tumor necrosis factor, prostaglandins, and histamine from mast cells (see Figure 42.3). Inflammation is an important nonspecific defense response because it isolates the damaged area to stop spread of damage, and recruits cells such as mast cells to kill pathogens. The process of inflammation also promotes healing. When the inflammation response is inappropriately strong can result in some allergic responses, autoimmunity or sepsis. Sepsis is usually the result of a severe bacterial infection in which the inflammation response does not remain

local. This condition causes all of the body's blood vessels to dilate and results in blood pressure to drop to lethal levels.

Cell signaling pathways stimulate the body's defenses.

- In general, an invading pathogen can be regarded as a signal that causes cells to produce proteins in the body's defense and stimulate other cells in response.
- The link between signal and response is known as a signal transduction pathway.
- The group of receptors that initiate this pathway is known as toll-like receptors. They are part of a protein kinase cascade that ultimately results in the activation of at least 40 genes in an immune response.
- Figure 42.2 shows one example of the interaction between CD14 ("cluster differentiation" 14), a toll receptor, and the activation of a transcription factor NF-κB.

Specific defenses of the immune response occur when immune cells (B and T cells) recognize and destroy nonself substances.

- Adaptive immunity has four main features:
  - An antigen is a molecule or organism that is reconized by specific antibodies produced by B cells or T cell receptors to provide specificity. Immune cells recognize antigenic determinants, or epitopes (i.e., specific amino acids of a protein) on invading pathogens. There can be many antigenic determinants on a pathogen surface.
  - The immune cells can distinguish self from nonself, which prevents them from attacking their own cells.
  - Immune cells respond to pathogens by activating lymphocytes of the appropriate specificity. Humans can respond to 10 million different antigenic determinants.
    - The immune system "remembers" a pathogen and can respond more rapidly and more effectively the next time it is exposed to it. This immunological memory is the basis of vaccination.
- There are two interactive immune responses against invaders. The humoral immune response involves antibodies from B cells that react with antigenic determinants. The cellular immune response involves T cells binding with antigens. These systems act in concert and share some of the same mechanisms.
  - With the humoral immune response, antigens are presented by antigen-presenting cells to a B cell receptor. This activates the B cell and it secretes multiple copies of the antibody.
  - T cells involved in the cellular immune response recognize and destroy cells that are expressing viral antigens or cells that are abnormal, such as cancer cells. These T cells express T cell receptors that recognize antigens, and are similar in structure and function to B cell–produced antibodies. One these T cells recognize and bind to the antigen, an immune response that results in the destruction of the antigen-containing cell is started.
  - T-helper (T_H) cells bind to antigens on an antigen-presenting cell and activate both the humoral and cellular immune systems. T_H cells are a key player in integrating the two responses (see Figure 42.6).
  - ***For review, go to Diagram Exercise 1.***

Genetic changes and clonal selection account for the characteristic features of the immune response.

- The genetic changes resulting in diversity are produced by DNA changes, including chromosomal rearrangements and other mutations that occur as B and T cells are produced in the bone marrow. Each B cell can make only one type of antibody, and the T cells involved in cellular immunity have a specific receptor to recognize an antigen. There are millions of different kinds of B cells and T cells with specific receptors. Essentially, the immune system has the machinery in place to recognize antigens before they are encountered.
- When an antigen that fits the surface antibody binds to the B cell, that cell is activated and divides to produce clonal B cells, which secrete antibodies (see Figure 42.7). T cells are clonally selected in a similar manner. This process is known as clonal selection.

Immunologic memory and immunity result from clonal selection.

- In the primary immune response, activated B and T cells produce effector cells and memory cells. The production of both cell types serves as a basis for the rapid recognition of antigens even years after the primary immune response.
  - Effector cells attack a pathogen by producing specific antibodies (in the case of B cells) or cytokines (in the case of T cells). Effector B cells are called plasma cells. During a first encounter with an antigen, the primary immune response occurs. Effector cells generally live for only a few days.
  - Following the primary immune response, long-lived memory cells remain and divide at a slow rate. The memory B and T cells divide rapidly to produce effector cells and more memory cells to serve as a more powerful immune response in the event of reexposure to a pathogen. This rapid response, the secondary immune response, provides a natural immunity to diseases caused by those pathogens.
- Vaccinations and immunizations are administered to give the recipient artificial immunity and are effective because of this secondary immune response. The vaccine initiates a primary immune response. If a pathogen carrying the same antigens subsequently attacks, vaccinations are produced by heat or chemical inactivation of the pathogen, attenuation of the pathogen by means of laboratory-generated mutants that are no longer capable of causing disease, or by recombinant DNA technology that produces nontoxic peptide fragments that can activate lymphocytes. A summary of some of the available human vaccines is provided in Table 42.2.

- Animals are able to tolerate their own antigens due to clonal deletion, a process that removes any B and T cells that recognize self-antigens during their differentiation. An immune cell that recognizes self-antigens undergoes programmed cell death (apoptosis).

B cells and antibodies are the basis of the humoral response.

- Activation of B cells results from the binding of an antibody to a particular antigenic determinant, and the arrival of a signal from a helper T cell that has bound the antigen on an antigen-presenting cell. Once this occurs, the plasma cell (effector B cell, see Figure 42.8) is able to secrete antibodies. As the effector B cells proliferate, they produce antibodies that are specific for the antigen that bound to the parent B cell.
- Antibodies (also called immunoglobulins) are proteins that are similar in structure, but they can be grouped into five different classes. Antibody molecules consist of two identical heavy polypeptide chains and two identical light polypeptide chains held together by disulfide bonds. Each polypeptide consists of a constant region that is similar in amino acid sequence from one immunoglobulin to another and a variable region that forms the antigen-binding site (see Figure 42.9).
  - The variable region produces the specificity of the millions of antibodies. The constant region determines whether the antibody will be secreted or remain on the cell surface of the B cell. It also determines what type of action will be taken to eliminate the antigen.
- The antigen binding sites on each immunoglobulin is identical, so antibodies are bivalent. Each antibody can recognize two antigen molecules, and since most antigens have multiple epitopes, antibodies can form large complexes with antigens. These complexes are easy targets for the rest of the immune system.
- The five immunoglobulin classes (IgG, IgM, IgD, IgA, and IgE) are based on differences in the constant region of the heavy chain (see Table 42.3). IgG is the most abundant, and its heavy chains can attach to receptors on macrophages that allows the macrophages to destroy antigens by phagocytosis.

Monoclonal antibodies are produced by hybridomas.

- Normal immune responses are polyclonal, i.e., generated by the activation of many different B cells and resulting in a complex mixture of antibodies.
- Monoclonal antibodies are produced from a clone of B cells that makes only one antibody that binds a unique determinant. Monoclonal antibodies are generated by the fusion of normal B lymphocytes with tumor cells of plasma cells to produce hybridomas (see Figure 42.10).
- Monoclonal antibodies have been used in immunoassays to detect proteins that are produced in small amounts such as hormones, in immunotherapy to target cancer cells with radioactive ligands or toxins, and for passive immunization.

T cells direct the cellular immune response.

- T cells have specific glycoprotein surface receptors that are made up of two different polypeptide chains, both with a variable and a constant region (see Figure 42.11). The variable region provides the specificity of the T cell receptor.
- T cell receptors bind antigen fragments that are displayed on antigen-presenting cells.
- When a T cell is activated, it proliferates and forms a clone. The resulting effector cells are of two different types:
  - Cytotoxic T (T_C) cells recognize virus-infected cells and kill them by causing them to lyse.
  - Helper T (T_H) cells assist both the cellular and the humoral immune systems. They send out signals to stimulate B cell and T cell proliferation.

The major histocompatibility complex (MHC) encodes proteins that present antigens to the immune system.

- MHC gene products are plasma membrane glycoproteins that present antigens on an antigen-binding site of the MHC protein. These antigens are inspected by T cell receptors to distinguish between self and nonself. MHC proteins can display antigenic peptides of about 10 to 20 amino acids.
- Class I MHC proteins are present on the surface of every nucleated cell in an animal. Degraded cellular proteins are presented to T_C cells on MHC I complexes. T_C cells have a surface protein called CD8 that can recognize MHC I proteins.
- Class II MHC proteins are found on the surfaces of antigen-presenting cells, including B cells and macrophages. When an antigen-presenting cell ingests a foreign antigen, it is broken down inside the cell. MHC II molecules bind to antigen fragments and carry it to the cell surface. T_H cells recognize class II MHC proteins by means of the CD4 cell surface protein (see Figure 42.12).
- T cell receptors recognize both the antigenic fragment and the MHC I or MHC II molecules that are bound to the antigen.
- MHC molecules play a critical role in selecting for T cells that can bind self MHC and in selecting against T cells that have self peptides bound to self MHC. Both types of selection occur in the thymus.

The humoral response and cellular immune response involve an interaction between T cells and MHC complexes.

- In the humoral response, both macrophages and B cells bind antigens and present antigen fragments on the surface of their cells on MHC II complexes. In the activation phase, a T_H cell binds to an antigen-presenting macrophage, resulting in the release of cytokines by the T_H cells. Those cytokines activate the T_H cells, causing it to divide. This pool of T_H cells is a clonal population with the same specificity. In the effector phase, the TH cells activate B cells that have the same specificity. The B cells then produce antibodies.

- In the next step of the effector phase, the B cells can also present antigens to the T_H cells on class II MHC proteins. When the T_H cell binds to this complex, it releases cytokines, and the B cell produces a clonal population of plasma cells. These plasma cells secrete the same antibodies as the B cell parent (see Figure 42.13A).
- In the cellular immune response, cytotoxic T cells destroy cells that are displaying antigens from virus or mutant proteins on their MHC I complexes, eliminating virally infected cells as well as tumor cells. As with the humoral response, the cellular response is divided into an activation phase and effector phase (see Figure 42.13B). T_C cells must receive a second signal for activation, which involves an interaction between another receptor on the T_C cell surface and a protein called B7 on the antigen-presenting cell. The T_C cell then proliferates and divides, releasing cytokines.

Humoral and cellular immune responses are suppressed by regulatory T cells.

- Regulatory T cells (Tregs) are made in the thymus, express the T cell receptor, and are activated if they bind to antigen–MHC complexes. However, Tregs recognize self-antigens.
- When Tregs are activated, they secrete a cytokine called interleukin 10. This blocks the activation of T_C and T_H cells and causes them to undergo apoptosis (see Figure 42.14).

MHC molecules play a key role in the tolerance of self.

- T cells are tested in the thymus to see if they bind to MHC displaying one of the body's own antigens, or if they are unable to recognize the body's MHC proteins. If T cells have either of these characteristics, they are destroyed.
- MHC proteins are specific to each individual; tissue transplants are thus recognized as nonself in the recipient. Suppression of rejection of organ transplants by the recipient's immune system is accomplished by treating a patient with drugs such as cyclosporin, which suppress the immune system.

DNA rearrangements and mutations are used to generate all the diverse antibodies within an organism.

- The genome of the B cell undergoes genetic rearrangement during differentiation, and the combination of different alleles allows a B cell's genome to encode for a unique immunoglobulin.
- DNA fragments are rearranged and joined during B cell development to generate antibody supergenes (see Figures 42.15 and 42.16A).
- The variable region of the light chain originates from two families of genes, and the variable region of the heavy chain originates from three families of genes.
- For the constant and variable regions of the heavy chain in mice, there are multiple genes coding for each of four kinds of segments: 100 *V*, 30 *D*, 6 *J*, and 8 *C* (see Figure 42.15). Each B cell randomly selects one gene from each of these multiple genes to make the final coding sequence, *VDJC* for the heavy chain. Light chains are similarly constructed from DNA segments.
- Light and heavy chains are combined to create billions of possible antibodies.
- Increased diversity is achieved by imprecise recombination during DNA rearrangements, addition of extra nucleotides to cut DNA fragments as insertion mutations from an enzyme called terminal transferase, and an increased mutation rate in immunoglobulin genes.
- Before it becomes a plasma cell, a B cell produces IgM antibodies that are responsible for the B cell's recognition of specific antigens. If the B cell becomes a plasma cell, class switching occurs by rearrangements in the DNA that position the *VDJ* segment (the same variable region on the antibody) adjacent to a constant region farther down the DNA (see Figure 42.17). This results in the production of an antibody with a different constant region. The plasma cell will produce IgA, IgD, IgE, or IgG depending on which constant region is placed next to the variable region genes. The antibody will still retain the same variable regions as the IgM produced by the parent B cell. Class switching in B cells is caused by cytokine release from T_H cells.
- ***For review, go to Diagram Exercise 2.***

Disorders of the immune system

- Hypersensitivity occurs when the immune system overreacts to a dose of antigen and may cause inflammation as seen in the allergic response. The antigen may not present a threat, but the inappropriate immune response may cause inflammation and other symptoms that sometimes can cause life-threatening illnesses.
- Immediate hypersensitivity in allergic reactions is caused by IgE binding the foreign antigen and causing mast cells to release histamines (see Figure 42.18). Pollen allergies can be treated by desensitization, in which a small amount of the allergen is injected in the skin to stimulate the production of IgG. If a person is exposed to the same allergen, IgG reacts before IgE can interact with it. This approach generally does not work for food allergens, as the response is often too strong for the therapy to be effective.
- Delayed hypersensitivity is a response that takes place hours after exposure to an antigen. Antigen-presenting cells process an antigen and initiate a T cell response.
- Autoimmunity occurs when T cells direct their response against self-antigens. Potential origins of autoimmunity include the failure to delete a clone that makes antibodies against a self-antigen, and molecular mimicry, in which a T cell that recognizes a nonself antigen from a virus also recognizes a self-antigen that has a similar structure. Examples of autoimmune diseases include systemic lupus erythematosis, rheumatoid arthritis, Hashimoto's thyroiditis, and insulin-dependent diabetes mellitus.

- AIDS is caused by HIV, a virus that eventually destroys T_H cells. HIV is transmitted through body fluids, including blood and semen.
  - HIV initially infects macrophages, T_H cells, and dendritic cells (one kind of antigen-presenting cell). These cells carry the virus to the lymph nodes, where they preferentially infect activated T_H cells. HIV itself kills T_H cells, and the infected T_H cells are also a target of T_C cells.
  - During the first phase of infection, up to 10 billion viruses are made every day, causing mononucleosis-like symptoms. During the second phase, the virus infects T_H cells, and an immune response is mounted, producing HIV antibodies in the blood. The viral load in the blood decreases, but the infection remains at a low level because of the depletion of T_H cells (see Figure 42.19). A person may remain at this "set point" for 8 to 10 years on average. Eventually, the T_H cells are destroyed, and the patient becomes susceptible to a wide variety of infections that normally would be eliminated by T_H cells, including Kaposi's sarcoma, *Pneumocystis carinii*, and virally caused tumors.
- The biology of HIV was discussed in Chapter 16. Drug treatments generally focus on HIV reverse transcriptase and protease. When used in combination, drugs targeted to both proteins can prolong survival.

Test Yourself

Diagram Exercises

1. Name the parent cell type and the two daughter cell types in the diagram. For the left-hand lineage, list the three different cell types that arise from this parent cell. For the right-hand lineage, list the nine different cells and cell fragments that arise from this parent cell. Indicate if the cell is granular or agranular if appropriate, and list the primary function of each cell.

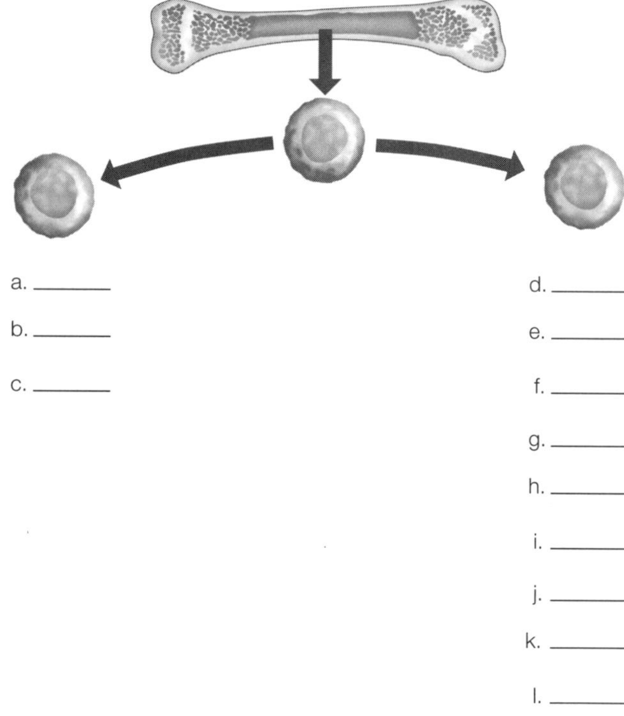

Textbook Reference: *42.1 What Are the Major Defense Systems of Animals? p. 876, Figure 42.2*

2. In the diagrams below, identify the cellular and humoral immune responses. Label each cell type and MHC protein with its correct class. Describe what is occurring in each step of the process for both diagrams.

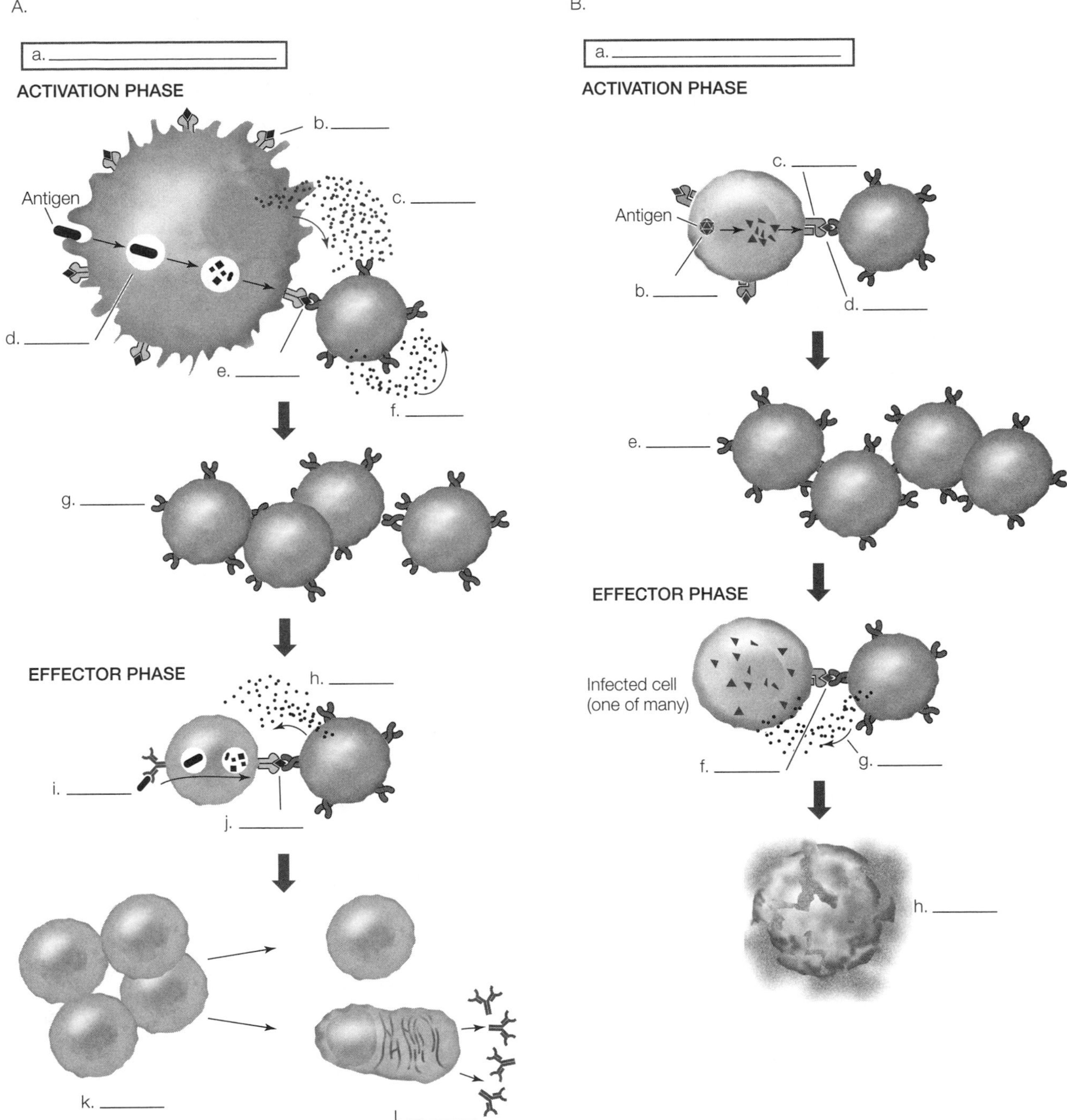

Textbook Reference: *42.5 What Is the Cellular Immune Response? pp. 887–891, Figure 42.13*

Knowledge and Synthesis Questions

1. Phagocytes
   a. are derived from T and B cells.
   b. present antigen on MHC I complexes.
   c. digest nonself materials.
   d. are a type of erythrocyte.
   e. secrete antibodies.
   Textbook Reference: *42.1 What Are the Major Defense Systems of Animals? p. 874*
2. B cells
   a. secrete antibodies.
   b. present antigen on MHC II.
   c. ingest antigens.
   d. are part of the humoral response.
   e. All of the above
   Textbook Reference: *42.3 How Does Specific Immunity Develop? p. 882*
3. Which of the following is a nonspecific defense of the immune system?
   a. Macrophages
   b. Natural killer cells
   c. Complement proteins
   d. Cilia on the mucous membranes
   e. All of the above
   Textbook Reference: *42.2 What Are the Characteristics of the Nonspecific Defenses? pp. 877–878 and Table 42.1*
4. When the receptor of a T_H cell binds to a pathogen presented on a macrophage, it
   a. inactivates itself.
   b. secretes cytokines.
   c. inactivates B cells.
   d. inactivates the macrophage.
   e. becomes a T_C cell.
   Textbook Reference: *42.5 What Is the Cellular Immune Response? pp. 882–883*
5. Part of the normal immune response includes
   a. the production of B memory cells.
   b. the production of memory macrophages.
   c. antibody secretion by eosinophils.
   d. the production of B cells that attack the individual's own cells.
   e. the production of complement proteins as a specified immune response.
   Textbook Reference: *42.3 How Does Specific Immunity Develop? p. 883*
6. Antibody molecules are
   a. produced by B cells and have only a constant region.
   b. secreted by B cells, once a signal (a cytokine) is received from a T cell.
   c. produced by T cells and have a variable and a constant region.
   d. proteins that are used to induce B cells to form memory cells.
   e. proteins with antibody binding sites.
   Textbook Reference: *42.4 What is the Humoral Immune Response? pp. 885–886*
7. Cytotoxic T cells
   a. release cytokines that activate B cells.
   b. attack pathogens by binding to cell surface antigens on those pathogens.
   c. destroy pathogens by lysing them.
   d. destroy host cells that are infected with virus.
   e. stimulate the classical complement pathway.
   Textbook Reference: *42.5 What Is the Cellular Immune Response? p. 889*
8. The humoral response
   a. is the secretion of antibodies.
   b. occurs when T cells bind antigen-presenting cells.
   c. is due to T cells secreting their receptors.
   d. occurs when natural killer cells engulf cancer cells.
   e. is initiated with macrophages phagocytozing bacteria.
   Textbook Reference: *42.5 What Is the Cellular Immune Response? pp. 889–891*
9. Autoimmunity is
   a. active when organ transplantation is successful.
   b. caused by viruses.
   c. a response in which the immune cells attack the body's own tissues.
   d. a result of the destruction of the immune system.
   e. a desirable immune response.
   Textbook Reference: *42.7 What Happens When the Immune System Malfunctions? p. 895*
10. Patients with HIV are susceptible to a variety of infections because
    a. the virus produces cell surface receptors that bind to pathogens, making it easier for those pathogens to be infective.
    b. when a DNA copy of the viral genome is synthesized it makes a person feel sick.
    c. HIV attacks and destroys the T helper cells, which are central to mounting an effective immune response.
    d. HIV destroys B cells so that antibodies cannot be made in response to invading pathogens.
    e. HIV mutates the B cells so they cannot make the huge array of antibodies needed for an effective immune response.
    Textbook Reference: *42.7 What Happens When the Immune System Malfunctions? pp. 895–896*
11. DNA rearrangements in the B cell
    a. are responsible for generating single B cells that can express many different antibodies.
    b. lead to mutations in T cells, resulting in the elimination of essential T cell genes.
    c. occur only in B memory cells.
    d. are responsible for generating many different antibodies, with each B cell expressing only one set of identical antibodies.
    e. occur only in fully mature B cells.
    Textbook Reference: *42.6 How Do Animals Make So Many Different Antibodies? p. 892*

12. Major histocompatibility proteins function in the immune system by
  a. presenting antigens to T_C cells.
  b. presenting antigens to T_H helper cells.
  c. generating antibodies to different pathogens.
  d. presenting antigen fragments to B cells.
  e. presenting macrophages to T_H cells.

  Textbook Reference: *42.5 What Is the Cellular Immune Response? p. 888*
13. Which of the following is *not* a granular cell?
  a. Basophil
  b. Neutrophil
  c. B cell
  d. Mast cell
  e. Eosinophyl

  Textbook Reference: *42.1 What Are the Major Defense Systems of Animals? p. 876*
14. Inflammation occurs when _______ release _______.
  a. mast cells; histamine
  b. neutrophils; toxins
  c. B cells; histamine
  d. mast cells; toxins
  e. neutrophils; antihistamine

  Textbook Reference: *42.3 How Does Specific Immunity Develop? p. 878*
15. The process in which an antigen binds to a specific B cell and it begins to divide is called
  a. a nonspecific defense.
  b. clonal selection.
  c. normal flora.
  d. meiosis.
  e. a secondary immune response.

  Textbook Reference: *42.3 How Does Specific Immunity Develop? p. 883*

Application Questions

1. Imagine that a particular bacterium has surface molecules that resemble membrane proteins found on the heart valves. What would happen to an individual infected with this bacterium if the infection was not treated?

  Textbook Reference: *42.7 What Happens When the Immune System Malfunctions? p. 895*
2. Organ transplants are more successful when the donor is a relative of the recipient. Why?

  Textbook Reference: *42.5 What Is the Cellular Immune Response? p. 891*
3. Vaccinations are effective against an array of different pathogens. How do they work? Why have vaccinations against the HIV virus not been effective in preventing the disease?

  Textbook Reference: *42.3 How Does Specific Immunity Develop? p. 884; 42.7 What Happens When the Immune System Malfunctions? pp. 895–896*
4. If the cytotoxic T cells were eliminated from a person's array of immune defenses, what kinds of disease would he or she be susceptible to?

  Textbook Reference: *42.5 What Is the Cellular Immune Response? p. 889*
5. The polio virus comes in two types: an inactivated, injectible form and an oral, attenuated form. Which would be the safer vaccine to give to a child with an immunodeficiency disorder? Explain your reasoning.

  Textbook Reference: *42.3 How Does Specific Immunity Develop? p. 884*

Answers

Diagram Exercise Answers

1.
  a. B lymphocytes
  b. T lymphocytes
  c. Natural killer cells
  d. Erythrocytes
  e. Platelets
  f. Basophils
  g. Eosinophils
  h. Neutrophils
  i. Mast cells
  j. Monocytes
  k. Macrophages
  l. Dendritic cells
2. A.
  a. Humoral immune response (B cells)
  b. MHC class II receptor presents fragments of antigen.
  c. Interleukin-1 is produced by the macrophage in response to phagocytosis of an antigen.
  d. An antigen is taken into the macrophage by phagocytosis and broken down in a lysosome.
  e. Antigen is presented to the CD4 receptor on the T helper cell
  f. Interleukin-1 from the macrophage stimulates the T helper cell to produce interleukin-2. Interleukin-2 tells T helper cells to proliferate.
  g. The T helper cell replicates itself multiple times.
  h. Interleukin-2 is produced by the T helper cell. This will stimulate the B cell to proliferate and generate both memory cells and plasma cells.
  i. An antigen binds to a B cell receptor, specifically IgM. This triggers phagocytosis of the antigen, followed by degradation and presentation of fragments on the MHC class I receptor.
  j. The B cell MHC Class II receptor presents the antigen fragment to the T helper cell's CD4 receptor.
  k. The proliferating B cells give rise to both memory cells and B plasma cells.
  l. The B plasma cell produces antibodies.

  B.
  a. Cell mediated immune response (T cells)
  b. The viral proteins produced in the infected cell are degraded.

c. The viral proteins are displayed on the MHC class I receptor.
d. The cytotoxic T cell's CD8 receptor binds to the MHC Class I receptor that is presenting the viral antigen.
e. The cytotoxic T cell, (demonstrating the clonal theory of selection) proliferates after binding to an antigen presented on the infected cell's MHC Class I receptor.
f. The activated cytotoxic T cell recognizes an infected cell by the antigen presented on the infected cells' MHC Class I receptor. This MHC receptor binds to the T_C cell's CD8 receptor.
g. The T_C releases perforin to destroy the infected cell.
h. The infected cell lyses, halting the viral replication process in the infected cell.

Knowledge and Synthesis Answers

1. **c.** Phagocytes are nonspecific cells (not B and T cells) that digest nonself materials and present protein fragments of those nonself materials on their surface via MHC class II complexes. They do not have antibodies.
2. **e.** B cells are antigen-presenting cells; they bind antigen, digest it, present fragments of that antigen on their MHC II proteins, and secrete antibodies as part of the humoral response.
3. **e.** Macrophages, natural killer cells, cilia on mucous membranes, and complement proteins are all part of the nonspecific response of the immune system. B and T cells are part of the specific response of the immune system.
4. **b.** When the T_H cell binds antigen being presented on a macrophage, it secretes cytokines, which activate the T_H cell and B cells.
5. **a.** In a normal immune response, B memory cells are produced, and the organism can mount a faster and more effective response to any subsequent encounter with the pathogen. Memory macrophages do not exist. Eosinophils do not secrete antibodies. In an abnormal immune response, B cells that attack the individual's own cells can be activated, as seen in autoimmune diseases. Complement proteins are part of the nonspecific immune response.
6. **b.** T cells have cell surface receptors with a variable and a constant region and do not produce antibody molecules. Antibodies are not produced by macrophages. Only B cells produce antibodies in response to cytokines released from the T_H cell. Each antibody molecule has two identical heavy chains and two identical light chains, and each of these chains has a variable and a constant region. Antigens have "antibody binding sites" called epitopes.
7. **d.** Cytotoxic cells bind to virus antigen being presented on MHC I protein and destroy those cells. T_H cells release cytokines to activate B cells. The antibodies of B cells bind cell-surface antigens on pathogens. Pathogens are lysed by complement proteins (and not by cytotoxic T cells), which are activated by antibodies bound to those pathogens. T cells have no involvement in the classical complement pathway.
8. **a.** The humoral response refers to that part of the specific response that releases antibodies (B cells) in the lymph and blood (the "humors" of the body). In the cellular response, T cell receptors binding antigen on antigen-presenting cells. T cells do not secrete their receptors. Macrophages are part of the nonspecific response.
9. **c.** Autoimmunity occurs when the immune cells attack the body's own cells. Transplant rejection can occur if the transplanted tissue is recognized by the body as nonself, so this is not an autoimmune response. The immune system can be destroyed by an immune deficiency disorder.
10. **c.** An HIV-infected individual is more susceptible to a variety of infections because the virus destroys T_H cells, which are essential for mounting an effective immune response. HIV does not bind to pathogens and does not destroy B cells or cause mutations in their DNA that alter antibody production.
11. **d.** DNA rearrangements occur in B cell precursors and result in each mature B cell expressing a unique kind of antibody. DNA rearrangements also occur in T cells to generate T cell receptors. This rearrangement does not destroy essential T cell genes. Memory B cells are cells in which the DNA rearrangement has already occurred.
12. **b.** MHC proteins present nonself antigens and self antigens to T cells. T cells that react with self antigens are destroyed, preventing autoimmunity. The cytokines of T cells (and not MHC proteins) activate B cells. MHC proteins do not generate antibodies. Macrophages are antigen presenting cells, but are not themselves presented as antigens to T_H cells.
13. **c.** The B cells are lymphocytes, not granular cells. The basophils, neutrophils, eosinophils, and mast cells are all examples of granular cells (see Figure 42.2).
14. **a.** Inflammation occurs in response to an infection or tissue injury. Once an infection or injury occurs, the local mast cells release histamine. The histamine in turn makes the capillaries leak, and fluid accumulating in the area produces the inflammation. Antihistamines block histamine activity in the body.
15. **b.** When a B cell attaches to an antigen, it begins to divide and make clone cells, which results in more B cells that recognize the antigen. This process is known as clonal selection. A secondary immune response occurs when memory cells encounter an antigen and begin to actively divide.

Application Answers

1. If the bacterial infection was not treated, the individual's immune response to this pathogen would include the secretion of antibodies. Such antibodies could cross-react with the individual's own membrane

proteins on the cells of the heart valve, leading to an autoimmune response and possibly heart disease.

2. Organ transplants are successful if the MHC proteins and other cell surface markers are similar in the donor and recipient, making the immune system of the recipient more tolerant of the foreign tissue. Since close relatives have more genes in common than unrelated people do, they are more likely to have surface antigens on their cells that are similar. Organ transplants from unrelated individuals require the administration of immunosuppressive drugs to the recipient so that the patient's own immune system does not reject the foreign tissue.
3. Vaccinations work against foreign pathogens because B and T memory cells are made during every immune response. If these memory cells are reexposed at a later time to the same antigen, they will mount a faster and more efficient immune response to that pathogen. One challenge in creating vaccines against HIV is that the virus mutates frequently (In its reproductive cycle, HIV uses reverse transcriptase, an enzyme prone to introducing genetic mutations). A vaccine that creates memory cells specific to a certain protein in one strain of HIV will not be effective against mutant HIV forms if the protein is modified in that form and the memory cells do not recognize it.
4. Cytotoxic T cells target virally infected cells and some cancer cells. If cytotoxic T cells were eliminated, the individual would be much more susceptible to viral infections and cancer.
5. The inactivated, injectible form would be safer for a patient with a compromised immune system. Because the pathogen (in this case the polio virus) is completely inactivated, it cannot cause the disease in the patient. In contrast, the attenuated vaccine contains a mutant form of the virus that will reproduce itself in the patient, although at a much slower rate than the normal. It is possible for the virus to mutate back to a virulent form (which does happen, but rarely) and cause disease.

43 Animal Reproduction

The Big Picture

- Animals reproduce both asexually and sexually. Asexual reproduction includes budding, regeneration, and parthenogenesis. Sexual reproduction has an advantage over asexual reproduction in that it helps increase genetic diversity. Sexual reproduction has three main stages: gametogenesis, mating (or spawning), and fertilization.
- Gametogenesis is the process by which the sex cells are produced. In males, haploid sperm develop from diploid spermatogonia by the process of spermatogenesis. In females, the haploid egg develops from diploid oogonia by the process of oogenesis.
- The human female has two interrelated cycles, the ovarian and the uterine cycles, during which an egg is produced and released from the ovary and the uterus is prepared for implantation and pregnancy. These two cycles occur in parallel and are coordinated by changes in hormone levels. Luteinizing hormone, follicle-stimulating hormone, estrogen, and progesterone all play a role in these cycles.
- External fertilization is typical of many aquatic animals. Internal fertilization, often with the help of accessory sex organs to ensure fertilization, occurs in terrestrial animals. The eggs and sperm of a given species recognize one another.
- Several methods are available for preventing pregnancy and for overcoming infertility.

Common Problem Areas

- Understanding the ploidy level of the developing egg and sperm can be very confusing, but it is essential in determining how genetic inheritance works.
- Following the hormonal changes and cues associated with human ovarian and menstrual cycles can be difficult. Create a chronological sequence of hormones and the events they trigger during both cycles.

Study Strategies

- Recognize that although sperm and egg formation are completely separate processes, each with its own characteristics, the same general events are happening in both. This will help you recognize and remember the stages of spermatogenesis and oogenesis. Be sure, though, to also understand the differences between spermatogenesis and oogenesis with respect to timing and number of gametes produced.
- Go to yourBioPortal.com to review the following tutorials and activities:

  Animated Tutorial 43.1 Fertilization in a Sea Urchin

  Animated Tutorial 43.2 The Ovarian and Uterine Cycles

  Web Activity 43.1 The Human Male Reproductive Tract

  Web Activity 43.2 Spermatogenesis

  Web Activity 43.3 The Human Female Reproductive Tract

Important Concepts

Some animals reproduce asexually.

- A variety of animals, mostly invertebrates, employ asexual reproduction, which produces offspring that are genetically identical to one another and to the parent. Individuals that reproduce asexually often are either sessile or members of a sparse population; both of these situations make searching for and finding a mate challenging.
- Asexual reproduction has the following two advantages: (1) time and energy are not wasted on mating; and (2) every member of the population can produce offspring. The main disadvantage of asexual reproduction is that it does not generate genetic diversity, which can be essential if the environment changes.
- New offspring can be produced either by budding off outgrowths to produce a new individual or by regenerating a new individual from pieces of an animal (see Figure 43.1). These methods rely on mitosis.
- Another way that animals can produce new individuals asexually is through parthenogenesis—the development of unfertilized eggs into new offspring.
- Sometimes sexual behavior is required for asexual reproduction; this is the case, for example, in a species of parthenogenetic whiptail lizard (see Figure 43.2).

Sexual reproduction produces genetic diversity.

- Sexual reproduction in animals produces genetic diversity when two haploid gametes join to form a diploid cell.
- Sexual reproduction has three stages: gametogenesis (producing gametes), mating or spawning (bringing gametes together), and fertilization (fusing gametes).
- Gametogenesis produces haploid gametes using meiotic cell division. During meiosis, crossing over between homologous chromosomes and independent assortment of chromosomes contribute to genetic diversity.
- Gametogenesis occurs in an animal's gonads. Male gametes—sperm—are produced in the testes, and female gametes—eggs or ova—are produced in the ovaries. Gametes are derived from germ cells. These germ cells migrate to the gonads of the embryo.

Males produce sperm by spermatogenesis.

- Spermatogenesis begins with a male germ cell ($2n$) that proliferates through mitosis to produce spermatogonia ($2n$; see Figure 43.3A). Spermatogonia mature into primary spermatocytes ($2n$) that enter meiosis. The first meiotic division produces secondary spermatocytes (n), and a second meiotic division produces four spermatids (n).
- Mammalian spermatids remain in cytoplasmic contact and share gene products of the X chromosome. These gene products are essential for development into mature sperm.
- The spermatids differentiate and mature into sperm cells in a process called spermiogenesis. Mature mammalian sperm are streamlined and have a flagellum; they do not remain in cytoplasmic contact with one another once mature.

Females produce eggs by oogenesis.

- Oogenesis begins with a female germ cell ($2n$) that proliferates through mitosis to produce oogonia ($2n$; see Figure 43.3B). Oogonia mature into primary oocytes ($2n$). The primary oocytes enter prophase of the first meiotic division and in human females remain arrested in prophase for at least 10 years (until puberty is reached); some remain arrested for up to 50 years. No new primary oocytes will be produced during the lifetime of the female.
- Upon exiting prophase, the primary oocyte completes the first meiotic division to produce a secondary oocyte (n) and the first polar body. The second meiotic division produces an ootid (n) and the second polar body. This second meiotic division may not be completed until fertilization in some species. Polar bodies degenerate.
- Two main differences between spermatogenesis and oogenesis are: (1) spermatogenesis continues to completion once the primary spermatocyte has differentiated, whereas oogenesis involves a period of arrest during the first meiotic division; and (2) cytoplasm is apportioned equally to resulting cells in spermatogenesis and unequally to resulting cells in oogenesis (polar bodies receive much less cytoplasm than do secondary oocytes or ootids).

Fertilization is the joining of a haploid sperm with a haploid egg to form a diploid zygote.

- For successful fertilization to occur, the sperm and egg must recognize one another. Specific recognition molecules on the gametes mediate recognition. These molecules prevent eggs from being fertilized by sperm from a different species. In sea urchins, the eggs release species-specific chemical attractants, and other recognition molecules are involved later in fertilization (see Figure 43.4B). In mammals, a glycoprotein in the zona pellucida (a layer surrounding the mammalian egg) binds to recognition molecules on the head of the sperm and triggers the acrosomal reaction (see Figure 43.5).
- During the acrosomal reaction, the plasma membrane covering the head of the sperm breaks down, as does the underlying acrosomal membrane, releasing enzymes that digest a path for the sperm through the protective layers surrounding the egg.
- Fusion of the plasma membranes of the sperm and egg triggers blocks to polyspermy that prevent more than one sperm from entering the egg. Entry by more than one sperm usually means death of the embryo.
- Soon after entry by a sperm, the egg is metabolically activated and stimulated to begin development. The diploid nucleus of the zygote is created when the haploid nucleus of the egg fuses with the haploid nucleus of the sperm.

Fertilization occurs differently in the water and on land.

- Many aquatic animals have external fertilization of eggs. The eggs and sperm are released into the water (in a process called spawning), where fertilization may occur.
- Many fishes and amphibians that display external fertilization employ special behaviors to bring gametes close to each other to increase the probability that fertilization will occur.
- Terrestrial animals employ internal fertilization, in which sperm and egg fuse within the female reproductive tract rather than in the external environment. Some aquatic animals (e.g., sharks) also have internal fertilization.
- Gonads (testes and ovaries) are primary sex organs. Organs other than gonads that are part of an animal's reproductive system are considered accessory sex organs. For example, in males of many species with internal fertilization, the penis is an accessory sex organ needed to deposit sperm in the female's reproductive tract. The vagina is an example of a female accessory sex organ. Glands, ducts, and tubules associated with reproduction are also accessory sex organs. The term *genitalia* describes external sex organs.

- Copulation is the physical joining of male and female accessory sex organs. Some male invertebrates (for example, mites and scorpions) and vertebrates (salamanders) produce sperm-filled spermatophores that they deposit in the environment. When a female comes across the spermatophore, she straddles it and takes the packet of sperm into her reproductive tract. Thus, these animals have internal fertilization without copulation.

Some animals have both male and female reproductive systems.

- Dioecious species have separate male and female individuals.
- In monoecious species, a single individual may produce both eggs and sperm; such individuals are called hermaphrodites. Hermaphrodites can be either simultaneous (individuals are male and female at the same time) or sequential (individuals change sex at some point during their life span).

The move to land required a reproductive system that functions in a dry environment.

- Amphibians were the first vertebrates to live on land, but they still require water to reproduce.
- Many reptiles and all birds have solved the problem of a dry environment by being oviparous; they lay protective shelled eggs (termed amniote eggs) that contain food (yolk) and a watery environment within which the embryo develops. Internal fertilization must occur before shell formation. Prior to hatching, all nourishment comes from nutrients stored in the egg.
- Among mammals, monotremes are oviparous and all others are viviparous, retaining the embryo within the uterus where it is nourished via the placenta. Most non-mammalian viviparous animals are ovoviviparous, with embryos developing in shelled eggs that are retained in females until hatching.

Human males produce sperm and deliver it to the female.

- Males produce semen containing sperm (n) and fluids that support the sperm and facilitate fertilization. In the testis, sperm cells are produced by spermatogenesis in the seminiferous tubules. Sertoli cells surround developing sperm cells and provide protection and nourishment (see Figure 43.9C). In most mammals, the testes are held in the scrotum; this pouch of skin holds the testes outside the body cavity where temperatures are slightly lower than normal body temperature and optimal for spermatogenesis.
- Sperm develop a flagellum at the back and an acrosome cap that contains enzymes at the front. (Recall that when released from the acrosome, these enzymes digest a path through the protective layers around the egg.) Mitochondria provide energy for the flagellum. Immature sperm move from the seminiferous tubules to the epididymis, where they mature, become motile, and are stored. The epididymis connects to the vas deferens and then to the urethra, which opens to the outside of the body at the tip of the penis.
- In addition to containing sperm, semen contains fluids from accessory glands. The bulbourethral glands produce an alkaline and mucoid secretion that helps control pH in the urethra and lubricates the tip of the penis. The seminal vesicles produce seminal fluid that contains fructose (an energy source for sperm), mucus, and proteins. The prostate gland produces the milky prostate fluid and a clotting enzyme that causes the protein in seminal fluid to turn semen into a gelatinous mass, which aids its retention in the female reproductive tract. Another enzyme activated shortly after semen enters the female reproductive tract dissolves the clotted semen and liberates sperm. Prostate secretions are alkaline, so they also help neutralize the acidity of the male and female reproductive tracts.
- During sexual arousal, the penis becomes engorged with blood, and contraction of smooth muscle in the ducts and at the base of the penis results in ejaculation of semen. At the cessation of sexual stimulation, the vessels at the base of the penis constrict, leading to loss of the erection. The dilation of blood vessels in the penis during an erection is initiated by the neurotransmitter nitric oxide (NO) and its effects on the second messenger cGMP. After ejaculation, decreases in NO and cGMP lead to the constriction of blood vessels associated with loss of an erection.
- Testosterone, produced in the Leydig cells (also called interstitial cells) of the testes, is the male sex hormone that controls sexual function and sperm production (see Figure 43.10). Testosterone also prompts development and maintenance of secondary sexual characteristics, such as facial hair and a deep voice in males. These characteristics are not directly involved in reproduction but are responsible for differences in the physical appearances of males and females.

Human females produce eggs, receive sperm, and gestate the embryo.

- The ovaries release eggs into the body cavity, where they enter the oviduct (also called the Fallopian tube) and move toward the uterus (see Figure 43.11A). Sperm swim up the vagina, through the opening to the uterus (cervix), through the uterus, and into the oviduct. Fertilization occurs in the oviduct; embryos develop in the uterus.
- The initial division of the zygote produces a blastocyst that moves into the uterus, where it burrows into the uterine wall (the endometrium); this process is called implantation. The blastocyst interacts with the endometrium to form the placenta, the organ that provides nutrients and removes wastes from the developing embryo.
- Mature eggs are produced during a 28-day ovarian cycle (see Figure 43.12). Initially, 6 to 12 primary oocytes begin to mature and are surrounded by follicle cells to produce individual follicles. After one week, one

primary oocyte continues to grow while the others shrink. Prior to ovulation, the primary oocyte undergoes a meiotic division to become a secondary oocyte and is expelled from the ovary. The remaining follicle cells become the corpus luteum, a glandular structure that produces progesterone and estrogen.

- The uterine cycle, in concert with the ovarian cycle, prepares the endometrium to receive the blastocyst. If a blastocyst does not implant by around day 14 of the uterine cycle, the endometrium is broken down and expelled in the process of menstruation. The uterine cycles of most mammals other than humans do not include menstruation. Other mammals resorb the endometrium; in these species, females display a period of sexual receptivity around ovulation termed estrus.
- Luteinizing hormone (LH) and follicle-stimulating hormone (FSH) control the timing of the ovarian and uterine cycles (see Figure 43.13). The first step of the uterine cycle is menstruation. Prior to day 10 of the uterine cycle, FSH and LH levels are kept low by negative feedback from estrogen. Around days 12 to 14, this changes, and estrogen exerts positive feedback, which causes a rise in FSH levels. Rising levels of FSH stimulate the follicles to begin to mature. LH levels also increase, triggering ovulation, the release of the egg from the mature follicle. Following ovulation, the follicle cells form the corpus luteum. Estrogen and progesterone produced by the corpus luteum stimulate growth and maintenance of the endometrium. In the absence of fertilization, the corpus luteum degenerates, estrogen and progesterone levels decrease, and menstruation occurs to start the cycle again.
- If the egg is fertilized and the blastocyst implants in the uterus, human chorionic gonadotropin is produced, which keeps the corpus luteum functional. Continued production of estrogen and progesterone by the corpus luteum supports development and maintenance of the endometrium, thereby preparing the uterus for pregnancy. The placenta develops, and eventually replaces the corpus luteum as the main source of estrogen and progesterone.
- Late in pregnancy, estrogen increases contractility of uterine muscle. The uterine contractions of childbirth are triggered by pressure from the fetal head on the maternal cervix. This mechanical stimulus triggers release of oxytocin by the mother and fetus. Oxytocin stimulates increased strength and frequency of uterine contractions in a positive feedback cycle (see Figure 43.15B). Stages of childbirth include dilation of the cervix, delivery of the baby, and delivery of the placenta.

Human sexual responses have four phases.

- There are four phases of response to sexual stimulation in men and women: excitement, plateau, orgasm, and resolution.
- The excitement phase is characterized by increases in heart rate, blood pressure, and muscle tension, and erection of the penis in males and clitoris in females. Breathing rate increases during the plateau phase, and further increases occur in heart rate and blood pressure. Ejaculation occurs during orgasm in males; females may have several orgasms in quick succession. During the resolution phase, blood drains from the genitals, and body physiology returns to resting conditions.
- In males, a refractory period occurs after orgasm. During this period, which may last for minutes or hours, a full erection and orgasm are not possible.

Humans use several methods to prevent pregnancy.

- Methods of contraception differ in their mode of action and failure rate (see Table 43.1).
- The rhythm method involves avoiding sexual intercourse during the time period when the female is most likely to be fertile and has a 15 to 35 percent failure rate. Coitus interruptus, another nontechnological method, involves the male withdrawing his penis before ejaculation; the failure rate of this method is 20 to 40 percent.
- Condoms, diaphragms, and cervical caps work by creating a barrier to sperm. They have a failure rate of 10 to 15 percent.
- Several methods employ hormones to prevent ovulation by the female. Oral contraceptives contain synthetic estrogen and progesterone that interfere with the development of the ova and suspend the ovarian cycle. The failure rate of oral contraceptives is less than 3 percent. Synthetic estrogen and progesterone can also be administered through injections, patches, and vaginal rings.
- Intrauterine devices work by blocking the implantation of a fertilized egg in the uterus and have a failure rate of less than 5 percent.
- Both males and females can be sterilized. Males can undergo a vasectomy in which the vasa deferentia are cut and tied off, whereas females can have their oviducts tied off in a procedure called tubal ligation.
- The termination of pregnancy is known as abortion. A spontaneous abortion occurring early in pregnancy is commonly called a miscarriage. RU-486 blocks progesterone receptors in the uterine lining; when used early in pregnancy it causes shedding of the endometrium and embryo.

Medical science has found ways to help couples overcome infertility.

- Several problems may prevent couples from conceiving. A male may not produce adequate numbers of sperm or the sperm may lack motility. In the female, the environment of the uterus may be poor for hosting sperm or the fertilized egg, or the oviducts may be blocked, preventing passage of gametes.
- Artificial insemination places sperm in the female reproductive tract where fertilization can occur. This technique is commonly used when males have a low

sperm count (sperm is collected and concentrated before the procedure, or donor sperm is used).

- Assisted reproductive technologies (ARTs) take eggs from a female and then either fertilize them outside the body (in vitro fertilization, IVF) or place a mixture of sperm and eggs in the female oviduct. Sperm that cannot gain access to the plasma membrane of an egg may be injected into an egg in a procedure called intracytoplasmic sperm injection (ICSI).
- Genetic testing may be performed on cells taken from early blastocysts in a procedure known as preimplantation genetic diagnosis (PGD). This procedure does not compromise the embryo's development and allows those produced through IVF to be tested for genetic defects.

Test Yourself

Diagram Exercise

To better understand the changes in ploidy that occur in gametes as they form, create a flowchart that shows how ploidy changes from germ cell to mature gamete in spermatogenesis and oogenesis. Be sure to label each stage of gamete formation (e.g., spermatogonium, primary spermatocyte, etc.) and include its ploidy.
Textbook Reference: *43.2 How Do Animals Reproduce Sexually? p. 903*

Knowledge and Synthesis Questions

1. Asexual reproduction is an effective strategy in stable environments because
   a. gametogenesis is most efficient under these conditions.
   b. the resulting offspring, genetically identical to their parents, are preadapted to their environment.
   c. asexual parthenogenesis produces a large amount of genetic diversity.
   d. animal cells tend to be more totipotent under stable conditions.
   e. sessile animals and sparse populations are more common under stable conditions.
   Textbook Reference: *43.1 How Do Animals Reproduce without Sex? p. 900*
2. An important difference between a sperm and an egg concerns
   a. their size.
   b. the amount of cytoplasm they contain.
   c. whether or not they are motile.
   d. whether or not protective layers exist outside the plasma membrane.
   e. All of the above
   Textbook Reference: *43.2 How Do Animals Reproduce Sexually? pp. 902–903, 905*
3. External fertilization
   a. is typical of terrestrial animals.
   b. requires a penis.
   c. can occur when a spermatophore is transferred.
   d. occurs in some, but not all, aquatic animals.
   e. is confined to invertebrates.
   Textbook Reference: *43.2 How Do Animals Reproduce Sexually? p. 906*
4. Which of the following best represents the normal path of a sperm cell as it makes its way from the point of entry into a female's reproductive tract to the location where fertilization typically occurs?
   a. Cervix, vagina, ovary, oviduct
   b. Vagina, cervix, uterus, oviduct
   c. Uterus, cervix, vagina, oviduct
   d. Vagina, uterus, cervix, oviduct
   e. Cervix, vagina, oviduct, ovary
   Textbook Reference: *43.3 How Do the Human Male and Female Reproductive Systems Work? p. 912*
5. The function of the seminal vesicle is to
   a. produce a solution of fructose to provide energy for the sperm.
   b. secrete alkaline fluids that neutralize the acidity of a female's reproductive tract.
   c. initiate the muscular contractions that lead to emission.
   d. produce lubrication for the tip of the penis.
   e. serve as the location for spermatogenesis.
   Textbook Reference: *43.3 How Do the Human Male and Female Reproductive Systems Work? p. 911*
6. Which of the following is an example of positive feedback control in the reproductive cycle of males or females?
   a. The increased response of the hypothalamus and anterior pituitary gland in response to estrogen
   b. The decreased response of the hypothalamus and anterior pituitary gland in response to estrogen
   c. The inhibition of luteinizing hormone by high levels of testosterone
   d. The stimulation of luteinizing hormone by low levels of testosterone
   e. The inhibition of oxytocin release by increased uterine contractions during childbirth
   Textbook Reference: *43.3 How Do the Human Male and Female Reproductive Systems Work? pp. 914–915*
7. Which of the following statements about oogenesis is *false*?
   a. The polar bodies degenerate after the second meiotic division.
   b. The ovum produced is haploid.
   c. The major growth phase of the primary oocyte occurs in prophase I.
   d. The primary oocyte is haploid.
   e. The secondary oocyte is haploid.
   Textbook Reference: *43.2 How Do Animals Reproduce Sexually? p. 903*
8. Which of the following is *not* an accessory sex organ?
   a. Penis
   b. Prostate gland
   c. Ovary

d. Vagina
e. Uterus
Textbook Reference: *43.2 How Do Animals Reproduce Sexually? p. 907*

9. For approximately how long during the human female's menstrual cycle are progesterone concentrations high enough to maintain the uterus in a proper condition for pregnancy?
a. The entire duration of the cycle
b. No portion of the cycle
c. During the first half of the cycle
d. During the second half of the cycle
e. For a 5-day window mid-cycle
Textbook Reference: *43.3 How Do the Human Male and Female Reproductive Systems Work? p. 914*

10. Which of the following statements about the birth control pill is *false*?
a. It works by preventing ovulation.
b. It works by preventing implantation.
c. The ovarian cycle is suspended by the birth control pill.
d. It contains low doses of estrogen and progesterone.
e. It allows the uterine cycle to continue because hormones are not taken for one week every 21–28 days.
Textbook Reference: *43.4 How Can Fertility Be Controlled? p. 918*

11. Among the methods of contraception listed below, which has the lowest failure rate?
a. Vaginal ring
b. Rhythm method
c. Diaphragm
d. Coitus interruptus
e. Condom
Textbook Reference: *43.4 How Can Fertility Be Controlled? p. 918*

12. If you compared the genetic makeup of a female animal produced by parthenogenesis with that of its mother, which of the following would you expect?
a. About 100 percent genetic similarity
b. About 50 percent genetic similarity
c. No genetic similarity
d. About 25 percent genetic similarity
e. Parthenogenetic animals do not have mothers.
Textbook Reference: *43.1 How Do Animals Reproduce without Sex? pp. 900–901*

13. Which of the following statements about oviparity is *false*?
a. Of terrestrial animals, only birds and reptiles are oviparous.
b. The large amount of yolk in the egg provides all the nutrients for the developing embryo prior to hatching.
c. The shell of the amniote egg protects the embryo from dehydrating.
d. Both oxygen and carbon dioxide can diffuse through the shell of the amniote egg.
e. In animals that produce amniote eggs, internal fertilization occurs before a shell develops.
Textbook Reference: *43.2 How Do Animals Reproduce Sexually? p. 908*

14. Spermatogenesis
a. results in two sperm cells from each primary spermatocyte.
b. involves a period of arrest during the first meiotic division.
c. results in diploid gametes.
d. apportions cytoplasm equally among resulting cells.
e. apportions cytoplasm unequally among resulting cells.
Textbook Reference: *43.2 How Do Animals Reproduce Sexually? pp. 902–903*

15. Which of the following statements about fertilization is *false*?
a. The egg permits several sperm to enter it.
b. The plasma membranes of sperm and egg fuse.
c. The egg is activated and stimulated to begin development.
d. Species-specific recognition occurs between egg and sperm.
e. During the acrosomal reaction, enzymes spill from a cap on the head of the sperm and digest a path through the protective layers surrounding the egg.
Textbook Reference: *43.2 How Do Animals Reproduce Sexually? pp. 905–906*

16. Which of the following statements is *false*?
a. Delivery of the placenta occurs after delivery of the baby.
b. If used correctly, condoms can prevent pregnancy and protect against sexually transmitted diseases (STDs).
c. RU-486 blocks estrogen receptors.
d. Men, but not women, have a refractory period following orgasm.
e. Breastfeeding immediately after birth helps shrink the uterus.
Textbook Reference: *43.3 How Do the Human Male and Female Reproductive Systems Work? p. 916; 43.4 How Can Fertility Be Controlled? pp. 917–918*

Application Questions

1. Some animals can reproduce sexually or asexually. When and why might these animals switch between sexual and asexual reproduction?
Textbook Reference: *43.1 How Do Animals Reproduce without Sex? p. 900*

2. What is the life history characteristic that animals that reproduce asexually and those that reproduce sexually as simultaneous hermaphrodites share?
Textbook Reference*: 43.1 How Do Animals Reproduce without Sex? p. 900; 43.2 How Do Animals Reproduce Sexually? p. 907*

3. Why do physicians recommend that men who want children but who have been diagnosed with low sperm counts avoid wearing tight-fitting shorts?
   Textbook Reference: *43.3 How Do the Human Male and Female Reproductive Systems Work? pp. 909–910*
4. Given what you know about the hormones that regulate the ovarian and uterine cycles in human females, explain the mechanism by which birth control pills containing synthetic estrogen and progesterone prevent pregnancy.
   Textbook Reference: *43.4 How Can Fertility Be Controlled? p. 918*
5. Why are multiple births (for example, septuplets) associated with assisted reproductive technologies (ARTs)?
   Textbook Reference: *43.4 How Can Fertility Be Controlled? p. 919*

Answers

Diagram Exercise Answer

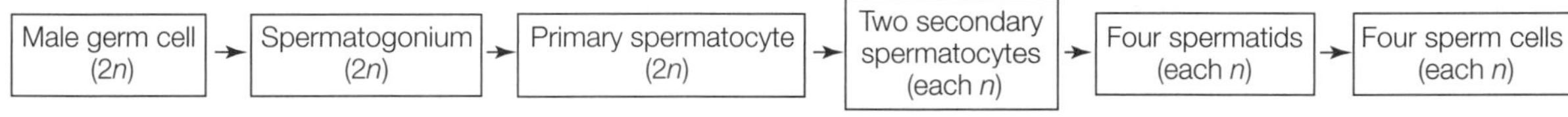

Oogenesis:

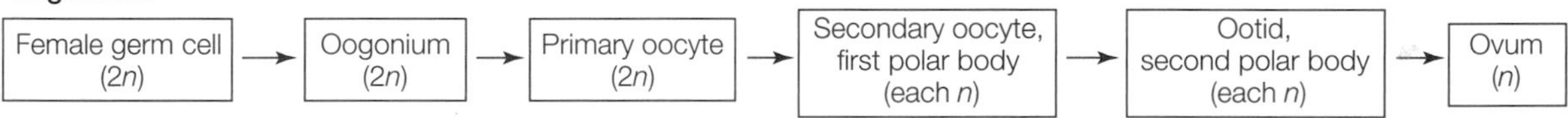

Knowledge and Synthesis Answers

1. **b.** The parents that have survived to reproduce asexually are able to survive in the current stable environment. Therefore, the offspring should be preadapted for this stable environment.
2. **e.** There are many differences between eggs and sperm. Eggs are larger than sperm and contain more cytoplasm and organelles. Whereas eggs are immotile, sperm have a flagellum and are motile. Eggs, but not sperm, have protective layers outside their plasma membranes.
3. **d.** External fertilization occurs in some, but not all, aquatic animals. It occurs in invertebrates (e.g., oysters and sea urchins) and vertebrates (e.g., lampreys and salmon), and does not require a penis. Transfer of a spermatophore (sperm packet) results in internal fertilization without copulation.
4. **b.** A sperm is ejected by the male into the vagina. From the vagina the sperm move through the cervix into the uterus and finally into the oviduct, where fertilization occurs.
5. **a.** The seminal vesicles contribute substances to semen. One of the components of seminal fluid is fructose, which serves as an energy source for the sperm.
6. **a.** During days 12 through 14 of the ovarian cycle, estrogen exerts positive feedback control on the pituitary, prompting it to release LH and FSH.
7. **d.** During oogenesis, the primary oocyte is diploid; after the first meiotic division into the secondary oocyte the cell becomes haploid.
8. **c.** The ovary is not an accessory sex organ. Gonads are not accessory sex organs; they are the primary sex organs producing the sperm and the egg.
9. **d.** High levels of progesterone are needed to maintain the uterus in the proper condition for pregnancy. The levels of progesterone are high only during the second half of the uterine cycle.
10. **b.** The birth control pill interferes with the maturation of the follicles and the ova, inhibiting the release of an egg. Thus, it blocks ovulation, not implantation.
11. **a.** Among those methods listed, the vaginal ring has the lowest failure rate (<1%).
12. **a.** Parthenogenesis is a form of asexual reproduction in which offspring develop from unfertilized eggs; there is no sexual recombination of genes, so a female animal born parthenogenetically will be nearly identical to its mother genetically. In many hymenopterans (bees, wasps, and ants), only males are born from unfertilized eggs. Such males are haploid, so they are not genetically identical to their diploid mothers.
13. **a.** Birds and reptiles are not the only oviparous taxa on land—a group of mammals, the monotremes, are also egg-layers.
14. **d.** During spermatogenesis cytoplasm is apportioned equally among resulting cells. This differs from

oogenesis, in which cytoplasm is apportioned unequally among daughter cells.

15. **a.** During fertilization, mechanisms in the egg that operate as blocks to polyspermy prevent more than one sperm from entering it.
16. **c.** RU-486 blocks progesterone (not estrogen) receptors and causes sloughing off of the endometrium and embryo.

Application Answers

1. Asexual reproduction does not waste time and energy on mating, and every member of the population can produce offspring. However, asexual reproduction does not yield genetic diversity, which can be needed when environmental conditions change. Sexual reproduction generates genetic diversity. Thus, we would expect animals that can reproduce by either method to use asexual reproduction when environmental conditions are stable and favorable, and to switch to sexual reproduction when environmental conditions are unstable and unfavorable.
2. Animals that reproduce asexually and those that reproduce sexually as simultaneous hermaphrodites often have a low probability of encountering a potential mate.
3. The optimal temperature for spermatogenesis is slightly below normal body temperature, which explains why the testes are located outside the body cavity in the scrotum. Tight clothing pulls the testes close to the body, exposing them to higher than optimal temperatures for spermatogenesis.
4. During normal cycling, estrogen and progesterone serve as negative feedback signals (except at very high levels) to the hypothalamus and pituitary gland and thus keep follicle-stimulating hormone (FSH) and luteinizing hormone (LH) at low levels. The low doses of estrogen and progesterone in birth control pills function in the same manner; these hormones prevent ovulation (and thus pregnancy) by keeping FSH and LH low.
5. Normally, a woman ovulates a single secondary oocyte each month; if this oocyte is fertilized, then a single birth may result. Many ARTs use hormones ("fertility drugs") to trigger superovulation—the release of several secondary oocytes from the ovary. If more than one of these oocytes is fertilized, then multiple births may result. Also, in some ARTs more than one embryo is introduced into the female reproductive tract in the hope that at least one will implant itself. If all become implanted, then multiple births can result.

44 Animal Development

The Big Picture

- Animal development involves several steps: fertilization, cleavage, gastrulation, and organogenesis (for example, neurulation).
- Fertilization is the joining of the egg and sperm to form a zygote. Following fertilization, the cells of the embryo begin to divide in a process known as cleavage. Eggs undergo either complete cleavage or incomplete cleavage, depending on the amount of yolk present. Those that undergo complete cleavage form a blastula, and those that undergo incomplete cleavage form a flat blastodisc.
- Cleavage in mammals is different from cleavage in other groups. It occurs more slowly, and the products of embryonic genes play a role in directing its course; in other animals, cleavage is directed entirely by molecules present in the egg prior to fertilization, not the products of embryonic genes.
- During gastrulation, the germ layers—endoderm, ectoderm, and mesoderm—form and move into specific positions. In the sea urchin, an archenteron that will become the gut forms, and specific cells form the three germ layers. In frogs, a dorsal lip is formed, and the germ cells move into place. The dorsal lip is considered the primary embryonic organizer in amphibians. The blastodisc of reptiles, birds, and mammals goes through a much different pattern of gastrulation.
- Neurulation occurs early in organogenesis. The notochord induces overlying ectoderm to form a neural plate. The neural plate forms a neural tube that will become the brain and spinal cord. Somites produce the repeating segments of the vertebrate body plan. Several genes play a large role in the differentiation of tissues along the body axes. For example, Hox genes control differentiation along the anterior–posterior axis.
- Reptiles and birds have four major extraembryonic membranes: yolk sac, amnion, chorion, and allantois. Mammals have a special placenta for the exchange of nutrients, gases, and wastes between the mother and the developing fetus.

Common Problem Areas

- All fertilized cells must divide to grow into adult animals. Sorting out the features of division that are common to all animals from those specific to certain taxonomic groups can be difficult and confusing until you begin to recognize the basic patterns.
- Many students find gastrulation difficult to picture and understand, especially given the different ways in which gastrulation occurs in different organisms.
- Students are often confused about the structure and role of the placenta. Is it embryonic? Is it maternal? And how do nutrients, gases, and wastes travel between the mother and embryo or fetus?

Study Strategies

- This chapter contains a great deal of terminology. Yet many terms are common to many animals. Create a list of new terms from this chapter. Then determine which are the more general terms applicable to numerous animals (e.g., "somites," "gastrulation") and learn these first. Then tackle those referring to development in specific animals (e.g., "dorsal lip of the blastopore," "primitive streak").
- A key point in understanding animal development is that all specialized tissues arise from the three germ layers: endoderm, ectoderm, and mesoderm. More important than just committing these three layers to memory is learning the specialized tissues that derive from them. This will enable you to make predictions about tissue functions and to compare body plans among animals.
- Go to yourBioPortal.com to review the following tutorials and activity:

  Animated Tutorial 44.1 Gastrulation

  Animated Tutorial 44.2 Tissue Transplants Reveal the Process of Determination

  Web Activity 44.1 Extraembryonic Membranes

Important Concepts

Fertilization is the joining of the sperm and the egg.

- Early development of animals has been studied extensively in a few model organisms: sea urchins, frogs, chickens, and humans.
- Development begins when a haploid sperm joins with a haploid egg. However, fertilization does more than produce a diploid zygote; fertilization activates development.

At fertilization, the egg and sperm contribute differently to the zygote.

- The egg contributes most of the cytoplasm and organelles to the zygote.
- The sperm contributes its haploid nucleus and a centriole that becomes the centrosome, which forms the mitotic spindles necessary for cell division.
- In an unfertilized frog egg, nutrients are concentrated in the lower half of the egg, called the vegetal hemisphere, and the haploid nucleus is found in the upper half of the egg, known as the animal hemisphere.
- Upon fertilization, the cytoplasm of the frog egg undergoes rearrangement (see Figure 44.1). A sperm enters in the animal hemisphere at what will become the ventral side of the frog and causes rotation of the cortical cytoplasm to create a gray crescent, which will then become the dorsal region of the frog. This helps establish the left-to-right axis and the anterior-to-posterior axis of the developing animal.
- During the movement of the cytoplasm, the transcription factor β-catenin is degraded in parts of the frog egg so that it becomes concentrated in the dorsal side of the embryo (see Figure 44.2). This happens because the protein GSK-3 moves to the ventral side of the zygote, where it targets β-catenin for degradation. On the dorsal side, proteins inhibit the action of GSK-3, and as a result, the concentration of β-catenin is higher there.

Cleavage is a rapid series of cell divisions early in development.

- In a process known as cleavage, the first cell divisions occur rapidly with little growth or differentiation. The embryo divides into smaller and smaller cells, producing first a solid ball of cells and then a blastula. The blastula has a central fluid-filled cavity called the blastocoel. Individual cells in the blastula are called blastomeres.
- The amount of yolk in the egg determines the plane along which cell division occurs (see Figure 44.3). Cleavage furrows are impeded by yolk. Frog eggs have small amounts of yolk and complete cleavage in which early cleavage furrows completely divide the egg. The eggs of fishes, reptiles, and birds have large amounts of yolk and incomplete cleavage. In these organisms the embryo develops from a disc of cells (known as the blastodisc) on top of the yolk mass. Some insects undergo superficial cleavage, a variation of incomplete cleavage in which a syncytium (single cell with many nuclei) is produced early in development when cycles of mitosis occur without cell division.
- The orientation of mitotic spindles determines the cleavage planes and thus the arrangement of daughter cells. In the frog, the mitotic spindles are parallel or perpendicular to the animal–vegetal axis, and the resulting cleavage pattern is radial. In mollusks, the mitotic spindles are at oblique angles to the animal–vegetal axis, and the resulting cleavage pattern is spiral.

In placental mammals, the early cell divisions are unique.

- The early cell divisions of placental mammals occur more slowly than in other animals, and are asynchronous.
- Mammals have a unique rotational cleavage pattern. In other animals, cleavage is directed by molecules already present in the egg. In placental mammals, genes are expressed during cleavage and play a large role in the process. The mammalian zygote produces both the embryo and extraembryonic structures that interact with the mother.
- During cleavage from the 16-cell to the 32-cell stage, the cells separate into an inner cell mass that will become the embryo proper, and the trophoblast that will become part of the placenta and attach to the uterus during implantation. At the 32-cell stage, the mass of mammalian cells is called a blastocyst (see Figure 44.4). Within the blastocyst is a blastocoel.
- Mammalian fertilization occurs in the oviduct. As the zygote undergoes cleavage, it moves down the oviduct and eventually enters the uterus as a blastocyst that implants itself in the uterine lining (endometrium). The zona pellucida inhibits early implantation. Once the blastocyst reaches the uterus, it hatches from the zona pellucida, making implantation possible. Implantation anywhere other than the uterus is a dangerous condition called an ectopic pregnancy.

Blastomeres become determined during late cleavage.

- During cleavage, cytoplasm is distributed to the cells of the blastula in such a way that cells in different regions have different levels of nutrients and informational molecules. Specific blastomeres will become specific tissues and organs, and this can be mapped out in the developing cells as a fate map of the blastula (see Figure 44.6). The blastomeres become determined (committed to specific fates) at different times for different species.
- Animals with mosaic development have blastomeres that are set very early to contribute to specific parts of the embryo. In animals with regulative development, the blastomeres can be removed, and other cells will compensate for the loss.

Gastrulation involves movement and differentiation of cells.

- During gastrulation, the germ layers of the tissues form and position themselves in the embryo. Gastrulation makes possible the inductive interactions between cells, which trigger differentiation and organ formation.
- Three germ layers develop during gastrulation. The inner layer, or endoderm, will become the lining of the digestive tract, the lining of the respiratory tract, and some organs, such as the pancreas and liver. The outer layer, or ectoderm, will become the epidermis (outer layer of the skin), hair, nails, glands (sweat and oil), and the nervous system. The middle layer, or mesoderm, will become the heart, blood vessels, muscle, and bone.

Gastrulation in sea urchins involves invagination of the vegetal pole.

- The vegetal pole in sea urchins first flattens. The flat portion of the vegetal pole then moves inward as an invagination to form the endoderm and primitive gut (archenteron). Some cells from the vegetal pole break free to become the mesenchyme cells that make up the mesoderm.
- The opening of the archenteron or blastopore will become the anus, and the place where the tip of the archenteron meets the ectoderm will become the mouth (see Figure 44.7).

The dorsal lip controls embryonic organization during gastrulation in the frog.

- The frog embryo has more yolk than the sea urchin, and gastrulation is more complex (see Figure 44.8). Initially, cells near the gray crescent begin to bulge into the blastocoel. These initial cells (bottle cells) move along the interior of the blastula and pull the outer surface of cells along with them, creating a dorsal lip; this process is called involution. The first cells moving in are prospective endoderm, and they form the archenteron. This rearrangement of cells, called convergent extension, involves changes in cell properties, including elongation of cells and intercalation between cells.
- As gastrulation continues, cells from the animal hemisphere move toward the site of involution in a process known as epiboly. As epiboly continues, the dorsal lip forms a complete circle around a plug of yolk-rich cells. At the end of gastrulation, the cells have become fate-determined and are layered as the ectoderm, endoderm, and mesoderm.
- Hans Spemann examined the timing and fate of cells during gastrulation in salamanders to determine if they were totipotent or able to direct development (see Figure 44.9). Spemann and Hilde Mangold, his student, found that when the dorsal lip was transplanted to another gastrula, the result was a second site of gastrulation and eventually two embryos attached at the belly. They concluded that the dorsal lip acts as the primary embryonic organizer (see Figure 44.10).
- β-catenin, a transcription factor, is a candidate for initiator of the signal cascade involved in organizer activity. When β-catenin activity is knocked out by antisense RNA, the embryo does not go through gastrulation; overexpression of β-catenin can induce a second axis of embryo formation.
- Presence of β-catenin and a complex series of interactions between growth and transcription factors create the organizer and lead to induction of the body plan. Primary embryonic organizer activity is started from the vegetal cells below the gray crescent. The transcription factors Goosecoid and Siamois are critical in the signal cascade. Tcf-3 proteins act as repressors for the *siamois* gene. When the Tcf-3 protein is blocked by β-catenin, expression of *siamois* occurs, and the goosecoid protein is produced (see Figure 44.11).
- As the organizer migrates from the dorsal lip, it inhibits various growth factors along the way to achieve different patterns of differentiation along the anterior–posterior axis.

In reptilian and avian gastrulation, the primitive groove is the blastopore, and Hensen's node corresponds to the amphibian dorsal lip.

- In birds and reptiles, cleavage produces a blastodisc on top of the large amount of yolk in the egg.
- The blastula is a circular layer of cells composed of an outer epiblast layer and an inner hypoblast layer. The epiblast will form the embryo proper and the hypoblast will form extraembryonic membranes. The fluid-filled space between the two layers is the blastocoel.
- The primitive streak is formed by the movement of cells in the epiblast toward the midline (see Figure 44.13). Along the primitive streak, a primitive groove forms and cells move through this groove, becoming endoderm and mesoderm in the blastocoel.
- In the chicken, there is no archenteron. Endoderm and mesoderm move forward and form gut structures. A group of cells at the anterior end of the primitive groove, known as Hensen's node, acts in the same manner as the dorsal lip of the frog blastopore. Cells that move over Hensen's node differentiate into the notochord and structures of the head.

Gastrulation in placental mammals is similar to that in reptiles and birds.

- In placental mammals, just as in birds and reptiles, the cells of the embryo segregate into an epiblast and a hypoblast with a blastocoel in between. Extraembryonic membranes, including that which contributes to the placenta, develop from the hypoblast. The embryo and some extraembryonic membranes develop from the epiblast.
- Gastrulation in mammals occurs just as in birds, with the formation of a primitive groove through which epiblast cells migrate to form endoderm and mesoderm.

The nervous system begins to develop during neurulation.

- After gastrulation, the organs begin to develop through the process of organogenesis. Neurulation—the initial development of the nervous system—begins early in organogenesis.
- During neurulation, a rod of connective tissue, called the notochord, develops in the blastocoel from chordamesoderm cells derived from the dorsal mesoderm. The notochord provides structural support to the developing embryo and induces overlying cells to form the nervous system. Eventually the notochord will be replaced by the vertebral column.
- A neural plate forms above the notochord from the ectoderm, which will begin to fold into a cylinder forming the neural tube (see Figure 44.14). The anterior end of the neural tube will become the brain, with the rest forming the spinal cord. Failure of this process leads to serious birth defects such as spina bifida and anencephaly.
- Neural crest cells break from the neural tube and migrate inward to become peripheral nerves and other structures.
- The repeating pattern of the vertebrate body plan forms from blocks of somite tissue derived from mesoderm located on both sides of the notochord (see Figure 44.15). The somites will become ribs, vertebrae, and trunk muscles.

Homeotic genes control differentiation in the anterior–posterior directions during development.

- Body segments differentiate as the embryo develops. Hox genes control this differentiation.
- Hox genes are found on different chromosomes and are expressed along the anterior–posterior axis in their order on the chromosomes (see Figure 44.16).
- Dorsal–ventral differentiation is controlled by a separate set of genes. For example, the *sonic hedgehog* gene, expressed in the notochord, prompts the most ventral cells in the overlying neural tube to become motor neurons.
- After segmentation occurs, organs and organ systems develop rapidly.

Birds and reptiles develop extraembryonic membranes.

- Extraembryonic membranes surround the embryos of reptiles, birds, and mammals.
- In birds, the yolk sac is derived from the hypoblast and nearby mesoderm, and it surrounds the entire yolk to help retrieve nutrients from the yolk and deliver them to the embryo via blood vessels (see Figure 44.17).
- Cells from the ectoderm and mesoderm form the amnion, which helps provide an aqueous environment, and the chorion, which develops just under the shell. The chorion regulates water, oxygen, and carbon dioxide exchanges across the shell.
- The allantois is derived from endoderm and is a membrane that forms a sac to store metabolic wastes.

The placenta of mammals provides exchanges between the mother and the embryo.

- In placental mammals, the first extraembryonic membrane to form, the trophoblast, interacts with the endometrium to attach to the uterine wall and begin implantation of the blastocyst.
- The hypoblast cells interact with the trophoblast to form the chorion. The chorion and tissues from the uterine wall form the placenta (see Figure 44.18).
- The amnion of the developing mammal surrounds the embryo to produce a closed, fluid-filled environment.
- The allantois of mammals has the function of removing nitrogenous wastes, but its function is relatively minor in many mammals, including humans. The tissues of the allantois help to form the umbilical cord, which carries major blood vessels that provide a route for exchanges of nutrients, wastes, carbon dioxide, and oxygen between the mother and fetus.

Human pregnancy can be divided into trimesters.

- In humans, pregnancy lasts about 266 days, compared with 21 days in mice and 600 days in elephants.
- During the first trimester of human development, cell division and tissue differentiation are rapid, and organ development begins. At this stage, the developing human is most sensitive to environmental disrupters. By the end of the first trimester, the embryo is considered a fetus.
- During the second trimester of development, organ systems continue to grow and mature. The first fetal movements are felt and the fetus reaches a size of about 600 grams.
- The third trimester is marked by extremely rapid growth of the fetus and further maturation of the internal organs.
- Development continues after birth, with changes in the brain particularly evident until adolescence.

Test Yourself

Diagram Exercises

1. In the diagram below of a sea urchin gastrula, label the ectoderm, endoderm, primary mesenchyme, secondary mesenchyme, blastopore, and archenteron.

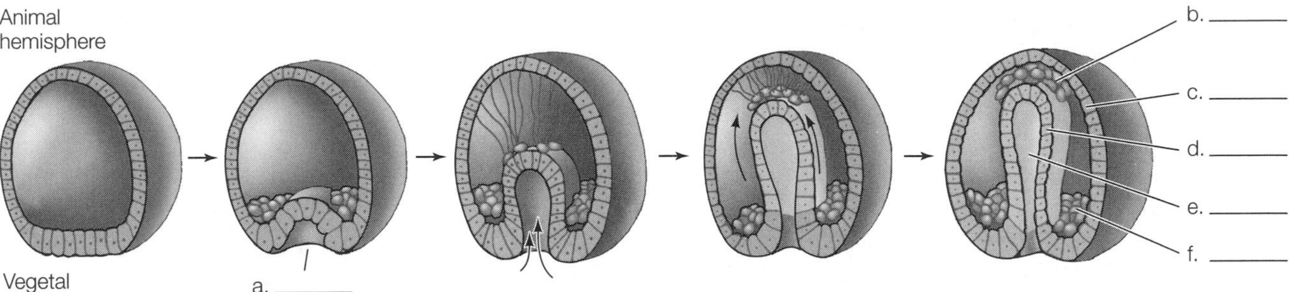

Textbook Reference: *44.2 How Does Gastrulation Generate Multiple Tissue Layers? p. 929*

2. In the diagram below of a 9-day chick embryo, label the embryo, amnion, chorion, yolk sac, allantois, and the allantoic membrane.

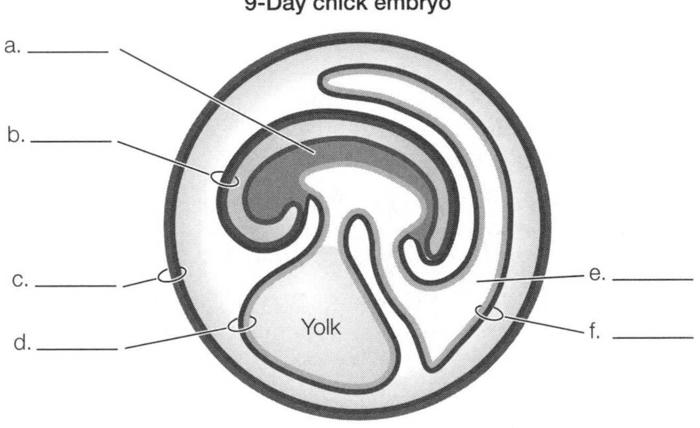

Textbook Reference: *44.4 How Is The Growing Embryo Sustained? p. 937*

Knowledge and Synthesis Questions

1. Which of the following statements about the respective contributions of the sperm and egg to the zygote is *false*?
   a. The sperm contributes most of the organelles.
   b. The egg contributes most of the cytoplasm.
   c. Both the sperm and the egg contribute a haploid nucleus.
   d. The sperm contributes a centriole, which becomes the centrosome of the zygote.
   e. All of the above are false.
   Textbook Reference: *44.1 How Does Fertilization Activate Development? p. 923*
2. In which of the following animals is the cleavage pattern rotational?
   a. Frog
   b. Mammal
   c. Sea urchin
   d. Mollusk
   e. Bird
   Textbook Reference: *44.1 How Does Fertilization Activate Development? p. 926*
3. Which of the following statements about complete cleavage versus incomplete cleavage is true?
   a. Incomplete cleavage occurs in species with small volumes of cytoplasm.
   b. Complete cleavage is found in mammals and is the more evolved characteristic.
   c. Incomplete cleavage occurs in species with large amounts of yolk.
   d. Complete cleavage occurs only in eggs that have been fertilized by two sperm.
   e. Incomplete cleavage occurs in species with small amounts of yolk.
   Textbook Reference: *44.1 How Does Fertilization Activate Development? pp. 925–926*
4. A fate map can be used to map out the tissues and organs that eventually will develop from specific germ layers. On the fate map, ectoderm will become
   a. the lining of the gut.
   b. the nervous system.
   c. muscle.
   d. the heart.
   e. the lining of the respiratory tract.
   Textbook Reference: *44.2 How Does Gastrulation Generate Multiple Tissue Layers? p. 928*
5. The location of the _______ determines the anterior–posterior axis of the embryo.

a. primitive streak
b. blastopore
c. vegetal hemisphere
d. hypoblast
e. tertiary mesenchyme
Textbook Reference: *44.2 How Does Gastrulation Generate Multiple Tissue Layers? p. 929*

6. Which of the following represents the correct order of the germ layers, from the inside to the outside?
a. Mesoderm, ectoderm, endoderm
b. Endoderm, ectoderm, mesoderm
c. Ectoderm, mesoderm, endoderm
d. Endoderm, mesoderm, ectoderm
e. Mesoderm, endoderm, ectoderm
Textbook Reference: *44.2 How Does Gastrulation Generate Multiple Tissue Layers? p. 928*

7. The dorsal lip of the blastopore organizes embryo formation in frogs. The equivalent structure in chickens is
a. the epiblast.
b. the hypoblast.
c. Hensen's node.
d. the bottle cell.
e. the gray crescent.
Textbook Reference: *44.2 How Does Gastrulation Generate Multiple Tissue Layers? pp. 933–934*

8. Birds develop extraembryonic membranes during development. Which of the following statements about avian extraembryonic membranes is *false*?
a. The yolk sac surrounds the yolk and provides nutrients.
b. The amnion and chorion are derived from ectoderm and mesoderm.
c. The allantois stores nutrients.
d. The chorion exchanges gases and water between the embryo and the environment.
e. The allantois is derived from endoderm and mesoderm.
Textbook Reference: *44.4 How Is The Growing Embryo Sustained? pp. 937–938*

9. Which of the following statements about the mammalian blastocyst is *false*?
a. The trophoblast gives rise to the embryo proper.
b. Maternal genes are expressed during cleavage.
c. The blastocyst implants itself in the mother's uterus.
d. Early mammalian cleavage is relatively slow.
e. The trophoblast contributes to formation of the chorion.
Textbook Reference: *44.1 How Does Fertilization Activate Development? pp. 926–927; 44.4 How Is The Growing Embryo Sustained? p. 938*

10. The primary embryonic organizer is most likely initiated by
a. the yolk.
b. Tcf-3 protein.
c. β-catenin.
d. cAMP.
e. None of the above
Textbook Reference: *44.2 How Does Gastrulation Generate Multiple Tissue Layers? pp. 931–932*

11. The _______ eventually develop into vertebrae, ribs, and trunk muscles, and are found along the sides of the _______.
a. somites; neural tube
b. neural tube cells; notochord
c. blastopore cells; dorsal lip
d. neural crest cells; dorsal lip
e. neural plate cells; archenteron
Textbook Reference: *44.3 How Do Organs and Organ Systems Develop? pp. 935–936*

12. Hans Spemann called the dorsal lip of the blastopore the embryonic organizer because it
a. is the point where gastrulation begins.
b. becomes part of the nervous system.
c. becomes part of the notochord.
d. leads to the establishment of the embryonic axes.
e. is the location at which the sperm enters the egg.
Textbook Reference: *44.2 How Does Gastrulation Generate Multiple Tissue Layers? pp. 930–931*

13. During its development, the human embryo is contained within a fluid-filled chamber enclosed by the extraembryonic membrane called the
a. yolk sac.
b. amnion.
c. chorion.
d. allantois.
e. trophoblast.
Textbook Reference: *44.4 How Is The Growing Embryo Sustained? p. 938*

14. The third trimester of human prenatal development is characterized by
a. extremely rapid growth.
b. the formation of major organs.
c. the greatest sensitivity to damage from drugs and radiation.
d. gastrulation.
e. the formation of the placenta.
Textbook Reference: *44.5 What Are the Stages of Human Development? p. 939*

15. Which of the following statements about human development is *false*?
a. At about 3 months gestation, the developing human is called a fetus.
b. Fetal movements are first felt by the mother during the second trimester.
c. Substantial developmental changes occur in the brain between birth and adolescence.
d. Humans have the longest gestation period among mammals because of the developmental requirements of their relatively large brains.
e. Development continues after postnatal growth has stopped.
Textbook Reference: *44.5 What Are the Stages of Human Development? pp. 939–940*

Application Questions

1. Explain how identical and non-identical twins form in humans. Are conjoined twins identical or non-identical twins?
   Textbook Reference: *44.1 How Does Fertilization Activate Development? p. 928*
2. Thalidomide is a drug known to disrupt limb development. Why might exposure to thalidomide during the first trimester cause more severe birth defects than a similar level of exposure in the third trimester?
   Textbook Reference: *44.5 What Are the Stages of Human Development? pp. 939–940*
3. Compare cleavage in birds and in mammals. How do differences in cleavage between these two groups influence gastrulation?
   Textbook Reference: *44.1 How Does Fertilization Activate Development? pp. 925–926; 44.2 How Does Gastrulation Generate Multiple Tissue Layers? p. 929, 933*
4. How is the development of the extraembryonic membranes different in birds and placental mammals and how is it similar? What accounts for the difference?
   Textbook Reference: *44.4 How Is The Growing Embryo Sustained? pp. 937–938*
5. Compare the trophoblast and inner cell mass of the human blastocyst with respect to function. From which of the two do embryonic stem cells come from?
   Textbook Reference: *44.1 How Does Fertilization Activate Development? pp. 926–928*

Answers

Diagram Exercise Answers

1.
   a. Blastopore
   b. Secondary mesenchyme
   c. Ectoderm
   d. Endoderm
   e. Archenteron
   f. Primary mesenchyme
2.
   a. Embryo
   b. Amnion
   c. Chorion
   d. Yolk sac
   e. Allantois
   f. Allantoic membrane

Knowledge and Synthesis Answers

1. **a.** The egg donates most of the organelles to the zygote. In most species, the sperm contributes a centriole.
2. **b.** Cleavage in mammals occurs in a rotational pattern.
3. **c.** Incomplete cleavage occurs because the cleavage furrows cannot completely penetrate the yolk. Incomplete cleavage occurs in animals, such as birds and reptiles, whose eggs have large amounts of yolk.
4. **b.** Ectoderm will eventually become the nervous system.
5. **b.** The blastopore, which will eventually become the anus, marks the posterior of the embryo.
6. **d.** The germ layers that are formed during gastrulation are the inner layer of endoderm, the middle layer of mesoderm, and the outer layer of ectoderm.
7. **c.** In chickens, Hensen's node is the equivalent of the dorsal lip of the frog blastopore.
8. **c.** During development, an avian embryo produces wastes, but because birds develop in a shell, the wastes must be stored in the egg. The allantois forms a sac that is used for the storage of metabolic wastes.
9. **a.** In the mammalian blastocyst, the trophoblast forms the fetal part of the placenta (chorion). A disc-shaped portion of the inner cell mass becomes the embryo.
10. **c.** β-catenin is thought to be the initiator of organizer activity during early development.
11. **a.** Somites are located along the neural tube in the developing vertebrate. These cells will develop into the vertebrae, ribs, and trunk muscles.
12. **d.** The dorsal lip leads to the establishment of the embryonic axes.
13. **b.** The amnion is the membrane that surrounds the developing mammalian fetus in its fluid-filled amniotic cavity.
14. **a.** Extremely rapid growth characterizes the third trimester. Gastrulation, formation of major organs, and formation of the placenta occur during the first trimester, which is also the period of greatest sensitivity to drugs and radiation.
15. **d.** The human gestation period of nine months is not the longest among mammals. Length of gestation is related to overall body size, not relative brain size.

Application Answers

1. Identical (monozygotic) twins are formed during an early stage of cleavage when the mass of cells splits. Non-identical twins form when two secondary oocytes are released from the ovaries and are fertilized by two different sperm. Conjoined twins are identical twins formed when splitting of the mass of cells is incomplete.
2. Exposure to thalidomide during the first trimester would likely cause more severe birth defects than exposure during the third trimester because tissues and organs are differentiating and forming during the first trimester and mainly growing during the third trimester.
3. Due to the large amount of yolk in the avian egg, cleavage in birds is incomplete, and the result is a blastodisc that sits on top of the yolk (see Figure 44.3B; although the egg pictured is from a zebrafish, bird eggs show a similar cleavage pattern). Unlike bird embryos, mammalian embryos undergo complete cleavage in a rota-

tional pattern, resulting in a structure called a blastocyst (see Figure 44.4C). The pace of cleavage is also slower in mammals than in birds, and embryonic genes play a role in directing cleavage in mammals, whereas in birds, molecules already present in the egg direct cleavage. Despite these differences, gastrulation is very similar in birds and mammals. The avian blastodisc develops into a flattened blastula with epiblast and hypoblast surrounding a fluid-filled blastocoel; the mammalian inner cell mass also splits into an epiblast and a hypoblast. Although the details vary, the developmental fates of these structures are fundamentally the same in both groups: the hypoblast will give rise to extraembryonic membranes, and the embryo proper will develop from the epiblast.

4. Both the birds and mammals develop extraembryonic membranes. The extraembryonic membranes of birds are used to obtain food from the large reserve of yolk (yolk sac), to exchange water and gases with the environment across the shell (chorion), to protect the embryo in a fluid-filled environment (amnion), and to store nitrogenous wastes (allantois). The eggs of placental mammals do not contain yolk; the developing embryo relies instead on the placenta. In placental mammals, extraembryonic membranes and the uterine lining of the mother form a placenta which exchanges nutrients, respiratory gases, and wastes between embryo and mother. The hypoblast cells that form the yolk sac in birds, along with the trophoblast cells and the chorion, all contribute to the placenta. An allantois also develops, but has a limited function in many mammals. Thus, the main reason for differences in the functions of extraembryonic membranes of birds and placental mammals relates to the placenta taking over many functions in mammals.

5. The trophoblast secretes enzymes that allow the blastocyst to burrow into the uterine lining (endometrium) of the mother and to implant there. The trophoblast then forms chorionic villi, the embryo's major contribution to the placenta. In contrast, the inner cell mass forms the embryo proper and contributes to extraembryonic membranes. Embryonic stem cells come from the inner cell mass.

45 Neurons and Nervous Systems

The Big Picture

- The neuron, with support from surrounding glia, is the functional unit of the nervous system. The neuron is composed of a cell body, dendrites, and an axon. The membrane of a neuron has a difference in voltage across it. Nerve impulses are passed down the axon of a neuron as action potentials. An action potential is a temporary disruption of the "battery-like" state of the resting neuron membrane due to the opening and closing of sodium and potassium voltage-gated channels. The action potential is an all-or-none response that occurs when the depolarization of an axon reaches a threshold level.
- At the end of the axon synapse, information flow is controlled through excitation or inhibition of synapses. There are many different neurotransmitters that transmit a nerve impulse from the presynaptic cell to the postsynaptic cell. The action of a specific neurotransmitter will depend on the receptor to which it binds.

Common Problem Areas

- One of the most difficult challenges when learning about the nervous system is understanding how the resting membrane potential and action potential are produced. Remember that the neuron at rest is like a battery and that the charge of the membrane is dependent on the ions that are on either side and moving across the membrane.
- Students often have difficulty with the function of a synapse and how the action potential is transmitted across the synapse. Remember that in many cases the signal goes from an electrical signal, to a chemical signal, and back to an electrical signal.

Study Strategies

- Students should understand the concept of "pre-" and "postsynaptic" neurons. Try to remember the sequence of events and relate them to the function of the neuron transmitting a stimulus from one cell to another.
- Students often have trouble understanding how the membrane potential is established. Students should learn what each important anion and cation does at rest and during the action potential. Then, piece by piece, put together a sense of how the neuron membrane is charged at rest, and what happens during an action potential.
- Go to yourBioPortal.com to review the following tutorials and activity:

  Animated Tutorial 45.1 The Resting Membrane Potential

  Animated Tutorial 45.2 The Action Potential

  Animated Tutorial 45.3 Synaptic Transmission

  Interactive Tutorial: Neurons: Electrical and Chemical Conduction

  Web Activity 45.1 Neurotransmitters

Important Concepts

The nervous system consists of neurons and glia.

- The nervous system is made up of two major types of cells: neurons and glia (also called glial cells).
- Neurons transmit information as electric signals (action potentials). Afferent neurons carry sensory information into the nervous system. Efferent neurons carry information from the nervous system to effectors, such as muscles or glands. Interneurons facilitate communication between afferent and efferent neurons.
- Simple animals, such as the sea anemone, possess a nerve net (see Figure 45.1A).
- The nervous system of more complex animals, such as earthworms and squid, is made up of clusters of neurons distributed throughout the body. These clusters are known as ganglia.
- Vertebrates have a central nervous system—the brain and spinal cord—and a peripheral nervous system—the neurons in the rest of the body. The brains of vertebrates vary in size (even among species with similar body masses) and complexity. The size of the cerebrum, the part of the brain responsible for complex behaviors, is especially variable across taxa.

- Information is passed from one neuron to another neuron at a synapse. The first neuron is the presynaptic neuron, and the second is the postsynaptic neuron.
- A neuron is composed of four parts: the cell body, dendrites, axons, and axon terminals (see Figure 45.3A). The cell body contains the nucleus, with dendrites coming off it and receiving information from other neurons. The axon conducts action potentials away from the cell body. Axon terminals interact with the neuron's target cells to form a synapse.
- Synapses can be either chemical or electrical. Chemical synapses are more common than electrical synapses in vertebrates. At a chemical synapse, an action potential arriving at the axon terminals of a presynaptic neuron causes the release of neurotransmitter, which travels across the synapse to bind with receptors on the postsynaptic neuron.
- Glia serve many functions in the nervous system, such as supplying nutrients to neurons, removing wastes, and helping neurons make the proper connections during development. Glia that insulate axons in the peripheral nervous system are Schwann cells, and those that insulate axons in the central nervous system are oligodendrocytes. The covering produced by Schwann cells and oligodendrocytes is called myelin; myelinated axons conduct action potentials more rapidly than do axons lacking myelin. Star-shaped glia, called astrocytes, contribute to the blood–brain barrier, which protects the brain from some toxins in the blood.

Neurons have an electrically charged cell membrane at rest.

- Resting neurons have a negative charge inside and a positive charge outside, resulting in a difference in electrical charge across the membrane known as the membrane potential. In an unstimulated neuron, this voltage difference is called a resting potential. Electrodes can be used to measure resting potentials; such measurements show that the resting potential is typically between –60 and –70 mV (see Figure 45.5).
- The electrical charge across the membrane at rest is due to differences in concentrations of the charged ions sodium (Na^+), chloride (Cl^-), potassium (K^+), and calcium (Ca^{2+}).
- The lipid bilayer of the plasma membrane is impermeable to ions. Ions move across the plasma membrane through channels or by ion pumps, both of which are formed by proteins.
- The sodium–potassium pump transports Na^+ out of the cell and K^+ into it, thereby maintaining higher concentrations of Na^+ ions outside the cell and higher concentrations of K^+ ions inside it (see Figure 45.6A). Because the pump is an enzyme complex that needs ATP to perform its work, it is sometimes called sodium–potassium ATPase.
- At rest, neurons have a specific charge due to K^+ movement to the outside of the cell, resulting in a resting potential. K^+ channels are the most common open (sometimes called leak) channels, allowing K^+ to diffuse out of the cell down the concentration gradient that has been set up by the Na^+–K^+ pump (see Figure 45.6B).
- Ion channels are selective pores in the plasma membrane that allow specific ions to diffuse across the membrane. Whereas some are always open (such as K^+ channels), others are gated (open under certain conditions and closed under other conditions). Ion channels can be voltage-gated (responding to changes in the voltage across the membrane), chemically gated (responding to the presence of a specific chemical), or mechanically gated (responding to mechanical force applied to the membrane).
- Patch clamping is a technique that allows the recording of voltage differences due to the movements of ions through channels in an isolated patch of plasma membrane (see Figure 45.8). Sakmann and Neher developed patch clamping in the late 1970s and received a Nobel Prize in 1991.

Neurons generate action potentials to conduct nerve impulses.

- Changes in resting membrane potential ("depolarization") produce action potentials or nerve impulses.
- Membranes can be depolarized or hyperpolarized (see Figure 45.9). Depolarization occurs when the inside of a neuron becomes less negative compared to the resting potential. Hyperpolarization occurs when the inside of a neuron becomes more negative compared to the resting potential.
- Changes in polarity of the plasma membrane due to opening and closing of ion channels are passed down along an axon to transmit a signal as an action potential.
- Action potentials are very short-lived, but large, changes in membrane potential (see Figure 45.10). Action potentials are generated when the membrane reaches a threshold potential, Na^+ voltage-gated channels open, and Na^+ enters the cell to make the inside of the axon positive. Voltage-gated K^+ channels then open, allowing positive-charged K^+ to leave the axons to help return the membrane potential back to the resting level. As the K^+ channels open, the Na^+ channels close and cannot be opened for about 1 to 2 milliseconds, which is the refractory period. The Na^+–K^+ pump helps return the concentration of ions back to the resting levels.
- Action potentials travel down an axon by a positive feedback mechanism that stimulates adjacent regions of an axon to generate the action potential. The Na^+ ions that enter during an action potential flow to adjoining regions of the axon, stimulating depolarization and the movement of the action potential along the axon (see Figure 45.11).

- The refractory period, during which the Na^+ channels cannot act, can be explained by the presence of two gates in the channel, an activation gate and an inactivation gate. The refractory period keeps an action potential moving in one direction, away from the cell body.
- An action potential is an all-or-none response; the depolarization must reach a threshold level for an action potential to occur. An action potential is also a self-regenerating response; once an action potential occurs at one location on an axon, it stimulates the adjacent area to generate an action potential.

Action potentials can jump down an axon.

- In the nervous systems of invertebrates, the conduction velocity of axons increases with the increasing diameters of axons. In the nervous systems of vertebrates, conduction velocity of axons is increased by myelination.
- Myelination occurs by glia wrapping themselves around some axons; nodes of Ranvier are gaps in the myelin wrapping at which depolarization can occur. Depolarization jumps from node to node, increasing the speed of transmission of an action potential along the axon in a process known as saltatory conduction (see Figure 45.12).

Neurons communicate across synapses.

- Transfer of information from one cell to another by either a chemical or electrical message occurs in a small junction called a synapse. Presynaptic nerve cells send a message across a synapse to postsynaptic cells.
- A chemical synapse uses a chemical messenger, or neurotransmitter, to communicate between the presynaptic and postsynaptic cells. The neurotransmitter is produced in the axon terminal and packaged into vesicles. The neurotransmitter is released into the synaptic cleft when the vesicle binds with the presynaptic membrane. The neurotransmitter crosses the synaptic cleft and binds with receptors on the surface of the postsynaptic cell. In electrical synapses, the action potential spreads directly from presynaptic to postsynaptic cell.

Acetylcholine is a common synaptic neurotransmitter.

- The neurotransmitter acetylcholine (ACh) is the chemical messenger carrying information between motor neurons and muscle cells at neuromuscular junctions (see Figure 45.13). Acetylcholine is enclosed in a vesicle at the presynaptic synapse, which fuses with the membrane to release the acetylcholine into the 20- to 40-nm-wide gap between the two cells, the synaptic cleft.
- Ca^{2+} channels open when the action potential reaches the axon terminal, causing Ca^{2+} to rush in and regulate the fusing of the acetylcholine-containing vesicles to the presynaptic membrane.
- Acetylcholine released into the synaptic cleft binds with receptors in the postsynaptic membrane, the motor end plate, which opens Na^+ channels, resulting in depolarization of the motor end plate.
- Numerous proteins are involved in the release of neurotransmitter from the presynaptic neuron. For example, some proteins transport neurotransmitter into vesicles, some anchor vesicles to cytoskeletal elements, and others are involved in the fusion of vesicular and cell membranes.
- Acetylcholinesterase breaks down acetylcholine in the synapse junction to halt the action of the released acetylcholine (see Figure 45.14).

Synapses can either depolarize or hyperpolarize postsynaptic membranes.

- Synapses in vertebrates can be excitatory and depolarize the postsynaptic membrane, or inhibitory and hyperpolarize the postsynaptic membrane.
- Neurons may receive synaptic inputs from many neurons. Excitatory and inhibitory postsynaptic potentials are summed by adding simultaneous potentials (spatial summation) or by summing rapid firing of one postsynaptic potential (temporal summation) (see Figure 45.15).
- The region of the cell body at the base of the axon, called the axon hillock, is the "decision-making" area of a neuron. If the axon hillock is depolarized, the axon will fire an action potential.

Two main types of receptors for neurotransmitters exist.

- The two general categories of neurotransmitter receptors are ionotropic and metabotropic. Ionotropic receptors are ion channels on the postsynaptic membrane that are activated by the binding of the neurotransmitter. They allow fast, short-lived responses.
- Metabotropic receptors are not ion channels. Metabotropic receptors act by initiating signaling cascades, which eventually cause changes in ion channels. When mediated by metabotropic receptors, postsynaptic cell responses are usually slower and longer-lived than those generated by ionotropic receptors.

Electrical synapses connect some neurons.

- Electrical synapses are formed by direct contact between adjacent neurons; these synapses contain numerous gap junctions.
- Two neurons forming an electrical synapse are joined by connexons, which are tunnels (pores) between the two neurons that allow ions to pass between the two cells.
- Electrical synapses are very fast connections and can transmit an action potential in either direction. These synapses are good for rapid communication.
- Electrical synapses are less common than chemical synapses in vertebrate nervous systems. With electrical synapses, there can be no temporal summation of

synaptic inputs. Electrical synapses also require large areas of contact and cannot be inhibitory.

The action of a neurotransmitter is determined by the receptor to which it binds.

- There are more than 50 neurotransmitters, including amino acids, peptides, gases (e.g., nitric oxide), purines, and monoamines. One neurotransmitter can act on several different receptors, and its particular action depends on the receptor to which it binds.
- Acetylcholine has nicotinic receptors and muscarinic receptors. Both types are found in the central nervous system. The nicotinic receptors are ionotropic and tend to be excitatory, whereas the muscarinic receptors are metabotropic and tend to be inhibitory. Receptors for acetylcholine also occur outside the central nervous system.

The neurotransmitter glutamate may have a role in learning and memory.

- Two types of ionotropic glutamate receptors—NMDA and AMPA—may play a role in learning. NMDA receptors allow for a slow, long-lasting influx of Na^+, whereas the AMPA receptors allow for rapid influx (see Figure 45.16). The NMDA receptors also let Ca^{2+} into the postsynaptic cell, resulting in long-term cellular changes.
- Both of these receptors are excited by glutamate.
- Long-term potentiation is an enhanced response in the postsynaptic neuron with repeated stimulation (see Figure 45.17). Long-term potentiation may be involved in memory.

Neurotransmitters must be removed from the synapse in order for their action to be turned off.

- The actions of neurotransmitters can be stopped in several ways. First, enzymes may destroy the neurotransmitter. Second, the neurotransmitter may simply diffuse away from the synaptic cleft. Third, nearby cell membranes may take up the neurotransmitter using active transport.
- Some gases used in chemical warfare cause death by paralysis by inhibiting acetylcholinesterase, the enzyme that destroys acetylcholine at the neuromuscular junction. The drug Prozac, used to treat depression, acts by slowing the reuptake of the neurotransmitter serotonin, thus prolonging serotonin's action at the synapse.

Test Yourself

Diagram Exercise

Label the following structures on the myelinated motor neuron shown below: axon, axon hillock, axon terminals, cell body, and dendrites. What is the direction and manner in which an action potential is conducted along the neuron.

Textbook Reference: *45.1 What Cells Are Unique to the Nervous System? pp. 946–947*

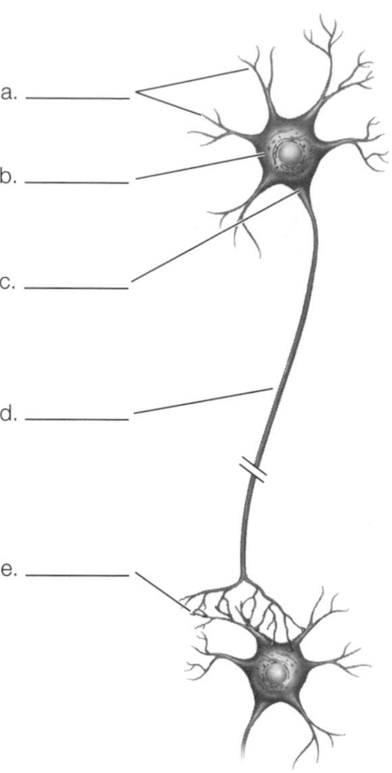

Knowledge and Synthesis Questions

1. The extensions of postsynaptic neurons that provide the main receptive surface for presynaptic neurons are the
   a. nuclei.
   b. somas.
   c. axons.
   d. dendrites.
   e. glia.
   Textbook Reference: *45.1 What Cells Are Unique to the Nervous System? p. 946*
2. The substance that wraps the axon of many neurons and provides for increased conduction speed is
   a. dendrase.
   b. histamine.
   c. acetylcholine.
   d. myelin.
   e. microglia.
   Textbook Reference: *45.1 What Cells Are Unique to the Nervous System? p. 947*
3. The long extension from the cell body of a neuron that provides the pathway for action potentials to the synapse is the
   a. dendrite.
   b. Schwann cell.
   c. axon.
   d. presynaptic membrane.
   e. nerve net.
   Textbook Reference: *45.1 What Cells Are Unique to the Nervous System? p. 946*
4. The threshold of a neuron is the
   a. amount of inhibitory neurotransmitter required to inhibit an action potential.

b. membrane voltage at which an axon potential will be suppressed.
c. amount of excitatory neurotransmitter required to elicit an action potential.
d. membrane voltage at which the membrane potential develops into an action potential.
e. closing of numerous sodium channels.

Textbook Reference: *45.2 How Do Neurons Generate and Transmit Electrical Signals? p. 954*

5. When a membrane is at the resting potential, the concentration of
   a. sodium and potassium ions is higher on the inside of its membrane than on the outside.
   b. sodium and potassium ions is higher on the outside of its membrane than on the inside.
   c. sodium ions is higher on the inside of its membrane and of potassium ions is higher on the outside.
   d. sodium ions is higher on the outside of its membrane and of potassium ions is higher on the inside.
   e. sodium equals the concentration of potassium inside the cell.

   Textbook Reference: *45.2 How Do Neurons Generate and Transmit Electrical Signals? pp. 951–952*

6. Glia are specialized to do all of the following *except*
   a. receive neural impulses.
   b. insulate axons.
   c. supply neurons with nutrients.
   d. help maintain a proper ionic environment for the neuron.
   e. guide neurons to make proper contacts during development.

   Textbook Reference: *45.1 What Cells Are Unique to the Nervous System? p. 947*

7. The cells that create the blood–brain barrier, keeping some toxic substances from entering the brain, are _______ and belong to a type of neural tissue called _______.
   a. endothelial cells; Schwann cells
   b. astrocytes; glia
   c. glial fibers; axons
   d. dendrites; synapses
   e. oligodendrocytes; glia

   Textbook Reference: *45.1 What Cells Are Unique to the Nervous System? p. 947*

8. A particular disease of the nervous system specifically involves the Ca^{2+} channels at the chemical synapses of motor neurons where neurotransmitter is stored and released. In other words, this disease affects the
   a. axon terminals of the presynaptic cell and the release of acetylcholine.
   b. axon terminals of the postsynaptic cell and the release of K^+.
   c. movement of Na^+ out of the postsynaptic cell.
   d. axon terminals of the presynaptic cell and the release of K^+.
   e. axon terminals of the postsynaptic cell and the release of Cl^-.

   Textbook Reference: *45.3 How Do Neurons Communicate with Other Cells? pp. 956–957*

9. Which of the following statements about electrical synapses is *false*?
   a. Connexons form molecular tunnels between two cells.
   b. Electrical synapses cannot be inhibitory.
   c. Electrical synapses do not allow for temporal summation.
   d. Their transmission capacity is very slow.
   e. They allow for transmission either toward or away from the cell.

   Textbook Reference: *45.3 How Do Neurons Communicate with Other Cells? p. 959*

10. Which of the following statements about neurotransmitter receptors is *false*?
    a. Ionotropic receptors are ion channels.
    b. The acetylcholine receptor of the motor end plate is a metabotropic receptor.
    c. Metabotropic receptors are not ion channels.
    d. Metabotropic receptors induce signaling cascades in the postsynaptic cell.
    e. Responses in the postsynaptic cell mediated by metabotropic receptors are usually slower than those mediated by ionotropic receptors.

    Textbook Reference: *45.3 How Do Neurons Communicate with Other Cells? p. 959*

11. The rapid depolarization of a neuron during the first half of an action potential is due to the
    a. exit of K^+ ions from the cell through gated potassium channels.
    b. rapid reversal of ion concentration caused by the action of the sodium–potassium pump.
    c. entry of Na^+ ions into the cell through gated sodium channels.
    d. movement of both Na^+ and K^+ ions through appropriate open channels.
    e. closing of sodium channels.

    Textbook Reference: *45.2 How Do Neurons Generate and Transmit Electrical Signals? pp. 953–954*

12. The refractory period of a neuron
    a. is the period when the sodium–potassium pump is nonfunctional.
    b. results from activation of voltage-gated chloride channels.
    c. results from closing of inactivated voltage-gated sodium channels.
    d. occurs when the action potential reaches the synapse.
    e. lasts about a minute.

    Textbook Reference: *45.2 How Do Neurons Generate and Transmit Electrical Signals? p. 954*

13. Which of the following statements about the process of summation in a neuron is *false*?

a. Slight perturbations of the membrane potential spread across the postsynaptic cell body.
b. Axons that terminate closer to the axon hillock have more influence on the summation process than those that do not.
c. It is essentially a comparison of all the excitatory and inhibitory postsynaptic inputs.
d. The concentration of voltage-gated sodium channels is highest in the dendrites of the postsynaptic cell.
e. Spatial summation adds up the simultaneous influences of synapses at different locations on the postsynaptic cell.

Textbook Reference: *45.3 How Do Neurons Communicate with Other Cells? pp. 958–959*

14. Which of the following statements about neurotransmitters is *false*?
a. Gases, such as nitric oxide, can act as neurotransmitters.
b. Each neurotransmitter has a single type of receptor.
c. Amino acids and their derivatives, monoamines, function as neurotransmitters.
d. Neurotransmitters have different effects in different tissues.
e. Some neurotransmitters are cleared from synapses by enzymes that destroy them.

Textbook Reference: *45.3 How Do Neurons Communicate with Other Cells? pp. 960–961*

15. The electrical events labeled as EPSPs are the result of _______ of the _______ membrane.
a. hyperpolarization; postsynaptic
b. depolarization; postsynaptic
c. hyperpolarization; presynaptic
d. depolarization; presynaptic
e. repolarization; presynaptic

Textbook Reference: *45.3 How Do Neurons Communicate with Other Cells? pp. 958*

Application Questions

1. In order to determine the role of the potassium channels in a neuron, a researcher has knocked out all the functional potassium channels and depolarized the membrane potential. What will happen to the membrane potential after depolarization?
Textbook Reference: *45.2 How Do Neurons Generate and Transmit Electrical Signals? pp. 948–954*

2. The active ingredients in many nerve gases belong to a class of chemicals called anticholinesterases (chemicals that block acetylcholinesterase). Suggest a possible synaptic mechanism to explain how these chemicals can damage an animal's nervous system.
Textbook Reference: *45.3 How Do Neurons Communicate with Other Cells? p. 961*

3. Clinical depression is thought to be due, in part, to insufficient levels of the neurotransmitter serotonin. Drugs known as selective serotonin reuptake inhibitors (SSRIs) can be used to treat depression. Taking the name of this class of drugs as a clue, propose a mechanism by which they might act.
Textbook Reference: *45.3 How Do Neurons Communicate with Other Cells? p. 961*

4. Explain how an action potential travels more quickly down an axon wrapped in myelin than it does down an unmyelinated axon.
Textbook Reference: *45.2 How Do Neurons Generate and Transmit Electrical Signals? pp. 955–956*

5. Why do certain substances, such as anesthetics and alcohol, have rapid effects on the brain, whereas others cannot reach the brain?
Textbook Reference: *45.1 What Cells Are Unique to the Nervous System? p. 947*

Answers

Diagram Exercise Answer

a. Dendrites
b. Cell body
c. Axon hillock
d. Axon
e. Axon terminals

The action potential travels down the axon, away from the axon hillock.

Knowledge and Synthesis Answers

1. **d.** The neuron is composed of a cell body, an axon, and dendrites. The dendrites form synapses with presynaptic cells to create the junction where information from one neuron is transferred to another neuron.
2. **d.** The glia that coat the axon of some neurons form myelin.
3. **c.** The neuron is composed of the cell body, the dendrite, and the axon. The axon carries action potentials away from the cell body to the synapses.
4. **d.** For an action potential to occur in an axon, the membrane must be depolarized above a certain level. This level is known as the threshold.
5. **d.** The resting potential of a neuron membrane occurs when the sodium ion concentration is higher on the outside and the potassium ion concentration is higher on the inside.
6. **a.** Glia perform many functions in the nervous system, but they do not receive neural impulses.
7. **b.** The blood–brain barrier is formed by astrocytes that wrap around the blood vessels traveling through the brain. Astrocytes are a special kind of glia.
8. **a.** If the disease acts on a chemical synapse where the neurotransmitter is stored and released, it is affecting the axon terminals of the presynaptic cell. Ca^{2+} channels are involved in regulating the release of acetylcholine by allowing Ca^{2+} to enter the presynaptic cell and promoting the fusing of acetylcholine-containing vesicles to the membrane.

9. **d.** Electrical synapses join two cells with protein tunnels known as connexons. These synapses provide for very fast transmission between cells.
10. **b.** The acetylcholine receptor of the motor end plate is an ionotropic receptor.
11. **c.** The first step in an action potential is the influx of Na^+ leading to a depolarization of the axon membrane. Na^+ rushes into the cell due to the higher concentration outside of the cell and the negative membrane potential.
12. **c.** After the spike of the depolarization, the sodium voltage-gated channels close. One of the properties of these channels is that they will open again only after a short delay. This short delay (about 1–2 milliseconds) is known as the refractory period when the sodium voltage-gated channels are inactive.
13. **d.** Dendrites, and most of the cell body, have few gated sodium channels. These channels mediate the action potentials that travel down the axon, where their levels are high.
14. **b.** Each neurotransmitter has multiple types of receptors.
15. **b.** Excitatory postsynaptic potentials (EPSPs) make it easier for an action potential to occur, so they depolarize the postsynaptic membrane.

Application Answers

1. The potassium voltage-gated channels are responsible for setting up the resting potential of a membrane. Potassium ions have a tendency to diffuse out of the cell, leaving a negative charge inside. Knocking out the function of the potassium voltage-gated channels would result in the cell's being unable to maintain resting potential. If the cell was depolarized by the opening of sodium voltage-gated channels, then it might not repolarize because the potassium channels that help repolarize the membrane would not be functioning.
2. Acetylcholine is the neurotransmitter used by all neuromuscular synapses in vertebrates. It transmits the action potential from a presynaptic cell to a postsynaptic cell. The enzyme acetylcholinesterase is found in the synaptic cleft, and it cleaves acetylcholine to help remove it from the synaptic cleft after an action potential. A nerve gas with components that block the action of acetylcholinesterase would cause acetylcholine to build up in the synaptic cleft. This buildup would mean that the receptors on the postsynaptic cell would remain bound with acetylcholine, resulting in prolonged muscle contraction.
3. Selective serotonin reuptake inhibitors such as Paxil, Zoloft, and Prozac increase the level of serotonin at the synapse by reducing its rate of removal.
4. The conduction of an action potential down a myelinated axon is called saltatory conduction. The myelin acts to insulate areas of the axon, preventing depolarization. The areas of the axon between the myelin sheaths are known as nodes of Ranvier. Depolarization can occur only at these nodes. As the action potential moves down a myelinated axon, the influx of sodium ions at one node diffuses down the axon. This results in the depolarization of the next node of Ranvier. Depolarization can occur only in the downstream nodes because the upstream nodes are in a refractory period. As a result, the action potential moves quickly down the axon to the synapse.
5. Astrocytes are glia that help form the blood–brain barrier by surrounding tiny, very permeable blood vessels in the brain. However, because the barrier is made of plasma membranes, lipid-soluble substances such as anesthetics and alcohol can pass through it.

46 Sensory Systems

The Big Picture

- Sensory structures work by converting some form of stimulus—mechanical, chemical, light—into action potentials in the nervous system, which are then interpreted by the central nervous system as a perceived sense.
- Receptors are named on the basis of their sensitivity. For example, chemoreceptors respond to chemical stimulation, mechanoreceptors respond to mechanical stimulation, and photoreceptors respond to light.
- Different animals have different sensitivities of senses, as well as different types of senses.

Common Problem Areas

- The action potentials produced in the neurons of the ear, eye, knee, or stomach are identical. The action potentials coming from the eye, for example, are interpreted as light because of the region of the brain that receives and analyzes them.
- Many of the receptors with complex structures have both neural and nonneural components. The nonneural components (e.g., the ear pinnae) help channel or otherwise alter or filter the stimulus that will arrive at the neural component of the receptor.

Study Strategies

- The route by which sound travels in the ear can be very confusing. View the cochlea in the uncoiled form as in Figure 46.9. This will help you visualize how pressure waves of different wavelengths produce different sounds.
- All of the senses have what appear to be very different mechanisms for the transmission of information to the brain. To help sort all of this out, remember that there are only a few types of receptors that respond to stimuli and that they all generate action potentials. The steps in sensory transduction are also similar in all the different sensory systems.
- Go to yourBioPortal.com to review the following tutorials and activities:

  Animated Tutorial 46.1 Sound Transduction in the Human Ear

  Animated Tutorial 46.2 Photosensitivity

  Interactive Tutorial: Sensory Receptors

  Web Activity 46.1 Structures of the Human Ear

  Web Activity 46.2 Structure of the Human Eye

  Web Activity 46.3 Structure of the Human Retina

Important Concepts

Sensory cells detect stimuli and transmit the information to the CNS.

- Sensory cells are modified neurons specialized to detect internal and external stimuli, and to transmit information concerning these stimuli as action potentials to different sites in the central nervous system (CNS).
- Sensory cells are involved in maintaining homeostasis by transmitting information about the status of the internal environment to the CNS. We may not be conscious of this information.
- Sensory transduction is the process by which a stimulus (e.g., mechanical, thermal, or chemical) to a sensory cell is transformed ("transduced") into an action potential. Sensory transduction can occur by direct activation of receptor proteins that open or close ion channels to change membrane potential. In this situation, the sensory receptor protein is ionotropic. Alternatively, the stimulus can activate a second messenger within the cell that couples with a G protein to eventually open ion channels. In this situation, the sensory receptor protein is metabotropic.
- A receptor potential is a change in the resting membrane potential of a sensory receptor cell in response to a stimulus. Receptor potentials produce action potentials either by causing the release of a neurotransmitter that induces an associated neuron to generate action potentials, or by generating action potentials within the sensory cell itself.
- Even though all sensory systems process information as action potentials, we perceive different sensations

(e.g., pain, light, sound) because messages from the different sensory systems go to different areas of the CNS. The frequency of action potentials encodes the intensity of sensation.

- Sensory organs are groups of sensory cells that, along with other cells, collect, filter, and amplify stimuli. Eyes, ears, and noses are examples of sensory organs. Sensory systems include the sensory cells, the associated structures, and the networks of neurons that process the information.
- Many sensory cells have diminished responses to a stimulus over time in a process known as adaptation. Adaptation allows organisms to ignore background conditions and focus on new information.

Chemoreceptors produce action potentials in response to a chemical stimulus.

- Chemical stimuli in the external and internal environments stimulate chemoreceptors. Chemoreceptors are responsible for smell and taste, and for monitoring levels of particular chemicals (e.g., carbon dioxide) inside the body.
- Pheromones are chemicals used in within-species communication. Insects use pheromones to attract mates by remotely stimulating their target's chemoreceptors (see Figure 46.3). The concentration of the pheromone released by a female creates a gradient that provides information about her specific location.
- Olfaction is the sense of smell. The olfactory sensors of vertebrates are neurons with axons extending to the olfactory bulb of the brain; the dendrites of these neurons are exposed as hairs to the environment within the epithelium of the nasal cavity (see Figure 46.4). Olfactory receptor proteins are found on the hairs, and each receptor binds with specific odorants. Binding of the odorant generates action potentials, which are transmitted to glomeruli in the olfactory bulb. The ability to discriminate many different odors is due to the large number of specific receptors. Binding of an odorant results in depolarization of the cell through a G protein that activates a second messenger, which then opens sodium channels. The strength of a smell is related to the number of odorant molecules that bind to receptors.
- Amphibians, reptiles, and some mammals have a vomeronasal organ, a paired structure located in the nasal epithelium. In mammals, the vomeronasal organ senses pheromones and conveys information to the accessory olfactory bulb in the brain. In snakes, the forked tongue presents odorant molecules from the environment to the chemoreceptors of the vomeronasal organ on the roof of the mouth; thus, in snakes, the tongue is used in smell and not in taste.
- Gustation, the sense of taste, relies on clusters of chemoreceptor cells called taste buds (see Figure 46.5). Binding of the stimulus to receptor proteins on the microvilli of sensory cells causes a change in membrane potential and the release of neurotransmitters that stimulate sensory neurons at the base of the taste bud.
- Humans can perceive five general tastes: sweet, sour, salty, bitter, and umami (meaty).

Mechanoreceptors detect mechanical force.

- Mechanical force causes distortion of the membranes of mechanoreceptors, causing ion channels to open and an action potential to be generated.
- The skin has several different types of mechanoreceptors (see Figure 46.6). Meissner's corpuscles are very sensitive but adapt rapidly; they detect light touches to the skin. Merkel's discs adapt slowly and provide information about objects touching the skin. Pacinian corpuscles and Ruffini endings are deeper in the skin and respond to vibrations. The density of tactile mechanoreceptor cells varies with the region of the body, and can be assessed using the two-point spatial discrimination test.
- Muscle spindles are mechanoreceptors (specifically, stretch receptors) in skeletal muscles that perceive muscle stretch. Golgi tendon organs are mechanoreceptors found in the tendons and ligaments that provide information about forces generated during muscle contraction. Collectively, these mechanoreceptors provide information on limb position, as well as stresses and strains on muscles and joints.

The auditory system contains mechanoreceptors specialized for sound reception.

- The auditory system takes in sound as pressure waves and transforms them into action potentials.
- The pinna is the outer portion of the mammalian ear that collects sound waves and directs them into the auditory canal. At the end of the auditory canal is the tympanic membrane, which vibrates and transmits sound waves to tiny bones (ossicles) in the middle ear. The ossicles are the malleus (hammer), incus (anvil), and stapes (stirrup), and together they transmit sound waves to the membrane called the oval window.
- Sound travels through the oval window into the fluid-filled cochlea (in the inner ear), where pressure waves are turned into action potentials. Movement of the oval window generates pressure waves in the cochlear fluid. The cochlea is a three-canal chamber with two membranes: Reissner's membrane and the basilar membrane. The pressure waves in the cochlear fluid cause the basilar membrane to vibrate. The organ of Corti is supported on the basilar membrane, and it contains hair cells with stereocilia that are in contact with the overhanging tectorial membrane. When the basilar membrane vibrates, the hair cell stereocilia of the organ of Corti are pushed against the tectorial membrane. Movements of the stereocilia are transduced into action potentials that are carried to the brain by the auditory nerve (see Figure 46.8).

- The round window functions to relieve pressure created by movements of the oval window.
- Different pitches of sound cause the basilar membrane to flex at different locations, stimulating different hair cells. The brain interprets input from hair cells in different areas as sounds of different pitch.

Displacement is detected by hair cells.

- Hair cells are mechanoreceptors that have stereocilia projecting from their surface. Bending the stereocilia causes changes in the ion channels of the hair cell plasma membrane (see Figure 46.10). Bending in one direction opens the ion channels, causing depolarization and the release of neurotransmitters. Bending in the other direction closes ion channels.
- In the inner ear of mammals, hair cells are present in the organs of hearing (organ of Corti) and equilibrium (vestibular system). The vestibular system consists of three semicircular canals and two chambers called the saccule and utricle; the entire system is filled with the fluid endolymph.
- Changes in position of the head cause shifts in the fluid within semicircular canals, which pushes on the gelatinous cupulae of hair cells, causing their stereocilia to bend (see Figure 46.11A). In the saccule and utricle, the stereocilia are bent by gravitational forces on otoliths (see Figure 46.11B). Otoliths are granules of calcium carbonate that sit on top of the gelatinous mass overlying the hair cells.
- The lateral line system of fishes is composed of hair cells that line canals located just under the surface of the skin. This system detects displacement of water around the fish (see Figure 46.12).

Light-sensitive pigments are used in photoreceptors.

- Rhodopsin is a family of pigments made up of the protein opsin and the light-absorbing nonprotein group 11-*cis*-retinal (see Figure 46.13). The 11-*cis*-retinal absorbs photons of light and changes conformation to all-*trans*-retinal, causing opsin to change conformation and become photoexcited rhodopsin.
- Photoexcited rhodopsin triggers a G protein cascade, which ultimately leads to changes in membrane potential and the photoreceptor's response to light.
- In vertebrate eyes, rod cells are photoreceptor cells that contain an inner segment, a synaptic terminal, and an outer segment with many rhodopsin molecules (see Figure 46.14). Rods are found in the retina, along with a layer that transduces visual information into action potentials.
- Rod cells become hyperpolarized in response to light and respond by decreasing the levels of neurotransmitter released (see Figure 46.15).

Invertebrates display a variety of visual systems.

- Flatworms have photoreceptor cells organized into eye cups, which are used to orient the animal away from light sources.
- Arthropods have compound eyes with hundreds or tens of thousands of ommatidia (optical units) per eye. The ommatidia contain light-sensitive photoreceptors called retinula cells (see Figure 46.16). The inner border of the retinula cells are covered by microvilli that contain rhodopsin. The compound eye communicates a low-resolution image to the CNS.
- Cephalopod mollusks (and vertebrates) have eyes that form detailed images.

The image-forming eyes of vertebrates have several components.

- Cephalopod mollusks and vertebrates independently evolved image-forming eyes with very similar structures (see Figure 46.17).
- The vertebrate eye is surrounded by the sclera, formed from connective tissue. The cornea is the transparent sclera through which light passes.
- The iris controls the amount of light entering the eye through the pupil; it also gives the eye its color.
- Mammals and birds focus on near and far objects by changing the shape of the lens (see Figure 46.18). Fishes, amphibians, and reptiles focus by moving their lenses closer to or farther from their retinas.

The vertebrate eye focuses light onto photoreceptors.

- The center of the retina is the fovea, an area with the highest density of photoreceptor cells.
- The human retina contains both rods (which are more light-sensitive) and cones (which absorb light of various wavelengths, allowing for color vision). Cones have different opsin molecules that absorb blue, green, yellow, or red (see Figure 46.20).
- There are five layers of neurons in the retina, with the photoreceptive rods and cones in the last layer, farthest from the lens (see Figure 46.21).
- The first layer of cells consists of ganglion cells (which create the action potential), the axons of which form the optic nerve. Bipolar cells are stimulated by neurotransmitters from the photoreceptors to transmit the signal from the photoreceptor to the ganglion cells by the release of a neurotransmitter. Thus, information flow in the retina is from photoreceptor cells at the back to bipolar cells to ganglion cells, which send the information to the brain.
- The two other layers in the retina are the horizontal cells (which connect adjoining groups of photoreceptors and bipolar cells) and amacrine cells (which connect adjoining groups of bipolar cells and ganglion cells). Horizontal and amacrine cells are interneurons responsible for lateral communication across the retina.

Test Yourself

Diagram Exercises

1. In the diagram of the human ear below, label each of the following structures: tympanic membrane, malleus,

incus, stapes, oval window, round window, cochlea, semicircular canal of the vestibular system, and auditory nerve.

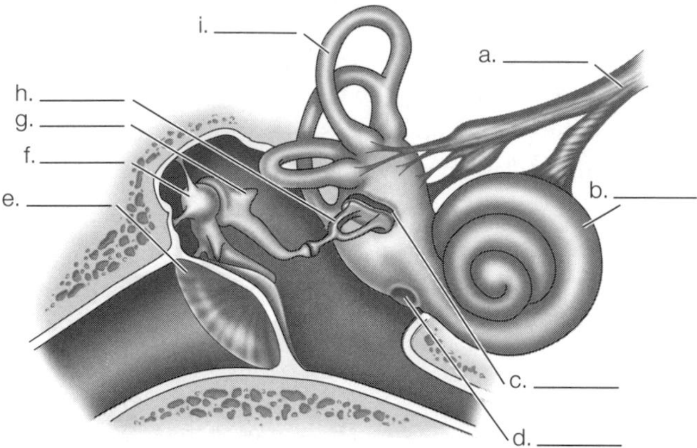

Textbook Reference: *46.3 How Do Sensory Systems Detect Mechanical Forces? p. 972*

2. In the diagram of the vertebrate eye below, label each of the following structures: sclera, cornea, iris, pupil, lens, retina, fovea, vitreous humor, and optic nerve.

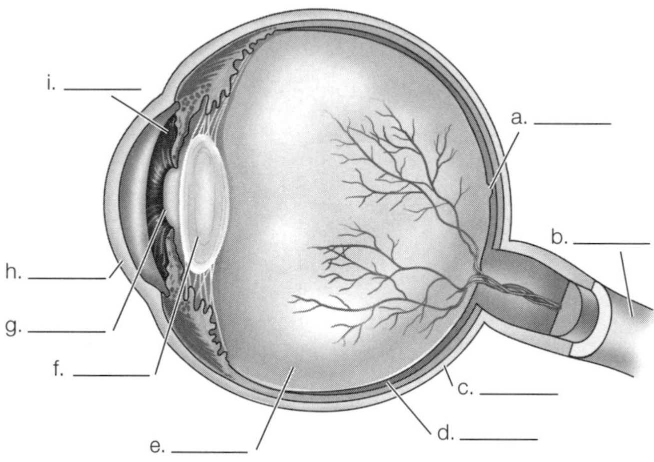

Textbook Reference: *46.4 How Do Sensory Systems Detect Light? p. 979*

Knowledge and Synthesis Questions

1. An electrode is inserted into a chemosensory nerve leading away from a taste bud in the mouth of a dog. A mild acid solution is then flushed continuously over the sensors associated with this nerve. Initially, the nerve responds to this stimulation, but over time it ceases to carry action potentials. Which of the following processes would best explain this observation?
   a. Translocation
   b. Adaptation of the sensory cells
   c. Depletion of neurotransmitter in the sensory nerve
   d. Second messenger influences that increase cell membrane potentials
   e. Action potentials arriving at the wrong area in the CNS.

   Textbook Reference: *46.1 How Do Sensory Cells Convert Stimuli into Action Potentials? p. 967*

2. Which of the following statements about sensory cells is *false*?
   a. Mechanoreceptors detect stimuli that distort membranes.
   b. Chemoreceptors monitor aspects of the internal environment.
   c. Chemoreceptors are involved in smell, taste, and hearing.
   d. Photoreceptors exhibit a conformational change when stimulated by light.
   e. Most chemoreceptors are metabotropic receptors.

   Textbook Reference: *46.1 How Do Sensory Cells Convert Stimuli into Action Potentials? pp. 965–966; 46.2 How Do Sensory Systems Detect Chemical Stimuli? pp. 967–968*

3. Silkworm moths use chemosensory signals known as _______ for mate attraction.
   a. rhodopsin
   b. hormones
   c. pheromones
   d. G proteins
   e. locally acting chemical messengers

   Textbook Reference: *46.2 How Do Sensory Systems Detect Chemical Stimuli? p. 968*

4. Stretch receptors in the aorta and carotid artery sense changes in arterial pressure. These receptors are therefore considered
   a. chemoreceptors.
   b. thermoreceptors.
   c. electroreceptors.
   d. mechanoreceptors.
   e. muscle spindles.

   Textbook Reference: *46.3 How Do Sensory Systems Detect Mechanical Forces? p. 970*

5. Which of the following does not employ hair cells as its transducer?
   a. Meissner's corpuscle
   b. Lateral line
   c. Organ of Corti
   d. Semicircular canal
   e. Saccule

   Textbook Reference: *46.3 How Do Sensory Systems Detect Mechanical Forces? pp. 970–976*

6. Which of the following statements about human gustation is *false*?
   a. Taste bud cells are relatively short lived because of the high degree of abrasion they encounter.
   b. Taste buds are confined to the oral cavity.
   c. Changes in the membrane potential of the taste bud sensory cells cause them to release neurotransmitter onto the dendrites of sensory neurons.
   d. Humans perceive only three categories of tastes: sweet, sour, and bitter.
   e. Most taste buds are found on the papillae of the tongue.

Textbook Reference: *46.2 How Do Sensory Systems Detect Chemical Stimuli? pp. 969–970*

7. Which of the following statements about the photosensitive molecule rhodopsin is *false*?
   a. Opsin is converted from the 11-*cis* to the all-*trans* form upon absorbing a photon of light.
   b. The retinal is the light-absorbing group.
   c. Photoexcited rhodopsin triggers a cascade of reactions that ultimately alters the membrane potential of a photoreceptor cell.
   d. Opsin is a protein; retinal is not a protein.
   e. 11-*cis*-retinal is covalently bonded to opsin.

   Textbook Reference: *46.4 How Do Sensory Systems Detect Light? p. 976*
8. In the human visual system, _______ send information directly to the brain.
   a. amacrine cells
   b. bipolar cells
   c. ganglion cells
   d. rods and cones
   e. horizontal cells

   Textbook Reference: *46.4 How Do Sensory Systems Detect Light? pp. 980–981*
9. Through which of the following cell layers must a photon of light pass before striking a cone cell in the eye of a human?
   a. Amacrine
   b. Bipolar
   c. Ganglion
   d. Horizontal
   e. All of the above

   Textbook Reference: *46.4 How Do Sensory Systems Detect Light? pp. 980–981*
10. Which of the following animals changes the shape of its lens to focus?
    a. Fishes
    b. Reptiles
    c. Amphibians
    d. Mammals
    e. All of the above

    Textbook Reference: *46.4 How Do Sensory Systems Detect Light? pp. 979–980*
11. Which of the following statements about sensory receptor proteins is *false*?
    a. Ionotropic receptor proteins are either ion channels themselves or they directly influence the opening of ion channels.
    b. Photoreceptors are ionotropic.
    c. Mechanoreceptors are ionotropic.
    d. Thermoreceptors are ionotropic.
    e. Metabotropic receptors influence ion channels indirectly through second messengers.

    Textbook Reference: *46.1 How Do Sensory Cells Convert Stimuli into Action Potentials? pp. 965–966*
12. Which of the following structures is *not* found in the inner ear?
    a. Reissner's membrane
    b. Tectorial membrane
    c. Tympanic membrane
    d. Basilar membrane
    e. Semicircular canal

    Textbook Reference: *46.3 How Do Sensory Systems Detect Mechanical Forces? pp. 972–973*
13. Which of the following statements about receptor potentials is *false*?
    a. They are changes in the resting membrane potential of a sensory cell in response to a stimulus.
    b. The receptor potential spreads from the cell body of a sensory cell to the axon hillock, where action potentials are generated.
    c. They must be converted to action potentials to travel long distances.
    d. A receptor potential always prompts the release of a neurotransmitter that induces an associated neuron to generate an action potential.
    e. They are graded membrane potentials.

    Textbook Reference: *46.1 How Do Sensory Cells Convert Stimuli into Action Potentials? pp. 965–966*
14. Which of the following statements about the detection of chemical stimuli is *false*?
    a. Snakes use their tongues to smell.
    b. Many mammals have a vomeronasal organ to detect pheromones.
    c. A greater frequency of action potentials is associated with perception of a more intense smell.
    d. Taste buds are confined to the oral cavity in aquatic animals.
    e. Chemoreceptors monitor aspects of the internal environment.

    Textbook Reference: *46.2 How Do Sensory Systems Detect Chemical Stimuli? pp. 967–970*
15. Which of the following statements about sensory systems is *false*?
    a. Rattlesnakes have pit organs to detect infrared wavelengths.
    b. Cephalopod mollusks and vertebrates independently evolved image-forming eyes.
    c. Nocturnal animals have a high percentage of cones in their retinas, whereas diurnal animals have a high percentage of rods.
    d. Fish detect water movements with their lateral lines.
    e. Bats have small muscles in their ears that contract to dampen sounds when they are emitting calls.

    Textbook Reference: *46.4 How Do Sensory Systems Detect Light? p. 981*

Application Questions

1. You had a cold for a number of weeks, during which time your sense of smell was diminished. What was the cause of your loss of smell?

   Textbook Reference: *46.2 How Do Sensory Systems Detect Chemical Stimuli? pp. 968–969*

2. After a loud rock concert your hearing appears to be dampened. What portion of your ear been altered during the concert, resulting in a dampening of your hearing?
   Textbook Reference: *46.3 How Do Sensory Systems Detect Mechanical Forces? p. 974*
3. You have just given a presentation in your biology class that had many elaborate red- and green-colored slides. Afterward, a male friend tells you that he could not see any of the differences you were reporting. Why could your friend not see the differences?
   Textbook Reference: *46.4 How Do Sensory Systems Detect Light? p. 981*
4. Gray squirrels are diurnal (active during the day) and southern flying squirrels are nocturnal (active at night). How would would you expect their retinas to differ?
   Textbook Reference: *46.4 How Do Sensory Systems Detect Light? p. 981*
5. An infection that causes vertigo (dizziness) would be located in which sensory organ and in which particular part of the organ?
   Textbook Reference: *46.3 How Do Sensory Systems Detect Mechanical Forces? p. 975*

Answers

Diagram Exercise Answers

1. a. Auditory nerve
   b. Cochlea
   c. Oval window (under stapes)
   d. Round window
   e. Tympanic membrane
   f. Malleus
   g. Incus
   h. Stapes
   i. Semicircular canal of the vestibular system

2. a. Fovea
   b. Optic nerve
   c. Sclera
   d. Retina
   e. Vitreous humor
   f. Lens
   g. Pupil
   h. Cornea
   i. Iris

Knowledge and Synthesis Answers

1. **b.** When a sensor cell is stimulated by an unchanging, steady-state stimulus, it will adapt to that stimulus. This allows the sensory system to ignore the unchanging stimulus while still being able to respond to new information.
2. **c.** Chemoreceptors are involved in smell and taste but not in hearing.
3. **c.** Pheromones are chemical signals used in communication within a species. The female silkworm moth releases a pheromone (bombykol) into the environment. The male uses chemoreceptors to follow the pheromone to the source.
4. **d.** The stretch receptors of the aorta, which detect changes in blood pressure, are examples of mechanoreceptors.
5. **a.** Meissner's corpuscle of the skin does not use hair cells to sense a stimulus. The cell membranes of the Meissner's corpuscle deform in response to light touching of the skin.
6. **d.** Humans can perceive five tastes: sweet, salty, sour, bitter, and umami (a meaty taste). The combination of taste and smell provides the complex subtle flavors of the food we eat.
7. **a.** Rhodopsin contains two groups: the protein opsin and the light-sensitive group retinal. Retinal, not opsin, is converted from the 11-*cis* to the all-*trans* form upon absorbing a photon of light. Opsin does change conformation in response to a change in the rhodopsin to signal the detection of light.
8. **c.** The ganglion cells transmit information from the bipolar cells to the brain. The axons of the ganglion cells connect with the optic nerve.
9. **e.** The photoreceptive cells are located at the back of the retina. Light must pass through a layer of ganglion cells, a layer of amacrine and bipolar cells, and a horizontal cell layer.
10. **d.** Mammals (and birds) change the shape of their lens to focus. Fishes, amphibians, and reptiles move their lens closer to or farther from their retinas to focus.
11. **b.** Photoreceptors are metabotropic because they influence ion channels indirectly, through G proteins and second messengers.
12. **c.** Although the tympanic membrane is found in the human ear, it is not found in the inner ear. The tympanic membrane is the membrane that transmits sounds from the auditory canal to the middle ear.
13. **d.** The receptor potential does not always prompt the release of a neurotransmitter to induce an associated neuron to generate an action potential. Sometimes the receptor potential generates action potentials within the sensory cell itself.
14. **d.** Taste buds are confined to the oral cavity in terrestrial animals. However, some aquatic animals, such as fish, have taste buds in their skin.
15. **c.** Nocturnal animals have a high percentage of rods in their retinas, and diurnal animals have a high percentage of cones.

Application Answers

1. Your sense of smell depends on olfactory cilia that line the surface of the nasal epithelium. The cilia's receptors bind with odorant molecules, triggering an action

potential that is sent to the olfactory bulb of the brain. Usually this epithelium is covered with a thin layer of protective mucus. However, when you have a cold, the production of mucus increases and mucus covers the epithelium and the olfactory cilia, making it more difficult for odorant molecules to reach the cilia. Thus your sense of smell is decreased.

2. Sounds that are too loud will eventually damage the hair cells of the organ of Corti, resulting in nerve deafness. This damage to the hair cells is cumulative and permanent.
3. Your friend has red-green color blindness. The cones in our eyes allow us to see color. We have cones for red, green, and blue. Your friend either has red and green cones that do not function properly, or lacks them altogether.
4. Cones are responsible for color vision. Rods are responsible for highly sensitive black-and-white vision. We would expect diurnal species (such as the gray squirrel) to have mostly cones in their retinas, and nocturnal species (such as the flying squirrel) to have mostly rods.
5. Vertigo is caused by infection in the ear, specifically the inner ear, which contains the organs of equilibrium (in addition to the cochlea).

47 The Mammalian Nervous System: Structure and Higher Function

The Big Picture

- The mammalian nervous system is divided both anatomically and functionally. The anatomical divisions are the central and peripheral nervous systems. The functional divisions are the sympathetic and parasympathetic nervous systems. Afferent nerves carry information toward the central nervous system from the peripheral nervous system, and efferent nerves carry information from the central nervous system to the peripheral nervous system.
- The brain can be divided anatomically and functionally into many different regions, each responsible for its many vital actions. The "higher" brain centers of the cerebrum are responsible for conscious thought and deliberate (voluntary) movements. The "lower" brain centers such as the cerebellum, pons, and medulla regulate involuntary movements and are involved primarily in maintaining homeostasis throughout the body.

Common Problem Areas

- It is easy to confuse"afferent"and "efferent," and it is important in your understanding of the nervous system to differentiate them. Think of *a*fferent as *a*rriving, and *e*fferent as *e*xiting the reference point—usually the CNS.
- It is important to recognize that although the neurotransmitters acetylcholine and norepinephrine always have antagonistic effects on each other, there is no universal pattern as to which stimulates tissue and which inhibits tissue. For example, acetylcholine causes the smooth muscle in blood vessels in many regions of the body to relax, but it causes the smooth muscle in the stomach and intestines to contract.
- It is often difficult for students to appreciate that, when dealing with sensory input to the brain, specific regions of the body "map onto" specific regions of the cerebrum. Similarly, with regard to control of movements, specific regions of the brain "map onto" specific regions of the body. Moreover, the amount of brain matter devoted to a particular body region depends on the amount of muscle control and sensors contained in that body region. Thus, a relatively small area of the cerebral hemispheres is devoted to the upper leg (which has a relatively limited range of movement and sensation), whereas the tongue, with its many sensory receptors and high degree of mobility, commands more of the tissue of the cerebrum.

Study Strategies

- Several key terms in this chapter occur repeatedly: parasympathetic, sympathetic, afferent, efferent, agonist, antagonist, preganglionic, and postganglionic. Until you master this vocabulary, it will be difficult to put together a comprehensive picture of the nervous system. Create your own list of the terms you see repeatedly, and make sure that you understand their meanings.
- Brain anatomy is complex. Start by dividing up the structures into forebrain, midbrain, and hindbrain regions. Then identify each of the subcomponents. Your understanding will be more complete if you learn the general functions of each brain section as you go along. That is, rather than learning the anatomy of the brain and then starting over to learn the functions, learn the two at the same time.
- Go to yourBioPortal.com to review the following tutorials and activities:

  Animated Tutorial 47.1 Information Processing in the Spinal Cord

  Animated Tutorial 47.2 Information Processing in the Retina

  Interactive Tutorial: Visual Receptive Fields

  Web Activity 47.1 The Human Cerebrum

  Web Activity 47.2 Language Areas of the Cortex

  Web Activity 47.3 Structures of the Human Brain

Important Concepts

The mammalian nervous system consists of the central nervous system and the peripheral nervous system.

- The brain and the spinal cord together constitute the central nervous system (CNS). The part of the nervous system outside the brain and spinal cord is the peripheral nervous system (PNS); it includes cranial and spinal nerves.
- The peripheral nervous system is composed of afferent nerves that carry information to the CNS and efferent nerves that carry information from the CNS to muscles and glands (see Figure 47.1). We are conscious of some information carried by afferent pathways (e.g., sounds, light), but not all of it (e.g., blood pressure). Efferent pathways can be classified as voluntary (executing our conscious movements) or involuntary (autonomic, controlling physiological functions).
- A nerve is a bundle of axons that carries information in both directions between the organs of the body and the CNS.
- The CNS also receives chemical information from circulating hormones, and releases neurohormones.

The CNS develops from the embryonic neural tube.

- The CNS develops from a tube of neural tissue running down the length of a vertebrate embryo in its early developmental stages (see Figure 47.2). The anterior end of the neural tube develops into the main parts of the brain—the hindbrain, the midbrain, and the forebrain—and the spinal cord develops from the rest of the neural tube.
- The hindbrain is made up of the medulla and pons, which control physiological functions, and the cerebellum, which coordinates commands leaving the brain with the actions of the muscles.
- The midbrain receives and processes auditory and visual information.
- The forebrain consists of a central diencephalon region and a surrounding telencephalon region. The diencephalon contains the thalamus, a relay station for sensory information, and the hypothalamus, which maintains homeostasis. The cerebral hemispheres make up the telencephalon (cerebrum).

The spinal cord is the main neural highway between the body and the brain.

- The spinal cord is a bidirectional neural pathway for information flow between the peripheral nervous system and the brain. Each spinal nerve has two roots: afferent (sensory) axons enter the spinal cord via the dorsal root, and efferent (motor) axons leave via the ventral root.
- Cell bodies are found in gray matter and axons are found in white matter of the nervous system. The white color is due to myelin.
- The spinal cord converts some afferent information from the peripheral nervous system into efferent information sent back to the peripheral nervous system in a process known as a spinal reflex (see Figure 47.3).
- A monosynaptic reflex, such as the knee-jerk reflex, involves only an afferent neuron, an efferent neuron, and one synapse. Leg muscle stretch receptors trigger the sensory neuron to conduct action potentials to the spinal cord upon stretching. The action potential is passed through the synapse of the sensory and motor neuron located in the central gray matter of the spinal cord. More complicated reflexes involve interneurons and additional synapses.
- Flexor muscles, which flex limbs, and extensor muscles, which straighten limbs, control limb movement. The motor neurons that stimulate these muscles are antagonistic and polysynaptic, with each sensory neuron stimulating one motor neuron while inhibiting the other motor neuron at two distinct synapses.

The reticular system and limbic system interact with the forebrain.

- A nucleus is an anatomically distinct group of neurons in the CNS.
- The nuclei of the reticular system are located in the brainstem. Many of the connections involved in functional control of the body occur in the reticular system, along with the control of sleep and waking.
- The limbic system is involved in instincts and emotions (see Figure 47.4). A portion of the limbic system, known as the hippocampus, helps transfer short-term memory to long-term memory. The amygdala, another part of the limbic system, is involved in fear and fear memory.

The cerebrum is the largest portion of the mammalian brain.

- The cerebral hemispheres are covered by a convoluted cerebral cortex, a sheet of gray matter that processes sensory information and higher-order information in the association areas (see Figure 47.5). The convolutions are ridges known as gyri and the valleys are sulci. Underneath the gray matter is white matter containing axons that connect cell bodies in the cortex and other areas of the brain. Each cerebral hemisphere is broken down into four main regions: the temporal lobe, the frontal lobe, the occipital lobe, and the parietal lobe. Each region has association areas that integrate information.
- The temporal lobe processes auditory information, and the association areas are involved with recognition, identification, and the naming of objects. Damage to the temporal lobe results in conditions (agnosias) in which individuals are unable to identify a stimulus (e.g., a face), even though they can perceive it. Integration of spoken language can also be diminished with damage to the temporal lobe.

- The frontal lobe contains the primary motor cortex made up of axons that project to muscles in the body. The primary motor cortex can be mapped according to the locations that control movements of various body parts (see Figure 47.7A). The association areas of the frontal lobe are involved in planning and personality.
- The central sulcus separates the parietal lobe from the frontal lobe. The primary somatosensory cortex is located in the parietal lobe. As with the motor cortex, the location of the body that is sensed by the somatosensory cortex can be mapped on the brain (see Figure 47.7B). Contralateral neglect syndrome results from damage to the right parietal lobe and is characterized by the inability to detect stimuli from the left side of the body.
- Visual information is processed by the occipital lobe of the cerebrum. The association areas for the occipital lobe translate visual stimuli into language.

Several features distinguish the brain of humans from those of other vertebrates.

- Among vertebrates, humans (and porpoises) stand out as having larger brains than would be predicted by their body sizes.
- Degree of convolution of the cerebral cortex (a measure of the area of cortex) is greatest in humans, as is the percentage of cortex that is association cortex (a measure of the area of cortex devoted to the integration of information).

The sympathetic and parasympathetic systems make up the autonomic nervous system.

- The sympathetic and parasympathetic divisions of the autonomic nervous system have antagonistic effects on the organ systems they innervate and play a large role in the maintenance of homeostasis.
- The sympathetic division is involved in the fight-or-flight response of increased heart rate, blood pressure, and cardiac output. The parasympathetic division slows down heart rate and decreases blood pressure and cardiac output; however, it accelerates digestive activities.
- Preganglionic neurons of the autonomic efferent pathway have cell bodies in the brain stem or spinal cord. They use the neurotransmitter acetylcholine at synapses where the ganglion lies outside the CNS. The axon that runs from the ganglion to the target cells is known as the postganglionic neuron.
- Norepinephrine is the neurotransmitter in postganglionic neurons of the sympathetic system, whereas acetylcholine is the neurotransmitter in postganglionic neurons of the parasympathetic system.
- Both acetylcholine and norepinephrine influence the pacemaker cells in the heart that control heart rate.
- The parasympathetic system has preganglionic neurons that come from the brain stem and the last segment of the spinal cord (the sacral region). In the sympathetic system, preganglionic neurons come from the middle (lumbar) and upper (thoracic) regions of the spinal cord (see Figure 47.10).

Visual images are constructed in the occipital cortex.

- In the retina, one ganglion cell receives and integrates the information from many groups of photoreceptors that make up a circular receptive field with a center and a surround (see Figure 47.11). Light falling on the center of an on-center receptive field excites ganglion cells; light falling on the surround of an on-center receptive field inhibits ganglion cells. The opposite pattern characterizes off-center receptive fields. Receptive fields of ganglion cells can overlap.
- Information from the retina is transferred by the optic nerve to the thalamus and then to the visual cortex located in the occipital lobe.
- The visual cortex is composed of cells with specific receptive fields associated with areas of the retina that respond to specific light patterns. Neurons in the visual cortex (called simple cells) are stimulated by static bars of light with specific orientations (see Figure 47.12). Complex cells receive input from several simple cells that have receptive fields in different locations on the retina. Mental images of the world are determined by analyzing the edges of the patterns of light falling on the retina.
- The optic nerve from the two eyes joins at the optic chiasm (see Figure 47.13). At the optic chiasm, half of the axons from one eye cross over to the opposite side of the brain. Binocular vision occurs because the overlap in the field of view for each eye is transmitted to the same location in the visual cortex, the binocular cells. Binocular cells receive overlapping visual information from both eyes and interpret the disparity between the overlapping information from the two eyes to produce a three-dimensional image.

Sleep involves complex cerebral function.

- An electroencephalogram (EEG) measures the electrical activity of neurons in the cerebral cortex; at any one time, large numbers of neurons are monitored in the regions below the electrodes. An electromyogram (EMG) records electrical activity of muscles, and an electroocculogram (EOG) records eye movements.
- There are two main states of sleep in humans: rapid-eye movement (REM) sleep and non-REM sleep (see Figure 47.14).
- Non-REM sleep has four states progressing from stage 1 to the restorative stages 3 and 4. During non-REM sleep the neurons in the thalamus and cerebral cortex become hyperpolarized due to the opening of K^+ channels. During hyperpolarization, the cells fire off in synchronized bursts due to Ca^{2+} channels deactivating, which produces a slow-wave pattern.
- During REM sleep, dreams and nightmares occur along with near complete paralysis of skeletal muscles. Neurons that were hyperpolarized during non-REM sleep return to waking levels, allowing information to be

processed. Afferent and efferent pathways are inhibited during REM sleep.

- During a night of sleep, the brain cycles between REM sleep and non-REM sleep, with 80 percent of sleep being non-REM sleep.

The language center is located in the left cerebral hemisphere.

- In most people, the ability to produce and interpret language occurs in the left hemisphere of the cerebrum.
- The two cerebral hemispheres are connected by the corpus callosum. This connection allows the two hemispheres to communicate. Severing an individual's corpus callosum results in the person's inability to express in language the knowledge that is in the right hemisphere.
- Several areas have been located that are important for language (see Figure 47.15). The frontal lobe contains Broca's area, which is involved in motor aspects of speech. Wernicke's area is located in the temporal lobe and influences sensory aspects of language. The angular gyrus is thought to integrate spoken and written language.
- Language involves the flow of information among these areas of the left cerebral cortex. Damage to any of the areas can result in aphasia, the condition of not being able to use or understand written or spoken words.

Learning and memory involve long-lasting synaptic changes.

- Learning occurs when behavior is modified as a result of experience. Long-lasting synaptic changes must occur for learning to take place. Long-term potentiation (LTP) occurs when high-frequency electrical stimulation makes certain circuits more sensitive to subsequent stimulation.
- Long-term depression (LTD) occurs when certain circuits become less sensitive as a result of repetitive, low-level stimulation.
- Associative learning in animals involves linking two unrelated stimuli, as in the conditioned reflex of Pavlov's dogs. The dogs were conditioned to associate eating with the ringing of a bell, so that even in the absence of food the mere ringing of a bell stimulated salivation. Other experiments have shown that the neural circuitry for a conditioned reflex is located in the cerebellum.
- Memory is the phenomenon whereby the nervous system retains what has been experienced. Immediate memories are very short-term vivid memories of what has just happened. Short-term memories last between 10 and 15 minutes and contain less information than immediate memories. Long-term memory can last for days, months, years, or even a lifetime. Repetition or reinforcement enhances the transfer of short-term memory to long-term memory.
- Declarative memory is memory of people, places, events, and things. Procedural memory is memory of how to perform motor tasks, such as riding a bicycle.
- The insular cortex (insula) of the forebrain is expanded in humans and the great apes, and may be related to self-awareness and conscious experience.

Test Yourself

Diagram Exercise

Label the following structures in the diagram of the knee-jerk reflex below: stretch receptors, sensory neuron, motor neuron(s), interneuron, dorsal root, ventral root, gray matter, and white matter. Indicate the direction of information flow along the neurons. Which pathway is polysynaptic?

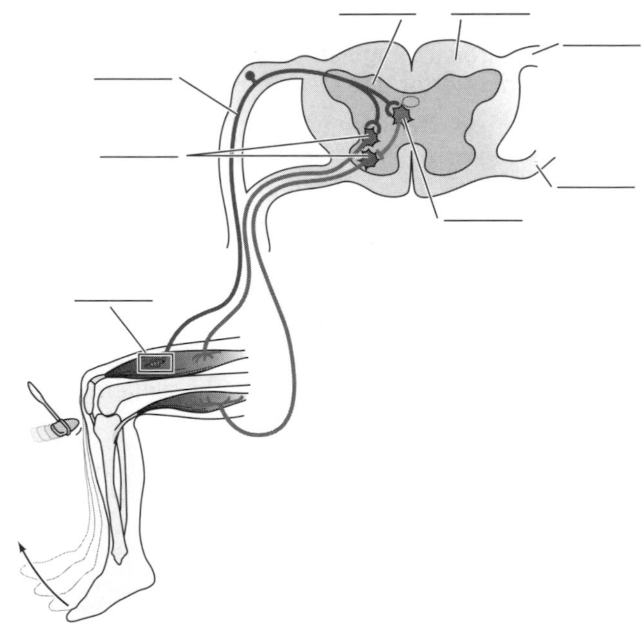

Textbook Reference: *47.1 How Is the Mammalian Nervous System Organized? p. 988*

Knowledge and Synthesis Questions

1. A man has damage to his brain that affects his ability to recognize the faces of people he knows. The damage must have occurred in the
   a. hypothalamus.
   b. temporal lobe.
   c. parietal lobe.
   d. frontal lobe.
   e. occipital lobe.
   Textbook Reference: *47.1 How Is the Mammalian Nervous System Organized? pp. 990–991*
2. The secretion of hormones from an endocrine gland is most directly under the control of which of the following components of the nervous system?
   a. Autonomic
   b. Voluntary

c. Afferent portion of the PNS carrying information of which we are conscious
d. Limbic
e. Afferent portion of the PNS carrying information of which we are unconscious

Textbook Reference: *47.2 How Is Information Processed by Neural Networks? pp. 993–994*

3. The primary motor cortex of the cerebrum
a. is mapped from the head region on the lower side of the cortex to the lower part of the body on the upper side of the cortex.
b. receives touch and pressure information from the body.
c. is located in the parietal lobe.
d. occurs behind the central sulcus.
e. is located in the temporal lobe.

Textbook Reference: *47.1 How Is the Mammalian Nervous System Organized? p. 991*

4. Which of the following statements about the sympathetic division of the autonomic nervous system is *false*?
a. It increases heart rate.
b. It relaxes the urinary bladder.
c. It stimulates digestion.
d. It increases blood pressure.
e. It relaxes airways.

Textbook Reference: *47.2 How Is Information Processed by Neural Networks? pp. 993–994*

5. Which of the following is *not* part of the central nervous system?
a. Brain stem
b. Spinal gray matter
c. Cerebellum
d. Neuronal cell body of a sensory afferent
e. Pons

Textbook Reference: *47.1 How Is the Mammalian Nervous System Organized? pp. 986–987*

6. Observations of people with aphasia indicate that
a. only Broca's area is essential for normal language skills.
b. language skills depend on proper flow of neural information between the temporal lobes and the motor cortex.
c. in humans the right hemisphere is dominant in the production and use of language.
d. Broca's area influences the sensory aspects of language and Wernicke's area influences the motor aspects.
e. lateralization does not pertain to language ability.

Textbook Reference: *47.3 Can Higher Functions Be Understood in Cellular Terms? p. 1000*

7. A friend wakes you from sleep, and you have the sensation of having just experienced a vivid dream. Which of the following statements about the state of sleep from which you were awakened would be *false*?
a. Your hands and feet were twitching slightly.
b. Your eyes were twitching.
c. Most of your voluntary body muscles were inactive.
d. Your cerebral cortex was not as active as it is when you are awake.
e. You were in the state of sleep that accounts for about 20 percent of your sleep.

Textbook Reference: *47.3 Can Higher Functions Be Understood in Cellular Terms? p. 1000*

8. To locate a complex cell in a cat's visual system one would look in
a. Broca's area.
b. the spinal cord.
c. the occipital cortex.
d. the reticular system.
e. the amygdala.

Textbook Reference: *47.1 How Is the Mammalian Nervous System Organized? p. 992*

9. Which of the following statements about the peripheral nervous system is *false*?
a. Afferent portions carry information to the CNS.
b. Efferent portions carry information from the CNS.
c. It communicates only with the circulatory and digestive systems.
d. Efferent portions contain voluntary and involuntary divisions.
e. It consists of cranial and spinal nerves.

Textbook Reference: *47.1 How Is the Mammalian Nervous System Organized? pp. 986–987*

10. Which of the following statements about the developing CNS is *false*?
a. The CNS develops from a solid neural cylinder.
b. The midbrain becomes part of the brain stem.
c. The forebrain develops into both the diencephalon and the telencephalon.
d. The hindbrain develops into the medulla, pons, and cerebellum.
e. In humans, the telencephalon develops into the largest part of the brain.

Textbook Reference: *47.1 How Is the Mammalian Nervous System Organized? pp. 987–988*

11. In which of the following cortical lobes can association areas be found?
a. Temporal
b. Parietal
c. Occipital
d. Frontal
e. All of the above

Textbook Reference: *47.1 How Is the Mammalian Nervous System Organized? p. 990*

12. Which of the following groups of vertebrates would likely have the smallest ratio of telencephalon size to body size?
a. Fishes
b. Amphibians
c. Mammals
d. Reptiles
e. Birds

Textbook Reference: *47.1 How Is the Mammalian Nervous System Organized? p. 988*

13. Which of the following statements about the knee-jerk reflex is *false*?
 a. It is a monosynaptic reflex.
 b. It causes the leg extensor muscle to contract.
 c. Chemoreceptors sense a physician's hammer tap.
 d. The afferent nerve travels from the receptor to the spinal cord.
 e. The motor neuron leaves via a ventral root of the spinal cord.

 Textbook Reference: *47.1 How Is the Mammalian Nervous System Organized? p. 988*

14. When compared with brains of other vertebrates, the human brain
 a. is larger than body size might lead one to predict.
 b. has proportionately more of the cortex devoted to information integration.
 c. has proportionately more cerebral cortex.
 d. has more convolutions in the cortex.
 e. All of the above

 Textbook Reference: *47.1 How Is the Mammalian Nervous System Organized? p. 992*

15. Which of the following describes the processing of visual information by the retina?
 a. Convergence of information
 b. Telencephalization
 c. Long-term depression
 d. Long-term potentiation
 e. Divergence of information

 Textbook Reference: *47.2 How Is Information Processed by Neural Networks? p. 995*

16. The insular cortex
 a. is located in the hindbrain.
 b. is greatly expanded in fishes.
 c. is most active during times of mild emotion in humans.
 d. is unrelated to perception of self.
 e. is greatly expanded in humans and great apes.

 Textbook Reference: *47.3 Can Higher Functions Be Understood in Cellular Terms? pp. 1002–1003*

Application Questions

1. What are the differences between the brain functions involved in reading a written sentence out loud and those involved in repeating a sentence one has just heard?

 Textbook Reference: *47.3 Can Higher Functions Be Understood in Cellular Terms? pp. 1000–1001*

2. Describe the neurological basis of the phenomenon of sleepwalking. What type of sleep is a person who is sleepwalking in?

 Textbook Reference: *47.3 Can Higher Functions Be Understood in Cellular Terms? p. 1000*

3. Imagine that you have been eyeing candy in a dish and finally decide to unwrap a piece and eat it. As you begin to suck on the candy, your salivary glands begin to secrete saliva. What parts (divisions or branches) of the nervous system are involved in this sequence of events?

 Textbook Reference: *47.1 How Is the Mammalian Nervous System Organized? pp. 986–987; 47.2 How Is Information Processed by Neural Networks? pp. 993–994*

4. Humans have the ability to see things in three dimensions. What allows us to have binocular vision?

 Textbook Reference: *47.2 How Is Information Processed by Neural Networks? pp. 997–998*

5. During a boxing match, a sharp punch to the jaw of a boxer may cause loss of consciousness. Which area of the brain is likely to have been affected by such a knockout punch?

 Textbook Reference: *47.1 How Is the Mammalian Nervous System Organized? p. 989*

Answers

Diagram Exercise Answer

The pathway that contains the neuron is polysynaptic.

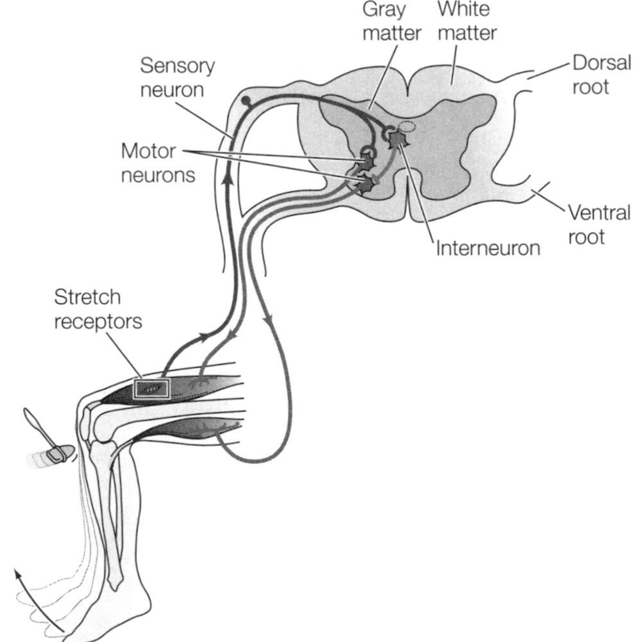

Knowledge and Synthesis Answers

1. **b.** The temporal lobe is involved in recognition of people and objects. A person who has had damage to the temporal lobe will not be able to identify someone by face and must use other cues.
2. **a.** Secretion of hormones by endocrine glands is involuntary. The autonomic nervous system consists of efferent pathways that link the CNS with many physiological functions.
3. **a.** The motor neurons in the head region control specific parts of the body. Parts of the body can be mapped on the primary motor cortex, from the head

region on the lower side to the lower part of the body at the top.

4. **c.** The sympathetic division of the autonomic nervous system inhibits digestion rather than stimulates it.
5. **d.** The cell bodies of the sensory neurons are located in the periphery and send their axons to the CNS.
6. **b.** Speaking a written or heard word requires neural flow from Wernicke's area in the temporal lobe to Broca's area in the frontal lobe. Therefore, language skills depend on proper flow of neural information between the temporal lobes and the motor cortex.
7. **d.** During REM sleep the body is paralyzed, except for the twitching of muscles. During this stage of sleep the brain is as active as when you are awake.
8. **c.** The occipital cortex is the higher brain region controlling vision.
9. **c.** The peripheral nervous system is in contact with every tissue in the body sending and receiving information to the CNS.
10. **a.** The CNS develops from a hollow tube composed of neural tissues.
11. **e.** Association areas are present in all of these regions of the cerebral cortex.
12. **a.** Fishes have the most undeveloped telencephalon of the vertebrates listed, resulting in the smallest ratio.
13. **c.** The knee-jerk reflex is an example of a monosynaptic reflex. Stretch receptors (not chemoreceptors) sense the hammer tap on the tendon.
14. **e.** Relative to the brains of other vertebrates, the human brain is larger than body size might lead one to predict, and it exhibits a greater degree of convolution of the cerebral cortex. The human brain also has proportionately more cerebral cortex and association areas within the cortex.
15. **a.** In the retina, the information from over 100 million photoreceptors is integrated by about 1 million ganglion cells; this type of processing is called convergence of information.
16. **e.** The insular cortex is greatly expanded in humans and great apes. This area of the forebrain may be involved with self-recognition and conscious experience.

Application Answers

1. Both speaking written language and repeating heard language involve similar pathways in the brain. The main difference has to do with the initial region of the brain perceiving the word. In reading a word, the area at the back of the cerebrum is used to visualize it. The spoken word uses an area of the cerebrum just behind the area used for speech. Once the word has been processed by the initial centers, the path used for speaking the word is the same. Wernicke's area is stimulated, followed by Broca's area, and then the motor area.
2. Sleepwalking takes place during non-REM sleep. During REM sleep the skeletal muscles of the body become paralyzed, and the sleeper is unable to move. A sleepwalker will not exhibit the eye movements typical of REM sleep.
3. The peripheral system contributed to both seeing the candy and the movements of the arms and legs that you used to pick it up. The parasympathetic branch of the autonomic nervous system stimulated salivation. The central nervous system was involved in the recognition and decision to unwrap and eat the candy.
4. The right side of your brain receives visual information from the left visual field, and the left side receives information from the right visual field. In the visual cortex, the cells are organized in columns that alternate between receiving information from the right and left eyes. At the borders of the columns, the inputs from the right and left eyes overlap. The cells that receive the overlap are called binocular cells, and they interpret the disparity between what the two eyes sense. This provides a three-dimensional image.
5. A sharp punch to the jaw will cause the head to turn sharply, and this is likely to twist the medulla and reticular activating system. The reticular system is a network of neurons in the brain stem that, unless inhibited by other regions of the brain, activates the cerebral cortex and causes consciousness. A sharp blow can affect the reticular system and cause temporary loss of consciousness.

48 Musculoskeletal Systems

The Big Picture

- Actin and myosin are the "universal" proteins for motion. Whether they are located in a unicellular animal or the leg muscle of a human, the molecular interactions of actin and myosin produce movement in the structures in which they reside.
- The sliding filament theory provides a model for how actin and myosin filaments slide past each other, resulting in the shortening (contraction) or lengthening (relaxation) of muscle cells. The whole process of actin and myosin interaction is tightly regulated by the movement of calcium ions into and out of the intracellular spaces of muscle cells, which in turn is activated by the arrival of action potentials in motor neurons. Muscle contraction requires the use of energy in the form of ATP.
- In vertebrates, muscles act in concert with an internal skeleton made of bone. Bone is living tissue that is constantly remodeled. Bones are articulated, forming joints that provide for specialized directional movements of the tissues supported by the bones. Muscles controlling joint movement are often located in pairs acting antagonistically, with one set of muscles causing bending, or flexion, of the joint and the other causing straightening, or extension, of the joint. Some invertebrates have hydrostatic skeletons (a fluid-containing body cavity surrounded by muscles), while others have exoskeletons (rigid outer coverings).

Common Problem Areas

- The sarcomere is the functional unit of the muscle cell, and until you understand its fine structure—Z lines, H lines, etc.—it will be difficult to appreciate how the sarcomere shortens through the actions of actin and myosin.
- Many students think of bone as tissue that is not living. Bone is a living tissue that is constantly remodeled. Become familiar with the three types of living cells in bone: osteocytes, osteoblasts, and osteoclasts.
- The way that movements of multiple muscle groups cause both flexion and extension of a joint can be confusing. Thinking of joints in terms of levers and pulleys may help you understand their actions.

Study Strategies

- The interactions of myosin, actin, troponin, tropomyosin, and calcium and the role of action potentials in stimulating muscle contraction make up a complex, multistep process.
  - First, break the process down into its constituents and learn their locations and general structures.
  - Second, determine how actin and myosin move relative to each other through a series of power strokes.
  - Finally, understand how calcium ions released from the sarcomeres by action potentials initiate and maintain the whole process of muscle contraction.
- Go to yourBioPortal.com to review the following tutorials and activities:

  Animated Tutorial 48.1 Molecular Mechanism of Muscle Contraction

  Animated Tutorial 48.2 Smooth Muscle Action

  Web Activity 48.1 The Structure of a Sarcomere

  Web Activity 48.2 The Neuromuscular Junction

  Web Activity 48.3 Joints

Important Concepts

Muscle cells are responsible for tissue contraction.

- Vertebrates have three types of muscle: skeletal, cardiac, and smooth. Skeletal muscle is responsible for voluntary movements and the unconscious movements associated with breathing. Cardiac muscle is responsible for the beating of the heart. Smooth muscle is controlled by the autonomic nervous system, and it is responsible for the contraction that occurs in many hollow organs, such as the bladder and the gut.
- In all three types of muscle tissue, contraction is due to the interaction between the contractile proteins actin and myosin.

Muscle contraction is described by the sliding filament theory.

- Skeletal muscles are striated voluntary muscles made of large muscle fibers (see Figure 48.1). Muscle fibers have many nuclei and are composed of actin and myosin. Molecules of actin are organized into thin filaments and those of myosin are organized into thick filaments. Bundles of actin and myosin filaments are arranged into myofibrils.
- The contracting unit of myofibrils is the sarcomere, which contains actin and myosin filaments and has very distinct repeating patterns.
- In a sarcomere, the actin filaments are anchored by the Z lines, and myosin filaments are found at the center in the A band (see Figure 48.1).
- In relaxed muscle, the H zone and I band are the regions in which there is no overlap of actin and myosin. Within the H zone is the M band, which contains proteins that help hold myosin filaments in their regular arrangement. The protein titin runs from Z line to Z line, and provides resistance to stretch in relaxed skeletal muscle. During muscle contraction, the Z lines move toward each other, and the H zone and I band shrink in size due to the sliding of actin filaments along the myosin filaments. This is the sliding filament theory of muscle contraction (see Figure 48.2).

The proteins myosin and actin are the key to muscle contraction.

- Myosin molecules are made of two polypeptide chains wrapped around each other, each with a globular head at one end, much like two twisted golf clubs. A myosin filament is composed of many myosin molecules (see Figure 48.3).
- Actin filaments are composed of two monomer chains in a helical arrangement and look like two linear strings of pearls wrapped around each other. The proteins tropomyosin and troponin are associated with actin (see Figure 48.3).
- Myosin heads change conformation when they bind to actin filaments at myosin binding sites, forming a cross-bridge connection. The conformational change in the myosin pulls the actin in toward the middle of the sarcomere. ATP then binds to an ATP binding site on myosin, resulting in the bound actin's release from the myosin and the myosin's return to the original conformation. Many myosin molecules cycle through binding with actin to shorten a sarcomere.

Calcium ions regulate the movements of myosin and actin filaments.

- All the fibers activated by a single motor neuron constitute a motor unit. Action potentials spread deep into the sarcoplasm (cytoplasm) of the muscle through transverse tubules (T tubules) that are in contact with the sarcoplasmic reticulum (endoplasmic reticulum) throughout the sarcoplasm (see Figure 48.5). The sarcoplasmic reticulum takes up and releases Ca^{2+} ions into the sarcoplasm, thereby controlling relaxation and contraction of the myofibrils.
- In relaxed muscle, tropomyosin and troponin cover the myosin binding sites on actin filaments, inhibiting muscle contraction. Calcium regulates contraction by binding with troponin, causing the tropomyosin to change conformation and expose the actin–myosin binding sites on the actin filaments (see Figure 48.6).

Cardiac muscle causes beating of the heart.

- Cardiac muscle is found in the heart and is composed of branched muscle cells that form a strong meshwork (see Figure 48.7). Cardiac muscle cells are smaller than skeletal muscle cells and each cell has only a single nucleus. Intercalated discs add additional strength by holding the cells together. Gap junctions within the intercalated discs allow cardiac muscle cells to be electrically coupled.
- Heartbeats originate at the pacemaker cardiac muscle cells and spread rapidly through gap junctions in the muscle. The heartbeat is described as myogenic because it is generated by the heart muscle itself.
- The mechanism of excitation–contraction coupling in cardiac muscle cells is called Ca^{2+} -induced Ca^{2+} release.

Smooth muscle causes contraction in many internal organs.

- Smooth muscles are involuntary muscles composed of long spindle-shaped cells, each with a single nucleus. Smooth muscles are controlled by acetylcholine and norepinephrine of the autonomic nervous system (see Figure 48.8). When smooth muscle is stretched, it contracts with strength that is proportional to the stretch of the muscle.
- Smooth muscle cells are arranged in sheets. Gap junctions allow electrical contact between the cells and promote coordinated contraction of cells in a sheet.
- In smooth muscle, contraction is controlled by a calmodulin–Ca^{2+} complex. This complex activates myosin kinase, an enzyme that phosphorylates the myosin head to cause contraction. Myosin phosphatase works in the opposite direction by dephosphorylating myosin and stopping interactions between actin and myosin (see Figure 48.9).

Skeletal muscles show graded contractions, ranging from twitches to tetanus.

- An action potential in a skeletal muscle fiber causes twitch or contraction of the muscle. Twitches can occur as discrete contractions, or if they occur frequently enough, they can be summed together (see Figure 48.10).
- The level of tension generated by a muscle depends on the number of motor units activated and the frequency with which the motor units fire. Maximum muscle tension, or tetanus, occurs when there is a high rate of stimulation by action potentials.

- During tetanic contraction, actin and myosin bonds cycle to help keep a muscle fiber from stretching. ATP levels control the length of a tetanic contraction because ATP provides the energy for myosin to break the bond with actin.
- Muscle tone reflects the small but changing number of motor units active in a muscle at any given time.

The strength and endurance of muscles depend on fiber type.

- Slow-twitch muscle fibers are highly resistant to fatigue because they are well supplied with myoglobin (an oxygen-binding protein similar to hemoglobin), mitochondria, and blood vessels. They are also called "red" or "oxidative" muscle (see Figure 48.11).
- Fast-twitch muscle fibers rapidly develop maximum tension but fatigue quickly. Compared with slow-twitch fibers, they have fewer mitochondria and blood vessels, little or no myoglobin, and are called "white" or "glycolytic" muscle (see Figure 48.11).

The strength of a muscle fiber is related to its length.

- The amount of force a sarcomere can generate depends on its resting length (see Figure 48.12).
- Stretching a muscle causes the sarcomeres to lengthen, resulting in less overlap between actin and myosin filaments and less force.

Exercise enhances the strength and endurance of muscle.

- Anaerobic exercise, such as weight lifting, increases strength. Such exercise induces the formation of new actin and myosin filaments in existing muscle fibers, thus producing bigger fibers and bigger muscles.
- Aerobic exercise, such as jogging, increases endurance. Such exercise increases myoglobin, number of mitochondria, density of capillaries, and enzymes involved in energy utilization.

Muscles have three systems for obtaining ATP.

- Muscles use the immediate, glycolytic, and oxidative systems to obtain ATP needed for contraction (see Figure 48.13).
  - The immediate system utilizes preformed ATP and creatine phosphate.
  - The glycolytic system follows the immediate system within seconds and metabolizes carbohydrates to lactate and pyruvate.
  - The oxidative system is fully activated within about one minute and completely metabolizes carbohydrates or fats to water and carbon dioxide.

The flight muscles of insects are asynchronous.

- The striated muscle of vertebrates and many invertebrates is described as synchronous because cycling of the contractile mechanism is tied to the firing of motor neurons.
- The flight muscles of insects are asynchronous because cycling of the contractile mechanism is not tied to the firing rate of the flight motor neurons; this allows for very high rates of cycling and wingbeat frequencies.

Invertebrates use hydrostatic skeletons and exoskeletons for support and movement.

- Many soft-bodied invertebrates have a fluid-filled body cavity that acts as a hydrostatic skeleton. Earthworms have circular muscles and longitudinal muscles that oppose each other to act on the hydrostatic skeleton and control elongation and shortening of body segments (see Figure 48.14).
- Arthropods have an exoskeleton, or cuticle, composed of chitin that offers protection and provides sites for muscle attachment.
- As an animal that has an exoskeleton grows, it undergoes the process of molting. With each molt the old exoskeleton is shed, revealing a new one that has developed underneath.

The vertebrate endoskeleton provides a frame for support and movement.

- Endoskeletons are growing, living tissue that provide sites for muscle attachment and support for the body.
- The human skeleton is composed of a central axial skeleton (skull, vertebral column, sternum, and ribs) and an appendicular skeleton (pectoral girdle, pelvic girdle, arms, hands, legs, and feet) (see Figure 48.15).
- The pliable portions of the endoskeleton, such as the framework of the nose and the surface of the joints of the endoskeleton, are composed of cartilage containing the protein collagen.
- The bone in the endoskeleton is strong and solid and is composed mainly of collagen and calcium phosphate. Bone is constantly being remodeled by two types of cells: osteoblasts, which lay down new bone, and osteoclasts, which break down bone (see Figure 48.16). When an osteoblast becomes enclosed by the matrix it is laying down, it stops forming matrix and exists within a lacuna; at this stage, the cell is called an osteocyte. Osteocytes communicate with one another and influence the activities of osteoblasts and osteoclasts.
- Developing bone is created as membranous bone growing on a scaffolding of connective tissue or as cartilage bone hardening from an initial cartilage model. The long bones of the arms and legs are cartilage bones that ossify first at the center and then at each end; elongation occurs at epiphyseal plates (see Figure 48.17). Compact bone is solid, whereas cancellous bone is lightweight, with many cavities. In mammals, most compact bone is known as Haversian bone; it is composed of concentric rings with blood vessels and nerves running through a central canal. These structural units are known as Haversian systems (see Figure 48.18).

Movable bones come together at joints.

- There are six types of joints where bones meet: ball-and-socket, pivot, saddle, ellipsoid, hinge, and plane (see Figure 48.19).

- The muscles attached to bones at joints work antagonistically (see Figure 48.20). The flexor muscles bend the joints and the extensor muscles straighten them.
- Two types of connective tissues hold joints and bones together. Ligaments hold bone to bone, and tendons hold muscle to bone.

Test Yourself

Diagram Exercises

1. Label the following structures in the diagram below: muscle, tendon, single muscle fiber, single myofibril, sarcomere, actin filament, myosin filament, Z line, A band, H zone, I band, M band. Also label as many of the last seven terms as you can on the myofibril and on the enlargement of the sarcomere.

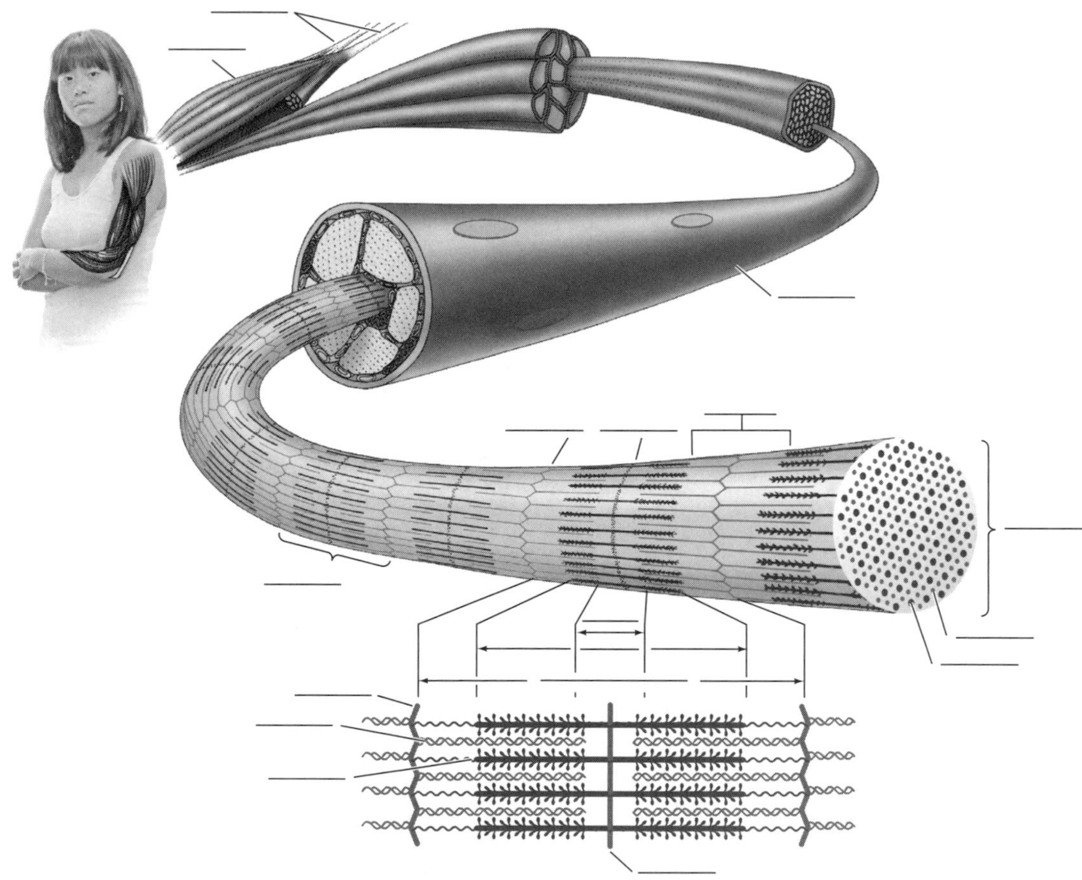

Textbook Reference: *48.1 How Do Muscles Contract? p. 1008*

2. On the diagram below, label the four main components of the axial skeletal system.

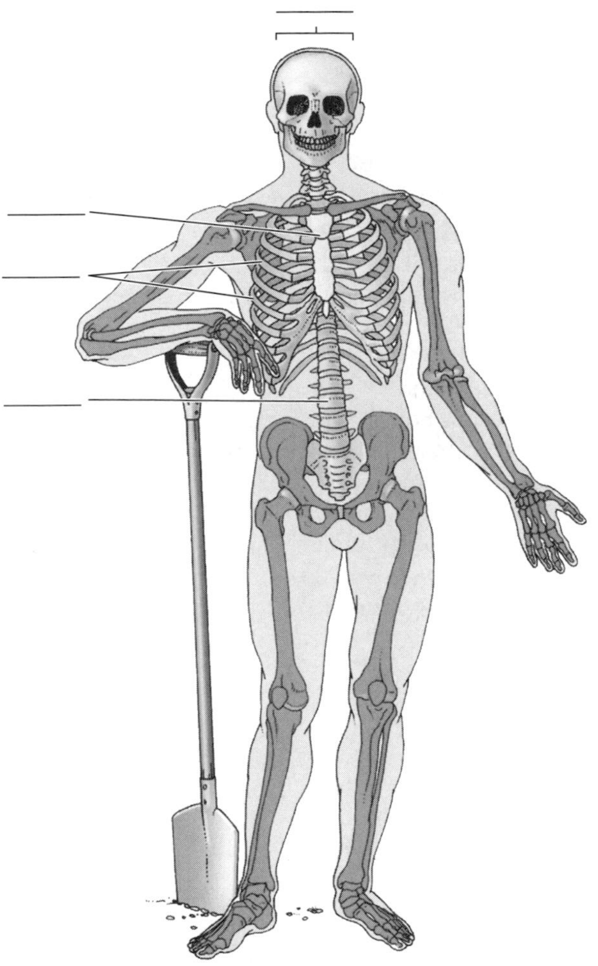

Textbook Reference: 48.3 *How Do Skeletal Systems and Muscles Work Together?* p. 1019

Knowledge and Synthesis Questions

1. Which of the following statements is *false*?
   a. Strength training induces the formation of new actin and myosin filaments in existing muscle fibers.
   b. Satellite cells are muscle stem cells.
   c. Aerobic exercise increases the number of mitochondria in muscle cells and the density of capillaries in muscle.
   d. A common effect of strength training is the production of more muscle fibers.
   e. Aerobic exercise increases myoglobin in skeletal muscle cells.

   Textbook Reference: *48.2 What Determines Muscle Performance? p. 1017*

2. A motor unit is best described as
   a. all the nerve fibers and muscle fibers in a single muscle bundle.
   b. one muscle fiber and its single nerve fiber.
   c. a single motor neuron and all the muscle fibers that it innervates.
   d. the neuron that provides the central nervous system with information about muscle contraction.
   e. a neuron that communicates information from sensory to motor neurons.

   Textbook Reference: *48.1 How Do Muscles Contract? p. 1010*

3. Which of the following statements is *false*?
   a. Cardiac muscle is striated.
   b. Smooth muscle does not contain actin.
   c. Skeletal muscle is considered voluntary.
   d. Smooth muscle is found in the digestive tract and the walls of the bladder.
   e. A single skeletal muscle cell has many nuclei.

   Textbook Reference: *48.1 How Do Muscles Contract? pp. 1007, 1012–1014*

4. The oxygen-binding molecule in skeletal muscle is
   a. myoglobin.
   b. hemoglobin.
   c. ATP.
   d. myokinase.
   e. creatine phosphate.

   Textbook Reference: *48.2 What Determines Muscle Performance? p. 1017*

5. The action potential that triggers a muscle contraction travels deep within the muscle cell by means of
   a. sarcoplasmic reticulum.
   b. transverse (or T) tubules.
   c. synapses.
   d. motor end plates.
   e. neuromuscular junctions.

   Textbook Reference: *48.1 How Do Muscles Contract? p. 1010*

6. A sarcomere is best described as a
   a. moveable structural unit within a myofibril bounded by H zones.
   b. fixed structural unit within a myofibril bounded by Z lines.
   c. fixed structural unit within a myofibril bounded by A bands.
   d. moveable structural unit within a myofibril bounded by Z lines.
   e. collection of myofibrils.

   Textbook Reference: *48.1 How Do Muscles Contract? p. 1008*

7. ATP provides the energy for muscle contraction by allowing for the
   a. formation of an action potential in the muscle cell.
   b. breaking of actin–myosin bonds.
   c. formation of actin–myosin bonds.
   d. release of calcium by the sarcoplasmic reticulum.
   e. formation of T tubules.

   Textbook Reference: *48.1 How Do Muscles Contract? pp. 1009–1010*

8. Ca^{2+} binds to _______ in skeletal muscle and leads to exposure of the binding site for _______ on the _______ filament.
   a. troponin; myosin; actin
   b. troponin; actin; myosin
   c. actin; myosin; troponin
   d. tropomyosin; myosin; actin
   e. myosin; actin; troponin
   Textbook Reference: *48.1 How Do Muscles Contract? p. 1011*
9. Tropomyosin is moved by which of the following proteins?
   a. Calmodulin
   b. Acetylcholine
   c. Actin
   d. Troponin
   e. Titin
   Textbook Reference: *48.1 How Do Muscles Contract? p. 1011*
10. Summation of frequent muscle twitches to give maximum contraction is called
    a. motor unit summation.
    b. twitch.
    c. facilitation.
    d. tetanus.
    e. muscle tone.
    Textbook Reference: *48.2 What Determines Muscle Performance? p. 1015*
11. _______ are responsible for the dynamic remodeling of bone that occurs continuously.
    a. Osteoblasts
    b. Osteoblasts and osteoclasts
    c. Osteoclasts and osteocytes
    d. Osteoblasts, osteoclasts, and osteocytes
    e. Myoblasts
    Textbook Reference: *48.3 What Roles Do Skeletal Systems Play in Movement? p. 1020*
12. Endoskeletons
    a. are characteristic of arthropods.
    b. are located on the inside of the body.
    c. lack joints.
    d. require molting as the animal grows.
    e. provide support to earthworms, along with their hydrostatic skeleton.
    Textbook Reference: *48.3 How Do Skeletal Systems and Muscles Work Together? p. 1019*
13. A soccer player who has suffered a knee injury that damages the tissue holding his upper and lower leg bones together has most likely damaged _______ tissue
    a. muscle
    b. tendon
    c. ligament
    d. cartilage
    e. membrane
    Textbook Reference: *48.3 How Do Skeletal Systems and Muscles Work Together? p. 1022*
14. Sites of elongation between the ossified regions of long bones are called
    a. glue lines.
    b. fulcrums.
    c. hinge joints.
    d. epiphyseal plates.
    e. Haversian systems.
    Textbook Reference: *48.3 How Do Skeletal Systems and Muscles Work Together? p. 1021*
15. Which of the following statements is *false*?
    a. In humans, cartilage is the principal component of the embryonic skeleton.
    b. Some vertebrates retain a cartilaginous endoskeleton into adulthood.
    c. Cancellous bone has numerous cavities.
    d. Physical stress on bones causes them to become thinner.
    e. Calcitonin and parathyroid hormone regulate the deposition of calcium in bone.
    Textbook Reference: *48.3 How Do Skeletal Systems and Muscles Work Together? p. 1020*

Application Questions

1. Smooth muscle contracts involuntarily, whereas skeletal muscle contraction is under voluntary control. How do the mechanisms that control smooth muscle and skeletal muscle contraction differ?
   Textbook Reference: *48.1 How Do Muscles Contract? pp. 1010–1011, 1012–1014*
2. White muscle and red muscle are found in different parts of the body and are used for different types of movement. What are the physiological and morphological characteristics that distinguish the two types of muscle?
   Textbook Reference: *48.2 What Determines Muscle Performance? pp. 1015–1016*
3. Smooth muscle contracts involuntarily in the digestive tract, blood vessels, and urinary bladder. Describe the two main ways that smooth muscle contraction and the membrane potential of smooth muscle are controlled.
   Textbook Reference: *48.1 How Do Muscles Contract? pp. 1012–1014*
4. Earthworms have a hydrostatic skeleton. Describe how a hydrostatic skeleton is used to move an earthworm through the soil.
   Textbook Reference: *48.3 How Do Skeletal Systems and Muscles Work Together? pp. 1018–1019*
5. Why does weight-bearing exercise help prevent osteoporosis?
   Textbook Reference: *48.3 How Do Skeletal Systems and Muscles Work Together? p. 1020*

Answers

Diagram Exercise Answers

1.

2.

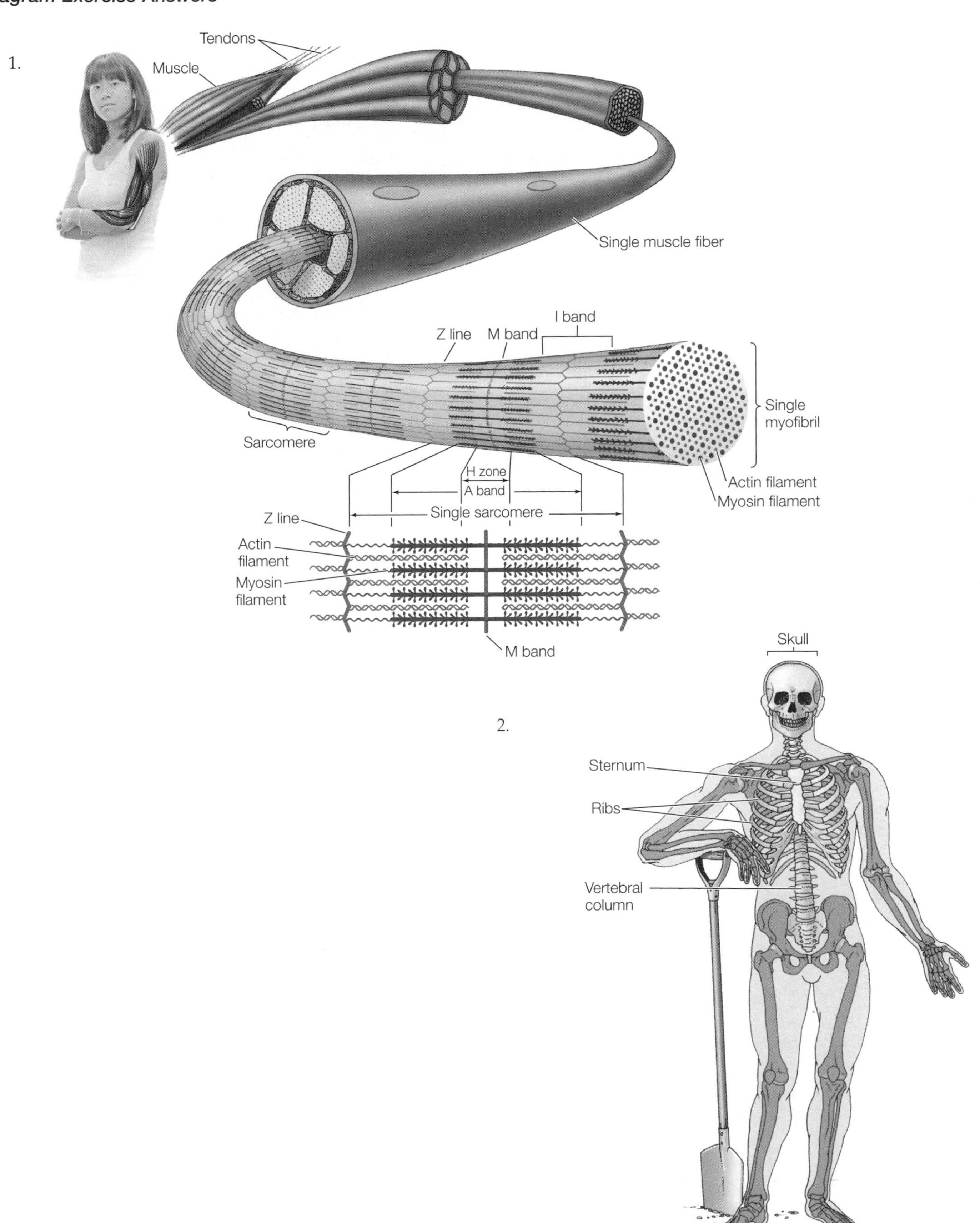

Knowledge and Synthesis Answers

1. **d.** Strength training typically produces bigger, rather than more, muscle fibers.
2. **c.** A motor unit is a single motor neuron and all the muscle fibers that it innervates.
3. **b.** Although contraction of smooth muscle is controlled differently from that of skeletal muscle, smooth muscle does contain actin and myosin.
4. **a.** Myoglobin is the main oxygen-carrying molecule in skeletal muscle.
5. **b.** The action potential arriving to the muscle travels into the muscle through the transverse (or T) tubules.
6. **d.** A sarcomere is a structural unit within a myofibril bounded by Z lines; it contains actin and myosin.
7. **b.** ATP provides energy that is used to break actin–myosin bonds.
8. **a.** Calcium is released from the sarcoplasmic reticulum and binds with troponin, resulting in exposure of the myosin-binding site on actin.
9. **d.** The binding of calcium with troponin causes a conformational change in tropomyosin.
10. **d.** Tetanus is the maximum level of muscle contraction.
11. **d.** Osteoblasts, osteoclasts, and osteocytes are all involved in remodeling bone.
12. **b.** Endoskeletons (such as those of mammals) are found inside the body, and exoskeletons (such as those of insects) are found outside the body.
13. **c.** Ligaments hold bones together.
14. **d.** Epiphyseal plates are the sites of elongation in long bones.
15. **d.** Physical stress on bones causes them to thicken.

Application Answers

1. Smooth muscle contraction and skeletal muscle contraction are regulated by the presence of Ca^{2+}. In smooth muscle, Ca^{2+} joins with the protein calmodulin to activate myosin kinase in the sarcoplasm. Myosin kinase then phosphorylates the myosin head, allowing the myosin head to bind with actin. In skeletal muscle, Ca^{2+} binds with troponin on the actin filaments. This binding of Ca^{2+} causes a conformational change in tropomyosin and uncovers the myosin binding sites on the actin, allowing myosin to bind with actin.
2. Fast-twitch fibers are known as white muscle and have few mitochondria, small amounts of myoglobin, and few blood vessels. Muscles with many fast-twitch fibers are good for short-term work that requires maximum strength. Slow-twitch fibers are known as red muscle. Red muscle has many mitochondria, large amounts of myoglobin, and many blood vessels. Muscles with many slow-twitch fibers function well in endurance activities.
3. Smooth muscle is sensitive to stretching. It will depolarize in response to any stretching and contract, with the strength of the contraction proportional to the amount of stretch in the muscle. Parasympathetic inputs of acetylcholine also cause depolarization of the muscle membrane resulting in contraction, whereas norepinephrine hyperpolarizes the muscle membranes.
4. The hydrostatic skeleton of the earthworm is an incompressible fluid-filled cavity surrounded by longitudinal and circular muscles. Contraction of the longitudinal muscles causes the segments to contract and the body to shorten. Contraction of the circular muscles causes the body segments to elongate and the body to lengthen. Alternating contractions between the longitudinal and the circular muscles move the animal in a push and pull manner. Bristles on the body help hold the animal in place after elongation, and the body is pulled forward during shortening.
5. Osteoporosis is a decrease in bone density that results when the destruction of bone by osteoclasts outpaces the formation of new bone by osteoblasts, leading to thin, brittle bones. Weight-bearing exercises place stress on bone, ultimately altering the interplay of osteoblast and osteoclast activity to induce thickening of bone.

49 Gas Exchange in Animals

The Big Picture

- Cellular metabolism requires O_2 and produces CO_2 as a waste product that must be eliminated. Animals have evolved a variety of structures and mechanisms for exchanging these gases with the environment. Tracheal systems in insects, gills in fishes, and lungs in terrestrial vertebrates are three examples of gas-exchange organs. Very short diffusion distances and very large surface areas—adaptations designed to maximize gas exchange, characterize all.
- O_2 is transported by the protein hemoglobin in vertebrates. The P_{O_2} levels surrounding hemoglobin determine O_2 binding. If P_{O_2} is high, hemoglobin accepts O_2, but it readily gives up its O_2 when the P_{O_2} falls. These properties allow hemoglobin to bind O_2 in the gas-exchange organ, transport it to the metabolizing tissues, and then release its O_2 for consumption in cellular metabolism.

Common Problem Areas

- The structure and pattern of air flow through the avian respiratory system are complex and very different from the mammalian pattern. Make sure to study Figure 49.8, and be sure to remember that avian air sacs are not sites of gas exchange.
- A common error is the conception of O_2 and CO_2 exchange as a "two-way street," with O_2 taking the immediate place of CO_2 in the lungs and CO_2 then taking the place of O_2 in the tissues. This is incorrect. In fact, the mechanisms for exchange of these two gases are completely different.
- Understanding how O_2 is picked up, transported by, and released from hemoglobin can be quite confusing. Think of the O_2 dissociation curve as a "tool." Think first of the situation at the top of the curve, where P_{O_2} is high (in the lungs or gills, for example). If the P_{O_2} is high, hemoglobin will maximally bind O_2. Think, then, of transporting that blood to a region with a given P_{O_2}, and picture on the curve what must happen to O_2 saturation under the new P_{O_2} level. The O_2 from hemoglobin is now available for tissue respiration.

Study Strategies

- Refer to Fick's law of diffusion as a guideline for understanding gas exchange. The various components of the law—distance for diffusion, partial pressure gradient, surface area over which gas exchange occurs—provide a good framework for understanding why gas-exchange organs have evolved with certain characteristics in common. For example, gas-exchange organs tend to have large surface areas, have very short diffusion distances, and experience large partial gradients for O_2 across their surfaces. All of this makes sense in the context of Fick's law of diffusion. Fick's law of diffusion also governs the movement of O_2 and CO_2 in the body, and it can help you remember where and why these gases are picked up and released in the body.
- Go to yourBioPortal.com to review the following tutorials and activities:

  Animated Tutorial 49.1 Airflow in Birds

  Animated Tutorial 49.2 Airflow in Mammals

  Interactive Tutorial: Hemoglobin: Loading and Unloading

  Web Activity 49.1 Human Respiratory System

  Web Activity 49.2 Oxygen-Binding Curves

  Web Activity 49.3 Concept Matching

Important Concepts

Respiratory gas exchange occurs by diffusion.

- The respiratory gas oxygen (O_2) is required by cells to produce energy in the form of ATP.
- The respiratory gas carbon dioxide (CO_2) is one of the waste by-products of ATP production and must be eliminated from an animal's body.
- The transfer of these gases occurs by simple diffusion in the respiratory systems of animals. Partial pressures are used to express the concentrations of gases in a mixture.

Fick's law of diffusion describes the exchange of gas in the respiratory system.

- The partial pressure gradient is the difference between the partial pressure of a gas at two locations.

- The partial pressure gradients of O_2 and CO_2 drive the movement of O_2 into the body and CO_2 out of the body.
- Rate of diffusion, described by Fick's law of diffusion, depends in part on a diffusion coefficient that varies according to temperature, the medium, and the diffusing molecules. Rate of diffusion also depends on the cross-sectional area over which the gas is diffusing, and the partial pressure gradient.

Several physical parameters affect respiratory gas exchange.

- There are major differences in the O_2 capacity of air and water.
- Water, because of its density, is more expensive to move over a respiratory surface than air is. Water also contains far less O_2 compared to air, and O_2 also diffuses far more slowly in water than in air.
- Temperature affects respiration of aquatic animals because there is less O_2 in warm water than in cold, and an animal's need for O_2 increases with temperature (see Figure 49.2).
- The tendency for a gas to move by diffusion across gills, skin, or lungs depends on its partial pressure. The sum of all the gases' partial pressures in a gas mixture is the total partial pressure, which in the atmosphere equals the atmospheric pressure.
- O_2 makes up 20.9 percent of the atmospheric pressure. As elevation increases and atmospheric pressure decreases, the total amount of O_2 in air decreases.
- CO_2 diffuses across the respiratory organs in both air and water. In air, diffusion is rapid because of the large partial pressure gradient between the blood and the atmosphere. The partial pressure gradient for CO_2, unlike that of O_2, does not change with altitude.
- Getting rid of CO_2 typically is not a problem for water-breathing animals. Lack of O_2 becomes a problem for water-breathing animals long before problems with CO_2 exchange occur.

Animals have adaptations to maximize gas exchange.

- Animals maximize respiration by decreasing the distance that gases must diffuse between the blood and the external environment. They achieve this with very thin respiratory surface membranes.
- Animals also actively move the environmental respiratory fluid over the gills and through the lungs and perfuse the internal side of the respiratory organ to carry the respiratory gases.
- Some aquatic amphibians and insects have external gills with large surface areas for exchange of respiratory gases with water (see Figure 49.3A).
- Internal gills have a large surface area and are protected from the environment by an animal's body cavity. They must be actively ventilated with water to achieve gas exchange.

Diverse types of respiratory organs have evolved in animals.

- Insects have tracheae—air tubes that end at the tissue cells as air capillaries and open to the environment through spiracles. Gases diffuse through the tracheae into the air capillaries, but are also moved by movement of the animal's body parts (see Figure 49.4).
- Fish have internal gills with a large surface area that they ventilate with unidirectional flowing water to maximize external P_{O_2} levels. Each gill has hundreds of gill filaments with folds, or lamellae, that act as the respiratory gas exchange surfaces. Countercurrent flow (blood in the lamellae flowing in the direction opposite to that of water flowing over the lamellae) maximizes the P_{O_2} gradient between the water and the blood (see Figures 49.5 and 49.6). As blood flows through the lamellae it is always in contact with water that has a higher O_2 level, resulting in a continuous O_2 gradient to maximize the uptake of O_2 into the blood.
- Birds have a unidirectional flow of air through their lungs. Gas exchange occurs in the parabronchi and air capillaries of the lungs; gas exchange does not occur in the associated air sacs (see Figure 49.7). Air requires two cycles of inhalation and exhalation to move from the posterior air sac, to the lung, to the anterior air sac, and out of the respiratory tract (see Figure 49.8).

The lungs of most vertebrates are ventilated by tidal ventilation.

- Lungs evolved in fishes as outpocketings of the digestive tract. In mammals, lungs are dead-end sacs in which ventilation is tidal. In tidal ventilation, air flows in and exhaled gases flow out by the same route.
- The tidal breathing of mammals can be described in terms of a series of lung volumes measured with a spirometer (see Figure 49.9). The normal volume of air moved during one cycle is known as the tidal volume. The vital capacity consists of the resting tidal volume, plus the additional volume of air that can be taken in with a large breath (inspiratory reserve volume) and the additional volume of air that can be forced out with a large exhale (expiratory reserve volume). Not all of the air can be forced out of the lungs. Some air remains in the bronchi and trachea, making up the dead space; this air is called the residual volume. The total lung capacity is the sum of the vital capacity and the residual volume. The residual volume cannot be measured directly with a spirometer, but it can be measured indirectly by means of the helium dilution method. Increases in the residual volume are associated with certain respiratory diseases, such as emphysema.
- Tidal breathing limits the partial pressure gradient needed to drive the diffusion of O_2 from air into blood. Fresh air is not moving into the lungs during some of the breathing cycle, and when it does enter the lungs, it mixes with stale air.

The lung is the respiratory organ in mammals.

- Mammalian lungs have a very large surface area and a very short gas diffusion pathway.
- Air enters at the oral cavity or nasal passage, travels down the pharynx through the larynx, and then through the trachea, which branches into two bronchi, one leading to each lung. In the lung, more branching occurs to produce bronchioles and finally the site of gas exchange—small air sacs, the alveoli (see Figure 49.10).
- O_2 diffuses across the thin-walled alveoli (less than 2 μm) into many surrounding capillaries.
- The alveoli have surface tension due to an aqueous layer that makes inflation difficult. This is overcome by a surfactant produced by the cells forming the alveolar walls. The aqueous layer is needed to ensure the diffusion of gases across the membrane.
- Mucus, produced along the lung's larger airways, catches and removes inhaled dust and microorganisms. Cilia along the airways move the mucus up the airways and into the throat, acting as a mucus escalator.
- Lungs are inflated by negative pressure created by contraction of the diaphragm muscle in the thoracic cavity. Closed pleural membranes line the thoracic cavity and allow for the development of negative pressure. During exhalation, the diaphragm relaxes and the thoracic cavity contracts (see Figure 49.11). At times of strenuous exercise, intercostal muscles also help change the volume of the thoracic cavity.

Oxygen is transported in the blood by hemoglobin in vertebrates.

- Gases diffuse between the respiratory organ and the circulatory system, where they are transported in blood. The liquid blood plasma transports only a small amount of dissolved O_2, with the vast majority of O_2 transported by the O_2-binding pigment hemoglobin found in the red blood cells of vertebrates.
- Hemoglobin contains four polypeptide subunits, each with a heme group that reversibly binds O_2. The amount of O_2 bound to hemoglobin depends on the partial pressure of O_2. At the high blood P_{O_2} that is characteristic of blood leaving the lungs, hemoglobin is almost 100 percent saturated with O_2. At the tissues, the P_{O_2} is lower and hemoglobin releases approximately 25 percent of its bound O_2. The relationship between the P_{O_2} of the environment and the percentage of hemoglobin bound with O_2 is known as the S-shaped O_2-binding curve (see Figure 49.12). At low levels of P_{O_2}, only one of the hemoglobin subunits can bind O_2. Once it binds a single O_2 molecule, the conformation of the hemoglobin changes, making it easier for the other three sites on the hemoglobin to bind O_2. This process is known as positive cooperativity.
- In muscle, myoglobin is the O_2-carrying molecule and functions primarily to store O_2 intracellularly. The O_2 affinity is higher in myoglobin than in hemoglobin, but myoglobin has only one unit to bind one O_2 molecule.
- Fetal hemoglobin has a different structure from the adult form, with a higher affinity for O_2. This allows the efficient movement of O_2 from maternal to fetal blood.
- The Bohr effect is the shifting of the hemoglobin binding curve to the right in response to a decrease in pH of the blood. This results in the release of more O_2 at a given environmental P_{O_2} at the lower pH.
- Hemoglobin's affinity for O_2 is lowered by 2,3-bisphosphoglyceric acid (BPG), a metabolite of glycolysis. BPG binds with hemoglobin to cause a conformational change in the hemoglobin, resulting in a right shift in the O_2-binding curve.

Carbon dioxide is transported as dissolved CO_2 and as bicarbonate ions.

- Tissues produce CO_2 as a by-product of metabolism. CO_2 must be moved from the tissues and excreted into the environment.
- Significant amounts of CO_2 dissolve in the plasma and are carried in this form to the lungs.
- Most of the CO_2 is transported as bicarbonate ions. CO_2 reacts with water to make bicarbonate ions and a proton. Carbonic anhydrase speeds up the conversion of CO_2 to carbonic acid, which then dissociates into bicarbonate ions.

Breathing is controlled by the brain and blood pH levels.

- Breathing is controlled by the autonomic nervous system from neurons in the medulla. Brain areas above the medulla modify breathing in response to speaking, eating, and emotional states.
- In air-breathing animals, chemoreceptors on the medulla are sensitive to CO_2 levels and the pH of cerebral spinal fluid. Breathing increases when CO_2 levels increase and pH decreases. Control of breathing is relatively insensitive to blood O_2 levels, but chemoreceptors in the arteries leaving the heart (carotid and aorta) can signal the medulla in response to low levels of O_2.
- In water-breathing animals, O_2 is the primary feedback stimulus for breathing.

Test Yourself

Diagram Exercise

The diagrams below highlight important aspects of the respiratory system of a bird. Label the following structures on the top diagram: anterior air sacs, posterior air sacs, trachea, lung, parabronchus, and bronchus. On each diagram, indicate with arrows the path of airflow at that stage, and note whether inhalation or exhalation is occurring.

Breath 1

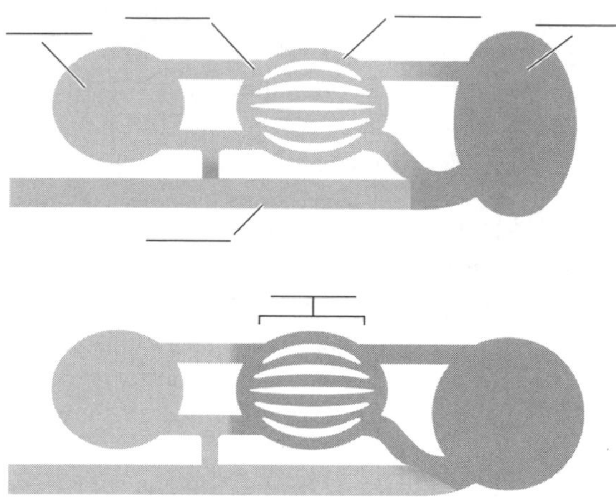

Breath 2

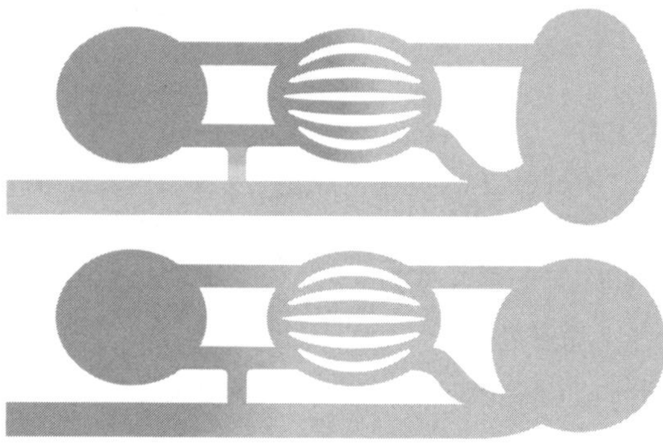

Textbook Reference: 49.2 What Adaptations Maximize Respiratory Gas Exchange? p. 1032

Knowledge and Synthesis Questions

1. Which of the following statements is *false*?
   a. Compared to given volume of water, the same volume of air is moved more easily across a respiratory surface.
   b. Compared to water, air holds more O_2 per unit volume.
   c. Water breathers have a difficult time ridding themselves of CO_2 because CO_2 does not dissolve well in water.
   d. Temperature increases affect the O_2 content of water more than they do that of air.
   e. O_2 diffuses more rapidly in air than in water.

   Textbook Reference: 49.1 What Physical Factors Govern Respiratory Gas Exchange? pp. 1027–1028

2. External gills, tracheae, and lungs share which of the following sets of characteristics?
   a. They are parts of gas-exchange systems, they exchange both CO_2 and O_2, and they increase surface area for diffusion.
   b. They are used by water breathers, their functioning is based on countercurrent exchange, and they make use of negative pressure for breathing.
   c. They exchange only O_2, they are associated with a circulatory system, and they are found in vertebrates.
   d. They are found in insects, they employ positive-pressure pumping, and their functioning is based on crosscurrent flow.
   e. They are parts of gas-exchange systems, they exchange both CO_2 and O_2, and they decrease surface area for diffusion.

   Textbook Reference: 49.2 What Adaptations Maximize Respiratory Gas Exchange? p. 1029

3. Which of the following represents a larger volume of air than is normally found in the resting tidal volume of a human lung?
   a. Residual volume
   b. Inspiratory reserve volume
   c. Expiratory reserve volume
   d. Vital capacity
   e. All of the above

   Textbook Reference: 49.2 What Adaptations Maximize Respiratory Gas Exchange? p. 1033

4. Both bird and mammal lungs
   a. need two cycles of inhalation and exhalation in order for air to move.
   b. contain alveoli at the terminal ends.
   c. have an anatomical dead space.
   d. exchange O_2 and CO_2 with blood in capillaries.
   e. have a unidirectional flow of air moving through them.

   Textbook Reference: 49.2 What Adaptations Maximize Respiratory Gas Exchange? pp. 1031–1032

5. According to Fick's law of diffusion, which of the following does *not* play a role in the diffusion of O_2 across a membrane?
   a. Surface area
   b. Volume
   c. Difference in concentration or partial pressure
   d. Diffusion distance
   e. Diffusion coefficient

   Textbook Reference: 49.1 What Physical Factors Govern Respiratory Gas Exchange? p. 1027

6. Because of the relatively high altitude of Antonito, Colorado, the town has a normal barometric pressure of about 600 mm Hg rather than 760 mm Hg as at sea level. The partial pressure of O_2 in Antonito's air is approximately _______ mm Hg.
   a. 75
   b. 126
   c. 160
   d. 76
   e. 21

   Textbook Reference: 49.1 What Physical Factors Govern Respiratory Gas Exchange? p. 1026

7. Air flows into the lungs of mammals during inhalation because the
   a. pressure in the lungs falls below atmospheric pressure.
   b. volume of the lungs decreases.

c. pressure in the lungs rises above atmospheric pressure.
d. diaphragm moves upward toward the lungs.
e. internal intercostal muscles contract.
Textbook Reference: *49.3 How Do Human Lungs Work? pp. 1035–1036*

8. The movement of O_2 and CO_2 between the blood in the tissue capillaries and the cells in tissues depends most directly upon
a. active transport of O_2 and CO_2.
b. total atmospheric (barometric) pressure differences across the cell membranes.
c. diffusion of O_2 and CO_2 down a concentration gradient.
d. diffusion of O_2 and CO_2 down a partial pressure gradient.
e. osmosis across cell membranes.
Textbook Reference: *49.1 What Physical Factors Govern Respiratory Gas Exchange? p. 1028*

9. The alveoli of the lungs do not contain air with 20.9% oxygen because
a. we normally do not ventilate our lungs at a high enough rate.
b. the lungs have too many alveoli to ventilate.
c. there is dead space in the trachea and bronchi.
d. the trachea and bronchi are too small in volume
e. some O_2 has been exchanged before reaching the alveoli.
Textbook Reference: *49.3 How Do Human Lungs Work? p. 1035*

10. Which of the following statements about hemoglobin is *false*?
a. Hemoglobin allows the blood to carry a large amount of O_2.
b. Hemoglobin contains a single polypeptide chain with very high affinity for O_2.
c. Hemoglobin is packaged inside red blood cells.
d. Fetal hemoglobin is structurally different from adult hemoglobin.
e. Compared to adult hemoglobin, fetal hemoglobin has a higher affinity for O_2.
Textbook Reference: *49.4 How Does Blood Transport Respiratory Gases? pp. 1037–1039*

11. As blood becomes fully O_2 saturated, hemoglobin is combining with _______ molecule(s) of O_2.
a. 1
b. 2
c. 4
d. 8
e. 6
Textbook Reference: *49.4 How Does Blood Transport Respiratory Gases? p. 1037*

12. The Bohr shift describes the
a. outward movement of Cl^- from the blood cell in exchange for HCO_3^- moving into the cell.
b. leftward shift of the entire O_2 equilibrium curve when temperature rises.
c. rightward shift of the entire O_2 equilibrium curve when pH rises.
d. rightward shift of the entire O_2 equilibrium curve when pH falls.
e. leftward shift of the entire O_2 equilibrium curve when BPG increases.
Textbook Reference: *49.4 How Does Blood Transport Respiratory Gases? p. 1039*

13. The presence of CO_2 in blood will lower pH because CO_2 combines with _______, with the rate of reaction increased by _______.
a. H_2O to form H^+ and HCO_3^-; carbonic anhydrase
b. H_2O to form only HCO_3^-; carbonic anhydrase
c. H_2O to form only H^+; carbonic ions
d. H^+ to form HCO_3^-; oxyhemoglobin
e. H_2O to form H^+ and HCO_3^-; fetal hemoglobin
Textbook Reference: *49.4 How Does Blood Transport Respiratory Gases? p. 1039*

14. The largest proportion of CO_2 carried by the blood is in the form of
a. bicarbonate ions (HCO_3^-) carried in the plasma.
b. molecular CO_2 dissolved in the plasma.
c. bicarbonate ions (HCO_3^-) carried within the red blood cells.
d. molecular CO_2 chemically bound to hemoglobin.
e. molecular CO_2 chemically bound to myoglobin.
Textbook Reference: *49.4 How Does Blood Transport Respiratory Gases? p. 1039*

15. Which of the following is involved in the control of breathing?
a. Neurons in the medulla
b. Chemoreceptors on the surface of the medulla
c. Chemoreceptors in the aorta
d. Chemoreceptors in the carotid arteries
e. All of the above
Textbook Reference: *49.5 How Is Breathing Regulated? pp. 1040–1042*

16. Which of the following statement is *false*?
a. Air-breathing animals are remarkably insensitive to falling levels of O_2.
b. In water-breathing animals, O_2 is the primary feedback stimulus for breathing.
c. In humans, chemoreceptors for CO_2 are located in the medulla.
d. Areas of the brain above the medulla have no impact on breathing rate.
e. Breathing is an involuntary function of the central nervous system.
Textbook Reference: *49.5 How Is Breathing Regulated? pp. 1040–1042*

Application Questions

1. How can an aquatic animal with no specialized respiratory surfaces get enough O_2 to all of its cells to survive?
Textbook Reference: *49.2 What Adaptations Maximize Respiratory Gas Exchange? p. 1029*

2. The gills of fish employ a countercurrent exchange mechanism to ensure that $O_2$is efficiently extracted from the water. Describe how the countercurrent exchange works.
Textbook Reference: 49.2 *What Adaptations Maximize Respiratory Gas Exchange? pp. 1030–1031*
3. What is the difference between the myoglobin O_2-binding curve and the hemoglobin O_2-binding curve?
Textbook Reference: 49.4 *How Does Blood Transport Respiratory Gases? p. 1038*
4. Why would you be more likely to find birds than mammals at high altitudes?
Textbook Reference: 49.2 *What Adaptations Maximize Respiratory Gas Exchange? pp. 1031–1032*
5. Why would a tropical fish in a fish tank face severe respiratory problems (and possibly death) if the heater in the tank malfunctioned and caused extremely high water temperatures?
Textbook Reference: 49.1 *What Physical Factors Govern Respiratory Gas Exchange? p. 1028*

Answers

Diagram Exercise Answer

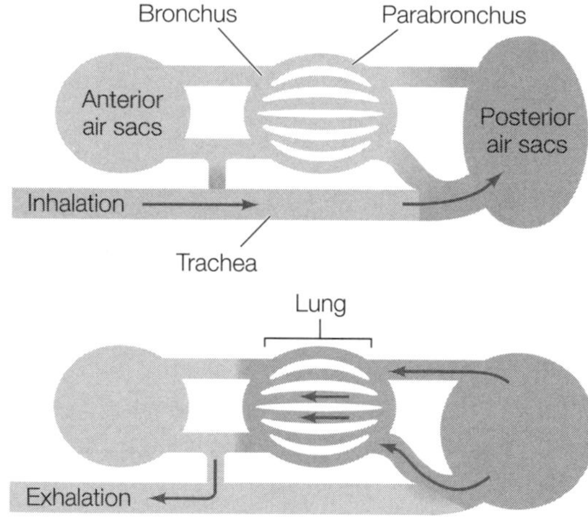

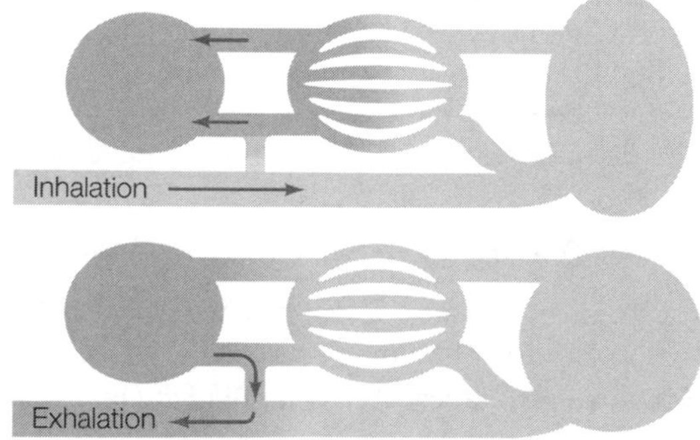

Knowledge and Synthesis Answers

1. **c.** Water breathing is more difficult than air breathing because the higher density of water makes it more expensive to move across the respiratory surfaces than air, it has less O_2 than air, and the O_2 content is very dependent on the temperature of the water. CO_2, however, dissolves easily in water and it is easily expelled, even in stagnant water.
2. **a.** The external gill, tracheae, and lungs are all examples of gas-exchange systems used by various animals to exchange both CO_2 and O_2. One of the special properties of all respiratory systems is that they increase surface area for diffusion.
3. **e.** All have a larger volume than the resting tidal volume.
4. **d.** The lungs of birds and mammals are both sites for the exchange of O_2 and CO_2 with blood in capillaries. The lungs of birds require two cycles in order for air to move through them, whereas the mammalian lung only requires one cycle. The mammalian lung is the only one that ends in alveoli and has an anatomical dead space. Flow of air is unidirectional in birds, but tidal in mammals.
5. **b.** The volume of the gas is not included in Fick's law of diffusion. All the other parameters are important in determining the rate of diffusion of gases in the respiratory system of animals.
6. **b.** Air at sea level with an atmospheric pressure of 760 mm Hg has a partial O_2 pressure of 160 mm Hg (21% of 760). Therefore, the partial pressure of O_2 at Antonito, Colorado, is 126 mm Hg (21% of 600).
7. **a.** Inhalation in the mammalian lung occurs by means of negative pressure produced by contraction of the diaphragm. Thus the pressure in the lungs falls below atmospheric pressure.
8. **d.** Movement of O_2 and CO_2 from the blood to the tissues always occurs by diffusion of O_2 and CO_2 down their partial pressure gradients.
9. **c.** The alveoli do not contain "air" with 20.9 percent O_2 because incoming air is mixed with air left in the dead space of the trachea and bronchi, which has had some of the O_2 removed by the lungs.
10. **b.** Hemoglobin is composed of four subunits, which act with positive cooperativity to bind O_2. O_2-binding myoglobin found in muscle has only one unit.
11. **c.** Hemoglobin has four subunits, each of which binds to one molecule of O_2 for a total of four molecules of O_2 bound to one hemoglobin molecule.
12. **d.** The Bohr effect describes the action of pH on the O_2-binding curve. Decreases in pH result in a net rightward shift of the entire O_2 equilibrium curve.
13. **a.** Carbon dioxide combines with H_2O in the plasma to form H^+ and HCO_3^-. The enzyme carbonic anhydrase catalyzes the reaction.

14. **a.** The majority of CO_2 is carried as bicarbonate ions (HCO_3^-) in the plasma.
15. **e.** All are involved in the control of breathing.
16. **d.** Areas of the brain above the medulla modify the breathing rate with respect to speech, eating, coughing, and emotional state.

Application Answers

1. Since it has no respiratory surfaces, an aquatic animal must rely on simple diffusion from the environment into its cells. For simple diffusion to be an effective means of O_2 transport, the animal must be very small, with all of its cells only a few cell layers from the environment. The metabolic rate of this animal is most likely very low in order to accommodate the lack of any special respiratory surfaces.
2. In the countercurrent exchange in the gills of fish, water and blood move in opposite directions. This results in blood with low O_2 content entering the gills and flowing past water with a higher O_2 content. O_2 diffuses down the partial pressure gradient into the blood. As blood moves along the lamella, it picks up O_2, but it always has less O_2 than the water moving in the opposite direction. In this way the blood is able to optimize O_2 uptake during the entire trip through the lamella.
3. The O_2-binding curve of myoglobin is shifted to the left in relationship to the O_2-binding curve of hemoglobin. This allows myoglobin to pick up O_2 and hold it at lower partial pressures of O_2.
4. Birds have an extremely efficient respiratory system characterized by the continuous unidirectional flow of air through the lungs. In contrast, mammals use tidal ventilation, a less-efficient method in which air flows in and exhaled gases flow out the same route. Tidal ventilation results in the mixing of fresh air with residual air in the lungs. O_2 availability decreases with altitude, and thus we would expect to see birds rather than mammals at very high elevations.
5. The tropical fish would be facing the double bind of needing more O_2 as the water temperature increased (due to its increasing metabolic rate with increasing water temperature), and yet facing lower levels of O_2 in the warmer water (since warm water holds less dissolved O_2 than cold water does).

50 Circulatory Systems

The Big Picture

- Cardiovascular systems are required when animals grow too big to acquire nutrients and eliminate waste by diffusion alone. Animals have evolved a wide variety of cardiovascular systems, which can be classified generally as open or closed. Open systems have relatively low pressures and have large sinuses that bathe tissues and organs in circulating fluid. Closed systems have an interconnected series of vessels that keep blood separated from interstitial fluid but allow gases, nutrients, and wastes to diffuse through their walls.
- During the evolution of vertebrates, the heart has changed from the simple two-chambered heart of fishes to the complex four-chambered hearts of crocodilians, birds, and mammals. Reptiles (turtles, lizards, snakes, and crocodilians) often breathe intermittently; their hearts allow the shunting of blood away from the lungs and into the systemic circuit. Amphibians pick up oxygen at their lungs and skin.
- The human heart is really two pumps in one. The left pump provides oxygenated blood under high pressure to the systemic tissues, and the right pump sends deoxygenated blood under low pressure to the lungs. A cardiac pacemaker sets the cardiac rhythm.
- The flow of blood through capillaries is highly regulated at the local level by changes in local surroundings; blood flow through arteries and veins is regulated by the combined action of neural activity and circulating hormones.

Common Problem Areas

- It is frequently thought that the hearts of amphibians and reptiles are primitive structures, awaiting "repair through evolution" to the more highly evolved bird and mammal heart. In fact, the hearts of amphibians and reptiles are highly adapted to suit their particular lifestyle, specifically their intermittent breathing and relatively low metabolic rate. By being able to shunt blood from pulmonary to systemic circuits, these animals can save the energy that would otherwise be wasted in sending blood to the temporarily nonventilated lungs.
- Students often fail to appreciate that the human heart is actually two entirely distinct pumps—the right heart pumping blood to the lungs and the left heart pumping blood to the body. The two pumps just happen to be packaged in the same structure—the heart. In fact, cardiologists refer to the "left heart" and "right heart," emphasizing the separate nature of these two pumps. Figuring out the flow of blood through the human heart becomes easier when it is considered from this perspective.

Study Strategies

- Determining the pattern of blood flow through the human heart will be easier if you mentally "unfold" the circulation into a great circle, in which there are two pumps at opposite sides of the circle with a vascular bed intervening between each pump. Once you have mastered this concept, you will see that the only way to place the two pumps in proximity in the same structure (the heart) is to twist the circle into a folded figure eight—the typical diagram of the circulation in the textbook.
- Following from the preceding, learn the various chambers and valves in the context of a sequence of structures in each of the right and left pumps of the heart, rather than as a list of terms printed on a complex, anatomically accurate diagram of the heart.
- Go to yourBioPortal.com to review the following tutorials and activities:

  Animated Tutorial 50.1 The Cardiac Cycle

  Interactive Tutorial: Blood Pressure and Heart Rate Regulation

  Web Activity 50.1 Vertebrate Circulatory System

  Web Activity 50.2 The Human Heart

  Web Activity 50.3 Structure of a Blood Vessel

Important Concepts

Small organisms do not need circulatory systems.

- An animal's circulatory system transports nutrients and wastes to and from tissues and cells.

- In the smallest multicellular animals, every cell within the body is no more than one to two cells from the environment. Such small animals do not require a circulatory system because materials can move efficiently into and out of cells by diffusion alone.
- Some aquatic invertebrates have a gastrovascular system, which consists of a highly branched central cavity where exchange of gases and nutrients occurs. Other organisms are flattened to increase the surface-to-volume ratio, ensuring that cells are close to the outside environment.
- Large active animals need a circulatory system to carry wastes, gases, and nutrients to and from all the cells in the body.

Circulatory systems may be classified structurally as open or closed.

- In an open circulatory system, blood and other tissue fluids are not separated; the fluid is called hemolymph (see Figure 50.1A and B). A pump may be present to help move hemolymph through the various large compartments, assisted by movements of the animal.
- In a closed circulatory system, blood is pumped through a series of interconnected closed vessels; it is kept separate from interstitial fluid (see Figure 50.1C). There are numerous advantages to closed systems compared to open ones: rapid transport of fluid in the closed vessels, the ability to divert the flow of blood between different tissues, and direct hormone and nutrient transport to the tissues.
- The closed system typical of vertebrates is composed of arteries and arterioles, which carry blood from the heart to the body; veins and venules, which carry blood from the body to the heart; and intervening capillary beds. In fishes, blood is pumped from the heart to the gills to the tissues and back to the heart. In birds and mammals, the systemic circuit flows from the heart to the body tissues, and the pulmonary circuit flows from the heart to the gas exchange organ.

The gills and systemic tissues play an important role in the fish circulatory system.

- Fishes pump blood with a heart that has one atrium and one ventricle. Blood leaves the heart and enters the gills, where gas exchange occurs. Some arterial blood pressure is lost in passing through the gills.
- Blood leaving the gills collects in the aorta and travels to the systemic organs and tissues, eventually returning to the heart.
- African lungfish have a well-vascularized lung for gas exchange that is formed embryonically from an outpocketing of the gut. A heart with a partially divided atrium helps separate oxygenated blood returning from the lung and deoxygenated blood returning from the tissues as they enter the single ventricle. A large proportion of oxygenated blood bypasses the gills and goes directly to the tissues. The lungfish heart and vascular system exhibit adaptations that partially separate systemic and pulmonary circuits.

Amphibians have partial separation of pulmonary and systemic blood.

- Amphibians have two atria: one that receives deoxygenated blood from the tissues and one that receives oxygenated blood from the lungs.
- Anatomical features of the ventricle help direct deoxygenated blood preferentially to the lungs and skin (which also acts as a gas exchange organ in amphibians) and oxygenated blood to the systemic body tissues. Even though the two circuits are not fully divided, blood tends to flow in parallel.

Reptiles can shunt blood away from the pulmonary circuit when they are not breathing.

- Turtles, snakes, and lizards have two atria and a ventricle that is internally divided into subchambers that allow for control of the distribution of blood to the lungs and body.
- During air breathing, blood from the right side of the ventricle is preferentially pumped to the lungs because there is a higher resistance in the systemic circuit than in the pulmonary circuit. Also, there is an asynchronous ejection from the heart because pumping occurs first from the right and then the left side of the ventricle. At the stage in the cardiac cycle in which blood from the left side leaves the heart, the resistance is higher in the pulmonary circuit. Thus, the last blood leaving the heart tends to be deoxygenated blood and is preferentially directed to the pulmonary circulation.
- When a reptile stops breathing, vessel constriction in the lungs causes pulmonary circuit resistance to increase, causing blood from the right side to largely bypass the pulmonary circulation and flow to the systemic circuit.
- Alligators and crocodiles are unique in that they have a four-chambered heart, resembling that of a bird or mammal. However, alligators and crocodiles retain a left aorta that arises from the right side of the heart. There is a small passageway between the left and right aortas through which blood can bypass the pulmonary circulation. Thus, crocodiles and alligators can operate their heart like that of a mammal, distributing blood equally to the pulmonary and systemic circulations, or they can partially bypass the lungs by directing blood from the left aorta into the right aorta and on to the systemic circulation.

Birds and mammals have a completely parallel circulatory system.

- Birds and mammals have a four-chambered heart that allows for complete separation and no mixing of systemic and pulmonary blood.
- This cardiovascular arrangement maximizes O_2 transport by eliminating mixing between pulmonary and systemic circulations.

- The pulmonary circulation operates at low pressure, protecting the delicate lung membranes, whereas the systemic circuit operates at high pressures, allowing the perfusion of tissues and organs distant from the heart.

The human heart contains separate right and left sides pumping to the lungs and tissues, respectively.

- The human heart has two atria and two ventricles (see Figure 50.2). Between each atrium and ventricle is an atrioventricular valve, which stops backflow of blood from the ventricle into the atrium. The pulmonary valve resides between the right ventricle and the pulmonary artery and the aortic valve resides between the left ventricle and the aorta; these two valves help maintain one-way flow through the heart.
- Deoxygenated blood flows from the body through the superior vena cava and inferior vena cava into the right atrium and into the right ventricle by passive filling during heart relaxation. During heart contraction, blood flows from the right ventricle to the pulmonary artery and the lungs, where blood becomes oxygenated.
- Oxygenated blood from the lungs flows back to the heart via pulmonary veins into the left atrium, then into the left ventricle, and finally out through the aorta to the systemic tissues.
- Resistance in the blood vessels of the body's systemic tissues is higher than in the lungs, so the left ventricle must be larger and more muscular than the right ventricle to generate more force while pumping the same volume of blood.
- Systole is the contraction of the ventricle, and diastole is the relaxation of the ventricle. At the very end of diastole, the atria contract. Together these two phases make up the cardiac cycle.
- During the cardiac cycle the pressure is highest during ventricle contraction (systole) and lowest during relaxation (diastole) (see Figure 50.3).
- Blood pressure can be measured using a sphygmomanometer (see Figure 50.4). A typical reading for a healthy young adult is 120 (systolic pressure) over 80 (diastolic pressure).

The human heartbeat is generated in the heart's pacemaker.

- The pacemaker, or sinoatrial node, is a small section of highly modified muscle tissue located at the junction of the superior vena cava and right atrium.
- The pacemaker rhythmically produces action potentials that pass on to the rest of the heart and stimulate a heartbeat (see Figure 50.6).
- Gap junctions connect cardiac muscle cells and thus allow the action potentials to spread rapidly from cell to cell.
- The action potential moves from the atria to the ventricles through the atrioventricular node (AV node). The AV node slows down the transmission of the action potential from the atria to the ventricles. The action potential passes rapidly down the bundle of His and to the Purkinje fibers, fanning out in the ventricular tissue. From the Purkinje fibers the action potential spreads through both ventricles, causing contraction to occur slightly after the contraction of the atria.
- The autonomic nervous system also plays a role in the control of the heartbeat. Sympathetic activity increases heart rate, and parasympathetic activity decreases heart rate (see Figure 50.5).
- The electrocardiogram (ECG or EKG) measures the electrical activity of the heart as a complex wave pattern (see Figure 50.8). The wave has five parts: P (depolarization and contraction of the atria), Q, R, and S (depolarization and contraction of the ventricles), and T (relaxation and repolarization of the ventricles).

Blood is a tissue composed of fluid plasma and many different cell types.

- The fluid matrix of blood is known as plasma and contains water, salts, and proteins. The hematocrit (also called packed-cell volume) is the percentage of blood volume composed of cells. A normal hematocrit is about 42 percent for women and 46 percent for men.
- Several different cell types in blood carry out a wide range of functions (see Figure 50.9). The cell types include the erythrocytes, basophils, eosinophils, neutrophils, lymphocytes, monocytes, and platelets.
- Erythrocytes, or red blood cells, are shaped like biconcave discs when mature. Their function is to transport respiratory gases. In mammals, but not other vertebrates, mature red blood cells lack a nucleus and other organelles.
- Erythrocytes are produced in stem cells located in the marrow of bone. Once they are released into the bloodstream they survive for about 120 days. The hormone erythropoietin is produced in the kidneys and regulates the production of red blood cells.
- The spleen serves as a reservoir for red blood cells.
- Platelets, which help to start blood clotting, are also found in blood and are produced by the breaking off of cell fragments from megakaryocytes in the bone marrow. Platelets bind to the edges of a break in a vessel wall, not only providing a mechanical plug but also starting a chemical chain reaction that leads to the conversion of the circulating protein prothrombin into thrombin. Thrombin stimulates the formation of sticky fibrin threads that form a meshwork and cover the hole in the vessel (see Figure 50.10).
- Plasma contains many gases, ions, nutrients, proteins, and other molecules. The plasma also contains the hormones that are circulated by the blood to target cells. The composition of plasma is very similar to that of interstitial fluid, except that proteins are present in higher concentrations in plasma.

The vascular system is composed of arteries, capillaries, and veins.

- Arteries carry blood away from the heart to the body under relatively high pressure. Veins carry blood toward the heart from the body under low pressure (see Figure 50.11). The very fine capillaries interconnect the arteries with the veins in the tissues, dissipate pressure, and are sites for all exchanges of gases, nutrients, and wastes.
- Arteries and the smaller arterioles are elastic, allowing them to stretch during systole and rebound during diastole. In vertebrates, the walls of vessels contain smooth muscle that controls the vessel's resistance by causing changes in its diameter. Arteries and arterioles are known as resistance vessels.

Capillaries are sites for exchange of gases, nutrients, and wastes, and the locations of considerable fluid loss and recovery.

- Capillaries connect the arterioles with the venules in the tissues. Capillaries have a large cross-sectional area compared with arteries and veins and extremely thin walls to allow the diffusion of gases, nutrients, and wastes (see Figure 50.11).
- Blood moves slowly as it enters the capillaries. In a process known as filtration, blood pressure from the arterial side forces fluids, ions, and small molecules out of the capillaries through small holes called fenestrations. Near the venule end of the capillaries, the difference in osmotic potential between the plasma inside the capillaries and the surrounding fluids creates an osmotic pressure resulting in fluid recovery back into the capillaries. Thus the net flow of water between the plasma and tissue fluid is determined by the difference in blood pressure and osmotic pressure (see Figure 50.13). The two opposing forces, blood pressure and osmotic pressure, are known as Starling's forces.
- Edema is tissue swelling that results from the accumulation of fluid in extracellular spaces. This condition is associated with protein starvation (which leads to low concentrations of proteins in the blood) and inflammation (which causes increased permeability of capillaries).
- Some capillaries are selective about which ions and molecules can pass through them, whereas others are less selective. The blood–brain barrier consists of highly selective brain capillaries.

Blood returns to the heart through veins.

- Blood returns to the heart at low pressure in expandable veins. Because of the low pressure, veins that act against gravity have valves, and blood flowing toward the heart is helped by the movement of the surrounding skeletal muscle (see Figure 50.14). Gravity can cause pooling of blood in the veins if a person is inactive for a long period of time.
- Veins are called capacitance vessels because of their ability to stretch and store blood.
- When veins return a greater volume of blood to the heart, the heart contracts more forcefully. This is explained by the Frank–Starling law, which states that if cardiac muscle cells are stretched (as happens when the volume of returning blood increases), they contract more forcefully.

The lymphatic system returns tissue fluid not reclaimed by the capillaries to the blood.

- Once interstitial fluid enters the lymphatic system, it is called lymph. In humans, lymph is moved through the lymphatic vessels up the body by surrounding skeletal muscle contractions. Lymph is returned to the blood through the thoracic ducts that empty into large veins at the base of the neck.
- Lymph nodes exist in mammals and birds and contribute to the defenses of the body. Foreign materials and microorganisms are removed by the phagocytic action of white blood cells in the lymph nodes.

Cardiovascular disease affects the arteries and veins.

- Hardening of the arteries is known as atherosclerosis.
- Plaque deposits begin to form at sites of endothelial damage, and lipids tend to accumulate on the deposits. Connective tissue invades the deposits and makes the wall of the artery less elastic. Arteries narrow as plaque deposits grow. Blood platelets stick to the plaque, and this can lead to the formation of a blood clot, or thrombus.
- The coronary arteries that supply the heart with blood are highly susceptible to blockage, resulting in a heart attack (myocardial infarction).
- An embolus is a piece of a thrombus that breaks off, travels in the vessels, and then lodges in a vessel. When this happens in the brain, it is known as a stroke. A stroke results in denial of blood flow to a particular region of the brain, causing paralysis, loss of cognitive abilities, or death.
- Genetics and age are the main factors that determine your risk of developing atherosclerosis. Environmental factors also contribute to risk and include high-fat and high-cholesterol diets, smoking, and lack of exercise.

The circulatory system is under hormonal, neural, and local control.

- Blood flow in the capillary beds of many tissues is controlled locally by autoregulatory mechanisms. Precapillary sphincters at the arterial end of the capillaries control the flow of blood, with contraction of these sphincters limiting or stopping blood flow through the capillary bed (see Figure 50.16). Low levels of O_2 and high levels of CO_2 can open the precapillary sphincters. The increased supply of blood brings in more O_2 and removes more CO_2; this response is called hyperemia (excess blood).
- Norepinephrine from sympathetic neurons and epinephrine released from the adrenal medulla cause

arteries and arterioles to contract, reducing blood flow and increasing central blood pressure.

- Autonomic control of heart rate and constriction of blood vessels occurs from the medulla. Stretch receptors (known as baroreceptors) in the aorta and carotid arteries help regulate blood pressure. With rising blood pressure they inhibit sympathetic output and cause the heart to slow and arterioles to dilate. When pressure is low, sympathetic output is stimulated, increasing heart rate and constriction of arterioles. Chemoreceptors in the aorta and carotid arteries sense changes in blood composition.
- In response to a fall in arterial pressure (signaled by decreased activity of baroreceptors), the posterior pituitary releases antidiuretic hormone (ADH, also known as vasopressin). ADH stimulates the kidneys to reabsorb more water, and this results in increased blood volume and pressure.

Test Yourself

Diagram Exercise

The diagrams below trace the course of the heartbeat. Label the following structures on the left diagram: sinoatrial node, atrioventricular node, bundle of His, Purkinje fibers, atria, and ventricles. On the middle and right diagram, indicate with arrows the path of the action potentials at the particular stage indicated at the top of each diagram.

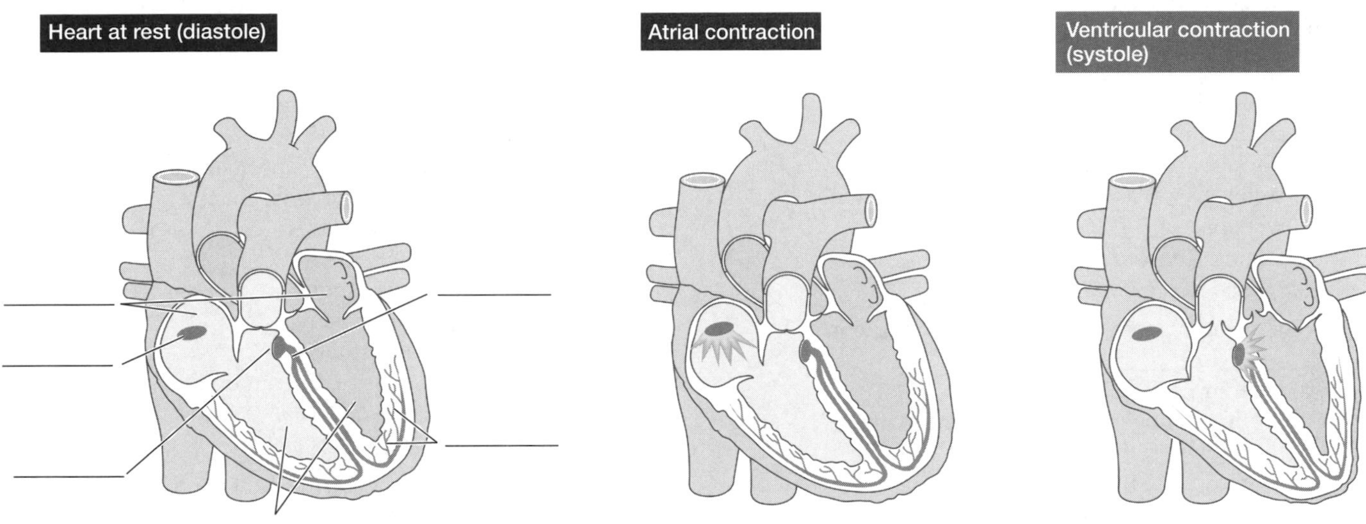

Textbook Reference: *50.3 How Does the Mammalian Heart Function? p. 1055*

Knowledge and Synthesis Questions

1. Which of the following is *not* one of the reasons that closed circulatory systems are more efficient than open circulatory systems?
   a. Closed systems rely exclusively on simple diffusion for transport, whereas open systems rely on pumping mechanisms.
   b. Transport within closed systems is more rapid than in open systems.
   c. Blood can easily be directed to specific areas in closed systems, but not in open systems.
   d. Compared to open systems, closed systems operate better under higher pressure.
   e. In closed systems molecules and cells that transport hormones and nutrients can be kept in vessels until they unload their goods at specific tissues.

   Textbook Reference: *50.1 Why Do Animals Need a Circulatory System? p. 1048*

2. When reptiles stop breathing, for example on extended dives,
   a. blood is shunted from the systemic circuit to the pulmonary circuit.
   b. blood is shunted specifically to the brain.
   c. blood is shunted from the pulmonary circuit to the systemic circuit.
   d. blood is shunted to the skin for gas exchange.
   e. the pattern does not change between the breathing and the nonbreathing periods.

   Textbook Reference: *50.2 How Have Vertebrate Circulatory Systems Evolved? p. 1050*

3. In which of the following would the highest blood pressure be recorded?
   a. In the ventricle supplying blood to the gills of a fish
   b. In the anterior dorsal artery of an ant
   c. In the pulmonary vein of a frog
   d. In the ventricle supplying blood to the systemic circuit of a bird
   e. In the vessels leaving the gills of a clam

   Textbook Reference: *50.2 How Have Vertebrate Circulatory Systems Evolved? pp. 1048–1051*
4. Blood consists of a fluid fraction consisting of _______ and a solid fraction consisting of _______.
   a. plasma; water, erythrocytes, and platelets
   b. erythrocytes; leukocytes, macrophages, and platelets
   c. plasma; erythrocytes, platelets, and leukocytes
   d. leukocytes; erythrocytes and platelets
   e. interstitial fluid; stem cells

   Textbook Reference: *50.4 What Are the Properties of Blood and Blood Vessels? pp. 1056–1057*
5. Red blood cells are
   a. biconcave cells containing hemoglobin.
   b. spherical cells containing hemoglobin.
   c. spherical cells capable of amoeboid motion and containing hemoglobin.
   d. biconcave cells containing platelets.
   e. biconcave cells that function in the immune response.

   Textbook Reference: *50.4 What Are the Properties of Blood and Blood Vessels? pp. 1056–1057*
6. Blood clotting pathways cause the conversion of _______ to _______.
   a. vitamin K; prothrombin
   b. fibrin; fibrinogen
   c. thrombin; prothrombin
   d. prothrombin; thrombin
   e. fibrinogen; thrombin

   Textbook Reference: *50.4 What Are the Properties of Blood and Blood Vessels? p. 1058*
7. Which of the following regions of the vascular bed is the actual site of gas exchange with surrounding tissue?
   a. Arteries
   b. Capillaries
   c. Lymphatic vessels
   d. Veins
   e. Venules

   Textbook Reference: *50.4 What Are the Properties of Blood and Blood Vessels? pp. 1058–1060*
8. Which is the correct sequence of parts through which cardiac action potentials pass?
   a. Purkinje fibers, AV node, SA node, bundle of His, atrial fibers
   b. AV node, atrial fibers, SA node, bundle of His, Purkinje fibers
   c. SA node, bundle of His, atrial fibers, AV node, Purkinje fibers
   d. Purkinje fibers, bundle of His, AV node, atrial fibers, SA node
   e. SA node, atrial fibers, AV node, bundle of His, Purkinje fibers

   Textbook Reference: *50.3 How Does the Mammalian Heart Function? pp. 1054–1055*
9. The purpose of the AV node is to _______, and the purpose of the Purkinje fibers is to _______.
   a. create simultaneous atrial and ventricular depolarization; speed up transmission of the cardiac impulse into the ventricle
   b. delay ventricular depolarization relative to atrial depolarization; insulate the cardiac impulse from the general ventricular fibers
   c. delay ventricular depolarization relative to atrial depolarization; ensure the cardiac impulse spreads rapidly and evenly throughout the ventricles
   d. delay atrial depolarization relative to ventricular depolarization; transmit the cardiac impulse to very small, localized groups of ventricular fibers
   e. initiate each heartbeat; transmit impulses to the atria.

   Textbook Reference: *50.3 How Does the Mammalian Heart Function? pp. 1053–1055*
10. The atrial walls are _______ the ventricular wall, and pressure generated in the atrial chambers is _______ the pressure in the ventricles.
    a. thinner than; higher than
    b. thinner than; lower than
    c. thicker than; higher than
    d. thicker than; lower than
    e. the same thickness as; the same as

    Textbook Reference: *50.3 How Does the Mammalian Heart Function? pp. 1051–1052*
11. The left ventricle exceeds the right ventricle in the
    a. amount of blood that enters during heart contraction.
    b. volume expelled during contraction.
    c. pressure developed during contraction.
    d. speed with which it contracts.
    e. All of the above

    Textbook Reference: *50.3 How Does the Mammalian Heart Function? p. 1052*
12. Which of the following structures of the human lymphatic system acts primarily as a filter for detecting and destroying microorganisms in lymph traveling through major lymph vessels?
    a. Lymph nodes
    b. Thymus
    c. Lymph capillaries
    d. Tonsils, but not the appendix
    e. Thoracic ducts

    Textbook Reference: *50.4 What Are the Properties of Blood and Blood Vessels? p. 1061*
13. The net loss of fluid from blood capillaries increases if
    a. plasma filtration decreases.

b. the osmotic pressure of plasma increases.
c. blood pressure increases in the capillaries.
d. the osmotic pressure of interstitial fluid decreases.
e. blood pressure drops below osmotic pressure.

Textbook Reference: *50.4 What Are the Properties of Blood and Blood Vessels? pp. 1059–1060*

14. Which of the following statements about the control of circulation in humans is *false*?
a. Sympathetic nerve input to skeletal muscle causes the blood vessels in the muscle to dilate.
b. Blood flow can be regulated by autonomic nerve signals emanating from cardiovascular control centers in the medulla of the brain.
c. Carotid artery chemoreceptors detect low O_2 levels in the blood and promote increased blood pressure.
d. Hormones such as angiotensin and vasopressin cause venules to constrict.
e. Autoregulatory mechanisms can cause constriction or dilation of arterioles.

Textbook Reference: *50.5 How Is the Circulatory System Controlled and Regulated? pp. 1062–1063*

15. Which of the following statements about vertebrate circulatory systems is *false*?
a. Crocodilians have a four-chambered heart.
b. In birds and mammals, pressures in the pulmonary circuit are lower than those in the systemic circuit.
c. The ventricle of turtles, snakes, and lizards is partly divided by a septum.
d. Amphibians have two atria and one ventricle.
e. In fishes, blood passes through the gills, returns to the heart, and then is pumped to the body.

Textbook Reference: *50.2 How Have Vertebrate Circulatory Systems Evolved? pp. 1048–1050*

Application Questions

1. What are the important differences between the circulatory system of a fish and that of a turtle?
Textbook Reference: *50.2 How Have Vertebrate Circulatory Systems Evolved? pp. 1048–1050*

2. The human heart pumps blood to the pulmonary and the systemic circuits. Describe the path that blood takes as it moves through the circulatory system, starting at the right ventricle.
Textbook Reference: *50.3 How Does the Mammalian Heart Function? pp. 1051–1052*

3. You are about to take a final exam in your freshman biology course and are very nervous. What responses will your circulatory system have to your nervousness?
Textbook Reference: *50.5 How Is the Circulatory System Controlled and Regulated? pp. 1063–1064*

4. Blood moves through the arteries and veins at a faster velocity than the blood moving through the capillaries. What is the cause and effect of blood slowing down as it moves through the capillaries?
Textbook Reference: *50.4 What Are the Properties of Blood and Blood Vessels? pp. 1058–1060*

5. Many animals with open circulatory systems are relatively inactive. Explain how insects, with their open circulatory systems, maintain their high levels of activity.
Textbook Reference: *50.1 Why Do Animals Need a Circulatory System? p. 1048*

Answers

Diagram Exercise Answer

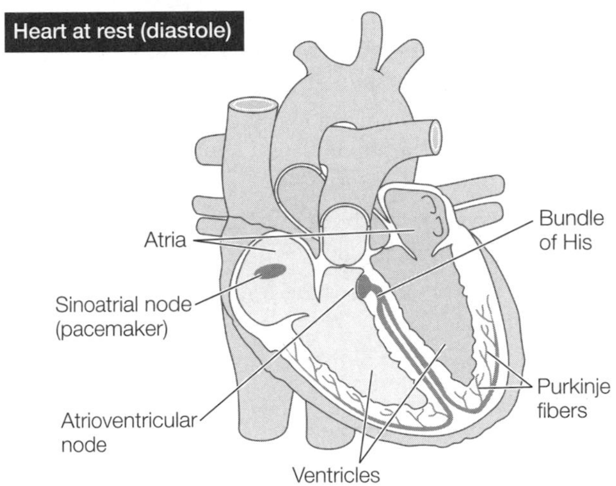

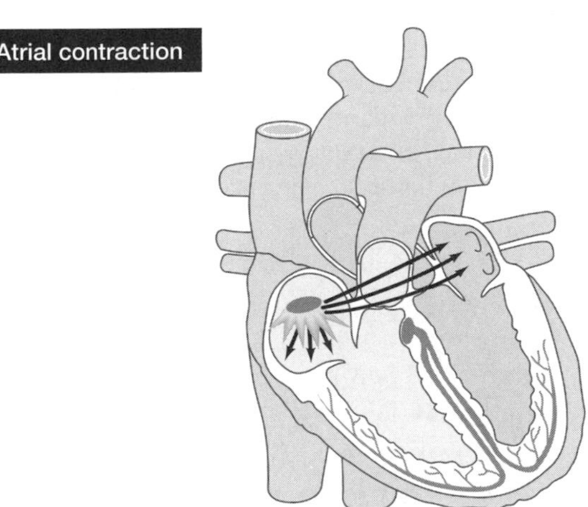

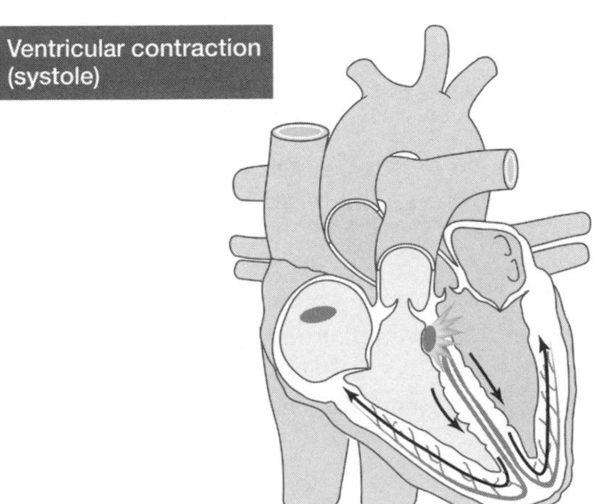

Knowledge and Synthesis Answers

1. **a.** Both closed systems and open systems rely on pumping mechanisms to distribute fluid throughout the body.
2. **c.** During non-breathing periods , the blood is shunted from the pulmonary circuit to the systemic circuit. The resistance in the pulmonary circuit increases, resulting in the shunt.
3. **d.** Of the vertebrate groups, mammals and birds tend to have the highest blood pressures. Therefore, the ventricle supplying blood to the systemic circuit of a bird has the highest pressure.
4. **c.** The fluid portion of blood is the plasma; three components of the solid fraction are erythrocytes, platelets, and leukocytes.
5. **a.** Red blood cells are biconcave cells that contain the oxygen-binding hemoglobin.
6. **d.** Blood clotting involves the conversion of prothrombin to thrombin.
7. **b.** Gas exchange at the tissues occurs across the capillaries.
8. **e.** The cardiac action potential passes through the SA node, atrial fibers, AV node, bundle of His, and finally the Purkinje fibers.
9. **c.** The AV node delays the ventricular depolarization relative to atrial depolarization, so atrial contraction occurs before ventricular contraction. The Purkinje fibers ensure that the action potential spreads rapidly and evenly throughout the ventricles.
10. **b.** The atrium has thinner walls and generates lower pressures than the ventricles.
11. **c.** The left ventricle generates a greater pressure in the blood flowing to the systemic circuit than the right ventricle with blood flowing to the pulmonary circuit.
12. **a.** The lymph nodes filter and destroy microorganisms that are traveling through the lymphatic system.
13. **c.** Changes in blood pressure will change the amount of filtration that occurs at the capillaries. Thus, increases in blood pressure will result in greater rates of filtration.
14. **d.** Angiotensin and vasopressin control constriction of the arterioles, not the venules.
15. **e.** In fishes, blood passes through the gills and directly out to the body; it does not return to the heart for additional pumping.

Application Answers

1. The fish have a two-chambered heart (one atrium and one ventricle), with blood flowing from the heart of the fish to the gills and then directly to the body. Turtles have a more complex three-chambered heart (two atria and one ventricle). The ventricle, however, is partially divided, so outflow can be directed to the pulmonary circuit and to the systemic circuit. Thus the circulatory system of a fish delivers blood to the respiratory organ and the body tissues in succession, whereas the circulatory system of a turtle delivers blood to the respiratory organ and the body tissues in parallel.
2. Blood from the right ventricle exits through the pulmonary valve into the pulmonary artery and then goes to the lungs. From the lungs, the blood returns to the left atrium via the pulmonary veins and passes through the atrioventricular valve into the left ventricle. Blood from the left ventricle is pumped to the systemic circuit through the aortic valve and into the aorta. From the aorta, blood moves to the tissues in arteries and arterioles, passes through tissues in the capillaries, and returns via the venules, veins, and ultimately the superior vena cava and inferior vena cava to the right atrium of the heart. From the right atrium, blood is pumped back where we started—into the right ventricle through the atrioventricular valve.
3. Stress and nervousness elicit the fight-or-flight response. Signals from the medullary cardiovascular control center stimulate sympathetic inputs from the autonomic nervous system. In response to sympathetic inputs, the adrenal gland releases epinephrine into the bloodstream. Epinephrine stimulates an increased heart rate and arterial pressure. The blood flow to the smooth muscles decreases due to constriction of blood vessels, and blood is diverted away from areas such as the digestive tract to the skeletal muscle needed for fight or flight.
4. Blood is pumped from the heart into the arteries, where it has a relatively fast velocity. Upon entering the capillaries, much of the velocity is lost. The capillaries have a much higher total cross-sectional area than the arteries, and blood flow velocity is inversely proportional to the cross-sectional area; thus, the more area through which the blood flows, the more slowly it will go. Just as water in a stream that widens to become a river flows more quickly upstream than it does downstream, blood moves slowly through the capillaries and more quickly through the arteries and veins. The slow velocity of blood through the capillaries allows for greater exchange of materials with the tissues.
5. Despite having an open circulatory system, insects are able to maintain their high levels of activity because they do not rely on their circulatory systems for the exchange of respiratory gases. Instead, they rely on a system of air-filled tubes, called tracheae, which serve as sites for gas exchange.

51 Nutrition, Digestion, and Absorption

The Big Picture

- Animals have evolved a wide variety of mechanisms for acquiring needed energy and raw materials for metabolism. All animals are heterotrophs; they can achieve energy and nutrient input by eating plants, animals, or both.
- The digestive system is specialized for efficient digestion of the type of food it must break down. The relatively short guts of carnivores have acid and protein-reducing enzymes. The longer guts of herbivores have numerous holding chambers for the laborious process of breaking down plant material. The teeth of carnivores and herbivores also reflect dietary differences.
- Food passing down the length of the gut is first broken down mechanically into small pieces and then broken down chemically into small molecules that can be easily absorbed across the gut. As nutrients and water are absorbed from ingested plant and animal material, the remaining material forms the feces, which are eventually eliminated from the body.
- The process of digestion is under both neural and hormonal control. The brain plays a role in regulating food intake, and the digestive tract has an intrinsic nervous system to regulate digestive functions. Numerous hormones control the peristaltic movements of the gut and the secretion of enzymes and other chemicals involved in digestion. Hormones are also involved in regulating food intake.

Common Problem Areas

- The study of digestive function may seem uninteresting or, worse, unpleasant to students. Yet, some of the best and clearest examples of negative feedback and homeostasis are to be found in the study of digestion.
- Sometimes students think that there is a single digestive mechanism for food and are surprised to learn of the complexity of the process. Be sure to recognize that different food components—proteins, carbohydrates, fats—are digested, absorbed, and transported by very different mechanisms.

Study Strategies

- Although the guts of higher animals are structurally complex, they can be divided functionally into a foregut, midgut, and hindgut. If you break down the structure of the gut into these sections and determine which physiological process occurs in each section, you will have divided the task into smaller pieces.
- Because different types of food are digested and absorbed by different mechanisms, learn how a particular type of food is digested and match the appropriate enzymes to that process. For example, it will be much easier to remember how proteins are digested, rather than to recount a long list of digestive enzymes and then try to pick the one that might break down protein. In short, learn the process, not the list!
- Go to yourBioPortal.com to review the following tutorials and activities:

  Animated Tutorial 51.1 The Digestion and Absorption of Fats

  Animated Tutorial 51.2 Insulin and Glucose Regulation

  Interactive Tutorial: Parabiotic Mice: Regulation of Food Intake

  Web Activity 51.1 Mineral Elements Required by Animals

  Web Activity 51.2 Vitamins in the Human Diet

  Web Activity 51.3 Mammalian Teeth

  Web Activity 51.4 The Human Digestive System

Important Concepts

All animals must eat to acquire energy for survival.

- Animals are heterotrophs, meaning that they must get their energy and nutrients from the fats, carbohydrates, and proteins of plants and other animals. The energy that animals acquire from food is used to make ATP, a high-energy phosphate compound that provides energy for cellular work. The conversion of ATP during cellular work creates heat as a by-product.
- The calorie and kilocalorie (Calorie) are measures of heat energy.

- Metabolic rate is a measure of the total energy used by an animal. Basal metabolic rate is the resting metabolic rate or the energy consumption needed for all essential physiological functions.
- Energy needed for metabolism can be obtained from food taken in or from stored food. Fats, carbohydrates, and proteins are the components of food that provide energy. Fat has the highest energy content and is the most important form of energy stored in an animal's body.
- Energy budgets compare calories consumed with calories expended and allow scientists to apply a cost–benefit approach when analyzing behaviors.

Fuel molecules are stored in the body.

- Carbohydrates are stored as glycogen in the liver and muscles.
- Fat has more energy per gram than carbohydrates and can be stored more compactly, making it the most important form of stored energy in animals.
- An animal is undernourished when it takes in too little food to meet its energy requirements. An animal is overnourished when it consistently takes in more food than it needs to meet its energy requirements.

Essential molecules must be obtained from a food source.

- The acetyl group supplies the carbon skeleton for larger organic molecules. Animals cannot produce acetyl groups through their own metabolism and so must obtain them from another source (see Figure 51.4).
- Amino acids needed to make proteins are obtained by breaking down proteins in food. Some amino acids can be synthesized, but the essential amino acids can only be obtained from outside sources.
- Essential amino acids are needed for proper protein production. In humans, the essential amino acids are isoleucine, leucine, lysine, methionine, phenylalanine, threonine, tryptophan, and valine. All eight essential amino acids can be found in meat, eggs, soybean products, and milk. Vegetarian diets should include complementary dietary mixtures, such as grains and legumes, to avoid protein malnutrition (see Figure 51.5). Human infants require four additional amino acids (histidine, tyrosine, cysteine, and arginine), and certain human populations require other amino acids, which are described as conditionally essential amino acids.
- Ingested proteins are not used in their ingested form because, as macromolecules, they are not absorbed by cells in the gut. Also, their structures and functions are species-specific. In addition, foreign proteins in the body would be recognized as invaders and would be attacked. Instead, proteins are broken down into peptides and amino acids.
- Essential fatty acids, such as linoleic acid, must also be acquired in food and are necessary for producing membrane phospholipids.
- The nutrients needed for survival are categorized as either macronutrients or micronutrients, depending on the quantity needed (see Table 51.1).
- Vitamins are carbon compounds that are required in small amounts and often function as coenzymes. Fat-soluble vitamins accumulate in the liver, whereas water-soluble vitamins are eliminated in urine. Vitamin D is needed to help in the uptake of calcium into the body.

Diseases can be caused by nutrient deficiency.

- Malnutrition is caused by the lack of any essential nutrient in the diet. Chronic malnutrition leads to a particular deficiency disease. For example, protein deficiency leads to kwashiorkor, which is characterized by swelling, distension of the abdomen and degeneration of the liver and other organs. When left untreated, protein deficiency will eventually cause death.
- Humans require 13 vitamins. Shortages in vitamins can lead to disease (see Table 51.2).
- Nutrient deficiencies may be due to the body's inability to absorb essential nutrients even when they are present in the diet. For example, pernicious anemia is caused by the inability to absorb vitamin B_{12} due to inadequate production of intrinsic factor, a peptide needed for absorption of the vitamin.

There are several different ways for animals to feed.

- Two types of organisms feed on dead matter: saprobes (also called decomposers or saprotrophs) absorb nutrients from decaying organic matter, whereas detritivores actively feed on dead matter. Predators feed on other organisms, and there are three main types: herbivores feed on plants, carnivores feed on animals, and omnivores feed on plants and animals.
- Teeth are composed of a hard outer surface (enamel), a bony layer (dentine), and a pulp cavity that contains blood vessels and nerves (see Figure 51.6A).
- Mammals have four types of teeth: incisors, canines, premolars, and molars. The shapes and organization of mammalian teeth reflect different diets (see Figure 51.6B). Because plant tissue is difficult to break down, herbivores have teeth that are good for clipping, shearing, crushing, and grinding. Because carnivorous animals must capture, hold, and eat other animals, their teeth are adapted for stabbing, holding, ripping, and shredding their prey. Omnivorous animals have teeth adapted for multiple purposes.

Digestion occurs in either a cavity or a tube in most animals.

- Food is generally digested extracellularly by enzymes that break it down. Most enzymes are produced in an inactive form called a zymogen and are converted to the active form at the site of use. Enzymes do not digest the gut because of a protective layer of mucus that lines it. There are several classes of digestive enzymes:

- Carbohydrases hydrolyze carbohydrates.
- Proteases hydrolyze proteins.
- Peptidases hydrolyze peptides.
- Lipases hydrolyze fats.
- Nucleases hydrolyze nucleic acids.

- In simple animals, a single opening that functions as both a mouth and an anus connects a gastrovascular cavity to the environment.
- More complex animals have tubular guts with a mouth that takes in food and an anus through which digestive wastes are eliminated. Food enters the mouth and, in some animals, is broken up by grinding mechanisms that may include teeth or mandibles; many birds grind their food using small stones or gravel in an upper region of the gut called the gizzard. After grinding, food moves to a storage chamber, such as the stomach or crop, where digestion may or may not occur. It then enters the midgut or intestine, where most of the nutrients are absorbed. The hindgut absorbs water and ions and stores waste until it is expelled from the anus during defecation.
- The parts of the gut that absorb nutrients have increased surface area in the form of fingerlike projections known as villi, which in turn have microscopic projections called microvilli (see Figure 51.8).

The gut wall is composed of the same layers of tissue types throughout its length.

- Four layers of different cell types form the wall of the gut (see Figure 51.10).
- The innermost layer is a layer of mucosa that surrounds the cavity of the gut, or lumen. Cells of the mucosal epithelium have secretory functions; some secrete mucus, while others secrete enzymes, hormones, or hydrochloric acid. In some regions of the gut, cells of this layer have absorptive functions.
- The submucosa, lying under the mucosa, contains blood and lymph vessels that transport the nutrients. This layer also contains a network of nerves that regulates gut activities.
- Smooth muscle surrounds the submucosa in both circular and longitudinal layers. These layers help move the contents of the digestive tract by peristalsis.
- The peritoneum is a membrane that lines the abdominal cavity and covers the gut and other abdominal organs.

Food passes through the gut by smooth muscle contraction.

- Food enters the mouth and is swallowed, passing through the pharynx into the esophagus (see Figure 51.11A).
- Food is then moved along by peristalsis, which is a wave of muscle contraction that moves food through the gut (see Figure 51.11B).
- The esophageal sphincter helps keep food from reentering the esophagus from the stomach, and the pyloric sphincter controls the movement of stomach contents into the intestines.

Several enzymes help break down food.

- The enzyme amylase, secreted by the salivary glands, begins the breakdown of starch in the mouth.
- In the stomach, the major enzyme is pepsin, which breaks down proteins into shorter peptides. Pepsin is initially secreted by gastric glands of the stomach in the zymogen form, pepsinogen (see Figure 51.12). Low pH activates the conversion of pepsinogen to pepsin through autocatalysis, a positive feedback process.
- The secretion of hydrochloric acid by the gastric glands in the wall of the stomach maintains a low pH.
- Ulcers of the stomach are locations where the mucosal lining has been damaged by the bacterium *Helicobacter pylori*.
- Small amounts of a few substances, such as alcohol, aspirin, and caffeine, are absorbed across the walls of the stomach.
- The acidic mixture of digesting food and gastric juices is known as chyme, which slowly passes into the intestines through the pyloric sphincter.

Most chemical digestion occurs in the small intestine.

- Carbohydrate and protein digestion continue in the small intestine.
- In humans, the small intestine can be more than six meters in length, with villi and microvilli contributing to a large surface area necessary for the absorption of nutrients. The small intestine is divided into the duodenum, jejunum, and ileum.
- Fat digestion begins in the small intestine with the help of secretions and enzymes from the liver and pancreas.
- The liver secretes bile through a side branch of the hepatic duct into the gallbladder, where the bile is stored. When lipids are present, bile is secreted by the gallbladder into the duodenum through the common bile duct (see Figure 51.14). Bile molecules have hydrophobic ends, which attach to the tiny fat droplets to stabilize particles of fat in the form of micelles. This increases the exposed surface area of the lipids for digestion by fat-digesting enzymes, the lipases (see Figure 51.15A).
- The zymogen trypsinogen is secreted by the pancreas and is converted into active trypsin by enterokinase. The pancreas produces several zymogens that are released through the pancreatic duct, which joins the common bile duct and empties into the duodenum. The zymogens are converted to the active form by trypsin. The pancreas also helps maintain a slightly

alkaline pH in the duodenum by secreting bicarbonate ions. This helps neutralize the acidic chyme entering from the stomach.

The small intestine is the site of most nutrient absorption.

- Final digestion of proteins and carbohydrates occurs among the microvilli of the small intestine. Proteins are broken down into amino acids by peptidases. The enzymes maltase, lactase, and sucrase break down disaccharides into monosaccharides. The mucosal cells secrete the peptidases and maltase, lactase, and sucrase.
- Absorption occurs via several methods, including diffusion, facilitated diffusion, active transport, osmosis, and cotransport.
- Fats move through the plasma membranes as fatty acids and monoglycerides (see Figure 51.15B). In the mucosal cells they are converted into triglycerides and combine with cholesterol and phospholipids to form chylomicrons that pass into the lymphatic system. Bile is not absorbed across the membrane of the duodenum, but is actively reabsorbed in the ileum and returned to the liver.
- The hepatic portal vein carries blood from the digestive organs to the liver, where nutrients are either stored or converted to needed molecules.

The large intestine is the site of water and ion absorption.

- Material entering the large intestine has had most of the nutrients removed, but still contains important ions and water. The water and ions are absorbed across the large intestine, leaving behind semisolid feces.
- In the colon, absorption of too much water can lead to constipation, whereas absorption of too little water can produce diarrhea.

Herbivores employ fermenting microorganisms to help break down plant matter.

- Cellulose is the main component of the food of herbivores. Nevertheless, most herbivores cannot produce cellulase enzymes, which break down plant cellulose. Microorganisms in the digestive tracts of herbivores produce the cellulase enzymes that break down cellulose into fatty acids.
- Ruminants have a large four-chambered stomach in which fermentation occurs (see Figure 51.16). The rumen and reticulum, the first two chambers, contain many microorganisms that break down cellulose. The contents of the rumen are periodically regurgitated into the mouth for rechewing; this process is often described as "chewing the cud."
- From the rumen, food passes through the omasum, where water absorption occurs, into the abomasum, or true stomach. In the abomasum, hydrochloric acid and proteases kill the microorganisms and digest them.
- Some herbivores, such as rabbits, have a microbial fermentation chamber called the cecum extending from the large intestine. Many of these species produce two types of feces, one that is pure waste and the other that still contains some nutrients. Coprophagy is the reingesting of this second type of feces to digest and absorb additional nutrients. In humans, the cecum has become the appendix, an organ with no known digestive function.

Unconscious reflexes and hormones regulate digestion.

- Salivation and swallowing are unconscious reflexes. The digestive tract has an intrinsic nervous system that coordinates motility and digestive activities.
- Many digestive functions are controlled by hormones (see Figure 51.17). The duodenum secretes secretin, a hormone that stimulates the pancreas to release bicarbonate ions into the small intestine to control pH.
- Fats and proteins stimulate the mucosa of the small intestine to secrete cholecystokinin, which stimulates the release of bile and pancreatic digestive enzymes.
- Gastrin is secreted by the stomach into the blood, where it is transported to the upper regions of the stomach to stimulate stomach secretions and movement.

The liver interconverts fuel molecules.

- When fuel molecules are abundant in the blood, the liver stores them as glycogen or fat. When fuel molecules are low in the blood, the liver delivers them back into the blood.
- The liver can convert monosaccharides into glycogen or fat, and vice versa. The liver can also convert amino acids and certain other molecules into glucose; this process is called gluconeogenesis.

Lipoproteins allow fats to be transported in the circulatory system.

- Lipoproteins are particles of fat and cholesterol covered with proteins. Chylomicrons are large lipoproteins produced by the mucosal cells of the intestine. The liver produces other lipoproteins.
- Very low-density lipoproteins (VLDLs) contain mostly triglyceride fats, which they transport to cells of adipose tissue; these lipoproteins are only 3 percent cholesterol. Low-density lipoproteins (LDLs) contain 50 percent cholesterol, which they carry around the body for storage or use. High-density lipoproteins (HDLs) remove cholesterol from tissues and carry it to the liver for bile synthesis; they are 15 percent cholesterol. LDLs and VLDLs are associated with cardiovascular disease.

Insulin and glucagon help regulate blood glucose levels.

- The absorptive period is the period after a meal when digestion occurs. The postabsorptive period is the period when the gut is empty and the body uses energy reserves.

- During the absorptive period, the liver takes up glucose from the blood and converts it to glycogen and fat, fat cells take up glucose, and the body preferentially uses glucose as the energy source.
- During the postabsorptive period, the liver breaks down glycogen into glucose, fatty acids are supplied to the blood by adipose tissues, and the body preferentially uses fatty acids as an energy source.
- The pancreatic hormones insulin and glucagon control metabolic fuel use (see Figure 51.18).
- Insulin is released in response to high blood glucose levels. Insulin stimulates cells to burn glucose, fat cells to use glucose to make fat, and liver cells to convert glucose to glycogen and fat.
- At postabsorptive blood glucose levels, insulin secretion is diminished, and the liver and fat cells break down glycogen.
- At low levels of blood glucose, the pancreas releases glucagon, which stimulates the liver to break down glycogen and produce glucose.

Regulation of food intake involves the hypothalamus and signals from hormones and enzymes.

- Hunger and satiety are influenced by the region of the brain known as the hypothalamus. Both the ventromedial hypothalamus and lateral hypothalamus play a role in food intake.
- Leptin, a hormone produced by fat cells, acts on the hypothalamus and is involved in control of hunger and satiety. Ghrelin, a hormone produced by the stomach, acts on the hypothalamus to stimulate appetite. The enzyme AMP-activated protein kinase (AMPK) is produced by cells deprived of nutrients, and also appears to play a role in regulating food intake.

Test Yourself

Diagram Exercise

Create a flow chart of the following organs, indicating the order in which they occur from the start to the end of the human gut: small intestine, anus, esophagus, large intestine, mouth, and stomach. Indicate in which of these organs chemical digestion occurs, and the particular nutrient (e.g., fat, protein) being broken down. Also indicate in which of the organs absorption occurs, and the particular nutrient or substance absorbed.
Textbook Reference: *51.3 How Does the Vertebrate Gastrointestinal System Function? pp. 1078–1083*

Knowledge and Synthesis Questions

1. Certain amino acids are essential to the diet of animals because they
   a. prevent overnourishment.
   b. are cofactors and coenzymes that are required for normal physiological function.
   c. cannot be directly synthesized by animals through the transfer of an amino group to an appropriate carbon skeleton.
   d. are needed to make stored fats that are used during hibernation and migration.
   e. are the most important source of stored energy.
   Textbook Reference: *51.1 What Do Animals Require from Food? p. 1070*
2. Which of the following statements about sheep digestion is true?
   a. Sheep are saprobes; they engulf food and perform intracellular digestion.
   b. Sheep are autotrophs; they synthesize organic nutrients and perform extracellular digestion.
   c. Sheep are herbivores; they ingest food and perform extracellular digestion.
   d. Sheep are detritivores; they ingest food and perform intracellular digestion.
   e. Sheep are omnivores; they ingest plants and animals and exhibit intracellular digestion.
   Textbook Reference: *51.2 How Do Animals Ingest and Digest Food? p. 1074*
3. Which of the following protects the walls of the stomach against the action of its own digestive juices?
   a. An antienzyme chemical formed by the gastric glands
   b. The nervous reactions of the lining of the stomach
   c. Control by a center in the medulla of the brain
   d. A mucus covering the stomach's inner surface
   e. Alkaline secretions of the gastric glands that neutralize acidic secretions
   Textbook Reference: *51.3 How Does the Vertebrate Gastrointestinal System Function? p. 1079*
4. Chylomicrons are produced in the
   a. mouth.
   b. stomach.
   c. lumen of the small intestines.
   d. epithelial cells of the small intestines.
   e. liver.
   Textbook Reference: *51.3 How Does the Vertebrate Gastrointestinal System Function? pp. 1082–1083*
5. The gallbladder
   a. produces bile.
   b. is part of the liver.
   c. stores bile produced by the liver.
   d. produces cholecystokinin.
   e. is vestigial in humans.
   Textbook Reference: *51.3 How Does the Vertebrate Gastrointestinal System Function? p. 1082*
6. The pancreas
   a. is exclusively an endocrine gland responsible for the production and release of insulin and glucagon.
   b. is exclusively an endocrine gland that produces salivary amylase.
   c. contains villi to increase surface area.
   d. produces urobiligen (a bile pigment).

e. produces exocrine products involved in chyme digestion.

Textbook Reference: *51.3 How Does the Vertebrate Gastrointestinal System Function? p. 1082*

7. Hydrochloric acid
   a. is secreted by the gastric glands of the liver.
   b. is secreted by the gastric glands of the stomach.
   c. produces a low pH in the small intestine.
   d. promotes the growth of microorganisms in the stomach.
   e. is secreted by salivary glands.

   Textbook Reference: *51.3 How Does the Vertebrate Gastrointestinal System Function? p. 1079*

8. Bile produced in the liver is associated with which of the following?
   a. Emulsification of fats into tiny globules in the small intestine
   b. Digestive action of pancreatic amylase
   c. Emulsification of fats into tiny globules in the stomach
   d. Digestion of proteins into amino acids
   e. Emulsification of fats into tiny globules in the large intestine

   Textbook Reference: *51.3 How Does the Vertebrate Gastrointestinal System Function? pp. 1081–1082*

9. Most of the chemical digestion of food in humans is completed in the
   a. mouth.
   b. appendix.
   c. ascending colon.
   d. stomach.
   e. small intestine.

   Textbook Reference: *51.3 How Does the Vertebrate Gastrointestinal System Function? p. 1081*

10. Which of the following does *not* contribute to the large surface area available for nutrient absorption in the small intestines?
    a. Villi
    b. Intestinal length
    c. Microvilli
    d. Bile duct
    e. Surface folds

    Textbook Reference: *51.3 How Does the Vertebrate Gastrointestinal System Function? p. 1081*

11. Waves of muscle contractions that move the intestinal contents are
    a. caused by contraction of skeletal muscle.
    b. regulated by liver secretions.
    c. called peristalsis.
    d. voluntary.
    e. caused by smooth muscles of the gut relaxing in response to being stretched.

    Textbook Reference: *51.3 How Does the Vertebrate Gastrointestinal System Function? pp. 1078–1079*

12. What is the function of enterokinase?
    a. It converts pepsinogen to pepsin.
    b. It converts trypsinogen to trypsin.
    c. It digests proteins.
    d. It activates HCl.
    e. It prompts opening of the lower esophageal sphincter.

    Textbook Reference: *51.3 How Does the Vertebrate Gastrointestinal System Function? p. 1082*

13. Digestive enzymes responsible for breaking down disaccharides include
    a. pepsin, trypsin, and trypsinogen.
    b. amylase, pepsin, and lipase.
    c. sucrase, lactase, and maltase.
    d. pepsin, trypsin, and chymotrypsin.
    e. nuclease and aminopeptidase.

    Textbook Reference: *51.3 How Does the Vertebrate Gastrointestinal System Function? p. 1082*

14. Which of the following is characteristic of the large intestine?
    a. It has almost no bacterial populations.
    b. It contains chyme.
    c. It absorbs much of the water remaining in waste materials.
    d. It is the site of most of digestion.
    e. It receives blood from the digestive tract via the hepatic portal vein.

    Textbook Reference: *51.3 How Does the Vertebrate Gastrointestinal System Function? p. 1083*

15. The innermost layer of the digestive tract is the
    a. peritoneum.
    b. mucosa membrane.
    c. submucosa membrane.
    d. lumen.
    e. circular muscle.

    Textbook Reference: *51.3 How Does the Vertebrate Gastrointestinal System Function? p. 1078*

16. Which of the following hormones stimulates gluconeogenesis?
    a. Glucagon
    b. Insulin
    c. Estrogen
    d. Secretin
    e. Gastrin

    Textbook Reference: *51.4 How Is the Flow of Nutrients Controlled and Regulated? p. 1086*

17. Which of the following statements about mammalian teeth is true?
    a. Enamel covers the crown and the root of a tooth.
    b. Teeth lack nerves and blood vessels.
    c. The teeth of omnivores are more specialized than the teeth of carnivores and herbivores.
    d. Carnivores have large canines, whereas herbivores have large premolars and molars.
    e. Dentine is restricted to the crown of a tooth.

    Textbook Reference: *51.2 How Do Animals Ingest and Digest Food? p. 1075*

18. Which of the following statements about lipoproteins is *false*?
    a. Lipoproteins move fats from storage sites to sites where they are used.
    b. Lipoproteins have a hydrophobic core of cholesterol and fat.
    c. High-density lipoproteins are "good" lipoproteins.
    d. Low-density lipoproteins are "bad" lipoproteins.
    e. The liver is solely responsible for synthesis of lipoproteins.

    Textbook Reference: *51.4 How Is the Flow of Nutrients Controlled and Regulated? pp. 1085–1086*

Application Questions

1. Suppose you have just eaten a small pizza with a lot of cheese. Because this food is high in lipids, your digestive system will have to perform specific tasks to digest and absorb the fats. What are the steps that your digestive system will take to digest and absorb the fats in this meal?

   Textbook Reference: *51.3 How Does the Vertebrate Gastrointestinal System Function? pp. 1081–1083*
2. During the absorptive period, the body digests and absorbs nutrients, whereas during the postabsorptive period, the body uses up stored fuel. What are the main differences in fuel use and control mechanisms during these two periods?

   Textbook Reference: *51.4 How Is the Flow of Nutrients Controlled and Regulated? pp. 1085–1086*
3. Ingested food typically contains many useful proteins. However, these proteins are not used in the form in which they are ingested, but are broken down into smaller peptides and amino acids. Why does the digestive system break down these ingested proteins?

   Textbook Reference: *51.1 What Do Animals Require from Food? p. 1071*
4. Some mammals have very high-crowned cheek teeth (premolars and molars). Would you expect these animals to be carnivores, herbivores, or omnivores?

   Textbook Reference: *51.2 How Do Animals Ingest and Digest Food? p. 1075*
5. In ruminants, such as cows and bison, microorganisms that break down cellulose are found in two chambers of the greatly enlarged stomach. In other herbivores, including rabbits, the microorganisms that break down cellulose are found in the cecum, a chamber off the large intestine. In which type of herbivore would the absorption of nutrients from cellulose be more efficient?

   Textbook Reference: *51.3 How Does the Vertebrate Gastrointestinal System Function? p. 1084*

Answers

Diagram Exercise Answer

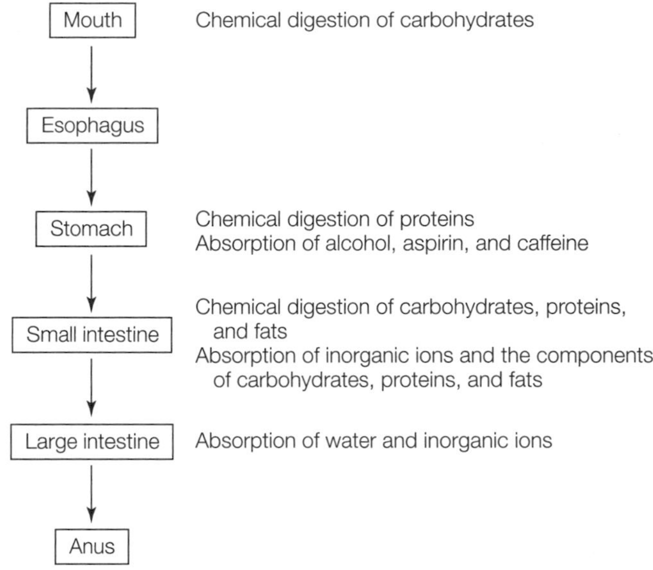

Knowledge and Synthesis Answers

1. **c.** Essential amino acids must be acquired through diet because an animal cannot directly synthesize all the amino acids needed for protein production.
2. **c.** Sheep are herbivores that ingest plant material for nutrition. Digestion occurs extracellularly in a sheep's digestive tract.
3. **d.** The stomach is protected from digestive enzymes and low pH by the mucus secreted over its inner surface.
4. **d.** The epithelial cells of the small intestine produce chylomicrons by combining triglycerides with cholesterol and phospholipids, allowing lipids to pass into the lymphatic system.
5. **c.** Bile is produced by the liver, stored in the gallbladder, and released into the small intestine to aid in lipid digestion.
6. **e.** The pancreas produces exocrine products such as lipases, nucleases, amylases, and trypsin, all of which are involved in chyme digestion.
7. **b.** Hydrochloric acid is a strong acid secreted by gastric glands in the lining of the stomach. It lowers the pH of the stomach fluid.
8. **a.** Bile aids in the digestion of lipids in the small intestine by changing fat droplets into small fat particles.
9. **e.** The small intestine is the main site for chemical digestion in humans.
10. **d.** The length of the intestines, surface folds, and the presence of microvilli and villi increase the surface area of the small intestine.
11. **c.** Food is moved through the digestive tract by wavelike contractions of smooth muscle called peristalsis.

12. **b.** Enterokinase converts the zymogen trypsinogen to trypsin.
13. **c.** The digestive enzymes sucrase, lactase, and maltase are involved in the breakdown of disaccharides into the monosaccharides glucose, galactose, and fructose.
14. **c.** The large intestine is the site of water and ion absorption. The large populations of bacteria found in the large intestine contribute useful vitamins to their hosts.
15. **b.** The membranes of the digestive tract are, from the inside to the outside: mucosa, submucosa, and circular and longitudinal muscles. The peritoneum surrounds the digestive tract.
16. **a.** Glucagon increases the rate of gluconeogenesis, and insulin reduces the rate of gluconeogenesis.
17. **d.** Carnivores have large canines for holding and killing prey animals. Herbivores have large premolars and molars for grinding plant material.
18. **e.** Chylomicrons are lipoproteins synthesized in the cells of the small intestine. Thus, the liver is not the sole producer of lipoproteins.

Application Answers

1. The digestion of fats begins in the small intestine. Lipases are water soluble, but lipids tend to aggregate into large droplets in an aqueous environment. Bile helps to increase the exposed surface area of the lipids by preventing the formation of large lipid droplets and promoting small micelles. This allows lipases to break down the fats into free fatty acids and monoglycerides. The fatty acids dissolve into the plasma membrane of the intestinal epithelial cells and will be absorbed across this surface. In mucosal cells, the free fatty acids and monoglycerides are turned into triglycerides, which are incorporated into chylomicrons. The chylomicrons can then pass out of the mucosal cells and into the lymphatic system (see Figure 51.15).
2. During the postabsorptive period, cells tend to metabolize fats, keeping the blood glucose reserves for the nervous system. Insulin levels are low during the postabsorptive period. During the absorptive period, the body attempts to store the nutrients and fuel that are being absorbed. The preferred fuel source at this time is glucose, and insulin is released to help facilitate glucose uptake.
3. Proteins in digested food are not absorbed and used in their original form for several reasons. First, the proteins are too large to be easily absorbed by the cells of the digestive tract, whereas amino acids are readily absorbed. Second, proteins are species-specific, and thus a protein that is optimal for a prey species may not be optimal for the predator species. Finally, the proteins of another species would be considered foreign and attacked by the immune system.
4. Mammals, such as horses, have very high-crowned cheek teeth. Even though enamel is very hard, the teeth of horses wear down over time because of their abrasive grass diet. Animals with high-crowned cheek teeth are herbivores. Carnivores and omnivores have low-crowned cheek teeth.
5. In vertebrates, most nutrients are absorbed in the small intestine. Because ruminants break down cellulose in chambers of their stomach (which comes before the small intestine in position along the gut), the nutrients from cellulose are available for absorption in the small intestine, and absorption is relatively efficient. In nonruminant herbivores, however, cellulose is broken down in the cecum, after it has passed through the small intestine, so nutrient absorption is less efficient. Some nonruminant species produce two types of feces and eat the one containing cecal material to gain greater access to nutrients.

52 Salt and Water Balance and Nitrogen Excretion

The Big Picture

- Different animals experience different types of stresses related to water and salt balance. Animals living in marine environments have to overcome problems with water loss due to osmosis, whereas animals living in freshwater environments have to overcome problems with water gain due to osmosis. Some invertebrates circumvent these problems by allowing their body fluid ion concentration to reflect that of their surroundings. Other invertebrates, and all vertebrates, retain their body fluids at more constant levels, expending energy to eliminate excess water and/or salts.
- All animals produce metabolic wastes (notably nitrogenous waste) as a by-product of metabolism. Invertebrates have evolved a variety of structures to eliminate wastes, whereas the vertebrates have evolved variations on the kidney. The nephron is the "functional unit" of the vertebrate kidney. By eliminating wastes without losing valuable nutrients and salts, the kidneys cleanse the blood of metabolic waste products.
- Kidneys work by differentially processing nitrogenous wastes, ions, and water. In the kidney, large volumes of plasma are filtered from the blood into the interior of the nephron. As the fluid passes through the various specialized regions of the nephron, the desirable components to be retained (specific ions, nutrients, and water) are pulled back into the tissues, leaving behind in the nephron the waste products in an ever more concentrated form of urine. As urine leaves the nephron, water content is adjusted in the collecting ducts through hormonal alteration of their water permeability. If the collecting ducts are highly water permeable, water will be drawn by osmosis out of the collecting ducts into the region of high salt concentration created by the action of the loop of Henle.

Common Problem Areas

- Depending on whether they are living in fresh water or salt water, fishes may excrete copious amounts of dilute urine or small amounts of more concentrated urine, or take up ions at their gills. Given that living in either of these environments creates osmoregulatory problems, think about whether the tissues have higher or lower solute concentration relative to the surrounding water, and then predict water and ion movements on that basis.
- Understanding how the nephron functions, and in particular how the loop of Henle works as a "countercurrent multiplier system," is one of the greater challenges in this chapter.

Study Strategies

- Nephron function can be understood more easily if you place yourself in the position of a single Na^+ ion in the blood entering the nephron. Follow your pathway—and your potential pathways—as you traverse the nephron, eventually ending up in the urine. Note especially that it may take you some time to pass through the loop of Henle as you leave the ascending limb, only to recycle back into the descending limb—a process that could be repeated many times before finally escaping the countercurrent multiplier. Repeat this imaginary journey again as a water molecule.
- Go to yourBioPortal.com to review the following tutorials and activites:

  Animated Tutorial 52.1 The Mammalian Kidney

  Interactive Tutorial: Glomerular Filtration Rate (GFR): Kidney Regulation

  Web Activity 52.1 Annelid Metanephridia

  Web Activity 52.2 The Vertebrate Nephron

  Web Activity 52.3 The Human Excretory System

  Web Activity 52.4 The Major Organ Systems

Important Concepts

Excretory organs control the solute concentrations of animals.

- Excretory organs control the level of solutes and the volume of tissue fluid and excrete urine. Urine composition depends on the environment surrounding an animal.
- Water movement in the body occurs either by pressure differences or differences in solute potential that

produce water movement by osmosis. There is no active transport of water in excretory systems.

- Excretory systems control water and solute balance by means of filtration across membranes, reabsorption, and secretion.
- Osmolarity is the number of moles of osmotically active solutes per liter of solvent.
- Osmoconformers do not actively regulate the osmolarity of their tissues; rather, they allow them to come to equilibration with their environment. Marine invertebrates tend to be osmoconformers unless the concentrations in the environment are extreme.
- Osmoregulators actively regulate the osmolarity of their tissues, even as environmental osmolarity changes. The brine shrimp is an osmoregulator that maintains its tissue fluid osmolarity below that of the environment in high osmolarity waters and above that of the environment in low osmolarity waters (see Figure 52.1).
- Ionic conformers allow the ionic composition of their extracellular fluid to match that of the environment. Ionic regulators employ active transport mechanisms to keep ions at optimal concentrations in their extracellular fluid. Birds excrete salt from consumed seawater through a nasal salt gland (see Figure 52.2).

Nitrogen is a potentially toxic metabolic waste.

- The end products of the metabolism of proteins and nucleic acids contain nitrogen and are called nitrogenous wastes.
- Ammonia (NH_3) is a highly toxic nitrogenous waste that either must be excreted or converted to the less toxic urea and uric acid (see Figure 52.3).
- Ammonotelic animals excrete ammonia to the aquatic environment, usually across gill membranes. Examples include most bony fishes and aquatic invertebrates.
- Ureotelic animals excrete nitrogenous waste as water-soluble urea. Examples include mammals, cartilaginous fishes, and most amphibians.
- To conserve water, uricotelic birds and reptiles excrete nitrogenous waste as insoluble uric acid. Insects are also uricotelic.
- Most species produce more than one nitrogenous waste.

Reflecting their diversity, invertebrates employ a number of excretory mechanisms.

- Flatworms have a protonephridium made up of tubules and flame cells that conserve ions and excrete water to produce dilute urine (see Figure 52.4).
- Annelid worms filter the blood across the capillaries into the coelom, where the fluid enters a metanephridium (a pair of metanephridia occurs in each segment) through a funnel-like opening called a nephrostome. Tubules of the metanephridia actively secrete and absorb various ions and end in nephridiopores, which open to the environment to excrete dilute urine containing nitrogenous wastes and other solutes (see Figure 52.5).
- Malpighian tubules are the excretory organ of insects. Malpighian tubules actively transport uric acid and sodium and potassium ions from the tissue into the tubules, with water following passively (see Figure 52.6). The tubules lead into the hindgut, where uric acid precipitates and water is reabsorbed. Insects eliminate semi-dry matter containing uric acid and other wastes.

The vertebrate excretory system is composed of nephrons in a kidney.

- Vertebrate excretory organs are kidneys, and nephrons are the functional units of kidneys.
- Vertebrates exhibit diverse excretory adaptations.

Marine and freshwater fishes employ two opposing mechanisms for controlling water and ion balance.

- Freshwater fishes live in an environment hypotonic to their own body fluids, with water tending to move into the body by osmosis. To counter this, they excrete large amounts of dilute urine.
- Marine bony fishes are osmoregulators, maintaining their extracellular fluids at one-third to one-half the osmolarity of seawater. They live in an environment hypertonic to their body fluids, so water tends to be drawn out of the body by osmosis. To cope with this environment, they produce a small amount of concentrated urine and actively secrete salts across the gills and the renal tubules.
- Cartilaginous fishes, almost all of which are marine, act as osmoconformers, allowing urea and trimethylamine oxide concentrations to increase, which increases the osmolarity of their tissues. Cartilaginous fishes are not ionic conformers; they have a rectal gland to excrete salts.

Terrestrial amphibians and reptiles must avoid desiccation.

- Amphibians living in or near fresh water have a large water influx, and they produce large amounts of dilute urine in response.
- In contrast, amphibians living in terrestrial environments have low metabolic rates and skin with reduced water permeability, resulting in low water turnover. Some also estivate (burrow underground and reduce metabolic rate) and use their bladders as canteens to store water for later use.
- Reptiles are amniotes; they lay shelled eggs, which allows them to live in a fully terrestrial environment (i.e., they do not have to return to water to reproduce). They have scaly, dry skin to decrease evaporative water loss, and they excrete nitrogenous wastes as uric acid with little water.

Mammals conserve water in their terrestrial habitat.

- Mammals display several adaptations for water conservation.

- A major adaptation of mammals is their ability to produce urine that is more concentrated than their blood.

Nephrons are the functional units of vertebrate kidneys.

- The nephron of the vertebrate kidneys is composed of the glomerulus, renal tubules, and peritubular capillaries, which are further broken down into vascular and tubule components (see Figure 52.7).
- Three main processes lead to the formation of urine: filtration, tubular reabsorption, and tubular secretion. Filtration occurs at the glomerulus and tubular reabsorption and secretion occur along the renal tubule.
- There are two sets of capillary beds in a nephron. The glomerulus is composed of the first set of capillary beds, with blood entering at the afferent arteriole and exiting at the efferent arteriole. From the efferent arteriole the peritubular capillaries of the second capillary bed emerge and surround the tubule component of the nephron (see Figure 52.7).
- The renal tubule begins with Bowman's capsule, which encloses the glomerulus. Within Bowman's capsule, podocyte cells wrap around the capillaries of the glomerulus. Blood is filtered by the glomerulus into Bowman's capsule. The endothelial walls of the capillaries in the glomerulus have pores that allow water and small molecules to leave the blood. Only small molecules and water can enter Bowman's capsule to form filtrate.
- Arterial blood pressure is the driving force for filtration at the glomerulus. The porous capillary beds in the glomerulus also contribute to the high filtration rate.
- The filtrate entering the tubules is similar to blood plasma, but as the fluid moves through the tubules, ions and molecules are actively reabsorbed and secreted. This controls the composition of the urine.

Mammalian kidneys produce concentrated urine.

- Humans have two kidneys that filter blood and produce urine (see Figure 52.9). Urine exits the kidney through the ureter and travels to the bladder, where it is stored until it is excreted through the urethra. The timing of urination is controlled by two sphincters in the urethra. One is made of smooth muscle and is involuntary, and the other is made of skeletal muscle and is voluntary.
- The human kidney is shaped like a kidney bean, with a central medulla and a surrounding cortex.
- The glomerulus and adjoining proximal convoluted tubules are located in the cortex. The descending limb of the renal tubule runs down into the medulla, and the ascending limb returns to the cortex in what is known as the loop of Henle. From the ascending limb, the tubules become the distal convoluted tubule, which connects to the collecting ducts and runs down the medulla.
- Nephrons with long loops of Henle that go deep into the medulla are important to the formation of concentrated urine.
- Blood vessels also run in a similar pattern to the tubules. An afferent arteriole carries blood to the glomerulus, where it is drained into the efferent arteriole, which gives rise to the peritubular capillaries. Capillaries run through the medulla, parallel to the loop of Henle, to form the vasa recta.

Filtration occurs at the glomerulus and reabsorption occurs largely at the proximal convoluted tubule.

- The typical glomerulus filters about 180 liters of blood per day in an adult human. However, 98 percent of this fluid is reabsorbed into the blood, so that only 2 percent of glomerular filtrate is excreted as urine.
- The majority of the filtrate is reabsorbed in the proximal convoluted tubules. Sodium ions and other solutes are actively transported out of the proximal convoluted tubule and water follows passively, all to be taken up by the peritubular capillaries.

The loop of Henle sets up a concentration gradient in the medulla.

- The loop of Henle produces a concentration gradient in the medulla by acting as a countercurrent multiplier system (see Figure 52.10). The purpose of this gradient is to move water across fluid compartments by osmosis, because there is no mechanism for active transport.
- The concentration of the fluids surrounding the loop of Henle is raised by the active reabsorption of Cl^- and the passive flow of Na^+ into the tissue from the thick ascending limb. The thick ascending limb is impermeable to water, so only solutes leave the tubules.
- The thin descending limb is permeable to water but not sodium and chloride ions. Water moves out of the tubules into the surrounding tissues due to the higher concentration of Na^+ and Cl^- in the surrounding tissues.
- In the thin ascending limb, the fluid is more concentrated than the fluid in the surrounding tissues, so Na^+ and Cl^- move out of the tubule. Water cannot move out because of the low permeability of the thin ascending limb.
- The fluid reaching the distal convoluted tubule is less concentrated than blood plasma.
- Aquaporins, a class of membrane proteins that form water channels, account for the high water permeability of some regions of the renal tubule (such as the proximal convoluted tubule) and the low permeability of others (such as the ascending limb of the loop of Henle).

Concentration of urine occurs in the collecting duct.

- The collecting duct passes through the medulla, with its high solute concentration set up by the loop of Henle.

- Water moves out of the collecting duct and into the tissue, resulting in concentrated urine. The permeability of the collecting ducts can be adjusted hormonally to regulate the amount of water excreted.
- Longer collecting ducts result in a greater ability to concentrate urine.

The kidneys help regulate the pH of the blood.

- Buffers can either absorb or release hydrogen ions.
- Bicarbonate ions are the major buffer in the blood. The kidneys control the level of H^+ and bicarbonate ions in the blood by varying their rate of secretion into the urine (see Figure 52.13).

Kidney failure can be treated by dialysis.

- Kidney (renal) failure severely disrupts homeostasis, causing high blood pressure (due to the retention of salts and water), uremic poisoning (due to the retention of urea), and acidosis (due to the retention of metabolic acids).
- A dialysis machine can replace kidney function until a kidney transplant can be arranged (see Figure 52.14).

Enzymes and hormones regulate glomerular filtration rate.

- Glomerular filtration rate is a function of the blood flow and pressure to the kidneys.
- The kidneys have autoregulatory mechanisms that help maintain their high filtration rate.
- In response to decreasing arteriole resistance or decreased arterial blood pressure, the afferent renal arterioles open, or dilate, to increase flow to the glomerulus. If this does not increase glomerular filtration rate sufficiently, the kidneys release the enzyme renin into the blood. Renin converts an angiotensin precursor into active angiotensin.
- Angiotensin constricts the efferent renal arterioles to increase pressure through the glomerulus. Angiotensin also constricts peripheral blood vessels, stimulates the release of aldosterone from the kidney, and stimulates thirst.
- Aldosterone stimulates the reabsorption of sodium by the kidney, which increases water reabsorption, and results in increased blood volume and pressure.

Antidiuretic hormone regulates blood pressure and blood osmolarity.

- Antidiuretic hormone (ADH) is produced in the hypothalamus and released from the posterior pituitary gland (see Figure 52.16). ADH controls the permeability of the collecting ducts to water by stimulating production of aquaporins over the long term, and insertion of aquaporins into the membrane over the short term.
- Stretch receptors in the aorta and carotid arteries inhibit the release of ADH in response to increased blood pressure, causing water to leave the body and blood volume to decrease.
- Osmosensors in the hypothalamus stimulate the release of ADH in response to increased osmolarity, resulting in increased water reabsorption and dilution of the blood.
- ADH also causes constriction of peripheral blood vessels.
- Atrial natriuretic peptide (ANP) is released by the heart in response to high blood pressure. At the kidneys, this hormone decreases sodium reabsorption, resulting in decreased water reabsorption and a lowered blood volume and pressure.

Test Yourself

Diagram Exercise

In the diagram below, label the following structures: loop of Henle, collecting duct, descending limb, proximal convoluted tubule, glomerulus, ascending limb, distal convoluted tubule, and Bowman's capsule. Also number the main locations where the following processes occur in the nephron: 1. filtration, 2. reabsorption, and 3. fine-tuning the filtrate.

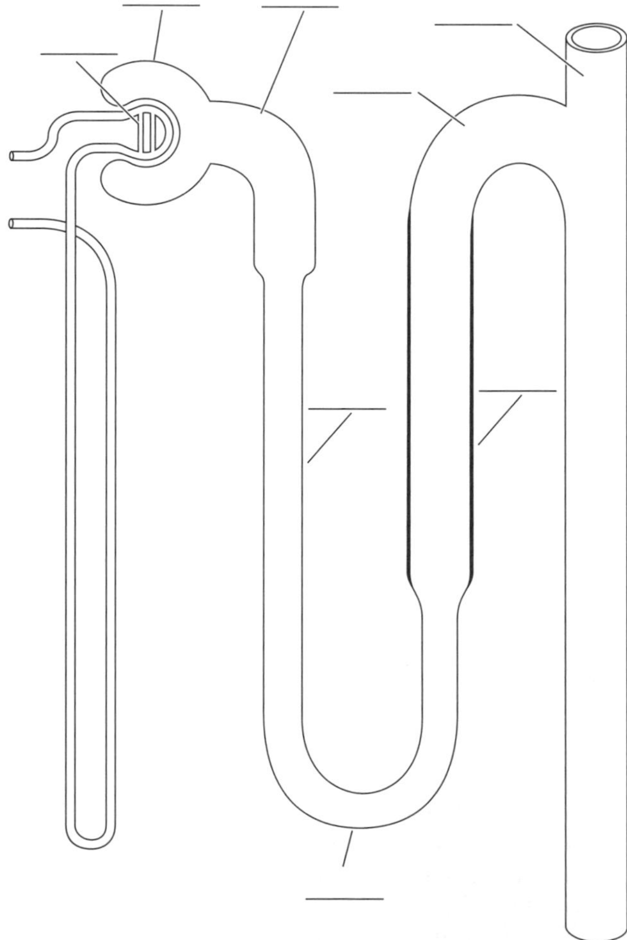

Textbook Reference: *52.4 How Do Vertebrates Maintain Salt and Water Balance? pp. 1099–1100; 52.5 How Does the Mammalian Kidney Produce Concentrated Urine? pp. 1101–1104*

Knowledge and Synthesis Questions

1. In order to maintain homeostasis, a marine bony fish
   a. excretes only small amounts of water and pumps sodium out of its body at the gills.
   b. excretes large amounts of water and pumps sodium into its body.
   c. converts nitrogenous wastes to urea.
   d. excretes uric acid.
   e. None of the above

   Textbook Reference: *52.4 How Do Vertebrates Maintain Salt and Water Balance? p. 1098*

2. Which of the following statements about the excretory system of insects is *false*?
   a. Active transport moves materials from the coelomic fluid into the Malpighian tubules.
   b. The Malpighian tubules can produce a highly concentrated urea solution, allowing insects to inhabit some of Earth's driest habitats.
   c. Reabsorption of salts takes place mostly in the gut.
   d. Water reabsorption takes place by osmotic movement only.
   e. Insects excrete urea.

   Textbook Reference: *52.3 How Do Invertebrate Excretory Systems Work? p. 1097*

3. Which of the following represents the correct pathway of water and solutes traveling through a nephron?
   a. Glomerulus, Bowman's capsule, proximal tubule, loop of Henle, distal tubule, collecting ducts
   b. Bowman's capsule, glomerulus, distal tubule, loop of Henle, proximal tubule, collecting ducts
   c. Glomerulus, Bowman's capsule, distal tubule, loop of Henle, proximal tubule, collecting ducts
   d. Glomerulus, Bowman's capsule, proximal tubule, collecting ducts, distal tubule, loop of Henle
   e. Glomerulus, Bowman's capsule, proximal tubule, distal tubule, collecting ducts, loop of Henle

   Textbook Reference: *52.4 How Do Vertebrates Maintain Salt and Water Balance? pp. 1098–1100*

4. Na^+ and Cl^- are actively transported out of the tubules to help set up the countercurrent multiplier. Sites of active Na^+ and Cl^- transport in the nephron are _______ and the _______.
   a. Bowman's capsule; descending limb of the loop of Henle
   b. the descending limb of the loop of Henle; ascending limb of the loop of Henle
   c. the ascending limb of the loop of Henle; proximal tubule
   d. the collecting duct; descending limb of the loop of Henle
   e. the proximal tubule; ascending limb of the loop of Henle

   Textbook Reference*: 52.5 How Does the Mammalian Kidney Produce Concentrated Urine? pp. 1102–1103*

5.–9. Refer to the diagram below to complete the statements about the mammalian nephron that follow.

5. The composition of the filtrate would be most like plasma in the tubule next to letter _______.

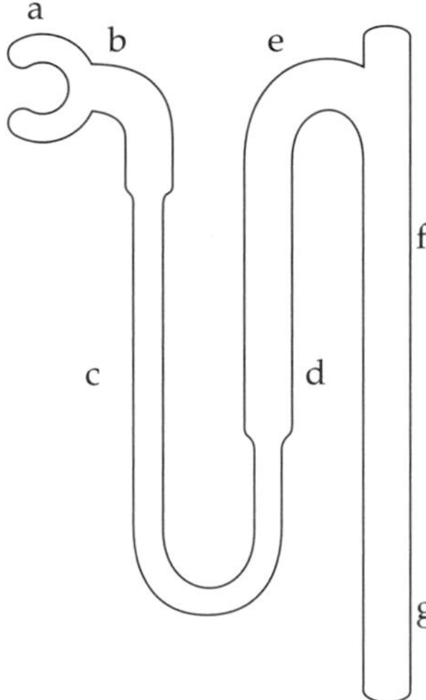

6. The NaCl concentration in the extracellular fluid would be greatest in the area of letter _______.
7. The osmolarity of the filtrate next to letters _______ is similar to the osmolarity of blood plasma.
8. The urine would be most concentrated in the collecting duct next to letter _______.
9. Most of the glomerular filtrate is reabsorbed into the blood in peritubular capillaries next to letter _______.

   Textbook Reference: *52.5 How Does the Mammalian Kidney Produce Concentrated Urine? pp. 1102–1103*

10. The sole mechanism for water reabsorption by the renal tubules is
    a. active transport.
    b. osmosis.
    c. cotransport with sodium ions.
    d. cotransport with bicarbonate ions.
    e. None of the above.

    Textbook Reference: *52.1 How Do Excretory Systems Maintain Homeostasis? p. 1093*

11. Several hormones help regulate water and solute uptake and release in the nephron. Antidiuretic hormone (ADH) promotes _______ in response to _______.
    a. active transport of Cl^-; increased solute concentration
    b. active transport of Na^+; increased blood pressure
    c. increased permeability of the collecting duct to water; decreased blood pressure
    d. decreased permeability of the collecting duct to water; increased solute concentration
    e. decreased permeability of the collecting duct to water; decreased blood pressure

    Textbook Reference: *52.6 How Are Kidney Functions Regulated? pp. 1107–1108*

12. Which of the following is *not* a normal constituent of the glomerular filtrate?

a. Red blood cells
b. Urea
c. Sodium ion
d. Glucose
e. Amino acids
Textbook Reference: *52.4 How Do Vertebrates Maintain Salt and Water Balance? pp. 1099–1100*

13. If the afferent arteriole that supplies blood to the glomerulus becomes dilated, then
a. the protein concentration of the filtrate decreases.
b. hydrostatic pressure in the glomerulus decreases.
c. the glomerular filtration rate increases.
d. the glomerular filtration rate decreases.
e. the water concentration of the filtrate decreases.
Textbook Reference: *52.6 How Are Kidney Functions Regulated? pp. 1107–1108*

14. Osmoreceptors in the brain detect changes in blood ion concentration that can result reflexively in _______ by the kidneys.
a. water retention
b. water loss
c. ion retention
d. All of the above
e. None of the above
Textbook Reference: *52.6 How Are Kidney Functions Regulated? pp. 1107–1108*

15. As the glomerular filtrate passes through the renal tubule, about _______ percent is returned to the blood.
a. 2
b. 10
c. 50
d. 75
e. 98
Textbook Reference: *52.5 How Does the Mammalian Kidney Produce Concentrated Urine? p. 1102*

Application Questions

1. Based on your knowledge of how the nephron works, explain why mammals produce concentrated urine whereas reptiles cannot.
Textbook Reference: *52.5 How Does the Mammalian Kidney Produce Concentrated Urine? pp. 1102–1103*

2. What would happen to the composition of excreted urine if all active transport processes in the nephron came to a halt?
Textbook Reference: *52.5 How Does the Mammalian Kidney Produce Concentrated Urine? pp. 1102–1103*

3. The drug urizadole inhibits antidiuretic hormone secretion from the pituitary. How does this change glomerular filtration rate, urine flow, and urine concentration in patients taking this medication?
Textbook Reference: *52.6 How Are Kidney Functions Regulated? pp. 1107–1108*

4. You have just eaten a large number of very salty potato chips, and the osmolarity of your blood has increased. What response will your body have in order to bring your blood osmolarity back to homeostasis?
Textbook Reference: *52.6 How Are Kidney Functions Regulated? pp. 1107–1108*

5. The excretory systems of earthworms and humans share many characteristics. List a few of the ways in which the excretory systems are similar in form and function.
Textbook Reference: *52.3 How Do Invertebrate Excretory Systems Work? p. 1096; 52.4 How Do Vertebrates Maintain Salt and Water Balance? pp. 1098–1100*

Answers

Diagram Exercise Answer

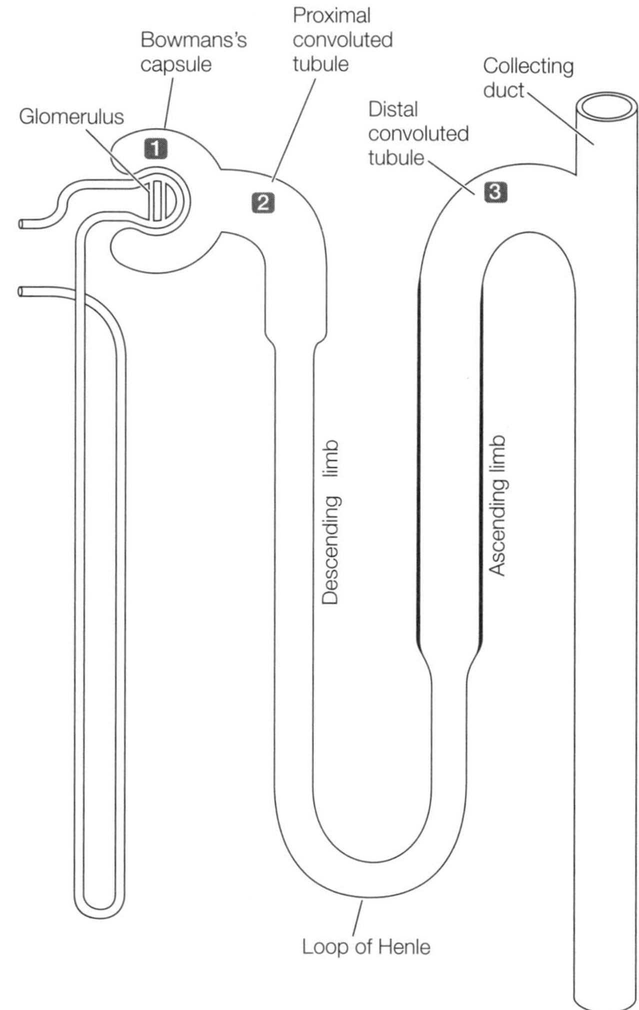

Knowledge and Synthesis Answers

1. **a.** Marine bony fishes live in an environment in which salts tend to be drawn into the body and water tends to be drawn out of the body. To counter these effects, these fishes excrete small amounts of water and actively pump sodium out of the body across the gills.
2. **e.** Insects excrete uric acid, not urea.

3. **a.** The route of water and solutes through the nephron is from the glomerulus, to Bowman's capsule, to proximal tubule, to loop of Henle, to distal tubule, to collecting ducts.
4. **e.** The proximal tubule is the site of active transport of Na^+ out of the tubule. Na^+ also moves out of the tubule at the ascending limb of the loop of Henle, but this is passive transport, with Cl^- being actively transported out.
5. **a.** At Bowman's capsule the filtrate is most similar to plasma.
6. **g.** The sodium concentration is the highest in the extracellular fluid near the middle of the medulla.
7. **a.**, **b.**, and **e.**The filtrate has a similar osmolarity to plasma in the cortex of the kidney, including the glomerulus and Bowman's capsule, the proximal convoluted tubule, and the distal convoluted tubule.
8. **g.** The highest concentration of the filtrate in the collecting ducts will be near their ends, deep in the medulla.
9. **b.** The bulk of the water and solute are reabsorbed at the proximal convoluted tubule.
10. **b.** The sole mechanism for water reabsorption in the renal tubules is by osmosis.
11. **c.** Antidiuretic hormone acts on the collecting ducts by increasing permeability to water. Antidiuretic hormone secretion is stimulated by a decrease in blood pressure.
12. **a.** Red blood cells are too large to be filtered out of the blood at the glomerulus and thus will not be found in the filtrate.
13. **c.** Changes in afferent arteriole pressure affect glomerular filtration rate. Dilation of the afferent arteriole increases pressure, which will increase filtration rate.
14. **d.** Osmoreceptors in the brain detect changes in blood ion concentration, resulting in reflexive increases or decreases in both ion and water retention, depending on the status of blood osmolarity.
15. **e.** About 98 percent of the filtrate is returned to the blood.

Application Answers

1. The loop of Henle acts as a countercurrent ion multiplier in the medulla of mammals. This allows mammalian kidneys to produce concentrated urine. Reptiles lack the countercurrent multiplier of the loop of Henle.
2. Active transport of Na^+ and Cl^- in the nephron provides the ions that set up the countercurrent multiplier, allowing for the production of concentrated urine in mammals. If active transport in the nephrons were to stop, the urine produced would eventually be isotonic with blood plasma because the countercurrent multiplier would disappear.
3. Inhibition of antidiuretic hormone by urizadole affects the permeability of the collecting ducts to water. This results in increased urine flow because water will not be reabsorbed across the collecting ducts. Blockage of ADH has no effect on the glomerular filtration rate.
4. In response to increased blood osmolarity, osmosensors in the hypothalamus stimulate the release of ADH from the pituitary into the blood. ADH acts to increase the permeability of the collecting ducts to water so that increased amounts of water can be resorbed to bring down the blood osmolarity. The osmosensors will also stimulate thirst, causing you to increase your water intake.
5. The metanephridia of the earthworm and the nephron of humans are both composed of tubules through which filtrate flows. In the tubules of both, certain molecules are actively transported out or secreted in. Both systems produce urine with an osmolarity different from that of the body fluid. The urine of the earthworm is hypotonic, whereas that of humans is usually hypertonic.

53 Animal Behavior

The Big Picture

- The distinction between proximate and ultimate explanations is of fundamental importance in the study of behavior and can be extended to many other fields of biology. Behaviorists concerned with proximate questions strive to unravel the underlying neural and hormonal mechanisms of behavior and to sort out the relative roles of genes and experience in shaping behavior. Behaviorists asking ultimate questions attempt to understand how the behavior that they are studying is related to the animal's survival and reproductive success. In this effort, they not only draw on concepts from evolution, ecology, and animal behavior, but also make important contributions to those fields.
- Animals exhibit a wide range of species-specific behaviors that can be either genetically based or environmentally determined. Genetically based behaviors may require triggers or releasers to stimulate an animal. Nevertheless, patterns of behavior typically have genetic and environmental influences.
- Hormones are important in controlling the behavior of animals. Sex steroids are frequently involved in stimulating the different behaviors of males and females of a species. Daily, seasonal, or annual rhythms are often carried out through the actions of hormones.
- Communication is an integral part of animal behavior. Communications can be chemical (involving pheromones), visual (involving displays of fins, feathers, fur, etc.), auditory (sounds created by either general or specialized structures and received by auditory sensors), tactile, or electric. Such communication is an important component of territory-marking and reproductive behavior.
- As you learned from Chapter 21 (Evidence and Mechanisms of Evolution), the cost–benefit approach applied in this chapter to the analysis of behavior can be extended to any kind of adaptation of organisms.
- Because standard Darwinian theory holds that only traits that contribute to individual fitness are favored by natural selection, explaining the evolution of altruistic behavior has presented a major challenge to evolutionary biologists. The concepts of inclusive fitness and kin selection have proved to be very fruitful in understanding altruism in animals and have also been applied to many aspects of human social behavior (though not without controversy).

Common Problem Areas

- It can sometimes be difficult to distinguish between behaviors founded in genetics (instinctive behaviors) and those that are learned during the course of an animal's lifetime. Further muddying the waters, the ability to learn new behavior is, of course, a heritable trait!
- Hormones control many behaviors, and hormone production is often a function of the time of day or the time of year. Thus, a particular hormone that stimulates (or inhibits) behaviors under one condition at one point in time may not have the same effect at another time. Be sure to learn the conditions and caveats that accompany hormone actions.
- The categories of animal orientation—piloting, distance-and-direction navigation, and bicoordinate navigation—can be confusing. Refer to Figure 53.18 to consolidate your understanding of how the sun is used as a time-compensated compass.

Study Strategies

- As you learn about various animal behaviors, it may be useful to design "mind experiments" to help you to distinguish between a genetically determined and an environmentally determined behavior. This also may help you to work out the nature of the behavior.
- Learn to distinguish the various behavior cycles associated with internal controls via "biological clocks" as opposed to behaviors that are directly stimulated by an immediate event in an animal's surrounding environment.
- To study the properties of the various sensory modes of communication (visual, chemical, etc.), make a table that compares them with respect to characteristics such as cost of production, effective signaling distance (and how signaling distance is affected by environmental conditions), durability, and information content.

- Go to yourBioPortal.com to review the following tutorials and activities:

  Animated Tutorial 53.1 The Costs of Defending a Territory

  Animated Tutorial 53.2 Foraging Behavior

  Animated Tutorial 53.3 Circadian Rhythms

  Animated Tutorial 53.4 Time-Compensated Solar Compass

  Interactive Tutorial: Circadian Rhythms: Time-Compensated Solar Compass

  Web Activity 53.1 Honeybee Dance Communication

  Web Activity 53.2 Concept Matching

Important Concepts

The study of animal behavior had its historical roots in behaviorism and ethology.

- Pavlov discovered the conditioned reflex through his experiments on the salivation response of dogs. Before conditioning, food is an unconditioned stimulus (UCS) that elicits salivation, an unconditioned response (UCR). If the UCS is presented immediately after a neutral stimulus, such as a particular sound, the dog eventually salivates solely in response to the sound. The sound has become a conditioned stimulus (CS) and the salivation response to sound is a conditioned response (CR). Salivation in response to sound is thus a learned response.
- Skinner showed that any random action of an animal could become a conditioned response to a stimulus if a reward was associated with the action and the stimulus. This type of learning is called operant conditioning.
- The work of Pavlov, Skinner, and their followers represents an approach to the study of animal behavior known as behaviorism, which emphasizes the study of learned behavior in a few model species (e.g., the albino rat) in a laboratory environment.
- Ethology arose at the same time as behaviorism, but largely in Europe. Ethologists were interested in many species, especially their evolutionary relationships and the ways in which their behaviors were adapted to their environments.
- The instinctive behaviors studied by ethologists are genetically determined fixed action patterns. They are performed without learning, are stereotypic (i.e., performed the same way each time), and cannot be modified by learning.
- Determination of inheritance of behaviors can be made by deprivation experiments, in which all relevant experience is denied. If the animal still exhibits the behavior, it is assumed that the behavior can develop without opportunities to learn it.
- A releaser is a stimulus that triggers a fixed action pattern. In general, releasers are very simple subsets of the information available in the environment (see Figure 53.2).
- Ethologists demonstrated the genetic basis of behavior by hybridizing closely related species with known behaviors and studying the behavior of the offspring.
- To ethologists, the fundamental questions about animal behavior fall into two categories: those concerning proximate causation and those referring to ultimate causation.
  - Questions concerning the proximate causes of behaviors relate to the immediate genetic, physiological, neurological, and developmental mechanisms that determine how an individual animal is behaving.
  - Questions about the ultimate causes of behavior relate to the evolutionary processes that have produced the animal's capacity and tendency to behave in particular ways.

Alterations in single genes can result in discrete behavior phenotypes.

- Behavioral geneticists conduct breeding experiments to analyze how particular behaviors (e.g., the hygienic behavior of honeybees) are inherited (see Figure 53.4).
- In a gene knockout experiment, biologists inactivate a particular gene (e.g., in the mouse) and investigate the behavioral effects of its loss (see Figure 53.5).
- Most behaviors are complex traits involving many genes that function in cascades and offer many points for a change in a single gene to influence behavior (see Figure 53.6).

An important proximate question is how behavior emerges as an animal develops and matures.

- Hormones can determine the development of a behavioral potential at an early age and the expression of that behavior at a later age. For example, in rats the sex steroids present during early development determine the sexual behavior pattern of the adult, but sex steroids present in the adult control the expression of those patterns.
- In imprinting, an animal learns a set of stimuli during a limited critical period. Imprinting of offspring on parents or of parents on offspring is a learned response that helps in recognition. The critical period for imprinting is often determined by the developmental or hormonal state of the animal.
- Birds use songs in territorial displays and in courtship. In some species, imprinting of the species-specific song is required in the nestling in order for it to sing the song as an adult, even though the juvenile bird never sings the song. Imprinting forms a memory of the song that is recalled as the bird approaches adulthood. The adults need both the initial imprinting of the song as a juvenile and the ability to match the song with auditory feedback in order to sing the correct song (see Figure 53.9).

- The influence of hormones on behavior has also been investigated through experiments on song development in birds. In male birds, testosterone levels control singing by inducing certain regions of the brain to grow during the breeding season. During the nonbreeding season, those regions of the brain associated with singing are reduced in size. Female birds that are treated with testosterone during the spring will also develop the species-specific song. In these birds, testosterone has stimulated growth in those areas of the brain associated with singing.

Behavioral responses to the environment influence fitness.

- The habitat of an animal is the environment in which it normally lives. In choosing their habitat, animals use cues that are good predictors of general conditions suitable for future survival and reproduction.
- The success of already-settled individuals of the same species (conspecifics) may indicate habitat quality (see Figure 53.10).
- Natural selection molds behavior in accordance with costs and benefits (see Figure 53.8). There are three aspects to the cost of behaving:
  1. Energetic cost—the amount of energy the animal expends during the behavior.
  2. Risk cost—the amount of risk the behavior entails for the animal.
  3. Opportunity cost—the benefits the animal forfeits by not engaging in other behaviors instead.
- The territory of an animal is an area from which other individuals of its own species (and sometimes individuals of other species) are excluded. By establishing a territory, an animal (usually a male) may improve its fitness by gaining exclusive use of the resources of part of its habitat.
- Cost–benefit analysis explains the diversity of territorial behaviors characteristic of different species (see Figure 53.11). Some animals defend all-purpose territories in which nesting, mating, and foraging take place. Other animals cannot establish feeding territories but defend nest sites or areas that provide access to females. Still other animals defend territories (leks) that are used only for mating.
- Cost–benefit analysis can also be applied to foraging behavior (i.e., what food an animal selects and when and where it searches for it). Optimal foraging theory predicts that the foraging choices of many animals result in their maximizing their rate of energy intake (see Figure 53.13).
- Because minerals and foods with medicinal value are important in the diets of many animals, they may sometimes forage in a way that deviates from the energy-maximization model.

Another important proximate question is what physiological mechanisms underlie the behavior of an animal.

- Control of behavior involves the nervous and endocrine systems. Execution of behavior involves effector mechanisms, not only the musculoskeletal system but also those that produce sound, secretions, color changes, etc.
- Responses to the environment must be timed appropriately. Circadian rhythms are daily cycles in activity, sleep, foraging, and other physiological processes and behaviors that are controlled by an endogenous clock (see Figure 53.15).
- The length of one cycle in a rhythm is defined as one period. Any point in the cycle is known as a phase. Two cycles can be in phase if the rhythms match or be phase-advanced or phase-delayed if they do not match.
- Circadian rhythms can be reset by environmental cues, such as the light–dark cycle, during entrainment. In constant conditions an animal's circadian clock is said to be free-running and will have a natural period that is different from the 24-hour period of the day.
- In mammals, the master circadian "clock" is located in the suprachiasmatic nuclei (SCN) of the brain. The molecular mechanism of the circadian clock involves negative feedback control of certain clock genes that are expressed in SCN cells.
- Day length (photoperiod) can trigger seasonal rhythms, which are important in controlling the timing of breeding and migrating. Animals such as hibernators and equatorial migrants, for whom changes in day length are not a reliable cue, rely instead on circannual rhythms—built-in neural calendars that keep track of the time of the year.
- Animals find their way around their environment by a variety of mechanisms, including piloting, distance-and-direction navigation, and bicoordinate navigation.
- Piloting is orientation by the recognition of landmarks. It is the mechanism used by some species that migrate or that are capable of homing (the ability to return to a specific location).
- Homing and migrating species that are able to take direct routes to their destinations through environments they have never experienced must use mechanisms of navigation other than piloting. Examples of such species are homing pigeons and the many species of migrating birds that are able to fly great distances and return to the same breeding ground each season.
- Many animals appear to have a compass sense, which allows them to use environmental cues to determine direction, and some appear to have a map sense, which allows them to determine their position.
- Distance-and-direction navigation involves knowledge of direction and distance to a destination. The position

of the sun and stars can be a source of directional information. Pigeons, for example, have the ability to determine direction by means of a time-compensated solar compass (see Figure 53.18). The stars offer two sources of information about direction: moving constellations and a fixed point (the point directly over the axis on which Earth turns).

- Bicoordinate navigation (also known as true navigation) requires knowing the map coordinates of both the current position and the destination. The behavior of many species (such as albatrosses) suggests that they are capable of this type of navigation.
- Communication is used by animals to transmit information. If the transmission of information benefits both the sender and the receiver, the behaviors of individuals may become elaborated through evolution into communication signals.
- Pheromones are chemical signals used to communicate among individuals of a species. Because of the diversity of their molecular structures, pheromones can communicate very specific, information-rich messages. Pheromones used in different types of communication vary in their volatility and diffusibility. Because pheromones remain in the environment for some time after they are released, they are useful for such functions as marking territories but unsuitable for the rapid exchange of information.
- Visual signals provide rapid, directional communication over considerable distances. One drawback to visual signals is the need for light, except in the case of species that have evolved their own light-emitting mechanisms. Visual signals can also be intercepted by other species.
- Sound has advantages over sight in that it can travel in complex environments, such as a forest, and over long distances, such as the sound produced by whales in the ocean. Sound also communicates directional information. The characteristics of acoustic signals are typically adjusted to their function and the environment of the animal.
- Tactile communication is used by animals in close contact with one another. Honey bees dance to communicate the location of a food source in the environment (see Figure 53.19). Round dances are used to communicate that food is less than about 80 meters from the hive. A waggle dance is used to communicate distance and direction to food that is more than about 100 meters from the hive. Speed of the waggle indicates distance to the food source; direction of the straight run of the waggle dance indicates the direction of the food source relative to the sun.
- The specificity of communication signals is enhanced by the use of multiple sensory modalities. Courtship behavior in fruit flies, for example, involves tactile, chemical, visual, and acoustic signals (see Figure 53.6A).

Sociobiology is the study of the evolution of social behavior.

- Social interactions range from very simple (a single male and a single female) to complex societies (such as honey bee colonies). Sociobiology proposes that the evolution of all variants of social behavior can be understood by asking how the behavior contributes to the fitness of the individuals involved.
- Mating systems evolve to maximize the fitness of both partners. Because males produce vast numbers of sperm that contain next to no resources, whereas females produce relatively few eggs that are rich in resources, the energetic and opportunity costs for reproduction are greater for the female than for the male. This disparity in the investment in the young is particularly large in mammals because females bear the costs associated with gestation. For a female, the best way to maximize her fitness is to make sure her young are healthy and survive to pass on her genes.
- Males have different options for maximizing their fitness, including monogamy (in which one male forms a pair bond with one female and both parents participate in rearing the young), polygamy (one male mates with many females), and polyandry (one female mates with multiple males).Which mating strategy maximizes his fitness depends on environmental factors.
- In species with polygamous mating systems, some males have high reproductive success while many males have none. As a rule, bigger, stronger males are the winners in the competition for females, and sexual dimorphism in body size evolves.
- Polyandry is a relatively rare mating systems that occurs in some birds and a few mammals in which paternal care for the young can have a large influence on fitness.
- Natural selection sometimes favors altruistic acts—behaviors that reduce the reproductive chances of the individual performing the act but increase the fitness of the helped individual. Typically, such behavior is directed toward a relative of the altruist, with whom it shares alleles. By helping its relatives, an individual can increase the representation of some of its own alleles in the population. An altruistic behavior pattern can evolve if it increases the inclusive fitness of the altruist: the fitness derived from an individual's own reproductive success (individual fitness) plus the fitness derived from the reproductive success of its relatives.
- The maximization of inclusive fitness underlies kin selection, which is selection for behaviors that increase the reproductive success of relatives even when they have some cost to the performer. According to Hamilton's rule, for an apparent altruistic act to be adaptive, the fitness benefit of that act to the recipient times the degree of relatedness of the performer and the recipient has to be greater than the cost to the performer.

- Eusocial species are those whose social groups include sterile individuals. In the Hymenoptera (ants, bees, and wasps), kin selection has probably facilitated the evolution of eusociality because of their sex determination system (haplodiploidy), in which males are haploid and females are diploid, with the result that sisters are genetically more similar to one another than to their own offspring. In eusocial species in which both sexes are diploid (termites, naked mole-rats), the difficulty of establishing independent colonies has favored the evolution of eusociality.
- Group living may benefit both predator and prey species. For predators, it may improve hunting efficiency or increase the size of the prey that can be captured. For prey, it may provide increased protection against predators.
- Living in a group imposes costs as well as benefits. Individuals in groups may compete for food, interfere with one another's foraging, injure one another's offspring, inhibit one another's reproduction, or transmit diseases to their associates.

Test Yourself

Diagram Exercise

Draw five diagrams showing the orientation of the straight run of the waggle dance that would be performed on the vertical surface of a honeycomb by a foraging honey bee that has discovered a food source located more than 100 meters away from the hive at the times and locations described below. Recall that in the northern hemisphere the direction (azimuth) of the sun is due south at noon; assume that the sun rises precisely in the east at 6:00 A.M. and sets precisely in the west at 6:00 P.M. As in Figure 53.19, let the top of the page indicate the "up" direction.

(a) Food location: due south of the hive; time: noon
(b) Food location: due north of the hive; time: 6:00 A.M.
(c) Food location: due north of the hive; time: noon
(d) Food location: due west of the hive; time: 9:00 A.M.
(e) Food location: due east of the hive; time: 6:00 P.M.

Textbook Reference: *53.5 What Physiological Mechanisms Underlie Behavior? pp. 1132–1133*

Knowledge and Synthesis Questions

1. On April mornings, you step outside your front door and notice the singing of a male robin. At that time you also observe that the female of the pair is building a nest in a nearby tree. A month later you observe the pair feeding their nestling offspring. Which of the following is a question about the *ultimate* cause of the behavior of these birds?
   a. Did the male bird begin to sing in April because the photoperiod had increased to a critical length?
   b. What were the relative roles of genes and experience in causing the female robin to build her nest out of particular building materials and to place it in a particular location?
   c. What combination of internal physiological factors and external cues stimulates the parents to feed their nestlings?
   d. Do robins that begin nesting in April raise more offspring than they would raise if they delayed nesting until June, and if so, why?
   e. None of the above

   Textbook Reference: *53.1 What Are the Origins of Behavioral Biology? pp. 1116–1117*

2. You train your pet dog to "sit" and "roll over" on command by rewarding it with a dog biscuit whenever it performs the desired behavior. This is an example of
   a. Pavlovian conditioning.
   b. imprinting.
   c. a releaser triggering a fixed action pattern.
   d. a conditioned reflex.
   e. operant conditioning.

   Textbook Reference: *53.1 What Are the Origins of Behavioral Biology? pp. 1114–1115*

3. Which of the following statements about the development of singing behavior in white-crowned sparrows is *false*?
   a. If a male is deafened after he sings his correct species-specific song, he will continue to sing like a normal bird.
   b. Female songbirds can be induced to sing by treatment with testosterone.
   c. Females learn their species song, but they do not normally express it under natural conditions.
   d. Testosterone is not necessary for singing in males once they have learned their song.
   e. To sing normally as an adult, a male must hear its species-specific song as a nestling.

   Textbook Reference: *53.3 How Does Behavior Develop? pp. 1121–1122*

4. Two species of mice live in the same geographical region, but Species 1 prefers open fields, whereas Species 2 lives in forests. When presented in an experiment with simulated "fields" and "forests," individuals of each species born in a laboratory preferred the environment in which they normally lived. This experiment illustrates the concept of
   a. habitat selection.
   b. optimal foraging strategy.
   c. territoriality.
   d. imprinting.
   e. Both a and b

   Textbook Reference: *53.4 How Does Behavior Evolve? p. 1123*

5. Birds spend some of their time scanning the horizon for predators. While scanning, they cannot be foraging for food. This situation illustrates the phenomenon of
   a. cooperative hunting.
   b. energetic cost.
   c. risk cost.
   d. optimal foraging strategy.
   e. opportunity cost.

   Textbook Reference: *53.4 How Does Behavior Evolve? pp. 1123–1124*

6. Suppose a predator has two different prey, Species 1 and Species 2. You perform a series of experiments in which the density of Species 1 is varied while the density of Species 2 is kept constant. At each Species 1 density, you determine how many prey of Species 1 and 2 are taken by the predator. The two curves (a and b) plotted in the following graph show the outcome of this study.

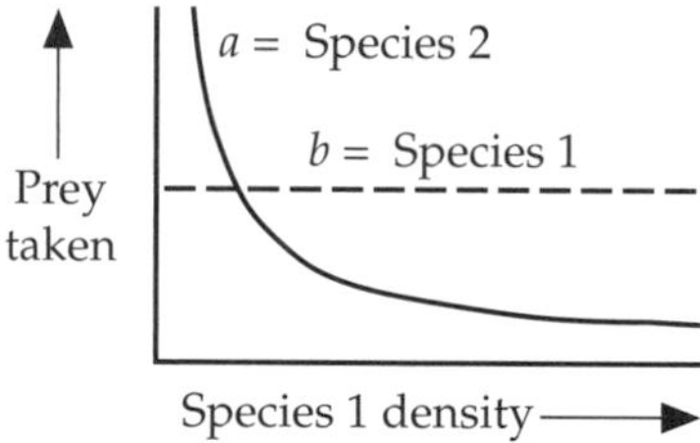

   Based on the graph, which of the following statements is true?
   a. Curve *a* indicates that Species 2 is the preferred prey.
   b. Curve *b* indicates that Species 1 is the preferred prey.
   c. Curve *b* indicates that Species 1 is not the preferred prey.
   d. Both a and c
   e. Neither curve provides insight into the prey preference of the predator.

   Textbook Reference: *53.4 How Does Behavior Evolve? pp. 1125–1126*

7. Hygienic behavior in honey bees
   a. is adaptive because it helps prevent bacterial infections in the hive.
   b. has two components, each controlled by a separate gene.
   c. has been studied by hybridizing two closely related species of honey bees.
   d. has been studied using gene knockout experiments.
   e. Both a and b

   Textbook Reference: *53.2 How Can Genes Influence Behavior? pp. 1117–1118*

8. Animals exhibit daily rhythms in their behavior and physiology. Which of the following statements about circadian rhythms is *false*?
   a. In mammals, the master circadian "clock" is located in the suprachiasmatic nuclei of the brain.
   b. A circadian clock can be entrained by environmental cues.
   c. Free-running circadian clocks are seldom exactly 24 hours long.
   d. Genetic mutations can cause changes in the length of the free-running circadian clock.
   e. Animals that are active at night do not display circadian rhythms.

   Textbook Reference: *53.5 What Physiological Mechanisms Underlie Behavior? pp. 1127–1128*

9. Which of the animals listed below would *not* be expected to have an endogenous circannual rhythm?
   a. A ground squirrel that hibernates
   b. A white-tailed deer living in eastern North America
   c. A cave salamander
   d. A migratory bird that summers in North America and winters in equatorial South America
   e. All of the above would have an endogenous circannual rhythm.

   Textbook Reference: *53.5 What Physiological Mechanisms Underlie Behavior? p. 1129*

10. Which of the following is an example of piloting?
    a. Silkworms following a trail of pheromones
    b. Marine migration over a featureless ocean
    c. Bees learning the direction of a food source from a dance
    d. Gray whale migration along the pacific coast of America
    e. None of the above

    Textbook Reference: *53.5 What Physiological Mechanisms Underlie Behavior? pp. 1129–1130*

11. There are a number of different navigational methods used by animals. Which of the following definitions of a navigational method is correct?
    a. Distance-and-direction navigation requires knowledge of latitude and longitude.
    b. Bicoordinate navigation requires knowledge of direction and of distance to a destination.
    c. The stars offer two sources of information about direction: a fixed point and moving constellations.
    d. Piloting involves the use of the sun as a compass.
    e. None of the above

    Textbook Reference: *53.5 What Physiological Mechanisms Underlie Behavior? pp. 1129–1130*

12. Which of the following statements about pheromones is true?
    a. Pheromones are used only for communication between individuals of the same species.
    b. Mammalian pheromones can communicate information about the size, sex, and reproductive status of the signaler.
    c. Because of their durability in the environment, pheromones are unsuitable for rapid exchange of information.
    d. Pheromones differ in their volatility and diffusibility depending on their function.
    e. All of the above

    Textbook Reference: *53.5 What Physiological Mechanisms Underlie Behavior? p. 1132*

13. You observe an example of an apparently altruistic act by an animal that seems to reduce its near-term likelihood of reproductive success. What would be the *least* plausible explanation for its behavior?
    a. The act aids the reproductive success of individuals sharing a high proportion of genes with the altruistic individual.
    b. The act is only apparently altruistic; over the long term, the behavior actually contributes to individual fitness.
    c. The act increases the inclusive fitness of the animal performing it.
    d. The act is advantageous because it helps the species to survive and reproduce, even if the altruist itself does not.
    e. From the information provided, it is impossible to determine the least plausible explanation.

    Textbook Reference: *53.6 How Does Social Behavior Evolve? p. 1133, 1135*
14. Some birds give a species-specific vocalization called an "alarm call" when they see a predator, although this call may direct the predator toward them. Other members of their species respond to these calls by taking cover. This would be an example of altruistic behavior that is beneficial to the calling bird if
    a. the bird giving the vocalization survives the attack.
    b. the inclusive fitness of the bird giving the vocalization is increased.
    c. all birds survive the attack.
    d. the birds benefiting are offspring of the bird giving the vocalization.
    e. All of the above

    Textbook Reference: *53.6 How Does Social Behavior Evolve? p. 1137*
15. Which of the following characteristics is shared by all eusocial species?
    a. A sex determination system in which males are haploid and females are diploid
    b. Queens that mate with a single male
    c. The presence of sterile classes
    d. Both a and c
    e. All of the above

    Textbook Reference: *53.6 How Does Social Behavior Evolve? p. 1136*
16. Which of the following statements about territoriality is true?
    a. A male defending a territory in a lek is defending neither food, a nest site, nor females.
    b. Territories do not always include foraging areas.
    c. Territorial defense is likely to impose three kinds of costs: energetic, risk, and opportunity.
    d. Both b and c
    e. All of the above

    Textbook Reference: *53.4 How Does Behavior Evolve? pp. 1123–1125*

Application Questions

1. Describe the experiments performed with bluegill sunfish preying on water fleas and the implications of these studies for foraging theory.

   Textbook Reference: *53.4 How Does Behavior Evolve? p. 1126*
2. A pigeon has been trained to feed at food bins at the eastern end of a circular cage from which it can see the sky. There are food bins at the N, NE, E, SE, S, SW, W, and NW ends of the cage. Based on this information, answer the following questions:

   a. If the cage is covered and a fixed light source is presented at the east end of the cage at the time of sunrise, where will the pigeon search for food at noon?

   b. If a mirror is used to shift the apparent position of the sun at noon from the south to the northwest, where will the pigeon search for food?

   c. The bird has been placed in a light-controlled environment for three weeks and phase-delayed by six hours. If the pigeon is returned to the cage under natural lighting conditions at sunset, where will it search for food?

   Textbook Reference: *53.5 What Physiological Mechanisms Underlie Behavior? p. 1131*
3. What is the function of a honey bee's dance, and what does it communicate to other bees in a hive?

   Textbook Reference: *53.5 What Physiological Mechanisms Underlie Behavior? p. 1132–1133*
4. For each of the two *y*-axes in the following graph, draw a labeled curve that correctly summarizes observations made on goshawks attacking wood pigeons, as described in the textbook.

   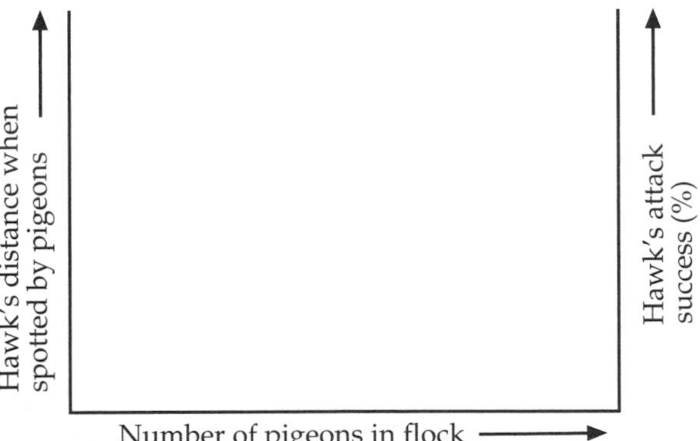

   Textbook Reference: *53.6 How Does Social Behavior Evolve? pp. 1136–1137*
5. Why does the unusual sex determination system in *Hymenoptera* predispose species in this group toward the evolution of eusociality?

   Textbook Reference: *53.6 How Does Social Behavior Evolve? p. 1136*

Answers

Diagram Exercise Answer

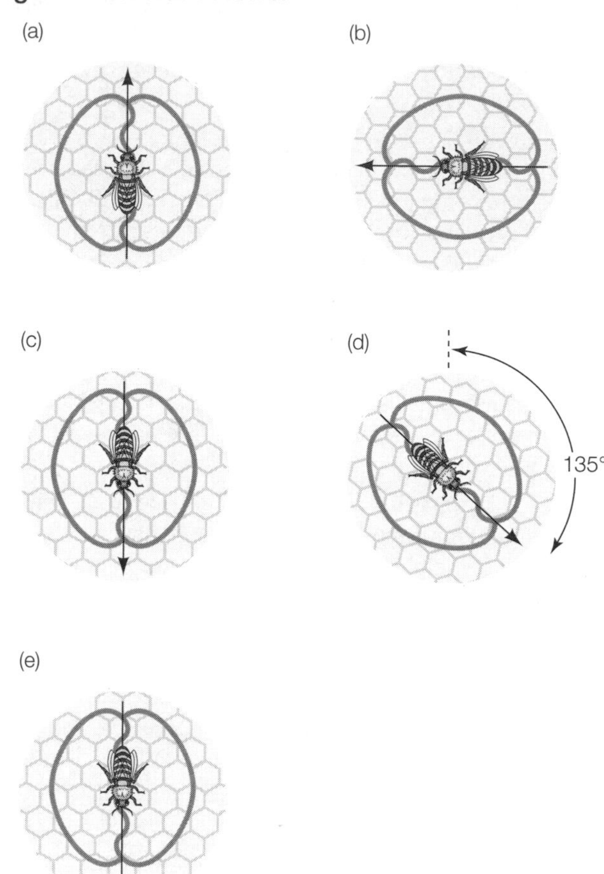

Knowledge and Synthesis Answers

1. **d.** The first three questions (answer choices a, b, and c) concern the *proximate* mechanisms that underlie a behavior. The fourth question is concerned with the *ultimate* cause of a behavior—the selection pressures that shaped its evolution.
2. **e.** The temporal pairing of a reward with a particular response to a stimulus is a feature of operant conditioning, which is the standard method of training an animal.
3. **d.** Young male songbirds must hear the species song and then later be able to hear themselves sing. In the adult, the presence of testosterone is needed to increase the size of regions in the brain associated with singing during the breeding season. Absence of testosterone will result in the inability to perform the correct song.
4. **a.** The environment in which a species normally lives is its habitat.
5. **e.** The forfeited benefits of behaviors that could not be achieved as a result of performing a different behavior, like scanning, constitute the opportunity cost of the performed behavior.
6. **b.** Because the number of Species 2 taken by the predator increases markedly at low Species 1 densities, you can conclude that the predator includes Species 2 in its diet only when insufficient Species 1 individuals are available. The number of Species 1 taken is independent of Species 1 density.
7. **e.** Crosses of honey bees exhibiting hygienic behavior with other non-hygienic bees of the same species revealed that two genes control this behavior. The behavior increases the resistance of the hive to a bacterium that is fatal to bee larvae.
8. **e.** All animals exposed to a daily cycle display circadian rhythms, regardless of when they are active.
9. **b.** Hibernators, cave-dwellers, and equatorial migratory species lack the necessary daily cues at some stage during the year, so they most likely have endogenous circannual rhythms. Thus the white-tailed deer would be the only animal that could use circadian cues.
10. **d.** Piloting is a means of navigation using visual cues and landmarks. Following the coast during migration is considered piloting.
11. **c.** Many animals use the stars for navigation. The stars present two types of information: a fixed reference point (the point directly above Earth's axis of rotation) and moving constellations, which appear to revolve around the fixed point.
12. **e.** By definition, pheromones are chemicals used for communication among individuals of a single species. Mammalian pheromones can communicate information about several characteristics of the signaler, including all those mentioned in answer b. Though pheromones differ in their durability (as well as their volatility and diffusibility) in the environment depending on their function, they are all much more durable than most visual, auditory, or tactile signals and are therefore relatively poorly suited for rapid exchange of information.
13. **d.** For an altruistic behavior to evolve by natural selection, it must increase the inclusive fitness of the individual performing the behavior. Any explanation of an altruistic act based on its purported benefit to the species as a whole violates this principle.
14. **b.** Altruistic behavior is beneficial to the performer when the improvement in the reproductive success of kin (not including offspring) exceeds the reduced reproductive success of the individual performing the act. If this condition is met, then the behavior has improved the inclusive fitness of the performer.
15. **c.** By definition, eusocial species live in social groups with sterile castes. The sex determination mechanism in which males are haploid and females are diploid is found in the *Hymenoptera* (ants, bees, and wasps) but not in termites and naked mole-rats. The textbook also mentions that naked mole-rat colonies include several reproductive males.

16. **e.** Territories of animals may not include foraging areas. In species in which males engage in communal displays at a lek, the territory is used only for mating. Maintaining a territory is costly in all the ways mentioned in answer c.

Application Answers

1. In studies in which bluegill sunfish were provided with small, medium, and large water fleas at three different prey densities, the fish took equal proportions of the three prey sizes at low densities, but mostly large water fleas at high prey densities. This agrees with predictions from foraging theory about how a predator should behave to maximize its energy intake.

2. a. The pigeon will eat from the northern bin at noon. Normally at noon the sun is in the south, and the bird would eat from the eastern bin that is 90° to the left of the sun. With the light in the east, the bird would eat in the north, which is 90° to the left of the light.

   b. The bird will go to the bin that is 90° from the sun at noon, and will feed from the southwestern food bin.

   c. At sunset the eastern bin with food would normally be 180° from the sun in the west. After the bird has been phase-delayed six hours, it will think that the setting sun is the noon sun. At noon the bird usually eats from the eastern bin, which is located 90° to the left of the sun. Therefore, the bird will eat from the southern bin.

3. Honey bees dance to inform other members of the hive about the distance and location of a food source. If food is less than about 80 meters from the hive, the bees will perform a round dance. If food is farther from the hive, the bees will perform a waggle dance. Only the waggle dance contains information about direction.

4. Your curves should show a positive relationship between the hawk's distance when spotted and pigeon flock size and a negative relationship between the hawk's attack success and pigeon flock size (see below).

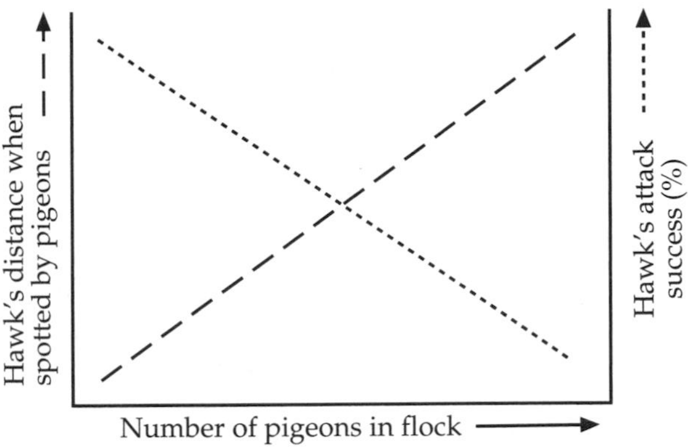

5. All species of the *Hymenoptera* (ants, bees, and wasps) have a sex determination system in which males are haploid and females are diploid; thus females share 75 percent of their genes with their sisters but only 50 percent of their genes with any offspring they could produce. In species whose sexes are determined in this way, females may increase their inclusive fitness by foregoing reproduction and helping to raise sisters. This explanation does not apply to the evolution of eusociality in species without this mode of sex determination.

54 Ecology and the Distribution of Life

The Big Picture

- Much evidence supporting the theory of continental drift (described in Chapter 25) is based on the distributions of organisms on Earth. By the same token, the acceptance by geologists and biologists of the reality of continental drift revolutionized the field of biogeography.
- The area phylogeny approach used by biogeographers to understand how the present distribution of groups of organisms came about is a good example of the broad applicability of the modern methods of phylogenetic analysis that were described in Chapters 22 and 24.

Common Problem Areas

- The relationship between a taxonomic phylogeny and an area phylogeny may be confusing. Study Figure 54.11 to understand how these two types of phylogeny can be used to reconstruct the evolutionary history of a taxon such as the Equidae.

Study Strategies

- In attempting to master the information concerning the many biomes described in this chapter, focus on the graphs that summarize the essential points about each one.
- You also should study Figure 54.5 to get a sense of the geographical location and extent of each biome. A useful exercise to help you learn the relative locations of biomes is to imagine a journey (e.g., from equatorial South America to Alaska, or from Maine to California), naming the biomes that you would pass through.
- Go to yourBioPortal.com to review the following tutorials and activity:

  Animated Tutorial 54.1 Biomes

  Animated Tutorial 54.2 Rain Shadow

  Web Activity 54.1 Major Biogeographic Regions

Important Concepts

Ecology is the scientific study of the interactions among organisms and between organisms and their physical environment.

- The science of ecology generates knowledge about interactions in the natural world. The greater our understanding of ecological interactions, the greater the likelihood that we can carry out activities such as growing food and managing pests without causing unintended consequences for ourselves and other organisms.
- Whereas ecology is a science, environmentalism uses ecological knowledge and other considerations to inform both personal decisions and public policy relating to the stewardship of natural resources and ecosystems.
- The environment comprises both abiotic factors (physical and chemical) and biotic factors (living organisms).
- Many new tools—such as mathematical models, molecular techniques, and satellite imaging—are available to ecologists as they strive to understand the biotic and abiotic forces that influence the distribution and abundance of organisms.

Climates vary geographically primarily because different places receive different amounts of solar energy.

- The climate of a region is the average of the atmospheric conditions found in that region over the long run. Weather is the short-term state of those conditions.
- Solar energy varies with latitude. Regions near the poles receive less energy per unit of ground area than regions near the equator because of the lower angle of the sun. At high latitudes there is also more variation over the course of a year in both day length and the angle of arriving solar energy than at latitudes closer to the equator (see Figure 54.1).
- Rising air expands and cools, releasing moisture, whereas descending air is compressed and warmed, taking up more moisture.
- Unequal heating of the atmosphere at low and high latitudes produces vertical and latitudinal movements of air masses. These movements cause very moist

climates to occur at the equator and at 60° north and south latitudes (where air rises) and arid climates at about 30°N and 30°S latitudes and near the poles (where air descends) (see Figure 54.2).

- The spinning of Earth on its axis causes air masses moving latitudinally to be deflected to the right in the Northern Hemisphere and to the left in the Southern Hemisphere. Thus, winds blowing toward the equator at low latitudes veer to become the northeast and southeast trade winds, whereas winds blowing away from the equator at mid-latitudes are deflected to become the prevailing Westerlies (see Figure 54.3).
- Ocean currents are driven primarily by prevailing winds but are deflected by continents. The poleward movement of ocean water warmed in the tropics is a major mechanism of heat transfer to high latitudes.
- Winds can also influence climate by causing upwellings: areas where colder water from depths below 50 meters rises to mix with and replace warmer surface water.
- Organisms must adapt to climatic conditions. Metabolic adaptations to an unfavorable environment include resting states characterized by greatly reduced metabolic activity and enhanced physiological resistance to adverse conditions. In mammals, such states may occur in summer (estivation) or winter (hibernation). In invertebrates, diapause is the equivalent of hibernation.
- To cope with extremely cold temperatures, many organisms produce antifreezes that lower the freezing point of their cell contents or body fluids.
- Morphological adaptations to climatic conditions include differences in body shape and pigmentation.
- Particularly among ectotherms, behavior mechanisms for temperature regulation are common. An example of a behavioral adaptation is moving to a more suitable microclimate—a subset of climatic conditions in a small, specific area that differ from those in the environment at large.
- Long-distance movements can be key adaptations to climatic challenges. In response to cyclic environmental changes, organisms may evolve life cycles that include migrations that appear to anticipate those changes.

A biome is a large, terrestrial environment defined by its climatic and geographic attributes and characterized by ecologically similar organisms.

- In biomes that occur in several widely separated areas of the globe, the species occurring in different locations are unlikely to be identical, but they are likely to share many adaptations to their environment as a result of convergent evolution.
- The distribution of biomes on Earth is strongly influenced by annual patterns of temperature and precipitation. Often the boundary between two biomes is somewhat arbitrary because one biome gradually merges into another.
- Tundra is found at high latitudes in the Arctic and in high mountains. It is a treeless biome dominated by short perennial plants. In the Arctic, permanently frozen soil called permafrost underlies tundra vegetation.
- Boreal forests are located at lower latitudes than Arctic tundra and at lower elevations on temperate-zone mountains. The short summers favor evergreen trees, which are the dominant vegetation. Temperate evergreen forests occur along the coasts of continents at middle to high latitudes, where winters are mild but wet and summers are cool and dry.
- Temperate deciduous forests are dominated by deciduous trees, which produce leaves that photosynthesize rapidly during the warm, moist summers and are lost during the cold winters.
- Temperate grasslands occur in areas that are relatively dry for much of the year. Grasses dominate this biome because they are well adapted to grazing and fire. Because of their rich topsoil, most temperate grasslands have been turned over to agriculture.
- Hot deserts, characterized by very warm and dry conditions year-round, are found in two belts around the 30°N and 30°S latitudes. Plants and animals of this biome are characterized by adaptations for conserving water.
- Cold deserts are found in dry regions at middle to high latitudes. Seasonal changes in temperature are great. They are dominated by a few species of low-growing shrubs.
- The chaparral biome, found on the west side of continents at mid-latitudes, is dominated by evergreen shrubs and low trees that are adapted to survive periodic fires. Winters are cool and wet; summers are hot and dry.
- Thorn forests and savannas are found in semiarid climates on the equatorial side of hot deserts. Savanna, a grassland punctuated by scattered trees, is maintained by grazing, browsing, and burning. In their absence, it reverts to dense thorn forest dominated by small, spiny shrubs and trees.
- Tropical deciduous forests are dominated by trees that lose their leaves during the long, hot dry season. Because their soils are less leached of nutrients than the soils of wetter areas, most of these forests have been cleared for grazing cattle and growing crops.
- Tropical evergreen forests (rainforests), found in equatorial regions where annual rainfall exceeds 250 cm annually, have the greatest species richness of all biomes and the highest productivity of all ecological communities. Nevertheless, their soils are poor and usually cannot support long-term agriculture unless massively fertilized. Epiphytes—plants that grow on other plants, deriving their nutrients and water from the atmosphere—thrive in tropical mountain forests. Rainforests are being deforested or converted to agriculture at a high rate.
- The movement of air over mountains often results in a rain shadow—a dry area on the leeward side of the range (see Figure 54.6).

- The distribution of biomes is determined not only by climate but also by other factors, particularly soil fertility and fire.

Biogeography is the scientific study of the distributional patterns of populations, species, and ecological communities.

- Biogeographers divide Earth into a number of biogeographic regions, each containing characteristic assemblages of species. The biotas of the biogeographic regions differ because barriers such as oceans restrict the dispersal of organisms (see Figure 54.8).
- Species found in only one place are said to be endemic to that location. Remote islands contain many endemic species.
- Two scientific advances changed the field of biogeography: the acceptance of the theory of continental drift and the development of phylogenetic taxonomy.
- Continental drift has influenced the evolution and mixing of species throughout the history of life on Earth. It explains some discontinuous distributions that include several biogeographic regions (see Figure 54.10).
- The development of phylogenetic taxonomy has given biogeographers the ability to convert a taxonomic phylogeny into an area phylogeny, a process that involves replacing the names of the taxa with the names of the places where those taxa live or lived.
  This method is used to help explain how the current distribution of particular groups of species came about (see Figure 54.11).
- In the process of biotic interchange, two different biota merge following the fusion of two formerly separated land masses. An example of this process was the formation of the Central American land bridge connecting North and South America. After it formed, many species of mammals that had evolved on one continent colonized the other (see Figure 54.12).
- A split distribution of a species can be accounted for in two ways.
  - A barrier may appear that splits a species' distribution (a vicariant event).
  - A species may cross an existing barrier and establish a new population (dispersal).
- Both vicariance and dispersal influence most biogeographic patterns. To determine which is more important in a particular case, scientists apply the parsimony principle. That is, they prefer the explanation that requires the smallest number of unobserved events to account for the pattern (see Figure 54.13).
- In some cases, vicariance cannot be the explanation for a split distribution because it is clear that no separation of continuous populations could ever have taken place. For example, the occurrence of related species on a continental landmass and on islands never connected to a continent (e.g., the Hawaiian Islands) can only be the result of dispersal.
- Humans have deliberately or inadvertently introduced many species to new regions, often with harmful results.

Aquatic ecosystems can be divided into a number of life zones.

- Earth's oceans form one interconnected water mass, with only partial barriers to dispersal. But though the oceans are connected, living successfully in a particular region requires that organisms have specific physiological tolerances and morphological characteristics. As a consequence, most marine organisms have restricted ranges.
- The ocean can be divided into several life zones (see Figure 54.14).
  - The photic zone (in both marine and freshwater environments) extends from the surface to the depth at which photosynthesis can no longer occur. Most aquatic life inhabits this zone. The dominant autotrophs in this zone are floating microscopic phytoplankton, which are fed upon principally by zooplankton.
  - The coastal zone extends from the shoreline to the edge of the continental shelf. The near-shore region of the coastal zone affected by wave action is the littoral zone, and the portion of this zone between high- and low-tide levels is the intertidal zone.
  - The pelagic zone is the open ocean beyond the coastal zone.
  - The benthic zone is the ocean bottom.
  - The aphotic zone is that portion of both the coastal and pelagic zones that is below the depth at which photosynthesis can occur. Organisms inhabiting this zone either subsist on decaying organic matter that descends from the photic zone or are sustained, directly or indirectly, by chemosynthetic microbes.
- Freshwater environments comprise running water (streams and rivers) and standing water (lakes and ponds). Although these environments contain less than 3 percent of Earth's water, they are the habitat for about 10 percent of all aquatic species. Like oceans, bodies of standing fresh water can be divided into zones based on depth and light penetration. An important reason for discontinuities in the ranges of aquatic animals is that most organisms that live in fresh water cannot survive in the oceans, and vice versa.
- Estuaries are bodies of water where salt and fresh water mix. High in species diversity and important as a resource for humans, estuarine environments are now threatened in many places by pollution and overfishing.

Test Yourself

Diagram Exercise

Construct a concept map whose theme is "Oceans." Include in your map the following terms: oceans, aphotic zone, coastal zone, currents, depth, direction of prevailing winds, distance from shore, intertidal zone,

latitudinal differences in solar energy input, life zones, littoral zone, location of continents, pelagic zone, photic zone, photosynthetic organisms, and rotation of Earth. Connect these concepts by verbs, phrases, or comparative terms to indicate the relationships among them.
Textbook Reference: *54.5 How Is Life Distributed in Aquatic Environments? pp. 1163–1164*

Knowledge and Synthesis Questions

1. Which of the following are *not* usually included within the domain of ecology?
   a. Interactions between conspecifics
   b. Modifications of the environment by organisms
   c. Interactions between humans and domesticated plants and animals
   d. Modifications of the environment by physical processes
   e. Both a and b
   Textbook Reference: *54.1 What Is Ecology? pp. 1141–1142*
2. Which of the following is *not* a difference between an acre of land in Colombia and an acre of land in Michigan?
   a. The angle of the sun reaching the ground in the month of July
   b. The solar energy flux in the month of July
   c. The annual solar energy flux
   d. The total hours of daylight per year
   e. Both b and c
   Textbook Reference: *54.2 Why Do Climates Vary Geographically? p. 1142*
3. If Earth did not spin on its axis, the northeast trade winds would blow from the
   a. northeast.
   b. south.
   c. north.
   d. east.
   e. southwest.
   Textbook Reference: *54.2 Why Do Climates Vary Geographically? pp. 1143–1144*
4. Which of the following does *not* influence ocean circulation patterns?
   a. Circulation of Earth's atmosphere
   b. Deflection by land masses
   c. Upwelling of deep water
   d. Rotation of Earth on its axis
   e. All of the above influence ocean circulation patterns.
   Textbook Reference: *54.2 Why Do Climates Vary Geographically? pp. 1143–1144*
5. The following diagram shows a mountain with a sea breeze blowing as indicated by the arrow. Circle the letter for the area with air that would be *both* relatively warm and dry.

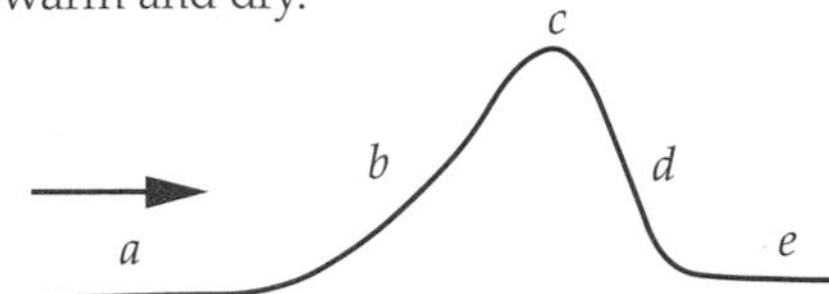

   Textbook Reference: *54.3 What Is a Biome? p. 1157*
6. In what area of the diagram in Question 5 is a process occurring that is similar to the process that occurs in the region surrounding the equator?
   Textbook Reference: *54.2 Why Do Climates Vary Geographically? p. 1143*
7. The biome that is maintained by browsers, grazers, or fire is the
   a. tundra.
   b. temperate grasslands
   c. cold desert.
   d. tropical deciduous forest.
   e. tropical savanna.
   Textbook Reference: *54.3 What Is a Biome? p. 1154*
8. Match the letters of the following biomes with the descriptions that follow.
   a. Tundra
   b. Boreal forest
   c. Hot desert
   d. Chaparral
   e. Tropical deciduous forest
   ____ Mostly coniferous, wind-pollinated and wind-dispersed tree species
   ____ Leaves lost during dry season; agriculturally desirable land
   ____ Cool winters, hot dry summers; maritime climate
   ____ Prominent succulent plants; found at 30°N and 30°S latitudes
   ____ Distribution altitudinally or latitudinally determined; permafrost present
   Textbook Reference: *54.3 What Is a Biome? pp. 1147–1156*
9. The following climograph shows yearly variation in rainfall and temperature for four biomes. Select the correct curve for each of the following biomes.
   ____ Tundra
   ____ Tropical evergreen forest
   ____ Temperate deciduous forest
   ____ Cold desert

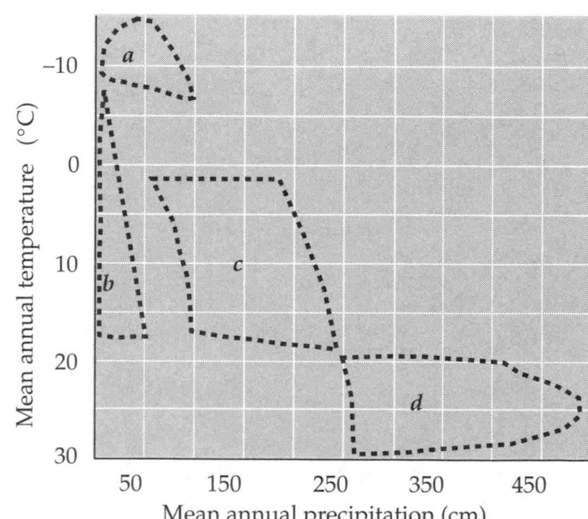

   Textbook Reference: *54.3 What Is a Biome? pp. 1147–1156*

10. The tropical deciduous forest biome has relatively constant _______, but _______ varies seasonally; the temperate deciduous forest biome has relatively constant _______, but _______ varies seasonally.
    a. temperature; rainfall; rainfall; temperature
    b. rainfall; temperature; temperature; rainfall
    c. rainfall; temperature; rainfall; temperature
    d. temperature; rainfall; temperature; rainfall
    e. None of the above

    Textbook Reference: *54.3 What Is a Biome? p. 1149, 1155*

11. Which of the following biogeographical regions represents the largest area?
    a. Nearctic
    b. Palearctic
    c. Neotropical
    d. Oriental
    e. Ethiopian

    Textbook Reference: *54.4 What Is a Biogeographic Region? p. 1158*

12. The following diagram shows three islands. Islands A and B were connected in the past, C was always separate. A species of land snail is found on all three islands.

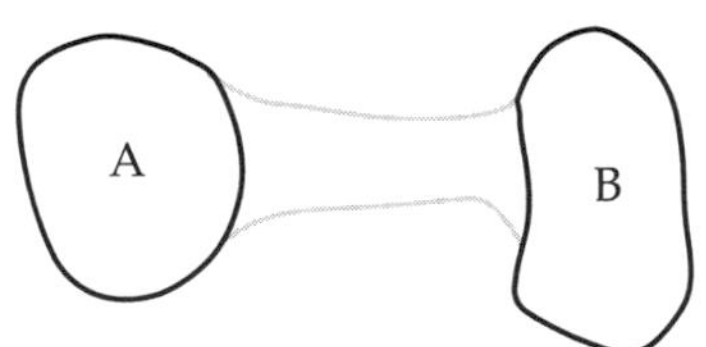

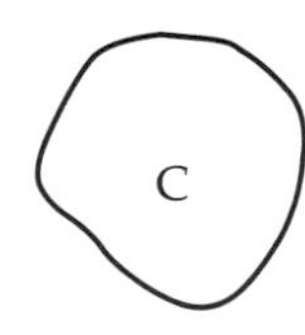

    Which of the following does *not* correctly describe the distribution of this snail relative to the three islands?
    a. Vicariant distribution relative to A and B
    b. Dispersal distribution relative to A and C
    c. Dispersal distribution relative to B and C
    d. Vicariant distribution relative to A and C
    e. All of the above are correct.

    Textbook Reference: *54.4 What Is a Biogeographic Region? pp. 1161–1162*

13. If the species of snail in the question above is found *only* on the three islands shown in the diagram, which of the following would correctly describe the endemism of this species?
    a. Endemic relative to A and B
    b. Endemic relative to C
    c. Endemic relative to A, B, and C
    d. All of the above
    e. None of the above

    Textbook Reference: *54.4 What Is a Biogeographic Region? pp. 1158–1159*

14. The region of the ocean that lies close enough to shore to be affected by wave action is the _________ zone.
    a. abyssal
    b. pelagic
    c. benthic
    d. aphotic
    e. littoral

    Textbook Reference: *54.5 How Is Life Distributed in Aquatic Environments? p. 1163*

15. Which of the following statements comparing freshwater and marine habitats is *false*?
    a. About 10 percent of all aquatic species live in freshwater habitats.
    b. Unlike the oceans, freshwater lakes and ponds cannot be divided into zones based on depth and light penetration.
    c. The global volume of marine habitats is much greater than that of freshwater habitats.
    d. Freshwater species richness is less than marine species richness in proportion to the relative extent of the two habitats.
    e. Both b and d

    Textbook Reference: *54.5 How Is Life Distributed in Aquatic Environments? p. 1164*

16. An aquatic life zone in which mixing of fresh water and salt water occurs is called
    a. an estuary.
    b. an intertidal zone.
    c. a littoral zone.
    d. a pelagic zone.
    e. a photic zone.

    Textbook Reference: *54.5 How Is Life Distributed in Aquatic Environments? p. 1164*

Application Questions

1. Why is an understanding of ecology essential for the future well-being of humanity?

   Textbook Reference: *54.1 What Is Ecology? p. 1141*

2. Many organisms move from place to place in an effort to locate and occupy the environment to which they are best adapted. Compare the movement of a bird flying back and forth between its tropical wintering area and its temperate breeding range with the movement of a lizard as it shifts between its burrow and the surface and between shaded and sunny areas on the surface. Is the term "migration" properly applied to both types of movement?

   Textbook Reference: *54.2 Why Do Climates Vary Geographically? p. 1145*

3. Tropical evergreen forests have greater overall productivity and species richness than tropical deciduous forests, yet the latter are more easily converted to productive agricultural land than the former. Why?

   Textbook Reference: *54.3 What Is a Biome? pp. 1155–1156*

4. More living species of the horse family live in Africa than in any other continent. Yet biogeographers do not believe that the horse family evolved in Africa. What kinds of evidence have biogeographers relied on to explain how the current distributions of horses came about?

   Textbook Reference: *54.4 What Is a Biogeographic Region? pp. 1160–1161*

5. About four million years ago, a land bridge formed between North and South America. Describe the results of the ensuing biotic interchange.
   Textbook Reference: *54.4 What Is a Biogeographic Region? pp. 1160–1161*

Answers

Diagram Exercise Answer

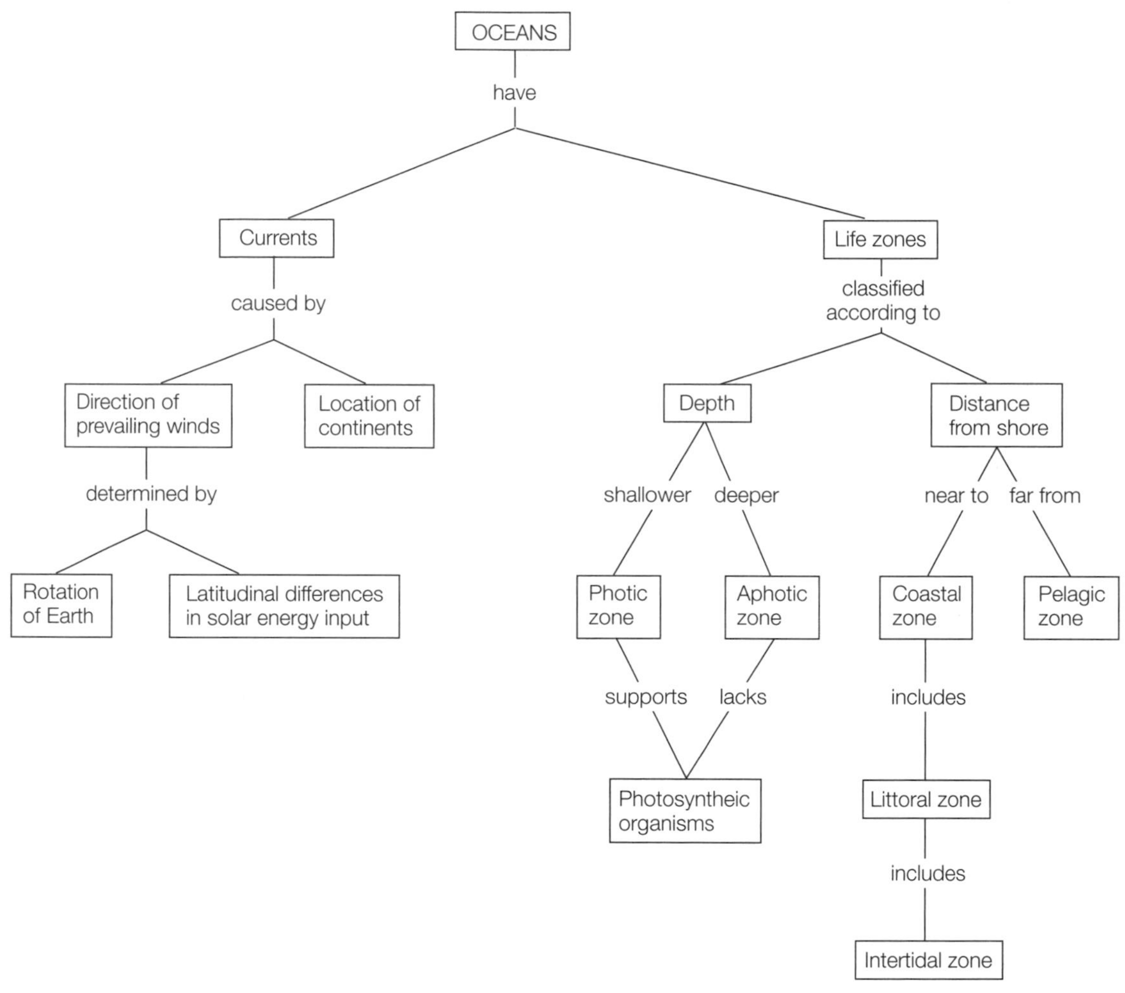

Knowledge and Synthesis Answers

1. **d.** All except the modification of the environment by physical factors are normally included within the domain of ecology.
2. **d.** All areas on Earth receive equal hours of daylight per year. The seasonal distribution of those hours, the solar energy input, and the sun angle do vary latitudinally.
3. **c.** If Earth did not spin on its axis, air flowing south toward the equator would not be deflected to the right. Therefore, the northeast trade winds would blow directly from the north instead of from the northeast.
4. **c.** The upwelling of deep water, especially along the coasts of continents, is an effect of the pattern of ocean current circulation and not a cause of the pattern.
5. **e.** The air would have lost most of its moisture while rising on the windward side of the mountain, and in descending to area *e* it would have warmed again.
6. **b.** In regions surrounding the equator, warm air rises and loses much of its moisture. These same events occur when moist air rises over a mountain.
7. **e.** If tropical savanna vegetation is not grazed, browsed, or burned, it typically reverts to dense thorn forest.
8. **b.** Mostly coniferous, wind-pollinated and wind-dispersed tree species

   e. Leaves lost during dry season; agriculturally desirable land

d. Cool winters, hot dry summers; maritime climate

c. Prominent succulent plants; found at 30°N and 30°S latitudes

a. Distribution altitudinally or latitudinally determined; permafrost present

9. **a.** Tundra

   d. Tropical evergreen forest

   c. Temperate deciduous forest

   b. Cold desert

10. **a.** The tropical deciduous forest biome has relatively constant temperature, but rainfall varies seasonally; the temperate deciduous forest biome has relatively constant rainfall (or precipitation, because snow may occur in winter) but temperature varies seasonally.
11. **b.** See Figure 54.8 in the textbook.
12. **d.** A vicariant distribution requires that the species once had a continuous distribution in what are now separate areas.
13. **c.** The snail is endemic to the group of islands, but not to any single one or pair of them because it is also found on the remaining islands or island.
14. **e.** See Figure 54.14 in the textbook.
15. **e.** Life zones occur in bodies of fresh water just as they do in oceans. Even though freshwater habitats occupy a very small global area, they have about 10 percent of all aquatic species, which, by proportion, makes them rich in species.
16. **a.** An estuary is the body of water found at the mouth of a river, where salt water mixes with fresh water.

Application Answers

1. A grasp of ecological science is essential for understanding (and hence for preserving) the ecosystems that provide essential services for human society (such as clean air and water). Also, as we alter or destroy natural ecosystems in the interest of providing food and other resources for the growing human population, an understanding of ecological principles may enable us to avoid many of the unanticipated and possibly disastrous consequences of our actions.
2. Only the bird is engaging in migration, which is the active movement of members of a population, often over long distances, in response to cyclical, predictable environmental events, such as seasonal changes. The lizard, in contrast, travels relatively short distances as it searches for its ideal microclimate, which is a subset of climatic conditions in a small, specific area. For the lizard, temperature is a particularly important aspect of the microclimate. The bird, of course, may also seek out particularly favorable microclimates not only within its wintering and breeding ranges but also during its migratory journeys.
3. The reason for the greater agricultural productivity of deciduous tropical forests is that their soils are richer in nutrients than the soils of tropical evergreen forests. The higher precipitation characteristic of tropical evergreen forests causes their soils to lose mineral nutrients very quickly if they are not recycled into the living vegetation, which is in fact where most of the nutrients are located in this type of forest.
4. The two main sources of evidence used by biogeographers to explain the present distributions of horses are the fossil record and the area phylogeny method. Fossils indicate clearly that the earliest ancestors of horses evolved in North America. Inasmuch as Przewalski's horse inhabits central Asia (as the area phylogeny in Figure 54.11 shows) and represents the earliest lineage to diverge from the lineages leading to the other living species, it is reasonable to assume that the ancestor of all living species of horses first dispersed from North America to Asia. Similar reasoning leads to the conclusion that further speciation events occurred as horses moved from Asia to Africa and that Africa was the site of the speciation of zebras.
5. The land bridge permitted the mammalian faunas of both North America and South America to disperse to the other continent. Many North American species were able to colonize South America, and subsequently, some of these invaders formed new species (such as the guanaco) found today only in South America. In contrast, only a few South American species (such as the nine-banded armadillo) became established in North America.

55 Population Ecology

The Big Picture

- An understanding of concepts of population ecology such as exponential growth and carrying capacity is fundamental to your ability to think intelligently about issues such as human population growth, the preservation of biodiversity, and the control of undesirable species.
- Darwin's realization that all populations have the inherent capacity for exponential growth was crucial to the development of his theory of natural selection.

Common Problem Areas

- Students frequently have problems interpreting graphs depicting concepts of population ecology. Look carefully at how the axes of a graph are labeled. Ask yourself questions such as: Are the scales arithmetic or logarithmic? Is the fate of a cohort of the population being traced, or is the growth pattern of the entire population being described?

Study Strategies

- Redrawing graphs from the textbook is a good way to reinforce your understanding of the concepts being presented.
- Take advantage of laboratory activities involving computer simulations of population growth or other aspects of population ecology.
- Go to yourBioPortal.com to review the following tutorials and activity:

  Animated Tutorial 55.1 Exponential Population Growth

  Animated Tutorial 55.2 Logistic Population Growth

  Animated Tutorial 55.3 Habitat Fragmentation

  Interactive Tutorial: Age Structure and Survivorship Relationships

  Web Activity 55.1 Logistic Population Growth

Important Concepts

Ecologists use a variety of methods to study populations.

- A population consists of the individuals of a species that interact with one another within a given area at a particular time. Demography is the study of populations.
- Populations have a characteristic age structure (distribution of individuals across age categories) and dispersion pattern (the way those individuals are spread over the environment).
- The density of a population is the number of individuals of the population per unit of area. It is a function of the processes that increase the size of a population (births and immigration, i.e., movements into the population) and processes that decrease its size (deaths and emigration, i.e., movements out of the population).
- Population dynamics refers to the patterns and processes of change in populations.
- For some species, especially those in which individuals are large and population size is small, investigators can perform a full census (complete count) of the population. The size of most populations, however, must be estimated from representative samples using statistical methods. For stationary organisms, ecologists count the number of individuals in sampling plots (quadrats) or along transects and extrapolate the counts to the entire range of the population. For mobile organisms, investigators often use the mark-recapture method.
- The age structure of a population describes the proportions of individuals in all age categories. Because reproductive capacity varies with age, the age structure of a population has a profound effect on its potential for growth; it also reveals much about the recent history of births and deaths in the population (see Figure 55.2).
- Dispersion refers to the spatial distribution of the individuals in a population. Dispersion patterns may be clumped, regular, or random. Abiotic environmental conditions and social interactions such as cooperation and competition are factors that may influence the dispersion pattern of a population.

- Ecologists use multiple estimates of population densities made over time to estimate the rate at which a population is growing or decreasing. Over a given interval of time, the number of individuals in a population increases by the number of individuals added to the population by birth and by immigration, and decreases by the number of individuals lost from the population by death and by emigration.
- A life table is a tool used by ecologists to keep track of demographic events (births, deaths, immigration, and emigration) in a population and to determine the rate at which these events occur.
  - A cohort life table (also called a horizontal life table) is constructed by determining for a cohort (a group of individuals born at the same time) the number still alive at specific times. From these data, investigators can calculate the mortality rate for each age class as well as survivorship (l_x), which is the likelihood of an individual member of the cohort's surviving to reach age *x* (see Table 55.1).
  - The fecundity schedule of a cohort life table tracks the number of offspring produced per female for each age class (see Table 55.2). Data concerning fecundity (m_x) are used to estimate a population's potential for growth.
  - A vertical life table is constructed by sampling a population at a single time.
- Survivorship curves show the number of individuals in a cohort still alive at different times over the life span. Graphs of survivorship in real populations often resemble one of three types (see Figure 55.4).
  - Type I: Most individuals survive for most of their potential life span and die at about the same age. Parental care and low fecundity are typical of Type I species.
  - Type II: Survivorship is about the same throughout most of the life span.
  - Type III: Survivorship of the young is very low, but is high for most of the remainder of the life span. Little or no parental care and high fecundity are typical of Type III species.

Environmental conditions affect life histories.

- The life history of an organism describes how it allocates its time and energy among growth, reproduction, and other activities. Life histories of different organisms vary dramatically. For example, organisms' life cycles differ in the timing of reproduction and number of offspring produced. Ecologists study life histories because they influence how populations grow and are distributed.
- The per capita growth rate of a population is called its intrinsic rate of increase (*r*). Leaving aside migration, *r* is the difference between the birth rate (*b*) and the death rate (*d*) per individual. It is expressed mathematically by the equation $r = b - d$.
- The intrinsic rate of increase of a population is dependent on life history traits that are influenced by environmental conditions. The traits most influenced by these conditions are generation time, number of broods per female, and number of offspring per brood. These factors vary not only between species, but also between populations of the same species.
- Iteroparous species are those that reproduce multiple times over the course of their lifetimes. Semelparous species reproduce only once.
- Interspecific interactions such as predation influence the evolution of life histories.

Many factors limit population densities.

- All populations have the potential to grow exponentially. In exponential growth, the per capita rate of increase (*r*) remains constant per unit of time, whereas the number of individuals added to the population accelerates. This occurs because, as the population grows, more and more individuals are alive to reproduce (see Figure 55.6).
- Exponential growth is expressed mathematically as $dN/dt = rN$, where dN/dt is the rate of change in the population size (dN = change in number of individuals; dt = change in time), *r* is the intrinsic rate of growth, and *N* is the population size.
- The biotic potential of a population is expressed when it is growing exponentially.
- The environmental carrying capacity (*K*) is the maximum population size that the environment can support indefinitely. Because growth tends to slow as the carrying capacity is approached, an S-shaped growth curve (logistic growth) is characteristic of many populations growing in environments with limited resources. This type of growth can be generated from the exponential growth equation by adding the term $K - (N/K)$, which represents environmental resistance—the reduction in population growth caused by preemption of available resources. Hence the equation for logistic growth is $dN/dt = r\,N((K - N)/K)$. Population growth stops when $N = K$ because then $(K - N)/K = 0$, and thus $dN/dt = 0$ (see Figure 55.7).
- Density-dependent regulation factors—such as resource depletion, predation, and disease—cause per capita birth or death rates of a population to change in response to changes in population density. As a result, population growth slows down as density increases.
- Factors (such as adverse weather) that change per capita birth or death rates of a population irrespective of its density are said to be density-independent. Abiotic factors tend to be density-independent in their effect on populations, whereas biotic factors are typically density-dependent.
- Different population regulation factors and different habitats are associated with different life history strategies (see Figure 55.8).

- Species termed r-strategists have life histories geared to achieve the maximum rate of population increase. They tend to live in a broad range of unpredictable environments and are characterized by a short life span, density-independent mortality, semelparity, and a Type III survivorship curve.
- Species termed K-strategists have life histories that allow them to persist at or near the carrying capacity of their environment. They tend to be specialized in their resource use and live in predictable environments. They are characterized by a long life span, density-dependent mortality, iteroparity, and a Type I or Type II survivorship curve.
- The life histories of most species combine elements of both strategies.

- Some species are more common than others for several reasons. Factors favoring high population density are utilization of abundant resources, small body size, complex social organization, and recent introduction into a new region (see Figure 55.9). In some cases, species have a small population because they have arisen only recently by polyploidy or a founder event, or they are declining toward extinction.

Spatially variable environments affect population dynamics.

- Most populations are divided into geographically separated subpopulations that live in habitat patches—areas of a particular kind of environment that are surrounded by other kinds. The larger population to which the subpopulations belong is called the metapopulation. Because the individual subpopulations are much smaller than the metapopulation, they are prone to fluctuations leading to extinction. In the rescue effect, immigrants from other subpopulations prevent declining subpopulations from becoming extinct (see Figure 55.10).
- Because corridors connecting habitat patches facilitate dispersal, they enhance the rescue effect (see Figure 55.11).

Humans use the principles of population dynamics to manage and control populations.

- Understanding the life history traits of a commercially valuable species is important for the successful management of its populations.
- To maximize the number of individuals that can be harvested from a population, the population should be held far enough below carrying capacity to have high birth and growth rates.
- Demographic traits, particularly reproductive capacity, determine how heavily a population can be exploited. For commercial reasons, many fish and whale populations have been overharvested in violation of basic principles of population management (see Figure 55.13).
- The populations of undesirable species are most effectively controlled by lowering the carrying capacity of the environment for those species. For non-native pest species, biological control—the introduction of predators and parasites from the pest's original habitat—is sometimes an effective means to reduce populations but is sometimes disastrous (see Figure 55.14).
- The size of the human population contributes to most of the environmental problems we are facing today, from pollution to extinctions of other species.
- Earth's carrying capacity for humans has drastically increased because of the development of social systems and communication, the domestication of plants and animals, and ongoing technological advances in food production and disease control. Today, the planet supports approximately 7 billion human beings.
- Human populations are growing at different rates in different parts of the world. Industrialized countries typically have an age structure that reflects their low birth rates and stable (or even declining) population size. In the developing world, many countries have a high rate of growth and populations skewed toward younger age categories, portending high rates of growth in the future (see Figure 55.15).
- Earth's carrying capacity for humans depends on our use of resources and the effects of our activities on other species.

Test Yourself

Diagram Exercise

Construct a concept map whose theme is "Population." Include in your map the following terms: population, abiotic, biotic, carrying capacity (K), clumped, density-dependent regulation factors, density-independent regulation factors, dispersion pattern, exponential growth pattern, intrinsic rate of increase (r), logistic growth, pathogens, population crash, predation, random, regular, resource limitation, and weather. Connect these concepts by verbs, phrases, or comparative terms to indicate the relationships among them.

Textbook Reference: *55.2 How Do Environmental Conditions Affect Life Histories? pp. 1173–1174; 55.3 What Factors Limit Population Densities? pp. 1174–1178*

Knowledge and Synthesis Questions

1. Which of the following is an aspect of population structure studied by ecologists?
   a. Distribution of genotypes
   b. Age distribution
   c. Spacing of population members
   d. Both b and c
   e. All of the above

   Textbook Reference: *55.1 How Do Ecologists Study Populations? p. 1168*
2. In an effort to measure the size of a bluegill population in a pond, you capture 40 bluegills and mark each with a tag on the tail fin and then release them. A week later you return to the pond and catch 40 more, of which 8

have a tag on the tail fin. What is the best estimate of the bluegill population in the pond?
a. 100
b. 160
c. 200
d. 320
e. 400
Textbook Reference: *55.1 How Do Ecologists Study Populations? p. 1169*

3. In the life table below, survivorship is greatest during which time interval?

Age (years)	Number Alive
0	800
1	770
2	550
3	125
4	75
5	0

a. 0–1 years
b. 1–2 years
c. 2–3 years
d. 3–4 years
e. 4–5 years
Textbook Reference: *55.1 How Do Ecologists Study Populations? pp. 1169–1171*

4. Refer to the graph of age distributions below.

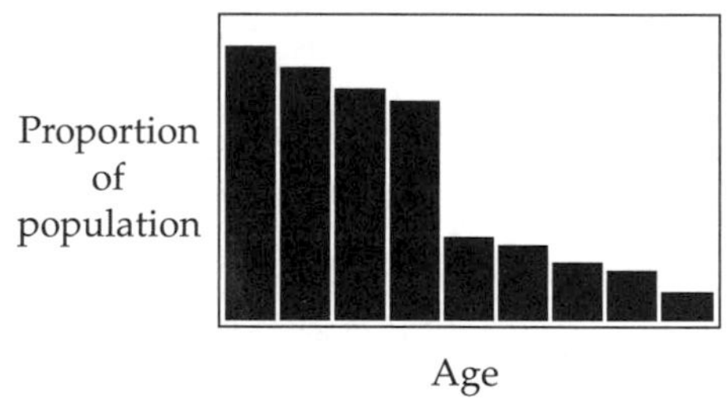

For a population that is stable in size, the age distribution of the graph would indicate
a. high birth and death rates.
b. low birth and death rates.
c. a low birth rate but a high death rate.
d. a high birth rate but a low death rate.
e. Either a or d
Textbook Reference: *55.5 How Can Populations Be Managed Scientifically? p. 1182*

5.–6. Refer to the graph below.

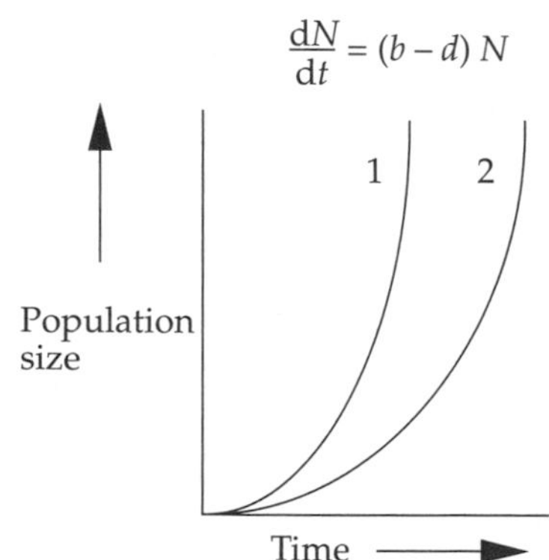

5. In the graph, which of the expressions of the exponential growth equation would need to be increased for curve 1 to become more like curve 2?
a. d*N*/d*t*
b. *d*
c. *b*
d. (*b* – *d*)
e. Either b or d
Textbook Reference: *55.3 What Factors Limit Population Densities? p. 1175*

6. The intrinsic rates of increase (*r*) of species A = 0.25 and of species B = 0.50. According to the graph, the population growth curve for A should be more like curve _______.
Textbook Reference: *55.3 What Factors Limit Population Densities? p. 1175*

7.–8. Refer to the graph below.

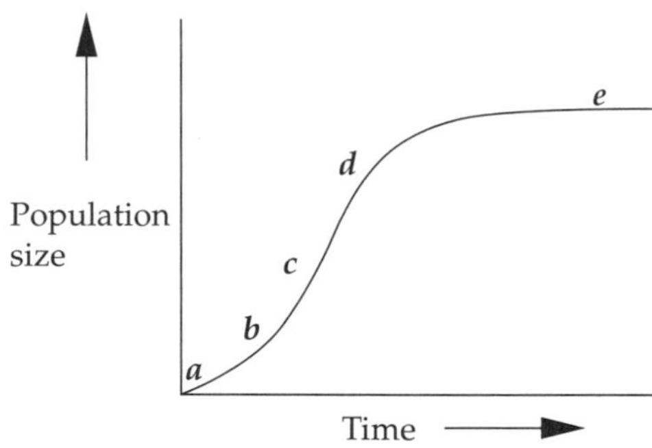

7. In the logistic population growth curve shown in the graph, the rate of growth is greatest at which point?
Textbook Reference: *55.3 What Factors Limit Population Densities? pp. 1175–1176*

8. At which point in the graph would there be zero population growth (d*N*/d*t* = 0)?
Textbook Reference: *55.3 What Factors Limit Population Densities? pp. 1175–1176*

9.–10. Refer to the graph below.

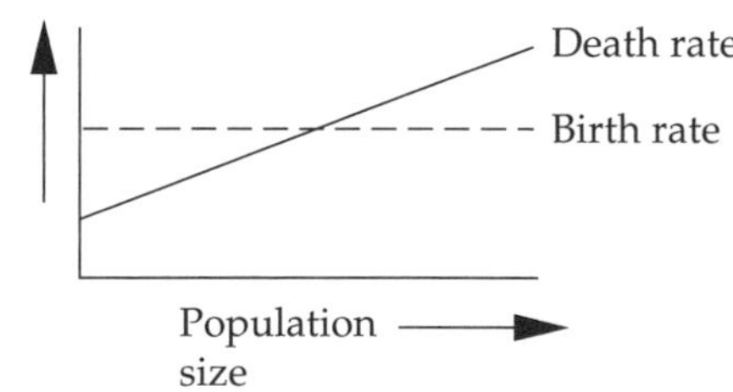

9. Which of the following events is *least* likely to be true of this population?
a. Parasites spread between population members.
b. Increased competition for food causes some individuals to delay reproduction.
c. The number of predators in the area varies with population density.
d. Territorial disputes lead to injury and death of some males.

e. Periods of extreme cold periodically reduce the population density.
Textbook Reference: *55.3 What Factors Limit Population Densities? p. 1176*

10. In the graph, the population size at which the curves for the birth and death rates intersect is
   a. an estimate of the environmental carrying capacity.
   b. an estimate of the intrinsic rate of increase.
   c. the point at which density-dependent regulation begins.
   d. the point at which density-independent regulation begins.
   e. an estimate of the biotic potential of the population.
   Textbook Reference: *55.3 What Factors Limit Population Densities? p. 1176*

11. Which of the following would probably *not* be true of a population whose dynamics are primarily influenced by density-independent factors?
   a. A growth pattern that is similar to the logistic growth curve
   b. A birth rate that is dependent on the nutritional status of its adult females
   c. The highest percentage of deaths caused by unfavorable weather conditions
   d. Both a and b
   e. All of the above
   Textbook Reference: *55.3 What Factors Limit Population Densities? pp. 1176–1177*

12. Which of the following would probably *not* be true of a *K*-strategist?
   a. A long life span
   b. A Type III survivorship curve
   c. Density-dependent mortality
   d. A specialized diet
   e. Iteroparity
   Textbook Reference: *55.1 How Do Ecologists Study Populations? pp. 1172–1173; 55.3 What Factors Limit Population Densities? pp. 1176–1177*

13. Which of the following tends to be associated with the ability of a species to attain high population densities?
   a. Small body size
   b. Utilization of abundant resources
   c. Complex social organization
   d. Recent introduction into a new region
   e. All of the above
   Textbook Reference: *55.3 What Factors Limit Population Densities? pp. 1176–1177*

14. The best way to decrease the population size of a pest species is to
   a. poison it.
   b. introduce additional predators.
   c. decrease the carrying capacity of its habitat.
   d. add competitors.
   e. introduce a pathogen.
   Textbook Reference: *55.5 How Can Populations Be Managed Scientifically? p. 1181*

15. Which of the following factors is significant in setting the present carrying capacity of Earth for the human population?
   a. Earth's ability to absorb by-products of human consumption of fossil fuel energy
   b. Availability of water suitable for drinking, irrigation, and other human uses
   c. Human willingness to cause the extinction of other species in order to expand the human share of Earth's resources
   d. Both a and c
   e. All of the above
   Textbook Reference: *55.5 How Can Populations Be Managed Scientifically? p. 1182*

Application Questions

1. Draw and explain survivorship curves characteristic of humans in a wealthy country, wild birds, and a typical plant.
   Textbook Reference: *55.1 How Do Ecologists Study Populations? pp. 1172–1173*

2. Some species of fish in the salmon family (called salmonids) are anadromous, meaning that they mature in salt water but migrate to bodies of fresh water to breed. These species are usually semelparous. Other nonanadromous species in this family live their entire lives in bodies of fresh water and are generally iteroparous. What might be the explanation for the difference in the life history patterns of these two groups of salmonids?
   Textbook Reference: *55.2 How Do Environmental Conditions Affect Life Histories? pp. 1173–1174*

3. The zebra mussel was not found in North America before 1985, yet it is now an abundant pest species in the Great Lakes region and Mississippi River drainage. What principles of species abundance would apply to the explosive increase of the zebra mussel?
   Textbook Reference: *55.3 What Factors Limit Population Densities? pp. 1177–1178*

4. How does the division of some populations into discrete subpopulations relate to the process known as the rescue effect? How do barriers to immigration influence the rescue effect?
   Textbook Reference: *55.4 How Does Habitat Variation Affect Population Dynamics? pp. 1178–1179*

5. How do the demographic traits of fish and whales influence how these marine resources respond to overexploitation?
   Textbook Reference: *55.5 How Can Populations Be Managed Scientifically? pp. 1180–1181*

6. As an alternative to chemical pesticides, ecologists have often recommended biological control: the introduction of predators or parasites of pests (especially exotic species) to bring an undesirable population under control. What are the risks of this strategy?
   Textbook Reference: *55.5 How Can Populations Be Managed Scientifically? p. 1181*

7. How does the age structure of a country's population affect its potential for future population growth? In your answer, contrast an industrialized country such as Germany with a developing country such as Uganda (see Figure 55.15).
Textbook Reference: *55.5 How Can Populations Be Managed Scientifically? p. 1182*

Answers

Diagram Exercise Answer

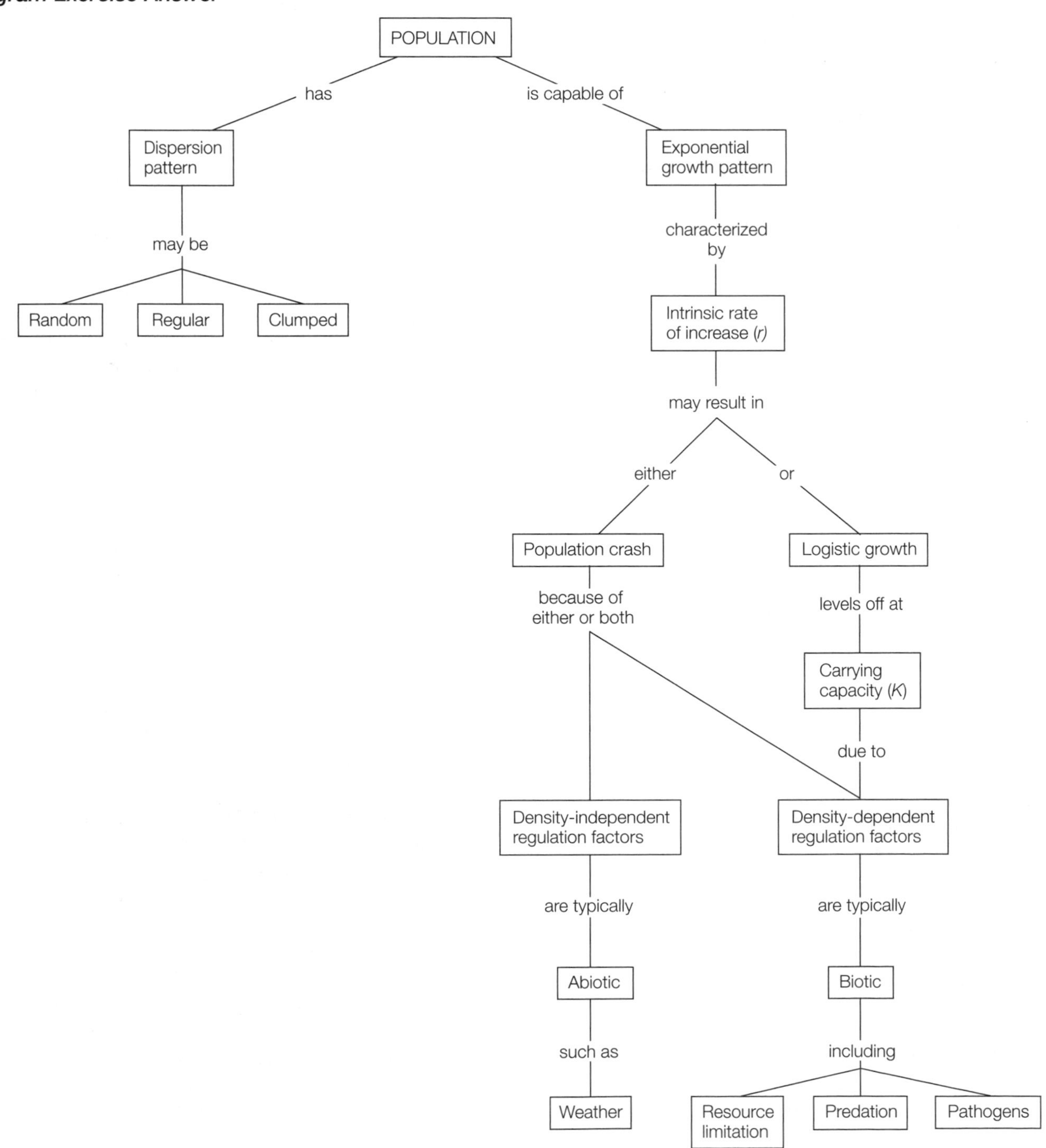

Knowledge and Synthesis Answers

1. **d.** The aspects of population structure studied by ecologists include the spacing and age distribution of the members of the population but not the distribution of genotypes.
2. **c.** If we assume that the population of bluegills randomly mixed in the interval between the capture of the first sample and the capture of the second sample, and if marked and unmarked individuals were equally likely to survive, then the proportion of tagged bluegills in the second sample can be used to estimate the total bluegill population. Because 8 of the 40 bluegills in the second sample were tagged (i.e., one-fifth of the sample), the best estimate of total population of bluegills is 5×40 (i.e., the size of the first sample) = 200.
3. **a.** The decrease in consecutive age classes is the least between 0 and 1 year, so the survivorship is greatest during that interval.
4. **a.** When a population's birth rate and death rate are both high, the age distribution is dominated by young individuals. A similar age distribution would occur in a population with a high birth rate and a low death rate, but such a population would not be stable in size.
5. **b.** Both curves show exponential growth, but the rate of exponential growth for curve 2 is less than the rate for curve 1. Because the rate of growth is determined by the expression $(b - d)$, increasing d, the death rate, would cause curve 1 to be more like curve 2.
6. **Curve 2.** Because $r = (b - d)$, you would expect the growth curve for species A to correspond to the curve with the lower rate of exponential growth.
7. **c.** The curve showing the relationship between population size and time is steepest at point c, so the growth rate would be greatest at that point.
8. **e.** The growth rate at e would be zero. This is the point called the environmental carrying capacity at which the birth and death rates are equal.
9. **b.** The graph indicates that the birth rate is independent of population size, so event b would not be true. Because the graph also shows that the death rate is density dependent, the population is unlikely to be limited by a density-independent factor such as extreme cold.
10. **a.** The environmental carrying capacity is the equilibrium population size when the birth rate equals the death rate. The birth rate equals the death rate at the intersection of the two curves.
11. **d.** The logistic growth curve describes density-dependent growth because the rate of increase in the size of the population decreases steadily as the carrying capacity is approached. The nutritional status of females in the population would depend on the availability of food and hence would be density dependent. By contrast, unfavorable weather generally occurs without any predictable relationship to the size of a population and thus is a density-independent factor.
12. **b.** Species with Type III survivorship curves produce many offspring, the vast majority of which die before reaching adulthood. *K*-strategists, in contrast, produce relatively few offspring, many of which are likely to survive to adulthood because of the large parental investment in each one.
13. **e.** All four factors favor the ability to achieve high population densities.
14. **c.** The best way to control a pest species is to reduce the carrying capacity of the habitat for the species by removing the resources it depends on.
15. **e.** All three of these factors contribute to setting the present carrying capacity of Earth for humanity, though water availability is not a factor everywhere.

Application Answers

1. See Figure 55.4 of the textbook. All of the graphs depict the survivorship of a group of individuals born at the same time (a cohort). In the human population in the United States, death rates are low until old age. In many species of wild birds, the probability of an individual's dying at a particular age is almost the same over most of its lifetime. In plants (and in other organisms that produce many offspring), poor survivorship of young individuals is followed by much lower death rates for the rest of the life span.
2. Semelparity is typical of organisms that experience very high rates of mortality as adults, whereas iteroparity is typical of organisms whose survival chances increase once they reach adulthood. Anadromous fish die in large numbers in the course of their breeding migration, so the chance that a salmon could make a second successful migration is extremely low. By comparison, nonanadromous salmonids live as adults in a comparatively stable environment in which they may have multiple opportunities to breed.
3. The high population density of zebra mussels in this continent illustrates that species introduced into a new region often reach levels of abundance above those found in their native ranges because their normal diseases and predators are absent (see Figure 55.9).
4. The rescue effect occurs if a subpopulation that has undergone a large decline is saved from extinction by the immigration of individuals from other subpopulations. As one would predict, barriers to immigration reduce the rescue effect (see Figure 55.11).
5. Although stocks of many fish and many whale species have been overexploited, the recovery of fish populations is potentially faster because most species of fish have high reproductive capacities. In contrast, whales, as marine mammals, reproduce at very low rates, and therefore recover more slowly from overexploitation (see Figure 55.13).

6. One risk of the biological control strategy is simply that the introduced parasite or predator will fail to control the pest. A more serious hazard is that the introduced species will attack or outcompete other species that are considered valuable. An instance of this outcome—the case of the toad *Bufo marinus* introduced into Australia—is described in this chapter (see Figure 55.14).

7. Because a high proportion of the German population is relatively old, the death rate in that country is likely to equal or exceed the birth rate in coming years and the population should remain stable or possibly decline. In contrast, the high proportion of pre-reproductive individuals in the age structure of Uganda portends a high rate of population growth in the future because the birth rate is likely to exceed the death rate in that country for many years to come.

56 Species Interactions and Coevolution

The Big Picture

- As discussed in Chapter 55, competition for resources, predation, and parasitism are all major density-dependent factors limiting population growth.
- The central role that the concept of competition played in Darwin's thinking as he formulated the theory of natural selection is indicated by his inclusion of the phrase "the preservation of favoured races in the struggle for life" in the full title of *The Origin of Species*.
- As you learned in this and previous chapters, some mutualistic relationships have extraordinary evolutionary or ecological significance. Examples described in previous chapters are the role of endosymbiosis in the evolution of eukaryotic cells and the association of nitrogen-fixing bacteria with their host plants.

Common Problem Areas

- The niche of a species is often confused with its habitat. The latter term refers to the preferred physical environment of a species (e.g., pond versus stream), whereas the former term refers to the entire set of physical and biological conditions a species needs in order to survive and reproduce. Thus the habitat of a species is just one of many attributes of its niche. In addition, the distinction between the fundamental niche of a species and its realized niche can be confusing. Reviewing Connell's classic experiment on barnacles will help clarify the distinction (see Figure 56.16).

Study Strategies

- This chapter describes many different categories of interspecific interactions and introduces some terms relating to these categories that may be new to you. To learn the basic types of interactions, study the chart presented in Figure 56.1. Then try creating your own charts or concept maps to organize the material in more detail and to master the terminology.

  Go to yourBioPortal.com to review the following tutorials and activity:

  Animated Tutorial 56.1 Mutualism

  Interactive Tutorial: Coevolution: Strategies for Survival

  Web Activity 56.1 Ecological Interactions

Important Concepts

Ecologists study several types of interactions between species.

- Interactions between species in a community fall into five general categories that reflect whether the outcome of the interactions is positive (+), negative (–), or neutral (0) for each of the species involved (see Figure 56.1).
  - Antagonistic interactions are those in which one species benefits and the other is harmed. Such interactions include (1) predation, in which an individual of one species kills and consumes multiple individuals of another species; (2) herbivory, in which an individual of a species consumes part or all of a plant; and (3) parasitism, in which an individual of one species consumes only certain tissues of one or a few individuals of another species (the host). Pathogens are parasites that cause symptoms of disease in their hosts.
  - Mutualism is a type of interaction between species that benefits both.
  - Competition in an interaction in which two or more species use the same resource. It is harmful to all species involved.
  - Commensalism is an interaction that is beneficial to one participant while leaving the other unaffected.
  - Amensalism is harmful to one participant while leaving the other unaffected.
- These five types of interaction are in reality part of a continuum, and their outcomes depend on ecological and evolutionary circumstances (see Figure 56.2).
- In the process known as reciprocal adaptation, or coevolution, an adaptation in one species leads to the evolution of an adaptation in a species with which it interacts.
- In a coevolutionary arms race, the evolution of traits that improve the fitness of a predator, herbivore, or parasite exerts selection pressure on its prey or host to counter the consumer's adaptation. The prey or host adaptation, in turn, exerts selection pressure on the consumer to improve its fitness, resulting in an escalating arms race.

Antagonistic interactions result in a wide range of adaptations.

- Predators invariably kill their prey, and over its lifetime an individual predator kills and eats many prey individuals. Predators tend to be less specialized than other consumers, and most are larger than their prey, though a few are smaller. Predators in the former category generally use their strength or swiftness to capture their prey, whereas those in the latter category employ strategies such as venom production (see Figure 56.3).
- Prey species have evolved several kinds of defenses against predators, including swiftness and protective exteriors, such as tough skins, shells, and spines.
- Animals that do not have features such as swiftness or morphological adaptations to protect them often rely on chemical defenses to escape or repel predators; these may, in turn, evolve to overcome their prey's defenses. In some cases, the predator may even ingest and sequester the prey's defensive chemicals and use them to protect itself against its own predators.
- Prey species protected by chemical defenses often advertise their toxicity by warning coloration, or aposematism. Visually hunting predators, especially among the vertebrates, can learn to avoid these species (see Figure 56.4).
- In Batesian mimicry, a nontoxic prey species (the mimic) evolves a resemblance to an aposematic toxic species (the model). In Müllerian mimicry, two or more aposematic species converge to resemble one another (see Figure 56.5).
- Prey species have evolved several methods of avoiding detection. These include crypsis (camouflage), homotypy (resemblance to an inedible object), ceasing to move, and even "playing dead" (see Figure 56.6).
- Herbivory is the most widespread interaction between species on Earth. Most of the world's herbivores are insects, and they are generally oligophagous—meaning that they feed on one or a few species of plants. In contrast, the great majority of vertebrate herbivores are polyphagous, feeding on many plant species. Though herbivores seldom kill their food plants, they can still reduce plant fitness if the plants they attack produce fewer offspring.
- Although most plants defend themselves against their consumers by producing defensive chemicals (secondary metabolites), many plants have additional defenses. These include thorns, spines, hooked hairs on leaf surfaces, cuticles, and silica—a defense found in grasses that wears down the teeth of herbivores.
- The great biochemical diversity of the flowering plants and the comparable diversity of herbivorous insects is the consequence of an evolutionary arms race in which plants that have evolved a novel secondary metabolite undergo an adaptive radiation that leads to an adaptive radiation of insects that have evolved resistance to the chemical. With sufficient selection pressure, a resistant herbivore can evolve to use the chemical against its own predators.
- Large, polyphagous herbivores such as deer and horses minimize their exposure to any particular defensive chemical by feeding on a wide variety of plant species and by learning to avoid plants with an unpleasant taste.
- Microparasites are many orders of magnitude smaller than their hosts and generally live and reproduce inside their hosts. Pathogenic microparasites may cause the death of the host, but because pathogens must continually infect new host individuals, natural selection may favor less deadly strains that can infect a larger number of new hosts. Thus, pathogen and host may reach a state of coexistence that involves increased host resistance and decreased pathogen virulence (ability to cause disease).
- The hosts of pathogens fall into three classes: (1) susceptible (capable of being infected), (2) infected, and (3) recovered (and thus, in many cases, immune to another infection). When a pathogen invades a population, rates of infection typically rise at first and then fall as the number of susceptible individuals decreases. The rates do not rise again until a sufficiently dense population of susceptible hosts has reappeared.
- Larger parasites, called macroparasites, rarely cause the kinds of disease symptoms that pathogenic microparasites cause, but they can affect host fitness nevertheless. Ectoparasites (external parasites) are often associated with their hosts only casually. They typically have a number of morphological adaptations that enable them to remain attached to their hosts (see Figure 56.8). Some biologists believe that hairlessness, bipedality, and the opposable thumb evolved in humans as a response to ectoparasites.

Many mutualistic interactions are tightly coevolved.

- Mutualisms can form virtually without regard to the taxonomic groups to which the participants belong. Many mutualisms involve an exchange of food for housing or defense and arise in environments in which resources are in short supply.
- Reciprocal adaptations, which ensure that both partners benefit from the exchange, are most likely to arise when an increase in dependency on a partner provides an increase in the benefits realized from the interaction.
- About three-quarters of all flowering plant species are animal-pollinated. A mutualistic pollination system requires that (1) the plant provide an attraction or reward to entice the pollinator; (2) the pollinator be able to transport the plant's pollen; and (3) the pollinator visit more than one individual of a plant species.
- Both plants and pollinators may take advantage of their partners. For example, some plant species have evolved flowers that deceive insects into attempting to copulate with them (see Figure 56.9), and thieves that do not transport pollen may consume nectar provided as a reward to pollinators.

- Floral characteristics and the timing of flowering can restrict the number of potential pollinators and encourage pollinator fidelity (see Figure 56.10). Nevertheless, most flowers can be pollinated by a number of animal species. Broad suites of floral characteristics that attract certain groups of pollinators exemplify diffuse coevolution: the evolution of similar traits in groups of species experiencing similar selection pressures (see Table 56.1). A few plant–pollinator relationships are much more exclusive and lead to highly specific coevolution (see Figure 56.11).
- Many animals that eat fruits (frugivores) disperse the seeds of the plants that produce them. Plants that depend on frugivores for plant dispersal must achieve a balance among three factors: (1) discouraging frugivores from eating fruits before the seeds are mature; (2) attracting frugivores when the seeds are mature; and (3) protecting the seeds from destruction in an animal's digestive tract.
- Because the mutualism between plants and frugivores is often asymmetric (i.e., one partner benefits more than the other), relatively few highly specialized frugivores exist.
- In fungus-"farming" mutualisms, the farmer species provides housing, nutrition, and care for the fungal partner, which in turn provides food for the host species by converting plant materials that the farmers cannot digest for themselves into an edible form (see Figure 56.13).
- In the mutualistic relationship between acacia ants and bull's horn acacia trees, the trees provide food and housing for the ants, and in exchange the ants provide protection against the trees' herbivores and competitors (see Figure 56.14).

Competition between individuals of different species can result in several evolutionary outcomes.

- Competition occurs whenever any resource is not sufficiently abundant to meet the needs of all the organisms with an interest in that resource. Intraspecific competition occurs among individuals of the same species and is a major density-dependent mechanism of population regulation. Interspecific competition occurs among individuals of different species; in addition to limiting population growth, it can influence the persistence and evolution of species.
- The principle of competitive exclusion states that no two species can coexist for long if they share the same limiting resource. In some cases, the inferior competitor may go extinct locally (competitive exclusion). In other cases, selection pressures resulting from competition may change the ways in which different species use a limiting resource (resource partitioning).
- Both intraspecific and interspecific competition occur by means of two major mechanisms.
  - In interference competition, an individual interferes with a competitor's access to a limiting resource. Interference competition may involve restriction of habitat use by a competitor or the production of chemicals that interfere with life processes of a competitor.
  - In exploitation competition, a limiting resource is available to all competitors and the outcome depends on the relative efficiency with which competitors use up the resource. One possible evolutionary outcome of exploitation competition is resource partitioning. As a result, guilds—groups of species that exploit the same resource in slightly different ways—may evolve. Another possible outcome is character displacement, in which morphological attributes of a species vary geographically depending on whether a competitor is present or absent (see Figure 56.15).
- Indirect competition may occur if (1) one species so alters the quality of a resource that it is rendered less usable by other species that encounter it afterward; or (2) the outcome of competition hinges on how the competitors interact with a shared predator.
- The niche of a species is the set of physical and biological conditions necessary for its persistence. Although a species may be able to persist under a wide range of resource conditions (i.e., in its "fundamental" niche), competitors may restrict its use of resources in a particular location, thus restricting it to its "realized" niche (see Figure 56.16).

Test Yourself

Diagram Exercise

Construct a concept map whose theme is "Predator–Prey Interactions." Include in your map the following terms: predator–prey interactions, antagonistic interaction, aposematism, armor, arms race, Batesian, chemical defenses, coevolution, crypsis, defenses, homotypy, mimics, Müllerian, predator, prey, and swiftness. Connect these terms by verbs or short phrases to indicate the relationships among them.
Textbook Reference: *56.1 What Types of Interactions Do Ecologists Study? pp. 1186–1188; 56.2 How Do Antagonistic Interactions Evolve? pp. 1188–1193; 56.3 How Do Mutualistic Interactions Evolve? pp. 1194–1197; 56.4 What Are the Outcomes of Competition? pp. 1198–1200*

Knowledge and Synthesis Questions

1. Which of the following does *not* describe an antagonistic interaction between two living organisms?
   a. Sheep grazing on grass in a pasture
   b. A robin catching and eating a worm
   c. A flea biting a dog to obtain a blood meal
   d. Lions skirmishing with hyenas to prevent them from feeding from a zebra carcass
   e. All of the above describe antagonistic interactions.
   Textbook Reference: *56.1 What Types of Interactions Do Ecologists Study? pp. 1186–1187*

2. Certain birds follow swarms of foraging army ants and prey upon the insects that the ants flush out. The relationship between these birds and the ants is an example of
   a. competition.
   b. commensalism.
   c. mutualism.
   d. amensalism.
   e. Both a and b

   Textbook Reference: *56.1 What Types of Interactions Do Ecologists Study? pp. 1186–1187*

3. Which of the following types of interaction would be *least* likely to lead to coevolution of the interacting species?
   a. Predator–prey
   b. Commensal
   c. Plant–herbivore
   d. Parasite–host
   e. Mutualistic

   Textbook Reference: *56.1 What Types of Interactions Do Ecologists Study? p. 1188*

4. If a species of plant that is trampled by an animal eventually evolves sharp spines that prevent trampling, we can say that its association with the animal has changed from _______ to _______.
   a. amensalism; competition
   b. amensalism; commensalism
   c. commensalism; competition
   d. commensalism; mutualism
   e. commensalism; amensalism

   Textbook Reference: *56.1 What Types of Interactions Do Ecologists Study? p. 1187*

5. As their name suggests, stick insects ("walking sticks") strongly resemble sticks or twigs. The most specific term for this method of avoiding detection by predators is
   a. mimicry.
   b. camouflage.
   c. aposematism.
   d. crypsis.
   e. homotypy.

   Textbook Reference: *56.2 How Do Antagonistic Interactions Evolve? p. 1190*

6. Skunks, which are noted for their ability to spray would-be predators with a foul-smelling chemical, typically have a coat that is mostly black with contrasting white stripes or spots. This color pattern is an example of
   a. mimicry.
   b. camouflage.
   c. aposematism.
   d. crypsis.
   e. homotypy.

   Textbook Reference: *56.2 How Do Antagonistic Interactions Evolve? pp. 1189–1190*

7. The most common defense of plants against herbivores is
   a. the production of secondary metabolites.
   b. a waxy cuticle covering the surfaces of the leaves.
   c. thorns or spines.
   d. hairy leaves.
   e. leaves containing silica, which wears down the teeth of herbivores.

   Textbook Reference: *56.2 How Do Antagonistic Interactions Evolve? p. 1191*

8. Which of the following statements about microparasites is true?
   a. Microparasite infection rates typically rise only when a sufficiently dense population of susceptible host individuals is present.
   b. Microparasites include viruses, bacteria, protists, and parasitic worms.
   c. Microparasites are always pathogens.
   d. Both a and c
   e. All of the above

   Textbook Reference: *56.2 How Do Antagonistic Interactions Evolve? pp. 1192–1193*

9. Which of the following statements about ectoparasites is *false*?
   a. Many ectoparasites have evolved adaptations to keep them attached to their hosts.
   b. Some ectoparasites have a relatively brief or casual association with their hosts.
   c. Most ectoparasitic insects are highly specialized, sometimes feeding on only a single host species.
   d. Many ectoparasitic insects have lost their wings during the course of their evolution.
   e. All of the above are true.

   Textbook Reference: *56.2 How Do Antagonistic Interactions Evolve? p. 1193*

10. Match the following pollinators with the descriptions of the flowers that follow.
    Bees
    Flesh flies
    Beetles
    Butterflies
    Moths
    Hummingbirds
    Bats
    a. Pendant white flowers with a heavy odor
    b. Irregularly shaped purplish flowers with the odor of carrion
    c. Tubular red flowers with no perceptible odor
    d. Irregularly shaped flowers with a sweet odor
    e. Bowl-shaped white flowers with a faint odor
    f. Cuplike white flowers with a musty odor
    g. Flowers with a tubular platform and a faint odor

    Textbook Reference: *56.3 How Do Mutualistic Interactions Evolve? p. 1194*

11. The suites of floral characteristics listed in the question above have evolved to attract various groups of pollinators through the process of
    a. specific coevolution.

b. interspecific competition.
c. resource partitioning.
d. character displacement.
e. diffuse coevolution.

Textbook Reference: *56.3 How Do Mutualistic Interactions Evolve? p. 1195*

12. A bird eats the fruit of a plant species. The seeds are not digested and germinate in the bird's excrement at some distance from the parent plant. This is an example of
a. predation.
b. competition.
c. commensalism.
d. mutualism.
e. amensalism.

Textbook Reference: *56.3 How Do Mutualistic Interactions Evolve? p. 1196*

13. In Eurasia, two closely related species of rock nuthatches (small birds that forage for insects and spiders on bare, rocky ground) occur either separately, or in zones of sympatry. Individuals of both species that exist in the former situation, with no interspecies competition, have bills that are similar in length. But in their zone of sympatry in Iran, one species has a longer bill than it has elsewhere and the other species has a shorter bill. This is an example of _______ competition that has resulted in _______.
a. exploitation; resource partitioning
b. exploitation; character displacement
c. interference; resource partitioning
d. interference; character displacement
e. None of the above

Textbook Reference: *56.4 What Are the Outcomes of Competition? p. 1199*

14. Which of the following is a possible outcome of competition involving two species?
a. Some individuals of both species experience reduced growth and reproductive rates.
b. One species excludes the other from a habitat it would occupy in the absence of competition.
c. One species restricts the geographic range of the other species.
d. The species evolve to divide up the limiting resource and continue to coexist in the same locality.
e. All of the above

Textbook Reference: *56.4 What Are the Outcomes of Competition? pp. 1198–1199*

15. Which of the following statements about the niche concept is *false*?
a. Compared to species with dissimilar niches, species with similar niches are likely to be stronger competitors with one another.
b. Two species can occupy the same habitat but have different niches.
c. Guilds are groups of species that occupy the same niche.
d. The realized niche of a species is never larger than its fundamental niche, though they may be identical in size.
e. All of the above are true.

Textbook Reference: *56.4 What Are the Outcomes of Competition? pp. 1198–1200*

Application Questions

1. Two of the main characters in the animated film *Finding Nemo* are anemonefish (also known as clown fish), which live inside of sea anemones and are unaffected by their stings. How is this relationship an example of the ways in which interactions between species are not always easily classifiable?

Textbook Reference: *56.1 What Types of Interactions Do Ecologists Study? pp. 1187–1188*

2. Some mimicry systems (e.g., those involving tropical butterflies) include several toxic species and several palatable species, all of which are aposematic and resemble one another closely. How would you classify these mimicry systems—Batesian, Müllerian, or both? Why might natural selection have favored the evolution of such complex mimicry systems?

Textbook Reference: *56.2 How Do Antagonistic Interactions Evolve? p. 1190*

3. After the European rabbit was introduced into Australia in 1859, it quickly became a devastating invasive species, causing the extinction or decline of many native animals and plants, major damage to crops, and soil erosion. To control the exponential growth of the population (which had grown to an estimated 600 million individuals from an original 24), biologists infected rabbits with the myxoma virus in 1950. It is estimated that in the first years after introduction of the virus, as many as 99 percent of infected rabbits died. Over a period of years, however, biologists observed an increase both in the average life span of infected rabbits and in the percentage of rabbits surviving infection. How would you explain these changes?

Textbook Reference: *56.2 How Do Antagonistic Interactions Evolve? p. 1193*

4. Review the description of Connell's classic experiment on competition between *Semibalanus* and *Chthamalus* barnacle populations (see Figure 56.16). Do you think that the interaction between these species is an example of interference competition or exploitation competition? Why?

Textbook Reference: *56.4 What Are the Outcomes of Competition? p. 1200*

5. Describe an experiment that would enable you to assess the degree of interspecific competition that exists between two rodent species living in the same habitat. How would you interpret the results?

Textbook Reference: *56.4 What Are the Outcomes of Competition? p. 1200*

Answers

Diagram Exercise Answer

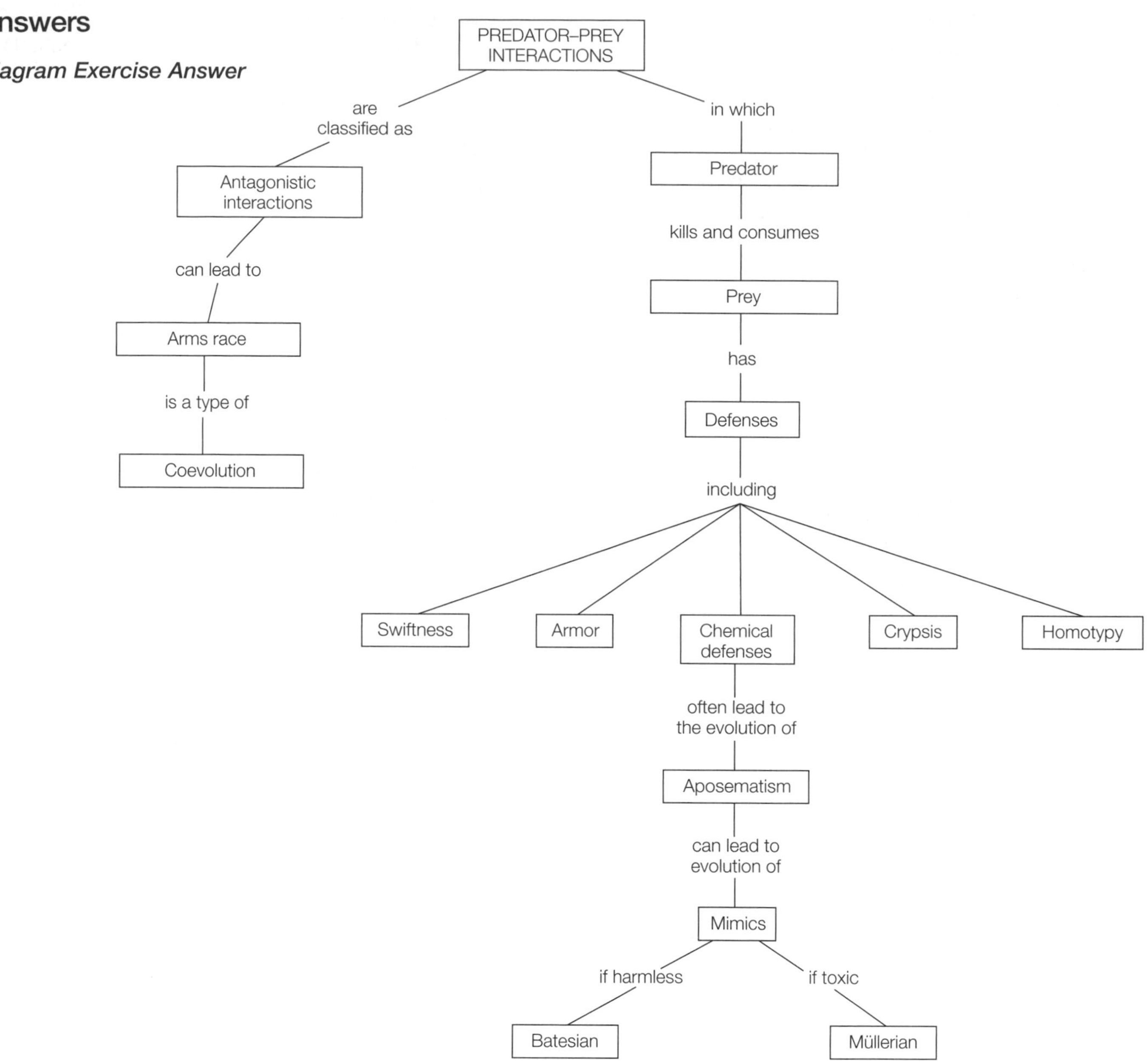

Knowledge and Synthesis Answers

1. **d.** Antagonistic interactions are those in which one species benefits and the other species is harmed. They include herbivory, predation, and parasitism. The competition between the lions and hyenas for food is harmful to both species and hence is not classified as an antagonistic interaction.
2. **e.** Both ants and birds are competing for the same food supply; the birds also benefit from the ants' activity in flushing out the insects, but the ants might not be affected.
3. **b.** The types of interactions most likely to lead to coevolution are those that have a strong effect on all of the interacting species. In commensalism, one species benefits from the interaction, but the other is unaffected by it.
4. **a.** Initially the plant is harmed and the animal is unaffected (amensalism). After the evolution of sharp spines, the two species are competitors for space.
5. **e.** Homotypy is the resemblance of a living organism to an object considered inedible by its predators.
6. **c.** Aposematism is warning coloration, frequently exhibited by toxic or otherwise noxious species.
7. **a.** The production of toxic secondary metabolites is by far the most widespread means by which plants defend themselves against herbivores.
8. **a.** Microparasites include viruses, bacteria, and protists—all of which are far smaller than their hosts—

but not multicellular parasites such as worms. Only those microparasites that cause disease symptoms are considered pathogens.

9. **e.** The mosquito is an example of an ectoparasite that remains with the host only long enough to feed. Lice and fleas are examples of ectoparasites that lack wings and that are highly specific with regard to their hosts (lice more so than fleas). Figure 56.8 illustrates the winglessness of lice as well as their adaptations for remaining attached to their host.

10. **a.** Moths

    b. Flesh flies

    c. Hummingbirds

    d. Bees

    e. Beetles

    f. Bats

    g. Butterflies

11. **e.** Diffuse coevolution is the evolution of similar traits in species experiencing similar selection pressures.

12. **d.** Both species have benefited; the bird dispersed and provided fertilizer for the plant's seed, and the plant provided food for the bird.

13. **b.** Because bill length in these species is clearly related to the type of food resources and foraging sites that the birds utilize, this case must be considered an example of exploitation competition. It is also an example of character displacement, because in both species bill length is dependent on the presence or absence of the competing species. The Iranian populations of the two species have evolved so as to minimize competition between them.

14. **e.** It is generally true that competition, whether intraspecific or interspecific, causes reduced growth and reproductive rates of some individuals participating in the competitive interaction. Interspecific competition in particular may lead to exclusion from a habitat of one of the competing species (i.e., competitive exclusion), or to range restriction, or to resource partitioning.

15. **c.** Guilds are defined as groups of species that exploit the same resource but in slightly different ways. This definition implies that their niches are similar but not identical. If their niches were identical, the principle of competitive exclusion would prevent them from coexisting in the same locality.

Application Answers

1. The anemonefish clearly benefit from the interaction with their host by gaining protection from their enemies. It is less clear that the sea anemones are unaffected (in which case the relationship would be a commensalism). The anemones may gain nutrients from the feces of the fish, or the fish may occasionally steal prey from the anemones, or both. Thus the interaction could be mutualistic or competitive, or both simultaneously.

2. This type of mimicry system is both Batesian and Müllerian. The palatable species are Batesian mimics of the toxic species (the models), and the toxic species are Müllerian mimics of one another (each species in a sense being both model and mimic). Natural selection may have favored the convergence of these aposematic species on a common color pattern because it is probably easier for predators to learn to avoid a single pattern than multiple patterns.

3. Natural selection favored the survival of those rabbits that by chance had relatively high genetic resistance to the virus. Thus, one would predict that over time, the rabbit population as a whole would evolve to become more resistant. One would also predict selection on the mxyoma virus favoring slower or lesser lethality: a mutant strain of the virus that allowed its host to survive for a longer period of time, or to not be killed at all, would be more likely to be transmitted to new hosts than the original strain that killed its host more quickly. Biologists studying this classic example of host–parasite coevolution have documented these evolutionary changes in both the rabbit and virus populations.

4. Ecologists regard the interaction between the barnacle populations as an instance of interference competition. *Semibalanus* individuals smother, crush, or dislodge *Chthamalus* individuals but do not directly reduce the food resources available to them.

5. A removal experiment could be done in which one or the other species would be removed in two similar areas. You then would look for changes in the abundance of the remaining species. The species with the largest increase in population density would be the one most affected by competition with the removed species.

57 Community Ecology

The Big Picture

- In urging a cautious approach to human manipulation of the natural environment, ecologists like to say that "you can never do just one thing." In this chapter, the descriptions of the effects of keystone species and of other complex community interactions illustrate the meaning of this saying.
- The island biogeographic model, which relates the area of an island to the equilibrium number of species inhabiting the island, has applications in conservation biology (see Chapter 59). For example, habitat destruction often leaves habitat islands that, if small, inevitably lose some of the species that existed in the previously more extensive habitat.
- As discussed in Chapter 59, if we wish to preserve a rare species, in many cases the most important step is to protect its habitat. This often requires an understanding of (and manipulation of) ecological succession to prevent changes in habitat unfavorable to the rare species.

Common Problem Areas

- Students sometimes have difficulty understanding the difference between energy pyramids and biomass pyramids. Energy pyramids can *never* be inverted, because there must always be a lower rate of energy flow through any given trophic level than through the trophic level below it. Biomass pyramids *can* be inverted, because they do not represent energy flow rates, but rather the amount of living matter present in the various trophic levels of the community at a given moment in time.
- The graphical representations of the concepts of island biogeographic theory are sometimes troublesome for students. Try redrawing the graphs for yourself, paying particular attention to how the axes are labeled.

Study Strategies

- This chapter contains descriptions of a number of studies of community structure, including several that relate species richness to other community characteristics. As you read about these studies, focus on the particular question that the study was designed to answer and how the results support the conclusions drawn from the study.
- Go to yourBioPortal.com to review the following tutorials and activities:

  Animated Tutorial 57.1 Biogeography Simulation

  Animated Tutorial 57.2 Primary Succession on a Glacial Moraine

  Web Activity 57.1 Energy Flow through an Ecological Community

  Web Activity 57.2 The Major Trophic Levels

Important Concepts

An ecological community consists of all the species that live and interact in a particular area.

- Communities vary greatly in species richness (i.e., the number of species that live in them).
- Gross primary productivity is defined as the rate at which plants assimilate energy through photosynthesis; the accumulated energy is called gross primary production.
- Net primary production is the portion of assimilated energy left over after the energy used by plants to power their own metabolism is subtracted. All of the other organisms in an ecosystem derive their energy either directly or indirectly from net primary production, which takes the form of plant growth and reproduction (see Figure 57.1).
- The organisms in a community use diverse sources of energy, and organisms are grouped into trophic levels according to the number of steps through which energy passes to reach them. Energy passes, in sequence, from autotrophic primary producers (photosynthesizers) to heterotrophs, beginning with primary consumers (herbivores) followed by secondary consumers (which feed on herbivores), and so on. Detritivores (decomposers) feed on the dead bodies and waste products of other organisms. Omnivores obtain their food from more than one trophic level (see Table 57.1).

- A food chain is a sequence of feeding linkages beginning with a primary producer. Food chains in a community are usually interconnected in a complex food web (see Figure 57.2).
- Ecological efficiency is the ratio of production of a trophic level to production of the trophic level immediately below it. On average, ecological efficiency is about 10 percent. Three factors account for such low ecological efficiency, and these are also the reasons that most communities have only three to five trophic levels.
  - Organisms use most of the energy they accumulate for respiration and other metabolic processes; this energy is ultimately dissipated as heat.
  - Consumers do not ingest all of the biomass available to them.
  - Consumers cannot assimilate (digest and absorb) all of the biomass they ingest.
- An energy pyramid shows how energy decreases in moving from lower to higher trophic levels. A biomass pyramid shows the mass of organisms existing at different trophic levels at a given moment of time. For the same community, these pyramids usually have similar shapes. Major exceptions are most aquatic communities, in which a small mass of rapidly dividing unicellular photosynthesizers can support a much larger biomass of herbivores (see Figure 57.3).
- Primary productivity and species richness depend not only on the supply of energy from the sun, but also on the availability of nutrients and water. Annual evapotranspiration—the amount of water released by evaporation from land and water surfaces and by transpiration from plants—is a measure of the availability of water to organisms.
- Species richness increases with primary productivity up to a point, after which it declines (see Figure 57.4). One hypothesis to explain this decline postulates that interspecific competition becomes more intense when productivity is higher, resulting in competitive exclusion of some species.

Species interactions can influence community structure.

- In a trophic cascade, predators or herbivores affect many different species, often at lower trophic levels (see Figure 57.5).
- Ecosystem engineers are organisms that build structures that create environments for other species.
- Keystone species exert an influence on ecological communities that is out of proportion to their abundance. Examples are predatory species that increase species richness by feeding on dominant competitors (see Figure 57.6) and plant species that serve as a food source for many different animals when food is otherwise scarce.

Ecologists attempt to measure and explain the species diversity of communities.

- Species diversity has many components depending on the scale at which it is measured.
  - Alpha diversity is a measure of diversity within a single community or habitat.
  - Beta diversity is a measure of the change in species composition from one community or habitat to another.
  - Gamma diversity is a measure of the regional diversity found over a range of communities or habitats in a geographic region.
- The alpha diversity of a community has two components: species richness and species evenness (see Figure 57.7). A widely used mathematical measure of alpha diversity that takes into account both of its components is the Shannon diversity index (H), which measures the certainty with which the next item sampled from a series can be predicted. The higher the H, the greater the uncertainty about the species composition of a sample is, and therefore the greater the diversity of the community.
- Beta diversity can be measured mathematically using Sorenson's index.
- Gamma diversity can be used to make comparisons among geographical regions. It is equivalent to the alpha diversity of each community combined with the beta diversity between the communities.
- Species diversity in many taxa decreases as distance from the equator increases (see Figure 57.8). Ecologists have proposed at least four hypotheses, each supported by some evidence, to account for latitudinal gradients in diversity.
  - The time hypothesis argues that organisms in tropical regions have had more time to diversify than those in less climatically stable temperate regions do.
  - The spatial heterogeneity hypothesis suggests that tropical regions have more habitat diversity and hence many more species than temperate regions do.
  - The specialization hypothesis proposes that greater competition among tropical species has led to the evolution of narrower niches and correspondingly more species.
  - The predation hypothesis, in contradiction to the specialization hypothesis, proposes that higher predation intensity in the tropics keeps prey populations low, allowing for the persistence of rare species that otherwise would be eliminated by interspecific competition.
- The species–area relationship states that within any latitudinal band, larger areas of habitat contain more species than smaller areas. Moreover, continental areas contain more species than oceanic islands of comparable size.

- According to the theory of island biogeography, equilibrium species richness (the number of species living in an area) on islands is determined by the rates of arrival of new species and extinction of species already present.
- The island biogeography theory predicts that the equilibrium number of species should increase with island size and decrease with distance from the mainland (see Figure 57.9). Biogeographers have confirmed the predictions of the theory both by observation and experiment.
- The theory of island biogeography can also be applied to habitat islands—isolated patches of suitable habitat surrounded by areas of unsuitable habitat.

All ecological communities are subjected to a variety of disturbances.

- A disturbance is an event that changes the survival rate of one or more species in a community. Small disturbances are much more common than large ones, but a few large events may cause most of the changes in a community. A community's history of disturbance may explain patterns of species diversity that otherwise would be puzzling.
- Ecological succession is a change in community composition following a disturbance. In the most common type of succession—directional succession—species come and go in a predictable pattern until a climax community capable of perpetuating itself under local environmental conditions persists for an extended period of time.
- Primary succession begins on sites that lack living organisms. Disturbances such as glaciers and volcanic activity can initiate this type of succession (see Figure 57.11).
- Secondary succession begins when some organisms survive a disturbance and reestablish themselves on a disturbed site (see Figure 57.12).
- Directional succession is characterized by several trends. In early stages, productivity is high and food webs are simple. As succession proceeds, food webs become more complex and nutrients become more concentrated in biomass than in biotic sources. Colonizing species in early succession tend to be *r*-strategists with good dispersal ability, whereas late successional species tend to be *K*-strategists.
- As succession progresses, species that have already become established may facilitate or inhibit colonization by other species.
- In cyclical succession, the climax community is maintained by periodic disturbance (see Figure 57.13).
- Heterotrophic succession occurs in detritus-based communities, which are found in dung, carrion, or dead plants. Because these communities cannot generate more energy through photosynthesis, energy resources are depleted in the course of succession, with the result that biomass and biodiversity also decline.
- In the decomposition of an animal body, three major stages occur: (1) autolysis, in which the body itself generates fermentative degradation of proteins and other organic molecules; (2) putrefaction, in which decomposition by microorganisms and saprobes takes place; (3) dry decay, which takes place after most fluids have evaporated.

Highly productive and stable communities tend to be species-rich.

- Communities with more species use resources more efficiently and hence tend to be both more productive and less variable in their productivity than communities with fewer species (see Figure 57.14). Population densities of individual species, however, may not be stable, regardless of species richness.
- Monocultures—plantings of only a single crop species in a particular area—are notoriously unstable because of their susceptibility to pest outbreaks. Agricultural systems in which two or more crops are grown on the same plot tend to be more stable.

Test Yourself

Diagram Exercise

Construct a concept map whose theme is "Succession." Include in your map the following terms: succession, burned area, cyclical succession, directional selection, disturbed sites, glacial moraine, *K*-strategists, persistent climax community, pioneer community, primary succession, *r*-strategists, and secondary succession. Connect these terms by verbs or short phrases to indicate the relationships among them.

Textbook Reference: *57.4 How Do Disturbances Affect Ecological Communities? pp. 1214–1217*

Knowledge and Synthesis Questions

1. The ecological community of a pond comprises
   a. all the species living in the pond and the abiotic environment (water, sunlight, etc.) that supports them.
   b. all the species living in the pond.
   c. the invertebrates and vertebrate animals living in the pond.
   d. all of the producers and consumers living in the pond, but not the decomposers.
   e. None of the above

   Textbook Reference: *57.1 What Are Ecological Communities? p. 1204*

2.–3. Refer to the food web below to answer the questions that follow.

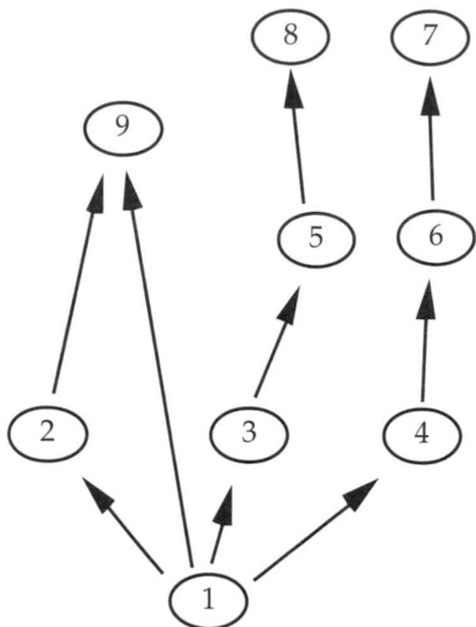

2. Organism 9 is a(n)
   a. herbivore.
   b. primary carnivore.
   c. secondary carnivore.
   d. primary producer.
   e. omnivore.

   Textbook Reference: *57.1 What Are Ecological Communities? pp. 1205–1206*

3. The food web has _______ trophic levels.
   a. two
   b. three
   c. four
   d. five
   e. six

   Textbook Reference: *57.1 What Are Ecological Communities? pp. 1205–1206*

4. In which of the following ecosystems would you expect to observe the lowest biomass of primary consumers relative to the biomass of the primary producers?
   a. Tropical evergreen forest
   b. Temperate grassland
   c. Freshwater lake
   d. Open ocean
   e. Both c and d

   Textbook Reference: *57.1 What Are Ecological Communities? pp. 1206–1207*

5. A plant in the dark uses 0.02 ml of O_2 per minute. The same plant in sunlight releases 0.14 ml of O_2 per minute. Its rate of gross primary production would be approximately _______ ml of O_2 per minute.
   a. 0.02
   b. 0.12
   c. 0.14
   d. 0.16
   e. 0.18

   Textbook Reference: *57.1 What Are Ecological Communities? p. 1205*

6. If the net primary productivity of a community is 2000 kcal/m^2/yr, what would be the best estimate of the productivity (in kcal/m^2/yr) of the secondary consumers in that community?
   a. 4000
   b. 2000
   c. 200
   d. 20
   e. 2

   Textbook Reference: *57.1 What Are Ecological Communities? p. 1206*

7. Which of the following statements about the relationship between species richness and productivity in ecological communities is true?
   a. Species richness is usually highest in the most productive communities.
   b. Species richness is usually highest in the least productive communities.
   c. Species richness is frequently highest at an intermediate level of productivity.
   d. Species richness is frequently lowest at an intermediate level of productivity.
   e. Species richness is not correlated with productivity.

   Textbook Reference: *57.1 What Are Ecological Communities? pp. 1206–1207*

8. Which of the following statements about keystone species is true?
   a. They are very abundant.
   b. Their removal has a great effect on community structure.
   c. They practice herbivory.
   d. They have regular dispersion patterns.
   e. They have clumped dispersion patterns.

   Textbook Reference: *57.2 How Do Interactions among Species Influence Community Structure? p. 1209*

9. Which of the following statements regarding species diversity is *false*?
   a. If two communities have the same species richness, the one with the more even distribution of species is considered more diverse.
   b. Beta diversity is a measure of the change in species composition from one community or habitat to another.
   c. For many taxa, species diversity decreases with distance from the equator.
   d. Gamma diversity is the regional diversity found over a range of communities or habitats in a geographic region.
   e. Sorenson's index is used to estimate the alpha diversity of a community.

   Textbook Reference: *57.3 What Patterns of Species Diversity Have Ecologists Observed? pp. 1210–1211*

10. Refer to the graph below showing the effect of species number on arrival and extinction rates.

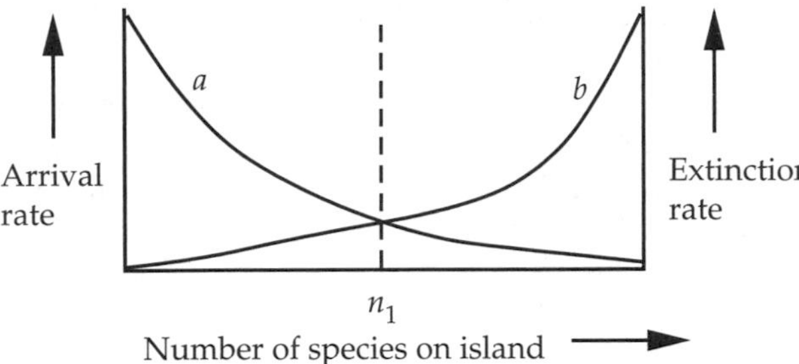

According to the theory of island biogeography, which of the following statements about the graph is true?
a. Curve *a* is the arrival rate curve.
b. Curve *b* is the extinction rate curve.
c. The arrival rate equals the extinction rate at n_1.
d. The extinction rate is zero at n_1.
e. Both a and b
Textbook Reference: *57.3 What Patterns of Species Diversity Have Ecologists Observed? p. 1212*

11. In the figure below, the species numbers of many islands of different sizes were plotted against their distance from the mainland, and data points for islands of similar size were connected to form the four curves.

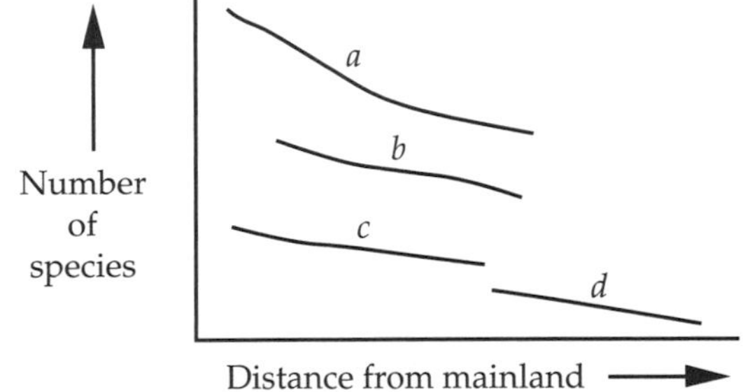

Circle the letter of the curve corresponding to the group of islands with the smallest size.
Textbook Reference: *57.3 What Patterns of Species Diversity Have Ecologists Observed? p. 1212*

12. The theory of island biogeography makes predictions about the effects of species number on the rate of extinction. Refer to the figure below, in which the solid curve shows this relationship for a large island.

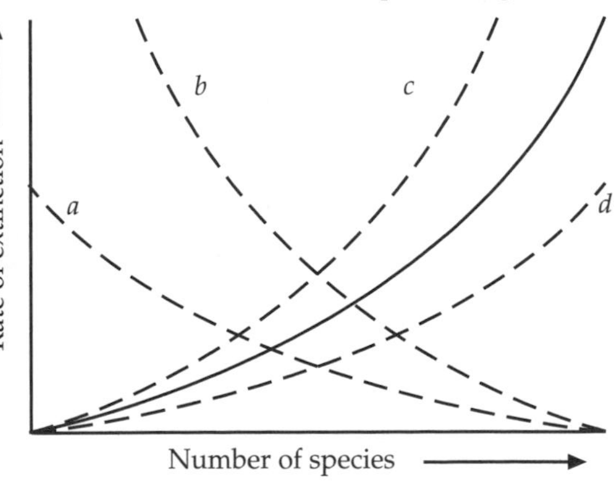

Which of the curves shows the expected relationship for a small island?
a. Curve *a*
b. Curve *b*
c. Curve *c*
d. Curve *d*
Textbook Reference: *57.3 What Patterns of Species Diversity Have Ecologists Observed? p. 1212*

13. Which of the following places stages of ecological succession occurring on glacial moraines in Alaska in correct chronological sequence, from earliest to latest?
a. Increase in soil nitrogen; arrival of lichens; arrival of willows and alders; arrival of conifers; arrival of bacteria, fungi, and photosynthetic microorganisms
b. Increase in soil nitrogen; arrival of bacteria, fungi, and photosynthetic microorganisms; arrival of lichens; arrival of willows and alders; arrival of conifers
c. Arrival of bacteria, fungi, and photosynthetic microorganisms; increase in soil nitrogen; arrival of lichens; arrival of willows and alders; arrival of conifers
d. Arrival of bacteria, fungi, and photosynthetic microorganisms; arrival of lichens; arrival of willows and alders; increase in soil nitrogen; arrival of conifers
e. Arrival of bacteria, fungi, and photosynthetic microorganisms; arrival of conifers; arrival of willows and alders; arrival of lichens; increase in soil nitrogen
Textbook Reference: *57.4 How Do Disturbances Affect Ecological Communities? pp. 1214–1215*

14. Which of the following represents primary succession?
a. The development of an aquatic community in a newly excavated farm pond, eventually followed by the filling in of the pond with sediment and its conversion to a forest
b. The development of a terrestrial community on lava produced by a volcanic eruption
c. The gradual establishment of a forest following a fire that destroyed the previous community
d. The invasion of weeds followed by shrubs and trees on an abandoned farm field
e. Both a and b
Textbook Reference: *57.4 How Do Disturbances Affect Ecological Communities? p. 1214*

15. Which of the following statements about succession is *false*?
a. Primary succession proceeds more rapidly than secondary succession.
b. Species that are early colonists may either facilitate or inhibit colonization by later-arriving species.
c. As succession proceeds, a lower proportion of the nutrients in a community are found in abiotic forms.
d. As succession proceeds, food webs become more complex.
e. The endpoint of directional succession is the establishment of a self-perpetuating climax community.
Textbook Reference: *57.4 How Do Disturbances Affect Ecological Communities? pp. 1214–1216*

16. In comparison with temperate regions at higher latitudes, tropical communities show greater species richness. Which of the following is a hypothesis that has been proposed to explain this difference?
    a. Tropical regions have had more time to diversify under stable climatic conditions.
    b. Tropical regions have more habitat types.
    c. Higher levels of interspecific competition in the tropics have led to narrower niches and hence a greater number of more specialized species.
    d. Higher predation intensity in the tropics prevents interspecific competition among prey species and thus allows rare species to persist.
    e. All of the above

    Textbook Reference: *57.3 What Patterns of Species Diversity Have Ecologists Observed? pp. 1211–1212*

Application Questions

1. Suppose that you are responsible for managing community succession to provide suitable habitat for an endangered animal species that is long-lived and has a low reproductive rate. Would you expect this species to be adapted to early or late successional stages?

   Textbook Reference: *57.4 How Do Disturbances Affect Ecological Communities? p. 1215*

2. Where do biomass pyramids like the one shown below occur, and what are the conditions that create them?

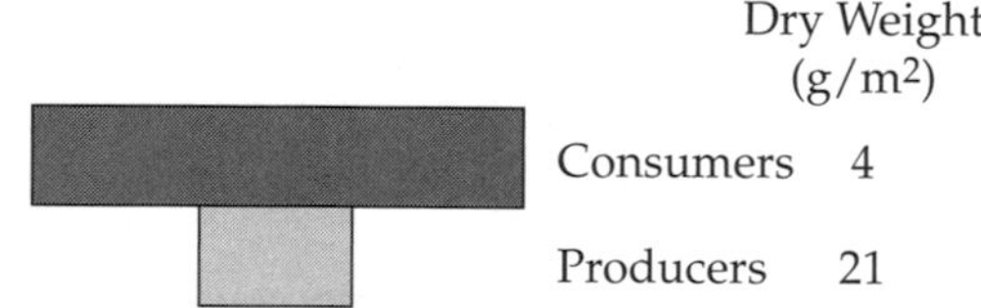

   Textbook Reference: *57.1 What Are Ecological Communities? pp. 1206–1207*

3. How is an understanding of heterotrophic succession useful in the field of law enforcement?

   Textbook Reference: *57.4 How Do Disturbances Affect Ecological Communities? pp. 1216–1217*

4. Generations of Americans have been instructed by Smokey Bear that fires are always harmful to the environment and should be prevented under all circumstances. What are some communities adapted to periodic fires? Explain why fires are beneficial to these communities.

   Textbook Reference: *57.4 How Do Disturbances Affect Ecological Communities? p. 1216*

5. Wolves were reintroduced to Yellowstone National Park in 1995. Describe the ecological effects within the park of this politically controversial decision.

   Textbook Reference: *57.2 How Do Interactions among Species Influence Community Structure? p. 1208*

6. Growing crops as monocultures is standard practice in modern agriculture. Why are monocultures unstable? What agricultural practices might result in more stable ecological communities?

   Textbook Reference: *57.5 How Does Species Richness Influence Community Stability? p. 1218*

Answers

Diagram Exercise Answer

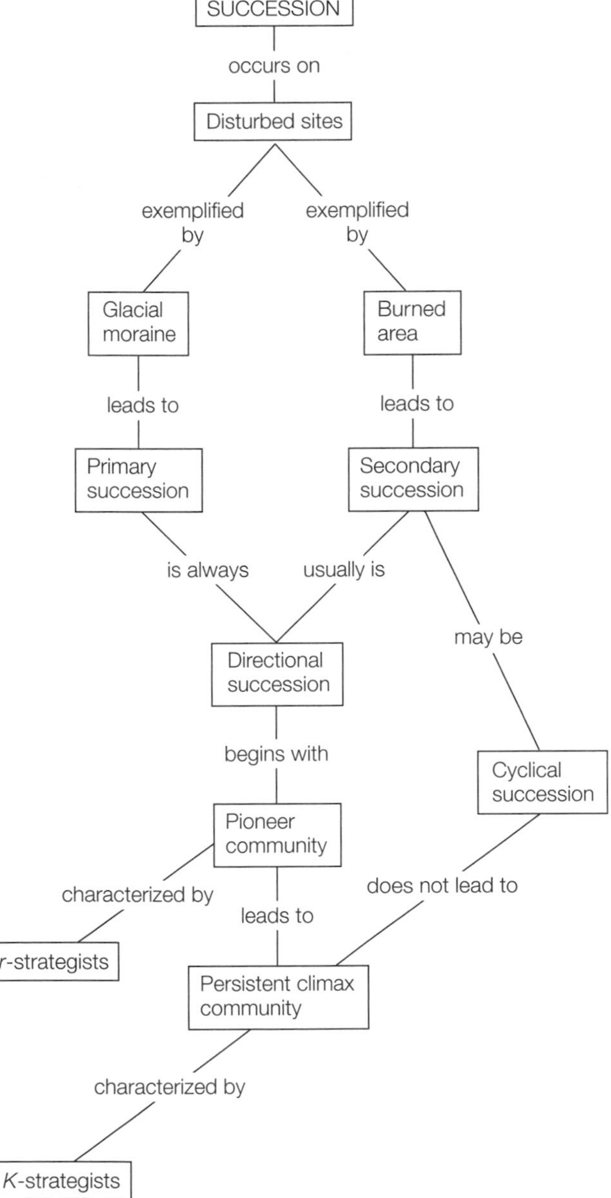

Knowledge and Synthesis Answers

1. **b.** An ecological community comprises all of the populations living in a defined area but excludes its nonliving components.
2. **e.** Because organism 9 eats from both the primary producer level (1) and the herbivore level (2), it is an omnivore.
3. **c.** Four trophic levels are depicted. The levels are primary producer (1), herbivore (2, 3, 4), primary carnivore (5, 6), and secondary carnivore (7, 8). The omnivore (9) occupies either the herbivore or primary carnivore level, depending on whether it is feeding on a primary producer or on an herbivore.
4. **a.** Compared to grasslands or aquatic ecosystems, forests typically have a lower ratio of primary consumer biomass to primary producer biomass because trees (the dominant primary producers) store energy for long periods of time in difficult-to-digest forms.
5. **d.** The amount of production used in maintenance and biosynthesis is added to net primary production to determine gross primary production. Photosynthetic release of O_2 in the light is an estimate of net production; O_2 use in the dark is an estimate of maintenance and biosynthesis costs. You would add 0.02 ml to 0.14 ml to get an estimate of gross primary production of 0.16 ml per minute.
6. **d.** Net primary productivity is the rate at which energy is incorporated into the bodies of primary producers through growth and reproduction. As a rule, only about 10 percent of the energy at one trophic level is transferred to the next. Because the secondary consumers are two trophic levels above the primary producers, their productivity would be expected to be only 1 percent of net primary productivity.
7. **c.** Both observational and experimental evidence indicate that species richness often peaks in communities of intermediate productivity.
8. **b.** By definition, removal of a keystone species has a larger effect on the community than would be expected based on its abundance.
9. **e.** The Shannon diversity index measures alpha diversity. Sorenson's index is used to estimate the change in species composition from one community to another (i.e., beta diversity).
10. **d.** The extinction rate is greater than zero and equals the arrival rate at n_1, which is therefore the equilibrium species number on the island.
11. **d.** Each curve corresponds to a group of similar-sized islands whose species number is plotted against their distance from the mainland. For each curve, species number decreases with distance from the mainland, and the curve that is lowest relative to the vertical axis would be the group of smallest islands.
12. **c.** With less space available, populations of the different species would be smaller. This would subject them to higher extinction rates than would be expected on larger islands.
13. **d.** Of the stages listed, the arrival of bacteria, fungi and photosynthetic microorganisms would be first. The arrival of lichens would be second, followed by the arrival of willows and alders, an increase in soil nitrogen content, and finally the arrival of conifers.
14. **e.** Primary succession begins on a newly available site where all preexisting living organisms have been stripped away, such as a freshly excavated pond or the surface of lava. Succession that occurs after disturbances such as fire or the conversion of a natural community to farmland is considered secondary.

15. **a.** Because primary succession begins on a site that is devoid of preexising organisms, it is a slower process than secondary succession.
16. **e.** Community ecologists have not been able to agree on the reasons for the greater species diversity characteristic of tropical communities. All four hypotheses have been proposed, and each has some evidence to support it.

Application Answers

1. As explained in Chapter 55, long life span and low reproductive rate are typical characteristics of *K*-strategists. Such species are typical of late successional stages. Pioneer species, in contrast, are likely to be replaced quickly and therefore tend to be *r*-strategists with a high reproductive rate, short life span, and good dispersal ability.
2. In most aquatic ecosystems, the dominant primary producers are unicellular algae. Their populations multiply so rapidly and are cropped so efficiently by slower-growing herbivores that they frequently support a larger herbivore mass than their own. Note that the diagram depicts an inverted pyramid of biomass. A pyramid of energy can never be inverted.
3. Heterotrophic succession occurs in communities whose energy source is detritus such as human corpses. The community of organisms found on a decomposing corpse can be used as evidence of the time since death.
4. The lodgepole pine forests discussed in this chapter are an example of a community that is maintained by periodic fires, which return nutrients to the soil and are required for seed germination. Other examples of fire-dependent or fire-adapted communities include tropical savanna and chaparral (described in Chapter 54). The use of fire as a management tool to protect biodiversity is discussed in Chapter 59.
5. In the Lamar Valley of the park, the reintroduction of wolves resulted in a reduction of the population of elk, the wolves' principle prey. In a trophic cascade, the diminution in the number of elk led to increased tree reproduction of aspens and willows, both of which had been heavily browsed by elk. In turn, the increase in the population of streamside willows resulted in an increase in the number of beaver colonies in the valley.
6. Monocultures are unstable because they are subject to attack by insect pests and pathogens that destroy or damage crops. Experiments have shown that more diverse agroecosystems in which two or more crops are grown on the same plot are less subject to pest outbreaks. In corn–sweet potato dicultures, for example, sweet potato pests were reduced because the corn provided a structural barrier, food for protective parasitoid wasps, and chemicals that interfered with the ability of pests to find their host plants.

58 Ecosystems and Global Ecology

The Big Picture

- Keep in mind the different ways in which energy and matter move through ecosystems. Energy flows unidirectionally from producers to higher trophic levels and is ultimately dissipated as heat, whereas elements and water continually cycle through ecosystems.
- The productivity of ecosystems, both terrestrial and aquatic, is relevant to the problem of providing food for the exploding human population (see Chapter 55). The increasing share of global primary production that is appropriated by the expanding human population is also relevant to the problem of maintaining biodiversity (see Chapter 59).
- Knowledge of how materials move through biogeochemical cycles is crucial for understanding and predicting the effects of human alterations of these cycles. Global warming and eutrophication of lakes are two important examples.
- The important roles of nitrogen fixers and other microorganisms in the nitrogen cycle illustrate the sometimes overlooked ways in which rather obscure members of the biotic community may be essential for the healthy functioning of ecosystems.

Common Problem Areas

- The reason that eutrophication—the "enrichment" of a body of water with nutrients—often results in a "dead zone" may be puzzling to you. The reason is that the additional growth of photosynthetic algae and bacteria caused by added nutrients such as phosphorus and nitrogen result in an increased rate of aerobic respiration by microorganisms as they decompose the dead photosynthesizers. This process may lead to oxygen depletion in part or all of a body of water, such as the deeper waters of a lake.

Study Strategies

- In studying the material on biogeochemical cycles, focus on the following basic questions: Where are the abiotic reserves of an element located? How does the element leave the reserve and enter living organisms? Why do living organisms need the element? How does the element return to its abiotic reserve?
- Go to yourBioPortal.com to review the following tutorials and activity:

  Animated Tutorial 58.1 The Global Hydrologic Cycle

  Animated Tutorial 58.2 The Global Carbon Cycle

  Animated Tutorial 58.3 The Ocean Conveyor Belt

  Animated Tutorial 58.4 The Global Nitrogen Cycle

  Web Activity 58.1 Concept Matching

Important Concepts

Earth's system is composed of cycles of materials, inputs of solar energy, and interactions between living organisms and the physical environment.

- An ecosystem is any ecological system within defined boundaries. It includes all the organisms within those boundaries as well as the chemical and physical factors that interact with those organisms.
- Earth is essentially a closed system with respect to atomic matter, but it is an open system with respect to energy.
- Energy from the sun, combined with the energy of radioactive decay in Earth's interior, drives the processes that move materials around the planet.
- The rate at which energy moves through a system is called its *flux*. An accumulation of an element in some component of an ecosystem is a *pool*; the term *flux* also applies to movements of elements between pools. Some pools are *sinks*, in which an element is locked up for a long period.
- Earth is an unusual planet in that it has (1) lithospheric plates that move continuously; (2) a moderate surface temperature; (3) a large quantity of liquid water on its surface; and (4) diverse forms of life. The moon (1) stabilizes the tilt of Earth's axis and thereby influences climate; (2) helps produce tides; and (3) slows Earth's rotation.
- The gases of Earth's atmosphere help moderate planetary temperatures and thus keep water in a liquid state. The chemical composition of the atmosphere is influenced by volcanic eruptions.

- The activities of living organisms have an enormous impact on the chemical composition not only of the atmosphere, but also of the water and the land.
- Energy flows through ecosystems unidirectionally and is dissipated as heat. The elements on which life depends cycle among the four compartments of Earth's physical environment: the atmosphere, the oceans, fresh waters, and land (see Figure 58.1).
- The atmosphere consists of two layers, the troposphere and the stratosphere.
  - The lowest layer is the troposphere, where almost all water vapor is located and where most global air circulation takes place.
  - Above the troposphere is the stratosphere, where incoming ultraviolet radiation is absorbed by a layer of ozone (O_3) (see Figure 58.2). The depletion of stratospheric ozone caused by the release of chlorinated fluorocarbons (CFCs), such as the refrigerant Freon, is a source of concern because increased ultraviolet radiation at Earth's surface is associated with increased rates of skin cancer, cataract formation, and crop damage.
- The most abundant gases in the atmosphere are N_2 (about 78 percent) and O_2 (about 21 percent). Although CO_2 constitutes only 0.03 percent of the atmosphere, it is the source of the carbon used by terrestrial photosynthetic organisms and of the dissolved carbonate used by marine producers. The greenhouse gases (such as CO_2, water vapor, methane, and nitrous oxide) trap outgoing infrared (heat) radiation emitted by Earth and thus raise its surface temperature (see Figure 58.3 and Table 58.1).
- Concentrations of mineral nutrients are very low in most ocean waters because most elements that enter the oceans gradually sink to the seafloor. Rates of photosynthesis in the oceans are highest in zones of upwelling adjacent to continents, where nutrient-rich cold bottom water rises to the surface (see Figure 58.4).
- Most mineral nutrients that enter fresh waters are released by the weathering of rocks and are carried to lakes and rivers via groundwater or by surface flow.
- In lakes, surface waters tend to become depleted of nutrients as organisms die and sink to the bottom, whereas deeper waters become depleted of O_2 as organic matter decomposes. Vertical movements called turnover bring nutrients and dissolved CO_2 to the surface and O_2 to deeper water. In deep lakes found in temperate climates, turnover occurs in the spring and fall when the water temperature is uniformly 4°C (see Figure 58.5).
- About one-fourth of Earth's surface is covered by land—the terrestrial compartment of the physical environment. Because elements move slowly on land and usually only over short distances, local variations in the supply of particular elements influence ecosystem processes in this compartment. The type of soil that forms in an area and the elements it contains depend on the underlying rock, as well as on climate, topography, the organisms living there, and the length of time that soil-forming processes have been acting.

Energy flows through the global ecosystem unidirectionally.

- With very few exceptions, energy flow in ecosystems originates with photosynthesis.
- Net primary production is the portion of assimilated energy left over after the energy used by primary producers for their own metabolism is subtracted (see Chapter 57). All of the other organisms in an ecosystem derive their energy directly or indirectly from net primary production, which takes the form of growth and reproduction of the producer organisms.
- The major limits on primary production in terrestrial ecosystems are low temperatures and lack of moisture. In aquatic systems, production is limited by light, nutrients, and temperature (see Figures 58.6 and 58.7).
- Human activities are having an increasing effect on the flow of energy through the global ecosystem. Humans are now appropriating about 24 percent of the annual net primary production on Earth, but the percentage varies widely among regions.

Elements cycle through the global ecosystem.

- The pattern of movement of a chemical element through organisms and compartments of the physical environment is called its biogeochemical cycle.
- The sun powers the hydrologic cycle by causing evaporation, most of it from ocean surfaces. The average residence time of a water molecule in a particular compartment varies greatly, from under a week in organisms to about 3000 years in the ocean. Water in underground aquifers (sedimentary rocks) has such a long residence time that it plays a small role in the hydrologic cycle (see Figure 58.8). Pumping of this groundwater for irrigation has altered the flux of water from the land to the oceans. If current water consumption trends continue, about half of the world's population will have an inadequate supply of water by 2025.
- Fires consume the energy of, and release chemical elements from, the vegetation they burn. Biomass burning also contributes a significant quantity of CO_2 and other greenhouse gases to the atmosphere.
- Nearly all the carbon in organisms comes from CO_2 in the atmosphere or dissolved carbonate (CO_3^{-2}) or bicarbonate (HCO_3^-) in water. Carbon is incorporated into organic molecules by photosynthesis; respiration and combustion break down these organic molecules and return carbon to the atmosphere and water (see Figure 58.9).
- Water near the ocean surface is becoming more acidic because of the absorption of quantities of CO_2 larger than at any time during the past 20 million years. The

combination of decreasing pH and increasing water temperature is endangering coral organisms and hence entire reef communities.

- Fossil fuels exist because in the remote past, large quantities of organic carbon were removed from the carbon cycle by burial of organisms in sediments lacking oxygen. The ever-increasing rate of burning of fossil fuels over the past 150 years has elevated the CO_2 concentration of the atmosphere (see Figure 58.10).
- Less than half of the CO_2 released into the atmosphere by human activities remains there; most of the rest dissolves in the oceans, which contain 50 times the amount of dissolved inorganic carbon as the atmosphere.
- Currently, the photosynthetic consumption of CO_2 exceeds the metabolic consumption of CO_2, so Earth's terrestrial vegetation is storing carbon that would otherwise be increasing atmospheric CO_2 concentrations. Nevertheless, it is unlikely that terrestrial vegetation can store the extra CO_2 that human activities are producing.
- Historical records and computer models indicate that Earth is warmer when atmospheric CO_2 levels are higher and cooler when they are lower (see Figure 58.11). The global warming caused by a doubling of the atmospheric CO_2 concentration would increase mean annual temperatures worldwide and disrupt current precipitation patterns. It might also melt the polar ice caps, thereby causing the flooding of coastal regions. Global warming has already caused an increase in insect infestations in some temperate forests and in the future may cause an increase in the incidence of some human diseases.
- Nitrogen gas (N_2) makes up 78 percent of Earth's atmosphere, but nitrogen can be converted into biologically useful forms by only a few species of microorganisms that can carry out nitrogen fixation. Thus, nitrogen is often in limited supply in ecosystems. Microorganisms also perform denitrification, which removes nitrogen from the biosphere and returns it to the atmosphere (see Figure 58.13).
- Total nitrogen fixation by humans as a result of the use of fertilizers and burning of fossil fuels equals global natural nitrogen fixation (see Figure 58.14). When nitrogen applied as fertilizer to cropland enters bodies of water, it can lead to eutrophication, a process in which high nutrient levels promote excessive algal growth, depleting the oxygen in the water and creating a deoxygenated "dead zone" (see Figure 58.15). Human-caused perturbations of the nitrogen cycle have a variety of other adverse effects, including increased air pollution, increased atmospheric concentrations of greenhouse gases, and reductions in species richness.
- Most of Earth's sulfur supply is located in rocks on land and as sulfate salts in deep-sea sediments. Sulfur in soil is taken up by plants and incorporated into proteins. Atmospheric sulfur (in the form of dimethyl sulfide, released into the atmosphere when phytoplankton decay) plays an important role in the global climate because it is a major component of particles that promote cloud formation.
- The burning of fossil fuels produces emissions that form sulfuric and nitric acid in the atmosphere. The resulting acid precipitation affects all industrialized countries and has been shown to have harmful effects on lake ecosystems and on terrestrial plants (see Figure 58.16).
- The phosphorus cycle differs from the cycles of carbon, nitrogen, and sulfur in that it lacks a gaseous phase. On land, most phosphorus becomes available through the weathering of phosphorus-containing rocks (see Figure 58.17).
- Because phosphorus is often a limiting nutrient for plant growth, it is a typical component of fertilizer. Because it is also frequently a limiting factor on algal growth in lakes, the addition of phosphorus (from fertilizer runoff, soil erosion, detergents, human wastes, animal manure, and industrial wastes) to freshwater bodies is a major cause of eutrophication.
- Recovery and recycling of phosphorus from sewage and animal wastes could reduce the amount of phosphorus entering lakes and rivers while supplying much of the needs of the fertilizer and detergent industries.
- Organisms may be deficient in any of several minerals that they require in very small amounts. Over much of the ocean, a scarcity of dissolved iron limits the rate of photosynthesis. In some terrestrial regions, iodine, cobalt, and selenium are not available in the concentrations necessary to meet the needs of endothermic vertebrates.
- Because biochemical cycles interact in significant ways, alterations in any one cycle affect the others. Scientists are constantly discovering previously unknown interactions. An example is the recent finding that elevated atmospheric CO_2 concentrations can cause a decrease in the rate of nitrogen fixation (see Figure 58.18).

Ecosystems provide human society with indispensable goods and services.

- Among the goods and services provided by ecosystems are food, clean water, clean air, fiber, building materials, fuel, flood control, soil stabilization, pollination, climate regulation, spiritual fulfillment, and aesthetic enjoyment. It would be either impossible or prohibitively expensive to replace these benefits.
- Modifications of Earth's ecosystems have contributed to human welfare, but the benefits have not been distributed equally. Moreover, short-term increases in some ecosystem goods and services have resulted in the long-term degradation of others.
- The most important cause of alterations in ecosystems has been changes in land use as natural ecosystems have been converted to more intensive uses. Though

such altered ecosystems provide many benefits (such as food production), they also have resulted in the degradation of other services (such as the ability to provide clean water, habitat for wildlife, and protection from floods).

- The ecological importance of coastal wetlands was illustrated by Hurricane Katrina, which would not have caused as much flooding in New Orleans if the wetlands surrounding the city had been intact.

Many options exist for the sustainable management of ecosystems.

- The economic value of a sustainably managed ecosystem is often higher than that of a converted or intensively exploited ecosystem (see Figure 58.19).
- Because ecosystem services are considered "public goods" that have no market value, government action and incentives may be required to encourage sustainable ecosystem management. In addition, education programs are needed to increase public awareness of how human activities affect ecosystem sustainability.

Test Yourself

Diagram Exercise

Draw a diagram showing the compartments of the global ecosystem and the processes by which water and nutrients move from one compartment to another.
Textbook Reference: *58.3 How Do Materials Cycle Through the Global Ecosystem? pp. 1229–1230*

Knowledge and Synthesis Questions

1. Zones of upwelling near the shores of land masses have high primary production because such zones
   a. bring nutrients from the seafloor up to the surface.
   b. are characterized by clear water that permits light to penetrate to an unusually great depth.
   c. bring warm water to the surface.
   d. trap nutrients washed into the ocean from nearby land masses.
   e. have water of lower salinity than other oceanic regions.
   Textbook Reference: *58.1 What Are the Compartments of the Global Ecosystem? pp. 1224–1225*
2. Which of the following statements is true of oceans but *not* true of freshwater ecosystems?
   a. They receive material from land mostly via groundwater.
   b. Elements are buried in bottom sediments for long periods of time.
   c. There is a seasonal mixing of materials.
   d. Bottom waters frequently lack oxygen.
   e. All of the above are characteristic of both oceans and freshwater ecosystems.
   Textbook Reference: *58.1 What Are the Compartments of the Global Ecosystem? pp. 1224–1226*
3. In the following graph, the temperature of a freshwater lake is plotted against its depth.

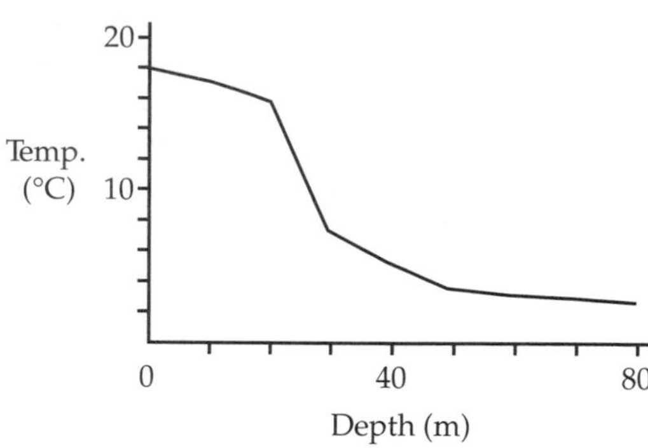

   This temperature profile of the graph would be most characteristic of the lake during the
   a. winter.
   b. spring.
   c. summer.
   d. fall.
   e. Both b and c
   Textbook Reference: *58.1 What Are the Compartments of the Global Ecosystem? p. 1226*
4. Based on data in the graph from Question 3, the thermocline for the lake is located between _______ and _______ meters.
   a. 0; 10
   b. 10; 20
   c. 20; 30
   d. 30; 50
   e. 50; 80
   Textbook Reference: *58.1 What Are the Compartments of the Global Ecosystem? p. 1226*
5. Which of the following statements is true of the stratosphere but *not* of the troposphere?
   a. Most greenhouse gases are concentrated here.
   b. Most water vapor resides here.
   c. Most of the mass of the atmosphere lies here.
   d. Circulation of this layer influences ocean currents.
   e. Most ultraviolet radiation is absorbed here.
   Textbook Reference: *58.1 What Are the Compartments of the Global Ecosystem? pp. 1223–1224*
6. Which of the following compartments of the global ecosystem is characterized by very slow movement of materials within it?
   a. Oceans
   b. Fresh waters
   c. Atmosphere
   d. Land
   e. All of the above
   Textbook Reference: *58.1 What Are the Compartments of the Global Ecosystem? p. 1226*

7. Which of the following environmental factors frequently limits primary production in *both* terrestrial and aquatic ecosystems?
   a. Temperature
   b. Moisture
   c. Light
   d. Nutrient supply
   e. None of the above

   Textbook Reference: *58.2 How Does Energy Flow through the Global Ecosystem? p. 1228*

8. Which of the following statements about biogeochemical cycles is *false*?
   a. Most elements remain longest in the living portion of their cycle.
   b. Gaseous elements cycle more quickly than elements without a gaseous phase.
   c. Some atoms in the human body may once have been part of a dinosaur.
   d. Biogeochemical cycles all include both organismal and nonliving components.
   e. Perturbations in one biogeochemical cycle can have major effects on other cycles.

   Textbook Reference: *58.3 How Do Materials Cycle through the Global Ecosystem? pp. 1229–1238*

9. Next to each of the following features, place a *c, n, p,* or *s* if it is characteristic of the biogeochemical cycles of *c*arbon, *n*itrogen, *p*hosphorus, or *s*ulfur. (Note: Some features may apply to more than one cycle.)

   _____ Major reservoir is atmospheric
   _____ Major reservoir is in sedimentary rock
   _____ Often in short supply in ecosystems
   _____ Fossil fuel reserve is part of this cycle
   _____ Involved in cloud formation
   _____ Lacks a gaseous phase
   _____ Includes a form that is a greenhouse gas
   _____ Most fluxes involve organisms

   Textbook Reference: *58.3 How Do Materials Cycle through the Global Ecosystem? pp. 1229–1237*

10. Which of the following is *not* one of the major threats to the climate caused by our alterations of the carbon cycle?
    a. The polar ice caps may melt if global warming continues.
    b. The burning of fossil fuels adds compounds of sulfur to the atmosphere.
    c. The increase in atmospheric CO_2 exceeds the ability of the oceans to absorb the increase.
    d. CO_2 is a gas that traps infrared radiation.
    e. Tropical storms are likely to increase in intensity if global warming continues.

    Textbook Reference: *58.3. How Do Materials Cycle through the Global Ecosystem? pp. 1230–1234*

11. Which of the following statements about acid precipitation is *false*?
    a. Acids that enter the atmosphere primarily affect ecosystems located less than 100 kilometers from the source of the pollution.
    b. Though lakes are very sensitive to acidification, their pH can return rapidly to normal values.
    c. Regulation of emission sources has raised the pH of precipitation in many parts of the eastern United States in the last two decades.
    d. Sulfuric acid and nitric acid from the burning of fossil fuels are the major causes of acid precipitation.
    e. Organisms lost from freshwater ecosystems because of acidification may not return even decades after pH values have returned to normal.

    Textbook Reference: *58.3 How Do Materials Cycle through the Global Ecosystem? pp. 1234–1235*

12. The process by which a lake ecosystem is altered by eutrophication involves several stages. Which of the following represents the correct ordering of stages in the chain of causation?
    a. Algal populations die off; algal blooms occur; oxygen levels drop in deeper water; respiratory demand from decomposers increases; phosphorus input from sewage and agricultural runoff increases
    b. Oxygen levels drop in deeper water; phosphorus input from sewage and agricultural runoff increases; respiratory demand from decomposers increases; algal blooms occur; algal populations die off
    c. Phosphorus input from sewage and agricultural runoff increases; algal blooms occur; algal populations die off; respiratory demand from decomposers increases; oxygen levels drop in deeper water
    d. Phosphorus input from sewage and agricultural runoff increases; algal blooms occur; oxygen levels drop in deeper water; respiratory demand from decomposers increases; algal populations die off
    e. Phosphorus input from sewage and agricultural runoff increases; respiratory demand from decomposers increases; oxygen levels drop in deeper water; algal blooms occur; algal populations die off

    Textbook Reference: *58.3 How Do Materials Cycle through the Global Ecosystem? p. 1235*

13. The availability of which of the following micronutrients limits the rate of photosynthesis over much of the ocean?
    a. Cobalt
    b. Iron
    c. Selenium
    d. Iodine
    e. Nitrogen

    Textbook Reference: *58.3 How Do Materials Cycle through the Global Ecosystem? p. 1236*

14. Which of the following ecosystem types is being converted to cropland most rapidly at the present time?
    a. Deserts
    b. Temperate forests
    c. Tropical and subtropical biomes
    d. Chaparral
    e. Temperate grasslands

    Textbook Reference: *58.4 What Services Do Ecosystems Provide? p. 1238*

15. Which of the following is a major obstacle to the sustainable management of ecosystems?
   a. Many ecosystem services are considered "public goods" that have no market value.
   b. The total economic value of a sustainably managed ecosystem is almost always lower than that of an intensively exploited ecosystem.
   c. Most people are not aware of the extent to which human activities affect the functioning of ecosystems.
   d. Both a and c
   e. All of the above

   Textbook Reference: *58.5 How Can Ecosystems Be Sustainably Managed? p. 1239*

Application Questions

1. How does the use of groundwater for irrigation alter the hydrological cycle?

   Textbook Reference: *58.3 How Do Materials Cycle through the Global Ecosystem? pp. 1229–1230*
2. Why do humans deliberately set fire to natural vegetation? In what ways are human-set fires contributing to global warming?

   Textbook Reference: *58.3 How Do Materials Cycle through the Global Ecosystem? p. 1230*
3. What is the connection between global warming and the spread of human disease?

   Textbook Reference: *58.3 How Do Materials Cycle through the Global Ecosystem? p. 1232*
4. In what ways are farming methods that require the large-scale use of manufactured fertilizers contributing to eutrophication and other adverse environmental effects?

   Textbook Reference: *58.3 How Do Materials Cycle through the Global Ecosystem? pp. 1233–1234*
5. Wetlands have frequently been regarded as "waste lands" that would have greater economic value if they were drained and converted to agricultural or commercial use. Describe what we learned from Hurricane Katrina, which struck the Gulf Coast in 2005, about the value of intact wetlands for flood protection? Describe two other ways in which the wetlands of this region are valuable.

   Textbook Reference: *58.4 What Services Do Ecosystems Provide? p. 1238*

Answers

Diagram Exercise Answer

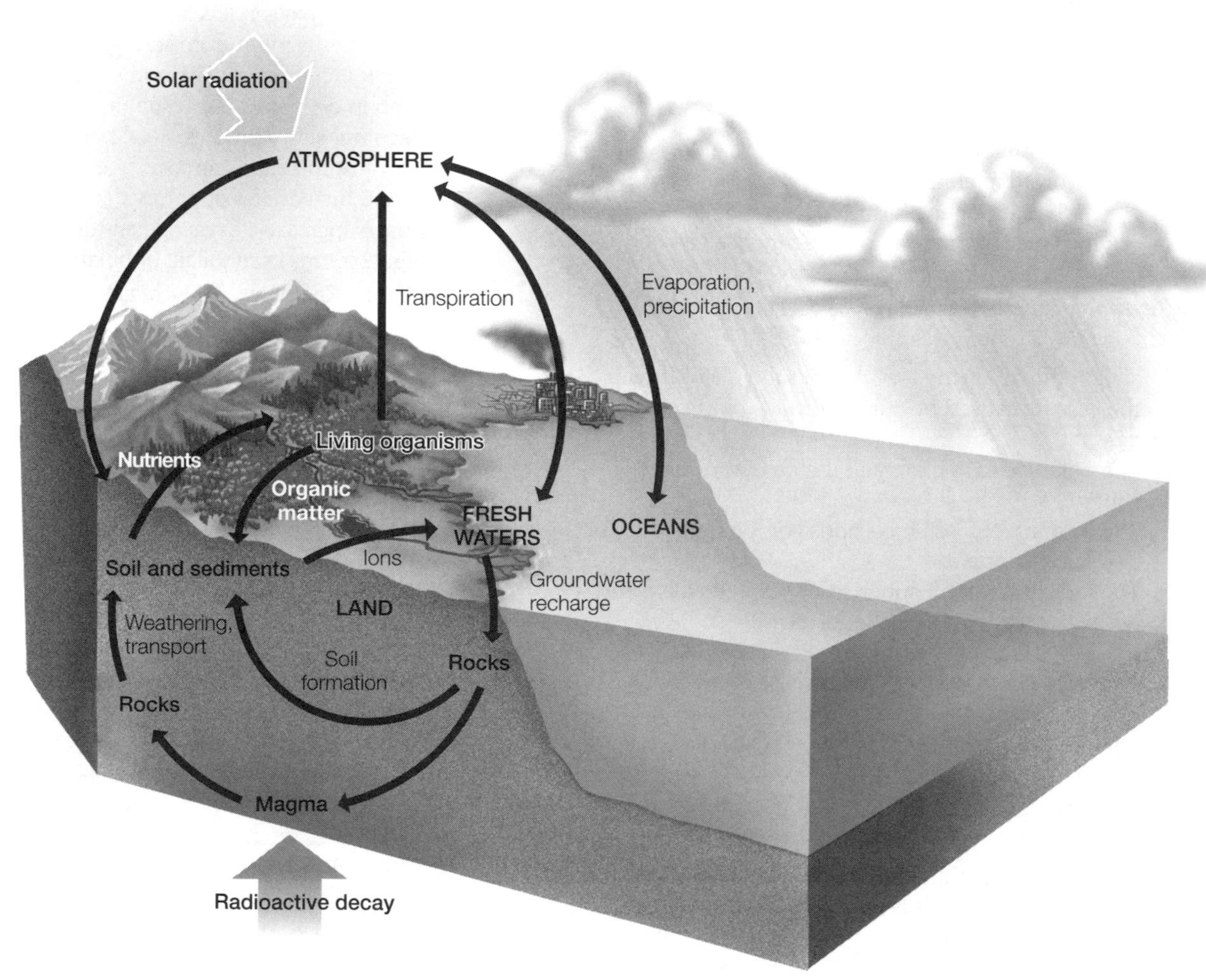

Knowledge and Synthesis Answers

1. **a.** Low concentrations of mineral nutrients limit the primary production of most ocean waters. In zones of upwelling, nutrient-rich water is brought to the surface from the ocean bottom, thereby enhancing the growth of photosynthesizing plankton.
2. **b.** Most elements remain in bottom sediments of the ocean for millions of years, until they are elevated above sea level by movements of Earth's crust.
3. **c.** The steep thermocline evident in this temperature profile would typically be established only by mid- to late-summer.
4. **c.** The range of depths over which the temperature changes most abruptly is 20 to 30 meters.
5. **e.** The ozone that shields the surface of Earth from most incoming ultraviolet radiation is located in the stratosphere.
6. **d.** Unlike their behavior in air and water, elements on land move slowly and usually only short distances.
7. **a.** Temperature, nutrient supply, and light can limit production in aquatic ecosystems; temperature and moisture most often limit production on land.
8. **a.** Most elements cycle through organisms more quickly than they cycle through the nonliving world.
9. **n.** Major reservoir is atmospheric
   c., **p.** Major reservoir is in sedimentary rock
   n., **p.** Often in short supply in ecosystems
   c. Fossil fuel reserve is part of this cycle
   s. Involved in cloud formation
   p. Lacks a gaseous phase
   n., **c.** Includes a form that is a greenhouse gas
   n. Most fluxes involve organisms
10. **b.** The addition of compounds of sulfur to the atmosphere has more to do with acid precipitation than it does with global warming.
11. **a.** Acid precipitation is a regional, not a local, environmental problem. Acids may travel hundreds of kilometers before they settle to Earth in precipitation or as dry particles.
12. **e.** The correct sequence is **e**.
13. **b.** Because iron is insoluble in oxygenated water, it sinks rapidly to the ocean floor and therefore may be very scarce in the surface waters in which photosynthesis occurs. Nitrogen is sometimes a limiting nutrient, but it is a macronutrient.
14. **c.** Conversion to cropland is causing rapid reductions of a number of tropical and subtropical biomes.
15. **d.** Two significant barriers to sustainable ecosystem management are the lack of market value of many ecosystem services and the lack of public understanding of how human activities may harm ecosystems.

Application Answers

1. Groundwater normally has a long residence time and plays a small part in the hydrologic cycle. Thus the extensive use of groundwater for irrigation may result in the depletion of this resource as the water is removed from aquifers more quickly than it can be replaced. Use of this groundwater has also increased flows of water to the ocean and has contributed to the rising sea level of the past century.
2. Most fires that occur in forests, savannas, and other biomes are deliberately set by humans to clear land for agriculture. These fires—together with other types of biomass burning, such as combustion of wood and alcohol—account for about 40 percent of the annual influx of CO_2 into the atmosphere and contribute to the production of other greenhouse gases.
3. Cold temperatures in winter kill many pathogens. The milder winters associated with global warming will allow increasing numbers of pathogens to survive; hence the diseases they cause will become more common. An example of this process is the spread of dengue fever, a tropical human disease that is expanding its range to higher latitudes. Some diseases of plants and other organisms are also expected to spread as a consequence of global warming.
4. If farmers apply fertilizers to croplands in quantities that exceed the plants' abilities to absorb them, some of the excess nitrates and phosphates leave the soil and enter bodies of water. This process contributes to eutrophication of lakes and the creation of "dead zones" in the sea, such as the one in the Gulf of Mexico around the mouth of the Mississippi River. Excessive use of nitrogen-containing fertilizers has also caused contamination of groundwater and has been linked to outbreaks of toxic dinoflagellates in estuaries on the Atlantic coast.
5. If coastal wetlands surrounding New Orleans had been intact, they would have absorbed the storm surge produced by the hurricane and thus protected the city from flooding. Furthermore, even in their reduced capacity, these wetlands remain important both as spawning grounds for many marine organisms, including some of commercial value, and as wintering habitat for migratory birds.

59 Conservation Biology

The Big Picture

- Conservation biology provides an excellent example of the importance of the synthetic approach to science. In this case, knowledge and concepts derived from different areas of biology and the social sciences are being applied to the solution of the problem of species and ecosystem preservation.
- Preserving biological diversity is considered by many informed individuals to be one of the most important challenges facing humanity in the twenty-first century.

Common Problem Areas

- It may not be apparent to you why a fragmented habitat is unlikely to support as many species as similar unbroken habitat of equal extent. One reason is that some species may require large expanses of a particular habitat in order to survive. A second reason is that smaller habitat patches are more susceptible to edge effects (see Figure 59.5).

Study Strategies

- This chapter refers to and relies on a number of concepts introduced in earlier chapters on ecology, such as: invasive species, habitat fragmentation, and the management of populations (Chapter 55); species richness, disturbance, species–area relationships, and island biogeography (Chapter 57). You may find it useful to review those concepts as you study the material in this chapter.
- Go to yourBioPortal.com to review the following tutorial and activity:

  Animated Tutorial 59.1 Edge Effects

  Web Activity 59.1 Concept Matching

Important Concepts

Conservation biology is the scientific study of how to preserve the diversity of life.

- The science of conservation biology draws on concepts and knowledge from ecology, ethology, and evolutionary biology. The development of this science has been spurred by the accelerating pace of human-caused extinctions of species.
- Conservation biology is guided by three basic principles: Evolution is the process that unites all of biology; the ecological world is dynamic; and humans are a part of ecosystems.
- Extinctions have occurred throughout Earth's history at a "background" rate as the environment has changed to favor some species and negatively affect others. Although human beings have probably caused extinctions for thousands of years (see Figure 59.1), today the human population is largely responsible for species extinctions that rival those of the five great mass extinctions of life's history.
- There are many reasons that people value biodiversity.
  - Species are necessary for the functioning of ecosystems and the many benefits and services those ecosystems provide.
  - Many individual species are important because they supply useful products such as food, fiber, and medicinal drugs.
  - Extinctions lessen the aesthetic pleasure many people derive from interacting with other organisms.
  - Extinctions deprive us of opportunities to study the structure and functioning of ecological communities and ecosystems.
  - Extinctions of other species caused by human activities raise ethical issues because those species are judged to have intrinsic value.

Biologists use several methods to predict changes in biodiversity.

- Accurate estimates of the number of extinctions that will occur in the next century are impossible for four reasons: (1) we do not know how many species exist on Earth today; (2) we do not know the ranges of most species; (3) it is difficult to determine when a species becomes extinct (see Figure 59.2); and (4) we rarely know how the extinction of one species may affect other species.
- Estimates of current rates of extinction worldwide are based primarily on species–area relationships, the

theory of island biogeography, and rates of tropical deforestation (see Figure 59.3).

- To estimate the risk of extinction of a particular population, biologists develop models that incorporate information about its size, its genetic variation, and the morphology, physiology, and behavior of its members.
- Endangered species are those that are in imminent danger of extinction over all or much of their range. Threatened species are those that are likely to become endangered in the near future. Species with rapidly shrinking populations are especially vulnerable to extinction because such reductions in population size can lead to genetic drift and loss of genetic variation.

Human activities that threaten the survival of species include habitat destruction, overexploitation, introduction of exotic species, and climate change.

- In the United States, the most important cause of species extinction today is habitat loss, degradation, and fragmentation, especially for species that live in fresh waters (see Figure 59.4). As habitats are fragmented into small patches, populations of species that require large areas cannot be maintained.
- The proportion of a habitat patch subject to detrimental edge effects increases as patch size decreases (see Figure 59.5). For example, species from surrounding habitats may invade the edges of the patch to compete with or prey upon the patch inhabitants. The persistence of species in small patches may be improved if the patches are connected by corridors of suitable habitat through which individuals can disperse.
- Overexploitation for commercial reasons (such as use in traditional medicine, and trade in pets, ornamental plants, and tropical hardwoods) continues to threaten many species (see Figure 59.7).
- Some exotic (non-native) species become invasive, reproducing rapidly and spreading widely to the detriment of native species. Exotic pests, predators, and competitors introduced by humans have driven many species to extinction and threaten many others (see Figure 59.8).
- As rapid global warming occurs, species with poor dispersal abilities may become extinct because of their inability to keep pace with climate change by shifting their ranges. Other effects of global warming may include the entire disappearance of some habitats (such as alpine tundra), and the development of new climates (especially at low elevations in the tropics).

Conservation biologists use many strategies to maintain biological diversity.

- Protected areas preserve habitat, prevent overexploitation, and may serve as nurseries for the replenishment of populations in unprotected areas. Using the criteria of species richness and endemism, biologists have identified a number of "hotspots" for biodiversity protection (see Figure 59.10).
- To further pinpoint sites with threatened species found nowhere else, conservation biologists have identified 595 "centers of imminent extinction." These are concentrated in tropical forests, on islands, and in mountainous regions (see Figure 59.11).
- Practitioners of restoration ecology attempt to return degraded habitats to their natural state. An example is the current project to restore an extensive prairie ecosystem in Montana (see Figure 59.12).
- Creating new wetlands to substitute for the ones being destroyed by development requires detailed ecological knowledge. Experiments have shown that restoration is most successful when wetlands are planted with a mixture of species instead of just one or two (see Figure 59.14).
- Humans often reduce the frequency and intensity of disturbances such as fires and thereby endanger species dependent on these disturbances. Reestablishment of historical disturbance patterns (for example, by controlled burning in forests), may help preserve such species (see Figure 59.15).
- Because most endangered species cannot survive further reductions in their breeding populations, it is important to prevent their exploitation. The Convention on International Trade in Endangered Species (CITES) determines which species are banned in international trade. For example, CITES is regulating the international trade in elephant ivory.
- Controlling invasions of exotic species is often important in preserving biodiversity. Using the traits found in most invasive species, conservation biologists have developed a decision tree to help them determine whether an exotic plant species should be introduced into North America (see Figure 59.16).
- Studies conducted in the field of ecological economics have demonstrated that conserving biodiversity is often economically valuable. Ecotourism, for example, is a major source of income for many developing nations (see Figure 59.17). Intact ecosystems may also be a reliable and relatively inexpensive source of water (see Figure 59.18).
- Reconciliation ecology is the practice of using land in which people live and extract resources in ways that sustain biodiversity. It is based on the principle that most ecosystem services are provided locally, and that people are more motivated to protect their local interests than they are to work on national or global issues.
- For a small fraction of endangered species, captive propagation can help prevent extinction by maintaining the species during critical periods and by providing individuals for reintroduction into the wild (see Figure 59.19).
- Science can supply information about the accelerating rate of species extinctions, but it is up to society as a whole to determine what rate of species loss due to human activities is acceptable.

Test Yourself

Diagram Exercise

Construct a concept map whose theme is "Biodiversity" Include in your map the following terms: Biodiversity, air purification, economic benefits, ecosystem services, ecotourism, edge effects, fiber, food, global warming, habitat fragmentation, habitat loss, invasive species, medicines, natural products, overexploitation, and water purification. Connect these terms by verbs or short phrases to indicate the relationships among them.
Textbook Reference: *59.1 What Is Conservation Biology? pp. 1242–1245; 59.2 How Do Biologists Predict Changes in Biodiversity? pp. 1245–1246; 59.3 What Factors Threaten Species Survival? pp. 1247–1250; 59.4 What Strategies Do Conservation Biologists Use? pp. 1251–1258*

Knowledge and Synthesis Questions

1. The concepts of conservation biology come mainly from all of the following fields *except*
   a. ecology.
   b. evolutionary biology.
   c. population genetics.
   d. immunology.
   e. ethology.
   Textbook Reference: *59.1 What Is Conservation Biology? p. 1243*
2. The most likely cause of the extinctions of many large mammals in North America in the last 14,000 years is
   a. rapid climate change associated with glaciation.
   b. overhunting by humans.
   c. the formation of land bridges to neighboring continents that allowed many new species of competitors and predators to invade these regions.
   d. massive volcanism that caused destruction of the food supply for these mammals.
   e. pathogens introduced by humans.
   Textbook Reference: *59.1 What Is Conservation Biology? p. 1244*
3. Based on species–area relationships, ecologists predict that
   a. about one million tropical evergreen forest species may become extinct in the next hundred years.
   b. a 90 percent loss of habitat will result in loss of about 9 percent of the species living there.
   c. the area required by most species will need to be reduced.
   d. extinction is inevitable.
   e. None of the above
   Textbook Reference: *59.2 How Do Biologists Predict Changes in Biodiversity? p. 1246*
4. In the United States, the groups of organisms with the highest proportion of endangered or extinct species live in
   a. grasslands.
   b. the deciduous forest biome.
   c. freshwater habitats.
   d. deserts.
   e. temperate evergreen forest.
   Textbook Reference: *59.3 What Factors Threaten Species Survival? p. 1247*
5. In the graph below, select the curve (*a, b,* or *c*) that correctly shows the expected relationship between habitat patch area and the proportion influenced by edge effects.

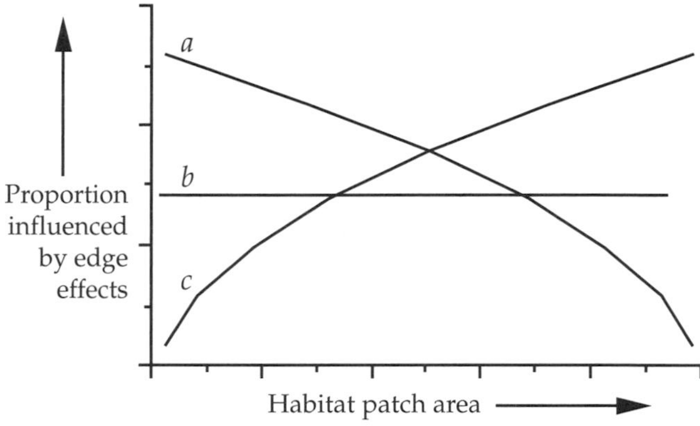

   Textbook Reference: *59.3 What Factors Threaten Species Survival? p. 1247*
6. Which of the following is *not* currently a major cause of the global reduction in biodiversity?
   a. Overexploitation
   b. Global warming
   c. Habitat destruction
   d. Introduction of foreign predators and disease
   e. All of the above are currently major causes of extinction.
   Textbook Reference: *59.3 What Factors Threaten Species Survival? pp. 1247–1250*
7. Why do conservation biologists believe that global warming may lead to extensive decimation of species?
   a. Because little change in plant community composition has occurred in the past, we cannot expect present communities to adapt to climate change.
   b. The magnitude of climate change will be much greater than past periods of climatic change.
   c. Many sedentary species may not be able to shift their ranges at the same pace as the northern movement of temperature zones.
   d. Both b and c
   e. All of the above
   Textbook Reference: *59.3 What Factors Threaten Species Survival? pp. 1249–1250*
8. A defect of the "hotspot" approach to conserving biodiversity is that it does not direct attention to
   a. regions of low species richness that may nevertheless contain unique species or ecological communities.
   b. marine regions.
   c. regions with a high number of endemic species.
   d. Both a and b
   e. All of the above

Textbook Reference: *59.4 What Strategies Do Conservation Biologists Use? p. 1251*

9. Which of the following efforts to preserve biodiversity best exemplifies restoration ecology?
   a. A campaign to discourage pesticide use on lawns
   b. A project to convert ranch land into a natural prairie
   c. Designation of the habitat of an endangered species as a protected area
   d. Elimination of commercial trade in products derived from endangered and threatened species
   e. A captive breeding program to maintain an endangered species

   Textbook Reference: *59.4 What Strategies Do Conservation Biologists Use? p. 1252*

10. Which of the following efforts to preserve biodiversity best exemplifies reconciliation ecology?
   a. A campaign to discourage pesticide use on lawns
   b. A project to convert ranch land into a natural prairie
   c. Designation of the habitat of an endangered species as a protected area
   d. Elimination of commercial trade in products derived from endangered and threatened species
   e. A captive breeding program to maintain an endangered species

   Textbook Reference: *59.4 What Strategies Do Conservation Biologists Use? pp. 1256–1257*

11. Experiments on wetland restoration have demonstrated that planting a richer mixture of species is associated with
   a. faster accumulation of belowground nitrogen.
   b. more complex vegetation structure.
   c. more rapid development of vegetation cover.
   d. Both b and c
   e. All of the above

   Textbook Reference: *59.4 What Strategies Do Conservation Biologists Use? p. 1253*

12. To help identify species of exotic plants that have the potential to become invasive if they were to be introduced into North America, conservation biologists make use of a "decision tree." Which of the following plant characteristics might cause biologists to deny approval for a proposed introduction of an exotic plant species?
   a. Seeds of the plant do not require pretreatment for germination.
   b. The plant spreads quickly by vegetative propagation.
   c. The plant is invasive outside of North America.
   d. The juvenile period of the plant is less than five years.
   e. All of the above

   Textbook Reference: *59.4 What Strategies Do Conservation Biologists Use? p. 1255*

13. The fynbos shrub community in South Africa
   a. is threatened by introduced species of taller, faster-growing plants.
   b. is a fire-adapted community.
   c. helps maintain a regional supply of high-quality water.
   d. includes thousands of plant species found nowhere else in the world.
   e. All of the above

   Textbook Reference: *59.4 What Strategies Do Conservation Biologists Use? p. 1256*

14. Which of the following statements about elephants and the ivory trade is *false*?
   a. African elephants are endangered throughout their range.
   b. The geographical source of ivory can be determined by the use of DNA markers.
   c. There is a strong demand for ivory in Japan and China because of its use in folk medicines.
   d. Online sales of ivory on eBay have been banned.
   e. The legal mechanism for prohibiting the ivory trade is the international agreement called the Convention on International Trade in Endangered Species.

   Textbook Reference: *59.4 What Strategies Do Conservation Biologists Use? p. 1254*

15. The California condor
   a. is being introduced into regions that were not part of its historic geographical range as part of the effort to save the species.
   b. became endangered partly because of high mortality that resulted from its eating carcasses containing lead shot.
   c. has increased in numbers because of captive propagation, though released captive-bred condors have not yet bred in the wild.
   d. has been reintroduced to the wild over the objections of cattle ranchers, who believe correctly that condors kill livestock.
   e. All of the above

   Textbook Reference: *59.4 What Strategies Do Conservation Biologists Use? p. 1257*

Application Questions

1. Aside from the aesthetic benefits of biodiversity, list three economic reasons people should care about species extinctions.

   Textbook Reference: *59.1 What Is Conservation Biology? pp. 1244–1245*

2. Under the Endangered Species Act of 1973, some species are listed as endangered, whereas others are deemed threatened. What is the distinction between these two categories?

   Textbook Reference: *59.2 How Do Biologists Predict Changes in Biodiversity? p. 1246*

3. What are wildlife corridors, and how have experiments demonstrated that they help species persist in patchy environments?

   Textbook Reference: *59.3 What Factors Threaten Species Survival? p. 1248*

4. Why are exotic species frequently a threat to the biological diversity of the region in which they are introduced?

Textbook Reference: *59.3 What Factors Threaten Species Survival? p. 1248*

5. In recent years, severe forest fires in many parts of the western United States have raised controversy concerning the use of controlled burning as a forest management tool. What are the benefits of this practice?
Textbook Reference: *59.4 What Strategies Do Conservation Biologists Use? pp. 1253–1254*

Answers

Diagram Exercise Answer

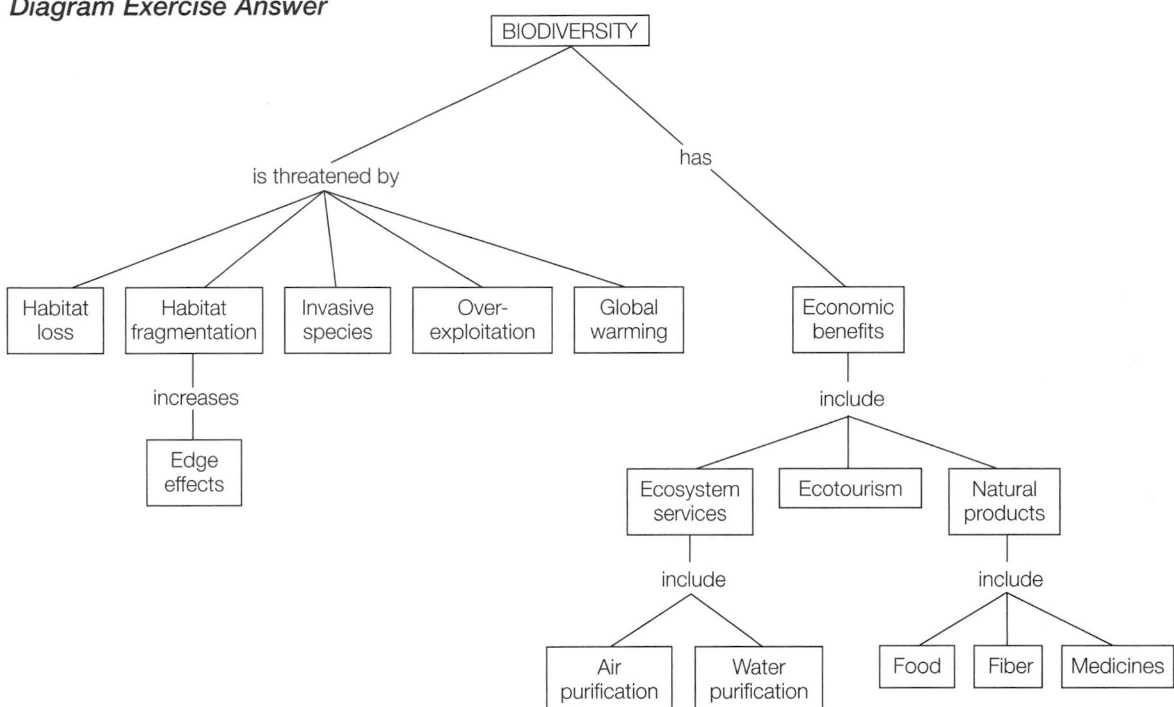

Knowledge and Synthesis Answers

1. **d.** Although immunology is sometimes a useful tool in assessing the amount of genetic variation that exists in a population, its importance is less than the other fields listed.
2. **b.** Because these extinctions followed human colonization of North America, most biologists favor the hypothesis that overhunting by humans was the most likely cause.
3. **a.** Based on estimates of current and future disappearance of tropical evergreen forests, about one million species that live in these communities could become extinct. This loss is not inevitable if we reduce the rate at which these forests are converted to pasture and cropland.
4. **c.** Habitat destruction and pollution have caused extinction or endangerment of a very high proportion of aquatic freshwater species.
5. **a.** Because the edge of a habitat patch equals the perimeter of the patch, it is proportionally greater for smaller habitat patches. Therefore, the relationship between edge effects and habitat patch areas is an inverse relationship, as shown by curve *a*.
6. **b.** Global warming is predicted to be a major cause of extinction in the future, but is not yet having the kind of impact on populations as overexploitation, habitat destruction and species introductions are.
7. **c.** Although the expected magnitude of the climate change due to global warming may be similar to past climate changes, the rate of warming will be greater. This may make it impossible for many species to extend their ranges at the same rate as the northward movement of the temperature zones.
8. **d.** The "hotspot" concept emphasizes the preservation of regions of high species richness and endemism. It overlooks marine regions and terrestrial regions of relatively low species richness.

9. **b.** Restoration ecology focuses on the reestablishment of entire ecosystems to their natural state.
10. **a.** Reconciliation ecology focuses on ways that biodiversity can be sustained in landscapes in which people live.
11. **e.** When a species-rich mixture is introduced, wetland restoration is more successful for all of the reasons listed.
12. **e.** See Figure 59.16 in the textbook.
13. **e.** See Figure 59.18 in the textbook.
14. **a.** A number of countries in southern Africa have so many elephants that some must be killed to prevent them from dispersing to populated regions and damaging crops.
15. **b.** The California condor's historic range included all of the regions where it is currently being reintroduced. Condors are now breeding in the wild and do not kill livestock.

Application Answers

1. A list of economic reasons people should care about species extinctions would include the importance of natural products (food, fiber, and medicine), services such as fermentation, ecosystems services, and nature tourism.
2. Endangered species are in imminent danger of extinction over all or a significant portion of their range. Threatened species are those likely to become endangered in the near future.
3. Habitat corridors are relatively thin strips of habitat of a particular type that connect larger patches of the same type of habitat. Their importance in permitting individuals to disperse from one patch to another was demonstrated in experiments described in Figure 55.11 of the textbook.
4. Native species that live in a particular community have evolved to cope successfully with the specific predators, competitors, and diseases that are part of its environment. Introduced species represent a change in the environment for which native species may not have been prepared by natural selection.
5. In ecosystems that naturally experience periodic fires, controlled burning prevents the excessive accumulation of fuel (dead branches, leaf litter, etc.) and thereby lessens the danger that catastrophic, tree-consuming canopy fires will occur. In addition, because many species in such ecosystems require periodic fires for successful establishment and reproduction, controlled burning may be necessary to maintain species richness.